AF581911

Computational Intelligence and Soft Computing: Recent Applications

Computational Intelligence and Soft Computing: Recent Applications

Editors

László T. Kóczy
István Á. Harmati

MDPI • Basel • Beijing • Wuhan • Barcelona • Belgrade • Manchester • Tokyo • Cluj • Tianjin

Editors
László T. Kóczy
Budapest University of Technology and Economics
Hungary

István Á. Harmati
Széchenyi István University
Hungary

Editorial Office
MDPI
St. Alban-Anlage 66
4052 Basel, Switzerland

This is a reprint of articles from the Special Issue published online in the open access journal *Symmetry* (ISSN 2073-8994) (available at: https://susy.mdpi.com/special_issue/process/379127/closed).

For citation purposes, cite each article independently as indicated on the article page online and as indicated below:

LastName, A.A.; LastName, B.B.; LastName, C.C. Article Title. *Journal Name* **Year**, *Volume Number*, Page Range.

ISBN 978-3-0365-6155-4 (Hbk)
ISBN 978-3-0365-6156-1 (PDF)

Contents

About the Editors

László T. Kóczy

Laszlo T. Koczy received his Ph.D. degree from the Technical University of Budapest (BME) in 1977, and his D.Sc. from the Hungarian Academy of Science in 1998. He spent his career at BME until 2001, and in 2002 moved to Szechenyi Istvan University (Gyor, SZE), where he was the Dean of Engineering. From 2013 to 2022 he was President of the University Research and University Ph.D. Councils. In March 2022 he received the Professor Emeritus title from both universities.

His main research activities have been in the field of computational intelligence, especially in fuzzy systems, evolutionary and memetic algorithms, and neural networks, as well as applications in engineering, logistics, management, etc. He has published close to 800 research articles, with over 3000 published fully independently and over 7700 Google Scholar citations. His h-index is 41. His main findings are: the concept of rule interpolation in sparse fuzzy models and hierarchical interpolative fuzzy systems; fuzzy Hough transform; fuzzy signatures, fuzzy situational maps, and fuzzy signature state machines; the node reduction algorithm in fuzzy cognitive maps; the Bacterial Memetic Evolutionary Algorithm; and the Discrete Bacterial Memetic Evolutionary Algorithm (for NP-hard continuous and discrete optimization and search). His research interests include applications of CI for telecommunication, transportation and logistics, vehicles and mobile robots, control, built environment evaluation, maintenance problems, information retrieval, employee attitude evaluation, management system investigation, etc.

István Á. Harmati

István Á. Harmati received his MSc from Eötvös Loránd University in 2001, in mathematics and physics. He received his PhD in mechanical engineering from the Technical University of Budapest in 2009, and his habilitation from Széchenyi István University in 2015. He is an associate professor at the Department of Mathematics and Computational Sciences, at Széchenyi István University. His main research interests are mathematical modelling and analysis of uncertain systems, including the classical stochastic view and non-standard approaches like fuzzy logic and possibility theory. Moreover, he is also interested in explainable neural networks, fuzzy mathematics and decision-making in uncertain environments.

Editorial

Symmetry or Asymmetry? Complex Problems and Solutions by Computational Intelligence and Soft Computing

László T. Kóczy [1,2]

[1] Department of Informatics, Széchenyi István University, 9026 Győr, Hungary; koczy@tmit.bme.hu
[2] Department of Telecommmnication and Media Informtics, Budapest University of Technology and Economics, 1111 Budapest, Hungary

Citation: Kóczy, L.T. Symmetry or Asymmetry? Complex Problems and Solutions by Computational Intelligence and Soft Computing. *Symmetry* **2022**, *14*, 1839. https://doi.org/10.3390/sym14091839

Received: 9 August 2022
Accepted: 19 August 2022
Published: 5 September 2022

1. Searching for Symmetry in the Solution of Complex Problems

What is the role of symmetry in the seemingly far away topics of solving complex applied problems by approaches offered by Soft Computing and Computational Intelligence? At first sight, it may be hard to give a direct answer. Nevertheless, there is a very important aspect that I try to explain in this little introductory study that forms a bridge between the two concepts.

When solving complex problems, setting up models for complex systems, and developing algorithms for search and optimization in such models and systems, it must be considered that such problems are intractable from the mathematical point of view. For the concept of intractability, see e.g., the very classic textbook [1]. This means that for problems of a given type, it is impossible to give an exact or optimal solution if the size of the problem (i.e., the number of components) exceeds a usually quite low value. Researchers unfamiliar with the theory of computational complexity may reply that it is a matter of computer speed and capacity, but it is absolutely not true. It is easy to show that the "Galactic Computer", a hypothetical computer consisting of all atoms of the Galaxy, operating with the speed of light, would not be able to exactly solve even problems of everyday life in the mathematical sense. It may be then surprising that such problems are often tackled rather efficiently by human experts, operators, or simple technical solutions. Is there any contradiction? Of course, there is none, just it must be realised that most complex problems do not need a really exact solution, but a "good enough" one, one which satisfies the expectations of the problem setter. In a general sense, such complex problems I will call "engineering problems", even though they often come from management, economics, social sciences, and the like. In a general sense, these problems have a common feature, namely, they reflect real phenomena, where the number of components is very high (this is why they are a priori intractable), and/or they contain elements that must be considered non-deterministic from the point of view of the problem solver, moreover, they often contain uncertainty in the formulation of the problem itself from the side of the problem setter. Let us assume there is an imaginary scale, where in one pan of the scale the expectations of the problem setter, while in the other, the resources offered by the problem solver are put. How can be this scale brought in balance, with other words, how can this approach made symmetric? Of course, the next question is, symmetric, but in what sense? Can highly complex and mathematically intractable problems somehow be weighed? Can solutions be weighed? Definitely, not in the ordinary sense, but there must be some measure found that connects the two sides and that may serve as the unit which helps balance the scale. This unit or measure is referred to in the literature as cost. As a matter of course, the cost is not considered a financial matter but the amount of resources, such as capacity and speed, used, as the loss of accuracy of the solution, and there may be other points.

A study of this issue considered from a specific point of view, applying rule-based Fuzzy Systems for modelling was given in [2]. This approach is definitely one of the key components of Computational Intelligence and it was the actual initial sub-discipline of

Soft Computing, both proposed by Zadeh (cf. [3]). There exists, however, an analytic mathematical approach that studies the question of whether an optimal solution for setting up a fuzzy rule-based model, where the cost is a combined function of the amount of needed resources (space and time complexity), and the efficiency of the method (say, the accuracy), both expressed in a merged way by a formula "total time needed", can be found. Such an optimal solution may be considered as an idealistically symmetric solution. In our paper [4], we found that for certain special cases, this optimum could be exactly found, while for some more complex cases, the existence of such an optimum could be proved [4].

Nowadays, meta-heuristic methods are widely applied, and in this Special Issue, we also published a paper on the successful application of a certain novel meta-heuristics for a rather difficult problem class. Meta-heuristic approaches offer probably the best solution when NP-hard or other highly complex problems have to be solved, and nowadays, evolutionary, memetic, and other population-based approaches have really produced sometimes marvelous results. This Special Issue presents a large variety of such papers, as I will show it in the second part of the Editorial. The mentioned meta-heuristics form another very important component of Computational Intelligence. In our experience, especially, memetic algorithms produce excellent results. The concept originates from [5], and a more recent overview was published in [6]. The main idea is that evolutionary or population-based algorithms may be used as a "wrapper", as a global search or optimisation technique, while a local search is conducted by some more traditional mathematical method, such as gradient-based search, for example. But the main point is in applying different algorithms for global and local search, thus speeding up and improving the accuracy of the algorithm: affecting the total cost in the pan of the computational resources of the scale. Various evolutionary algorithms have often rather different costs in this sense. An earlier comparison of some widely applied techniques was given in [7], another one can be found in [8].

The highly complex and mathematically intractable flow shop scheduling problem is definitely one of the rather sophisticated and complex problems, where a feasible solution needs a good meta-heuristic. In our paper, we proposed to use the new modified discrete bacterial memetic evolutionary algorithm (DBMA) where in a stricter sense, both the global and the local search are conducted by a certain meta-heuristic, the latter, namely, by Simulated Annealing. The results thus obtained proved to be better than any other approach for this problem. Here, it can be nicely pointed out that by the proper weighing of the costs, namely, the error in the accuracy of the optimization in one pan and the need for resources, especially, the running time of the optimization meta-heuristics in the other one must be brought to equilibrium, this way generating a symmetry in the solution. The exact position of the symmetrical (balanced) solution can, however, be calibrated by the designer of the solution, thus it may fit the application context of the concrete problem, considering the available resources and the expected quality of the quasi-optimum found. Thus, the asymmetric role played by the problem to solve and the model/algorithm for the solution must be balanced and, that way, the whole problem–solution complex must be brought into a symmetrical configuration. It is worthwhile mentioning that a very recently published article in the same journal tackles a similar type of highly complex problem, and proposes a rather different meta-heuristic for a solution, with some promising results [9]. There is a certain symmetry in the problem solution itself, but the general concept of targeting symmetry of costs in the optimal solution is applicable here, too. Finally, one more closely related paper may be mentioned here [10], where a logistic type path planning problem is tackled, although it is obvious that the solution method is easily applicable in other related fields, very likely, in VLSI design, among others. Here, the well-known Greedy Algorithm is proposed for path optimisation. Let me refer to our earlier paper where the same algorithm is applied, although, in a more complex embedding [11]. In this paper, the problem of symmetry and balance is clearly occurring twice, hierarchically, at the level of the costs of resources and accuracy loss that was mentioned above, and at the level of balancing the adaptive scheduler costs (runtime overhead) with the overall optimisation costs of the basic problem on hand.

It is possible to introduce other types of costs when analysing CI approaches in solving complex applied problems. Once Mamdani established his fuzzy model and control algorithm [12], an amazing explosion of applications followedThe most striking success was first observable in Japan [13], where an incredible number of successful commercial applications emerged within less than a decade—after a decade of stagnation of fuzzy applications, following the first few attempts. What was the reason for this success? Japanese scientists agree that the transparency of the fuzzy rule-based models, the possibility of directly tuning the parameters of fuzzy controllers based on expert domain knowledge, and the lack of necessity of using complicated analytical calculations such as, e.g., Lagrange transform, opened the door to efficient control of highly non-linear and partly uncertain systems, even for engineers with a modest knowledge of control theory. Thus, transparency is one of the most important features of SC and CI approaches, which may weigh in the pan of the wage representing the symmetric approach from our point of view.

And last but not least, there is another pair of cost components that heavily weigh in the desire to establish symmetry in the intertwined system of the problem and solution complex. This pair is what we referred to as predictability, and "universality", more precisely, general applicability of an approach, which is especially important when problems of similar type (e.g., various NP-hard discrete problems, cf. [1]) are in the focus of the solution. Sometimes, if there is a "guarantee" for obtaining a reasonably good solution for a particular concrete problem, where the expected time and space costs can be well estimated for arbitrary size, the applier is happier than when having a sometimes more efficient, but for some topologies, problem sub-classes, or certain large sizes inapplicable algorithm. This is rather typical for the metaheuristics which occur in a considerable number in this Special Issue. Our very paper here, optimising job scheduling came to life from the starting assumption that Discrete Bacterial Memetic Evolutionary Algorithm had its "universal" applicability. There are plenty of references in the paper showing evidence for this approach being quite well applicable in many different discrete problem groups. So, it was worthwhile to try—and we obtained good results, better than other authors so far!

Summarising the above thoughts, in one pan we collect the costs of space and time complexity (resource intensity), the overhead in the case of hierarchically built-up algorithms, and the lack of transparency, predictability, and general applicability, while in the other pan, there is the expectation of the applier, the accuracy, the feasibility, and similar components. As well, the solution provider attempts to find a well-balanced, in other words, symmetrical solution.

2. Let Us Quickly Overview Now the Contents of This Special Issue

The three main pillars of CI/SC are Fuzzy Systems (FS), Artificial Neural Networks/ Connectionist Systems (NN), and Evolutionary/Population-Based Algorithms (EA), which the latter of which includes Swarm Intelligence as well.

Although recently, the number of papers published in the fuzzy field seems to be braked a little, in this Special Issue, there are five papers, roughly one quarter, applying this by now well-established branch of non-conventional mathematics. Cruz-Aguilar et al. propose a combined Failure Mode and Effect Analysis and fuzzy method for the non-invasive measurement of methane and carbon dioxide. A. Łyczkowska-Hanćkowiak applies trapezoidal fuzzy numbers in portfolio analysis. A connected topic is K. Piasecki et al.'s paper on present value evaluation by oriented fuzzy numbers. One of the guest editors, I. Harmati, discusses the dynamics of fuzzy-rough cognitive networks. Finally, M. Holčapek et al. deal with fuzzy interpolation using extensional fuzzy numbers.

The situation is similar with Artificial Neural Networks. The number of related papers is five or even seven (counting two connected to the EA pillar as well) In a broader sense, the connectionist hybrid approach by Y. Zhao et al. on Key performance indicator (KPI) anomaly detection applies bi-directional long short-term memory replacing a traditional feedforward NN. H. Achicanoy et al. discuss the matter of generating synthetic images by applying StyleGAN, which latter reliably attributes every generated image to a particular network. E. Jeczmionek et al. deal with layer pruning in Convolutional Neural Networks. P. Li et al. discuss text summarisation based on the Dynamic Memory Network. Z. Xiao et al. apply a special NN for image processing: lung segmentation. S. Zeybek et al. combine the population-based Bees Algorithm for training recurrent NNs. At last, E. Kaya et al.'s work hybridises ANN and EA in their nonlinear system identification approach.

The last main group of papers deals with some version of the EA approach. In addition to the above-mentioned two hybrid NN and EA methods, eight further articles fall into this category. Z. H. Chin et al. apply a Genetic Algorithm (GA) for calculating Proof-of-Work blockchains. L. Wang et al. deal with Android malware detection, deploying a self-variant GA. A. H. G. Ruiz with co-authors also apply GA, for energy saving in an air-conditioning system. S. Nantogma and co-authors propose the use of artificial immune-based algorithms for learning in air-defence systems. A. Agárdi et al., including the present Guest Editor, offer a hybrid, Bacterial Evolutionary and Simulated Annealing memetic algorithm for the so far most efficient solution to the Job Scheduling Problem. Another hybrid approach is proposed by M. Zhang and co-authors, the combination of Butterfly and Particle Swarm Optimisation in the presence of high dimensionality. The paper by H. El Raoui et al. discusses the very important general problem of using meta-heuristics in problem-solving. This topic reflects the thoughts in the first part of this Editorial.

At last, a paper authored by K. K. Sharma et al. applying modified spectral clustering for the prediction of customer churn may be mentioned, the latter being an alternative technique for machine learning.

I am convinced the Reader will find a number of extremely interesting and thought-provoking ideas in this rich collection of articles.

Funding: This research was funded by the Hungarian Office for Research Innovation and Development (NKFIH), grant number K-124055.

Conflicts of Interest: The authors declare no conflict of interest.

References

1. Aho, A.V.; Hopcroft, J.E.; Ullman, J.D. *The Design and Analysis of Computer Algorithms*; Addison-Wesley: Amsterdam, The Netherland, 1983; 436p.
2. Kóczy, L.T. Fuzzy Rule Based Systems as Tools towards Solving the "Key Problem of Engineering". In *On Fuzziness, Studies in Fuzziness and Soft Computing 298*; Springer: Berlin/Heidelberg, Germany, 2013; pp. 311–323.
3. Zadeh, L.A. Fuzzy Logic, Neural Networks, and Soft Computing. *Commun. ACM* **1994**, *37*, 77–84. [CrossRef]
4. Gedeon, T.D.; Kóczy, L.T.; Zorat, A. Optimal Size Fuzzy Models. In *JACIII 11*; Fuji Technology Press: Tokyo, Japan, 2007; pp. 335–341.
5. Moscato, P.; Mathieson, L. Memetic Algorithms for Business Analytics and Data Science: A Brief Survey. In *Business and Consumer Analytics: New Ideas*; Springer: Berlin/Heidelberg, Germany, 2019; pp. 545–608.
6. Neri, F.; Cotta, C.; Moscato, P. *Handbook of Memetic Algorithms, Studies in Computational Intelligence*; Springer: Berlin/Heidelberg, Germany, 2019; 309p.
7. Balázs, K.; Botzheim, J.; Kóczy, L.T. Comparison of various evolutionary and memetic algorithms. In *Integrated Uncertainty Management and Applications*; Springer: Berlin/Heidelberg, Germany, 2010; pp. 431–442.
8. Nguyen, P.T.H.; Sudholt, D. Memetic algorithms beat evolutionary algorithms on the class of hurdle problems, GECCO' 18. In Proceedings of the Genetic and Evolutionary Computation Conference, Kyoto Japan, 15–19 July 2018; pp. 1071–1078.
9. Lei, D.; Yi, T.A. Novel Shuffled Frog-Leaping Algorithm for Unrelated Parallel Machine Scheduling with Deteriorating Maintenance and Setup Time. *Symmetry* **2021**, *13*, 1574. [CrossRef]
10. Li, D.; Wang, L.; Geng, S.; Jiang, B. Path Planning of AS/RS Based on Cost Matrix and Improved Greedy Algorithm. *Symmetry* **2021**, *13*, 1483. [CrossRef]
11. Balázs, K.; Kóczy, L.T. Adaptive scheduling of optimization algorithms in the construction of interpolative fuzzy systems. In Proceedings of the 2013 IEEE International Conference on Fuzzy Systems (FUZZ-IEEE), Hyderabad, India, 7–10 July 2013; pp. 1–7.
12. Mamdani, E.H.; Assilian, S. An experiment in linguistic synthesis with a fuzzy logic controller. *Int. J. Man-Mach. Stud.* **1975**, *7*, 1–13. [CrossRef]
13. Bellon, C.; Bosc, P.; Prade, H. Fuzzy boom in Japan. *Int. J. Intell. Syst.* **1992**, *7*, 293–316. [CrossRef]

Article

Segmentation of Lung Nodules Using Improved 3D-UNet Neural Network

Zhitao Xiao [1,2], Bowen Liu [1,2], Lei Geng [1,2,*], Fang Zhang [1,2] and Yanbei Liu [1,2]

1 School of Electronics and Information Engineering, Tiangong University, Tianjin 300387, China; xiaozhitao@tiangong.edu.cn (Z.X.); 1830091229@tiangong.edu.cn (B.L.); zhangfang@tiangong.edu.cn (F.Z.); liuyanbei@tiangong.edu.cn (Y.L.)

2 Tianjin Key Laboratory of Optoelectronic Detection Technology and Systems, Tiangong University, Tianjin 300387, China

* Correspondence: genglei@tiangong.edu.cn

Received: 28 September 2020; Accepted: 27 October 2020; Published: 28 October 2020

Abstract: Lung cancer has one of the highest morbidity and mortality rates in the world. Lung nodules are an early indicator of lung cancer. Therefore, accurate detection and image segmentation of lung nodules is of great significance to the early diagnosis of lung cancer. This paper proposes a CT (Computed Tomography) image lung nodule segmentation method based on 3D-UNet and Res2Net, and establishes a new convolutional neural network called 3D-Res2UNet. 3D-Res2Net has a symmetrical hierarchical connection network with strong multi-scale feature extraction capabilities. It enables the network to express multi-scale features with a finer granularity, while increasing the receptive field of each layer of the network. This structure solves the deep level problem. The network is not prone to gradient disappearance and gradient explosion problems, which improves the accuracy of detection and segmentation. The U-shaped network ensures the size of the feature map while effectively repairing the lost features. The method in this paper was tested on the LUNA16 public dataset, where the dice coefficient index reached 95.30% and the recall rate reached 99.1%, indicating that this method has good performance in lung nodule image segmentation.

Keywords: lung nodule segmentation; 3D-UNet; 3D-Res2UNet; multi-scale features; deep learning

1. Introduction

Lung cancer is one of the most common cancers worldwide and the main cause of death for cancer patients. According to "Global Cancer Statistics" [1], in 2018, there were approximately 2.1 million new cases of lung cancer worldwide and 1.77 million lung cancer-related deaths. Since lung cancer has no obvious symptoms in the early stages and is difficult to detect, it is often discovered in the middle and late stages of cancer, and the best treatment time is missed. On the one hand, studies have found that most of the lung cancer is in the form of lung nodules. The lung nodules are divided into benign and malignant. The probability of malignant lung nodules becoming cancerous are greatly increased. An accurate early identification of benign and malignant lung nodules is essential for the prevention of lung cancer. On the other hand, lung nodules are present as three-dimensional spherical shapes in images and are affected by factors such as variable shapes, different sizes, and complex surrounding tissues. The shape of pulmonary nodules is especially critical for doctors to screen. The process of using CT images to detect lung cancer is actually a process of identifying lung nodules.

The current research on lung nodules mainly focuses on two kinds of methods.

The first method is based on traditional image processing segmentation methods. Carvalh et al. [2] combined traditional image processing methods with machine learning methods. First, the lung parenchyma was segmented from the CT image as the region of interest. Then, the candidate nodules are detected in terms of shape, texture, and extraction of expression features (such as size) through segmentation methods. Messay et al. [3] designed an algorithm for the segmentation-based detection of nodules using candidate nodule shape, location, brightness, and gradient characteristics. Although the methods in [2,3] have achieved significant results, the edges of the lung nodules in the CT image are blurred due to the small size of the lung nodules (usually between 3 and 30 mm) and low contrast; additionally, the image gray scale is uneven due to adhesions. The influence of noise and artifacts also limits the accuracy of such methods in detecting and segmenting lung nodules.

The second segmentation method used to detect lung nodules is based on machine learning or deep learning. With the continuous development of deep learning, more and more deep learning-based convolutional neural networks have played a vital role in lung nodule research, mainly focusing on methods based on 2D convolutional neural networks and 3D-based methods.

(1) Methods based on the 2D convolutional neural network. Ding et al. [4] borrowed from the successful application of deep convolutional neural networks (DCNNs) in natural image recognition and proposed a lung nodule detection method based on DCNNs. In the faster R-CNN neural network, a deconvolution structure is introduced for candidate detection of axis slices. Aiming at the problem that the size of lung nodules is too small, and it is easy to lose features, a deconvolution is added after the VGG16 network to restore the size of the feature map so that the network can capture features more accurately. Deng Zhonghao et al. [5] aimed at solving the problem of low detection sensitivity of traditional algorithms and reducing the large number of false positives. The UNet network was improved to reduce the complexity of a deep neural network while maintaining its sensitivity. Although the two-dimensional detection and segmentation methods have made great progress compared with the traditional methods, a CT image is a three-dimensional image sequence, and the lung organs of the human body are not based on a two-dimensional plane, so making inferences about a three-dimensional object from single plane cuts is often not objective or specific enough. The artificial reduction of one dimension of information often results in low recall rates and high false positives. This undoubtedly brings a lot of unnecessary work to physicians. Therefore, because lung organs and lung nodules are three-dimensional objects, a three-dimensional convolutional neural network is required to further improve the detection and classification accuracy.
(2) Methods based on the 3D convolutional neural network. Aiming at the characteristics of lung nodules in a three-dimensional space and its variability in shape, Zhu et al. [6] considered the three-dimensionality of lung CT data and the compactness of the dual-path network, and designed two deep three-dimensional DPNs for the nodules: Detection and classification. Specifically, a three-dimensional fast convolutional neural network area (R-CNN) is designed for nodule detection. This modified R-CNN uses a three-dimensional dual-path block and UNet encoding and decoding structure to effectively learn nodule characteristics. This method makes full use of a lung nodule's spatial information and integrates feature maps with different abstract levels to repair the lost features, making the detection accuracy higher. Gong et al. [7] proposed an automatic computer-aided detection scheme for lung nodules based on deep convolutional neural networks (DCNNs). A three-dimensional dynamic neural network (SE-ResNet) based on a compressed excitation network and residual network was used to detect lung nodules and reduce false positives. Specifically, by fusing the 3D-SE-ResNet module to design a three-dimensional area suggestion network, with a UNet network structure to detect candidate lung nodules, the 3D-SE-ResNet module recalibrates the residual characteristic response of the channel to enhance the network. The model uses the 3D-SE-ResNet module to effectively learn the characteristics of nodules and improve nodule detection performance. Although this method detects lung nodules in three dimensions and can make full use of the spatiality and completeness of CT image sequences, it does not fully consider the adhesion of lung nodules and surrounding tissues, resulting in inaccurate segmentation.

Considering the influence of the surrounding tissues of lung nodules on the segmentation of lung nodules, as well as the diversification of lung nodules, this paper proposes a CT image lung nodule segmentation method based on the 3D convolutional neural network. Compared with other methods, the main difference is the 3D feature extraction method. By making full use of three-dimensional spatial information, our network can learn more feature information than two-dimensional networks. We also transform the Res2Net module into a 3D-Res2Net module and integrate it into the 3D-UNet network and design a 3D-Res2UNet network for candidate nodule detection and segmentation. Therefore, the network takes CT image sequences as input, can make full use of the spatial information of lung nodules, and minimizes artificial loss of dimensionality. First of all, in the designed network, the improved Res2Net network module can make the overall network deeper, effectively solving the problems of gradient explosion and gradient disappearance that are prone to deep networks. Secondly, the 3D-UNet network acts as the basic network, which can ensure its effectiveness while restricting the size of the feature map and repairing the lost features. Experiments have shown that, compared with the existing methods, the method in this paper can detect lung nodules more accurately and segment them effectively.

This article combines the Res2Net module with 3D-UNet and has achieved good results on small targets. This structure is not only suitable for the detection of lung nodules, but also for the detection of small particles and irregular objects in other fields. Thus, this research has laid the foundation for future applied work.

2. Related Work

2.1. UNet Segmentation Network

UNet [8] is a semantic segmentation network developed based on a fully convolutional neural network. The network has a total of 23 layers, and the number of layers is far less than the other networks while ensuring accuracy. The UNet network mainly includes two parts, down-sampling and up-sampling. Down-sampling is also called the feature extraction part, which mainly uses the convolutional and pooling layer to extract features of the input image. Up-sampling uses a deconvolution operation to up-sample the feature map. This structure of down-sampling and up-sampling is also called a decoder-encoder structure. In the down-sampling part, the input image passes through the convolutional and pooling layer to obtain feature maps of different levels. These feature maps contain image features with different levels of abstraction. In the up-sampling part, the deconvolution layer is used to gradually restore the size of the feature map, and the down-sampled feature map is merged to repair the less abstract detail information lost in the training process and improve the segmentation accuracy of the network.

However, because the lung is a three-dimensional structure, the UNet network uses two-dimensional convolution and pooling operations to extract the features of lung nodules in lung CT images, which will cause a lot of spatial information to be lost. Thus, a lot of contextual information is lost in the down-sampling process. It cannot be fully restored during up-sampling, which leads to fuzzy up-sampling results and insensitiveness to the details of the image. Combined with the above problems, it requires a three-dimensional network for further optimization.

2.2. 3D-UNet Segmentation Network

3D-UNet [9] is an improved semantic segmentation network based on UNet, since a lot of data in the field of medical imaging are three-dimensional data. Therefore, if you directly use the UNet network to process images, you need to perform slice preprocessing on the 3D data first. That is to divide the three-dimensional image into multiple layers of two-dimensional data, and then divide the two-dimensional data. This method is not only cumbersome and inefficient, but also loses the dimensional information of the three-dimensional data. This leads to the loss of correlation between adjacent slices, making the network accuracy low. The 3D-UNet network mainly includes an encoder part and a decoder part. Additionally, in the encoder part, a 3D convolution layer and a 3D pooling layer are used to extract the expressive features of the input image, and in the decoder part, 3D deconvolution is used to restore the feature map size. Moreover, the feature maps of the decoder and the encoder are merged through the cascade operation, which brings richer semantic information to the segmentation. This makes the segmentation accuracy higher.

3. Method

Since the lung CT image is a three-dimensional tomogram, and lung nodules are small in size, have a variable morphology, and have rich semantic information around the lung nodules (such as blood vessels and bronchus), the 3D segmentation of lung nodules is an extremely difficult research problem. However, these disadvantages can be mitigated by the Res2Net module because of its ability to capture the characteristics of tiny particles. In this work, the module is converted from 2D to 3D, which optimizes the 3D-UNet network to establish a 3D-Res2UNet network. This ensures that the entire network can be more accurate to improve image segmentation of lung nodules.

3.1. 3D-Res2Net

On the one hand, with the development of deep learning, deep neural networks have made a series of achievements in image classification tasks. This relies on the ability of deep neural networks to capture features more effectively. The number of these features can be increased by stacking more layers, which shows that this method is effective to establish connections within the network through a multi-layer end-to-end approach. On the other hand, in order to make the neural network achieve better learning results, the increasing number of network layers has brought some problems to the training of the network, of which gradient disappearance and gradient explosion are most important.

A residual network (ResNet) [10] effectively solves this problem. The residual learning unit introduces identity mapping to establish a direct correlation channel between the output and input; therefore, the parameterized layer can learn the input and output, and the residual difference between them achieves the purpose of protecting the integrity of the information and simplifies the goal and speed of learning.

For visual tasks, different scale information contains different features, so it is very important to be able to express features on multiple scales. Considering the general applicability of the 3D network in the future, this paper proposes to add a channel to the original Res2Net [11] module to upgrade it to the 3D level. The 3D-Res2Net module is shown in Figure 1.

3D-Res2Net changes the internal structure of the basic residual module. Compared with the basic residual unit, the new module replaces the original filter with multiple sets of $3 \times 3 \times 3$ filters and combines different filter groups. The residual cascade is connected to construct a new layered residual connection in a single residual block. A 3D-SE block is added, after the last 1×1 convolution, to each channel to reallocate weight. As shown in Figure 1, the 3D-Res2Net module uses a new size. It is the number of feature groups with scale = 4 so that the feature map sent to this structure is converted to eight channels after the $1 \times 1 \times 1$ convolution. In addition, x_1 represents the feature map with channel numbers 1, 2, and x_2 represents the feature map with channel numbers 3 and 4; in this way, if the channels are grouped and then trained, the weight of the convolution kernel trained by each grouped channel is also different. Compared with the weight of each channel convolution kernel caused by non-group training, this technique shows a greater advantage. Although the computational load of the Res2Net module is similar to the residual network architecture, the hierarchically connected network has strong multi-scale feature extraction capabilities, which enables the network to express multi-scale features at a finer granularity, and increases the network capacity of each layer. The output formula is as follows:

$$y\begin{cases} x_i \; i = 1 \\ K_i(x_i) \; i = 2 \\ K_i(x_i + y_{i-1}) \; 2 < i \leq s \end{cases} \quad (1)$$

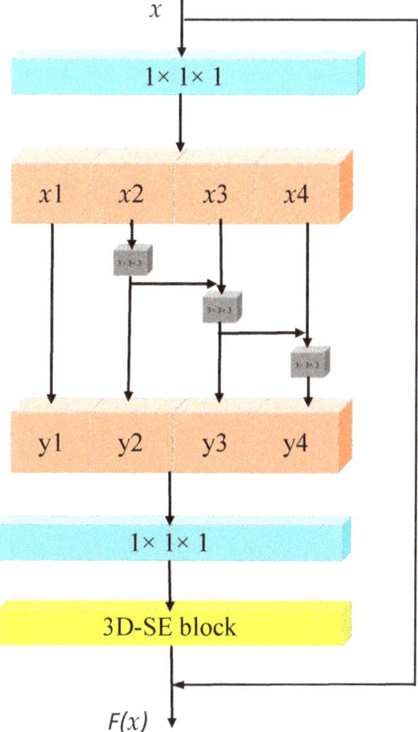

Figure 1. Three-dimensional residual network (3D-Res2Net) module.

Here, s is the number of groups that Res2Net integrates the 3×3 volumes of the original residual module into groups. Although a similar external jump connection is performed inside a bottleneck, the concat operation is performed on y_1, y_2, y_3, and y_4, so that the channel remains unchanged. The direct mapping of the first x_1 to y_1 is based on two considerations. The first is to reduce network parameters, and the second is to reuse features.

3.2. Network Design

This article combines the 3D-Res2Net module with the 3D-UNet network to form a new network structure, namely 3D-Res2UNet. The network structure diagram is shown in Figure 2.

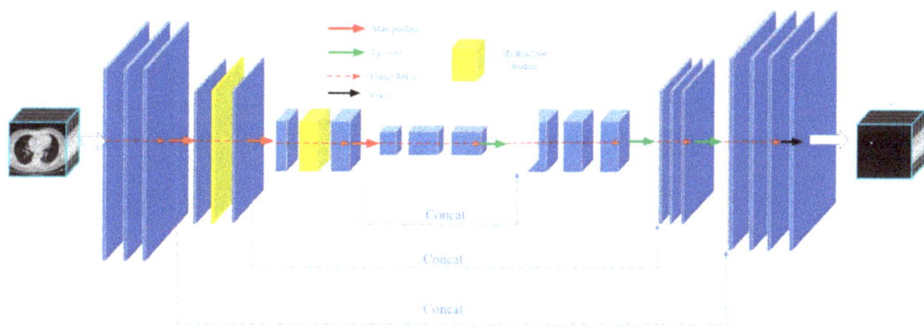

Figure 2. Three-dimensional residual UNet (3D-Res2UNet) network structure diagram.

Down-sampling is the most important process of extracting features in a network, and it can extract feature details well, which has an important significance for subsequent segmentation operations. Compared with the 3D-UNet convolutional neural network, the 3D-Res2UNet convolutional neural network changes the original CBR module in the two down-samplings, namely Conv + BN + ReLU, and replaces the 3D basic convolution module with the 3D-Res2Net module as a whole. Therefore, the method of integrating the residual network into the entire network can make the network deeper, and because of the strong feature extraction ability of the 3D-Res2Net module and the phased update of the convolution kernel weights of each channel, the overall network is able to capture very small features. With an improved performance, the connection method where x_1 is directly mapped to y_1 in the module effectively reduces network parameters and the resource consumption. Therefore, the network will not affect the network speed due to excessive redundancy.

4. Experience and Results

4.1. Dataset

This method is validated on the public dataset of LUNA16 [12]. The dataset contains a total of 888 sets of CT images with a total of 1186 lung nodules. During the experiment, the data was randomly divided into three parts, 70% of the data was used for training, 20% of the data was used for testing, and 10% of the data was used for verification. The criterion for judging a nodule in the LUNA16 dataset is that at least three of the four radiologists determine that the radius of the nodule is greater than 3 mm. Therefore, in the annotations of the dataset, non-nodules, nodules with a radius less than 3 mm, and nodules with a radius greater than 3 mm are considered by one or two radiologists as unrelated findings.

4.2. Data Preprocessing

4.2.1. Data Format Conversion

In the dataset, each chest CT sample mainly contains two parts: (1) The raw data of chest CT, including two files with the suffix zraw and mhd. The zraw file saves the original CT data and the mhd file saves the header file information of the CT data. The most important information is establishing an axis origin and pixel spacing. The origin represents the coordinates of the origin of the CT data in the world coordinate system, while the spacing subdivides the CT data to a length of a pixel in the world coordinates. (2) A CSV (Comma Separated Values) file with nodule annotation information. The content of the file is shown in Table 1.

Table 1. Labeling information of pulmonary nodules in a CSV file.

Seriesuid	CoordX	CoordY	CoordZ	Diameter (mm)
LKDS-00539	−70.7218023	−87.5886319	36.5	9.4028638
LKDS-00540	−50.2048319	−90.0324701	50.5	7.9106295

Since the CT data are expressed in a natural coordinate system, the world coordinates of the lung nodules need to be converted to the natural coordinates of the CT voxels to correspond to the CT data before the actual operation. The conversion formulas are as follows:

$$voxelcoord = \frac{(coord - origin)}{spacing} \quad (2)$$

$$voxeldiam = \frac{daimeter\_mm}{spacing} \quad (3)$$

The terms voxelcoord and voxeldiam, respectively represent the coordinates and diameter of the lung nodule in the CT pixel coordinate system, the terms coord and diameter_mm, respectively represent the coordinates and diameter of the nodule in the world coordinate system.

4.2.2. Lung Parenchymal Segmentation

Since the lung tissue is relatively complex, there are a large number of lung trachea, pulmonary blood vessels, tissue mucosa, and other structures around the lung nodules. As a result, there are some inevitable errors in the direct use of the original slice image to segment the lung nodules. If the data are preprocessed first, the lung parenchyma are segmented before the detection step. The lung nodules are detected in the lung parenchyma, which can avoid the interference of the external tissues and organs of the lung parenchyma on the detection task, thereby improving the detection accuracy. The lung CT image is shown in Figure 3.

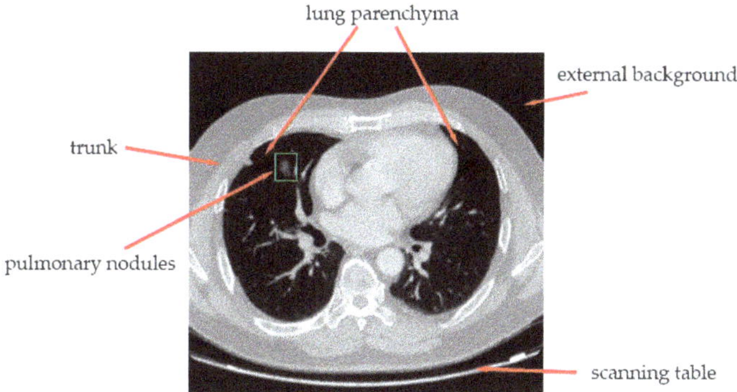

Figure 3. CT image of lung.

In this paper, the method proposed by Mansoor et al. [13] is used to segment the lung parenchyma. After the lung is segmented, the lung parenchyma is accurately segmented to remove the influence of surrounding tissues. This has a positive effect on lung nodule segmentation. Since the lung is a 3D model, the segmentation is divided into multiple horizontal layers, and the overall segmentation result is shown in Figure 4.

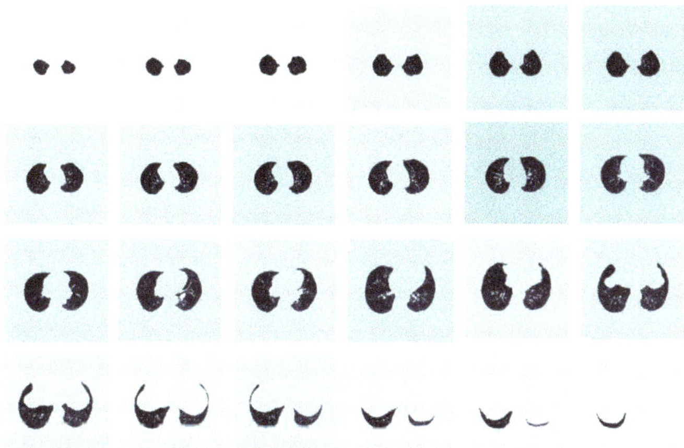

Figure 4. Results of lung parenchymal segmentation.

4.2.3. Voxel Value Normalization

X-rays are used to scan the lungs when collecting data in CT. Since different tissues of the human body have different X-ray absorption characteristics, different ray attenuations will occur when X-rays pass by, and the remaining X-rays after attenuation will be converted into a digital signal to acquire the desired CT value. The CT values of different human tissues are shown in Table 2.

Table 2. CT values of different human tissues.

Name of Organization	HU Values
Air	−1000
Lung	−500
Fat	−100~−50
Water	0
Aorta	35~50
Kidney	40~60
Bones	150~3000

Table 2 shows that the CT value of the lung is at −500, so the HU value of the CT sample is intercepted as [−1000, +400]. The [−1000, +400] is normalized to a [0, 1] range. This helps reduce the influence of other scalings and enhances the ability of neural networks to capture features.

4.2.4. Data Enhancement

Medical image data are essential for training the model. The severe imbalance of positive and negative samples in the dataset will affect the network performance. As a result, the network weight cannot reach the optimal value. Therefore, the AugGAN network [14] proposed by Huang SW et al. is used for data enhancement. It can solve the problem of insufficient positive samples to a certain extent. This also helps slow down the occurrence of model overfitting. In addition, the detection accuracy of lung nodules can also be improved. The resulting lung nodules are shown in Figure 5.

  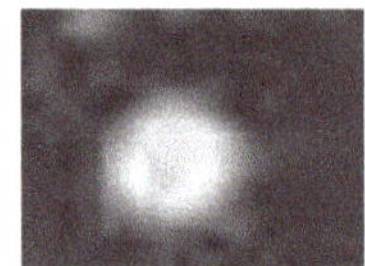 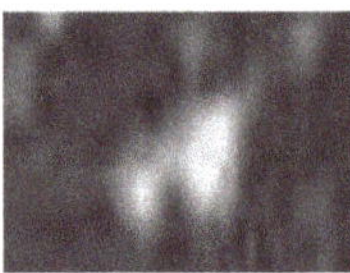

Figure 5. Lung nodules generated by the adversarial network.

4.3. *Evaluation Standard*

This article evaluates the article method from two perspectives, namely the dice index recall rate (Recall) and the average number of false positives per sample (FP/scan).

The dice index [15] refers to the degree of fit between the original target and the segmented target. The more the two objects fit, the higher the dice value and the lower the loss function value, indicating that the network segmentation of the lung nodules is more complete. The dice coefficient formula is as follows:

$$dice = \frac{2N_{truepositive}}{2N_{truepositive} + N_{falsepositive} + N_{falsenegative}} \tag{4}$$

The relationship between the dice coefficient and loss coefficient is as follows:

$$loss = 1 - dice \tag{5}$$

Here, $N_{truepositive}$ represents the area where the lung nodule exists and is correctly segmented, $N_{falsepositive}$ represents the area where the lung nodule exists but is not correctly segmented, and $N_{falsenegative}$ represents the area where the lung nodule does not exist and is not segmented. When the invalid dice coefficient is close to 1, the loss function loss is infinitely close to 0. At this time, the model segmentation result matches the real result.

The Recall indicator refers to the sensitivity of the network to lung nodules, describing the ability to segment lung nodules. The larger the value, the more the lung nodules found through the network are complete. Among them, N'_{real} is the number of real nodules detected by the network, and N_{nodule} is the number of real nodules in the sample. The formula is as follows:

$$Recall = \frac{N'_{real}}{N_{nodule}} \tag{6}$$

The average number of false positives (FP/scan) describes the network's ability to judge lung nodules, whether it can effectively avoid vascular tomography or lung tissue, and accurately distinguish lung nodules, where N'_{no} is the network detected. The number of non-nodules, N_{sample} is the total number of training samples, and the formula is as follows:

$$FP/scan = \frac{N'_{no}}{N_{sample}} \tag{7}$$

4.4. Experimental Results

The method in this paper is tested on the LUNA16 public dataset. There are many types of lung nodules in the sample, which are divided into three categories according to the nodule's density, e.g., solid nodules, mixed nodules, and ground glass nodules. The variable density structure can fully detect the ability of network segmentation and avoid the occurrence of contingency.

Figure 6 shows the results of the detection of lung nodules by the 3D-Res2UNet neural network and the overall effect of segmentation, and Figure 7 shows the local effect of lung nodule segmentation. In the local renderings, it can be found that the lung nodules are completely segmented. Either ellipse-like smooth edges or prominent feature edges can be accurately segmented by the 3D-Res2UNet neural network. This is not only helpful to assist doctors in the diagnosis and treatment, but also lays the foundation for the follow-up study of false positive detection of lung nodules.

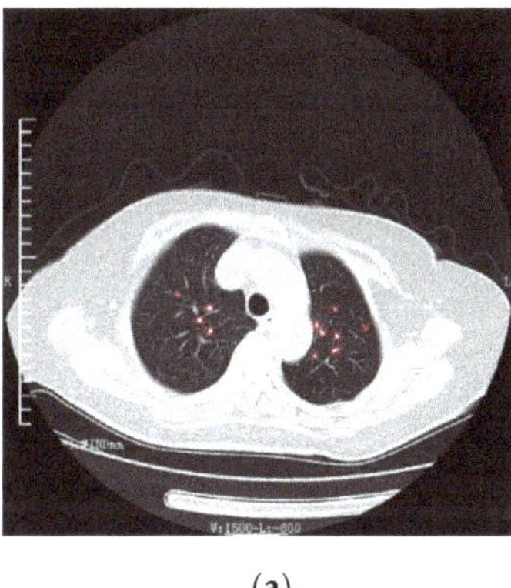

(a)

(b)

Figure 6. The detection results of lung nodules. (**a**) Results of lung nodule detection. (**b**) The overall effect of lung nodule segmentation.

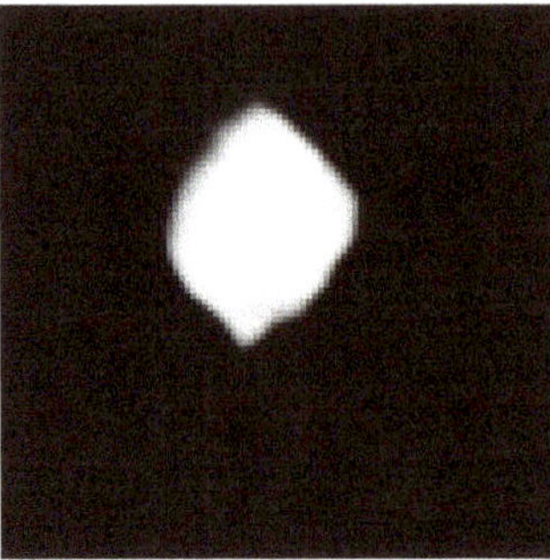

Figure 7. Local effect diagram of lung nodule segmentation.

In addition, Figure 8 lists some images whose segmentation is not very accurate. This is because the shape of such lung nodules is different from most lung nodules. Its edges are rough and jagged. Such lung nodules require a deeper network structure when segmenting the edges. However, a deeper network will inevitably lead to an increase in neural network training time and waste of resources. In fact, the ultimate purpose of segmentation of lung nodules is to assist doctors in the diagnosis, but the false positive probability of such nodules is relatively high. Therefore, the practical value of deepening the network for a more precise segmentation of such lung nodules is very small.

Figure 8. Image of lung nodules whose segmentation effect is not very accurate.

Model Comparison

The experiment first compared the ability of 3D-Res2UNet and the original network to segment and fit lung nodules. The comparison parameters are shown in Table 3.

Table 3. The dice coefficient of different networks.

Network Name	Dice (%)
UNet	81.32
3D-UNet	89.12
3D-UNet+fully CRF [16]	93.25
3D-Res2UNet (Ours)	95.30

Due to the three-dimensional CT tomogram of the lungs, as shown in the table, the 3D network has obvious advantages in capturing lung nodules compared with a 2D network. Since the 3D-Res2net module fused in this paper uses multiple sets of 3×3 filters, and different filter sets are connected by the residual cascade, a new structure is constructed in a single residual block. The hierarchical residual class is connected and finally a 3D-SE block is added to re-assign a weight to each channel. Such hierarchical filtering and gradual fusion of details make the 3D-Res2Unet network more sensitive to the small edges often found in lung nodules; thus, this method can more accurately restore the original shape of lung nodules during segmentation. The final dice coefficient is significantly better than the other networks by 95.30%, as shown in Figure 9, where the abscissa is the number of epochs and the ordinate is the dice coefficient.

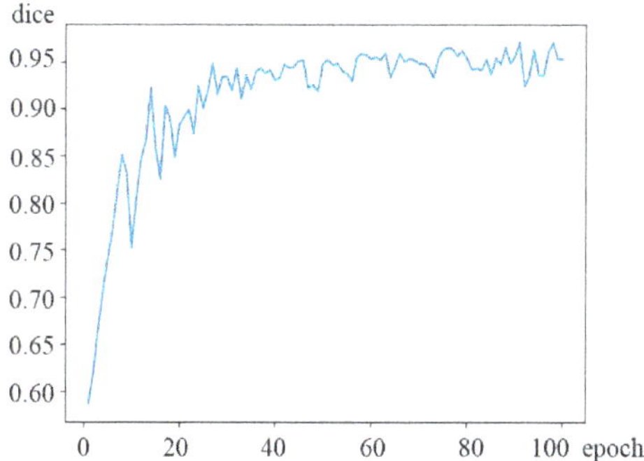

Figure 9. 3D-Res2UNet network dice coefficient.

Secondly, in terms of recall rate and the average number of false positive lung nodules, this article compares traditional and existing methods, as shown in Table 4.

Table 4. Comparison of detection algorithms.

Algorithm	Recall (%)	Number of False Positive Lung Nodules/CT
ISICAD [17]	85.7	329.3
ETROCAD [18]	92.2	333.0
DIAG_CONVENT [19]	93.3	269.0
LUNA16_V1 [20]	94.4	622.0
Dou [21]	97.1	219.1
LUNA16_V2 [20]	98.3	850.2
3D-Res2UNet (Ours)	99.1	276.3

In Table 4, ISICAD [17] is a traditional image processing method that uses artificially designed features to detect lung nodules and processes them according to the edge shape of lung nodules. This method mainly focuses on the main characteristics of large nodules. The eature extraction of mixed nodules and ground glass nodules is insufficient, so the overall recall rate is relatively low. LUNA16_V1 [20], as the official method, is representative of traditional image processing methods. It integrates the advantages of all previous traditional methods. This method has a high sensitivity to high-density nodules. The method works best on solid nodules and other large nodules. LUNA16_V2 [20] is an upgraded version based on LUNA16_V1. In addition to focusing on high-density large nodules, the network also has targeted designs for low-density mixed nodules and ground glass nodules. However, the impact on other lung tissues and the indiscriminate detection of vascular cross-sections has led to a sharp increase in the number of false positive lung nodules, which has caused great trouble for doctors when reading the film. A three-dimensional fully convolutional neural network designed by Dou et al. [21] can effectively detect the lung nodule memory. However, because the network does not have an up-sampling process, it cannot effectively repair the degree of abstraction lost in the training process due to the low detail information. The method in this paper can effectively restore the lost information. At the same time, this method can also extract detailed information many times to significantly improve the recall rate.

Finally, since the down-sampling process is a gradual amplification process, this article adds the 3D-Res2Net module to the second and third down-sampling of the 3D-UNet network to form symmetry in order to capture more target information. This effectively prevents the network from not capturing enough detail because the target is too large in the first down-sampling. What is more, it also avoids the target area of being too limited in the last down-sampling. In order to prove its rationality, the following comparative experiments were done. The comparison of experimental correlation coefficients is shown in Table 5. What is more, the dice coefficient variation curve is shown in Figure 10. Among them, the upper corner is marked as the position where the 3D-Res2Net module was added to the down-sampling.

Table 5. Comparison of the parameters of the 3D-Res2Net module in different positions.

Name	Dice	Recall	Number of False Positive Lung Nodules/CT
3D-Res2UNet ^{1,2}	92.4	97.2	320.7
3D-Res2UNet ^{1,3}	92.7	96.6	351.5
3D-Res2UNet ^{2,4}	93.50	98.5	300.1
3D-Res2UNet ^{3,4}	94.52	98.8	330.4
3D-Res2UNet ^{2,3}	95.30	99.1	276.3

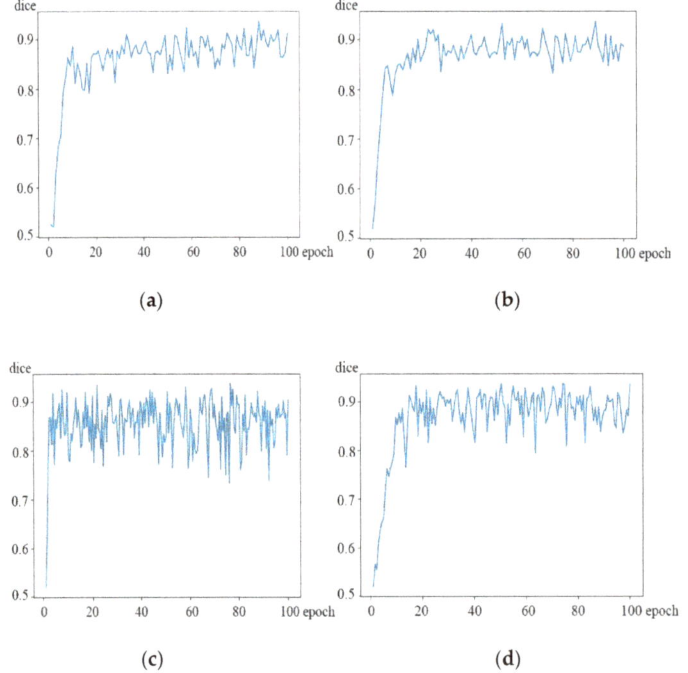

Figure 10. 3D-Res2UNet contrast network dice coefficient. (**a**) Describe the dice coefficient of 3D-Res2UNet ^{1,2}, (**b**) describe the dice coefficient of 3D-Res2UNet ^{1,3}, (**c**) describe the dice coefficient of 3D-Res2UNet ^{2,4}, (**d**) describe the dice coefficient of 3D-Res2UNet ^{3,4}.

Figure 11 shows the changes in the recall rate caused by the 3D-Res2Net module that is being added to different positions. Among them, the recall rate is the best when the module is added to the second and third down-sampling.

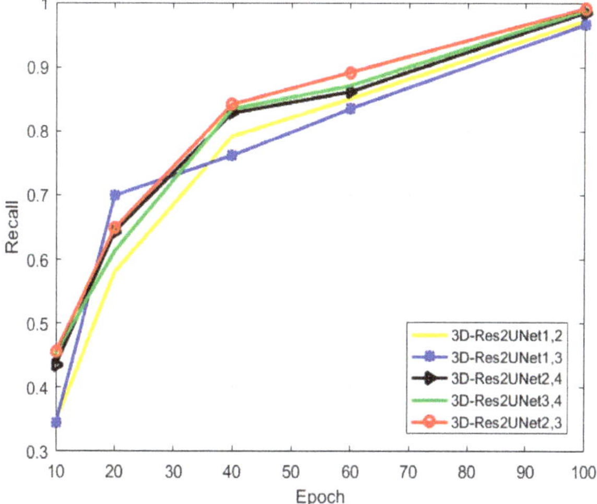

Figure 11. Comparison of recall rates of the 3D-Res2Net module added to different positions.

The method proposed in this paper has a dice coefficient of 95.30%, a recall rate of 99.1%, and a false positive nodule: 276.3 false positive nodules appear on each CT sample. These results are significantly better than the other design methods in terms of segmentation fit and recall.

5. Discussion

This article quantitatively analyzes the performance of the network through experiments. Firstly, for the situation of pulmonary nodules with variable shapes and different sizes, we chose the Res2Net module for improvement. The module is upgraded from two- to three-dimensional. This is because the 3D-Res2UNet network uses the means of decomposing and then fusing the input to effectively perform the multi-layer feature extraction. It helps improve the segmentation accuracy of the network, and it can more accurately distinguish the edges of lung nodules from other irrelevant tissues. Secondly, adding the 3D-Res2Net module twice during down-sampling is to be able to complement information at multiple levels. With the continuous enlargement of the feature map, the multi-layer detection network can progressively capture the missing target points in the previous layer to achieve better results.

The network proposed in this paper has achieved good results in the segmentation of lung nodules, and has made great progress in the segmentation of small nodules. This method can be effectively applied to other fields related to small detection and segmentation tasks. However, this method also has limitations, including high training costs and high false positives. This is also the direction of continued efforts in the future. It can be expected that 3D detection and segmentation methods will have better development in the future.

6. Conclusions

This paper proposes a CT image lung nodule segmentation method based on 3D-UNet and Res2Net. The 3D-Res2UNet neural network improves the training speed of the model while making the segmentation method more complete. Before the experiment, the preparation of and preprocessing methodology applied to the experimental data are introduced. After the experiment, the experimental results of the method in this paper are shown. The experimental results are compared with an original basic network, an existing same-task network, and the method presented in this paper. After testing, the method in this paper demonstrated better results than the other methods in terms of the dice coefficient and recall rate. The method in this paper also has a good overall performance in lung nodule segmentation.

Author Contributions: Conceptualization, Z.X.; validation, Z.X. and B.L.; investigation, L.G.; resources, F.Z.; writing—original draft preparation, B.L.; writing—review and editing, Y.L.; supervision, Y.L.; project administration, F.Z. All authors have read and agreed to the published version of the manuscript.

Funding: This work was supported by Tianjin Science and Technology Major Projects and Engineering (No. 17ZXSCSY00060, No. 17ZXHLSY00040) and the Program for Innovative Research Team in the University of Tianjin (No. TD13-5034).

Conflicts of Interest: The authors declare no conflict of interest.

References

1. Siegel, R.L.; Miller, K.D.; Jemal, A. Cancer statistics, 2019. *CA Cancer J. Clin.* **2019**, *69*, 7–34. [CrossRef] [PubMed]
2. Carvalho, A.O.; Sampaio, W.B.; Silva, A.C.; de Paiva, A.C.; Nunes, R.A.; Gattass, M. Automatic detection of solitary lung nodules using quality threshold clustering, genetic algorithm and diversity index. *Artif. Intell. Med.* **2014**, *60*, 165–177. [CrossRef] [PubMed]
3. Messay, T.; Hardie, R.C.; Rogers, S.K. A new computationally efficient CAD system for pulmonary nodule detection in CT imagery. *Med Image Anal.* **2010**, *14*, 390–406. [CrossRef] [PubMed]
4. Ding, J.; Li, A.; Hu, Z. Accurate pulmonary nodule detection in computed tomography images using deep convolutional neural networks. In Proceedings of the International Conference on Medical Image Computing and Computer-Assisted Intervention (MICCAI 2017), Quebec City, QC, Canda, 10–14 September 2017; pp. 559–567.
5. Deng, Z.; Chen, X. Pulmonary nodule detection algorithm based on deep convolutional neural network. *J. Comput. Appl.* **2019**, *39*, 2019–2115.
6. Zhu, W.; Liu, C.; Fan, W.; Xie, X. Deeplung: Deep 3D dual path nets for automated pulmonary nodule detection and classification. In Proceedings of the 18th IEEE Winter Conference on Applications of Computer Vision, Lake Tahoe, Stateline, NV, USA; 2017; pp. 673–681.
7. Gong, L.; Jiang, S.; Yang, Z.Y.; Zhang, G.; Wang, L. Automated pulmonary nodule detection in CT images using 3D deep squeeze-and-excitation networks. *Int. J. Comput. Assist. Radiol. Surg.* **2019**, *14*, 1969–1979. [CrossRef]
8. Ronneberger, O.; Fischer, P.; Brox, T. U-Net: Convolutional Networks for Biomedical Image Segmentation. In Proceedings of the International Conference on Medical Image Computing and Computer-Assisted Intervention, Munich, Germany, 5–9 October 2015; 9351, pp. 234–241. Available online: https://arxiv.org/abs/1505.04597v1 (accessed on 11 September 2020).
9. Çiçek, Ö.; Abdulkadir, A.; Lienkamp, S.S.; Brox, T.; Ronneberger, O. 3D U-Net: Learning Dense Volumetric Segmentation from Sparse Annotation. In Proceedings of the International Conference on Medical Image Computing and Computer Assisted Intervention, Athens, Greece, 17–21 October 2016; pp. 424–432. [CrossRef]
10. He, K.; Zhang, X.; Ren, S.; Sun, J. Deep Residual Learning for Image Recognition. In *Proceedings of the IEEE Conference on Computer Vision and Pattern Recognition*; IEEE: New York, NY, USA, 2016; pp. 770–778. Available online: http://arXiv.org/pdf/1512.03385 (accessed on 15 September 2020).
11. Gao, S.; Cheng, M.M.; Zhao, K.; Zhang, X.Y.; Yang, M.H.; Torr, P.H. Res2Net: A New Multi-scale Backbone Architecture. In *IEEE Transactions on Pattern Analysis and Machine Intelligence*; IEEE: New York, NY, USA, 2019. [CrossRef]
12. Armato, S.G.; Mclennan, G.; Bidautl, L.; McNitt-Gray, M.F.; Meyer, C.R.; Reeves, A.P.; Zhao, B.; Aberle, D.R.; Henschke, C.I.; Hoffman, E.A.; et al. The lung image database consortium (LIDC) and image database resource initiative (IDRI): A completed reference database of lung nodules on CT scans. *Med. Phys.* **2011**, *38*, 915–931. [CrossRef] [PubMed]
13. Mansoor, A.; Bagci, U.; Xu, Z.; Foster, B.; Olivier, K.N.; Elinoff, J.M.; Suffredini, A.F.; Jayaram, K.U.; Mollura, D.J. A generic approach to pathological lung segmentation. *IEEE Trans. Med. Imaging* **2014**, *33*, 2293–2310. [CrossRef] [PubMed]

14. Huang, S.W.; Lin, C.T.; Chen, S.P.; Wu, Y.-Y.; Hsu, P.-H.; Lai, S.-H. AugGAN: Cross Domain Adaptation with GAN-based Data Augmentation. In *ECCV 2018: European Conference on Computer Vision*; Springer: Cham, UK, 2018; pp. 731–744. [CrossRef]
15. Dice, L.R.; Dice, L.R. Measures of the Amount of Ecologic Association Between Species. *Ecology* **1944**, *26*, 297–302. [CrossRef]
16. Hou, T.X.; Zhao, J.J.; Qiang, Y. Pulmonary nodules segmentation based on CRF 3D-UNet structure. *Comput. Eng. Des.* **2020**, *41*, 1663–1669.
17. Murphy, K.; Ginneken, B.V.; Schilham, A.M.R.; de Hoop, B.J.; Gietema, H.A.; Prokop, M. A large-scale evaluation of automatic pulmonary nodule detection in chest CT using local image features and k-nearest-neighbour classification. *Med. Image Anal.* **2009**, *13*, 757–770. [CrossRef] [PubMed]
18. Tan, M.; Deklerck, R.; Jansen, B.; Bister, M.; Corneli, J. A novel computer-aided lung nodule detection system for CT images. *Med. Phys.* **2011**, *38*, 5630–5645. [CrossRef] [PubMed]
19. Setio, A.A.A.; Ciompi, F.; Litjens, G.; Gerke, P.; Jacobs, C.; van Riel, S.J.; Winkler, M.M.; Naqibullah, M.; Sánchez, C.I.; van Ginneken, B. Pulmonary nodule detection in CT images: False positive reduction using multi-view convolutional networks. *IEEE Trans. Med. Imaging* **2016**, *35*, 1160–1169. [CrossRef] [PubMed]
20. Lung Nodule Analysis 2016. Available online: https://luna16.grand-challenge.org (accessed on 18 January 2016).
21. Dou, Q.; Chen, H.; Jin, Y. Automated pulmonary nodule detection via 3D convnets with online sample filtering and hybrid-loss residual learning. In Proceedings of the International Conference on Medical Image Computing and Computer-Assisted Intervention (MICCAI 2017), Quebec City, QC, Canada, 10–14 September 2017; pp. 630–638.

Article

A Chaotic Hybrid Butterfly Optimization Algorithm with Particle Swarm Optimization for High-Dimensional Optimization Problems

Mengjian Zhang [1], Daoyin Long [2], Tao Qin [1] and Jing Yang [1,3,*]

1 Electrical Engineering College, Guizhou University, Guiyang 550025, China; gs.mjzhang18@gzu.edu.cn (M.Z.); tqin@gzu.edu.cn (T.Q.)
2 Power China Guizhou Engineering CO. LTD, Guiyang 550001, China; yj.china@163.com
3 Guizhou Provincial Key Laboratory of Internet + Intelligent Manufacturing, Guiyang 550025, China
* Correspondence: jyang7@gzu.edu.cn

Received: 23 September 2020; Accepted: 27 October 2020; Published: 30 October 2020

Abstract: In order to solve the problem that the butterfly optimization algorithm (BOA) is prone to low accuracy and slow convergence, the trend of study is to hybridize two or more algorithms to obtain a superior solution in the field of optimization problems. A novel hybrid algorithm is proposed, namely HPSOBOA, and three methods are introduced to improve the basic BOA. Therefore, the initialization of BOA using a cubic one-dimensional map is introduced, and a nonlinear parameter control strategy is also performed. In addition, the particle swarm optimization (PSO) algorithm is hybridized with BOA in order to improve the basic BOA for global optimization. There are two experiments (including 26 well-known benchmark functions) that were conducted to verify the effectiveness of the proposed algorithm. The comparison results of experiments show that the hybrid HPSOBOA converges quickly and has better stability in numerical optimization problems with a high dimension compared with the PSO, BOA, and other kinds of well-known swarm optimization algorithms.

Keywords: butterfly optimization algorithm (BOA); particle swarm optimization (PSO); cubic map; nonlinear; high dimension

1. Introduction

The butterfly optimization algorithm (BOA) was proposed by Arora and Singh in 2018 [1]. The method and concept of this algorithm was proposed [2] firstly at the 2015 International Conference on Signal Processing, Computing and Control (2015 ISPCC). After the algorithm was proposed, the authors have performed many studies on BOA. Arora and Singh [3] proposed an improved butterfly optimization algorithm with ten chaotic maps for solving three engineering optimization problems. Arora and Singh [4] proposed a new hybrid optimization algorithm which combines the standard BOA with Artificial Bee Colony (ABC) algorithm. Arora and Singh [5] used the BOA to solve the node localization in wireless sensor networks and compared the results with the particle swarm optimization (PSO) algorithm and firefly algorithm (FA). Arora et al. [6] proposed a modified butterfly optimization algorithm for solving the mechanical design optimization problems. Singh and Anand [7] proposed a novel adaptive butterfly optimization algorithm, which a novel phenomenon of changing the sensory modality of the basic BOA. Sharma and Saha [8] proposed a novel hybrid algorithm (m-MBOA) to enhance the exploitation ability of BOA with the help of the mutualism phase of symbiosis organisms search (SOS). Yuan et al. [9] proposed an improved butterfly optimization algorithm, which is employed for optimizing the system performance that is analyzed based on annual cost, exergy and energy efficiencies, and pollutant emission reduction. Li et al. [10] proposed an improved BOA for engineering design problems using the cross-entropy method. A hybrid intelligent predicting model was proposed for exploring household CO^2 emission mitigation strategies derived from BOA [11]. Tan et al. [12] proposed an improved BOA to solve the wavelet neural networks problem based on solutions for elliptic partial differential equations. Malisetti and Pamula [13] proposed a novel BOA based on quasi opposition for the problem of cluster head selection in wireless sensor network (WSNs). Sharma et al. [14] proposed a bidirectional butterfly optimization algorithm for solving the engineering optimization problems. Above the studies of BOA, which are improvement research or applied research, there is only one paper for a hybrid algorithm with ABC and BOA.

In addition, concerning the optimization algorithms that were proposed, these are mainly divided into three categories according to their principles, and the famous meta-heuristic algorithm mainly including evolutionary algorithms: Genetic Algorithm (GA) [15,16], Differential Evolution (DE) [17]; swarm intelligence algorithms: Particle Swarm Optimization (PSO) [18], Ant Colony Optimization (ACO) [19], and Artificial Bee Colony (ABC) algorithm [20]; physics-based algorithms: Gravitational Search Algorithm (GSA) [21], Sine Cosine Algorithm (SCA) [22], and Henry Gas Solubility Optimization (HGSO) algorithm [23]. In the past ten years, scholars have proposed many new swarm intelligence optimization algorithms, which are based on the behavior of animals in nature and also named nature-inspired heuristic algorithms, such as Bat-Inspired Algorithm (BA) [24], Krill herd (KH) [25], Fruit Fly Optimization Algorithm (FOA) [26], Grey Wolf Optimizer (GWO) [27], Moth-flame optimization (MFO) algorithm [28], Whale Optimization Algorithm (WOA) [29], Salp Swarm Algorithm (SSA) [30], Grasshopper Optimization Algorithm (GOA) [31], and Marine Predators Algorithm (MPA) [32]. For more details, the reader can refer to the papers [33–35], where the recent and popular algorithms are well reviewed.

The research status of the hybrid algorithm of different intelligent optimization algorithms and PSO algorithm are introduced. Zhen et al. [36] proposed a new memetic algorithm called shuffled particle swarm optimization (SPSO), which combines the PSO with the shuffled frog leaping algorithm (SFLA). Niu and Li [37] proposed a new hybrid global optimization algorithm PSODE combining PSO with DE. Lai and Zhang [38] proposed a novel hybrid algorithm, which combines PSO and GA, and the experiment for 23 benchmark problems was also presented. Mirjalili and Hashim [39] proposed a new hybrid PSOGSA algorithm for function optimization. Wang et al. [40] proposed a hybrid algorithm based on krill herd and quantum-behaved particle swarm optimization (QPSO) for benchmark and engineering optimization. Trivedi et al. [41] proposed a novel hybrid PSO-DA algorithm, which combined the PSO algorithm with the dragonfly algorithm (DA) for global numerical optimization. Trivedi et al. [42] proposed a novel PSOWOA for the global numerical optimization problems. Laskar et al. [43] proposed a new hybrid HWPSO algorithm for electronic design optimization problems according to the studies of hybrid algorithms with PSO and other meta-heuristic algorithms. In addition, the structure of PSO algorithm and BOA has certain similarities, and it is meaningful for a novel hybrid algorithm of PSO with BOA to be studied.

For the research of chaotic theory and chaotic attractors of nonlinear control systems, a general polynomial function was derived for Hopf controlling bifurcations using nonlinear state feedback by Xu and Chen [44]. Xu et al. [45] analyzed the n-scroll chaotic attractors of modified Chua's circuit and proved the chaos of the Chua system. Yu and Lü [46] studied three-dimensional chaotic systems for Hopf controlling bifurcations in detail. In addition, Yu et al. [47] used the inverse trigonometric function, $\tan^{-1}(x)$, to obtain one-, two-, and three-directional multiscroll integer and fractional order chaotic attractors, and they analyzed stabilization of the chaotic system with the application of chaos theory in the improvement of swarm intelligent optimization algorithms [48,49], and it has been recognized by scholars in the field.

In order to improve the ability of the algorithm for high-dimensional optimization problems that we proposed, the method for hybrid the meta-heuristic algorithms, which combines the basic PSO and BOA, and the chaotic theory, is also used in the improved method. In addition, the control parameter of the power exponent a in BOA is also analyzed in detail, and a nonlinear control strategy is proposed for adjusting the ability of the global search and local search capabilities of the improved algorithm.

The rest of this paper is organized as follows: Section 2 presents the basic BOA model. The basic PSO model is presented in Section 3. In Section 4, a novel HPSOBOA algorithm is proposed, and three improved strategies are also introduced in detail. Section 5 illustrates the experimental results on 26 high-dimensional optimization problems and the comparison results of two experiments are also introduced in detail. Finally, conclusion and future studies are summarized in Section 6.

2. The Basic Butterfly Optimization Algorithm (BOA)

The nature-inspired meta-heuristic algorithm is proposed, named BOA [1,2], which simulates the foraging and mating behavior of the butterfly. One of the main characteristics of BOA different from other meta-heuristics is that each butterfly has its own unique scent. The fragrance can be formulated as follows:

$$f_i = cI^a \tag{1}$$

where f_i is the perceived magnitude of fragrance, c represents the sensory modality, and I is the stimulus intensity, and a represents the power exponent based on the degree of fragrance absorption.

Theoretically any value of the sensory morphology coefficient c in the range $[0,\infty]$ can be taken. However, its value is determined by the particularity of the optimization problem in the iterative process of the BOA. The sensory modality c in the optimal search phase of the algorithm can be formulated as follows:

$$c_{t+1} = c_t + [0.025/(c_t \cdot T_{max})] \tag{2}$$

where T_{max} is the maximum number of iterations of the algorithm, and the initial value of parameter c is set to 0.01.

In addition, there are two key steps in the algorithm, they are, respectively, global search phase and local search phase. The mathematical model of the butterflies' global search movements can be formulated as follows:

$$x_i^{t+1} = x_i^t + \left(r^2 \times g_{best} - x_i^t\right) \times f_i \tag{3}$$

where x_i^t denotes the solution vector x_i of the ith butterfly in t iteration and r means a random number in [0,1]. Here, g_{best} is the current best solution found among all the solutions in the current stage. Particularly, f_i represents the fragrance of the ith butterfly. The local search phase can be formulated as follows:

$$x_i^{t+1} = x_i^t + \left(r^2 \times x_i^k - x_j^t\right) \times f_i \tag{4}$$

where x_j^t and x_i^k are jth and kth butterflies chosen randomly from the solution space. If x_j^t and x_i^k belong to the same iteration, it means that the butterfly becomes a local random walk. If not, this kind of random movement will diversify the solution.

Both global and local searches for food and a mating partner by the butterfly in nature can occur. Therefore, a switch probability p is set to convert the normal global search and the intensive local search. In each iteration, the BOA randomly generates a number in [0,1], which is compared with switch probability p to decide whether to conduct a global search or local search.

3. The Basic Particle Swarm Optimization (PSO) Model

PSO algorithm [18] is based on the swarm of birds moving for searching food in a multidimensional search space. The position and velocity are the important characteristics of PSO, which are used to find the optimal value.

Each individual is called a particle, and each particle is first initialized with random position and velocity within the search space. The position of the best global particle in the optimal solution is as follows:

$$v_i^{t+1} = w \cdot v_i^t + c_1 \cdot rand_1 \times \left(p_{best} - x_i^t\right) + c_2 \cdot rand_2 \times \left(g_{best} - x_i^t\right) \tag{5}$$

$$x_i^{t+1} = x_i^t + v^{t+1} \tag{6}$$

where v_i^{t+1} and v^{t+1} represent the velocity of ith particle at iteration number (t) and (t + 1). Usually, $c_1 = c_2 = 2$, $rand_1$, and $rand_2$ are the random numbers in (0, 1). The w can be calculated as:

$$w(t) = w^{max} - \frac{\left(w^{max} - w^{min}\right) \cdot T_i}{T_{max}} \tag{7}$$

where $w^{max} = 0.9$, and $w^{min} = 0.2$, and T_{max} represents the maximum number of iterations.

4. The Proposed Algorithm

In this section, a novel hybrid algorithm is proposed, and the initialization of BOA by a cubic one-dimensional map is introduced, and a nonlinear parameter control strategy is also performed. In addition, the PSO algorithm is hybridized with BOA in order to improve the basic BOA for global optimization.

4.1. Cubic Map

Chaos is a relatively common phenomenon in nonlinear systems. The basic cubic map [50] can be calculated as follows:

$$z_{n+1} = \alpha z_n^3 - \beta z_n \tag{8}$$

where α and β represent the chaos factors, and when β in (2.3, 3), the cubic map is chaotic. When $\alpha = 1$, the cubic map is in the interval $(-2, 2)$, and the sequence in $(-1, 1)$ with $\alpha = 4$. The cubic map can also be:

$$z_{n+1} = \rho z_n\left(1 - z_n^2\right) \tag{9}$$

where the ρ is control parameter. In Equation (8), the sequence of the cubic map is in (0, 1), and when $\rho = 2.595$, the chaotic variable z_n generated at this time has better ergodicity. A graphical presentation of the cubic map for 1000 iterations is in Figure 1.

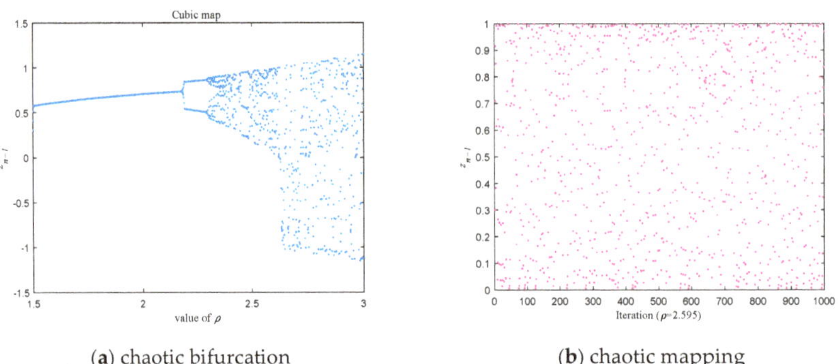

(**a**) chaotic bifurcation (**b**) chaotic mapping

Figure 1. Visualization of implemented cubic map with ρ in (1.5, 3) and $\rho = 2.595$, respectively.

In Figure 1, it can be seen that the chaotic map can distribute the population of butterflies to the random value in the interval (0, 1) during the search phase.

We propose the cubic map to initialize the position of the algorithm, and in order to ensure that the initialized interval is in (0, 1), the z (0) of cubic map is set to 0.315 in the proposed algorithm.

4.2. Nonlinear Parameter Control Strategy

From Equations (1), (3), and (4), we can see that the power exponent a plays an important role in BOA's ability to find the best optimization. When $a = 1$, it means that no scent is absorbed—that is, the scent emitted by a specific butterfly is perceived by other butterflies—which means that the search range will be narrowed and the local exploration ability of the algorithm will be improved. When $a = 0$, it means that the fragrance emitted by any butterfly cannot be perceived by other butterflies, so the group will expand the search range—that is, improve the global exploration ability of the algorithm. However, $a = 0.1$ in basic BOA, and taking a as a fixed value cannot effectively balance the global and local search capabilities. Therefore, we propose a nonlinear parameter control strategy as:

$$a(t) = a_{first} - \left(a_{first} - a_{final}\right) \cdot \sin\left(\frac{\pi}{\mu}\left(\frac{t}{T_{max}}\right)^2\right) \tag{10}$$

where a_{first} and a_{final} represent the initial value and final value of parameter a, μ is tuning parameter, and T_{max} represents the maximum number of iterations. In this paper, $\mu = 2$, $T_{max} = 500$, $a_{first} = 0.1$, and $a_{final} = 0.3$.

It can be seen from Figure 2a that for the intensity indicator coefficient a, the nonlinear control strategy based on the sine function proposed in this paper has a larger slope in the early stage, which can speed up the algorithm's global search ability. The mid-term slope is reduced, which is convenient for entering a local search. The later slope is gentle to allow the algorithm to search for the optimal solution. Therefore, it can effectively balance the global search and local search capabilities of the algorithm.

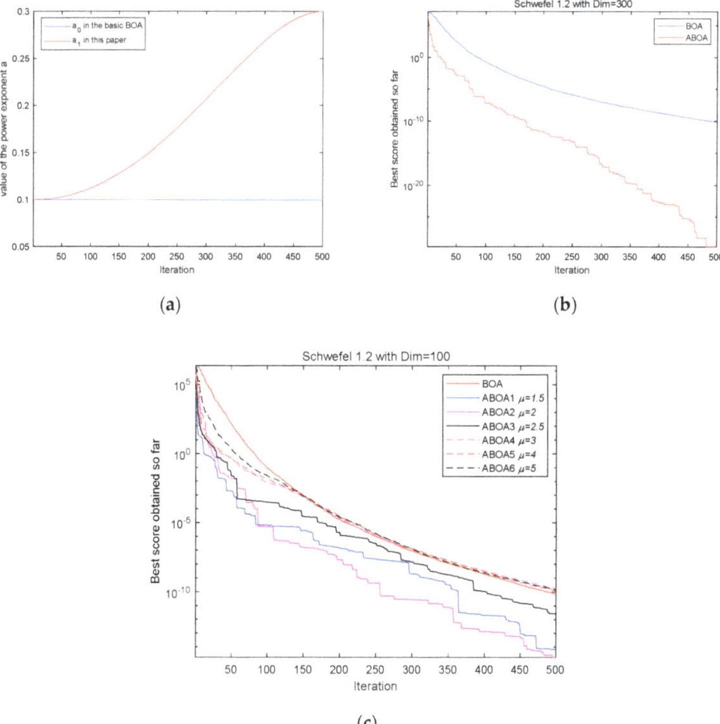

Figure 2. Variation curve of different intensity coefficients and convergence curve of test function. (**a**) Two control parameter strategies, (**b**) Convergence curve of Schwefel 1.2, (**c**) Convergence curve of Schwefel 1.2 with Dim = 100 for different parameter values setting.

From Figure 2, It can be seen from (b) that the convergence curve of improved BOA with the nonlinear parameter control strategy is better than the basic BOA in the optimal test of Schwefel 1.2 function. The curve has many turning points, indicating that the improved algorithm has the ability to jump out of the global optimum from Figure 2b.

The results of the main controlling parameter μ values of parameter a are shown in Figure 2c. As the value of parameter μ increases, the effect of the improvement strategy gradually worsens. It can be seen from (c) that the convergence curve of improved BOA with $\mu = 2$ is best in the seven convergence curves. When $\mu \geq 4$, the convergence curve is worse than the original BOA.

4.3. Hybrid BOA with PSO

In this section, a novel hybrid PSOBOA is proposed, which is a combination of separate PSO and BOA. The major difference between PSO and BOA is how new individuals are generated. The drawback of the PSO algorithm is the limitation to cover a small space for solving high-dimensional optimization problems.

In order to combine the advantages of the two algorithms, we combine the functionality of both algorithms and do not use both algorithm one after another. In other words, it is heterogeneous because of the method involved to produce the final results of the two algorithms. The hybrid is proposed as follow:

$$V_i^{t+1} = w \cdot V_i^t + C_1 \cdot r_1 \times \left(p_{best} - X_i^t\right) + C_2 \cdot r_2 \times \left(g_{best} - X_i^t\right) \tag{11}$$

where $C_1 = C_2 = 0.5$, and w can be also calculated by Equation (7), r_1 and r_2 are the random number in $(0, 1)$.

$$X_i^{t+1} = X_i^t + V^{t+1} \tag{12}$$

In addition, the mathematical model of the global search phase and local search phase in the basic BOA, which can be calculated by Equations (3) and (4). However, the global search phase of the hybrid PSOBOA can be formulated as follows:

$$X_i^{t+1} = w \cdot X_i^t + \left(r^2 \times g_{best} - w \cdot X_i^t\right) \times f_i \tag{13}$$

The local search phase of the hybrid PSOBOA can be formulated as follows:

$$X_i^{t+1} = w \cdot X_i^t + \left(r^2 \times X_i^k - w \cdot X_j^t\right) \times f_i \tag{14}$$

where X_i^k and X_j^t are jth and kth butterflies chosen randomly from the solution space, respectively.

The pseudo-code of hybrid PSOBOA is shown in Algorithm 1.

Algorithm 1. Pseudo-code of hybrid PSO with BOA (PSOBOA)

```
Generate the initialize population of the butterflies X_i (i = 1, 2, ..., n) randomly
Initialize the parameter r1, r2, C1 and C2
Define senser modality c, power exponent a and switch probability p
Calculate the fitness value of each butterflies
While t = 1: the max iterations
    For each search agent
        Update the fragrance of current search agent by Equation (1)
    End for
    Find the best f
    For each search agent
        Set a random number r in [0,1]
        If r < p then
            Move towards best position by Equation (13)
        Else
            Move randomly by Equation (14)
        End if
    End for
    Update the velocity using Equation (11)
    Calculate the new fitness value of each butterflies
        If f_new < best f
```

Update the position of best f using Equation (12)
End if
Update the value of power exponent a
$t = t + 1$
End while
Return the best solution and its fitness value

4.4. The Proposed HPSOBOA

In order to combine the advantages of the three improvement strategies proposed in this paper, a novel hybrid HPSOBOA is proposed in this section, which is a combination of the cubic map for the initial population, nonlinear parameter control strategy of power exponent a, PSO algorithm, and BOA.

The pseudo-code of novel HPSOBOA is shown in Algorithm 2.

Algorithm 2. Pseudo-code of novel HPSOBOA

Generate the initialize population of the butterflies X_i $(i = 1, 2, \ldots, n)$ using cubic map
Initialize the parameter r_1, r_2, C_1 and C_2 and switch probability p
Define senser modality c and the initial value of power exponent a
Calculate the fitness value of each butterflies
While $t = 1$: the max iterations
For each search agent
Update the fragrance of current search agent by Equation (1)
End for
Find the best f
For each search agent
Set a random number r in [0,1]
If $r < p$ then
Move towards best position by Equation (13)
Else
Move randomly by Equation (14)
End if
End for
Update the velocity using Equation (11)
Calculate the new fitness value of each butterflies
If $f_{new} <$ best f
Update the position of best f using Equation (12)
End if
Update the value of power exponent a using Equation (10)
$t = t + 1$
End while
Output the best solution

5. Experiments and Comparison Results

In this section, we choose the 26 high-dimensional test functions from CEC benchmark functions, and the name, range, type, and theoretical optimal value of the test functions are shown in Table 1. Then, two experiments are performed with ten algorithms, including improved BOA, novel BOAs in this paper, and other swarm algorithms or natural science-based algorithms. The performance of experiment 1 was compared through experimental data, which were compared with six algorithms by six benchmark functions in dimensions 100 and 300, respectively. Then, the performance of experiment 2 was, respectively, compared with the ten algorithms by 26 high-dimensional test functions in $Dim = 30$. Finally, the statistical methods were conducted, and the boxplots for the 30 times fitness of 26 test functions were also compared.

Table 1. High-dimensional test functions.

Name	Formula of Functions	*Dim*	Range	Type	$f_{\min}$
Sphere	$F_1(x) = \sum_{i=1}^{Dim} x_i^2$	30	[−100,100]	U	0
Schwefel 2.22	$F_2(x) = \sum_{i=1}^{Dim} \lvert x_i \rvert + \prod_{i=1}^{Dim} \lvert x_i \rvert$	30	[−10,10]	U	0
Schwefel 1.2	$F_3(x) = \sum_{i=1}^{Dim} \left(\sum_{j=i}^{i} x_j \right)^2$	30	[−100,100]	U	0
Schwefel 2.21	$F_4 = \max\{\lvert x_i \rvert, 1 \le i \le Dim\}$	30	[−10,10]	U	0
Step	$F_5 = \sum_{i=1}^{Dim} (x_i + 0.5)^2$	30	[−10,10]	U	0
Quartic	$F_6(x) = \sum_{i=1}^{Dim} Dim \cdot x_i^2 + rand(0,1)$	30	[−1.28,1.28]	U	0
Exponential	$F_7 = exp\left(0.5 \sum_{i=1}^{Dim} x_i\right)$	30	[−10,10]	U	0
Sum power	$F_8 = \sum_{i=1}^{Dim} \lvert x_i \rvert^{(i+1)}$	30	[−1,1]	U	0
Sum square	$F_9(x) = \sum_{i=1}^{Dim} \left(Dim \cdot x_i^2\right)$	30	[−10,10]	U	0
Rosenbrock	$F_{10}(x) = \sum_{i=1}^{Dim} \left(100\left(x_{i+1} - x_i^2\right) + (x_i - 1)^2\right)$	30	[−5,10]	U	0
Zakharov	$F_{11}(x) = \sum_{i=1}^{Dim} x_i^2 + \left(\sum_{i=1}^{Dim} 0.5ix_i\right)^2 + \left(\sum_{i=1}^{Dim} 0.5ix_i\right)^4$	30	[−5,10]	U	0
Trid	$F_{12}(x) = (x_i - 1)^2 + \sum_{i=1}^{Dim} i \cdot \left(2x_i^2 - x_{i-1}\right)^2$	30	[−10,10]	U	0
Elliptic	$F_{13}(x) = \sum_{i=1}^{Dim} \left(10^6\right)^{(i-1)/(Dim-1)} \cdot x_i^2$	30	[−100,100]	U	0
Cigar	$F_{14}(x) = x_1^2 + 10^6 \sum_{i=2}^{Dim} x_i^2$	30	[−100,100]	U	0
Rastrigin	$F_{15}(x) = \sum_{i=1}^{Dim} \left[x_i^2 - 10\cos(2\pi x_i) + 10\right]$	30	[−5.12,5.12]	M	0
NCRastrigin	$F_{16}(x) = \sum_{i=1}^{Dim} \left[y_i^2 - 10\cos(2\pi y_i) + 10\right]$, $y_i = \begin{cases} x_i, \lvert x_i \rvert < 0.5 \\ round(2x_i)/2, \lvert x_i \rvert > 0.5 \end{cases}$	30	[−5.12,5.12]	M	0

Table 1. *Cont.*

Name	Formula of Functions	Dim	Range	Type	$f_{\min}$
Ackley	$F_{17}(x) = -20\exp\left(-0.2\sqrt{\frac{1}{Dim}\sum_{i=1}^{Dim} x_i^2}\right) + exp\left(\frac{1}{Dim}\sum_{i=1}^{Dim}\cos(2\pi x_i)\right) + 20 + exp(1)$	30	[−50,50]	M	0
Griewank	$F_{18}(x) = \frac{1}{4000}\sum_{i=1}^{Dim} x_i^2 - \prod_{i=1}^{Dim}\cos\left(\frac{x_i}{\sqrt{i}}\right) + 1$	30	[−600,600]	M	0
Alpine	$F_{19}(x) = \sum_{i=1}^{Dim}\left\lvert x_i \cdot \sin(x_i) + 0.1x_i\right\rvert$	30	[−10,10]	M	0
Penalized 1	$F_{20}(x) = \frac{\pi}{Dim}\left\{\sum_{i=1}^{Dim-1}(y_i-1)^2\left[1+10sin^2(\pi y_{i+1})\right] + (y_{Dim-1})^2 + 10sin^2(\pi y_1)\right\} + \sum_{i=1}^{Dim} u(x_i, 10, 100, 4)$ $y_i = 1 + (x_i+1)/4, u_{y_i,a,k,m} = \begin{cases} k(x_i-a)^m, x_i > a \\ 0, -a \le x_i \le a \\ k(-x_i-a)^m, x_i < a \end{cases}$	30	[−100,100]	M	0
Penalized 2	$F_{21}(x) = \frac{1}{10}\left\{sin^2(\pi x_1) + \sum_{i=1}^{Dim-1}(x_i-1)^2\left[1+sin^2(3\pi x_{i+1})\right] + (x_{Dim-1})^2\left(1+sin^2(2\pi x_{i+1})\right)\right\}$ $+\sum_{i=1}^{Dim} u(x_i, 5, 100, 4)$	30	[−100,100]	M	0
Schwefel	$F_{22}(x) = \sum_{i=1}^{Dim}\left\lvert x_i \cdot \sin\left(\sqrt{\lvert x_i\rvert}\right)\right\rvert$	30	[−100,100]	M	0
Levy	$F_{23}(x) = sin^2(3\pi x_i) + \sum_{i=1}^{Dim-1}(x_i-1)^2\left[1+sin^2(3\pi x_{i+1})\right] + \lvert x_{Dim}-1\rvert \cdot \left[1+sin^2(2\pi x_{Dim})\right]$	30	[−10,10]	M	0
Weierstrass	$F_{24}(x) = \sum_{i=1}^{Dim}\left(\sum_{k=0}^{k_{max}}\left[a^k\cos\left(2\pi b^k(x_i+0.5)\right)\right]\right) - Dim\cdot\sum_{k=0}^{k_{max}}\left[a^k\cos\left(2\pi b^k \cdot 0.5\right)\right], a = 0.5, b = 3, k_{max} = 20$	30	[−1,1]	M	0
Solomon	$F_{25}(x) = 1 - cos\left(2\pi\sqrt{\sum_{i=1}^{Dim} x_i^2}\right) + 0.1\sqrt{\sum_{i=1}^{Dim} x_i^2}$	30	[−100,100]	M	0
Bohachevsky	$F_{26}(x) = \sum_{i=1}^{Dim}\left[x_i^2 + 2x_{i+1}^2 - 0.3\cdot cos(3\pi x_i)\right]$	30	[−10,10]	M	0

where U represents unimodal, and M represents multimodal.

5.1. Numerical Optimization Funtions and Experiments

The experiments were carried out on the same experimental platform. The results of all the algorithms were compared using MATLAB 2018a installed over Windows 10 (64 bit), Intel (R) Core (TM) i5-10210U, and @2.11G with 16.0GB of RAM.

5.1.1. The 26 Test Functions

The properties of unimodal and multimodal benchmark functions for numerical optimization, which are also high-dimensional test functions, are listed in Table 1, where *Dim* indicates the dimension of the function, and Range is the boundary of the function's search space. These functions are used to test the performance of the algorithms.

5.1.2. Experiment 1: Comparison with BOA, CBOA, PSOBOA, HPSOBOA, LBOA, and IBOA

In order to analyze the effectiveness of the improvement strategies proposed in this paper, the comparison experiment for BOA [1], CBOA, PSOBOA, HPSOBOA, LBOA [5], and IBOA [9] was designed for six high-dimensional functions from Table 1 with *Dim* = 100 and *Dim* = 100 as experiment 1. Additionally, there are three unimodal problems and three multimodal problems. The CBOA combines the basic BOA with the cubic map and nonlinear control strategy of the power exponent a. The hybrid PSOBOA just combines the basic BOA with PSO algorithm, the novel HPSOBOA is a combination of three improvement strategies in Section 4. In addition, two improved BOAs are also compared in this experiment; LBOA [5] was proposed by Arora and Singh, which was used in the improved algorithm to solve the node localization in wireless sensor networks in 2017. The IBOA [9] was proposed by Yuan et al., which was employed for optimizing the system performance that was analyzed based on annual cost, exergy and energy efficiencies, and pollutant emission reduction in 2019.

5.1.3. Experiment 2: Comparison with Other Swarm Algorithms

In order to prove the novel hybrid algorithm superior to other swarm algorithms, the experiment 2 was designed for 26 benchmark functions with *Dim* = 30. There are ten algorithms in this experiment, and we chose four swarm intelligence optimization algorithms besides the six algorithms in the experimental one. The four swarm algorithms including PSO [18], GWO [27], SCA [22], and MAP [32] were proposed in different years, and their principles are also different. The PSO and GWO algorithms simulate the behavior of animals in nature. The SCA is a physics-based algorithm, which moves towards the best solution using a mathematical model based on sine and cosine functions. The MPA is based on the widespread foraging strategy, namely Lévy and Brownian movements in ocean predators along with optimal encounter rate policy in the bio-logical interaction between predator and prey.

5.1.4. Performance Measures

In order to analyze the performances of the algorithms, three criteria of different swarm algorithms are considered, including the Mean (*Avg*), the Standard deviation (*Std*), and the Success Rate (*SR*). Here, we will use the Mean which is defined as:

$$Avg = \frac{1}{m}\sum_{i=1}^{m} F_i \tag{15}$$

where m is the number of optimization test runs, and F_i is the best fitness value.

The Standard deviation (*Std*) is defined as follows:

$$Std = \sqrt{\frac{1}{m}\sum_{i=1}^{m}(F_i - Avg)^2} \tag{16}$$

The Success Rate (*SR*) is defined as follows:

$$SR = \frac{m_{su}}{m_{all}} \times 100\% \tag{17}$$

where m_{all} is the total number of optimization test runs, and m_{su} is the times of the algorithm successfully reached to the specified value that $\varepsilon < 10^{-15}$ is called the specified value.

5.2. Comparison of the Parameter Settings of Ten Algorithms

In the experiments, ten comparison algorithms were selected, namely, BOA [1], CBOA, PSOBOA, HPSOBOA, LBOA [5], IBOA [9], PSO [18], GWO [27], SCA [22], and MPA [32]. The parameter settings of the ten algorithms are shown in Table 2. In addition, the population number of each algorithm is set to 30, and the max iteration is set to 500. Each algorithm is run for 30 times, and the Mean (Avg), Standard deviation (Std), Success Rate (SR), and Friedman rank [51] of the results are all taken in the two experiments.

Table 2. Parameter settings for algorithms.

NO.	Algorithms	Population Size	Parameter Settings
1	Butterfly Optimization Algorithm (BOA)	30	$a = 0.1, c(0) = 0.01, p = 0.6$
2	Butterfly Optimization Algorithm with Cubic map (CBOA)	30	$a_{first} = 0.1, a_{final} = 0.3, c(0) = 0.01, p = 0.6,$ $x(0) = 0.315, \rho = 0.295$
3	PSOBOA	30	$a = 0.1, c(0) = 0.01, p = 0.6, c_1 = c_2 = 0.5$
4	Hybrid PSO with BOA and Cubic map (HPSOBOA)	30	$a_{first} = 0.1, a_{final} = 0.3, c(0) = 0.01, p = 0.6,$ $x(0) = 0.315, \rho = 0.295, c_1 = c_2 = 0.5$
5	Butterfly Optimization Algorithm with Lévy flights (LBOA)	30	$a = 0.1, c(0) = 0.01, p = 0.6, \lambda = 1.5$
6	Improved Butterfly Optimization Algorithm (IBOA)	30	$a(0) = 0.1, c(0) = 0.01, p = 0.6, r(0) = 0.33,$ $\mu = 4$
7	Particle Swarm Optimization (PSO)	30	$c_1 = c_2 = 2, V_{max} = 1, V_{min} = -1,$ $\omega_{max} = 0.9, \omega_{min} = 0.2$
8	Grey Wolf Optimizer (GWO)	30	$a_{first} = 2, a_{final} = 0$
9	Sine Cosine Algorithm (SCA)	30	$a = 2, r_1(0) = 2$
10	Marine Predators Algorithm (MPA)	30	$a = 0.1, c(0) = 0.01, p = 0.6$

5.3. Results of Experiment 1

For the results of experiment 1, in order to analyze the robustness of the hybrid algorithm by three improved control strategies with other swarm intelligence algorithms, the convergence curves for six benchmark functions (*Dim* = 100) plots are shown in Figure 3.

It can be verified from the convergence curve that the proposed HPSOBOA converges faster than the other algorithms from Figure 3. The results show that the improved algorithm based on the three improvement strategies in this paper can effectively improve the convergence trend of the basic BOA when *Dim* = 100. From Figures 3 and 4a–f, it can be seen that the proposed HPSOBOA for those functions has a better convergence than the original BOA except the Schwefel 1.2 function when *Dim* = 300.

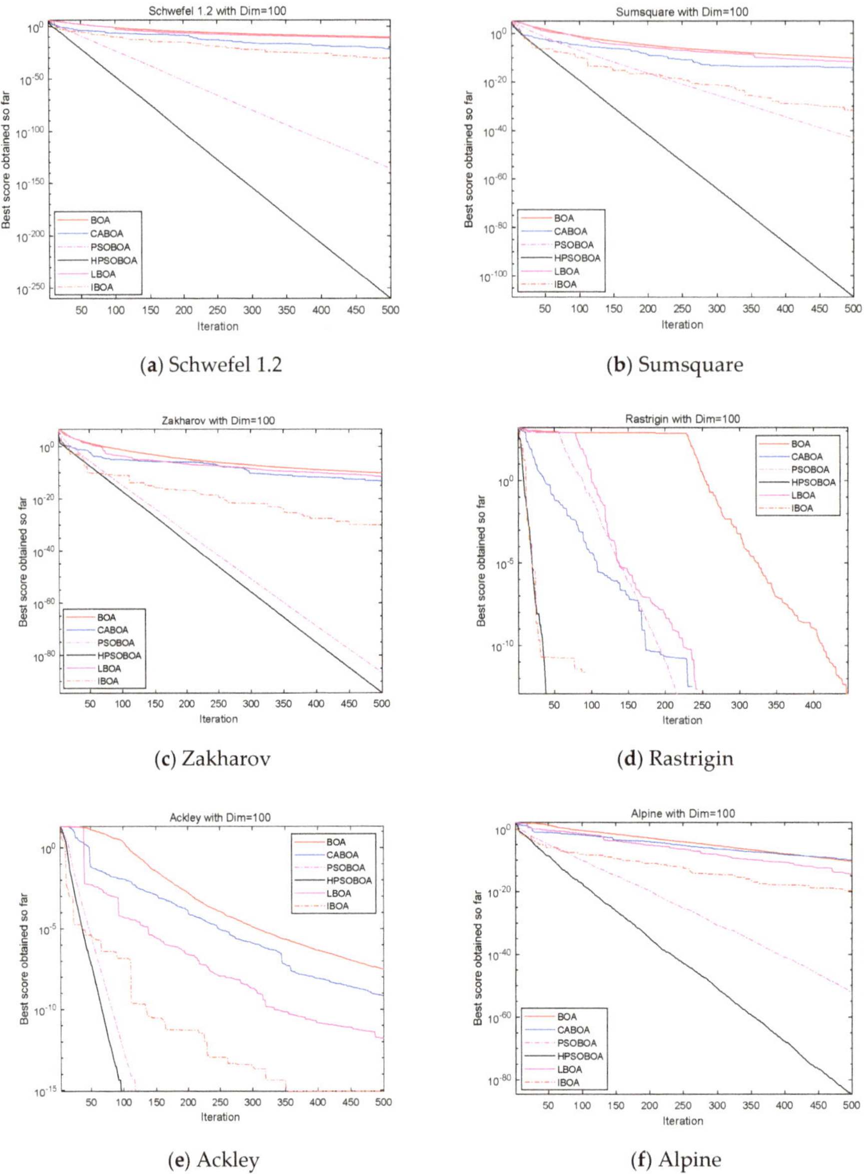

(**a**) Schwefel 1.2 (**b**) Sumsquare

(**c**) Zakharov (**d**) Rastrigin

(**e**) Ackley (**f**) Alpine

Figure 3. Convergence curve for six algorithms with Dim = 100; the six test functions' names are Schwefel 1.2, Sumsquare, Zakharov, Rastrigin, Ackley, and Alpine, respectively.

In order to analyze the robustness of the hybrid HPSOBOA by three improved control strategies with other five algorithms, the dimension of the six optimization problems is set to 300, and the convergence curves for six benchmark functions plots are shown in Figure 4.

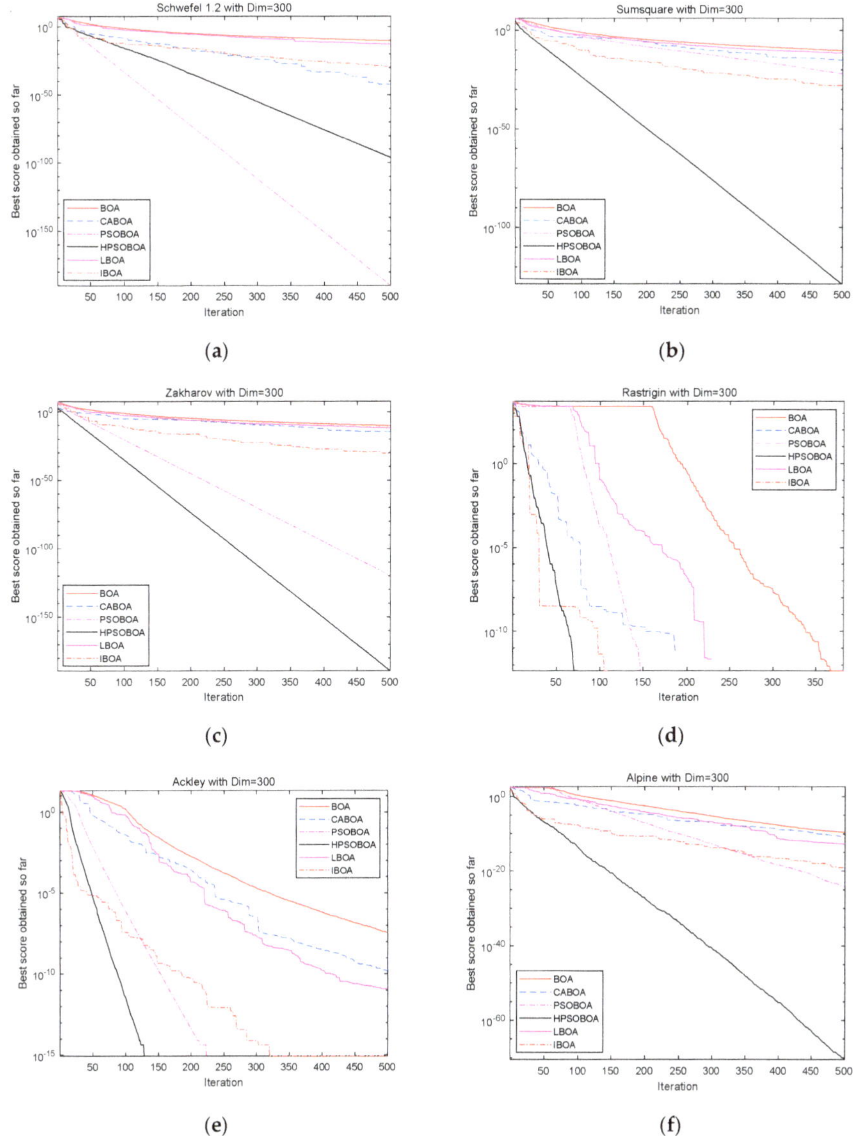

Figure 4. Convergence curve for six algorithms with $Dim = 300$; the six test functions' names are Schwefel 1.2, Sumsquare, Zakharov, Rastrigin, Ackley, and Alpine, respectively. (**a**) Schwefel 1.2, (**b**) Sumsquare, (**c**) Zakharov, (**d**) Rastrigin, (**e**) Ackley, (**f**) Alpine.

Figure 5 shows the box plots of optimization results of six high-dimensional problems by the six algorithms. The optimization result of the hybrid HPSOBOA is better than other algorithms from Figures 3–5.

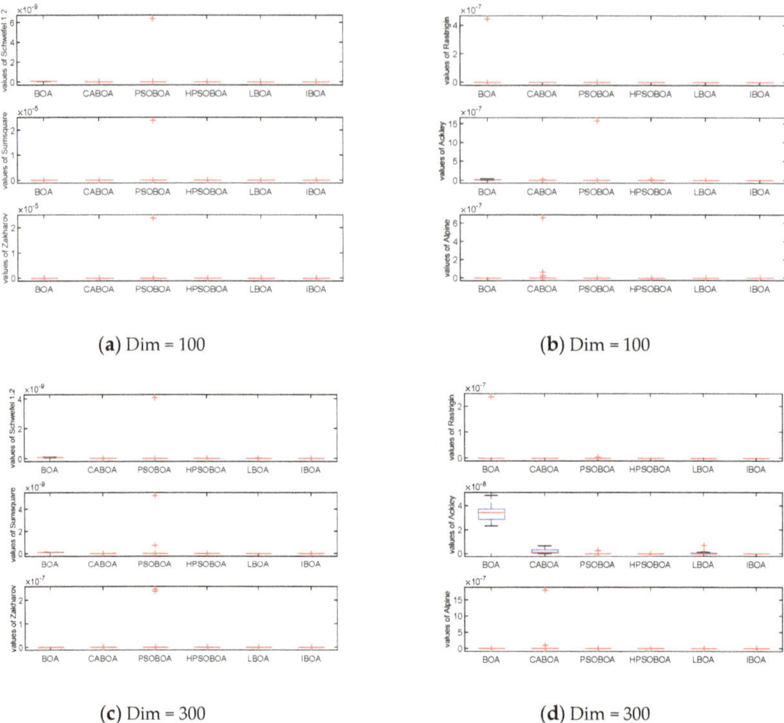

(**a**) Dim = 100 (**b**) Dim = 100

(**c**) Dim = 300 (**d**) Dim = 300

Figure 5. Boxplot for the 30 times fitness of six test functions with $Dim = 100$ and $Dim = 300$. (**a**) the three functions' names are Schwefel 1.2, Sumsquare and Zakharov with $Dim = 100$; (**b**) the three functions' names are Rastrigin, Ackley, and Alpine $Dim = 100$; (**c**) the three functions' names are Schwefel 1.2, Sumsquare, and Zakharov with $Dim = 300$; (**d**) the three functions' names are Rastrigin, Ackley, and Alpine $Dim = 300$.

In addition, statistical tests are essential to check significant improvements by novel algorithms over others, which were proposed. The Friedman rank test [51] was applied on the mean solutions, we used this method to compare the improved algorithms by different control strategies. The Avg-rank and overall rank are shown in Table 3. From the Friedman rank, the HPSOBOA outperforms all the comparison algorithms on six numerical optimization problems (Schwefel 1.2, Sumsquare, Zakharov, Rastrigin, Ackley, and Alpine), and the order of six algorithms with $Dim = 100$ is HPSOBOA > PSOBOA > IBOA > LBOA > CABOA > BOA. However, when the $Dim = 300$, the order of six algorithms is that HPSOBOA > PSOBOA > IBOA > CABOA > LBOA > BOA.

From the results of the analysis, we can see that although the order of HPSOBOA is better than others, the IBOA with chaotic theory for improving the control parameters also performed well. Thus, different one-dimensional chaotic maps can also have a good performance for improving the basic BOA.

5.4. Results of Experiment 2

In experiment 2, the performance of the proposed algorithm was compared with the other optimization algorithms using the 26 test functions with $Dim = 30$. The statistical results include the Mean (Avg), the Standard deviation (Std), the Success Rate (SR), Friedman rank test [51], and Wilcoxon rank-sum test [52] because the statistical test is a significance method to analyze the improved algorithm, and these comparison results are presented in Tables 4–8.

Table 3. Optimization comparison results for 100-dimensional and 300-dimensional functions.

Functions		BOA	CABOA	PSOBOA	HBOAPSO	LBOA	IBOA	BOA	CABOA	PSOBOA	HBOAPSO	LBOA	IBOA
		Dim = 100						Dim = 300					
Schwefel 1.2	Worst	8.23×10^{-11}	3.16×10^{-18}	6.40×10^{-9}	$\mathbf{2.89 \times 10^{-207}}$	1.67×10^{-11}	9.22×10^{-29}	9.21×10^{-11}	3.95×10^{-27}	4.06×10^{-9}	$\mathbf{2.65 \times 10^{-76}}$	2.56×10^{-11}	4.53×10^{-29}
	Best	5.68×10^{-11}	1.48×10^{-30}	4.03×10^{-287}	$\mathbf{7.12 \times 10^{-218}}$	6.56×10^{-14}	9.39×10^{-34}	6.14×10^{-11}	6.44×10^{-41}	4.93×10^{-285}	$\mathbf{4.57 \times 10^{-274}}$	3.27×10^{-13}	2.82×10^{-32}
	Avg	6.95×10^{-11}	1.12×10^{-19}	2.13×10^{-10}	$\mathbf{2.32 \times 10^{-207}}$	4.43×10^{-12}	4.34×10^{-30}	7.49×10^{-11}	1.32×10^{-28}	1.35×10^{-10}	$\mathbf{8.85 \times 10^{-78}}$	3.46×10^{-12}	2.73×10^{-30}
	Std	6.15×10^{-12}	5.76×10^{-19}	1.17×10^{-9}	$\mathbf{0.00 \times 10^{0}}$	4.29×10^{-12}	1.68×10^{-29}	7.44×10^{-12}	7.20×10^{-28}	7.41×10^{-10}	$\mathbf{4.85 \times 10^{-77}}$	4.86×10^{-12}	8.16×10^{-30}
	rank	5.97	3.97	1.97	**1.17**	4.97	2.97	5.97	2.80	1.87	**1.63**	4.97	3.77
	SR/%	0.00	100.00	96.67	**100.00**	0.00	100.00	0.00	100.00	96.67	**100.00**	0.00	100.00
Sumsquare	Worst	1.07×10^{-10}	2.33×10^{-12}	2.98×10^{-9}	$\mathbf{5.82 \times 10^{-20}}$	1.16×10^{-11}	1.35×10^{-29}	1.06×10^{-10}	2.50×10^{-12}	5.21×10^{-9}	$\mathbf{8.92 \times 10^{-24}}$	1.18×10^{-11}	4.98×10^{-30}
	Best	6.71×10^{-11}	4.45×10^{-19}	3.42×10^{-294}	$\mathbf{3.47 \times 10^{-294}}$	1.34×10^{-14}	9.55×10^{-34}	7.33×10^{-11}	1.00×10^{-16}	9.50×10^{-272}	$\mathbf{1.35 \times 10^{-292}}$	2.87×10^{-15}	5.72×10^{-33}
	Avg	8.63×10^{-11}	2.14×10^{-13}	1.01×10^{-10}	$\mathbf{1.94 \times 10^{-21}}$	3.20×10^{-12}	1.35×10^{-30}	8.95×10^{-11}	1.93×10^{-13}	1.98×10^{-10}	$\mathbf{2.97 \times 10^{-25}}$	3.03×10^{-12}	1.01×10^{-30}
	Std	8.78×10^{-12}	4.62×10^{-13}	5.43×10^{-10}	$\mathbf{1.06 \times 10^{-20}}$	2.80×10^{-12}	2.71×10^{-30}	8.84×10^{-12}	4.82×10^{-13}	9.56×10^{-10}	$\mathbf{1.63 \times 10^{-24}}$	3.64×10^{-12}	1.20×10^{-30}
	rank	5.97	3.93	1.90	**1.43**	4.93	2.83	5.93	3.90	1.70	**1.77**	4.90	2.80
	SR/%	0.00	43.33	93.33	**100.00**	0.00	100.00	0.00	46.67	90.00	**100.00**	3.33	100.00
Zakharov	Worst	1.11×10^{-10}	5.43×10^{-12}	2.38×10^{-5}	$\mathbf{2.28 \times 10^{-71}}$	1.95×10^{-11}	2.64×10^{-29}	1.03×10^{-10}	6.45×10^{-13}	2.45×10^{-7}	$\mathbf{2.74 \times 10^{-72}}$	2.04×10^{-11}	2.84×10^{-29}
	Best	5.70×10^{-11}	3.06×10^{-17}	2.14×10^{-294}	$\mathbf{4.25 \times 10^{-289}}$	7.01×10^{-15}	4.38×10^{-33}	6.88×10^{-11}	4.94×10^{-16}	7.09×10^{-293}	$\mathbf{4.43 \times 10^{-287}}$	4.80×10^{-14}	8.50×10^{-33}
	Avg	8.18×10^{-11}	5.01×10^{-13}	7.95×10^{-7}	$\mathbf{7.59 \times 10^{-73}}$	4.42×10^{-12}	1.57×10^{-30}	8.41×10^{-11}	9.96×10^{-14}	1.60×10^{-8}	$\mathbf{9.12 \times 10^{-74}}$	4.75×10^{-12}	3.43×10^{-30}

Table 3. *Cont.*

Functions		BOA	CABOA	PSOBOA	HBOAPSO	LBOA	IBOA	BOA	CABOA	PSOBOA	HBOAPSO	LBOA	IBOA
		Dim = 100						Dim = 300					
	Std	1.13×10^{-11}	1.21×10^{-12}	4.35×10^{-6}	$\mathbf{4.16 \times 10^{-72}}$	4.89×10^{-12}	4.85×10^{-30}	8.57×10^{-12}	1.61×10^{-13}	6.11×10^{-8}	$\mathbf{5.00 \times 10^{-73}}$	5.45×10^{-12}	6.09×10^{-30}
	rank	5.97	3.93	2.30	**1.07**	4.93	2.80	5.93	3.93	2.07	**1.47**	4.93	2.67
	SR/%	0.00	43.33	86.67	**100.00**	3.33	100.00	0.00	36.67	86.67	**100.00**	0.00	100.00
Rastrigin	Worst	4.44×10^{-7}	$\mathbf{0.00 \times 10^{0}}$	1.39×10^{-9}	$\mathbf{0.00 \times 10^{0}}$	$\mathbf{0.00 \times 10^{0}}$	$\mathbf{0.00 \times 10^{0}}$	2.36×10^{-7}	$\mathbf{0.00 \times 10^{0}}$	3.65×10^{-9}	$\mathbf{0.00 \times 10^{0}}$	$\mathbf{0.00 \times 10^{0}}$	$\mathbf{0.00 \times 10^{0}}$
	Best	0.00×10^{0}	$\mathbf{0.00 \times 10^{0}}$	0.00×10^{0}	$\mathbf{0.00 \times 10^{0}}$	$\mathbf{0.00 \times 10^{0}}$	$\mathbf{0.00 \times 10^{0}}$	0.00×10^{0}	$\mathbf{0.00 \times 10^{0}}$	0.00×10^{0}	$\mathbf{0.00 \times 10^{0}}$	$\mathbf{0.00 \times 10^{0}}$	$\mathbf{0.00 \times 10^{0}}$
	Avg	1.48×10^{-8}	$\mathbf{0.00 \times 10^{0}}$	4.65×10^{-11}	$\mathbf{0.00 \times 10^{0}}$	$\mathbf{0.00 \times 10^{0}}$	$\mathbf{0.00 \times 10^{0}}$	7.88×10^{-9}	$\mathbf{0.00 \times 10^{0}}$	1.22×10^{-10}	$\mathbf{0.00 \times 10^{0}}$	$\mathbf{0.00 \times 10^{0}}$	$\mathbf{0.00 \times 10^{0}}$
	Std	8.11×10^{-8}	$\mathbf{0.00 \times 10^{0}}$	2.55×10^{-10}	$\mathbf{0.00 \times 10^{0}}$	$\mathbf{0.00 \times 10^{0}}$	$\mathbf{0.00 \times 10^{0}}$	4.32×10^{-8}	$\mathbf{0.00 \times 10^{0}}$	6.67×10^{-10}	$\mathbf{0.00 \times 10^{0}}$	$\mathbf{0.00 \times 10^{0}}$	$\mathbf{0.00 \times 10^{0}}$
	rank	3.92	**3.40**	3.48	**3.40**	**3.40**	**3.40**	3.58	**3.47**	3.55	**3.47**	**3.47**	**3.47**
	SR/%	83.33	**100.00**	96.67	**100.00**	**100.00**	**100.00**	96.67	**100.00**	96.67	**100.00**	**100.00**	**100.00**
Ackley	Worst	3.23×10^{-8}	1.62×10^{-8}	1.58×10^{-6}	1.85×10^{-8}	5.46×10^{-10}	$\mathbf{8.88 \times 10^{-16}}$	4.86×10^{-8}	6.44×10^{-9}	2.67×10^{-9}	2.51×10^{-12}	7.15×10^{-9}	$\mathbf{8.88 \times 10^{-16}}$
	Best	1.59×10^{-9}	3.02×10^{-10}	8.88×10^{-16}	8.88×10^{-16}	4.44×10^{-15}	$\mathbf{8.88 \times 10^{-16}}$	2.30×10^{-8}	1.80×10^{-10}	8.88×10^{-16}	8.88×10^{-16}	4.49×10^{-13}	$\mathbf{8.88 \times 10^{-16}}$
	Avg	1.34×10^{-8}	3.17×10^{-9}	5.26×10^{-8}	6.38×10^{-10}	4.02×10^{-11}	$\mathbf{8.88 \times 10^{-16}}$	3.38×10^{-8}	2.09×10^{-9}	1.75×10^{-10}	1.33×10^{-13}	5.19×10^{-10}	$\mathbf{8.88 \times 10^{-16}}$
	Std	8.14×10^{-9}	3.96×10^{-9}	2.88×10^{-7}	3.38×10^{-9}	1.20×10^{-10}	$\mathbf{0.00 \times 10^{0}}$	5.63×10^{-9}	1.97×10^{-9}	6.66×10^{-10}	5.21×10^{-13}	1.33×10^{-9}	$\mathbf{0.00 \times 10^{0}}$
	rank	5.97	4.93	2.10	2.20	3.90	**1.90**	6.00	4.97	2.08	2.05	4.00	**1.90**
	SR/%	0.00	0.00	96.67	86.67	13.33	**100.00**	0.00	0.00	90.00	93.33	0.00	**100.00**

Table 3. *Cont.*

Functions		BOA	CABOA	PSOBOA	HBOAPSO	LBOA	IBOA	BOA	CABOA	PSOBOA	HBOAPSO	LBOA	IBOA
		Dim = 100						Dim = 300					
Alpine	Worst	1.30×10^{-10}	6.56×10^{-7}	1.80×10^{-12}	$\mathbf{9.13 \times 10^{-41}}$	9.60×10^{-12}	4.41×10^{-19}	7.31×10^{-10}	1.80×10^{-6}	1.04×10^{-11}	$\mathbf{8.54 \times 10^{-23}}$	6.14×10^{-11}	4.97×10^{-19}
	Best	2.31×10^{-11}	3.36×10^{-11}	3.11×10^{-146}	$\mathbf{9.85 \times 10^{-136}}$	2.17×10^{-17}	3.81×10^{-21}	5.21×10^{-11}	5.68×10^{-12}	8.46×10^{-147}	$\mathbf{3.37 \times 10^{-131}}$	9.80×10^{-18}	4.96×10^{-21}
	Avg	6.94×10^{-11}	2.57×10^{-8}	6.01×10^{-14}	$\mathbf{3.18 \times 10^{-42}}$	9.82×10^{-13}	7.90×10^{-20}	2.67×10^{-10}	6.34×10^{-8}	3.46×10^{-13}	$\mathbf{3.00 \times 10^{-24}}$	7.53×10^{-12}	1.33×10^{-19}
	Std	3.01×10^{-11}	1.20×10^{-7}	3.29×10^{-13}	$\mathbf{1.67 \times 10^{-41}}$	2.21×10^{-12}	9.09×10^{-20}	1.76×10^{-10}	3.28×10^{-7}	1.89×10^{-12}	$\mathbf{1.56 \times 10^{-23}}$	1.22×10^{-11}	1.48×10^{-19}
	rank	5.00	6.00	1.73	**1.37**	4.00	2.90	5.77	5.23	1.43	**1.73**	4.00	2.83
	SR/%	0.00	0.00	96.67	**100.00**	43.33	100.00	0.00	0.00	96.67	**100.00**	10.00	100.00
Avg.rank		5.464	4.361	2.247	**1.772**	4.356	2.800	5.531	4.050	2.117	**2.019**	4.378	2.906
Final rank		6	5	2	**1**	4	3	6	4	2	**1**	5	3

Table 4. Comparison results for 26 test functions with $Dim = 30$ for ten algorithms.

Functions		BOA	CABOA	PSOBOA	HBOAPSO	LBOA	IBOA	PSO	GWO	SCA	MPA
F_1	Avg	7.78×10^{-11}	1.01×10^{-13}	1.68×10^{-10}	**3.74×10^{-104}**	3.92×10^{-12}	1.61×10^{-30}	1.11×10^{-5}	6.20×10^{-28}	1.39×10^{1}	4.93×10^{-23}
	Std	7.67×10^{-12}	2.11×10^{-13}	9.17×10^{-10}	**2.05×10^{-103}**	4.46×10^{-12}	3.90×10^{-30}	2.12×10^{-5}	7.68×10^{-28}	2.88×10^{1}	7.29×10^{-23}
F_2	Avg	2.23×10^{-8}	1.25×10^{-14}	4.14×10^{-10}	**2.63×10^{-22}**	1.38×10^{-9}	5.11×10^{-19}	3.35×10^{-3}	1.04×10^{-16}	1.87×10^{-2}	2.99×10^{-13}
	Std	7.12×10^{-9}	2.15×10^{-14}	2.27×10^{-9}	**1.44×10^{-21}**	2.08×10^{-9}	1.73×10^{-18}	2.18×10^{-3}	8.66×10^{-17}	3.66×10^{-2}	2.56×10^{-13}
F_3	Avg	6.34×10^{-11}	6.30×10^{-13}	8.05×10^{-17}	**3.04×10^{-71}**	2.74×10^{-12}	6.15×10^{-31}	1.23×10^{2}	7.24×10^{-6}	8.03×10^{3}	1.52×10^{-4}
	Std	5.70×10^{-12}	1.37×10^{-12}	4.41×10^{-16}	**1.67×10^{-70}**	2.44×10^{-12}	1.16×10^{-30}	5.98×10^{2}	1.51×10^{-5}	6.30×10^{3}	3.15×10^{-4}
F_4	Avg	2.59×10^{-8}	2.77×10^{-10}	9.39×10^{-8}	**3.61×10^{-46}**	2.30×10^{-9}	1.36×10^{-19}	1.85×10^{-1}	8.57×10^{-8}	3.77×10^{0}	3.29×10^{-10}
	Std	2.58×10^{-9}	2.96×10^{-10}	5.14×10^{-7}	**1.97×10^{-45}**	2.36×10^{-9}	1.97×10^{-19}	4.62×10^{-2}	8.56×10^{-8}	1.30×10^{0}	2.23×10^{-10}
F_5	Avg	5.17×10^{0}	8.50×10^{-6}	6.47×10^{0}	**4.17×10^{-2}**	3.52×10^{0}	4.44×10^{0}	3.69×10^{-6}	6.84×10^{-1}	4.85×10^{0}	1.25×10^{-7}
	Std	6.09×10^{-1}	1.06×10^{-5}	3.90×10^{-1}	**6.41×10^{-2}**	8.50×10^{-1}	8.70×10^{-1}	4.74×10^{-6}	4.38×10^{-1}	7.32×10^{-1}	4.78×10^{-7}
F_6	Avg	2.03×10^{-3}	2.00×10^{-3}	**2.53×10^{-4}**	2.55×10^{-4}	2.10×10^{-3}	1.22×10^{-4}	7.98×10^{-2}	1.69×10^{-3}	1.19×10^{-1}	1.31×10^{-3}
	Std	8.70×10^{-4}	7.89×10^{-4}	**3.21×10^{-4}**	4.00×10^{-4}	9.63×10^{-4}	8.06×10^{-5}	3.14×10^{-2}	8.21×10^{-4}	1.04×10^{-1}	5.47×10^{-4}
F_7	Avg	1.05×10^{-11}	1.48×10^{-62}	8.41×10^{-11}	1.48×10^{-62}	5.23×10^{-20}	1.19×10^{-19}	0.00×10^{0}	5.10×10^{-58}	1.38×10^{-40}	**7.18×10^{-66}**
	Std	4.21×10^{-11}	6.67×10^{-63}	2.94×10^{-10}	6.70×10^{-63}	1.41×10^{-19}	5.94×10^{-19}	0.00×10^{0}	1.71×10^{-57}	7.24×10^{-40}	**7.74×10^{-70}**
F_8	Avg	6.33×10^{-14}	6.58×10^{-15}	1.42×10^{-17}	**3.19×10^{-118}**	7.51×10^{-16}	1.32×10^{-36}	1.37×10^{-14}	2.21×10^{-95}	7.27×10^{-5}	1.41×10^{-60}
	Std	3.60×10^{-14}	1.19×10^{-14}	7.78×10^{-17}	**1.68×10^{-117}**	9.49×10^{-16}	4.59×10^{-36}	4.69×10^{-14}	1.20×10^{-94}	2.25×10^{-4}	5.28×10^{-60}
F_9	Avg	7.01×10^{-11}	2.91×10^{-13}	1.87×10^{-16}	**2.72×10^{-99}**	2.36×10^{-12}	5.60×10^{-31}	1.67×10^{-4}	1.50×10^{-28}	7.67×10^{-1}	1.07×10^{-23}
	Std	7.91×10^{-12}	7.08×10^{-13}	1.02×10^{-15}	**1.31×10^{-98}**	2.76×10^{-12}	1.87×10^{-30}	3.97×10^{-4}	1.96×10^{-28}	1.13×10^{0}	1.36×10^{-23}
F_{10}	Avg	2.89×10^{1}	2.87×10^{1}	2.90×10^{1}	2.89×10^{1}	2.88×10^{1}	2.89×10^{1}	2.67×10^{1}	2.68×10^{1}	4.19×10^{1}	**2.53×10^{1}**
	Std	2.54×10^{-2}	1.39×10^{-5}	2.16×10^{-2}	8.18×10^{-2}	3.18×10^{-2}	3.40×10^{-2}	1.34×10^{0}	7.02×10^{-1}	4.32×10^{1}	**3.86×10^{-1}**
F_{11}	Avg	6.72×10^{-11}	2.37×10^{-14}	1.32×10^{-8}	**3.64×10^{-78}**	2.78×10^{-12}	1.10×10^{-30}	9.02×10^{-5}	2.24×10^{-28}	8.79×10^{0}	1.09×10^{-23}
	Std	6.90×10^{-12}	4.24×10^{-14}	6.84×10^{-8}	**1.99×10^{-77}**	2.63×10^{-12}	2.90×10^{-30}	1.05×10^{-4}	3.10×10^{-28}	1.74×10^{1}	2.38×10^{-23}
F_{12}	Avg	9.72×10^{-1}	**4.77×10^{-1}**	9.91×10^{-1}	9.75×10^{-1}	9.34×10^{-1}	9.71×10^{-1}	7.66×10^{-1}	6.67×10^{-1}	5.86×10^{2}	6.67×10^{-1}
	Std	1.14×10^{-2}	**3.45×10^{-1}**	5.24×10^{-3}	8.45×10^{-2}	2.11×10^{-2}	7.36×10^{-3}	3.39×10^{-1}	2.62×10^{-6}	2.29×10^{3}	5.38×10^{-8}
F_{13}	Avg	1.16×10^{-20}	4.17×10^{-30}	6.44×10^{-24}	5.73×10^{-92}	1.20×10^{-24}	3.03×10^{-35}	7.70×10^{-77}	0.00×10^{0}	2.43×10^{-96}	**1.97×10^{-162}**
	Std	6.14×10^{-20}	2.18×10^{-29}	3.00×10^{-23}	3.14×10^{-91}	5.29×10^{-24}	6.42×10^{-35}	3.27×10^{-76}	0.00×10^{0}	1.31×10^{-95}	**1.09×10^{-161}**
F_{14}	Avg	6.51×10^{-17}	2.24×10^{-23}	7.15×10^{-15}	1.28×10^{-63}	4.71×10^{-18}	8.45×10^{-31}	7.38×10^{-61}	**8.59×10^{-201}**	6.47×10^{-67}	1.07×10^{-63}
	Std	1.39×10^{-16}	7.51×10^{-23}	3.92×10^{-14}	7.04×10^{-63}	7.50×10^{-18}	2.52×10^{-30}	3.81×10^{-60}	**0.00×10^{0}**	3.37×10^{-66}	5.84×10^{-63}
F_{15}	Avg	2.51×10^{1}	**0.00×10^{0}**	1.10×10^{1}	**0.00×10^{0}**	**0.00×10^{0}**	**0.00×10^{0}**	4.56×10^{1}	3.36×10^{0}	4.42×10^{1}	**0.00×10^{0}**
	Std	6.52×10^{1}	**0.00×10^{0}**	4.22×10^{1}	**0.00×10^{0}**	**0.00×10^{0}**	**0.00×10^{0}**	1.11×10^{1}	4.42×10^{0}	3.71×10^{1}	**0.00×10^{0}**
F_{16}	Avg	9.36×10^{1}	**0.00×10^{0}**	2.00×10^{1}	**0.00×10^{0}**	**0.00×10^{0}**	**0.00×10^{0}**	4.48×10^{1}	8.20×10^{0}	7.08×10^{1}	1.01×10^{-8}
	Std	8.04×10^{1}	**0.00×10^{0}**	5.23×10^{1}	**0.00×10^{0}**	**0.00×10^{0}**	**0.00×10^{0}**	9.04×10^{0}	5.18×10^{0}	4.45×10^{1}	4.72×10^{-8}
F_{17}	Avg	1.09×10^{-9}	1.84×10^{-9}	5.63×10^{-8}	8.96×10^{-11}	2.34×10^{-12}	**8.88×10^{-16}**	1.69×10^{-3}	2.79×10^{0}	2.03×10^{1}	1.06×10^{-3}
	Std	8.16×10^{-10}	1.76×10^{-9}	3.06×10^{-7}	4.73×10^{-10}	7.87×10^{-12}	**0.00×10^{0}**	1.32×10^{-3}	7.22×10^{0}	5.27×10^{-2}	5.83×10^{-3}

Table 4. *Cont.*

Functions		BOA	CABOA	PSOBOA	HBOAPSO	LBOA	IBOA	PSO	GWO	SCA	MPA
F_{18}	Avg	7.64×10^{-12}	1.70×10^{-14}	2.61×10^{-8}	**0.00×10^{0}**	3.48×10^{-13}	**0.00×10^{0}**	5.33×10^{-3}	1.31×10^{-3}	2.17×10^{-1}	**0.00×10^{0}**
	Std	6.94×10^{-12}	1.82×10^{-14}	1.35×10^{-7}	**0.00×10^{0}**	8.78×10^{-13}	**0.00×10^{0}**	7.48×10^{-3}	4.99×10^{-3}	2.13×10^{-1}	**0.00×10^{0}**
F_{19}	Avg	1.90×10^{-10}	6.76×10^{-6}	4.77×10^{-7}	**2.54×10^{-45}**	6.32×10^{-14}	8.93×10^{-20}	1.15×10^{-3}	5.15×10^{-4}	3.02×10^{-1}	2.12×10^{-14}
	Std	1.00×10^{-10}	3.13×10^{-5}	1.78×10^{-6}	**1.39×10^{-44}**	1.73×10^{-13}	1.19×10^{-19}	9.36×10^{-4}	7.29×10^{-4}	5.38×10^{-1}	1.52×10^{-14}
F_{20}	Avg	5.56×10^{-1}	1.90×10^{-4}	8.75×10^{-1}	2.84×10^{-3}	3.06×10^{-1}	4.97×10^{-1}	4.44×10^{0}	4.74×10^{-2}	1.17×10^{6}	**5.79×10^{-5}**
	Std	1.40×10^{-1}	4.90×10^{-4}	2.11×10^{-1}	3.79×10^{-3}	9.96×10^{-2}	1.37×10^{-1}	2.62×10^{0}	2.27×10^{-2}	2.83×10^{6}	**3.17×10^{-4}**
F_{21}	Avg	3.52×10^{0}	1.36×10^{-2}	4.42×10^{0}	3.93×10^{-2}	2.41×10^{0}	3.15×10^{0}	**1.89×10^{-6}**	9.27×10^{-1}	3.48×10^{6}	1.38×10^{-2}
	Std	5.92×10^{-1}	4.57×10^{-2}	7.32×10^{-1}	4.57×10^{-2}	5.15×10^{-1}	4.40×10^{-1}	**3.64×10^{-6}**	2.80×10^{-1}	6.37×10^{6}	3.69×10^{-2}
F_{22}	Avg	9.76×10^{0}	5.52×10^{-16}	3.42×10^{-7}	**8.38×10^{-77}**	1.56×10^{-3}	3.56×10^{-26}	6.71×10^{-4}	5.34×10^{-1}	1.42×10^{1}	1.11×10^{-1}
	Std	7.84×10^{0}	4.56×10^{-16}	1.88×10^{-6}	**4.30×10^{-76}**	8.54×10^{-3}	7.84×10^{-26}	2.74×10^{-3}	4.25×10^{-1}	1.87×10^{0}	1.72×10^{-1}
F_{23}	Avg	1.17×10^{1}	**4.35×10^{-4}**	1.75×10^{1}	7.28×10^{-2}	8.42×10^{0}	9.83×10^{0}	4.77×10^{-2}	1.42×10^{0}	1.74×10^{1}	1.35×10^{-1}
	Std	2.66×10^{0}	**4.66×10^{-4}**	3.79×10^{0}	1.87×10^{-1}	2.56×10^{0}	2.47×10^{0}	6.58×10^{-2}	1.13×10^{0}	3.58×10^{0}	1.13×10^{-1}
F_{24}	Avg	6.21×10^{-1}	**0.00×10^{0}**	3.36×10^{-11}	**0.00×10^{0}**	**0.00×10^{0}**	**0.00×10^{0}**	9.15×10^{-1}	5.05×10^{0}	9.91×10^{0}	**0.00×10^{0}**
	Std	1.95×10^{0}	**0.00×10^{0}**	1.84×10^{-10}	**0.00×10^{0}**	**0.00×10^{0}**	**0.00×10^{0}**	1.40×10^{0}	2.35×10^{0}	2.00×10^{0}	**0.00×10^{0}**
F_{25}	Avg	7.65×10^{-1}	7.30×10^{-2}	1.41×10^{-1}	2.53×10^{-8}	3.65×10^{-2}	**2.25×10^{-32}**	1.15×10^{0}	3.48×10^{-1}	1.66×10^{0}	9.95×10^{-2}
	Std	2.21×10^{-1}	4.47×10^{-2}	1.21×10^{-1}	1.38×10^{-7}	4.88×10^{-2}	**5.88×10^{-32}**	3.41×10^{-1}	1.13×10^{-1}	2.15×10^{0}	8.20×10^{-17}
F_{26}	Avg	7.96×10^{-11}	6.54×10^{-15}	6.50×10^{-12}	**0.00×10^{0}**	3.22×10^{-12}	**0.00×10^{0}**	1.00×10^{-1}	**0.00×10^{0}**	7.60×10^{-1}	**0.00×10^{0}**
	Std	8.77×10^{-12}	1.52×10^{-14}	3.54×10^{-11}	**0.00×10^{0}**	2.68×10^{-12}	**0.00×10^{0}**	3.29×10^{-1}	**0.00×10^{0}**	1.27×10^{0}	**0.00×10^{0}**

Table 5. The Success Rate for 26 benchmark functions.

Functions	BOA	CABOA	PSOBOA	HBOAPSO	LBOA	IBOA	PSO	GWO	SCA	MPA
F_1	0.00	43.33	93.33	100.00	0.00	100.00	0.00	100.00	0.00	100.00
F_2	0.00	76.67	86.67	100.00	0.00	100.00	0.00	100.00	0.00	0.00
F_3	0.00	30.00	100.00	100.00	3.33	100.00	0.00	0.00	0.00	0.00
F_4	0.00	0.00	90.00	100.00	0.00	100.00	0.00	0.00	0.00	0.00
F_5	0.00	0.00	0.00	0.00	0.00	0.00	0.00	0.00	0.00	0.00
F_6	0.00	0.00	0.00	0.00	0.00	0.00	0.00	0.00	0.00	0.00
F_7	0.00	100.00	56.67	100.00	100.00	100.00	100.00	100.00	100.00	100.00
F_8	0.00	80.00	100.00	100.00	100.00	100.00	80.00	100.00	0.00	100.00
F_9	0.00	56.67	100.00	100.00	3.33	100.00	0.00	100.00	0.00	100.00
F_{10}	0.00	0.00	0.00	0.00	0.00	0.00	0.00	0.00	0.00	0.00
F_{11}	0.00	63.33	90.00	100.00	0.00	100.00	0.00	100.00	0.00	100.00
F_{12}	0.00	0.00	0.00	0.00	0.00	0.00	0.00	0.00	0.00	0.00
F_{13}	100.00	100.00	100.00	100.00	100.00	100.00	100.00	100.00	100.00	100.00
F_{14}	100.00	100.00	96.67	100.00	100.00	100.00	100.00	100.00	100.00	100.00
F_{15}	50.00	100.00	86.67	100.00	100.00	100.00	0.00	0.00	0.00	100.00
F_{16}	3.33	100.00	90.00	100.00	100.00	100.00	0.00	0.00	0.00	50.00
F_{17}	0.00	0.00	83.33	93.33	30.00	100.00	0.00	0.00	0.00	0.00
F_{18}	0.00	46.67	93.33	100.00	23.33	100.00	0.00	93.33	0.00	100.00
F_{19}	0.00	0.00	86.67	100.00	56.67	100.00	0.00	33.33	0.00	23.33
F_{20}	0.00	0.00	0.00	0.00	0.00	0.00	0.00	0.00	0.00	0.00
F_{21}	0.00	0.00	0.00	0.00	0.00	0.00	0.00	0.00	0.00	0.00
F_{22}	16.67	100.00	90.00	100.00	96.67	100.00	0.00	0.00	0.00	13.33
F_{23}	0.00	0.00	0.00	0.00	0.00	0.00	0.00	0.00	0.00	0.00
F_{24}	23.33	100.00	96.67	100.00	100.00	100.00	0.00	0.00	0.00	100.00
F_{25}	0.00	0.00	10.00	100.00	30.00	100.00	0.00	0.00	0.00	0.00
F_{26}	0.00	86.67	93.33	100.00	3.33	100.00	0.00	100.00	0.00	100.00
times	2	7	4	18	7	19	3	9	3	11
SR rank	8	5	6	2	5	1	7	4	7	3

Table 6. The rank test for 26 benchmark functions.

Rank	BOA	CABOA	PSOBOA	HBOAPSO	LBOA	IBOA	PSO	GWO	SCA	MPA
F_1	7.97	5.93	2.10	1.53	6.97	2.83	9.00	3.83	10.00	4.83
F_2	8.00	4.87	1.93	1.63	6.97	2.83	9.03	3.87	9.97	5.90
F_3	6.00	4.00	1.70	1.33	5.00	2.97	9.00	7.03	10.00	7.97
F_4	7.17	4.07	1.83	1.47	5.97	2.87	9.00	7.77	10.00	4.87
F_5	8.83	2.93	10.00	4.07	6.00	7.13	2.07	4.93	8.03	1.00
F_6	6.87	6.67	2.43	1.77	7.07	1.80	9.47	5.27	9.53	4.13
F_7	9.53	3.68	9.43	3.65	8.00	7.03	1.00	4.53	6.00	2.13
F_8	8.97	6.97	1.90	1.60	6.20	4.93	7.87	2.73	10.00	3.83
F_9	8.00	6.00	1.90	1.30	7.00	2.93	9.00	3.93	10.00	4.93
F_{10}	7.70	3.90	9.13	6.97	5.30	6.30	2.97	2.63	9.07	1.03
F_{11}	7.90	5.90	2.67	1.03	6.90	2.87	9.00	3.87	10.00	4.87
F_{12}	6.23	2.97	7.83	8.60	4.70	6.17	4.00	2.77	10.00	1.73
F_{13}	10.00	7.20	5.03	3.13	8.93	7.40	5.70	1.00	4.57	2.03
F_{14}	9.97	7.57	3.83	2.07	8.97	7.30	5.67	1.50	4.33	3.80
F_{15}	5.90	3.68	4.28	3.68	3.68	3.68	9.47	7.80	9.13	3.68
F_{16}	8.90	3.18	4.02	3.18	3.18	3.18	8.53	7.13	9.07	4.62
F_{17}	6.60	7.57	2.47	2.12	3.57	1.98	8.83	6.10	9.87	5.90
F_{18}	7.87	5.82	3.30	2.97	6.75	2.97	9.00	3.37	10.00	2.97
F_{19}	6.33	7.33	2.23	1.37	4.17	2.80	9.00	6.77	10.00	5.00
F_{20}	7.03	2.00	8.03	3.00	5.03	6.03	8.87	4.00	10.00	1.00
F_{21}	8.00	2.83	9.00	3.97	6.00	7.00	1.50	5.00	10.00	1.70
F_{22}	8.40	4.93	2.30	1.03	4.03	2.87	6.53	8.33	9.73	6.83
F_{23}	8.00	1.10	9.70	2.27	6.07	6.93	2.80	4.97	9.30	3.87
F_{24}	6.43	3.58	3.68	3.58	3.58	3.58	7.97	9.00	10.00	3.58
F_{25}	8.77	4.80	5.90	1.07	3.47	2.00	9.03	6.77	9.20	4.00
F_{26}	7.97	4.60	3.53	3.23	6.97	3.23	9.00	3.23	10.00	3.23
Avg-rank	7.82	4.77	4.62	2.75	5.79	4.29	7.05	4.93	9.15	3.83
Final rank	9	5	4	1	7	3	8	6	10	2

Table 7. The p-value of Wilcoxon rank-sum (WRS) test for 26 benchmark functions.

Ranksum	BOA	CABOA	PSOBOA	LBOA	IBOA	PSO	GWO	SCA	MPA
F_1	3.02×10^{-11}	3.02×10^{-11}	0.035137	3.02×10^{-11}	3.02×10^{-11}	3.02×10^{-11}	3.02×10^{-11}	3.02×10^{-11}	3.02×10^{-11}
F_2	3.02×10^{-11}	3.02×10^{-11}	**0.325527**	3.02×10^{-11}	3.02×10^{-11}	3.02×10^{-11}	3.02×10^{-11}	3.02×10^{-11}	3.02×10^{-11}
F_3	3.02×10^{-11}	3.02×10^{-11}	0.001597	3.02×10^{-11}	3.02×10^{-11}	3.02×10^{-11}	3.02×10^{-11}	3.02×10^{-11}	3.02×10^{-11}
F_4	3.02×10^{-11}	3.02×10^{-11}	0.014412	3.02×10^{-11}	3.02×10^{-11}	3.02×10^{-11}	3.02×10^{-11}	3.02×10^{-11}	3.02×10^{-11}
F_5	3.02×10^{-11}	3.02×10^{-11}	3.02×10^{-11}	3.02×10^{-11}	3.02×10^{-11}	3.02×10^{-11}	1.31×10^{-8}	3.02×10^{-11}	3.02×10^{-11}
F_6	4.20×10^{-10}	2.15×10^{-10}	**0.200949**	1.33×10^{-10}	**0.520145**	3.02×10^{-11}	7.38×10^{-10}	3.02×10^{-11}	2.44×10^{-9}
F_7	5.18×10^{-12}	$\mathbf{1.00 \times 10^{0}}$	5.18×10^{-12}	5.18×10^{-12}	5.16×10^{-12}	1.19×10^{-13}	0.009689	5.18×10^{-12}	9.85×10^{-11}
F_8	3.02×10^{-11}	3.02×10^{-11}	**0.122353**	3.02×10^{-11}	3.02×10^{-11}	3.02×10^{-11}	3.02×10^{-11}	3.02×10^{-11}	3.02×10^{-11}
F_9	3.02×10^{-11}	3.02×10^{-11}	0.001302	3.02×10^{-11}	3.02×10^{-11}	3.02×10^{-11}	3.02×10^{-11}	3.02×10^{-11}	3.02×10^{-11}
F_{10}	**0.340288**	3.02×10^{-11}	3.16×10^{-5}	0.003671	**0.200949**	2.20×10^{-7}	4.50×10^{-11}	1.34×10^{-5}	3.02×10^{-11}
F_{11}	**0.340288**	3.02×10^{-11}	3.16×10^{-5}	0.003671	**0.200949**	2.20×10^{-7}	4.50×10^{-11}	1.34×10^{-5}	3.02×10^{-11}
F_{12}	8.48×10^{-9}	4.69×10^{-8}	4.12×10^{-6}	8.48×10^{-9}	8.48×10^{-9}	1.43×10^{-8}	5.57×10^{-10}	3.02×10^{-11}	5.57×10^{-10}
F_{13}	3.02×10^{-11}	3.02×10^{-11}	0.00557	3.02×10^{-11}	3.02×10^{-11}	3.02×10^{-11}	1.21×10^{-12}	1.29×10^{-9}	2.53×10^{-4}
F_{14}	3.02×10^{-11}	3.02×10^{-11}	0.001953	3.02×10^{-11}	3.02×10^{-11}	4.62×10^{-10}	**0.09049**	7.69×10^{-8}	9.06×10^{-8}
F_{15}	1.27×10^{-5}	**NaN**	0.041926	**NaN**	**NaN**	1.21×10^{-12}	1.19×10^{-12}	1.21×10^{-12}	**NaN**
F_{16}	1.21×10^{-12}	**NaN**	0.041926	**NaN**	**NaN**	1.21×10^{-12}	1.21×10^{-12}	1.21×10^{-12}	1.27×10^{-5}
F_{17}	6.03×10^{-11}	2.89×10^{-11}	**0.248673**	5.93×10^{-7}	**0.160802**	2.37×10^{-12}	2.80×10^{-10}	2.37×10^{-12}	6.24×10^{-10}
F_{18}	1.21×10^{-12}	4.57×10^{-12}	**0.160802**	4.57×10^{-12}	**NaN**	1.21×10^{-12}	**0.160802**	1.21×10^{-12}	**NaN**
F_{19}	3.02×10^{-11}	3.02×10^{-11}	0.003671	3.02×10^{-11}	3.02×10^{-11}	3.02×10^{-11}	3.02×10^{-11}	3.02×10^{-11}	3.02×10^{-11}
F_{20}	3.02×10^{-11}	3.08×10^{-8}	3.02×10^{-11}	3.02×10^{-11}	3.02×10^{-11}	3.02×10^{-11}	3.34×10^{-11}	3.02×10^{-11}	1.09×10^{-10}
F_{21}	3.02×10^{-11}	9.51×10^{-6}	3.02×10^{-11}	3.02×10^{-11}	3.02×10^{-11}	3.02×10^{-11}	3.02×10^{-11}	3.02×10^{-11}	4.12×10^{-6}
F_{22}	3.02×10^{-11}	3.02×10^{-11}	1.58×10^{-4}	3.02×10^{-11}	3.02×10^{-11}	3.02×10^{-11}	3.02×10^{-11}	3.02×10^{-11}	3.02×10^{-11}
F_{23}	3.02×10^{-11}	1.39×10^{-6}	3.02×10^{-11}	3.02×10^{-11}	3.02×10^{-11}	1.11×10^{-4}	2.37×10^{-10}	3.02×10^{-11}	1.86×10^{-6}
F_{24}	1.95×10^{-9}	**NaN**	**0.333711**	**NaN**	**NaN**	1.21×10^{-12}	1.21×10^{-12}	1.21×10^{-12}	**NaN**
F_{25}	3.02×10^{-11}	5.49×10^{-11}	1.09×10^{-10}	8.89×10^{-10}	8.48×10^{-9}	1.90×10^{-11}	3.02×10^{-11}	3.02×10^{-11}	1.55×10^{-11}
F_{26}	1.21×10^{-12}	2.93×10^{-5}	**0.160802**	1.21×10^{-12}	**NaN**	1.21×10^{-12}	**NaN**	1.21×10^{-12}	**NaN**

Table 8. The hypothesis (H) of WSR test for 26 benchmark functions.

H	BOA	CABOA	PSOBOA	LBOA	IBOA	PSO	GWO	SCA	MPA
F_1	1	1	1	1	1	1	1	1	1
F_2	1	1	0	1	1	1	1	1	1
F_3	1	1	1	1	1	1	1	1	1
F_4	1	1	1	1	1	1	1	1	1
F_5	1	1	1	1	0	1	1	1	1
F_6	1	0	0	1	1	1	1	1	1
F_7	1	1	1	1	1	1	1	1	1
F_8	1	1	0	1	1	1	1	1	1
F_9	1	1	1	1	1	1	1	1	1
F_{10}	0	1	1	1	0	1	1	1	1
F_{11}	0	1	1	1	0	1	1	1	1
F_{12}	1	1	1	1	1	1	1	1	1
F_{13}	1	1	1	1	1	1	1	1	1
F_{14}	1	1	1	1	1	1	0	1	1
F_{15}	1	0	1	0	0	1	1	1	0
F_{16}	1	0	1	0	0	1	1	1	1
F_{17}	1	1	0	1	0	1	1	1	1
F_{18}	1	1	0	1	0	1	0	1	0
F_{19}	1	1	1	1	1	1	1	1	1
F_{20}	1	1	1	1	1	1	1	1	1
F_{21}	1	1	1	1	1	1	1	1	1
F_{22}	1	1	1	1	1	1	1	1	1
F_{23}	1	1	1	1	1	1	1	1	1
F_{24}	1	0	0	0	0	1	1	1	0
F_{25}	1	1	1	1	1	1	1	1	1
F_{26}	1	1	0	1	0	1	0	1	0

The alpha is set to 0.05 in the Wilcoxon rank-sum (WRS) test and Friedman rank test, and there are two hypotheses called the null and alternative. The null hypothesis is a significant difference from the proposed algorithm and the others. According to the statistical value, the null is accepted if this statistical value is greater than the value of alpha; otherwise, the alternative is accepted. The p-value and the Friedman rank depicted that this supremacy is statistically significant. Note, the last row in Table 4, Table 5, Table 6 represents the rank of each algorithm with the number of the best solutions. The p-value and the Friedman rank depicted that this supremacy is statistically significant.

From the comparison results of Table 4, it is proved that the HPSOBOA yields the best results on the 26 test functions with $Dim = 30$ except F_6, F_7, F_{10}, F_{12}, F_{13}, F_{14}, F_{17}, F_{20}, F_{21}, F_{23}, and F_{25}. For functions F_6, F_7, F_{10}, F_{12}, and F_{23}, the hybrid HPSOBOA can obtain the optimal fitness value, which is close to other algorithms but slightly worse. However, for F_{13}, F_{14}, F_{17}, F_{20}, F_{21}, and F_{25}, the best solutions of these functions are searched by the other algorithms, such as GWO, PSO, MPA, and IBOA, and MPA obtains the best solution twice. Additionally, the IBOA also obtains the best solution twice, which is improved by the logistic map for the control parameters. Combining the comparison results in Tables 5 and 6, we can see that the IBOA is better than others in the SR rank, which is set to $\varepsilon < 10^{-15}$, and is called the specified value, and the order of ten algorithms is IBOA > HPSOBOA > MPA > GWO > CABOA = LBOA > PSOBOA > PSO = SCA > BOA. The order of HPSOBOA and IBOA is only different once on the function F_{17}, and the SR of HPSOBOA is 93.33%, but the SR of IBOA is 100% for searching the global optimization value, which is set to $\varepsilon < 10^{-15}$, and is accepted in this paper. Therefore, the performance of the proposed algorithm needs to be improved in future work.

In addition, the comparison results of the Friedman rank test are shown in Table 6; from the Avg-rank, we can obtain that the final order of the rank means of the Friedman rank test—the ten swarm algorithms—is HPSOBOA > MPA > IBOA > PSOBOA > CABOA > GWO > LBOA > PSO > BOA > SCA. The WRS test values are given in Tables 7 and 8 for the 26 high-dimensional test functions of HPSOBOA vs. the others, respectively, where N/A means not applicable in Table 7. It can be seen from these tables that there is a significance different between the proposed hybrid HPSOBOA and the other algorithms for the 26 test functions with Dim = 30. In Table 8, if H=1, this indicates rejection of the null hypothesis at the 5% significance level. If H=0, this indicates a failure to reject the null hypothesis at the 5% significance level. In addition, Table 9 shows the comparison results of t-test for 26 benchmark functions of the proposed HPSOBOA with the other algorithms.

Table 9. The t-test for 26 benchmark functions.

t-tset	BOA	CABOA	PSOBOA	LBOA	IBOA	PSO	GWO	SCA	MPA
F_1	6.6456	6.6456	2.1068	6.6456	6.6456	6.6456	6.6456	6.6456	6.6456
F_2	6.6456	6.6456	0.9832	6.6456	6.6456	6.6456	6.6456	6.6456	6.6456
F_3	6.6456	6.6456	3.1565	6.6456	6.6456	6.6456	6.6456	6.6456	6.6456
F_4	6.6456	6.6456	2.4468	6.6456	6.6456	6.6456	6.6456	6.6456	6.6456
F_5	6.6456	−6.6456	6.6456	6.6456	6.6456	−6.6456	5.6846	6.6456	−6.6456
F_6	6.2464	6.3499	1.2789	6.4238	0.6431	6.6456	6.1577	6.6456	5.9655
F_7	6.9005	0.0000	6.9005	6.9005	6.9010	−7.4180	2.5867	6.9005	−6.4692
F_8	6.6456	6.6456	1.5450	6.6456	6.6456	6.6456	6.6456	6.6456	6.6456
F_9	6.6456	6.6456	3.2156	6.6456	6.6456	6.6456	6.6456	6.6456	6.6456
F_{10}	0.9536	−6.6456	4.1618	−2.9051	−1.2789	−5.1819	−6.5865	4.3540	−6.6456
F_{11}	6.6456	6.6456	3.7183	6.6456	6.6456	6.6456	−6.5865	6.6456	6.6456
F_{12}	−5.7585	−5.4628	−4.6053	−5.7585	−5.7585	−5.6698	−6.2021	6.6456	−6.2021
F_{13}	6.6456	6.6456	2.7721	6.6456	6.6456	6.6456	−7.1040	6.0690	−3.6591
F_{14}	6.6456	6.6456	3.0973	6.6456	6.6456	6.2316	−1.6928	5.3741	5.3446
F_{15}	4.3649	NaN	2.0343	NaN	NaN	7.1040	7.1063	7.1040	NaN
F_{16}	7.1040	NaN	2.0343	NaN	NaN	7.1040	7.1040	7.1040	4.3650
F_{17}	6.5431	6.6523	1.1536	4.9936	−1.4024	7.0110	6.3094	7.0110	6.1844
F_{18}	7.1040	6.9183	1.4024	6.9182	NaN	7.1040	1.4024	7.1040	NaN
F_{19}	6.6456	6.6456	2.9051	6.6456	6.6456	6.6456	6.6456	6.6456	6.6456
F_{20}	6.6456	−5.5368	6.6456	6.6456	6.6456	6.6456	6.6308	6.6456	−6.4534
F_{21}	6.6456	−4.4279	6.6456	6.6456	6.6456	−6.6456	6.6456	6.6456	−4.6053
F_{22}	6.6456	6.6456	3.7774	6.6456	6.6456	6.6456	6.6456	6.6456	6.6456
F_{23}	6.6456	−4.8271	6.6456	6.6456	6.6456	3.8661	6.3351	6.6456	4.7680
F_{24}	6.0023	NaN	0.9667	NaN	NaN	7.1040	7.1040	7.1040	NaN
F_{25}	6.6456	6.5569	6.4534	6.1281	5.7585	6.7136	6.6456	6.6456	6.7434
F_{26}	7.1040	4.1785	1.4024	7.1040	NaN	7.1040	NaN	7.1040	NaN

Figure 6 shows the box plots of the optimization results of 26 high-dimensional problems by the ten algorithms. It is clear from Figure 6 that the outcomes of the average of the fitness function are not normally distributed, in which each algorithm is run for 30 times for the 26 test functions. The values of SCA are relatively poor in the ten algorithms.

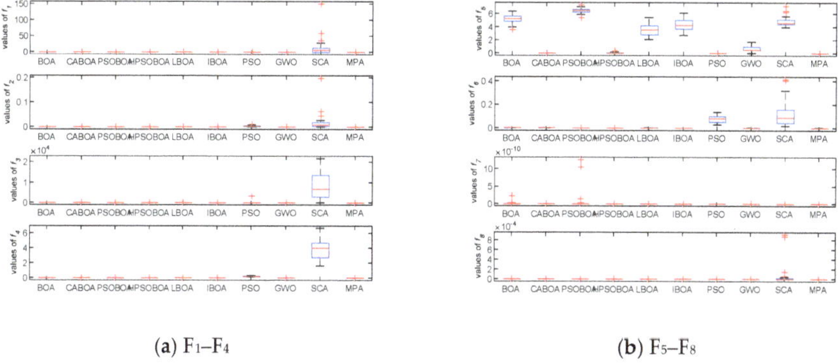

(a) F_1–F_4 (b) F_5–F_8

Figure 6. *Cont.*

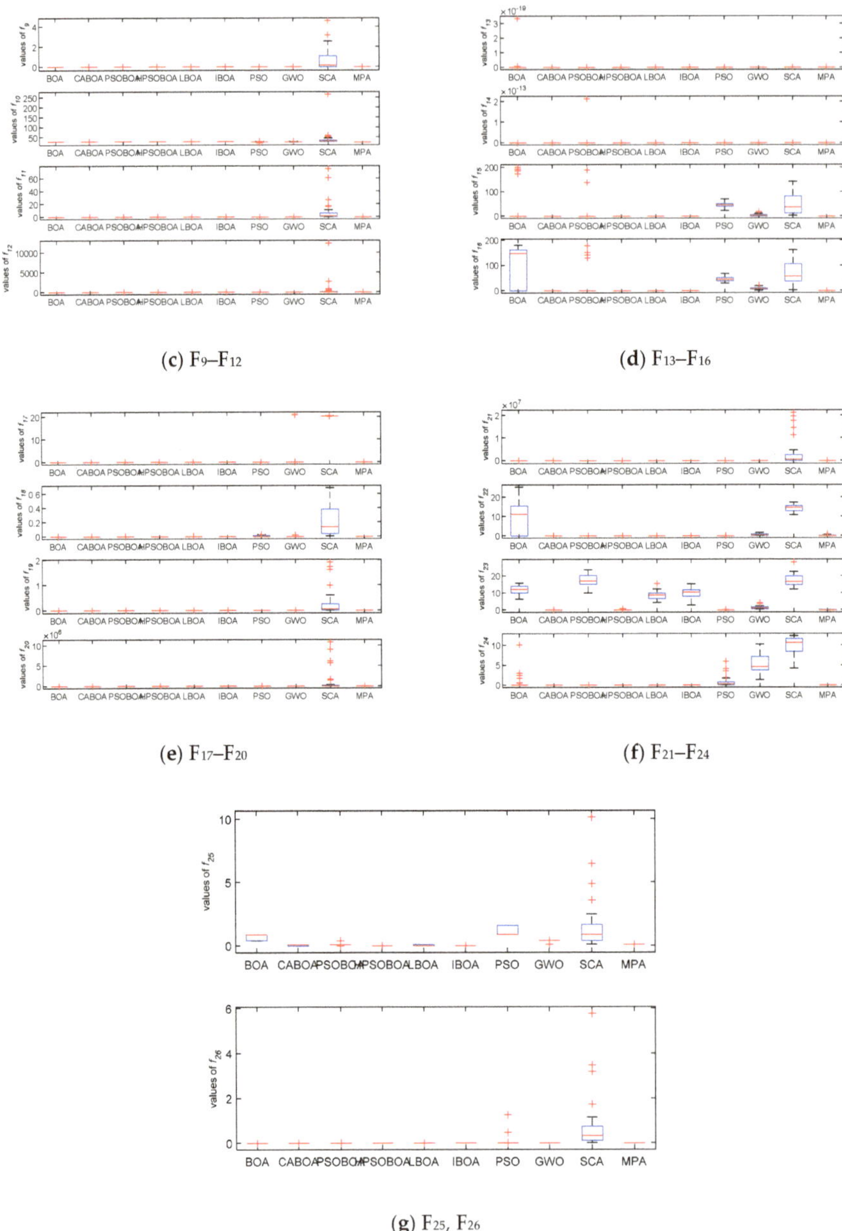

(c) F9–F12

(d) F13–F16

(e) F17–F20

(f) F21–F24

(g) F25, F26

Figure 6. Boxplot for the algorithms run 30 times for the fitness of 26 test functions with *Dim* = 30.

6. Conclusions and Future Work

In this paper, we proposed three improvement strategies, and they are as follows: (1) the initialization of BOA by cubic map; (2) a nonlinear parameter control strategy for the power exponent *a*; (3) hybrid PSO algorithm with BOA. These strategies all aim to improve the ability for global optimization of the basic BOA.

In order to analyze the effectiveness of the improvement strategies, a novel hybrid algorithm was compared with other swarm algorithms, and two experiments were designed. To deal with 26 high-dimensional optimization problems, a cubic map was employed for the initial population of HPSOBOA, and the experimental results show that the initial fitness value is superior to the BOA and other algorithms. In addition, the experimental results show that the one-dimensional chaotic maps may also have a good performance for improving the basic BOA. The MPA proposed in 2020 will be applied in more fields.

In future work, the performance of the proposed algorithm needs to be improved, and the improved BOA includes adjusting its control parameters to optimize algorithm performance. The two-dimensional and three-dimensional chaotic systems can also improve the BOA or other swarm intelligence algorithms in theory. The improved algorithm can also solve real-world problems, such as engineering problems, wireless sensor network (WSNs) deployment problems, proportional-integral-derivative (PID) control problems, and analysis of regional economic activity [53].

Author Contributions: M.Z.: Conceptualization, Methodology, Software, Writing, and Language modification; J.Y.: Data curation, Writing—original draft preparation, and Programming calculation; D.L.: Data curation; T.Q.: Example analysis. All authors have read and agreed to the published version of the manuscript.

Funding: This work is supported by the National Natural Science Foundation of China (under Grant No. 61640014 and 61861007), Industrial Project of Guizhou province (Grant No. Qiankehe Zhicheng (2019) 2152), Science and Technology Fund of Guizhou Province under Grant Qiankehe (No. (2020)1Y266), Science and Technology Plan of Guizhou province Qiankehepingtai (No. (2017) 5788), Guizhou Graduate Innovation Fund (No. YJSCXJH (2019) 005), the platform of IOT personnel form Guiyang hi tech Development Zone under Grant 2015, postgraduate case library (under Grant No. KCALK201708), and Innovation group 2020 of Guizhou Provincial Education Department.

Acknowledgments: The authors are grateful for the support provided by the Guizhou Provincial Key Laboratory of Internet + Intelligent Manufacturing, Guiyang 550025, China.

Conflicts of Interest: The authors declare no conflict of interest of this paper.

References

1. Arora, S.; Singh, S. Butterfly optimization algorithm: A novel approach for global optimization. *Soft Comput.* **2019**, *23*, 715–734. [CrossRef]
2. Arora, S.; Singh, S. Butterfly Algorithm with Lévy Flights for Global Optimization. In Proceedings of the 2015 International Conference on Signal Processing, Computing and Control (2015 ISPCC), Waknaghat, India, 24–26 September 2015; pp. 220–224.
3. Arora, S.; Singh, S. An improved butterfly optimization algorithm with chaos. *J. Intell. Fuzzy Syst.* **2017**, *32*, 1079–1088. [CrossRef]
4. Arora, S.; Singh, S. An Effective Hybrid Butterfly Optimization Algorithm with Artificial Bee Colony for Numerical Optimization. *Int. J. Interact. Multimed. Artif. Intell.* **2017**, *4*, 14–21. [CrossRef]
5. Arora, S.; Singh, S. Node Localization in Wireless Sensor Networks Using Butterfly Optimization Algorithm. *Arab. J. Sci. Eng.* **2017**, *42*, 3325–3335. [CrossRef]
6. Arora, S.; Singh, S.; Yetilmezsoy, K. A modified butterfly optimization algorithm for mechanical design optimization problems. *J. Braz. Soc. Mech. Sci.* **2018**, *40*, 1–17. [CrossRef]
7. Singh, B.; Anand, P. A novel adaptive butterfly optimization algorithm. *Int. J. Comput. Mater. Sci. Eng.* **2019**, *7*, 1850026. [CrossRef]
8. Sharma, S.; Saha, A.K. m-MBOA: A novel butterfly optimization algorithm enhanced with mutualism scheme. *Soft Comput.* **2019**, *24*, 4809–4827. [CrossRef]
9. Yuan, Z.; Wang, W.; Wang, H.; Khodaei, H. Improved Butterfly Optimization Algorithm for CCHP Driven by PEMFC. *Appl. Therm. Eng.* **2019**, *173*, 114766.
10. Li, G.; Shuang, F.; Zhao, P.; Le, C. An Improved Butterfly Optimization Algorithm for Engineering Design Problems Using the Cross-Entropy Method. *Symmetry* **2019**, *11*, 1049. [CrossRef]
11. Wen, L.; Cao, Y. A hybrid intelligent predicting model for exploring household CO_2 emissions mitigation strategies derived from butterfly optimization algorithm. *Sci. Total Environ.* **2020**, *727*, 138572. [CrossRef]

12. Tan, L.S.; Zainuddin, Z.; Ong, P. Wavelet neural networks based solutions for elliptic partial differential equations with improved butterfly optimization algorithm training. *Appl. Soft Comput.* **2020**, *95*, 106518. [CrossRef]
13. Malisetti, N.R.; Pamula, V.K. Performance of Quasi Oppositional Butterfly Optimization Algorithm for Cluster Head Selection in WSNs. *Procedia Comput. Sci.* **2020**, *171*, 1953–1960. [CrossRef]
14. Sharma, T.K.; Kumar Sahoo, A.; Goyal, P. Bidirectional butterfly optimization algorithm and engineering applications. *Mater. Today Proc.* **2020**, in press. [CrossRef]
15. Holland, J.H. Genetic Algorithms. *Sci. Am.* **1992**, *267*, 66–72. [CrossRef]
16. McCall, J. Genetic algorithms for modelling and optimisation. *J. Comput. Appl. Math.* **2005**, *184*, 205–222. [CrossRef]
17. Storn, R.; Price, K. Differential Evolution—A Simple and Efficient Heuristic for global Optimization over Continuous Spaces. *J. Glob. Optim.* **1997**, *11*, 341–359. [CrossRef]
18. Kennedy, J.; Eberhart, R. Particle Swarm Optimization. In Proceedings of the IEEE International Conference on Neural Networks, Perth, Australia, 27 November–1 December 1995.
19. Dorigo, M.; Di Car, G. Ant Colony Optimization: A New Meta-Heuristic. In Proceedings of the 1999 Congress on Evolutionary Computation-CEC99 (Cat. No. 99TH8406), Washington, DC, USA, 6–9 July 1999; pp. 1470–1477.
20. Karaboga, D.; Basturk, B. A powerful and efficient algorithm for numerical function optimization: Artificial bee colony (ABC) algorithm. *J. Glob. Optim.* **2007**, *39*, 459–471. [CrossRef]
21. Rashedi, E.; Nezamabadi-pour, H.; Saryazdi, S. GSA: A Gravitational Search Algorithm. *Inf. Sci.* **2009**, *179*, 2232–2248. [CrossRef]
22. Mirjalili, S. SCA: A Sine Cosine Algorithm for solving optimization problems. *Knowl. Based Syst.* **2016**, *96*, 120–133. [CrossRef]
23. Hashim, F.A.; Houssein, E.H.; Mabrouk, M.S.; Al-Atabany, W.; Mirjalili, S. Henry gas solubility optimization: A novel physics-based algorithm. *Future Gener. Comput. Syst.* **2019**, *101*, 646–667. [CrossRef]
24. Yang, X. A New Metaheuristic Bat-Inspired Algorithm. In *Nature Inspired Cooperative Strategies for Optimization*; Springer: Berlin/Heidelberg, Germany, 2010; pp. 65–74. [CrossRef]
25. Gandomi, A.H.; Alavi, A.H. Krill herd: A new bio-inspired optimization algorithm. *Commun. Nonlinear Sci.* **2012**, *17*, 4831–4845. [CrossRef]
26. Pan, W. A new Fruit Fly Optimization Algorithm: Taking the financial distress model as an example. *Knowl. Based Syst.* **2012**, *26*, 69–74. [CrossRef]
27. Mirjalili, S.; Mirjalili, S.M.; Lewis, A. Grey Wolf Optimizer. *Adv. Eng. Softw.* **2014**, *69*, 46–61. [CrossRef]
28. Mirjalili, S. Moth-flame optimization algorithm: A novel nature-inspired heuristic paradigm. *Knowl. Based Syst.* **2015**, *89*, 228–249. [CrossRef]
29. Mirjalili, S.; Lewis, A. The Whale Optimization Algorithm. *Adv. Eng. Softw.* **2016**, *95*, 51–67. [CrossRef]
30. Mirjalili, S.; Gandomi, A.H.; Mirjalili, S.Z.; Saremi, S.; Faris, H.; Mirjalili, S.M. Salp Swarm Algorithm: A bio-inspired optimizer for engineering design problems. *Adv. Eng. Softw.* **2017**, *114*, 163–191. [CrossRef]
31. Saremi, S.; Mirjalili, S.; Lewis, A. Grasshopper Optimisation Algorithm: Theory and application. *Adv. Eng. Softw.* **2017**, *105*, 30–47. [CrossRef]
32. Faramarzi, A.; Heidarinejad, M.; Mirjalili, S.; Gandomi, A.H. Marine Predators Algorithm: A nature-inspired metaheuristic. *Expert Syst. Appl.* **2020**, *152*, 113377. [CrossRef]
33. Li, M.D.; Zhao, H.; Weng, X.W.; Han, T. A novel nature-inspired algorithm for optimization: Virus colony search. *Adv. Eng. Softw.* **2016**, *92*, 65–88. [CrossRef]
34. Mirjalili, S.; Gandomi, A.H. Chaotic gravitational constants for the gravitational search algorithm. *Appl. Soft Comput.* **2017**, *53*, 407–419. [CrossRef]
35. Faramarzi, A.; Heidarinejad, M.; Stephens, B.; Mirjalili, S. Equilibrium optimizer: A novel optimization algorithm. *Knowl. Based Syst.* **2020**, *191*, 105190. [CrossRef]
36. Zhen, Z.; Wang, Z.; Gu, Z.; Liu, Y. A Novel Memetic Algorithm for Global Optimization Based on PSO and SFLA. *Lecture Notes Comput. Sci.* **2007**, *4683*, 127–136.
37. Niu, B.; Li, L. A Novel PSO-DE-Based Hybrid Algorithm for Global Optimization. *Lecture Notes Comput. Sci.* **2008**, *5227*, 156–163.

38. Lai, X.; Zhang, M. An efficient ensemble of GA and PSO for real function optimization. In Proceedings of the 2009 2nd IEEE International Conference on Computer Science and Information Technology, Beijing, China, 8–11 August 2009; pp. 651–655.
39. Mirjalili, S.; Hashim, S.Z.M. A new hybrid PSOGSA algorithm for function optimization. In Proceedings of the International Conference on Computer and Information Application (ICCIA 2010), Tianjin, China, 3–5 December 2010; pp. 374–377.
40. Wang, G.; Gandomi, A.H.; Alavi, A.H.; Deb, S. A hybrid method based on krill herd and quantum-behaved particle swarm optimization. *Neural Comput. Appl.* **2016**, *27*, 989–1006. [CrossRef]
41. Trivedi, I.N.; Jangir, P.; Kumar, A.; Jangir, N.; Totlani, R.H.; Totlani, R. A Novel Hybrid PSO-DA Algorithm for Global Numerical Optimization. In *Networking Communication and Data Knowledge Engineering*; Springer: Singapore, 2017; pp. 287–298. [CrossRef]
42. Trivedi, I.N.; Jangir, P.; Kumar, A.; Jangir, N.; Totlani, A.R. A Novel Hybrid PSO–WOA Algorithm for Global Numerical Functions Optimization. In *Advances in Computer and Computational Sciences*; Springer: Singapore, 2017; pp. 53–60. [CrossRef]
43. Laskar, N.M.; Guha, K.; Chatterjee, I.; Chanda, S.; Baishnab, K.L.; Paul, P.K. HWPSO: A new hybrid whale-particle swarm optimization algorithm and its application in electronic design optimization problems. *Appl. Intell.* **2019**, *49*, 265–291. [CrossRef]
44. Yu, P.; Chen, G. Hopf bifurcation control using nonlinear feedback with polynomial functions. *Int. J. Bifurcat. Chaos* **2004**, *14*, 1683–1704. [CrossRef]
45. Xu, F.; Yu, P.; Liao, X. Global analysis on n-scroll chaotic attractors of modified Chua's circuit. *Int. J. Bifurcat. Chaos* **2009**, *19*, 135–157. [CrossRef]
46. Yu, P.; Lü, J. Bifurcation control for a class of Lorenz-like systems. *Int. J. Bifurcat. Chaos* **2011**, *21*, 2647–2664. [CrossRef]
47. Xu, F.; Yu, P.; Liao, X. Synchronization and stabilization of multi-scroll integer and fractional order chaotic attractors generated using trigonometric functions. *Int. J. Bifurcat. Chaos* **2013**, *23*, 1350145. [CrossRef]
48. Wang, G.; Guo, L.; Gandomi, A.H.; Hao, G.; Wang, H. Chaotic Krill Herd algorithm. *Inf. Sci.* **2014**, *274*, 17–34. [CrossRef]
49. Yousri, D.; AbdelAty, A.M.; Said, L.A.; Elwakil, A.S.; Maundy, B.; Radwan, A.G. Chaotic Flower Pollination and Grey Wolf Algorithms for parameter extraction of bio-impedance models. *Appl. Soft Comput.* **2019**, *75*, 750–774. [CrossRef]
50. Palacios, A. Cycling chaos in one-dimensional coupled iterated maps. *Int. J. Bifurcat. Chaos* **2002**, *12*, 1859–1868. [CrossRef]
51. Meddis, R. Unified analysis of variance by ranks. *Br. J. Math. Stat. Psychol.* **1980**, *33*, 84–98. [CrossRef]
52. Wilcoxon, F. Individual Comparisons by Ranking Methods. *Biometr. Bull.* **1945**, *1*, 80–83. [CrossRef]
53. Caruso, G.; Di Battista, T.; Gattone, S.A. A Micro-level Analysis of Regional Economic Activity through a PCA Approach. *Adv. Intell. Syst. Comput.* **2020**, *1009*, 227–234.

Publisher's Note: MDPI stays neutral with regard to jurisdictional claims in published maps and institutional affiliations.

MDPI

Article

Fuzzy Interpolation with Extensional Fuzzy Numbers

Michal Holčapek, Nicole Škorupová and Martin Štěpnička *

CE IT4Innovations–IRAFM, University of Ostrava, 70103 Ostrava, Czech Republic; michal.holcapek@osu.cz (M.H.); nicole.skorupova@osu.cz (N.Š.)
* Correspondence: martin.stepnicka@osu.cz; Tel.: +42-0553-346-1403

Abstract: The article develops further directions stemming from the arithmetic of extensional fuzzy numbers. It presents the existing knowledge of the relationship between the arithmetic and the proposed orderings of extensional fuzzy numbers—so-called $\mathcal{S}$-orderings—and investigates distinct properties of such orderings. The desirable investigation of the $\mathcal{S}$-orderings of extensional fuzzy numbers is directly used in the concept of $\mathcal{S}$-function—a natural extension of the notion of a function that, in its arguments as well as results, uses extensional fuzzy numbers. One of the immediate subsequent applications is fuzzy interpolation. The article provides readers with the basic fuzzy interpolation method, investigation of its properties and an illustrative experimental example on real data. The goal of the paper is, however, much deeper than presenting a single fuzzy interpolation method. It determines direction to a wide variety of fuzzy interpolation as well as other analytical methods stemming from the concept of $\mathcal{S}$-function and from the arithmetic of extensional fuzzy numbers in general.

Keywords: extensional fuzzy numbers; MI-algebras; similarity; arithmetics of fuzzy numbers; orderings; fuzzy interpolation; economic data

Citation: Holčapek, M.; Škorupová, N.; Štěpnička, M. Fuzzy Interpolation with Extensional Fuzzy Numbers. *Symmetry* **2021**, *13*, 170. https://doi.org/10.3390/sym13020170

Received: 27 December 2020
Accepted: 20 January 2021
Published: 22 January 2021

1. Introduction

Fuzzy interpolation is a generally well-established area with crucial results. Indeed, it is a natural area that has a deep connection to distinct areas. We can consider, e.g., the interpolation of fuzzy rule bases with logical motivations stemming from approximate reasoning where the pairs of antecedent and consequent fuzzy sets of a fuzzy rule base establish "pairs of fuzzy points" and the task is to infer an appropriate output fuzzy set for an input fuzzy sets lying in between of the antecedents. In this logical setting, such an interpolation task leads to the solvability of fuzzy relational equations—an area initiated by Sanchez [1] and followed by many others, see [2–5]. The interest of the community in this area does not seem to disappear [1] as it moves to the investigation of distinct extensions [6–8].

Another direction of fuzzy interpolation was rather geometrically motivated although the connection to fuzzy rule-based systems was declared too. The geometrical approaches often stem from usual interpolations (linear [9] or spline [10]) extended to fuzzy numbers and allow the interpolation of sparse rule bases [11,12]. Due to the importance of the interpolation for the approximate reasoning, decision-making or optimization, we may observe the interest of the community into this direction too [13]. For the comparison of some fundamentals, we refer readers to [14] and we also highlight a work that up to some degree combines both approaches and serves us one of our motivations [15].

Our work follows this attempt of joining both views. It works with pairs of fuzzy numbers representing ague quantities [16] and interpolates even a sparse case. However, the fuzzy numbers as well as the manipulation with them will be deeply stemming from fuzzy relational calculus [17], similarity relations, extensional hulls and other concepts related to the approximate reasoning [18].

In particular, we will deal with the *MI-algebra*-based arithmetics based on *extensional fuzzy numbers*, i.e., extensional hulls of crisp numbers constructed with respect to some similarity relations [19]. Such arithmetic serves as an alternative to other technical approaches [20,21] allowing the formalization of the ideas of Mareš [22] in a way that splits the often time-consuming process of calculating operations on fuzzy numbers into a two-step procedure: one calculating the standard operation on crisp points, the other one on processing the similarity relations forming fuzzy numbers around the crisp points. Indeed, as the extensional fuzzy numbers are nothing else but extensional hulls of crisp numbers, we get genuine representations of vague quantities—collections of points similar to the given crisp point. And operations on such objects, e.g., when summing up *"about 3"* and *"about 5"*, are intuitively performed by a human brain by adding the crisp numbers 3 and 5 to a partial result 8 which is later viewed with a tolerance measure respecting that both entries were imprecise. This tolerance measure is nothing else but a similarity relation that is a result of the operation on similarities used in the construction of both extensional fuzzy numbers.

The difference in the used arithmetic, of course, also changes the interpolation method and its results. Any classical interpolation can be extended to a fuzzy interpolation method, assuming that the used arithmetic is functional and equipped with all the necessary operations. Apart from the arithmetical operations, also other fundamental mathematical concepts, such as ordering of fuzzy numbers, must be developed. Thus, we adopt the orderings of extensional fuzzy numbers introduced in [23] and studied in [24]. It is important to note that such an ordering is tightly connected to the arithmetic itself, which we find essential.

The structure of the article is as follows. Section 2 recalls the basic preliminaries about the arithmetics of extensional fuzzy numbers. Section 3 recalls the approach to the orderings of extensional fuzzy numbers and Section 4 sets up their crucial properties. Section 5 leads the investigation towards fuzzy interpolation that is, later, in Section 6 experimentally demonstrated on an illustrative example from the real practice. Section 7 closes the article with a brief discussion.

2. Preliminaries—Arithmetical Operations on Extensional Fuzzy Numbers

2.1. Motivation and the Main Concepts

The arithmetic of extensional fuzzy numbers has been originally proposed in [25,26] as an alternative to other functional arithmetics that are often stemming from Zadeh's extension principle. Such arithmetics often consider the α-cut calculus and they are more technically oriented based on distinct parametric representations of fuzzy numbers [21,27,28].

The proposed alternative (detail description in [19]), was motivated from the non-preservation of some algebraic properties by the existing approaches, especially the non-existence of inverse elements. This is nothing surprising as the models of vague quantities provided by fuzzy numbers are rather fuzzy intervals than fuzzy numbers, and the absence of an inverse element is a feature of the interval calculus, see [29]. However, as pointed out by Mareš [16], instead of considering only a single crisp neutral element, we may observe infinitely many "neutral-like" elements so, when solving equations with vague quantities, we get the equality of both sides up to "nearly zero". Therefore, there is no absence of the inverse, the operations on inverses only do not lead to the genuine neutral (identity) element but to any of the identity-like elements. This idea of Mareš is then developed in [19] and it leads to the *MI-algebras* (MI stands for many identities). Algebraically, it drops the standard field structure of real numbers $(\mathbb{R}, +, \cdot)$ however, it does not harm its ideas, just incorporates them with a certain vagueness.

Above-recalled alternative approach [29] correctly stems from the fact that the lack of the preserved algebraic properties is caused by the interval character of fuzzy numbers and the non-preservation inherited from interval calculus, is "healed" by using the concept of *gradual numbers* that are conceptually sort of inverse to the usual fuzzy numbers (intervals). Indeed, a gradual fuzzy number $\tilde{a}$ is a mapping from the interval $(0, 1]$ to the real line and the (crisp) real number a is its special case such that $\tilde{a}(\alpha) = a$ for arbitrary $\alpha \in (0,1]$. Then the arithmetic operating on such functions preserves the algebraic properties such as existence of an inverse to any element. Note that the concept is so general that it does not assume any restriction on the "shape" (e.g., on monotonicity or continuity) which is, on one hand, an advantage, on the other hand, it allows dealing with or obtaining functions hardly interpreting any vaguely given quantity.

The arithmetic of extensional fuzzy numbers stemmed from a natural model of a vague quantity and formalized human-style calculus with them. It models a vaguely given quantity *"about a"* as an extensional hull of the crisp value (singleton) a with respect to a given similarity relation. Thus, it has a genuine construction determining a fuzzy set of values similar to a. On one hand, as we stated, it is a fuzzy set of elements so, it leads to a set-like concept that will never possess the arithmetical properties of numbers however, it was not constructed as an (fuzzy) interval and the generating crisp number $a \in \mathbb{R}$ is available for further purposes including the calculus.

The price we pay for such a restriction to extensional hulls of crisp numbers is not high as we do not see other than technical reasons stemming from the usual arithmetics themselves (e.g., results of limits) to consider technical definitions allowing dealing with e.g., upper semi-continuous fuzzy sets as models of fuzzy numbers. In our opinion, the restriction pays off in the subsequent advantages.

Let us fix the notation for a set of fuzzy sets defined on a universe X: $\mathcal{F}(X) = \{A \mid A\colon X \to [0,1]\}$; and of the support of a fuzzy set A: $\mathrm{Supp}(A) = \{x \in \mathbb{R} \mid A(x) > 0\}$. Now, we recall the well-known fundamental definitions adopted to the universe of real numbers.

Definition 1. *[30,31] Let $\otimes$ be a left-continuous t-norm. A binary fuzzy relation $S \in \mathcal{F}(X)$ is called* $\otimes$-similarity *if the following holds*

(i) $S(x,x) = 1$ *(reflexivity)*
(ii) $S(x,y) = S(y,x)$ *(symmetry)*
(iii) $S(x,y) \otimes S(y,z) \le S(x,z)$ *($\otimes$-transitive)*

for all $x,y,z \in \mathbb{R}$.

Definition 2. *[19] Let S be a $\otimes$-similarity relation. It is called* separable *if $S(x,y) = 1$ implies that $x = y$.*

Definition 3. *[19] Let S be a $\otimes$-similarity relation. It is called* shift-invariant *if $S(x,y) = S(x+z, x+z)$ for any $x,y,z \in \mathbb{R}$.*

Convention 1. *Due to the practical importance of the separability property, we assume that all $\otimes$-similarity relations considered in this paper are separable.*

Definition 4. *[30] A fuzzy set $A \in \mathcal{F}(\mathbb{R})$ is called* extensional *regarding a $\otimes$-similarity relation S if the following holds:*

$$A(x) \otimes S(x,y) \le A(y)\,, \quad x,y \in \mathbb{R}\,.$$

Example 1. *Binary fuzzy relations on $\mathbb{R}$ given by*

$$S_p(x,y) = (1 - p|x-y|) \vee 0\,, \quad p > 0 \tag{1}$$

are the shift-invariant $\otimes$-similarity relations where $\otimes$ is the Łukasiewicz t-norm.

Example 2. *Binary fuzzy relations on $\mathbb{R}$ given by*

$$S_p(x,y) = e^{-p|x-y|}\,, \quad p > 0 \tag{2}$$

are the shift-invariant $\otimes$-similarity relations where $\otimes$ is the product t-norm.

Remark 1. *The shift-invariance is a useful property for connecting similarities and arithmetics, see Proposition 6. An example of similarity relation that is not shift-invariant can be found in Remark 2.2 in [19].*

An *extensional hull* of a fuzzy set A is the least fuzzy superset of A that is extensional.

Theorem 1. *[31] Let S be a $\otimes$-similarity relation and let $A \in \mathcal{F}(\mathbb{R})$. The extensional hull of A, denoted by $\mathrm{EXT}(A) \in \mathcal{F}(\mathbb{R})$, is given by*

$$\mathrm{EXT}_S(A)(x) = \bigvee_{y \in \mathbb{R}} (A(y) \otimes S(x,y)). \tag{3}$$

If we consider a singleton representation of a real number $a \in \mathbb{R}$ that is a fuzzy set $\tilde{a} \in \mathcal{F}(\mathbb{R})$ such that $\tilde{a}(a) = 1$, and $\tilde{a}(x) = 0$ for any $x \neq a$ we can construct an extensional hull of a real number. Such an object is called *extensional fuzzy number* or a fuzzy point, see [18,31,32].

Definition 5. *[19,25] Let $a \in \mathbb{R}$ and let S be a $\otimes$-similarity. Then $a_S \in \mathcal{F}(\mathbb{R})$ given as follows:*

$$a_S(x) = \mathrm{EXT}_S(a)(x) \tag{4}$$

is called extensional fuzzy number.

A correct version of Formula (4) would consider the extensional hull of the singleton $\mathrm{EXT}_S(\tilde{a})(x)$ but as there is no danger of confusion, we will neglect the difference between a and $\tilde{a}$ and stick to the simpler denotation. The extensional fuzzy number may be easily calculated by a simple substitution.

Lemma 1. *[19] Let $a \in \mathbb{R}$ and let S be a $\otimes$-similarity. Then $a_S(x) = S(a,x)$, for any $y \in \mathbb{R}$.*

Convention 2. *We restrict our choice of similarities to those ensuring that the fuzzy numbers a_S are formed by α-cuts $(a_S)_\alpha$ that are closed intervals in $\mathbb{R}$.*

2.2. Arithmetic of Extensional Fuzzy Numbers

The construction as well as the denotation of the extensional fuzzy number emphasizes its semantics—number a with close numbers around it—where the closeness is determined by similarity S. Such a restriction to a subclass of fuzzy numbers is reasonable and justified by the most natural representation of a vague quantity reflecting the semantics already in its construction. Furthermore, the construction allows the derivation of appropriate models of the arithmetic [25,26]—algebraically leading to MI-algebraic structures. We avoid going into algebraic details and only refer interested readers to the relevant sources, mainly to [19] and furthermore, to [33,34] for the quotient MI-groups, and to [35] for the topological MI-groups. Here, we continue only with the computational part of the arithmetic of extensional fuzzy numbers.

Consider a system of nested $\otimes$-similarities $\mathcal{S}$ with the bottom element $\perp_{\mathcal{S}}$ and the set of all fuzzy numbers that are extensional with respect to a similarity from the given system:

$$\mathcal{F}_{\mathcal{S}}(\mathbb{R}) = \{a_S \mid a \in \mathbb{R} \text{ and } S \in \mathcal{S}\}$$

preserving the Convention 2. Then the arithmetic operations on $\mathcal{F}_{\mathcal{S}}(\mathbb{R})$ are given as

$$a_S + b_T = (a+b)_{\max(S,T)}\,, \quad a_S \cdot b_T = (a \cdot b)_{\max(S,T)}\,, \quad S,T \in \mathcal{S} \tag{5}$$

with the maximum operating on the set of similarities $\mathcal{S}$ is defined via the inclusion:

$$S \subseteq T \quad \text{if} \quad S(x,y) \leq T(x,y)\,, \quad x,y \in \mathbb{R}\,.$$

The operations given by (5) are motivated by calculus performed by humans. Indeed, when one is asked to sum up "about 20" and "about 30", she or he ends up with "about 50" without any technical (α-cut-based) calculus simply by summing up 20 and 30 to 50, and by adding some neighborhood. The arithmetic (5) acts analogously, it sums up a and b to $a+b$ in the first phase, and, in the second phase, performs some operation on similarities to determine what are the acceptably close numbers. This approach is very fast, and it does not result inevitably wide fuzzy numbers after a few operations.

Remark 2. *The proposed calculus that does not widen the resulting fuzzy sets may be for some readers unusual, yet we do not view it unintuitive. Let us recall the argumentation used in [36] cit. "Suppose that we have two factors that affect the accuracy of a measuring instrument. One factor leads to errors ±10%—meaning that the resulting error component can take any value from* -10% *to* $+10\%$*. The second factor leads to errors of ±0.1%. What is the overall error? From the purely mathematical viewpoint, the largest possible error is 10.1%. However, from the common sense viewpoint, an engineer would say: 10%." These arguments in favor of common-sense in summing up errors (in our case similarities) are not left only on the motivation level but elaborated also mathematically based on the Hurwicz criterion [37], for details see [36].*

The bottom element $\perp_{\mathcal{S}} \in \mathcal{S}$ (the "narrowest" similarity) is used in the construction of the so-called *strong identity* elements for the arithmetic operations, in particular:

$$\mathbf{0} = 0_{\perp_{\mathcal{S}}} = \perp_{\mathcal{S}}(0,\cdot), \quad \mathbf{1} = 1_{\perp_{\mathcal{S}}} = \perp_{\mathcal{S}}(1,\cdot)\,.$$

Inverses are obtained using the arithmetic operations, i.e., $-(a_S) = (-a)_S$ and $(a_S)^{-1} = (a^{-1})_S$, where the division by "zero" must be omitted. The (non-strong) identities (*pseudoidentities*) are elements of $\mathcal{F}_{\mathcal{S}}(\mathbb{R})$ obtained as results of operations applied to elements and their inverses:

$$I^0_{\mathcal{S}} = \{a_S + (-a)_S \mid a_S \in \mathcal{F}_{\mathcal{S}}(\mathbb{R})\}\,, \quad I^1_{\mathcal{S}} = \{a_S \cdot (a^{-1})_S \mid a_S \in \mathcal{F}_{\mathcal{S}}(\mathbb{R}) \setminus I^0_{\mathcal{S}}\}\,.$$

Example 3. *The set* $\mathcal{S} = \{S_p \mid p \in [\ell, r]\}$ *where* S_p *is given by (1) or (2) and is a parametric system of nested similarities with the bottom element* $\perp_{\mathcal{S}} = S_r$*. The arithmetic operations are then given as follows:*

$$a_{S_p} + b_{S_{p'}} = (a+b)_{S_{p''}}\,, \quad a_{S_p} \cdot b_{S_{p'}} = (a \cdot b)_{S_{p''}}\,, \quad \text{where } p'' = \min\{p, p'\}\,.$$

Remark 3. *As the classical equality "=" is a* $\otimes$*-similarity for arbitrary t-norm* $\otimes$*, we can consider the systems of similarities from Example 3 with the parametrization* $p \in [\ell, +\infty)$ *and additionally defined* S_∞ *as the crisp equality. Then by adding this element to* $\mathcal{S}$*, we obtain a system* $\mathcal{S}_\infty = \mathcal{S} \cup S_\infty$ *allowing us to deal with crisp numbers as well. The bottom element will be then formed by the crisp equality* $\perp_{\mathcal{S}_\infty} = S_\infty$*.*

Let us fix the denotation $\mathcal{S}_\infty$ for a system containing the crisp equality and let us call them *systems of nested similarities with the crisp bottom element*.

For the purposes of our work, it is sufficient to recall a single particular structure with properties sufficient for the further studies. It is an *MI-prefield* $(\mathcal{F}_{\mathcal{S}}(\mathbb{R}), +, -, \cdot, ^{-1})$ preserving: the existence of the sets of pseudoidentities $I^0_{\mathcal{S}}$ and $I^1_{\mathcal{S}}$; associativity and commutativity of both arithmetic operations; closeness of the sets of pseudoidentitites with respect to the arithmetic operations (e.g., $a_S + b_T \in I^0_{\mathcal{S}}$ for any $a_S, b_T \in I^0_{\mathcal{S}}$); existence of inverse elements; and the distributive law:

$$a_R \cdot (b_S + c_T) = a_R \cdot b_S + a_R \cdot c_T\,.$$

As the MI-prefield structure possess most of the appropriate properties, though we often do not need such a rich structure in the subsequent sections of the article, we will always assume we are given an MI-prefield. This assumption will help the brevity and readability of the article.

3. Orderings of Extensional Fuzzy Numbers

3.1. Motivation

The necessity of defining the ordering (also ranking) of fuzzy numbers (or generally even fuzzy sets) has been pointed out already by Zadeh [38] and then in many other works, e.g., in [39–41] leading to distinct applications [42]. Moreover, the ordering is essential for defining monotonous fuzzy rule bases [43–45] and reasoning with them which is a topic extremely closely connected to the area of fuzzy interpolation [7,8].

The simplest yet maybe most natural example of an ordering is the ordering of intervals

$$[a,b] \leq_i [c,d] \;\Leftrightarrow\; a \leq c \text{ and } b \leq d$$

applied to all α-cuts of the fuzzy numbers:

$$A \leq_i B \;\Leftrightarrow\; A_\alpha \leq_i B_\alpha \quad \forall \alpha \in (0,1]\,. \tag{6}$$

In the following lemma, we show two equivalent definition of the ordering of fuzzy numbers using the orderings of their α-cuts.

Lemma 2. *Let A, B be fuzzy numbers. Then the following statements are equivalent.*

(1) $A \leq_i B$,

(2) *for any $\alpha \in (0,1]$ and $x \in A_\alpha$, there exists $y \in B_\alpha$ such that $x \leq y$, and vice versa, for any $y \in B_\alpha$, there exists $x \in A_\alpha$ such that $x \leq y$.*

(3) *for any $x \in \mathbb{R}$, there exists $y \in \mathbb{R}$ such that $x \leq y$ and $A(x) \leq B(y)$, and vice versa, for any $y \in \mathbb{R}$, there exists $x \in \mathbb{R}$ such that $x \leq y$ and $B(y) \leq A(x)$.*

Proof. (1) $\Rightarrow$ (2) Let $\alpha \in (0,1]$. Assume that $x \in A_\alpha$. Since $A_\alpha \leq_i B_\alpha$, it is sufficient to put $y = \max B_\alpha$, and we obtain that $y \in B_\alpha$ such that $x \leq y$. Similarly, if $y \in B_\alpha$, from $A_\alpha \leq_i B_\alpha$, it is sufficient to put $x = \min A_\alpha$, and we obtain that $x \in A_\alpha$ such that $x \leq y$.

(2) $\Rightarrow$ (3) Let $x \in \mathbb{R}$. If $A(x) = 0$, then it is sufficient to consider any $y \in \mathbb{R}$ such that $x \leq y$. Assume that $A(x) = \alpha > 0$. By (2), there exists $y \in B_\alpha$ such that $x \leq y$ and $A(x) \leq B(y)$ holds. Let $y \in \mathbb{R}$. Again, if $B(y) = 0$, then it is sufficient to consider any $x \in \mathbb{R}$ such that $x \leq y$. If $\alpha = B(y)$, then by (2), we find that there exists $x \in A_\alpha$ such that $x \leq y$ and $B(y) \leq A(x)$ holds.

(3) $\Rightarrow$ (1) Let $\alpha \in (0,1]$, and consider $A_\alpha = [a,b]$ and $B_\alpha = [c,d]$. By (3), for $b \in A_\alpha$, there exists $y \in \mathbb{R}$ such that $\alpha \leq A(b) \leq B(y)$ and $b \leq y$. Since $y \in B_\alpha$, we find that $b \leq y \leq d$. Similarly, for $c \in B_\alpha$, there exists $x \in \mathbb{R}$ such that $\alpha \leq B(c) \leq A(x)$ and $x \leq c$. Since $x \in A_\alpha$, we find that $a \leq x \leq b$. Hence, we obtain that $A_\alpha \leq_i B_\alpha$. □

This approach, however, does not preserve any information about the vagueness so, we will not apply it to extensional fuzzy numbers $\mathcal{F}_S(\mathbb{R})$. Indeed, general fuzzy sets or technically defined fuzzy numbers A, B do not necessarily encode the information on the position and the neighborhood unlike the extensional numbers that are generically created from these two sources of the information.

Another flaw of the α-cut-based ordering is that it is not a total ordering and, for example, two fuzzy numbers depicted in Figure 1 are not comparable. The totality in the strict sense, as with the one we know from classical mathematics, is not an unavoidable property. However, cases such as the one in Figure 1 are so intuitive that anyone would expect that A is smaller than B. Incidentally, appropriate arithmetic confirms this intuition as it leads to a positive result of the difference $(B - A)$.

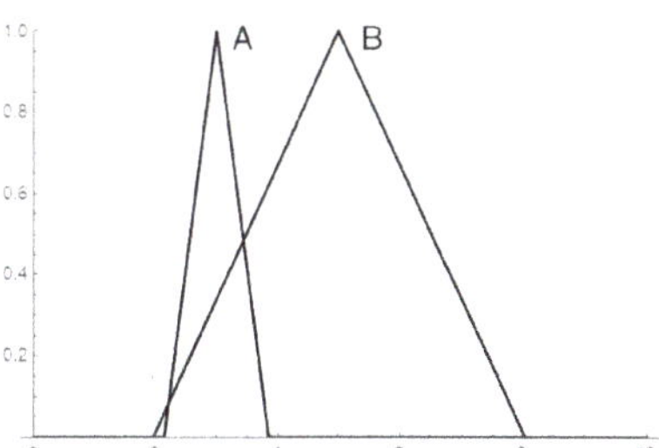

Figure 1. Fuzzy numbers modeling the vague quantity *"about 3"* (A), and the vague quantity *"about 5"* (B).

There are different approaches attempting to avoid the problem of the missing totality, mostly they consist of determining a certain ranking index (a crisp number representing the fuzzy number). On the other hand, this leads to a reduction of information. One of the exceptions that we found very elegant and formally founded was defined by Bodenhofer in [46,47]. It works with extending the original fuzzy numbers by constructing their extensional hulls up to the situation when the extended fuzzy numbers can be already ordered regarding their α-cuts. Naturally, this approach is very closely connected to the arithmetic of extensional fuzzy numbers and thus, it served for us a motivation, yet our approach comes from a bit more abstract setting and then lands to a particular type of Bodenhofer's ordering.

3.2. Definitions and Examples

With the goal to encode the necessary vagueness and preserve it as an essential property over the whole computational process, we extend the standard binary ordering relation (between crisp numbers) to a ternary relation $\leq$ operating on the Cartesian product $\mathcal{F}_{\mathcal{S}}(\mathbb{R}) \times \mathcal{F}_{\mathcal{S}}(\mathbb{R}) \times \mathcal{S}$. As there is never a danger of confusion between the order operating on real numbers and the newly defined order operating on extensional fuzzy sets, the difference will be always clear from the context, we will use the same symbol $\leq$.

Definition 6. *Let $(\mathcal{F}_{\mathcal{S}}(\mathbb{R}), +, -, \cdot, ^{-1})$ be an MI-prefield of extensional fuzzy numbers with respect to a system $\mathcal{S}$ of nested $\otimes$-similarities on $\mathbb{R}$. A ternary relation $\leq$ on $\mathcal{F}_{\mathcal{S}}(\mathbb{R}) \times \mathcal{F}_{\mathcal{S}}(\mathbb{R}) \times \mathcal{S}$ is called an* $\mathcal{S}$-ordering *on $\mathcal{F}_{\mathcal{S}}(\mathbb{R})$ provided that*

(O1) $\exists P \in \mathcal{S} : (a_S, a_T, P) \in \leq$, *(reflexivity)*
(O2) $(a_S, b_T, Q) \in \leq \ \& \ (b_T, a_S, R) \in \leq \Rightarrow a_S - b_T \in I^0_{\mathcal{S}}$, *(anti-symmetry)*
(O3) $(a_S, b_T, Q) \in \leq \ \& \ (b_T, c_U, R) \in \leq \Rightarrow \exists P \in \mathcal{S} : (a_S, c_U, P) \in \leq$, *(transitivity)*

holds for any $a, b, c \in \mathbb{R}$ and for any $Q, R, S, T, U \in \mathcal{S}$.

Notation 1. *For the case of brevity and clarity, let us introduce the following notation $(a_S \leq b_T)_U$ denoting the fact that $(a_S, b_T, U) \in \leq$. Furthermore, let $a_S \leq b_T$ denotes the fact that there exists some $U \in \mathcal{S}$ such that $(a_S \leq b_T)_U$. This second notation losses the information on which particular $U \in \mathcal{S}$ must be used to obtain a triplet $(a_S, b_T, U) \in \leq$ however, in certain situations, such a Boolean information about the order of two fuzzy numbers will be sufficient.*

Using Notation 1, the axioms of the $\mathcal{S}$-ordering on $\mathcal{F}_{\mathcal{S}}(\mathbb{R})$ can be rewritten to a very convenient form:

(O1) $\exists P \in \mathcal{S} : (a_S \leq a_T)_P$, (reflexivity)
(O2) $(a_S \leq b_T)_Q \ \& \ (b_T \leq a_S)_R \Rightarrow a_S - b_T \in I^0_{\mathcal{S}}$, (anti-symmetry)
(O3) $(a_S < b_T)_Q \ \& \ (b_T \leq c_U)_R \Rightarrow \exists P \in \mathcal{S} : (a_S \leq c_U)_P$ (transitivity) .

Let us recall some examples that first appeared in a bit different formalism in [23,48].

Example 4. *Let* $(\mathcal{F}_{\mathcal{S}}(\mathbb{R}), +, -, \cdot, ^{-1})$ *be an MI-prefield. Then* $\leq_{\max}$ *defined as*

$$\leq_{\max} = \{(a_S, b_T, \max(S, T)) \mid a, b \in \mathbb{R} \;\&\; S, T \in \mathcal{S}\}$$

is an $\mathcal{S}$*-ordering on* $\mathcal{F}_{\mathcal{S}}(\mathbb{R})$.

The simple denotation of the fact $(a_S, b_T, \max(S, T)) \in \leq_{\max}$ is according to Notation 1 very convenient, in particular, we may write $(a_S \leq_{\max} b_T)_{\max(S,T)}$ or in general $a_S \leq_{\max} b_T$.

Relation $\leq_{\max}$ demonstrates an $\mathcal{S}$-ordering reflecting the arithmetic operations: the width of the result of the arithmetic operations determines also the "width of the fuzzy truth" of the proposition stating that a_S *is smaller or equal to* b_T that is encoded in the third component of the triplet $(a_S, b_T, \max(S, T))$. If we mirror this idea into particular arithmetics from Example 3, we get the following ternary relation:

$$\leq_{\max} = \{(a_{S_p}, b_{S_{p'}}, \min(p, p')) \mid a, b \in \mathbb{R} \;\&\; S_p, S_{p'} \in \mathcal{S}\}$$

where $\mathcal{S} = \{S_p \mid p \in [\ell, r] \mid\}$ and S_p is given by (1). The demonstration of such an ordering is visualized in Figure 2.

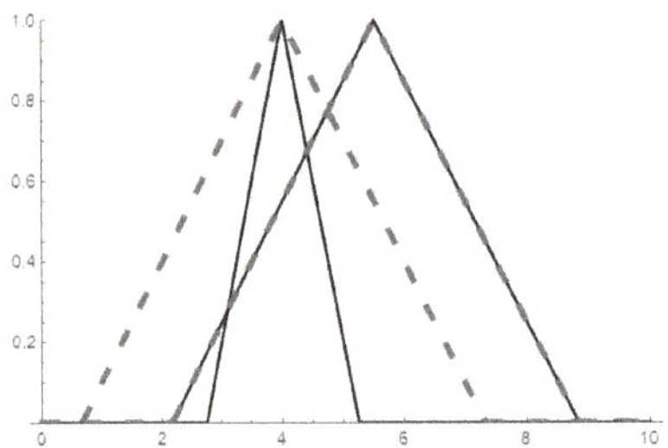

Figure 2. Fuzzy numbers $x_{S_{p'}}$ and $y_{S_{p''}}$ determined by the Łukasiewicz similarities from Example 3 based on values $x = 4$ and $p' = 0.8$ ("left" solid fuzzy set), $y = 5.5$ and $p'' = 0.3$ ("right" solid fuzzy set). Fuzzy sets $x_{S_{p'}}$ and $y_{S_{p''}}$ cannot be ordered by $\leq_i$ of α-cuts however, if we use $\leq_{\max}$, we obtain the dashed fuzzy sets $\mathrm{EXT}_{S_{p''}}(x_{S_{p'}})$ and $\mathrm{EXT}_{S_{p''}}(y_{S_{p''}})$, respectively, and thus $x_{S_{p'}} \leq_{\max} y_{S_{p''}}$.

Let us present another example of an $\mathcal{S}$-ordering. It assumes that set of similarities $\mathcal{S}$ has the greatest element.

Example 5. *Let* $\top_{\mathcal{S}}$ *be the greatest element of* $\mathcal{S}$*, i.e.,* $R \subseteq \top_{\mathcal{S}}$ *for all* $R \in \mathcal{S}$*. Then the ternary relation* $\leq_{\top_{\mathcal{S}}}$ *given by:*

$$\leq_{\top_{\mathcal{S}}} = \{(a_S, b_T, \top_{\mathcal{S}}) \mid a_S, b_T \in \mathcal{F}_{\mathcal{S}}(\mathbb{R}) \;\&\; \exists R \in \mathcal{S} : \mathrm{EXT}_R(a_S) \leq_i \mathrm{EXT}_R(b_T)\}$$

is an $\mathcal{S}$*-ordering.*

The order $\leq_{\top_{\mathcal{S}}}$ orders two fuzzy numbers if there exists a similarity such that the extensional hulls of both fuzzy numbers with respect to this similarity are α-cut ordered. Here we see a direct link to the works of Bodenhofer [46,47] that will not be here commented more detail but will turn to be very important in the subsequent sections.

Remark 4. *Note that the particular similarity relation R that was used to order the extensional hulls of two fuzzy numbers via their* α*-cuts is not mirrored in the information* $(a_S \leq_{\top_{\mathcal{S}}} b_T)_{\top_{\mathcal{S}}}$*. This feature, which is not present here, will also turn to be important in the subsequent sections. It actually tells us that* $\leq_{\top_{\mathcal{S}}}$ *is not that appropriate* $\mathcal{S}$*-ordering as it does not provide meaningful all three components of the triplet. But it is important to mention that even such structures meet the axiomatic definition of the* $\mathcal{S}$*-ordering.*

Let us consider, e.g., the product t-norm and

$$\mathcal{S} = \{S_p \mid p \in [1,5] \text{ and } S_p(x,y) = e^{-p|x-y|}\}.$$

The ordering $\leq_{\top_\mathcal{S}}$ can be visually demonstrated by Figure 3 where one can see two extensional fuzzy numbers $4_{S_{2.5}}$ and $5.5_{S_{1.2}}$ (displayed by solid lines) and their extensional hulls $\text{EXT}_{S_5}(4_{S_2.5}) = 4_{S_5}$ (left dashed fuzzy set) and $\text{EXT}_{S_5}(5.5_{S_{1.2}}) = 5.5_{S_5}$ (right dashed fuzzy set) that due to their interval-ordering $4_{S_5} \leq_i 5.5_{S_5}$ allow the ordering of the original fuzzy numbers $(4_{S_{2.5}} \leq_{\top_\mathcal{S}} 5.5_{S_{1.2}})_{\top_\mathcal{S}}$.

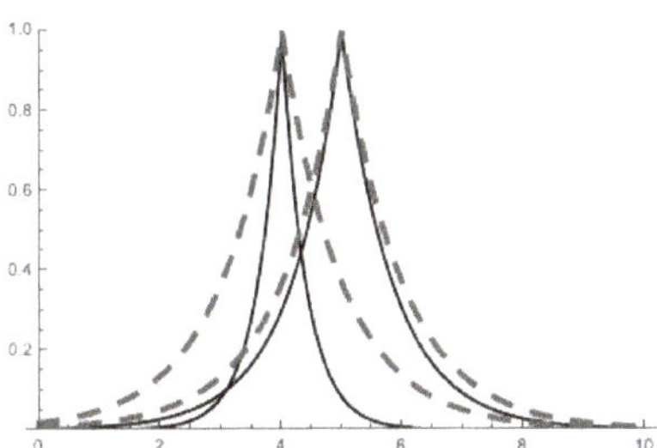

Figure 3. Demonstration of the $\mathcal{S}$-ordering $\leq_{\top_\mathcal{S}}$.

Let us equip the system of similarities $\mathcal{S}$ with its bottom element $\perp_\mathcal{S}$. Then, as $\mathcal{S}$ is the system of nested similarities, it has all infima, i.e., $\inf\{S \in \mathcal{C} \mid \mathcal{C} \subseteq \mathcal{S}\}$ exists for any subset $\mathcal{C}$. Then, we can define the ordering that determines the third component of the triplet (the tolerance added to the ordering) as the "narrowest" similarity relation that is sufficient to get ordered α-cut of extensional hulls (with respect to this similarity) of the given fuzzy numbers. Mathematically, such a $\mathcal{S}$-ordering $\leq_{\text{inf}}$ will be given by the intersection of all such similarities that lead to the interval-ordering of α-cuts of the extensional hulls:

$$\leq_\cap = \{(a_S, b_T, E) \mid a_S, b_T \in \mathcal{F}_\mathcal{S}(\mathbb{R}) \;\&\; E = \cap\{R \in \mathcal{S} \mid \text{EXT}_R(a_S) \leq_i \text{EXT}_R(b_T)\}\}.$$

The visualization of the ordering $\leq_\cap$ is provided in Figure 4.

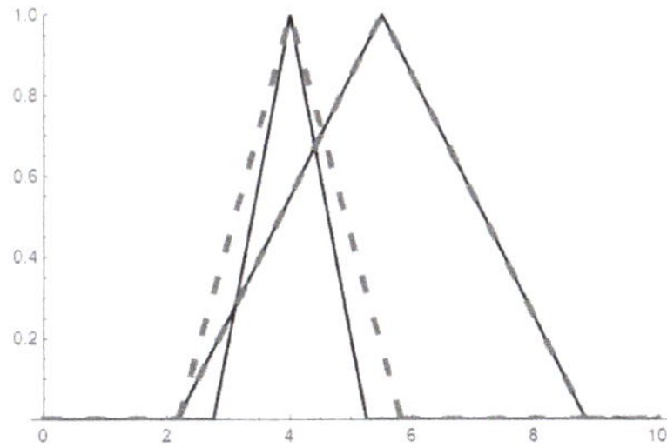

Figure 4. Example of $\leq_\cap$ applied to fuzzy sets from Figure 2.

Let us now come back to the properties of the $\mathcal{S} - orderings$ that were foreshadowed above. In particular, we will recall the so-called *pre-order compatibility* [23,24] and the *real-order* compatibility [24].

Definition 7. *Let $\leq$ be an $\mathcal{S}$-ordering on an MI-prefield $(\mathcal{F}_\mathcal{S}(\mathbb{R}), +, -, \cdot, ^{-1})$. Then $\leq$ is called* pre-order compatible *if for any $a_S, b_T \in \mathcal{F}_\mathcal{S}(\mathbb{R})$ and $R \in \mathcal{S}$,*

$$(a_S \leq b_T)_R \Rightarrow \text{EXT}_R(a_S) \leq_i \text{EXT}_R(b_T). \tag{7}$$

The name of the pre-order compatibility comes from the close connection to the works of Bodenhofer where the author dealt with pre-orderings of fuzzy sets, see [46,47]. The nature of the property is indeed such that it puts the given $\mathcal{S}$-ordering into the perspective of the above-mentioned pre-orderings. In other words, an $\mathcal{S}$-ordering that is pre-order compatible reflects in the third component of the triplet the width (similarity) of the extension that is necessary to obtain α-cut ordered extensional hulls of given fuzzy numbers.

The provided examples of $\mathcal{S}$-orderings $\leq_{\max}$, $\leq_{T_S}$, and $\leq_{\cap}$ are examples of pre-order compatible orderings. Now, we will show how easy it is to construct $\mathcal{S}$-orderings that will meet the axioms of Definition 6 but will not be pre-order compatible. There will be two such examples, each of them harming the pre-order compatibility from a different perspective. The first example harms the reflection of the necessary extension.

Example 6. *Let $\mathcal{S}$ be equipped with the bottom element $\bot_{\mathcal{S}}$. Then the ternary relation $\leq_{\bot_{\mathcal{S}}}$ given by*

$$\leq_{\bot_{\mathcal{S}}}=\{(a_S,b_T,\bot_{\mathcal{S}})\mid a_S,b_T\in\mathcal{F}_{\mathcal{S}}(\mathbb{R})\ \&\ a\leq b\}$$

is an $\mathcal{S}$-ordering that is not pre-order compatible.

The construction of $\leq_{\bot_{\mathcal{S}}}$ focuses only on the core elements—crisp numbers $a,b\in\mathbb{R}$ and ignores the widths of the extensional fuzzy numbers $a_S,b_T\in\mathcal{F}_{\mathcal{S}}(\mathbb{R})$ constructed above them. In other words, the fuzzy truth (similarity, tolerance) of the proposition a_S *is smaller or equal to* b_T is thus meaningless, it does not store any information and we get only binary information on the mutual position of a and b.

The other way how the pre-order compatibility can be easily harmed is presented in the following example that reverses the linear order on $\mathbb{R}$.

Example 7. *The ternary relation $\leq^{rev}_{\max}$ defined as follows*

$$\leq^{rev}_{\max}=\{(a_S,b_T,\max(S,T))\mid a_S,b_T\in\mathcal{F}_{\mathcal{S}}(\mathbb{R})\ \&\ b\leq a\}$$

is an $\mathcal{S}$-ordering that is not pre-order compatible.

Unlike the $\mathcal{S}$-ordering $\leq_{\bot_{\mathcal{S}}}$, the construction of $\leq^{rev}_{\max}$ reflects the necessary extension and the third component of the triplet would not be meaningful; however, it reverses the order. Thus, we get the information that "about 4" below "about 2". Please note that reversed order is still order so, nothing wrong in a sense that $\leq^{rev}_{\max}$ would harm Definition 6.

Let us present a lemma which demonstrates the practical impact of the pre-order compatibility as the crucial property of $\mathcal{S}$-orderings.

Lemma 3. *Let $\leq$ be a pre-order compatible $\mathcal{S}$-ordering on an MI-prefield $(\mathcal{F}_{\mathcal{S}}(\mathbb{R}),+,-,\cdot,^{-1})$. If $a_S\leq b_T$ for certain $S,T\in\mathcal{S}$ then $a\leq b$.*

Proof. Let $\leq$ be pre-order compatible. Then $a_S\leq b_T$ (according to Notation 1) means that there exists $U\in\mathcal{S}$ such that $\mathrm{EXT}_R(a_S)\leq_i \mathrm{EXT}_R(b_T)$. Let us expand the formula for the extensional hull of a_S

$$\mathrm{EXT}_R(a_S)(z)=\bigvee_{y\in\mathbb{R}} a_S(y)\otimes R(y,z)$$

and let us determine its core, i.e., let us put it equal to 1:

$$1=\bigvee_{y\in\mathbb{R}} a_S(y)\otimes R(y,z)=\bigvee_{y\in\mathbb{R}} S(a,y)\otimes R(y,z)\leq$$
$$\bigvee_{y\in\mathbb{R}}\max(S,R)(a,y)\otimes\max(S,R)(y,z)\leq\max(S,R)(a,z),$$

Due to the separability property assumed in Convention 1, we get $\max(S,R)(a,z) = 1$ if and only if $a = z$. Thus, the sought core (α-cut for $\alpha = 1$) is $\mathrm{EXT}_R(a_S)_1 = \{a\}$. Analogously, we get $\mathrm{EXT}_R(b_T)_1 = \{b\}$. Finally, $\mathrm{EXT}_R(a_S)_1 \leq \mathrm{EXT}_R(b_T)_1$ can be directly rewritten as $a \leq b$, which concludes the proof. □

Example 8. *Let us demonstrate Lemma 3 on examples. Consider the case depicted in Figure 2. For the pre-order compatible $\mathcal{S}$-ordering $\leq_{\max}$, the assumptions of the lemma are met, and it can be seen that $x_{S_{p'}} \leq_{\max} y_{S_{p''}}$ as well as $x \leq y$ holds. For the $\mathcal{S}$-ordering $\leq^{rev}_{\max}$ the assumption on the pre-order compatibility is dropped and from $y_{S_{p''}} \leq^{rev}_{\max} x_{S_{p'}}$ we cannot deduce that $y \leq x$ as the opposite is true.*

However, Lemma 3 claims only one direction implication, not the equivalence, and so, it does not exclude the possibility of a pre-order $\mathcal{S}$-ordering that would preserve the implication. Indeed, consider $\leq_{\perp_S}$ that is defined in such a way that from $x_{S_{p'}} \leq_{\perp_S} y_{S_{p''}}$ we can easily deduce $x \leq y$.

Naturally, one may revisit the pre-order compatibility defining Formula (7) and consider it with the equivalence instead of the implication. However, such a condition would be too restrictive and the only order that would in general meet such a requirement would be $\leq_\cap$. However, we can weaken the opposite implication and require only weaken version that ensures some dependence on the order of the real line but without reflecting the particular similarity. The reflection of the necessarily used similarity mirrored only in one of the implications—the one in Formula (7)—will turn to be sufficient, see Section 4.

Definition 8. *Let $\leq$ be an $\mathcal{S}$-ordering on an MI-prefield $(\mathcal{F}_\mathcal{S}(\mathbb{R}), +, -, \cdot, ^{-1})$. Then $\leq$ is called* real-order compatible *if for any $a_S, b_T \in \mathcal{F}_\mathcal{S}(\mathbb{R})$ and $R \in \mathcal{S}$ the following holds*

$$\mathrm{EXT}_R(a_S) \leq_i \mathrm{EXT}_R(b_T) \Rightarrow a_S \leq b_T\,. \tag{8}$$

The $\mathcal{S}$-orderings $\leq_{\max}$, $\leq_{\mathrm{T}_\mathcal{S}}$, $\leq_\cap$, and $\leq_{\perp_S}$ presented above are real-order compatible however, for example, $\leq^{rev}_{\max}$ is not real-order compatible. Can we construct an $\mathcal{S}$-ordering that is unlike $\leq^{rev}_{\max}$ pre-order compatible however, it is not real-order compatible? Indeed, the following example focusing only on the left points of the supports provides such an example.

Example 9. *The ternary relation $\leq^L_{\max}$ defined as follows*

$$\leq^L_{\max} = \{(a_S, b_T, \max(S,T)) \mid a_S, b_T \in \mathcal{F}_\mathcal{S}(\mathbb{R})\ \&\ \inf \mathrm{Supp}(a_S) \leq \inf \mathrm{Supp}(b_T)\}$$

is an $\mathcal{S}$-ordering that is pre-order compatible but it is not real-order compatible.

Now, putting both compatibilities together gives us an $\mathcal{S}$-ordering with properties sufficient for proving further results.

Definition 9. *If an $\mathcal{S}$-ordering is pre-order compatible and real-order compatible, it is said to be* strongly compatible.

Remark 5. *The $\mathcal{S}$-orderings $\leq_{\max}$, $\leq_{\mathrm{T}_\mathcal{S}}$, and $\leq_\cap$ presented above are strongly compatible.*

Another crucial property is the linearity or totality of the order. Naturally, also this property deserves to be incorporated into the context of $\mathcal{S}$-orderings.

Definition 10. *Consider an MI-prefield $(\mathcal{F}_\mathcal{S}(\mathbb{R}), +, -, \cdot, ^{-1})$ and an $\mathcal{S}$-ordering $\leq$ on $\mathcal{F}_\mathcal{S}(\mathbb{R})$. Then $\leq$ is called* total *if the following holds for arbitrary $a_S, b_T \in \mathcal{F}_\mathcal{S}(\mathbb{R})$:*

$$a_S \leq b_T \quad or \quad b_T \leq a_S\,. \tag{totality}$$

Let us focus on the relationship of the real-order compatibility and the totality.

Proposition 1. *Let* $(\mathcal{F}_{\mathcal{S}}(\mathbb{R}), +, -, \cdot, ^{-1})$ *be an MI-prefield and let* $\leq$ *be an* $\mathcal{S}$*-ordering on* $\mathcal{F}_{\mathcal{S}}(\mathbb{R})$*. If* $\leq$ *is real-order compatible then it is also total.*

Proof. We prove the proposition by contradiction. Let us assume that for some $a_S, b_T \in \mathcal{F}_{\mathcal{S}}(\mathbb{R})$ it holds that $a_S \not\leq b_T$ & $b_T \not\leq a_S$.

The assumption $a_S \not\leq b_T$ means that there does not exist any $R \in \mathcal{S}$ such that $\mathrm{EXT}_R(a_S) \leq_i \mathrm{EXT}_R(b_T)$. Indeed, if it existed, the real-order compatibility would lead to the contradiction with the assumption. As for any α, the α-cuts $\mathrm{EXT}_R(a_S)_\alpha$ and $\mathrm{EXT}_R(b_T)_\alpha$ are of the same width and $\mathrm{EXT}_R(a_S) \not\leq_i \mathrm{EXT}_R(b_T)$ holds then necessarily $\mathrm{EXT}_R(b_T) \leq_i \mathrm{EXT}_R(a_S)$ which, due to the real-order compatibility, means that $b_T \leq a_S$. □

4. Properties of $\mathcal{S}$-Orderings

4.1. Wang–Kerre Properties of Orderings of Fuzzy Numbers

As the orderings and rankings of fuzzy numbers attracted significant interest of the researcher, it is not surprising that some of them aimed at their properties. Many ordering techniques rely on comparing certain crisp numbers (indices or representative numbers) instead of comparing the original fuzzy numbers [49]. The natural consequence is a partiality of such orderings that is viewed as a problem by some authors. This is sometimes heuristically overcome by using multiple indices. This article does not focus on an exhaustive survey of all such methods and we are convinced that every single approach has its motivation justifying its existence. However, if we intend to apply the proposed arithmetic of extensional fuzzy numbers to the problem of fuzzy interpolation where orderings of fuzzy numbers play a crucial role, we need to know more about it. In particular, we should investigate under which conditions our approach preserves the most natural properties.

We have adopted the investigations of Wang and Kerre [49,50] as the starting point. This step is rather natural as the authors set up the most intuitive properties of orderings of fuzzy numbers and our goal is not to mimic such works but to investigate our technique in such a generally accepted framework.

Remark 6. *No matter if using indices or not, several techniques for ordering/ranking of fuzzy numbers operate on a finite subset of (reference) fuzzy numbers to which the ordering can be applied. Therefore, the ordering of two fuzzy numbers compares both to the reference set first. This fact is mirrored in the Wang–Kerre properties (axioms) recalled below. When mimicking the properties in our formalism of extensional fuzzy numbers, the properties will need to be slightly modified to drop the reference set(s), one of the properties will become totally redundant without considering the reference set.*

Wang–Kerre properties: Let us consider $\mathcal{F}(\mathbb{R})$ and let H, H_1, and H_2 be arbitrary finite subsets of $\mathcal{F}(\mathbb{R})$ (the reference sets). Furthermore, let $h \in \mathbb{R}$. Then

A1: $A \preceq A$, for $A \in H$;
A2: $A \preceq B$ and $B \preceq A \Rightarrow A \sim B$, for $A, B \in H$;
A3: $A \preceq B$ and $B \preceq C \Rightarrow A \preceq C$, for $A, B, C \in H$;
A4: $\sup \mathrm{Supp}(A) \leq \inf \mathrm{Supp}(B) \Rightarrow A \preceq B$, for $A, B \in H$;
A5: Let $A, B \in H_1 \cap H_2$. Then $A \preceq B$ in $H_1 \Leftrightarrow A \preceq B$ in H_2;
A6: Let $A \preceq B$ in $H_1 = \{A, B\}$. Then $A + C \preceq B + C$ in $H_2 = \{A + C, B + C\}$;
A7: Let $0 \leq h$ and $A, B, h \cdot A, h \cdot B \in H$.
Then $A \preceq B$ in $H_1 = \{A, B\}$ implies $h \cdot A \preceq h \cdot B$ in $H_2 = \{h \cdot A, h \cdot B\}$.

For the sake of brevity, when recalling the Wang–Kerre properties A1–A7, we dared to relax the precise definitions of the used symbols $\preceq$ and $\sim$ as the main motivation is to provide the readers with the ideas that must be accommodated to a different formalism in the framework of extensional fuzzy numbers anyhow.

4.2. Wang–Kerre Properties for $\mathcal{S}$-Orderings and Their Preservation

The implementation of the above-recalled Wang–Kerre properties to the framework of $\mathcal{S}$-orderings and extensional fuzzy numbers requires some modifications, for example, any use of reference sets is meaningless, $\mathcal{S}$-orderings of not use them to order objects from $\mathcal{F}_{\mathcal{S}}(\mathbb{R})$ at all. Consequently, axiom A5 is omitted from our consideration as whenever no reference set is used, it is always automatically fulfilled. Otherwise, one can see that the modification is only formal and preserving the same ideas and motivations.

Let $a_S, b_T, c_R \in \mathcal{F}_{\mathcal{S}}(\mathbb{R})$ and let $h \in \mathbb{R}$. Then the modified Wang–Kerre properties are given as follows:

A1′: $a_S \preceq a_S$;
A2′: $a_S \preceq b_T$ and $b_T \preceq a_S \Rightarrow (a_S - b_T) \in I_S^0$;
A3′: $a_S \preceq b_T$ and $b_T \preceq c_R \Rightarrow a_S \preceq c_R$;
A4′: $\sup \mathrm{Supp}(a_S) \leq \inf \mathrm{Supp}(b_T) \Rightarrow a_S \preceq b_T$;
A6′: $a_S \preceq b_T \Rightarrow a_S + c_R \preceq b_T + c_R$;
A7′: $0 \leq h$ and $a_S \preceq b_T \Rightarrow h \cdot a_S \preceq h \cdot b_T$;

The goal of the subsequent parts of the article is to investigate under which conditions an $\mathcal{S}$-ordering preserves the properties A1′–A7′.

Proposition 2. *Let $\preceq$ be an $\mathcal{S}$-ordering. Then it preserves properties A1′, A2′, and A3′.*

Proof. Using the reflexivity axiom (O1) from Definition 6, for any $S, T \in \mathcal{S}$ there must exist $P \in \mathcal{S}$ such that $(a_S \preceq a_T)_P$ which implies $a_S \preceq a_T$. It suffices to put $S = T$ to complete the proof of A1′.

The anti-symmetry axiom (O2) from Definition 6 states that $(a_S \preceq b_T)_Q$ and $(b_T \preceq a_T)_R$ implies $a_S - b_T \in I_S^0$. As it holds for any for any $S, T, Q, R \in \mathcal{S}$, we directly obtain the proof of A2′.

Finally, the transitivity axiom (O3) from Definition 6 states that $(a_S \preceq b_T)_Q$ and $(b_T \preceq c_U)_R$ implies $(a_S \preceq c_U)_P$ for some $P \in \mathcal{S}$. As it holds for any for any $S, T, U, Q, R \in \mathcal{S}$, we directly obtain the proof of A3′. □

The preservation of A4′ is not as automatic as the preservation of A1′-A3′ that stemmed directly from the axioms (O1)-(O3). An additional property must be assumed.

Proposition 3. *Let $\preceq$ be a real-order compatible $\mathcal{S}$-ordering. Then it preserves the property A4′.*

Proof. If $\sup \mathrm{Supp}(a_S) \leq \inf \mathrm{Supp}(b_T)$ holds, we immediately get the interval-ordering for all α-cuts of both fuzzy numbers: $a_S \leq_i b_T$ or equivalently $\mathrm{EXT}_S(a) \leq_i \mathrm{EXT}_T(b)$ by Lemma 1. Put $R = \min(S, T)$. Recall that the composition of similarities U and V is defined as $U \circ V(x,y) = \bigvee_{z \in \mathbb{R}} U(x,z) \otimes V(z,y)$. Obviously, we have $S = S \circ R$ and $T = T \circ R$, where $R = \min(S, T)$. Indeed, assume that $R = S$. Then $S = S \circ S$ follows immediately from the transitivity of S. In addition, we have

$$T \circ S(x,y) = \bigvee_{z \in \mathbb{R}} T(x,z) \otimes S(z,y) \geq T(x,y) \otimes S(y,y) = T(x,y) \otimes 1 = T(x,y), \quad x, y \in \mathbb{R},$$

therefore, $T \subseteq T \circ S$. On the other side, from $S \subseteq T$ and the transitivity of T, we have

$$T \circ S(x,y) = \bigvee_{z \in \mathbb{R}} T(x,z) \otimes S(z,y) \leq \bigvee_{z \in \mathbb{R}} T(x,z) \otimes T(z,y) = T(x,y), \quad x, y \in \mathbb{R},$$

therefore, $T \circ S \subseteq T$, and hence, we obtain $T = T \circ S$. Similarly, one can show the equalities for $R = T$. Since $\mathrm{EXT}_S(a) \leq_i \mathrm{EXT}_T(b)$, using the previous equalities, we simply find that $\mathrm{EXT}_R(a_S) = \mathrm{EXT}_{S \circ R}(a) = \mathrm{EXT}_S(a) \leq_i \mathrm{EXT}_T(b) = \mathrm{EXT}_{T \circ R}(b) = \mathrm{EXT}_R(b_T)$ which jointly with the real-order compatibility leads to $a_S \preceq b_T$. □

Example 10. *Let us again demonstrate Proposition 3 on examples. Let us consider two (disjoint) extensional fuzzy numbers such that* $\sup \mathrm{Supp}(a_S) \leq \inf \mathrm{Supp}(b_T)$. *For the real-order compatible* $\mathcal{S}$*-ordering* $\preceq_{\max}$*, we easily obtain* $a_S \preceq_{\max} b_T$. *However, for the* $\mathcal{S}$*-ordering* $\preceq_{\max}^{rev}$ *that is not real-order compatible, we get the opposite, in particular* $b_T \preceq_{\max}^{rev} a_S$.

The real-order compatibility turned to be useful in proving the preservation of A4′ however, it is not sufficient to prove the preservation of A5′. As it will be shown, we will need to use "the other direction" encoded in the pre-order compatibility and thus, the strong compatibility will be assumed.

Proposition 4. *Let* $\preceq$ *be a strongly compatible* $\mathcal{S}$*-ordering, and let any similarity from* $\mathcal{S}$ *be shift-invariant. Then it preserves the property A6′.*

Proof. Let $a_S, b_T, c_R \in \mathcal{F}_{\mathcal{S}}(\mathbb{R})$ and let $a_S \preceq b_T$ where $\preceq$ is strongly compatible. Put $U = \max(S, T, R)$. From the reflexivity axiom (O1) we have $(a_U \preceq a_S)_P$ and $(b_T \preceq b_U)_Q$ for certain similarities $P, Q \in \mathcal{S}$, i.e., $a_U \preceq a_S$ and $b_T \preceq b_U$, and by the transitivity axiom (O3), we find that $(a_U \preceq b_U)_V$ for a certain similarity $V \in \mathcal{S}$, i.e., $a_U \preceq b_U$. In what follows, for simplicity, we do not mention the existence of similarities P and Q in $\mathcal{S}$, and we will directly write $\mathcal{S}$-ordering. Since $\preceq$ is strong compatible, there exists $W \in \mathcal{S}$ such that $\mathrm{EXT}_W(a_U) \leq_i \mathrm{EXT}_W(b_U)$. We show that $\mathrm{EXT}_W((a+c)_U) \leq_i \mathrm{EXT}_W((b+c)_U)$ also holds. By Lemma 2, one can simply check that for any $x \in \mathbb{R}$, there exists $x' \in \mathbb{R}$ such that $x \leq x'$ and

$$\mathrm{EXT}_W(a_U)(x-c) = \bigvee_{y\in\mathbb{R}} U(a,y) \otimes W(y,x-c) \leq \bigvee_{z\in\mathbb{R}} U(b,z) \otimes W(z,x'-c) = \mathrm{EXT}_W(b_U)(x'-c) \quad (9)$$

and vice versa, for any $x' \in \mathbb{R}$, there exists $x \in \mathbb{R}$ such that $x \leq x'$ and

$$\mathrm{EXT}_W(b_U)(x'-c) = \bigvee_{z\in\mathbb{R}} U(b,z) \otimes W(z,x'-c) \leq \bigvee_{y\in\mathbb{R}} U(a,y) \otimes W(y,x-c) = \mathrm{EXT}_W(a_U)(x-c). \quad (10)$$

Hence, and using the shift-invariance of U and W, e.g., $U(a+c,y) = U(a,y-c)$ or $W(y,x) = W(y-c,x-c)$, if $x \in \mathbb{R}$, then there exists $x' \in \mathbb{R}$ such that $x \leq x'$ and (9) holds, which implies

$$\begin{aligned}
\mathrm{EXT}_W((a+c)_U)(x) &= \bigvee_{y\in\mathbb{R}} U(a+c,y) \otimes W(y,x) = \bigvee_{y\in\mathbb{R}} U(a,y-c) \otimes W(y-c,x-c) \\
&= \mathrm{EXT}_W(a_U)(x-c) \leq \mathrm{EXT}_W(b_U)(x'-c) = \bigvee_{z\in\mathbb{R}} U(b,z-c) \otimes W(z-c,x'-c) \\
&= \bigvee_{z\in\mathbb{R}} U(b+c,z) \otimes W(z,x') = \mathrm{EXT}_W((b+c)_U)(x').
\end{aligned}$$

Similarly, if $x' \in \mathbb{R}$, then there exists $x \in \mathbb{R}$ such that $x \leq x'$ and (10) holds, which implies

$$\begin{aligned}
\mathrm{EXT}_W((b+c)_U)(x') &= \bigvee_{z\in\mathbb{R}} U(b+c,z) \otimes W(z,x') = \bigvee_{z\in\mathbb{R}} U(b,z-c) \otimes W(z-c,x'-c) \\
&= \mathrm{EXT}_W(b_U)(x'-c) \leq \mathrm{EXT}_W(a_U)(x-c) = \bigvee_{y\in\mathbb{R}} U(b,y-c) \otimes W(y-c,x-c) \\
&= \bigvee_{y\in\mathbb{R}} U(b+c,y) \otimes W(y,x) = \mathrm{EXT}_W((b+c)_U)(x).
\end{aligned}$$

By Lemma 2, we find that $\mathrm{EXT}_W((a+c)_U) \leq_i \mathrm{EXT}_W((b+c)_U)$, and using the real-order compatibility, we obtain $(a+c)_U \preceq (b+c)_U$. Using reflexivity (O1), it holds that $(a+c)_{\max(S,R)} \preceq (a+c)_U$ and $(b+c)_U \preceq (b+c)_{\max(T,R)}$, and by transitivity (O3), we obtain $a_S + c_R = (a+c)_{\max(S,R)} \preceq (b+c)_{\max(T,R)} = b_T + c_R$. □

Example 11. *Let us demonstrate Proposition 4 on some examples. Let us again consider the case depicted in Figure 2 with $x_{S_{p'}}$ and $y_{S_{p''}}$ that are shift-invariant. Furthermore, let us consider $z_{S_{p'''}}$ such that z is an arbitrary number from $\mathbb{R}$ and $S_{p'''} \geq \max(S_{p'}, S_{p''})$ (i.e., $p''' < \min(p', p'')$).*

If we consider a strongly compatible $\mathcal{S}$-ordering $\leq_{\max}$ then we have $x_{S_{p'}} \leq_{\max} y_{S_{p''}}$ and easily, we can check that also $(x+z)_{S_{p'''}} \leq_{\max} (y+z)_{S_{p'''}}$ holds, where the left-hand side is equal to $x_{S_{p'}} + z_{S_{p'''}}$ and the right-hand side is equal to $y_{S_{p''}} + z_{S_{p'''}}$.

Now, let us consider the $\mathcal{S}$-ordering $\leq^{L}_{\max}$ and that is not strongly compatible, see Example 9. Due to the position of the left-hand sides of the supports, the following $y_{S_{p''}} \leq^{L}_{\max} x_{S_{p'}}$ holds. However, $x_{S_{p'}} + z_{S_{p'''}} = (x+z)_{S_{p'''}}$ and $y_{S_{p''}} + z_{S_{p'''}} = (y+z)_{S_{p'''}}$ so, we get two extensional fuzzy number of the same width and as $(x+y) \leq (y+z)$, also the left-hand sides of the supports are ordered in this way. Consequently, $(x+z)_{S_{p'''}} \leq^{L}_{\max} (y+z)_{S_{p'''}}$ which harms the property A6′.

Again, note that the proposition states only one implication but the opposite implication does not hold. Therefore, if we consider the $\mathcal{S}$-ordering that reverses the order $\leq^{rev}_{\max}$, we get $y_{S_{p''}} \leq^{L}_{\max} x_{S_{p'}}$ as well as $(y+z)_{S_{p'''}} \leq^{L}_{\max} (x+z)_{S_{p'''}}$ and thus, A6′ is preserved even though $\leq^{rev}_{\max}$ is not strongly compatible.

The last axiom that must be checked will be preserved again under the assumption of the strong compatibility. Additionally, the presence of the crisp bottom element in the considered MI-pregroup will be assumed.

Proposition 5. *Let $\leq$ be a strongly compatible $\mathcal{S}$-ordering on an MI-prefield $(\mathcal{F}_{S_\infty}(\mathbb{R}), +, -, \cdot, ^{-1})$ with the crisp bottom element. Then it preserves the property A7′.*

Proof. As S_∞ contains the crisp bottom element $\perp_S$ (classical equality), we may incorporate the representation of the scalar multiplication in $\mathcal{F}_{S_\infty}(\mathbb{R})$:

$$h \cdot a_S = h_{\perp_S} \cdot a_S = (h \cdot a)_{\max(S, \perp_S)} = (h \cdot a)_S \, .$$

If $a_S \leq b_T$ and $\leq$ is pre-order compatible, then $a \leq b$ and necessarily $h \cdot a \leq h \cdot b$ for any $h > 0$. Since $(\mathrm{EXT}_{\perp_S}((h \cdot a)_{\perp_S}))_\alpha = \{h \cdot a\}$ and $(\mathrm{EXT}_{\perp_S}((h \cdot b)_{\perp_S}))_\alpha = \{h \cdot b\}$ for any $\alpha \in (0, 1]$, we find that $\mathrm{EXT}_{\perp_S}((h \cdot a)_{\perp_S}) \leq_i \mathrm{EXT}_{\perp_S}(h \cdot b)_{\perp_S})$, which implies $(h \cdot a)_{\perp_S} \leq (h \cdot b)_{\perp_S}$ as a consequence of the real-order compatibility. Using the reflexivity axiom (O1), we obtain $(h \cdot a)_S \leq (h \cdot a)_{\perp_S}$ and $(h \cdot b)_{\perp_S} \leq (h \cdot b)_T$, and from the transitivity axiom (O3), we find that $(h \cdot a)_S \leq (h \cdot b)_T$. □

Corollary 1. *Let $\leq$ be a strongly compatible $\mathcal{S}$-ordering on an MI-prefield $(\mathcal{F}_{S_\infty}(\mathbb{R}), +, -, \cdot, ^{-1})$ with the crisp bottom element. Then it simultaneously preserves axioms A1′-A4′ and A6′-A7.*

Proof. We have to prove that A6′ remains true if the shift-invariance is omitted in $\mathcal{F}_{S_\infty}(\mathbb{R})$ (cf. Proposition 4). Let $a_S, b_T, c_R \in \mathcal{F}_{S_\infty}(\mathbb{R})$. As with the proof of Proposition 5, if $a_S \leq b_T$ and $\leq$ is pre-order compatible, then $a \leq b$ and necessarily $a + c \leq b + c$. Since $\mathrm{EXT}_{\perp_S}((a+c)_{\perp_S}) \leq_i \mathrm{EXT}_{\perp_S}(b+c)_{\perp_S})$, we find that $(a+c)_{\perp_S} \leq (b+c)_{\perp_S}$ because of the real-order compatibility. Using the reflexivity axiom (O1), we obtain $a_S + c_R = (a+c)_{\max(S,R)} \leq (a+c)_{\perp_S}$ and $(b+c)_{\perp_S} \leq (b+c)_{\max(T,R)} = b_T + c_R$, and from the transitivity axiom (O3), we obtain $a_S + c_R \leq b_T + c_R$. □

Corollary 1 directly gathers the results from Propositions 2–5 and shows an elegant and safe way how to construct a system of similarities and an $\mathcal{S}$-ordering that is safe in the sense of meeting all the preferable properties. Indeed, the list of assumptions may seem high, but it is not restrictive at all. As we showed above, the most natural and intuitive $\mathcal{S}$-orderings are strongly compatible and analogously, including the crisp equality as the crisp bottom element to the system of nested similarities is more natural than starting from a wider bottom element.

Definition 11. *Let $(\mathcal{F}_{\mathcal{S}}(\mathbb{R}), +, -, \cdot, ^{-1})$ be an MI-prefield, and let $\leq$ be an $\mathcal{S}$-ordering on $\mathcal{F}_{\mathcal{S}}(\mathbb{R})$. The structure $(\mathcal{F}_{\mathcal{S}}(\mathbb{R}), +, -, \cdot, ^{-1}, \leq)$ will be called an $\mathcal{S}$-*ordered MI-prefield *provided that for any $a_S, b_T, c_R \in \mathcal{F}_{\mathcal{S}}(\mathbb{R})$ and $U, V \in \mathcal{S}$, it holds that*

(P1) $a_S \leq b_T \Rightarrow a_S + c_U \leq b_T + b_U$,
(P2) $0_U \leq a_S$ *and* $0_V \leq b_T \Rightarrow \exists W \in \mathcal{S} : 0_W \leq a_S \cdot b_T$.

The following statement shows that the shift-invariance of similarities and strong compatibility of $\mathcal{S}$-ordering are sufficient conditions to get a $\mathcal{S}$-ordered MI-prefield.

Proposition 6. *Let $(\mathcal{F}_{\mathcal{S}}(\mathbb{R}), +, -, \cdot, ^{-1})$ be an MI-prefield such that any similarity from $\mathcal{S}$ is shift-invariant, and let $\leq$ be a strongly compatible $\mathcal{S}$-ordering on $\mathcal{F}_{\mathcal{S}}(\mathbb{R})$. Then $(\mathcal{F}_{\mathcal{S}}(\mathbb{R}), +, -, \cdot, ^{-1}, \leq)$ is an $\mathcal{S}$-ordered MI-prefield.*

Proof. (P1) follows from Proposition 4. Let $0_U \leq a_S$ and $0_V \leq b_T$ for $a_S, b_T \in \mathcal{F}_{\mathcal{S}}(\mathbb{R})$ and $U, V \in \mathcal{S}$. According to reflexivity of $\mathcal{S}$-ordering, we obtain $0_S \leq 0_U$ and $0_T \leq 0_V$, and by transitivity, we find that $0_S \leq a_S$ and $0_T \leq b_T$. Denote $R = \max(S, T)$. Then we have $0_R = 0_S \cdot 0_T$ and $(ab)_R = a_S \cdot b_T$. We will show that $0_R \leq (ab)_S$. Since $\leq$ is strongly compatible, as a consequence of the real-order compatibility, it is sufficient to show that there exists $W \in \mathcal{S}$ such that

$$\mathrm{EXT}_W(0_R) \leq_i \mathrm{EXT}_W((ab)_R). \tag{11}$$

By Lemma 2, we will show that for any $x \in \mathbb{R}$ there exists $x' \in \mathbb{R}$ such that $x \leq x'$ and

$$\bigvee_{y \in \mathbb{R}} R(0, y) \otimes W(y, x) \leq \bigvee_{z \in \mathbb{R}} R(ab, z) \otimes W(z, x') \tag{12}$$

and vice versa, for any $x' \in \mathbb{R}$ there exists $x \in \mathbb{R}$ such that $x \leq x'$ and

$$\bigvee_{z \in \mathbb{R}} R(ab, z) \otimes W(z, x') \leq \bigvee_{y \in \mathbb{R}} R(0, y) \otimes W(y, x). \tag{13}$$

Let $W \in \mathcal{S}$ be an arbitrary similarity. Since R and W are shift-invariant, for any $x \in \mathbb{R}$, we obtain

$$\bigvee_{y \in \mathbb{R}} R(0, y) \otimes W(y, x) = \bigvee_{y \in \mathbb{R}} R(ab, y + ab) \otimes W(y + ab, x + ab) = \bigvee_{y \in \mathbb{R}} R(ab, z) \otimes W(z, x'),$$

where we put $x' = x + ab$. Since R is strongly compatible, by Lemma 3 we find that $0 \leq a$ and $0 \leq b$, which implies $x \leq x + ab = x'$. Similarly, for any $x' \in \mathbb{R}$, we obtain

$$\bigvee_{z \in \mathbb{R}} R(ab, z) \otimes W(z, x') = \bigvee_{z \in \mathbb{R}} R(0, z - ab) \otimes W(z - ab, x' - ab) = \bigvee_{y \in \mathbb{R}} R(0, y) \otimes W(y, x), \tag{14}$$

where we put $x = x' - ab$, and thus $x \leq x'$. By Lemma 2, we obtain (11) and thus $0_R \leq (ab)_R = a_S \cdot b_T$, which concludes the proof of (P2). □

Example 12. *Let $(\mathcal{F}_{\mathcal{S}_\infty}(\mathbb{R}), +, -, \cdot, ^{-1})$ be the MI-prefield with $\mathcal{S}_\infty = \{S_p \mid p \in (0, +\infty]\}$, where S_p is given by (1) for any $0 < p < +\infty$ and S_∞ is the crisp equality (see, Remark 3), which is enriched by a strong compatible $\mathcal{S}_\infty$-ordering $\leq$. Since any similarity from $\mathcal{S}_\infty$ is shift-invariant (see, Example 1), by the previous proposition, $(\mathcal{F}_{\mathcal{S}_\infty}(\mathbb{R}), +, -, \cdot, ^{-1}, \leq)$ is an ordered MI-prefield.*

Since $\mathcal{S}$-ordering on an MI-prefield $\mathcal{F}_{\mathcal{S}}(\mathbb{R})$ is determined by the set $\mathcal{S}$ of similarities that is referred in the index of the MI-prefield, for convenience, we will omit $\mathcal{S}$ in "$\mathcal{S}$-ordered MI-prefield" and write only "ordered MI-prefield". In the following definition, we introduce the isomorphism of ordered MI-prefields which is used in our demonstration of fuzzy interpolation. As with algebraic literature, for convenience, we do not distinguish the operations and orderings of ordered MI-prefields using indexes if no confusion can appear.

Definition 12. *Let* $(\mathcal{F}_{\mathcal{S}_1}(\mathbb{R}), +, -, \cdot, ^{-1}, \leq)$ *and* $(\mathcal{F}_{\mathcal{S}_2}(\mathbb{R}), +, -, \cdot, ^{-1}, \leq)$ *be ordered MI-prefields. A map* $h\colon \mathcal{F}_{\mathcal{S}_1}(\mathbb{R}) \to \mathcal{F}_{\mathcal{S}_2}(\mathbb{R})$ *is a homomorphism of ordered MI-prefields provided that*

(H1) $h(a_S + b_T) = h(a_S) + h(b_T)$, $h(0_{\perp_{\mathcal{S}_1}}) = 0_{\perp_{\mathcal{S}_2}}$ *and* $h(-a_S) = -h(a_S)$ *for any* $a_S, b_T \in \mathcal{F}_{\mathcal{S}_1}(\mathbb{R})$;

(H2) $h(a_S \cdot b_T) = h(a_S) \cdot h(b_T)$, $h(1_{\perp_{\mathcal{S}_1}}) = 1_{\perp_{\mathcal{S}_2}}$ *and* $h((a_S)^{-1}) = h^{-1}(a_S)$ *for any* $a_S, b_T \in \mathcal{F}_{\mathcal{S}_1}(\mathbb{R}) \setminus I^0_{\mathcal{S}_1}$;

(H3) $a_S \leq b_T \Rightarrow h(a_S) \leq h(b_T)$ *for any* $a_S, b_T \in \mathcal{F}_{\mathcal{S}_1}(\mathbb{R})$.

A homomorphism h of ordered MI-prefields is called an isomorphism *provided that h is a bijective map.*

In the following example we show an isomorphism using the ordered MI-prefield $\mathcal{F}_{\mathcal{S}_\infty}(\mathbb{R})$ from Example 12.

Example 13. *Let* $(\mathcal{F}_{\mathcal{S}_\infty}(\mathbb{R}), +, -, ^{-1}, \leq)$ *be an ordered MI-prefield with a strong compatible* $\mathcal{S}$*-ordering* $\leq$. *Let* $r \in \mathbb{R}$ *such that* $0 < r$. *The the map* $h_r\colon \mathcal{F}_{\mathcal{S}_\infty}(\mathbb{R}) \to \mathcal{F}_{\mathcal{S}_\infty}(\mathbb{R})$ *given by* $h_r(a_{S_p}) = a_{S_{r\cdot p}}$ *for any* $a_{S_p} \in \mathcal{F}_{\mathcal{S}_\infty}(\mathbb{R})$ *is an isomorphism of ordered MI-prefields. It is easy to see that* h_r *is a bijective map, since* $f(p) = r \cdot p$ *for* $p \in (0, \infty]$ *is a bijection from* $(0, \infty]$ *onto* $(0, \infty]$, *where we naturally define* $r \cdot \infty = \infty$. *Let us show that* h_r *satisfies (H1) and (H2). Since the proofs are analogous for* $+$ *and* $\cdot$, *we verify only (H1). For any* $a_{S_{p_1}}, b_{S_{p_2}} \in \mathcal{F}_{\mathcal{S}_\infty}(\mathbb{R})$, *we have*

$$h_r(a_{S_p} + b_{S_{p_2}}) = h_r((a+b)_{\max(S_{p_1}, S_{p_2})}) = h_r((a+b)_{S_{\min(p_1,p_2)}}) = (a+b)_{S_{r\cdot\min(p_1,p_2)}} = (a+b)_{S_{\min(r\cdot p_1, r\cdot p_2)}} = (a+b)_{\max(S_{r\cdot p_1}, S_{r\cdot p_2})} = a_{S_{r\cdot p_1}} + b_{S_{r\cdot p_2}} = h_r(a_{S_{p_1}}) + h_r(b_{S_{p_2}}).$$

For the zero element 0_{S_∞} *in* $\mathcal{F}_{\mathcal{S}_\infty}(\mathbb{R})$, *we obtain* $h_r(0_{S_\infty}) = 0_{S_{r\cdot\infty}} = 0_{S_\infty}$, *where we use* $r \cdot \infty = \infty$. *Finally, for any* $a_{S_p} \in \mathcal{F}_{\mathcal{S}_\infty}(\mathbb{R})$, *we have* $h_r((a_{S_p})^{-1}) = h_r((a^{-1})_{S_p}) = (a^{-1})_{S_{r\cdot p}} = (a_{S_{r\cdot p}})^{-1} = h_r(a_{S_p})^{-1}$. *Now, we show that (H3) also holds. Let* $a_{S_{p_1}} b_{S_{p_2}} \in \mathcal{F}_{\mathcal{S}_\infty}(\mathbb{R})$ *such that* $a_{S_{p_1}} \leq b_{S_{p_2}}$. *Since* $\leq$ *is strong compatible, to prove that* $h_r(a_{S_{p_1}}) \leq h_r(b_{S_{p_2}})$, *it is sufficient to show that there exists* $S_q \in \mathcal{S}_\infty$ *such that* $\mathrm{EXT}_{S_q}(a_{S_{rp_1}}) \leq_i \mathrm{EXT}_{S_q}(b_{S_{rp_2}})$; *the desired inequality is a straightforward consequence of the real-order compatibility of* $\leq$. *Recall that* $S_p \circ S_q = S_q$ *whenever* $S_p \subseteq S_q$ *or equivalently* $p \geq q$ *(see the proof of Proposition 3). Since* $a_{S_{p_1}} \leq b_{S_{p_2}}$ *and* $\leq$ *is pre-order compatible, we find that* $a \leq b$ *according to Lemma 3. Put* $q = \min(rp_1, rp_2)$. *Then we obtain* $\mathrm{EXT}_{S_q}(a_{S_{rp_1}}) = a_{S_q}$ *and* $\mathrm{EXT}_{S_q}(b_{S_{rp_2}}) = b_{S_q}$. *Indeed, for any* $y \in \mathbb{R}$, *we have*

$$\mathrm{EXT}_{S_q}(a_{S_{rp_1}})(y) = \bigvee_{z\in\mathbb{R}} S_{rp_1}(a,z) \otimes S_q(z,y) = (S_{rp_1} \circ S_q)(a,y) = S_q(a,y) = a_{S_q}(y),$$

where $\otimes$ *is the Łukasiewicz multiplication, and similarly for the second equality. Now, for any* $\alpha \in (0,1]$, *we can simply calculate that*

$$(\mathrm{EXT}_{S_q}(a_{S_{rp_1}}))_\alpha = \left[a - \frac{1-\alpha}{q}, a + \frac{1-\alpha}{q}\right] \text{ and } (\mathrm{EXT}_{S_q}(b_{S_{rp_2}}))_\alpha = \left[b - \frac{1-\alpha}{q}, b + \frac{1-\alpha}{q}\right],$$

which implies that $\mathrm{EXT}_{S_q}(a_{S_{rp_1}}) \leq_i \mathrm{EXT}_{S_q}(b_{S_{rp_2}})$ *as a consequence of* $a \leq b$.

5. Application to Fuzzy Interpolation

5.1. The Concept of $\mathcal{S}$-Function

We will approach fuzzy interpolation from the analytical perspective rather than from the logical one. Therefore, we will always have in mind some function that is partially given (in a finite number of points) and our goal is to complete it to a total mapping that assigns an output to an arbitrary input. Therefore, first, we must start from defining such fundamental objects as a function operating on extensional fuzzy numbers.

Remark 7. *Deliberately, we avoid the notion of a fuzzy function as it has been used by many authors for distinct objects with distinct definitions. Usually, the ideas and motivations are similar, but the technical aspects differ. As the natural name involving the "extensional fuzzy numbers" might be too long, for the sake of brevity we adopt the name $\mathcal{S}$-function.*

Definition 13. *Let $\leq$ be an $\mathcal{S}$-ordering on an MI-prefield $(\mathcal{F}_S(\mathbb{R}),+,-,\cdot,^{-1})$ and let D be a non-empty subset of $\mathcal{F}_S(\mathbb{R})$. A mapping $f\colon D \to \mathcal{F}_S(\mathbb{R})$ is an* $\mathcal{S}$-function *if there exists a function $g\colon \mathbb{R} \to \mathbb{R}$ such that*

$$\forall x_S \in D\ \exists S' \in \mathcal{S}:\ f(x_S) = g(x)_{S'}\,. \tag{15}$$

Equation (15) states that each $\mathcal{S}$-function is determined by a real-valued function. This allows the introduction of fundamental properties known from mathematical analysis, such as monotonicity, injectivity, or surjectivity of functions, in an analogous way.

For example, an $\mathcal{S}$-function $f\colon D \to \mathcal{F}_S(\mathbb{R})$ is called increasing if

$$\forall u_R, v_S \in D:\ u_R \leq v_S \Rightarrow f(u_R) \leq f(v_S)\,. \tag{16}$$

Analogously, we could define monotonically decreasing $\mathcal{S}$-function and of course, also strictly increasing and strictly decreasing $\mathcal{S}$-function. Note the importance of the choice of the $\mathcal{S}$-ordering and the fact that the same ordering is used on both sides of the definition. Indeed, for distinct choices of the $\mathcal{S}$-ordering, the given $\mathcal{S}$-function may be or must not be monotonous. The generalization of the definition that would allow the use of different ordering on the left-hand side than the one used on the right-hand side is technically possible yet, would need to be strongly motivated by the modeled problem.

Using the concept of $\mathcal{S}$-function, we can very naturally define extensions of classical functions using the arithmetic operations, for example:

$$f(x_S) = a_T \cdot x_S + b_R\,, \quad \text{for } a_T, b_R \in \mathcal{F}_S(\mathbb{R})$$

is a linear $\mathcal{S}$-function;

$$f(x_S) = a_{T_0} + a_{T_1} \cdot x_S + \cdots + a_{T_n} \cdot x_S^n\,, \quad \text{for } a_{T_0}, \ldots, a_{T_n} \in \mathcal{F}_S(\mathbb{R})$$

is a polynomial $\mathcal{S}$-function (of the n-th order).

Moreover, we can define natural extensions of other fundamental functions, for example

$$e^{(x_S)} = (e^x)_{h(s)}\,, \quad \log(x_S) = \log(x)_{h(s)}\,, \quad h\colon \mathcal{S} \to \mathcal{S}$$

is an exponential (logarithm) $\mathcal{S}$-function;

$$\sin(x_S) = \sin(x)_{h(S)}\,, \quad \cos(x_S) = \cos(x)_{h(S)}\,, \quad h\colon \mathcal{S} \to \mathcal{S}$$

is a sine (cosine) $\mathcal{S}$-function.

The crucial for transferring the vagueness will be the choice and appropriateness of the mapping h. The most trivial would be to put h equal to the identity mapping however, such a choice would not be probably adequate for managing the uncertainty.

5.2. $\mathcal{S}$-interpolation

The fuzzy interpolation task can be in general formally define as follows. Let us be given two finite equal sized sets of fuzzy sets $A_1, \ldots, A_n$ and $B_1, \ldots, B_n$ on the universe $\mathbb{R}$ (or its non-empty subsets). We search for a mapping $F\colon \mathcal{F}(\mathbb{R}) \to \mathcal{F}(\mathbb{R})$ such that the condition $F(A_i) = B_i$ holds for any $i = 1, \dots, n$. Let us reformulate the problem into the formalism of $\mathcal{S}$-functions.

Definition 14. *Let $(\mathcal{F}_\mathcal{S}(\mathbb{R}), +, -, \cdot, ^{-1})$ be an MI-prefield and let $\leq$ be an $\mathcal{S}$-ordering on $\mathcal{F}_\mathcal{S}(\mathbb{R})$. Let $x^i, y^i \in \mathbb{R}$ for $i = 1, \dots, n$ and let $(x^1_{S_1}, y^1_{T_1}), \dots, (x^n_{S_n}, y^n_{T_n}) \in \mathcal{F}_\mathcal{S}(\mathbb{R})^2$ such that $x^i_{S_i} \leq x^{i+1}_{S_{i+1}}$ for any $i = 1, \dots, n-1$. We say that an $\mathcal{S}$-function f $\mathcal{S}$-*interpolates *above-given pairs of extensional fuzzy numbers if*

$$f(x^i_{S_i}) = y^i_{T_i}, \quad i = 1, \dots, n.$$

Naturally, we may mimic the classical interpolation methods that are well-founded. For example, the linear $\mathcal{S}$-interpolation will be defined for an arbitrary $x_S \in \mathcal{F}_\mathcal{S}(\mathbb{R})$ such that $x^i_{S_i} \leq x_S \leq x^{i+1}_{S_{i+1}}$ as follows:

$$f(x_S) = y^i_{T_i} + (x_S - x^i_{S_i}) \cdot \frac{y^{i+1}_{T_{i+1}} - y^i_{T_i}}{x^{i+1}_{S_{i+1}} - x^i_{S_i}}. \tag{17}$$

Remark 8. *The linear interpolation in the crisp sense can be computed by several formulas that are equivalent. Therefore, a natural question is whether this would hold also for the case of $\mathcal{S}$-interpolations stemming from these crisp equivalent formulas. The answer is positive. Due to the particular used arithmetic, the computation runs over the "centers" and the similarities are, in the second phase, calculated over their maximum. Therefore, as long as we use extensions of formulas that are equivalent in the crisp case, and the inputs are the same, the results of the extended formulas must be equal too. However, note that if we use another type of MI-algebra with different calculus over the similarities (see [19] for some examples), the results could differ.*

We will present a short lemma that will be used in the latter. The basically states that if an input is between two input nodes, the output obtained by Formula (17) is between the respective output nodes.

Lemma 4. *Let $(\mathcal{F}_\mathcal{S}(\mathbb{R}), +, -, \cdot, ^{-1})$ be an MI-prefield such that any similarity from $\mathcal{S}$ is shift-invariant and let $\leq$ be a strongly compatible $\mathcal{S}$-ordering on $\mathcal{F}_\mathcal{S}(\mathbb{R})$. Let $x^i, y^i \in \mathbb{R}$ for $i = 1, \dots, n$ and let $(x^1_{S_1}, y^1_{T_1}), \dots, (x^n_{S_n}, y^n_{T_n}) \in \mathcal{F}_\mathcal{S}(\mathbb{R})^2$ such that $x^i_{S_i} \leq x^{i+1}_{S_{i+1}}$ for any $i = 1, \dots, n-1$. For $x^i_{S_i} \leq x_S \leq x^{i+1}_{S_{i+1}}$ and for the $\mathcal{S}$-function f given by (17) it holds that:*

$$y^i_{T_i} \leq f(x_S) \leq y^{i+1}_{T_{i+1}}$$

Proof. Using the fact that $x^i_{S_i} \leq x_S \leq x^{i+1}_{S_{i+1}}$ and Lemma 3 we obtain $x^i \leq x \leq x^{i+1}$. Using Formula (17), we obtain $f(x_S) = y_T$ such that

$$y = y^i + (x - x^i) \cdot \frac{y^{i+1} - y^i}{x^{i+1} - x^i}$$

and $T = \max(S_i, S_{i+1}, T_i, T_{i+1}, S)$. Hence, we find that $y^i \leq y \leq y^{i+1}$. We show that $\mathrm{EXT}_T(y^i) \leq_i \mathrm{EXT}_T(y)$. Put $d = y - y^i$. Since T is shift-invariant, for any $x \in \mathbb{R}$, we obtain $\mathrm{EXT}_T(y_i)(x) = T(y^i, x) = T(y^i + d, x + d) = T(y, x + d) = \mathrm{EXT}_T(y)(x + d)$, where $x \leq x + d$, and vice versa, for any $x' \in \mathbb{R}$, we obtain $\mathrm{EXT}_T(y)(x') = T(y, x') = T(y - d, x' - d) = T(y^i, x' - d) = \mathrm{EXT}_T(y^i)(x' - d)$, where $x' - d \leq x'$. By Lemma 9, we find that $\mathrm{EXT}_T(y^i) \leq_i \mathrm{EXT}_T(y)$. Similarly, one can show that $\mathrm{EXT}_T(y) \leq_i \mathrm{EXT}_T(y^{i+1})$. Since $T_i \subseteq T$ and $T_{i+1} \subseteq T$ and from the transitivity of similarities, we have $T = T_i \circ T$, $T = T_{i+1} \circ T$ and $T = T \circ T$ as has been verified in the proof of Proposition 3. Hence, we have $\mathrm{EXT}_T(y^i) = \mathrm{EXT}_{T_i \circ T}(y^i) = \mathrm{EXT}_T(y^i_{T_i})$, $\mathrm{EXT}_T(y) = \mathrm{EXT}_{T \circ T}(y) = \mathrm{EXT}_T(y_T)$, and $\mathrm{EXT}_T(y^{i+1}) = \mathrm{EXT}_{T_{i+1}}(y^{i+1}) =$

$\mathrm{EXT}_T(y^{i+1}_{T_{i+1}})$, which implies $\mathrm{EXT}_T(y^i_{T_i}) \leq_i \mathrm{EXT}_T(y_T) \leq_i \mathrm{EXT}_T(y^{i+1}_{T_{i+1}})$, and due to the real-order compatibility of $\leq$, we obtain $y^i_{T_i} \leq f(x_S) \leq y^{i+1}_{T_{i+1}}$. □

There are two important observations coming from the definition of the $\mathcal{S}$-interpolation and the formula for the linear $\mathcal{S}$-interpolation.

First, the choice $\mathcal{S}$-ordering is absolutely crucial as the technique assumes all fuzzy points in the input space to be linearly ordered and moreover, every input needs to lie in between of two neighboring fuzzy points. Thus, the totality of the $\mathcal{S}$-ordering seems to be a natural requirement of the highest importance.

Secondly, the technique may determine reliably a unique solution for an extensional fuzzy number x_S that possesses the following position $x^i_{S_i} < x_S < x^{i+1}_{S_{i+1}}$ however, if the input point x_S is one of the given points, say, $x_S = x^i_{S_i}$ then there are two approaches leading to two results that need not be equal. In particular, the first approach (*interpolation from the right*) considers x_S to lie between $x^i_{S_i}$ and $x^{i+1}_{S_{i+1}}$:

$$f(x_S) = y^i_{T_i} + (x^i_{S_i} - x^i_{S_i}) \cdot \frac{y^{i+1}_{T_{i+1}} - y^i_{T_i}}{x^{i+1}_{S_{i+1}} - x^i_{S_i}} = y^i_{T_i} + 0_{\max(S_i,S_{i+1},T_i,T_{i+1})} = y^i_{\max(S_i,S_{i+1},T_i,T_{i+1})}\,,$$

while the other approach (*interpolation from the left*) considers x_S to be positioned between $x^{i-1}_{S_{i-1}}$ and $x^i_{S_i}$:

$$f(x_S) = y^{i-1}_{T_{i-1}} + (x^i_{S_i} - x^{i-1}_{S_{i-1}}) \cdot \frac{y^i_{T_i} - y^{i-1}_{T_{i-1}}}{x^i_{S_i} - x^{i-1}_{S_{i-1}}} = y^{i-1}_{T_{i-1}} + (y^i - y^{i-1})_{\max(S_{i-1},S_i,T_{i-1},T_i)} = y^i_{\max(S_{i-1},S_i,T_{i-1},T_i)}\,.$$

Moreover, none of the two results is necessarily equal to $y^i_{T_i}$ which means that the interpolation condition would not be met. This problem is not critical at all as then word interpolation actually assumes that it focuses on points between the given nodes so, nothing prevents us from defining the outputs for the given nodes $f(x^i_{S_i}) = y^i_{T_i}$.

However, the problem arises when the input is not equivalent to the node only due to a different width, i.e., we consider x^i_S such that $S \neq S_i$. In such a case, the interpolation from the right would give us $f(x^i_S) = y^i_{\max(S_i,S_{i+1},T_i,T_{i+1},S)}$ while the interpolation from the left would lead to $f(x^i_S) = y^i_{\max(S_{i-1},S_i,T_{i-1},T_i,S)}$. As it is not clear which result should be adopted and both "directions" can be justified, we can choose the one that lowers the uncertainty by picking

$$f(x^i_S) = y^i_T \text{ where } T = \min(\max(S_i, S_{i+1}, T_i, T_{i+1}, S), \max(S_{i-1}, S_i, T_{i-1}, T_i, S))\,. \tag{18}$$

However, we should check whether if we impose some reasonable properties, the proposed approach meets them. The most natural property one would expect is the monotonicity with respect to the uncertainty, i.e., less uncertain inputs cannot lead to more uncertain outputs:

$$x_S, x_T, y_U, y_V \in \mathcal{F}_\mathcal{S}(\mathbb{R}) \text{ s.t. } y_U = f(x_S), y_V = f(x_T) \text{ it holds that: } S \leq T \Rightarrow U \leq V\,. \tag{19}$$

It is easy to assume conditions to preserve the uncertainty monotonicity condition (19) for any $x \neq x^i$, $i = 1, \ldots, n$ but its preservation for $x = x^i$ must be ensured. Indeed, Formula (18) can easily lead to $T_i < T$ although $S \leq S_i$. Therefore, let us introduce a linear $\mathcal{S}$-interpolation that meets the constraint.

Definition 15. *Let $(\mathcal{F}_\mathcal{S}(\mathbb{R}), +, -)$ and $(\mathcal{F}_\mathcal{S}(\mathbb{R}), \cdot, ^{-1})$ be MI-pregroups and let $\leq$ be an $\mathcal{S}$-ordering on $\mathcal{F}_\mathcal{S}(\mathbb{R})$ that is strongly compatible. Let $x^i, y^i \in \mathbb{R}$ for $i = 1, \ldots, n$ and let $(x^1_{S_1}, y^1_{T_1}), \ldots, (x^n_{S_n}, y^n_{T_n}) \in \mathcal{F}_\mathcal{S}(\mathbb{R})^2$ such that $x^i_{S_i} \leq x^{i+1}_{S_{i+1}}$ for any $i = 1, \ldots, n-1$.* The linear $\mathcal{S}$-interpolation *of the given pairs of extensional fuzzy numbers is an* $\mathcal{S}$*-function* $f\colon D \to \mathcal{F}_{\mathcal{S}}(\mathbb{R})$*, where* $D \subseteq \mathcal{F}_{\mathcal{S}}(\mathbb{R})$ *such that* $D = \{x_S \mid x \in [x_1, x_n] \;\&\; S \in \mathcal{S}\}$*, i.e., for any* $x_S \in \mathcal{F}_{\mathcal{S}}(\mathbb{R})$ *given as follows:*

$$f(x_S) = \begin{cases} (17)\,, & \text{if } x^i_{S_i} \le x_S \le x^{i+1}_{S_{i+1}} \;\&\; x \neq x_i, x_{i+1}\,, \\ (18)\,, & \text{if } x = x_i \;\&\; S_i < S\,, \\ y^i_{T_i}\,, & \text{if } x = x_i \;\&\; S \le S_i\,. \end{cases} \tag{20}$$

Remark 9. *Let us note that as the basis for the above-introduced* $\mathcal{S}$*-interpolation, we have chosen the simplest crisp interpolation—the linear one. However, the concept of* $\mathcal{S}$*-function allows much more, basically, any standard interpolation method can be extended in this way, for example, Lagrange polynomial interpolation or Newton polynomial interpolation. On the other hand, one must take into account that such interpolations do not consider only neighboring points but basically all nodes which, using the arithmetic, causes inevitable widening of the results, which is exactly the problem that was aimed when introducing the arithmetic of extensional fuzzy numbers.*

Indeed, when calculating with two extensional fuzzy numbers, the result does not get wider than the widest one which is a positive aspect compared to arithmetics using Zadeh's extension principle. On the other hand, if we consider all nodes, the result will always be as wide as the widest one, which is a positive result when the used similarities are more or less of the same widths. But it is not very positive when we deal with significantly different values (e.g., daily COVID-19 increases in a particular country ranging from individuals to thousands per day) and the interpolation between two vaguely defined small (and thus also narrow) quantities would result into something as wide as the width of the highest values in the set of nodes. Such cases must be treated by special similarity transformation techniques or by using "relative" similarities flexibly changing the widths of the fuzzy numbers.

5.3. Properties

Proposition 7. *The linear* $\mathcal{S}$*-interpolation:*

(a) $\mathcal{S}$*-interpolates the given pairs of extensional fuzzy numbers;*
(b) preserves uncertainty monotonicity condition (19).

Proof. The proof is constructed directly. Let $x_S = x^i_{S_i}$ then (20) directly assigns $y^i_{T_i}$ which meets the interpolation conditions and thus, *a)* holds.

Let us consider x_S and x_R such that $S \le R$ and let us split the situation. First, let $x = x_i$. If both S and R are narrower than S_i then the result is $y^i_{T_i}$ and the monotonicity condition is preserved. If $S \le S_i$ and $S_i \le R$ then according to (18):

$$f(x_R) = y^i_T \;\text{ where }\; T = \min(\max(S_i, S_{i+1}, T_i, T_{i+1}, R), \max(S_{i-1}, S_i, T_{i-1}, T_i, R))$$

and clearly, $T_i \le T$ which again preserves the monotonicity.

Let $x \neq x_i$ then Formula (17) applies to $f(x_S)$ and $f(x_R)$ leads to $y_{\max(S_i,S_{i+1},T_i,T_{i+1},S)}$ and $y_{\max(S_i,S_{i+1},T_i,T_{i+1},R)}$, respectively, with $y \in \mathbb{R}$ being the solution of the crisp linear interpolation problem. As $S \le R$ and all the other similarities are present in both solutions, the uncertainty monotonicity condition (19) is preserved and thus, *(a)* holds. □

Finally, we investigate another important property of the interpolating $\mathcal{S}$-function that is, the preservation of the monotonicity with respect to the inputs. This property, naturally, makes sense only if the described dependency that is being modeled is assumed to be monotone. This assumption, though might be viewed as too restrictive, belongs to crucial and very usual properties. In the context of the approximate reasoning, the first studies on monotonicity of fuzzy rule bases were introduced by Van Broekhoven and De Baets [44,51,52]. Later, the original studies focusing mainly (not exclusively) on Mamdani–Assilian type of fuzzy rules, were followed by works aiming at the preservation of the monotonicity of implicative fuzzy rules [45,53,54]. Such research starts always by assuming that the given fuzzy rule base is monotone, i.e., abstracting from the mathematical formalism, considering any two antecedents, the bigger one is connected to a bigger consequent. Interestingly, the resulting function after applying the defuzzification is not always preserving the monotonicity, see [43]. Therefore, the other works aimed at determining the conditions under which the monotonicity is preserved.

Other two approximate reasoning tasks related to the above-given monotonicity problem are: preservation of the interpolation condition and monotonicity regarding fuzzy inputs.

The very first one became a natural consequence of the works [7,8] that, in order to preserve the monotonicity of the resulting function, modifies the original fuzzy rule base by specific at-least and at-most modifiers [55,56]. Then naturally, one must ask whether an input equivalent with one of the original antecedents still leads to the derived output equivalent to the original consequent, although the fuzzy rule base used for the inference contains already modified antecedents as well as consequents. The solution presented in [7,8] investigated the conditions under which the interpolativity remains preserved.

The other problem arises whenever we do not intend to involve a defuzzification and the inputs are fuzzy as well. Then the monotonicity preservation is investigated on a higher layer, in particular, assuming that we are given a partial mapping from fuzzy sets to fuzzy sets, e.g., by a fuzzy rule base, such that for any two fuzzy inputs (antecedents) that are ordered, the respective fuzzy outputs (consequents) reflect the order. Moreover, as in the above cases, the fuzzy inputs (are supposed to be ordered) and thus, the monotonic description is provided, although in a discrete form. Now, the task is to complete this partial mapping from fuzzy sets to fuzzy sets to a total one, but preserving the monotonicity. Therefore, if two order fuzzy inputs are processed, the generated fuzzy outputs must preserve the same order. Following the previous works, the whole description of the monotonicity is provided for the increasing case as the decreasing case is directly obtained by inverse steps.

This naturally leads to the investigation of the monotonicity of the $\mathcal{S}$-interpolation. Let us assume that the nodes are ordered $x^i_{S_i} \leq x^{i+1}_{S_{i+1}}$ and the respective consequents are reflecting the order, i.e., let $y^i_{T_i} \leq y^{i+1}_{T_{i+1}}$, for any $i = 1, \ldots, n-1$. Let us apply the linear $\mathcal{S}$-interpolation. Do we obtain a monotone $\mathcal{S}$-function f? The following proposition provides us with the answer.

Proposition 8. *Let $(\mathcal{F}_{\mathcal{S}}(\mathbb{R}), +, -, \cdot, ^{-1})$ be an MI-prefield such that any similarity from $\mathcal{S}$ is shift-invariant and let $\leq$ be a strongly compatible $\mathcal{S}$-ordering on $\mathcal{F}_{\mathcal{S}}(\mathbb{R})$. Let $(x^1_{S_1}, y^1_{T_1}), \ldots, (x^n_{S_n}, y^n_{T_n}) \in \mathcal{F}_{\mathcal{S}}(\mathbb{R})^2$ such that $x^i_{S_i} \leq x^{i+1}_{S_{i+1}}$ and $y^i_{T_i} \leq y^{i+1}_{T_{i+1}}$ for any $i = 1, \ldots, n-1$ and let f be given by the linear $\mathcal{S}$-interpolation of the given pairs of extensional fuzzy numbers. Then f is increasing.*

Proof. As the $\mathcal{S}$-order is strongly compatible, it is also total and thus, the problem is correctly defined. Let us consider inputs $u_R, v_S \in \mathcal{F}_{\mathcal{S}}(\mathbb{R})$ such that $u_R \leq v_S$. The situation splits into several cases.

First, let us assume that both inputs are indistinguishable from the nodes up to a certain similarity, i.e., $u_R - x^i_{S_i} \in I^0_{\mathcal{S}}$ and $v_S - x^{i+k}_{S_{i+k}} \in I^0_{\mathcal{S}}$ for some i and some k such that $i + k \leq n$. Then necessarily $u = x^i$ and $v = x^{i+k}$ and the respective outputs $f(u_R) = y^i_M$ and $f(v_S) = y^{i+k}_N$ for some $M, N \in \mathcal{S}$. Due to the assumption of the theorem, we know that $y^i \leq y^{i+k}$. Put $U = \max(M, N)$. Since $M \subseteq U$ and $N \subseteq U$, we obtain $U = M \circ U$ and $U = N \circ U$ as has been verified in the proof of Proposition 3. As with the proof of Lemma 4, from the shift-invariance of U, we simply obtain $\mathrm{EXT}_U(y^i) \leq_i \mathrm{EXT}_U(y^{i+1})$, and hence, $\mathrm{EXT}_{\max(M,N)}(y^i_M) = \mathrm{EXT}_{M \circ U}(y^i) = \mathrm{EXT}_U(y^i) \leq_i \mathrm{EXT}_U(y^{i+k}) = \mathrm{EXT}_{N \circ U}(y^{i+k}) = \mathrm{EXT}_{\max(M,N)}(y^{i+k}_N)$. Due to the real-order compatibility we get $y^i_M \leq y^{i+k}_N$ that is $f(u_R) \leq f(v_S)$.

Second, let us consider u_R, v_S such that they lie in different "intervals", i.e., let $x^i_{S_i} \leq u_R \leq x^{i+1}_{S_{i+1}}$ and let $x^{i+k}_{S_{i+k}} \leq v_S \leq x^{i+k+1}_{S_{i+k+1}}$ for some i, k such that $k \geq 1$ and $i + k \leq n - 1$.

Then using Formula (20) for the $\mathcal{S}$-interpolation and Lemma 4 we obtain $f(u_R) \leq y^{i+1}_{T_{i+1}}$ and $y^{i+k}_{T_{i+k}} \leq f(v_S)$ which, using the transitivity axiom (O3), proves that $f(u_R) \leq f(v_S)$.

Finally, we assume that both inputs belong to the same "interval", i.e., $x^i_{S_i} \leq u_R \leq v_S \leq x^{i+1}_{S_{i+1}}$. Using an analogous technique, we used in the second case, we obtain $f(u_R) = w^1_M$ and $f(v_S) = w^2_N$ such that $w^1 \leq w^2$. Then, we proceed analogously to the first case and we obtain $\mathrm{EXT}_{\max(M,N)}(w^1_M) \leq_i \mathrm{EXT}_{\max(M,N)}(w^2_N)$ that jointly with the pre-order compatibility leads to the desirable conclusion that $f(u_R) \leq f(v_S)$. □

Proposition 7 jointly with Proposition 8 provides us with the crucial answers. In particular, they confirm that the proposed linear $\mathcal{S}$-interpolation interpolates the given pairs of extensional fuzzy numbers and moreover, it is monotone in "both directions", with respect to the uncertainty as well as with respect to the ordering of vague quantities. Especially the latter has a strong impact on modeling the approximate reasoning in distinct schemes, namely when fuzzy interpolating a sparse fuzzy rule base.

6. Experimental Demonstration

The purpose of this experiment is not to demonstrate an exhaustive comparison of the suggested approach with the existing ones. First, such a comparison would need to have given a justified cost function on which we could evaluate some fair score. And the fairness of such a cost function would be a nutshell from the objectivity point of view. Indeed, for any interpolation function, there might be a cost function on which the particular method is optimal; however, this only creates biased information that we want to avoid deliberately. Secondly, the aim of this article is not centrally focused on the fuzzy interpolation problem. It provides a very general concept coming from the arithmetics of extensional fuzzy numbers and allowing the construction of concepts such as $\mathcal{S}$-ordering, $\mathcal{S}$-function, and/or $\mathcal{S}$-interpolation that can be either directly used or further extended to additional concepts known from the classical mathematical analysis. It is clear that a particular comparison would rather give information on the particular setting. Not only parametric but mainly conceptual, see Section 7.

However, for the sake of clarity, comprehensibility, and mainly for demonstrative purposes, we find the experimental example as a crucial part of the work. Below, we provide one such example that comes from the real-world problem, in particular, from our long-term grant supported co-operation with economists.

6.1. Ice-Cream Sales Modeling

The world ice-cream market is dominated by several well-established players who saturate a large market share. Their products are very similar and often differ only in marketing and seasonal innovations. The market basically consists of three types of products, namely artisanal ice-cream, pulse ice-cream, and take-home ice-cream, where the latter dominates the global market, with hypermarkets and supermarkets being the leading distribution channel. The largest market in terms of revenues is the US market, although the highest growth rate is currently observed in Asia–Pacific market.

The experimental demonstration is focused on the Czech Republic with the average per capital consumption close to 5 kg with low but stable annual growth of about 2% over the last decade and the high sensitivity to weather conditions leading to large differences between summer and winter sales, which is a feature observed in all countries outside the (sub)tropics. Analysis of ice-cream market data (see e.g., www.statista.com) shows that atypical weather conditions in seasons such as warm winter and early spring or colder summer can cause a drop in the ice-cream market revenue growth. Therefore, it is desirable to model the relationship between weather conditions in individual seasons and ice-cream sales, which will be used to predict the ice-cream market revenue growth when weather conditions are estimated.

For this purpose, we use dataset that consists of adjusted and normalized quarterly sales in the Ice category (various kinds of, mostly, take-home ice-cream) of one of the largest market players over 2008–2014, i.e., 28 observations are available, in the Czech Republic. It should be noted that only quarterly sales data are available; however, the production process cannot be rescheduled every day and the product has a relatively long use, allowing it to be stored relatively long in shops and households; therefore, the quarterly sales prediction appears to be sufficient to optimize the production process. To illustrate the $\mathcal{S}$-interpolation also for quarterly ice-cream sales that are inaccurate for some reason, we extend the crisp sales to extensional fuzzy numbers by adding a randomly determined standard deviation to each sale (see, Table 1). The quarterly sale expressed as an extension fuzzy number is then determined by the parameters (s, σ_s) and described as a fuzzy number $s_T : \mathbb{R} \to [0,1]$, where the similarity of the relation T on $\mathbb{R}$ is given (1) in Example 1, i.e.,

$$T(x,y) = 1 - \frac{|x-y|}{\sigma_s} \vee 0, \quad x,y \in \mathbb{R}, \tag{21}$$

and $s_T(x) = T(s,x)$. Unlike the quarterly ice-cream sale s expressed by a crisp number, the extensional fuzzy number s_T models the fact that the quarterly sale is "about s"; for example, we have "about 1.53" in spring 2010 as could be seen in Figure 5a.

Table 1. Quarterly sales of ice-cream rounded to two decimals and standard deviations rounded to three decimals over 2008–2014 (Q_1 = winter, Q_2 = spring, Q_3 = summer, Q_4 = fall).

	2008	2009	2010	2011	2012	2013	2014
Q_1	(0.23, 0.029)	(0.24, 0.053)	(0.25, 0.030)	(0.24, 0.048)	(0.27, 0.032)	(0.19, 0.047)	(0.31, 0.18)
Q_2	(1.54, 0.091)	(1.77, 0.094)	(1.53, 0.072)	(1.72, 0.082)	(1.66, 0.101)	(1.51, 0.118)	(1.60, 0.073)
Q_3	(2.53, 0.171)	(3.25, 0.299)	(2.59, 0.260)	(2.62, 0.433)	(2.57, 0.204)	(2.50, 0.399)	(2.22, 0.244)
Q_4	(0.32, 0.072)	(0.17, 0.106)	(0.35, 0.042)	(0.35, 0.075)	(0.22, 0.087)	(0.36, 0.064)	(0.27, 0.079)

To express weather conditions, for demonstration purposes, we will use the daily average temperatures (in degree Celsius) at Václav Havel Airport in Prague. Because each quarter contains approximately 90 records of temperature, and similarly to ice-cream sales, we summarize the temperature into an extensional fuzzy number determined by two parameters, namely the average temperature and the standard deviation in the quarter (see Table 2), which contains more information than only average temperature. Please note that the standard modeling of the relationship between temperatures and ice-cream sales (e.g., based on the interpolation or regression) use the average temperature in the quarters, while the random fluctuation of temperatures in the quarters expressed by the standard deviation is not considered (cf. [57]).

Table 2. Quarterly temperatures and standard deviations rounded to two decimals over 2008–2014.

	2008	2009	2010	2011	2012	2013	2014
Q_1	(3.11, 3.55)	(0.09, 4.44)	(−0.75, 5.63)	(0.94, 5.30)	(1.16, 6.95)	(−0.97, 4.04)	(3.44, 4.16)
Q_2	(13.37, 5.09)	(14.09, 3.22)	(12.70, 4.74)	(14.46, 4.37)	(13.94, 5.95)	(12.32, 5.59)	(13.19, 4.34)
Q_3	(16.53, 4.60)	(18.12, 2.96)	(17.02, 4.85)	(17.05, 3.26)	(17.40, 3.99)	(17.02, 4.64)	(17.05, 3.60)
Q_4	(4.75, 4.68)	(4.53, 8.81)	(2.13, 6.67)	(4.84, 4.54)	(3.96, 4.80)	(4.96, 4.53)	(6.35, 5.01)

An extensional fuzzy number expressing the temperature in a quarter and determined by parameters (t, σ_t) is the fuzzy set $t_S\colon \mathbb{R} \to [0,1]$, which is defined in the same way as the similarity R for the quarterly sales. An example of the extensional fuzzy number expressing the temperature "about 12.7" in spring 2010, which is determined by parameters $(12.70, 4.74)$ in Table 2, is displayed in Figure 5b.

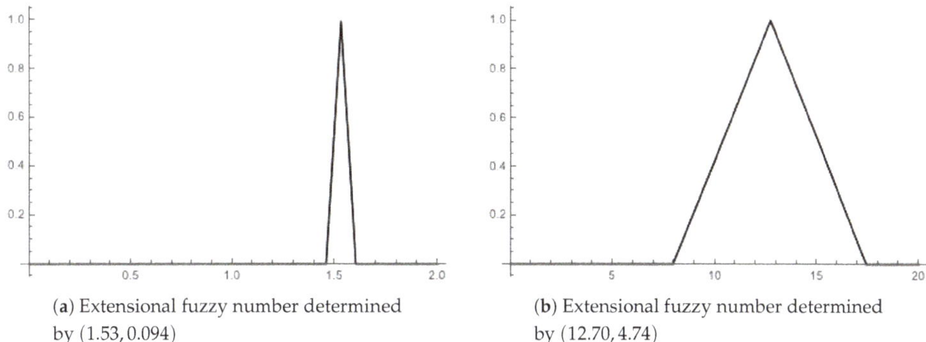

(**a**) Extensional fuzzy number determined by $(1.53, 0.094)$

(**b**) Extensional fuzzy number determined by $(12.70, 4.74)$

Figure 5. Extensional fuzzy numbers expressing the ice-cream sale "about 1.53" and temperature "about 12.70" in spring 2010.

In the following subsection we will describe the procedure how to $\mathcal{S}$-interpolate extensional fuzzy numbers expressing temperatures and ice-cream sales in quarters in the years 2008–2014.

6.2. General Description of $\mathcal{S}$-Interpolation Procedure

Let $(x^1_{S_1}, y^1_{T_1}), \ldots, (x^n_{S_n}, y^n_{T_n})$ be pairs of extensional fuzzy numbers such that $x^i_{S_i} \leq x^{i+1}_{S_{i+1}}$ for any $i = 1, \ldots, n$. We call these pairs of extensional fuzzy numbers as *extensional fuzzy data*. In contrast to the interpolation of real data, we must be more careful in the direct use of Formula (20) for extensional fuzzy data. The reason is that the fuzziness of extensional fuzzy numbers on input, which is expressed by a similarity relationship, is distributed through Formula (20) to another fuzziness specifying the similarity relation of extensional fuzzy number on the output of $\mathcal{S}$-interpolation. Since the input and output of extensional fuzzy data are in practice obtained by the measurement on different scales, the meanings of fuzziness are different for input and output fuzzy data, which leads to incorrect results by direct application of (20). In other words, unless the same measurement scale is used to determine the input and output fuzzy data, we cannot assume that extensional fuzzy data can be handled in one ordered MI-prefield, and the $\mathcal{S}$-interpolation introduced in Definition 14 cannot be applied for them. This problem can be overcome by a normalization procedure, which ensures that the normalized extensional fuzzy numbers will belong to the same ordered MI-prefield.

In what follows, we assume that extensional fuzzy data generally belong to $\mathcal{F}_{\mathcal{S}_I}(\mathbb{R}) \times \mathcal{F}_{\mathcal{S}_O}(\mathbb{R})$, where $\mathcal{F}_{\mathcal{S}_I}(\mathbb{R})$ and $\mathcal{F}_{\mathcal{S}_O}(\mathbb{R})$ are ordered MI-prefields with a strongly compatible $\mathcal{S}_I$-ordering $\leq_I$ and a strongly compatible $\mathcal{S}_O$-ordering $\leq_O$. respectively. We use I and O in the indexes to refer to the input and output, respectively. Let $\mathcal{F}_{\mathcal{S}}(\mathbb{R})$ be an ordered MI-prefield with a strongly compatible $\mathcal{S}$-orderings $\leq$, and let $N_I \colon \mathcal{F}_{\mathcal{S}_I}(\mathbb{R}) \to \mathcal{F}_{\mathcal{S}}(\mathbb{R})$ and $N_O \colon \mathcal{F}_{\mathcal{S}_O}(\mathbb{R}) \to \mathcal{F}_{\mathcal{S}}(\mathbb{R})$ be isomorphisms of ordered MI-prefields. We say that a pair (N_I, N_O) of isomorphisms is a *normalization* of extensional fuzzy numbers if the composition $N_I^{-1} \circ N_O$ and $N_O^{-1} \circ N_I$ introduce natural correspondences between the ordered MI-prefields $\mathcal{F}_{\mathcal{S}_I}(\mathbb{R})$ and $\mathcal{F}_{\mathcal{S}_O}(\mathbb{R})$, i.e., $N_I^{-1} \circ N_0(x_S)$ is a natural image of x_S in $\mathcal{F}_{\mathcal{S}_O}(\mathbb{R})$ and vice versa $N_O^{-1} \circ N_I(y_T)$ is a natural image of y_T in $\mathcal{F}_{\mathcal{S}_I}(\mathbb{R})$. Please note that the decision if the compositions $N_I^{-1} \circ N_0$ and $N_I^{-1} \circ N_0$ are natural correspondences between the ordered MI-prefields $\mathcal{F}_{\mathcal{S}_I}(\mathbb{R})$ and $\mathcal{F}_{\mathcal{S}_O}(\mathbb{R})$ depends on the user. If $\mathcal{F}_{\mathcal{S}_I}(\mathbb{R}) = \mathcal{F}_{\mathcal{S}_O}(\mathbb{R})$, i.e., $\mathcal{S}_I = \mathcal{S}_O = \mathcal{S}$, the pair (N_I, N_O) given by $N_I = id = N_O$, where id is the identity map on $\mathcal{F}_{\mathcal{S}}(\mathbb{R})$, defines the trivial normalization of extensional fuzzy numbers, nevertheless, the user must agree that extensional fuzzy data are determined by the measurement on the same scale and thus can be processed in the same ordered MI-prefield $\mathcal{F}_{\mathcal{S}}(\mathbb{R})$. Now, we can introduce the procedure of $\mathcal{S}$-interpolation of extensional fuzzy data.

Let $(x^1_{S_1}, y^1_{T_1}), \ldots, (x^n_{S_n}, y^n_{T_n})$ be extensional fuzzy data, and let $N_I\colon \mathcal{F}_{\mathcal{S}_I}(\mathbb{R}) \to \mathcal{F}_{\mathcal{S}}(\mathbb{R})$ and $N_O\colon \mathcal{F}_{\mathcal{S}_O}(\mathbb{R}) \to \mathcal{F}_{\mathcal{S}}(\mathbb{R})$ define a normalization of extensional fuzzy numbers. We put $x_S^{N_I} = N_I(x_S)$ and similarly $y_T^{N_O} = N_O(y_T)$. The procedure of linear $\mathcal{S}$-interpolation of extensional fuzzy data with respect to the normalization given by (N_I, N_O) is as follows:

Input: extensional fuzzy number x_S from $\mathcal{F}_{\mathcal{S}_I}(\mathbb{R})$

Step 1: normalize x_S with respect to N_I to get $x_S^{N_I}$

Step 2: find $x^{iN_I}_{S_i}, x^{i+1N_I}_{S_{i+1}}$ such that $x^{iN_I}_{S_i} \le x_S^{N_I} \le x^{i+1N_I}_{S_{i+1}}$

Step 3: compute z_R from $(x^{iN_I}_{S_i}, y^{iN_O}_{T_i})$ and $(x^{i+1N_I}_{S_{i+1}}, y^{i+1N_O}_{T_{i+1}})$ by Formula (20)

Step 4: compute $y_T = N_O^{-1}(z_R)$ (a denormalization procedure)

Output: extensional fuzzy number y_T from $\mathcal{F}_{\mathcal{S}_O}(\mathbb{R})$

It is easy to see that if $\mathcal{F}_{\mathcal{S}_I}(\mathbb{R}) = \mathcal{F}_{\mathcal{S}_O}(\mathbb{R})$ and the input and output extensional fuzzy numbers were determined by the measurement on the same scale, one can set (N_I, N_O) as the trivial normalization, and the output extensional fuzzy number coincides with the extensional fuzzy number obtained directly by Formula (20).

6.3. Experimental Results

Our extensional fuzzy data consist of 32 pairs $(t^i_{S_i}, s^i_{T_i})$, where we put $i = 1$ for winter 2008, $i = 2$ for spring 2008, etc. Recall that each input extensional fuzzy number $t^i_{S_i}$ is determined by parameters (t^i, σ^i_t) displayed in Table 2, where t_i is the average temperature, σ_i the standard deviation in the i-th quarter and S_i is given by (21). For the output values, we will consider two situations:

(a) the ice-cream sales are crisp fuzzy numbers, i.e., extensional fuzzy numbers $s^i_{T_i}$, where T_i is the equality relation and only the first parameter in each pair (s^i, σ^i_s) is considered in Table 1;

(b) the ice-cream sales are "true" fuzzy numbers, i.e., extensional fuzzy numbers $s^i_{T_i}$, where the similarity T_i is given by (21) and both parameters in each pair (s^i, σ^i_s) are considered in Table 1.

The first situation is very common in practice and occurs when output data (ice-cream sales) are available as real numbers, and we have no further information. The second situation requires additional information, for example, quarterly sales are average ice-cream sales that are calculated from weekly sales in quarters, where the standard deviation provides additional information, or the expert expresses an inaccuracy in the total sale in the quarter.

Let $\mathcal{F}_{\mathcal{S}_\infty}(\mathbb{R})$ be the ordered MI-prefield introduced in Example 12. Obviously, our extensional fuzzy data belong to $\mathcal{F}_{\mathcal{S}_\infty}(\mathbb{R})^2$, where p_i is the inverse element to the standard deviation σ_i, i.e., $p_i = \frac{1}{\sigma_i}$, but the input and output extensional fuzzy numbers cannot be reasonably processed in this ordered MI-prefield without normalization. Indeed, the scale for measuring temperature is different from the scale used for sales of ice-cream, and even scales within different quarters are different as could be seen from Table 3, where we display the standard deviations for the temperatures and ice-cream sales.

Table 3. Data for normalization computed in quarters 2008–2017.

Quarters	Temperature st. Deviation $(\sigma_{t,j})$	Ice-Cream Sale st. Deviation $(\sigma_{s,j})$
Q_1	8.110	0.036
Q_2	8.063	0.010
Q_3	8.161	0.313
Q_4	8.034	0.072

In contrast to the standard deviations of temperature in quarters (the standard deviation for all temperatures in 2008–2014 is 8.17), which are almost the same, there are large differences between the standard deviations of ice-cream sales in quarters (the standard deviation for all sales in 2008–2014 is 1.02). The variability of the standard deviations of ice-cream sales forces us to model the relationship between temperature and ice-cream sales in each quarter separately, otherwise the resulting sales uncertainty may be affected by inaccurate normalization. This would certainly happen if we applied overall normalization to all sales. Therefore, before we use the linear $\mathcal{S}$-interpolation procedure introduced in the previous subsection, we normalize the standard deviations with respect to the standard deviations of the temperatures and the sales in the respective quarters of the years 2008–2014, see Table 3. Please note that the normalized values (i.e., standard deviations) are measured in the unit of the standard deviation.

For any extensional fuzzy number t_{S_p} expressing a temperature in a quarter Q_j ($j = 1,\ldots,4$), we define $N_{I,j}(t_{S_p}) = t_{S_{p'}}$ with $p' = p \cdot \sigma_{t,j}$, where $\sigma_{t,j}$ is the standard deviation of temperatures in Q_j. Similarly, for any extensional fuzzy number s_{T_q} expressing an ice-cream sale in a quarter Q_j, we define $N_{O,j}(s_{T_q}) = s_{T_{q'}}$ with $q' = q \cdot \sigma_{s,j}$, where $\sigma_{s,j}$ is the standard deviation of sales in Q_j. For $p = q = \infty$, we put $p' = q' = \infty \cdot \sigma_t = \infty \cdot \sigma_s = \infty$, hence, $N_{I,j}(t_{S_\infty}) = t_{S_\infty}$ and $N_{O,j}(s_{T_\infty}) = s_{T_\infty}$. Please note that $p' = p \cdot \sigma_{t,j}$ (and similarly $q' = a \cdot \sigma_{s,j}$) is derived from the following formula

$$\frac{1}{p'} = \frac{\frac{1}{p}}{\sigma_t^j}$$

where $\frac{1}{p}$ represents a standard deviation for p and $\frac{1}{p'}$ a standard deviation for p', and thus the standard deviation $\frac{1}{p'}$ is a normalization of the standard deviation $\frac{1}{p}$ with respect to $\sigma_{t,j}$. According to Example 13, the maps $N_{I,j}$ and $N_{O,j}$ are isomorphisms from $\mathcal{F}_{S_\infty}(\mathbb{R})$ onto $\mathcal{F}_{S_\infty}(\mathbb{R})$. In addition, we agree that these isomorphisms are appropriately selected for normalization, where we do not measure in the original units, which are clearly different, but in the units of standard deviations in quarters, which are abstract and the same for input and output.

Unlike the standard linear interpolation, we cannot display the linear $\mathcal{S}$-interpolation in a clear way, because we have infinitely many extensional fuzzy numbers for any real number. On Figure 6, we illustrate the linear $\mathcal{S}$-interpolation on extensional fuzzy data observed in summer that are crisp and fuzzy, and we show separately the use of each rule of Formula (20) from Definition 15, where the computational algorithm is designed according to the procedure established in the previous subsection. In Figure 6a-left, the temperature measured as "about 17.7" and expressed by the extensional fuzzy number t_{S_p} with $t = 17.7$ and $p = 1/4$ (or $\sigma_t = 4$) gives the estimation of ice-cream sale s_{T_q} with $s = 2.857$ and $q = 1/0.153$ (or $\sigma_s = 0.153$). Since the sales of ice-cream in output data are crisp numbers, the uncertainty of the estimated sale is only affected by the uncertainty of the input data temperatures and the uncertainty of the temperature for which the sale is estimated. The standard deviation $\sigma_{t,3} = 8.161$ for normalization and $\sigma_{s,3} = 0.313$ for denormalization are used in our procedure to get an appropriate estimation of the uncertainty of the ice-cream sale. In Figure 6a-right, we consider the same input data as above, but the sales of ice-cream in output data are "true" extensional fuzzy numbers. It can be seen that in both situations the same estimate of ice-cream sales is obtained, which indicates a dominant similarity among the used temperatures. In other words, the uncertainty present in the sales nodes is lower than the uncertainty present in a used temperature. In Figure 6b-left, the input temperature determined as "about 17.4" and expressed by t_{S_p} with $t = 17.4$ and $p = 1/3$ (or $\sigma_t = 3$) gives the estimation of ice-cream sale that coincides with the crisp sale output 2.57 for the temperature node $t^{19}_{S_{p_{19}}}$ with $t^{19} = 17.4$ and $p_{19} = 1/3.99$ (or $\sigma_{t^{19}} = 3.99$). The reason is that $t = t^{19}$ and simultaneously $S_p \leq S_{p_{19}}$ (i.e., $p = 1/3 > 1/3.99 = p_{19}$), and due to the uncertainty monotonicity condition (19), the uncertainty of ice-cream sale corresponding to temperature t_{S_p} cannot be higher than the uncertainty of output corresponding to temperature node $t^{19}_{S_{p_{19}}}$. Similarly, in Figure 6b-right, the estimation of ice-cream sale coincides with the sale output $s^{19}_{T_{q_{19}}}$ with $s^{19} = 2.57$ and $q_{19} = 1/0.204$ (or $\sigma_{s^{19}} = 0.204$). On the other side, in Figure 6c-left, the input temperature determined as "about 17.4" and expressed by t_{S_p} with $t = 17.4$ and $p = 1/4.3$ (or $\sigma_t = 4.3$) gives the estimation of ice-cream sale s_{T_q} with $s = s^{19} = 2.57$ and $q = 1/0.165$ (or $\sigma_q = 0.165$). The reason is that $t = t^{19}$ and simultaneously $S_p > S_{p_{19}}$ and therefore, the uncertainty of ice-cream sale output corresponding to temperature t_{S_p} can be higher than the uncertain of output corresponding to temperature node $t^{19}_{S_{p_{19}}}$ (i.e., $p_{19} = 1/3.99 > 1/4.3 = p$), which results in the application of Formula (18). In Figure 6c-right, we display the resulting estimation of ice-cream sale in the situation when the input data are the same as above and the ice-cream sales in the output data are expressed by extensional fuzzy numbers. One can see that the estimation is the same as for sales expressed in real numbers, which indicates a dominant similarity among the used temperatures similarly as in Figure 6a.

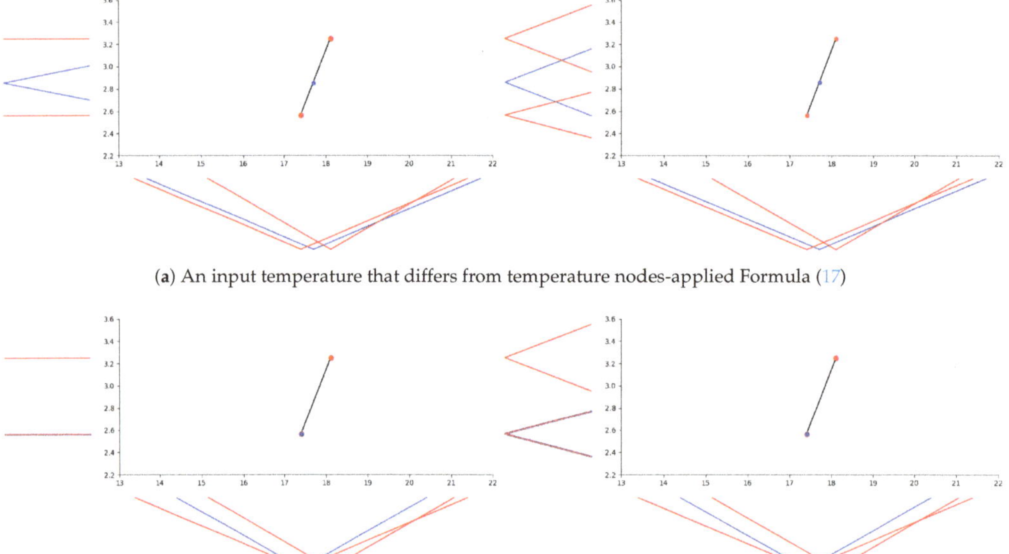

(**a**) An input temperature that differs from temperature nodes-applied Formula (17)

(**b**) An input temperature that differs from a temperature node up to similarity with lower similarity-applied the third rule of Formula (20)

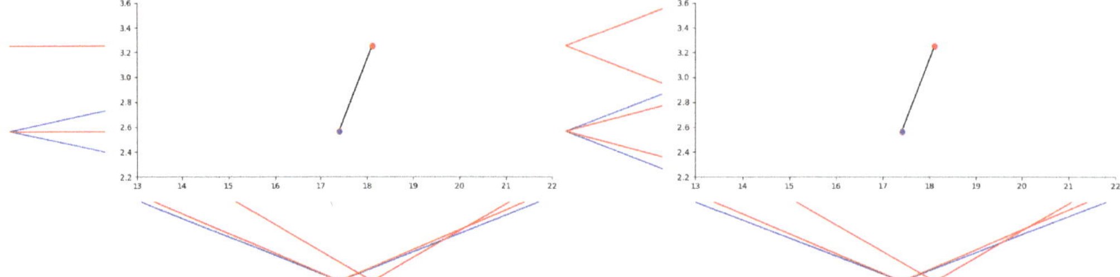

(**c**) An input temperature that differs from a temperature node up to similarity with higher similarity-applied Formula (18)

Figure 6. Estimation of ice-cream sales based on the linear $\mathcal{S}$-interpolation for selected temperatures in summer, where sale data are crisp (**left**) and fuzzy (**right**); temperature and ice-cream sale nodes are colored red, the selected (input) temperature and estimated (output) sale blue.

In Figure 7, we illustrate the linear $\mathcal{S}$-interpolation between three pairs of temperatures and ice-cream sales observed in winter, where sales are expressed by extensional fuzzy numbers, and show the estimation of ice-cream sales for three selected temperatures. Two of them, namely "about 1.7" ($\sigma_{t_1} = 3$) and "about 3.3" ($\sigma_{t_2} = 4.5$), lie between temperature nodes and "about 1.16" ($\sigma_{t_3} = 7.35$) coincides with the first node up to similarity with a higher similarity. The estimation of ice-cream sales is "about 0.257" ($\sigma_{s_1} = 0.032$), "about 0.277" ($\sigma_{s_2} = 0.029$) and "about 0.266" ($\sigma_{s_3} = 0.033$).

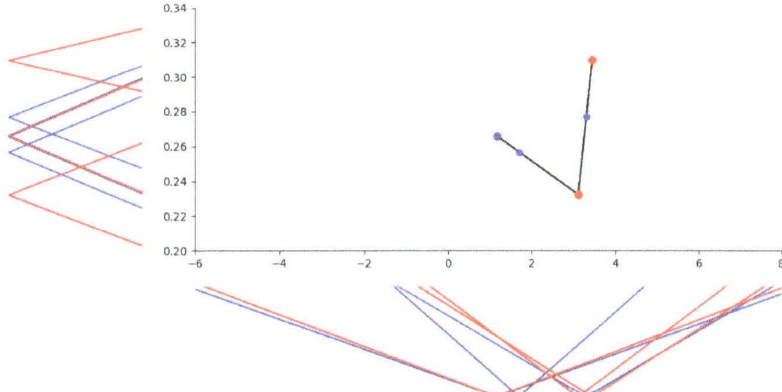

Figure 7. Estimation of ice-cream sales based on the linear $\mathcal{S}$-interpolation for selected temperatures in winter.

It should be noted that in practice it is very difficult to predict the average temperature in the coming quarter as a single number and it is more appropriate to use a confidence interval or an extensional fuzzy number as in our case. The linear $\mathcal{S}$-interpolation can be then used to predict ice-cream sales in the coming quarter, providing important information for production planning.

6.4. Comparing with Other Approaches—Discussion

As stated in the Introduction, fuzzy interpolation is a rather old problem with many well-developed techniques, no matter we talk about the geometrically motivated or rather logically motivated approaches. Thus, the formal apparatus designed in this article should naturally face some comparison with them. In data sciences (e.g., in pattern recognition or classification), we usually encounter an exhaustive experimental comparison with some objective evaluation on a given cost function. However, up to our best knowledge, there is no generally accepted quality criterion determining for fuzzy interpolation tasks. Indeed, there is always a way to define such a criterion, but it would be purely subjectively designed.

In such a case, we opt for a demonstrative comparison on the above-presented ice-cream sales data with two prototypical representatives of existing methods employing different paradigms. The first one that stems from the geometric motivation is the well-known Kóczy–Hirota (abbr. KH) methods introduced in [9]. Although this method is rather old and several improvements have been employed soon after its publication, it is a fundamental basis for many other geometrically oriented techniques, see [58]. Moreover, this method possesses the advantage of low computational costs and it is driven by the same crisp technique—the classical linear interpolation. For observing the differences, we find this choice very appropriate.

The second method stems from the logical (approximate reasoning) motivation and it is the standard fuzzy relational inference. Such method models the existing rules by a fuzzy relation that is either conjunctive (often called Mamdani–Assilian [59]) or implicative [60] and connects the fuzzy relation to a composition-based inference, i.e., to an appropriate image of a fuzzy set under the fuzzy relation. For more details on the distinct types of rules, their properties and the connection to the compositions/images as inference mechanisms, we refer to [61,62]. Briefly, the appropriate combinations are: (1) the implicative fuzzy rule base (antecedents and consequents are connected via a residated implication, rules are aggregated with help of the minimum) together with the basic composition, which is the well-known *Compositional Rule of Inference* (CRI); and (2) the conjunctive fuzzy rule base (antecedents and consequents are connected via a t-norm, rules are aggregated with help of the maximum) together with the Bandler–Kohout subproduct. The opposite combinations may cause complications, see [63].

Remark 10. *Many articles use the combination of the (Mamdani–Assilian) conjunctive fuzzy rules base and the CRI. This is often totally correct as the vast majority of cases deals with crisp inputs while the danger appears in the case of fuzzy inputs. Indeed, in the case of crisp inputs, no matter we deal with CRI or the Bandler–Kohout subproduct, the inference degenerates into a simple substitution of the crisp input into the given fuzzy relation so, we even cannot distinguish between these two compositional inferences, they coincide. But for the fuzzy interpolation task with fuzzy inputs, the choice of the combination is essential.*

Now, let us focus on several examples from the above elaborated ice-cream sales dataset. The first example considers "the simplest" case when the antecedents as well as consequents are ordered with respect to the interval-ordering of α-cuts and the fuzzy input lies "between" the antecedents, and the antecedents as well as consequents are overlapping. This case is demonstrated on a part of the autumn data from the ice-cream sales problem, see Figure 8. As we can see, the KH method and the proposed $\mathcal{S}$-interpolation method provides very comparable results. The proposed method generated a bit narrower output however, this advantage is caused by a well-chosen relativization determined by the standard deviation transformation caused and indeed, it could be easily lost. Therefore, we do not dare to claim it as a generally valid principle. The third method determined by fuzzy relational inference determined something that is logically correct however, not meeting the expectations of the analytical interpolation that is used under the implicit assumption of the satisfaction of the functionality axiom. Please note that the choice of the algebra has no significant influence, for example, using the Łukasiewicz implications would lead to a higher output fuzzy set but still reaching values around 0.6 at maximum and moreover, with an unlimited support. Here, we would like to emphasize that this is nothing against fuzzy relational inference. The problem lies in the fact that the antecedents are closer to each other than the consequents and thus, the fuzzy rule base is not coherent [64] which is a sort of consistency (non-conflictness) of the fuzzy rules. In other words, this means nothing wrong about fuzzy relational inference but rather about its non-applicability on such interpolative setting. Finally, the use of the other preferred combination (Mamdani–Assilian rules with Bandler–Kohout subproduct) would lead to a subnormal and non-convex output fuzzy set—again not meeting the implicit functionality axiom.

Now, let us consider the cases depicted on the right-hand side of Figure 6a. There, the antecedents are not ordered with respect to the interval-ordering of α-cuts which makes impossible to use the KH method (at least in its original form) while the proposed $\mathcal{S}$-interpolation works well due to the robustness of the $\mathcal{S}$-ordering. The fuzzy relational inference here can easily infer the outputs too as the ordering of antecedents does not play any role. However, the consequent part makes the inferred outputs inappropriate, in particular, the disjointness of the consequents is an extreme version of the incoherence and thus, the implicative rules would infer a fuzzy set equal to zero on the whole universe. For the completeness, let us mention that the Mamdani–Assilian rules would result again subnormal and non-convex outputs without any functional meaning.

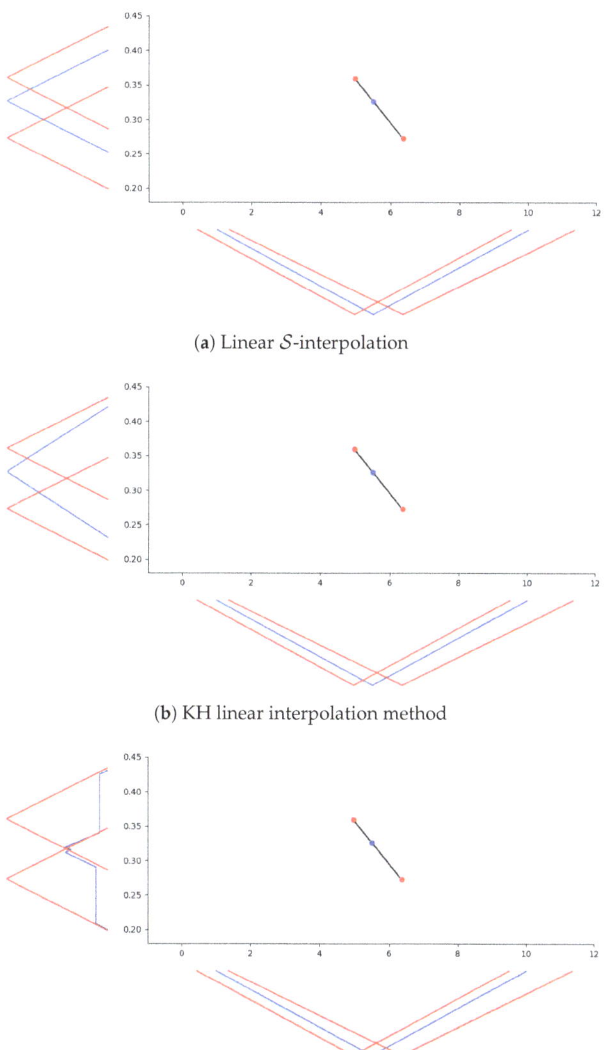

(c) Fuzzy relational inference—implicative rules with the CRI inference, the Gödel implication used.

Figure 8. Comparison of our method with two fundamental interpolation techniques on particular part of autumn ice-cream sales.

Another similar example is depicted on the right-hand side of Figure 6b. It is the same situation as in the previous case with the only difference that the fuzzy input is located in one of the nodes for the antecedents; however, the input is narrower than the antecedent. The figure shows how the proposed $\mathcal{S}$-interpolation generates the outputs while up to our best knowledge, KH method would not be applicable. But again, after some adjustments, KH method would deal with this input too and infer some reasonable output. The fuzzy relational inference would lead to the same as in the previous case.

The last case we consider is the situation of a sparse rule base where fuzzy relational inference would not again meet the expectations imposed by the assumed functionality axiom. In particular, if we were given two disjoint antecedents and the fuzzy input would be placed in between of them in such a way it would be disjoint with both, the inferred output would be equal to 1 on the whole output universe (or to 0, in the case of the Mamdani–Assilian rules with the Bandler–Kohout subproduct). Here, we would like to again emphasize that this not wrong at all from the logical perspective. If the rules are logical constraints stating that the consequent should be inferred under the assumption that the antecedent holds, and none of the antecedent is met by the input, we may deduce nothing, which is mirrored by the outputs 1 or 0, depending on the semantics of the rules. This is only again a case where, without further improvement like the one proposed in [15], relational inference is not appropriate for approximate reasoning schemes of this type of sparse cases. Here, the KH method performs a practical advantage as it is deliberately designed for the sparse rule bases. This case is not demonstrated on the real data as such a case has not occurred in the ice-cream sales dataset at all. However, from the principal point of view, the results obtained from the KH method and from the proposed $\mathcal{S}$-interpolation would be very comparable—mimicking the cases depicted in Figure 8.

7. Conclusions and Future Work

The presented approach stemmed from the particular fuzzy arithmetic, namely from the arithmetic of extensional fuzzy numbers [19,25]. The first steps from the arithmetic towards the ordering were published in [23,24] with the aim to construct metrics of fuzzy numbers [48]. The original motivation came from the need for modeling the triangle inequality in the metric-like structures. However, it is clear that the importance of $\mathcal{S}$-orderings is much only higher and mainly more general. Apart from the clear directions coming from the mathematical analysis based on the extensional fuzzy numbers (fuzzy approximation, mathematical analysis, construction of limits, convergences), already [24] mentioned approximate reasoning as one of the related areas where this research can have an importance.

Indeed, the fuzzy interpolation is one of the crucial problems in approximate reasoning and it is not possible to study it from the analytical point of view if we cannot order the used fuzzy sets (fuzzy numbers) in such a way to be able to state that an input lies "between two given fuzzy nodes". This article copes with this topic and incorporates the technique of the $\mathcal{S}$-interpolation generically from the formal roots (arithmetic, ordering, $\mathcal{S}$-functions) to the problem formalization ($\mathcal{S}$-interpolation, uncertainty monotonicity, and standard monotonicity) towards the practical implementation and the demonstrative example. It is clear that the proposed calculation is just one of the rich pool of possibilities and as mentioned in Remark 9 other techniques can be introduced analogously. The remark also contains a discussion pointing the potential drawbacks of more sophisticated approaches as well as the direction towards their elimination.

One can view the proposed approach as a single instance of more general classes of $\mathcal{S}$-interpolations not only form the linear, polynomial or other basis point of view, but also form the point of view of processing the vagueness. Indeed, following the arithmetic of extensional fuzzy numbers, also the $\mathcal{S}$-interpolation is a sort of two-step procedure. In the first one, the "crisp" part of the task is solved using the standard technique. In the second one, the resulting vagueness (uncertainty, tolerance, etc.) is calculated as a result of the computation with the similarities entering the interpolation problem. And as one can modify the first step of the procedure and consider, e.g., polynomial interpolations, the second step can be modified as well, and the resulting similarity may be deduced differently. So far, the procedure coming from the arithmetic of extensional numbers was naturally inherited as the linear $\mathcal{S}$-interpolation is derived from the arithmetic operations. However, considering the definition of the $\mathcal{S}$-function that is more general and the above-introduced examples (exponential, logarithmic, trigonometric), we are equipped with tools that allow modeling of the spreading of the vagueness in distinct ways according to the application or theoretical needs.

In our future research we plan to continue in the development of $\mathcal{S}$-interpolation for other types of functions and using metrics introduced in [48] we want to design more advanced interpolation approaches. For example, motivated by the Gaussian process regression (kriging) that has become popular in Machine Learning [65], where the interpolated values are modeled by Gaussian processes and are basically obtained as weighted averages of known neighbor nodes (see, e.g., [66]). In addition, it is a very challenging task to design an interpolation (regression) model that together admits probability and fuzziness of the data. The proposed theory seems to provide promising tools for successful solutions of this task.

Author Contributions: The conceptualization was contributed mainly by M.H. and M.Š.; methodology was mainly done by M.H.; software implementation was done by N.Š.; experimental design was proposed by M.H. and contributed by the other two authors as well; formal analysis was mainly done by M.Š. and contributed by the other authors as well; writing was done mainly by M.Š. and reviewed by the other two co-authors; visualization was done by N.Š. and reviewed by M.H.; the piloting ideas were created in joint brainstormings and discussions as a joint work. All authors have read and agreed to the published version of the manuscript.

Funding: The 1-st author announces support of Czech Science Foundation through the grant 18-13951S. The 2-nd author acknowledges funding from the project "Support of talented PhD students at the University of Ostrava" from the program RRC/10/2018 "Support for Science and Research in the Moravian-Silesian Region 2018". The 3-rd author announces support of Czech Science Foundation through the grant 20-07851S.

Institutional Review Board Statement: Not applicable.

Informed Consent Statement: Not applicable.

Data Availability Statement: Not applicable.

Acknowledgments: The authors would like to express their thanks to László T. Kóczy, Kaoru Hirota, and Ulrich Bodenhofer for their inspiring works that brought our attention to this field. Furthermore, we would like to express our thanks to the anonymous reviewers who helped to improve our article and directed our future research.

Conflicts of Interest: The authors declare no conflict of interest.

References

1. Perfilieva, I. Finitary Solvability Conditions for Systems of Fuzzy Relation Equations. *Inf. Sci.* **2013**, *234*, 29–43. [CrossRef]
2. Di Nola, A.; Sessa, S.; Pedrycz, W.; Sanchez, E. *Fuzzy Relation Equations and Their Applications to Knowledge Engineering*; Kluwer: Boston, MA, USA, 1989.
3. Pedrycz, W. Applications of fuzzy relational equations for methods of reasoning in presence of fuzzy data. *Fuzzy Sets Syst.* **1985**, *16*, 163–175. [CrossRef]
4. De Baets, B. Analytical solution methods for fuzzy relational equations. In *The Handbook of Fuzzy Set Series Vol. 1*; Dubois, D., Prade, H., Eds.; Academic Kluwer Publ.: Boston, MA, USA, 2000; pp. 291–340.
5. Perfilieva, I.; Lehmke, S. Correct models of fuzzy IF-THEN rules are continuous. *Fuzzy Sets Syst.* **2006**, *157*, 3188–3197. [CrossRef]
6. Cornejo, M.E.; Lobo, D.; Medina, J. On the solvability of bipolar max-product fuzzy relation equations with the product negation. *J. Comput. Appl. Math.* **2018**, *354*, 520–532. [CrossRef]
7. Štěpnička, M.; De Baets, B. Interpolativity of at-least and at-most models of monotone single-input single-output fuzzy rule bases. *Inf. Sci.* **2013**, *234*, 16–28. [CrossRef]
8. Štěpnička, M.; Jayaram, B. Interpolativity of at-least and at-most models of monotone fuzzy rule bases with multiple antecedent variables. *Fuzzy Sets Syst.* **2016**, *297*, 26–45. [CrossRef]
9. Kóczy, L.; Hirota, K. Approximate reasoning by linear rule interpolation and general approximation. *Int. J. Approx. Reason.* **1993**, *9*, 197–225. [CrossRef]
10. Vetterlein, T.; Štěpnička, M. Completing fuzzy if-then rule bases by means of smoothing. *Int. J. Uncertain. Fuzziness Knowl.-Based Syst.* **2006**, *14*, 235–244. [CrossRef]
11. Kóczy, L.; Hirota, K. Interpolative reasoning with insufficient evidence in sparse fuzzy rule bases. *Inf. Sci.* **1993**, *71*, 169–201. [CrossRef]
12. Baranyi, P.; Kóczy, L.; Gedeon, T. A generalized concept for fuzzy rule interpolation. *IEEE Trans. Fuzzy Syst.* **2004**, *12*, 820–837. [CrossRef]
13. Li, F.; Shang, C.; Li, Y.; Yang, J.; Shen, Q. Interpolation with Just Two Nearest Neighboring Weighted Fuzzy Rules. *IEEE Trans. Fuzzy Syst.* **2020**, *28*, 2255–2262. [CrossRef]

14. Bouchon-Meunier, B.; Dubois, D.; Marsala, C.; Prade, H.; Ughetto, L. A comparative view of interpolation methods between sparse fuzzy rules. In Proceedings of the Annual Conference of the North American Fuzzy Information Processing Society-NAFIPS, Vancouver, BC, Canada, 25–28 July 2001; Volume 5, pp. 2499–2504.
15. Perfilieva, I.; Dubois, D.; Prade, H.; Esteva, F.; Godo, L.; Hod'áková, P. Interpolation of fuzzy data: Analytical approach and overview. *Fuzzy Sets Syst.* **2012**, *192*, 134–158. [CrossRef]
16. Mareš, M. *Computation over Fuzzy Quantities*; CRC Press: Boca Raton, FL, USA, 1994.
17. Bělohlávek, R. *Fuzzy Relational Systems: Foundations and Principles*; Kluwer Academic, Plenum Press: Dordrecht, New York, NY, USA, 2002.
18. Klawonn, F. Fuzzy Points, Fuzzy Relations and Fuzzy Functions. In *Discovering the World with Fuzzy Logic*; Novák, V., Perfilieva, I., Eds.; Springer: Berlin, Germany, 2000; pp. 431–453.
19. Holčapek, M.; Štěpnička, M. MI-algebras: A new framework for arithmetics of (extensional) fuzzy numbers. *Fuzzy Sets Syst.* **2014**, *257*, 102–131. [CrossRef]
20. Dubois, D.; Prade, H. Operations on fuzzy numbers. *Int. J. Syst. Sci.* **1978**, *9*, 613–626. [CrossRef]
21. Stefanini, L.; Sorini, L.; Guerra, M. Parametric representation of fuzzy numbers and application to fuzzy calculus. *Fuzzy Sets Syst.* **2006**, *157*, 2423–2455. [CrossRef]
22. Mareš, M. Weak arithmetics of fuzzy numbers. *Fuzzy Sets Syst.* **1997**, *91*, 143–153. [CrossRef]
23. Štěpnička, M.; Holčapek, M.; Škorupová, N. Orderings of Extensional Fuzzy Numbers. In Proceedings of the IEEE International Conference on Fuzzy Systems (FUZZ-IEEE) 2019, New Orleans, LA, USA, 23–26 June 2019.
24. Štěpnička, M.; Škorupová, N.; Holčapek, M. On the properties of orderings of extensional fuzzy numbers. In Proceedings of the IEEE International Conference on Fuzzy Systems (FUZZ-IEEE) 2020, Glasgow, UK, 19–24 July 2020.
25. Holčapek, M.; Štěpnička, M. Arithmetics of extensional fuzzy numbers–part I: Introduction. In Proceedings of the IEEE International Conference on Fuzzy Systems, Brisbane, Australia, 10–15 June 2012.
26. Holčapek, M.; Štěpnička, M. Arithmetics of extensional fuzzy numbers–part II: Algebraic framework. In Proceedings of the IEEE International Conference on Fuzzy Systems, Brisbane, Australia, 10–15 June 2012.
27. Ma, M.; Friedman, M.; Kandel, A. A new fuzzy arithmetic. *Fuzzy Sets Syst.* **1999**, *108*, 83–90. [CrossRef]
28. Stefanini, L.; Sorini, L.; Guerra, M. Simulation of fuzzy dynamical systems using the LU-representation of fuzzy numbers. *Chaos Solitons Fractals* **2006**, *29*, 638–652. [CrossRef]
29. Fortin, J.; Dubois, D.; Fargier, H. Gradual Numbers and Their Application to Fuzzy Interval Analysis. *IEEE Trans. Fuzzy Syst.* **2008**, *16*, 388–402. [CrossRef]
30. Boixader, D.; Jacas, J. Extensionality based approximate reasoning. *Int. J. Approx. Reason.* **1998**, *19*, 221–230. [CrossRef]
31. Klawonn, F.; Castro, J.L. Similarity in Fuzzy Reasoning. *Mathw. Soft Comput.* **1995**, *2*, 197–228.
32. Klawonn, F.; Kruse, R. Equality relations as a basis for fuzzy control. *Fuzzy Sets Syst.* **1993**, *54*, 147–156. [CrossRef]
33. Holčapek, M.; Wrublová, M.; Bacovský, M. Quotient MI-groups. *Fuzzy Sets Syst.* **2016**, *283*, 1–25. [CrossRef]
34. Holčapek, M. On generalized quotient MI-groups. *Fuzzy Sets Syst.* **2017**, *326*, 3–23. [CrossRef]
35. Holčapek, M.; Škorupová, N. Topological MI-Groups: Initial Study. In *IPMU 2018: Information Processing and Management of Uncertainty in Knowledge-Based Systems, Proceedings of the Applications-17th International Conference, Cádiz, Spain, 11–15 June 2018*; Proceedings, Part III; Medina, J., Ojeda-Aciego, M., Galdeano, J.L.V., Perfilieva, I., Bouchon-Meunier, B., Yager, R.R., Eds.; Springer: Berlin/Heidelberg, Germany, 2018; Volume 855, Communications in Computer and Information Science, pp. 603–615.
36. Aryal, B.; Bokati, L.; Godinez, K.; Ibarra, S.; Liu, H.; Wang, B.; Kreinovich, V. Common Sense Addition Explained by Hurwicz Optimism-Pessimism Criterion. *J. Uncertain Syst.* **2019**, *13*, 171–174.
37. Hurwicz, L. *Optimality Criteria for Decision Making Under Ignorance*; Cowles Commission Discussion Paper, Statistics; Yale University, New Haven, 1951; pp. 171–174.
38. Zadeh, L. Similarity relations and fuzzy orderings. *Inf. Sci.* **1971**, *3*, 177–200. [CrossRef]
39. Bodenhofer, U. A Similarity-Based Generalization of Fuzzy Orderings Preserving the Classical Axioms. *Int. J. Uncertain. Fuzziness Knowl.-Based Syst.* **2000**, *8*, 593–610. [CrossRef]
40. Wang, W.; Wang, Z. Total orderings defined on the set of all fuzzy numbers. *Fuzzy Sets Syst.* **2014**, *243*, 131–141. [CrossRef]
41. Asady, B.; Zendehnam, A. Ranking fuzzy numbers by distance minimization. *Appl. Math. Model.* **2007**, *31*, 2589–2598. [CrossRef]
42. De Hierro, A.F.R.L.; Roldán, C.; Herrera, F. On a new methodology for ranking fuzzy numbers and its application to real economic data. *Fuzzy Sets Syst.* **2018**, *353*, 86–110. [CrossRef]
43. Van Broekhoven, E.; De Baets, B. A linguistic fuzzy model with a monotone rule base is not always monotone. In Proceedings of the EUSFLAT-LFA'05, Barcelona, Spain, 7–9 September 2005; pp. 530–535.
44. Van Broekhoven, E.; De Baets, B. On the monotonicity of Mamdani-Assilian models with two input variables. In Proceedings of the NAFIPS'06, Montreal, QC, Canada, 3–6 June 2006; pp. 102–107.
45. Štěpnička, M.; De Baets, B. Implication-based models of monotone fuzzy rule bases. *Fuzzy Sets Syst.* **2013**, *232*, 134–155. [CrossRef]
46. Bodenhofer, U. Orderings of Fuzzy Sets Based on Fuzzy Orderings Part I: The Basic Approach. *Mathw. Soft Comput.* **2008**, *15*, 201–218.
47. Bodenhofer, U. Orderings of Fuzzy Sets Based on Fuzzy Orderings Part II: Generalizations. *Mathw. Soft Comput.* **2008**, *15*, 219–249.

48. Štěpnička, M.; Škorupová, N.; Holčapek, M. From arithmetics of extensional fuzzy numbers to their distances. In Proceedings of the IEEE International Conference on Fuzzy Systems (FUZZ-IEEE) 2020, Glasgow, UK, 19–24 July 2020.
49. Wang, X.; Kerre, E.E. Reasonable properties for the ordering of fuzzy quantities (I). *Fuzzy Sets Syst.* **2001**, *118*, 375–385. [CrossRef]
50. Wang, X.; Kerre, E.E. Reasonable properties for the ordering of fuzzy quantities (II). *Fuzzy Sets Syst.* **2001**, *118*, 387–405. [CrossRef]
51. Van Broekhoven, E.; De Baets, B. Monotone Mamdani-Assilian models under mean of maxima defuzzification. *Fuzzy Sets Syst.* **2008**, *159*, 2819–2844. [CrossRef]
52. Van Broekhoven, E.; De Baets, B. Only smooth rule bases can generate monotone Mamdani-Assilian models under Center-of-gravity defuzzification. *IEEE Trans. Fuzzy Syst.* **2009**, *17*, 1157–1174. [CrossRef]
53. Štěpnička, M.; De Baets, B. Monotonicity of implicative fuzzy models. In Proceedings of the FUZZ-IEEE 2010, Barcelona, Spain, 18–23 July 2010; pp. 2334–2340.
54. Mandal, S.; Jayaram, B. Monotonicity of SISO Fuzzy Relational Inference With an Implicative Rule Base. *IEEE Trans. Fuzzy Syst.* **2016**, *24*, 1475–1487. [CrossRef]
55. Bodenhofer, U.; De Cock, M.; Kerre, E.E. Openings and closures of fuzzy preorderings: Theoretical basics and applications to fuzzy rule-based systems. *Int. J. Gen. Syst.* **2003**, *32*, 343–360. [CrossRef]
56. Bodenhofer, U.; Bauer, P. Interpretability of Linguistic Variables: A Formal Account. *Kybernetika* **2005**, *41*, 227–248.
57. Dvořáčková.; H..; A., K.; Tichý, T. Sales Prediction in the Ice Category Applying Fuzzy Sets Theory. *Ekon. Rev.–Cent. Eur. Rev. Econ. Issues* **2018**, *21*, 35–41.
58. Das, S.; Chakraborty, D.; Kóczy, L. Linear fuzzy rule base interpolation using fuzzy geometry. *Int. J. Approx. Reason.* **2019**, *112*, 105–118. [CrossRef]
59. Mamdani, E.H.; Assilian, S. An experiment in linguistic synthesis with a fuzzy logic controller. *Int. J. Man-Mach. Stud.* **1975**, *7*, 1–13. [CrossRef]
60. Jones, H.; Charnomordic, B.; Dubois, D.; Guillaume, S. Practical inference with systems of gradual implicative rules. *IEEE Trans. Fuzzy Syst.* **2009**, *17*, 61–78. [CrossRef]
61. Dubois, D.; Prade, H. What are fuzzy rules and how to use them. *Fuzzy Sets Syst.* **1996**, *84*, 169–185. [CrossRef]
62. Štěpnička, M.; Mandal, S. Fuzzy inference systems preserving Moser–Navara axioms. *Fuzzy Sets Syst.* **2018**, *338*, 97–116. [CrossRef]
63. Štěpnička, M.; Jayaram, B.; Su, Y. A short note on fuzzy relational inference systems. *Fuzzy Sets Syst.* **2018**, *338*, 90–96. [CrossRef]
64. Dubois, D.; Prade, H.; Ughetto, L. Checking the Coherence and Redundancy of Fuzzy Knowledge Bases. *IEEE Trans. Fuzzy Syst.* **1997**, *5*, 398–417. [CrossRef]
65. Rasmussen, C.; Williams, C. *Gaussian Processes for Machine Learning*; The MIT Press: Cambridge, MA, USA, 2006.
66. Le, N.D.; Zidek, J.V. *Statistical Analysis of Environmental Space-Time Processes*; Springer: New York, NY, USA, 2006.

Article

A Coordinated Air Defense Learning System Based on Immunized Classifier Systems

Sulemana Nantogma, Yang Xu * and Weizhi Ran

School of Computer Science and Engineering, University of Electronic Science and Technology of China, Chengdu 611731, China; 201814080012@std.uestc.edu.cn (S.N.); 202012081713@std.uestc.edu.cn (W.R.)
* Correspondence: xuyang@uestc.edu.cn; Tel.: +86-189-809-478-76

Abstract: Autonomous (unmanned) combat systems will become an integral part of modern defense systems. However, limited operational capabilities, the need for coordination, and dynamic battlefield environments with the requirement of timeless in decision-making are peculiar difficulties to be solved in order to realize intelligent systems control. In this paper, we explore the application of Learning Classifier System and Artificial Immune models for coordinated self-learning air defense systems. In particular, this paper presents a scheme that implements an autonomous cooperative threat evaluation and weapon assignment learning approach. Taking into account uncertainties in a successful interception, target characteristics, weapon type and characteristics, closed-loop coordinated behaviors, we adopt a hierarchical multi-agent approach to coordinate multiple combat platforms to achieve optimal performance. Based on the combined strengths of learning classifier system and artificial immune-based algorithms, the proposed scheme consists of two categories of agents; a strategy generation agent inspired by learning classifier system, and strategy coordination inspired by Artificial Immune System mechanisms. An experiment in a realistic environment shows that the adopted hybrid approach can be used to learn weapon-target assignment for multiple unmanned combat systems to successfully defend against coordinated attacks. The presented results show the potential for hybrid approaches for an intelligent system enabling adaptable and collaborative systems.

Keywords: air defense system; artificial immune system; command and control; learning classifier system; multi-agent systems; weapon-target assignment

Citation: Nantogma, S.; Xu, Y.; Ran, W. A Coordinated Air Defense Learning System Based on Immunized Classifier Systems. *Symmetry* **2021**, *13*, 271. https://doi.org/10.3390/sym13020271

Academic Editor: Kóczy László
Received: 15 January 2021
Accepted: 28 January 2021
Published: 5 February 2021

1. Introduction

In recent years, the challenges arising from the deployment of robotic systems in defense and security applications have been addressed from several different perspectives. As a typical multi-agent system, the complexity does not only emanate from the dynamics of the application domain, but also the limited and uncertainty in perceived information. In an attempt to handle these uncertainties, the research community has proposed several theories [1–3].

In the domains of defense application, unmanned combat platforms may be deployed to provide defense against targets and coordinated attacks by enemy adversaries. In these regards, an important functionality required by the multi-agent system is the threat evaluation and weapon assignment to targets that pose threats to its survival and/or the assets being protected. The threat evaluation and weapon assignment functionality is one of the fundamental problems in Command and Control Systems [4]. Threat evaluation and weapon assignment includes the evaluation and assignment of defensive weapons to intercept detected hostile threats with the primary objective of minimizing the total expected survivability of the threat [5,6]. Threat evaluation and weapon assignment functionalities are a fundamental part of achieving successful outcomes in naval combat scenarios.

To assist with the ever more complex and rapidly changing combat environment, combat systems require access to real-time adaptable control systems that can handle uncertain information to provide intelligent decision-making when executing combat tasks. Research work in this area seeks to realize the effective utilization of combat resources by minimizing defensive assets loss, amount of resource utilized during interception, and maximizing the effective kill probability while operating within the multiple constraints of the operating platforms [5,7–10]. Effective combat systems in a multi-agent setting must be able to solve conflicts of interest and optimize resource allocation.

Conflicts of interest and optimized resource allocation in general have received extensive attention in recent decades through the application of various approaches [11–13]. However, the limited operational capabilities and sophisticated battlefield environments with the requirement of timeless in decision-making are peculiar difficulties to be solved by Unmanned Combat Systems in modern battles. The dynamic and sophistication in missions and situations encountered in the battlefield makes it challenging to design intelligent and self-adaptive decision-making agents that are robust and fit for the battlefield since the urgency associated with combat decision-making has many consequences. Hence, the need to design methods and models to enable self-learning and coordinated decision-making that is capable of adapting online by employing learning algorithms and simulation-based training.

Multiple unmanned combat systems performing air defense operation is a typical multi-agent system. In these type of missions, multiple entities are connected in a network, with each system capable of computing and communicating autonomously and concurrently. Decision-making is based on the agents' knowledge of the environment and various entities in their locality of the network. The actions of individual agents in a multi-agent system are suppressed or stimulated during the lifetime of the agents. This suppression and stimulation are key characteristics of the artificial immune system.

The Artificial Immune System is an example of multi-agent and decentralized information processing system that is capable of learning and remembering [14]. The artificial immune system is based on working mechanisms exhibited by the Biological Immune System. The complexity exhibited by the immune system has inspired various theories and models which represent the different aspects proposed under the artificial Immune system such as the immune network [15], Clonal Selection [16], Negative Selection, and Danger Theory [17]. Several applications have been demonstrated based on these theories [18]. Despite the distributed nature of the immune system, their many engineering implementations are centralized.

On the other hand, the learning classifier system is one of the machine learning techniques introduced by Holland in 1976. It combines evolution and learning in an integrated way. A Learning classifier system evolves a group of $IF - THEN$ rules called classifiers [19] to find a suitable response of a learning system (agent) to incoming sensory inputs. As opposed to other learning approaches such as neural networks, rules evolved by a learning classifier system and its variants are easily understandable and general enough for a wide range of learning tasks such as traffic control problems [20,21], multi-agent control problems [22–25], and robotic control [26–28] etc. In the learning classifier system approach, agents acquire appropriate strategies while interacting with an environment by updating their classifier sets. The adaptation of classifiers is mostly done through evolutionary computing based on genetic algorithms.

In this paper, a hybrid of artificial immune and learning classifier system is utilized to design a self-learning system that implements autonomous cooperative threat evaluation and weapon assignment learning. Based on the combined strengths of the two approaches, we seek to realize an intelligent and adaptive system for a team of combat surface platforms to successfully perform air defense weapon-target allocation against a coordinated attack of heterogeneous targets in a realistic maritime environment. By considering the air defense operation from a force coordination perspective, heterogeneous distributed combat systems are required to make effective weapon-target assignment decisions to achieve interception of fast and multi-batch targets. To meet the requirement of real-time decision-making, combat systems are controlled by intelligent agents to coordinate and engage threats in stages based on the observed outcome of previous engagements. Threat evaluation and weapon assignment is done in a distributed manner to reduce communication complexity. Each agent only communicates and receives initial action strategies it generates with its neighbors. This approach can increase the adaptability and success rate of the overall mission. An experiment in a realistic environment shows that the method can be utilized to learn weapon-target assignment for groups of combat platforms to successfully defend against coordinated attacks. This research presents the following general contributions:

1. We investigate and demonstrate the applicability of a hybrid artificial immune and Learning Classifiers System for realizing air defense intelligence.
2. A hierarchical self-learning scheme for multiple unmanned combat systems air defense weapon-target allocation that integrates artificial Immune based algorithms with learning classifier system is presented.
3. We propose an approach to facilitate learning by applying a negative selection concept to filter out and condense situations from individual decision units.

The rest of this paper is presented as follows. In Section 2, the related work of this study and a brief background of the learning classifier system and the artificial immune system algorithms are discussed briefly. In Section 3, a description, analysis, and requirement of the air defense system are described and the air defense problem formulated. Based on this analysis and requirement, the framework allowing the multiple unmanned combat systems to learn an air defense strategy using artificial Immune system and learning classifier system is presented in Section 4. Having presented the framework for the autonomous air defense strategy learning, experimental results in a realistic combat environment are presented in Section 5. Finally, we present our concluding remarks by summarizing our contribution and experimental results.

2. Background and Related Work

In this section, we first present some related works in unmanned combat vehicles autonomous decision-making and learning approaches. Next, the biological principles of the artificial immune system that are relevant to this work are presented. Finally, the background knowledge of learning classifier system and it's working mechanism is presented.

2.1. Related Work

It is expected that unmanned combat systems such as unmanned combat aerial vehicles, unmanned combat surface vehicles, unmanned combat ground vehicles, etc., will be deployed in diverse missions in the near future. However, the insufficient and noisy information available for decision-making and control, simultaneous control variables, several conflicting objectives toppled with a dynamic and complex operating environment presents a challenge.

Several algorithms and approaches have been applied to realize control systems that enable autonomous systems to execute complex missions [29–35]. For instance, an autonomous decision-making for unmanned combat aerial vehicles (UCAVs) using genetic fuzzy trees is presented by authors in [29]. The proposed system shows the capability of obtaining strategies that are robust, aggressive and responsive against opponents. In [30], an intelligent air combat learning system based on the learning dynamics of the human brain is designed and its strategy acquisition without prior rules is demonstrated. The authors in [34] used deep q-neural networks to obtain combat strategies in an attack-defense pursuit-warfare of multiple UCAVs in a simplified environment. Alpha C2, an intelligent Air Defense Commander operations using a deep reinforcement learning framework was presented in [10]. The proposed system makes combat decision independent of human decision-making. A multi-agent-based training system for USVs was presented in [35] and a combat task for combat USVs was used to evaluate the system performance. Other methods such as optimizations and game theoretic approaches have been reported [31–33].

This work contributes to these body of literature by hybridising two known control and decision-making approaches; Learning Classifier Systems and Artificial Immune Systems.

2.2. Learning Classifier Systems

In addition to the different variations proposed since its first introduction by Holland, learning classifier systems have witnessed some improvement and real-world applications in recent years [36–41]. One of the most main-stream and widely applied learning classifier systems is the extended classifier system [42]. In the extended classifier system, classifiers that always receive the same reward are sort after. As in a standard learning classifier system, the extended classifier system consists of three main components; Performance, Reinforcement, and Discovery components.

$$P(a_i) = \frac{\Sigma_{cl_k \in [M]|a_i} cl_k.p \times cl_k.F}{\Sigma_{cl_l \in [M]|a_i} cl_l.F} \tag{1}$$

In the Performance Component, actions are selected and executed. To do this, a matched set $[M]$ of classifiers that match the current input messages in a population $[P]$ are formed and the acquisition reward for each action a_i is calculated to obtain a prediction array $P(a_i)$ using Equation (1). An action inferred from the prediction array is selected and performed in the environment. Meanwhile, classifiers in $P(a_i)$ that have the selected action are stored in an action set $[A]$. After the execution of the selected action, execution is passed on to the Reinforcement component.

In the Reinforcement Component, the parameters of the classifiers in $[M]$ are updated. These parameters include the prediction p, the error ϵ which is the difference between the reward received R and classifier prediction $cl.p$ and the fitness of classifiers. To update the p, the reward received from the environment is used as shown in Equation (2).

$$cl.p = cl.p + \beta(R - cl.p); \tag{2}$$

where β is the learning rate. Consequently, the prediction error, ϵ is updated using Equation (3)

$$cl.\epsilon = cl.\epsilon + \beta(|R - cl.p| - cl.\epsilon); \tag{3}$$

$$cl.F = cl.F + \beta(\hat{\lambda}(cl) - cl.F); \tag{4}$$

$$\lambda(cl) = \begin{cases} 1 & if\ \epsilon < \epsilon_0 \\ \alpha(\frac{\epsilon}{\epsilon_0})^{-v} & otherwise \end{cases}; \tag{5}$$

$$\hat{\lambda} = \frac{cl.n \times \lambda cl}{\sigma_{cl_b \in [A]} \lambda(b) \times b.n} \tag{6}$$

On the other hand, the fitness of a classifier is updated based on its accuracy λ of the classifier cl as represented in Equation (4), where $\hat{\lambda}$ is the relative accuracy of the classifier. And $\hat{\lambda}$ is computed using Equation (6) by first computing the accuracy, λ of the classifier with Equation (5). In Equation (5), ϵ_0 is an accuracy criterion constant. A classifier cl is deemed accurate when $cl.\epsilon$ is less than ϵ_0. The variables α and v are used to control the rate of reduction of accuracy.

Finally, the Discovery component applies Genetic Algorithms (GA) to generate new rules. This is usually done by selecting two parents from $[A]$ using a selection probability to produce new offspring through a crossover process. The new off-springs further undergo some mutation with a probability before they are added back to $[P]$. To keep the population fixed, classifiers with lower fitness are deleted if the number of classifiers in $[P]$ exceeds a certain value. A subsumption process can be applied to replace classifiers whose condition parts are included in more accurate and experienced classifiers.

A learning classifier system largely employs genetic algorithm for rules discovery. Genetic Algorithm is a gradient-free and parallel optimization algorithms. Genetic Algorithm searches for global optimum using performance criteria for evaluation and population of possible solutions. Due to its ability to handle complex and irregular solution spaces, various difficult optimization problems have been studied with genetic algorithm. Generally, in a genetic algorithm, parameters of a possible solution are encoded as chromosomes which are manipulated using genetic operators. In its simplest form, selection, crossover, and mutation are the main operators applied in a genetic algorithm.

During Selection operation, higher fitted chromosomes in the population are selected for reproduction. Thus, the fitter a chromosome, the more chances for it to be selected to reproduce.

Crossover on the other hand, enables the creation of new offspring. By biologically mimicking recombination between two single chromosome organisms, this process leads to the exchange of sub-sequences between two chromosomes to create two offspring.

Lastly, Mutation operator changes the values of genes in the chromosome. In the case of binary encoded chromosomes, some of the bits are randomly flipped. Mutation can occur on any of the genes of the chromosome. In this paper, genetic algorithm is applied as in standard learning classifier system for rules discovery at the strategy generation level while the clonal selection and immune network theory are applied at the strategy coordination level.

2.3. *Artificial Immune System*

The domain of artificial immune system involves the study and application of various computational techniques based on the biological immune system [43]. The biological immune system provides defense mechanisms that protect a host organism against harmful foreign agents such as viruses and bacteria. The two main biological defense mechanisms influencing each other include the Innate Immunity and the Adaptive Immunity that utilizes a set of cells referred to as T-cell to recognize harmful foreign agents (antigens) and B-cells to eliminate these harmful foreign agents. The acquired immune system located in the vertebrates of the organism retains a memory of encountered exposures which is recalled when reinfection occurs to cure the infection.

The dynamism exhibited by the biological immune system has inspired various theories and models which represent the different aspects proposed under the artificial Immune system such as the Immune Network [15], Clonal Selection [16], and the Danger Theory [17]. These under-listed theories which are relevant to our proposed approach are explained below.

2.3.1. Clonal Selection

The Clonal Selection Based Algorithms are inspired by the working mechanisms of antigen-antibody recognition, binding, cell propagation, and separation into memory cell [16]. The clonal selection algorithm, CLONALG [44] based on clonal selection and affinity maturation principles with engineering applications are one such algorithm. According to the clonal selection theory, the presence of affinity between stimulated antigen's epitope and B-cell receptors causes the B-cells to divide. These divided cells then undergo maturity to become plasma cells that secrete antibodies. Subsequently, the Antibodies with higher affinities undergo reproduction through hyper-mutation of B-cells. Some matching B-cells are retained by the Immune System. The adaptation feature emanates from the mechanism of building up of concentrations of the B-cells and the diversity in mutation of the cells in the bone marrow. The interaction in the clonal selection theory is between the antigens and the antibodies only.

2.3.2. Danger Theory

Among the mechanisms that inspire the immune system theories is its ability to identify events that signify danger and those that do not. The concept of responding to these harmful events that cause damage form the hypothesis of the danger theory [17]. The idea of danger theory is that the immune systems does not just respond to foreign antigens but harmful ones. Which introduces the concept of discrimination on which non-self/antigens to respond to. Hence, the immune system chooses to respond to danger instead. Danger in this case, is measured by cells' damage, indicated by distress signals emanating from unnatural deaths of cell. These danger signals essentially establish a danger zone. Consequently, only B-cells that match antigens within this danger zone are stimulated and undergo the clonal expansion process [45].

Similarly, in an immunized air defense system, determining which threat to prioritize is key to a successful defense. Since defense systems employ other mechanisms such as jamming, weapon assignment decision should be made taking into concentration the threatening level of the target. For instance, successfully jammed weapons should not be intercepted by the defense system.

2.3.3. Immune Network Concepts

The network theory of the immune system proposed by Jerne in [15] presents a generic view of the working mechanism of the biological immune system. It is premised on the concept that the immune system is constructed as a large-scale closed system of lymphocytes through mutual interaction between different species of lymphocytes. The interaction of cells happens regardless of the presents of harmful foreign agents or not. Jerne's theory stipulates that, the antibody of an immune cell possess paratope and idiotope. The idiotope of one antibody can be recognized by the paratope of another antibody with or without the presence of an antigen that posses an epitope (analogous to an idiotope). This recognition and interaction results in a network that is dynamic and leads to stimulation and suppression. The recognised antibody is suppressed while the recognizer antibody is simulated. This process of antibodies being able to recognize each other drives the adoption of this theory to heterogeneous robotic applications.

3. Air Defense System Model

In this paper, the problem of threat evaluation and weapon assignment decision-making for air defense operations is considered. In particular, this work considers the problem where a team of unmanned combat systems is deployed to protect a high-value target or contested region. The group consists of a heterogeneous set of unmanned combat systems with different capabilities and multiple types of air-to-surface weapon systems. In this scenario, high valued asset is to be protected from enemy air units trying to penetrate the defense area and destroy the defended asset. This problem is challenging as the characteristics of the scenario can scale to an unlimited number of stages based on the following.

- The number of detected targets at the decision time.
- The types of targets (attackers),
- The weapon capability vector of attackers,
- The number of air defense platforms,
- The weapon capability vector of defenders
- The state of high valued assets being protected, etc.

In all the different stages that the mission can be modified, the objective is to destroy enemy targets and protect the asset from being destroyed by enemy units. The objective is in conflict with the enemy targets who seek to penetrate the defense area. Hence, an efficient threat evaluation and weapon assignment system is required by both sides to achieve their mission objectives. For every detected target, the control system should make its decision based on the following.

- The defense system evaluation of the threat.
- The high valued asset state if any.
- Weapon capability and state of the unit.
- Available weapons and teammates.

Taking these into consideration, the control system decides on whether to deploy a weapon against the threat or not. If yes,

- the type of weapon to use against the target and
- quantity to be fired.

Formally, at a certain time T, the multi-agent system ag detects N targets H which can be a weapon (e.g., missile) or hostile flying unit. Meanwhile, each agent has C categories of weapons. Each category C_p has j quantity of ammunition, $C_p = \{c_{p1}, c_{p2}, ...c_{pj}\}$ to intercept targets at different ranges and efficiency. Here, the objective is to survive the attack with minimal loss and cost of resource utilization by maximizing targets being destroyed while minimizing the cost of operations in other to be able to defend against subsequent engagements. Hence, the objective can be formulated as:

$$O_b = \begin{cases} min \sum_{j=1}^{H'(t)} \sum_{i=1}^{C} \beta_i q_{ij} \delta_{ij} \\ max \sum_{j=1}^{H'(t)} v_j (1 - \prod_{i=1}^{C} (1 - p_{ij})^{\delta_{ij}(s)}) \end{cases} \quad (7)$$

1. W is the total remaining number of employable weapons of the combat platform. i.e., weapon that can be used against the contact by any member of the group.
2. H' is the remaining number of unassigned detected targets
3. δ_{ij} is the minimum delay before weapon i can be deployed against target j based on ready time and current allocation of the weapon of a unit.
4. q_{ij} is the quantity of ammunition of weapon of type i allocated to target j,
5. β_i represents the unit cost of the ammunition of weapon i
6. v_j means the threat value of target j
7. p_{ij} is the weapon kill probability

In this mission, the two main problems to be solved are: action strategy selection and cooperation. Action strategy selection in this case, involves deciding the type of weapon to deploy, the salvo or number of weapons. Cooperation is the coordination of the multi-agent system so that the assignment is distributed as much as possible among several agents to minimize the time between target detection and interception.

4. Approach

In this section, we present the approach and scheme adopted in this paper to realize a robust control system. The focus is to design a self-learning system to dynamically acquire threat evaluation and weapon assignment and coordination strategies. In this section, the details of system model are presented.

4.1. System Architecture

The framework which is an integration of artificial Immune system and learning classifier system proposed in this work is shown in Figure 1. The general framework is a typical multi-agent system consisting of two main agents; the Strategy Generation Agent and Strategy Coordination Agent. The Strategy Generation Agent is a low level agent that learns the appropriate weapons and quantity to deploys against a given target using learning classifier mechanism. The Strategy Coordination Agent receives proposed assignments from other agents and takes as input the generate classifiers as antigens from the Strategy Generation Agent perform immune network dynamics to determine the actions of the agent.

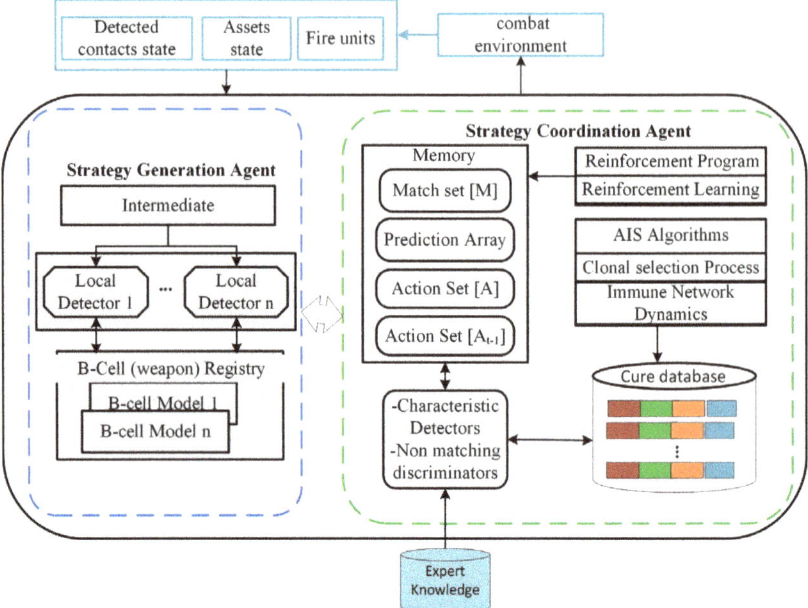

Figure 1. Immunized air Learning Classifier System Model.

In the Strategy Generation Agent, each weapon type is an abstract B-cell (decision unit) which is a vessel for carrying antibodies. Antibodies are classifiers that contain the control decisions as its consequent. While detected contacts are modeled as antigens, whose epitopes are expressed in terms of their intent, capability, and opportunity. The characteristic mapping based on the local detectors of antigen and individual B-cells and their activation leads to the generation of possible classifier sets that forms a sub-population of classifiers according to the specific targets. The competition between these sub-populations of classifiers produces a single classifier whose action part is applied against a target. The Strategy Generation Agent interfaces with the environment through the intermediate module which provides local detectors for B-cells.

Strategy Coordination Agent performs the final actions selection based on the concentration of classifiers in the Immune network by applying the AI algorithms. The Immune network models a collection of classifiers of the various B-cells. Each classifier has a connections part that is used to determine its appropriateness in a given situation with respect to the other classifiers. Strategy Coordination Agent receives classifiers from the B-cells, established their connections to form a network, and apply the immune network dynamics. The individual classifiers that undergo genetic operations and updates are based on the output classifiers of the immune network dynamics.

4.2. Classifier Representation and Encoding

In order to be able to apply immune network dynamics and coordinate classifiers generated by different B-cells, the form of classifiers is changed. In this case, classifiers generated by a b-cell are partitioned into 4 parts that corresponds to the paratope and idiotope of a cell. These connections are synonymous with idiotope and indicate which other classifiers it is connected to. The Identifier is used to indicate the particular b-cell that the classifier belongs. The antecedent indicates when the B-cell and weapon type is applicable while the consequent specifies quantity to fire. The parameters of standard extended classifier still applies. The antecedent part is encoded as singletons of fuzzy sets that represent the states of the variables of decision elements. The consequence part of each classifier is the number of ammunition to fire. The weapon to employ can directly be inferred from the posted classifiers since classifiers are identified by B-cells which represents the agent and weapon type. Figure 2 shows a schema of a classifier in our system. Both unit and target states are vectors of the individual variable representations of the classifier.

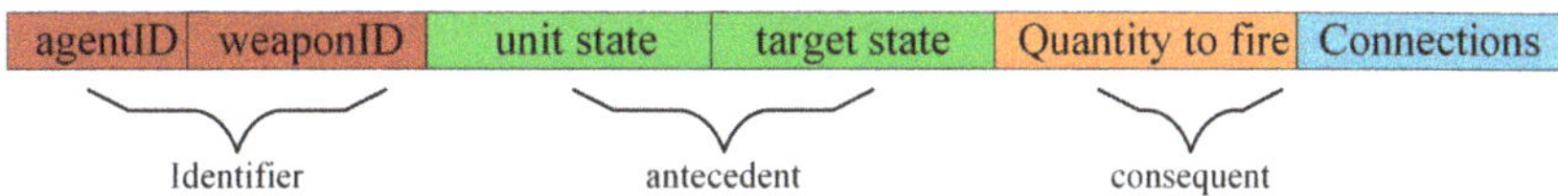

Figure 2. A schematic representation of a classifiers.

We design an intermediate module to further quantize all variable and apply binary coding scheme for the antecedent with size equal to the sum of the states of quantization in each input variable. The consequent part employs an integer coding scheme where each gene contains the index of the state used for the corresponding output variable. The levels of quantization of decision variables and continuous variables' value ranges are mapped to seven states from 1 to 7 of the corresponding value bits. For instance, consider the continues variable of 'velocity, distance, heading' of a detected target. If we are to quantize them to 3 states (fuzzy sets) each as shown in Equations (8)–(10). The indexes are 1, 2, and 3 from *low* to *high*. By this quantization, the velocity variable state can be represented by 3-bits only, where allele *'1'* means that the corresponding state/index is used in the velocity variable. That is, the encoding *'001'* means the velocity of a contact as read from the environment is *high*.

$$velocity = \begin{cases} high & when\ velocity\ > 500\ m/s \\ medium & when\ \ 250 < velocity \leq 500\ m/s \\ low & when\ \ 0 \leq velocity \leq 250\ m/s \end{cases} \tag{8}$$

$$distance = \begin{cases} high & when\ distance\ > 1000\ km \\ medium & when\ \ 500 < velocity \leq 1000\ km \\ low & when\ \ 0 \leq distance \leq 500\ km \end{cases} \tag{9}$$

$$heading = \begin{cases} high & when\ heading\ > 90\,\text{deg} \\ medium & when\ \ 45 < heading \leq 90\,\text{deg} \\ low & when\ \ 0 \leq heading \leq 45\,\text{deg} \end{cases} \tag{10}$$

Hence, a B-cell with sub characteristic detectors of distance, velocity and heading respectively for a target as an input, can have a classifier as *'1;2 | 100;010;001 | 2 | 1;3;5'* conforming schema as shown in Figure 2. The individual parts are separated by ' | ' and can be interpreted as; "IF *distance* IS *low* AND *velocity* IS *medium* AND *heading* IS *high* THEN fire 2 ammunition USING weapon 2". Also, in this particular example, the classifier is connected to classifiers 1, 3, 5.

It is worth noting that each hypothetical B-cell (decision unit) implement an extended classifier system for action selection. However, contrary to the many approach, the entire prediction array from individual classifiers are synthesized with respect to other B-cells predictions using the artificial immune network dynamics to obtain an optimal weapon-target assignment for the current environment state.

4.3. Coping with Multiple Targets

Learning to prioritizing contacts to counter first by an autonomous defense system when multiple threats are detected is key to a successful defense. Since each weapon is a B-cell with specific characteristics with respect to the detected threats, the concept of danger signals and affinity can be modeled. Based on the hypothesis of the danger theory, when a target is detected, a danger zone is constructed based on the trajectory of the detected target.

The value of the combat units within the danger zone signals how much priority should be given in intercepting the target. This is important since combat systems with jamming capability will succeed in jamming some incoming targets. Therefore it is unnecessary to intercepts those targets after it has been jammed.

4.4. Action Strategy Selection

At time t during an engagement, an agent receives observation from the environment. The trajectories of the targets are use to construct danger zone and the danger level of all detected targets determined. The intermediate module process the targets information and encode as antigens to the individual B-cells in range of the targets. The input information is of two types: (1) The external observation from the battlefield which include the observed state of detected targets, high value asset and teammate agents information. This information includes basic information such as distance, heading, speed, and contact type, etc. for detected targets as observed by a unit; (2) The characteristic internal state of decision units, including weapon types and states, range, quantity, etc.

Based on the quantized variables and characteristics of each B-cell, a matched set is generated by each B-cell for each target. To form the prediction array $[P_a]$, the expected action payoff for every action strategy of B-cells is calculated using Equation (11). ϖ is the affinity between B-cell of classifiers and antigens.

$$P(a_i) = \frac{\Sigma_{cl_k \in [M]|a_i} cl_k.p \times cl_k.F * \varpi}{\Sigma_{cl_l \in [M]|a_i} cl_l.F * \varpi} \tag{11}$$

The affinity between B-cell's generated classifiers and antigens is estimated base on Equation (12). In Equation (12), q is the quantity of ammo suggested by the classifier, d_g is normalized distance between target c and combat unit (B-cell), p_g is the speed advantage of fire unit weapon against target. w_e is the effectiveness (kill probability) of the weapon suggested, r_t is the ready time of weapon if it were to be deployed and v_c is the quantified target type that estimates the value of the target.

$$\varpi = \left[1 - \prod_{i=1}^{q} (1 - d_g * p_g * w_e * r_t)\right] * v_c \tag{12}$$

The classifiers are condensed by applying the negative selection described in the next section. Next, the condensed classifier set is transmitted to its neighbors and execution is passed to the Strategy Coordination Agent.

In the Strategy Coordination Agent, the connections of the classifiers are established based on the classifiers from the Strategy Generation Agent and neighbor units. Classifier i is said to be connected to classifier j if they produce antibodies for the same target. Based on the classifiers' connections the immune network dynamics are executed to obtain the classifiers whose action will be posted to the environment by a combat unit. The classifiers with the higher concentrations for each target are allowed to post their actions.

Since classifiers must cooperate and compete with other classifiers, each classifier has a concentration level based on it connectivity and affinity to the observed environment state. This concentration is determined by a system of ordinary differential equations, which corresponds to the immune network dynamics introduced in [46]. Equation (13) is used to control the network model's dynamics in this work.

$$\frac{da_i(t+1)}{dt} = \left(\alpha \sum_{j=1}^{N} m_{ji} a_j(t) - \beta \sum_{j=1}^{N} m_{ik} a_k(t) + \gamma m_i - k\right) a_i(t) \tag{13}$$

where:

- N is the number of classifiers (antibodies) that composes the sub-population of classifiers dealing with a target.

- m_i is the affinity between classifier i and current stimuli (antigen).
- m_{ji} is the mutual stimulus coefficient of antibody j into classifier i.
- m_{ki} represents the inhibitory effect of classifier k into classifier i.
- k is the rate of natural death rate of classifier i;
- $a(t)$ is the bounded concentrations imposed on classifiers;
- the coefficients α, β and γ is weight factor that determines the significance of the individual terms.

4.5. Methods for Coordinated Learning and Knowledge Sharing

The training of classifiers is done through the conventional extended classifier system mechanisms and facilitated by the negative selection and immune network mechanism of the artificial immune system. Figure 3 shows the top-level view of the negative selection mechanism. The training and learning procedure are as follows.

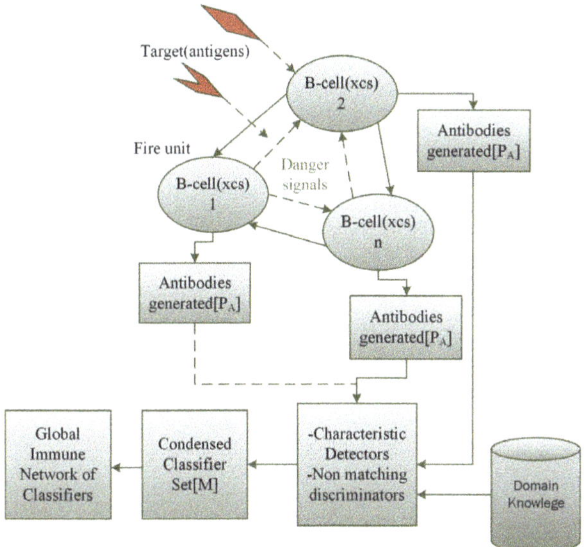

Figure 3. Negative selection Mechanism during action strategy selection.

1. Before the training begins each B-cells' population of individuals are initialized randomly. In this work, partial Pittsburgh-Style is adopted. That is, classifiers are treated both at the classifier level and as individuals. An individual is a collection of classifiers. Each classifier connection part is initialized as empty since the connections of a classifier are dynamically established during action selection. For each generation, an individual from each agent B-cells' population is selected and used to control the agent's actions. By initializing each population independently, a diverse population of classifier are generated collectively. Based on the characteristics detectors and quantized variables using the classifier encoding and representations above, the size of classifiers in a population can be determined a prior. However, the initial number of classifiers of an individual is chosen at the beginning of the training.
2. When individual B-cells generate their classifier sets based on the current environment, Non-matching and characteristics detectors in a form of rules obtained from an expert are used to filter out redundant classifiers in the matched set of the individual B-cells after merging to produce a condensed set of classifiers that undergo further processing for actions selection. In this case, manual rules are encoded to discriminates certain classifiers in the match set from being processed further. For example, if the status of a weapon is damage or a weapon has no ammunition remaining, all classifiers of that

particular weapon are filtered out. This forms a first phase of matching and message processing the system.

3. Next, the action selection mechanism in Section 4.4 is applied to obtain the weapon-target assignment of the agent. After executing the actions in the environment, the agent receives a reward based on the targets that were successfully intercepted as against the resources utilized. The reward obtained and concentration value of the classifier is used to update the fitness of classifiers within the individual under evaluation whose actions resulted in the reward. Also, in order to properly evaluate the classifiers of an individuals and their connections, each individual is simulated a predetermined number of times in each generation. While the immune network is applied each episode, genetic algorithm is applied on each extended classifier system at the individual level.
4. Finally, classifiers of individual with high accuracy are cloned and merged with other individuals of the same type. Two types of cloning are adopted: whole classifier cloning and merging and classifier action cloning, and replacement with higher accuracy classifiers' actions after all the individuals in the populations are evaluated.

5. Experimental Setup

In order to realize autonomous combat platforms operations, intelligent agents are required to interact with the battlefield environment during training. This is unrealistic in the physical environment as the process can be dangerous and costly. Hence, for facilitating the development of military intelligence, the physical environments are normally recreated in a virtual environment through modeling and simulation. To provide a realistic environment for training and evaluation, a scenario of digital battlefield is established in a real-time wargame, "Command: Modern Operations". The system generates battlefield data in real-time and log the damages of combat units after every engagement both in stages and end of engagement. In this scenario, there are two parties involved the ground combat units herein referred to as the ally (blue) faction who are the defending side and the air combat units herein referred to us the enemy (red) faction which are the invaders. The experiments setup for this work are described as follows.

5.1. Configuration of Ally Faction

The Ally faction consists of 7 combat ships. These include two (2) class A combat ships with 4 different weapons of ranges and kill probabilities, four (4) class B combat ships with two weapons types on board and one (1) carrier ship with 2 short ranges weapons on board. The number of weapon types of all the units, $M = 5$, the main difference between these weapons is the different interception range and probability of kill which is shown in Table 1. The per salvo is the number that can be fired in succession without delay. The subscripts for weapon types is the ID of the weapon used in classifiers.

Table 1. Information on weapons deployed on combat units.

Weapon Type	Min. Range (nm)	Max. Range (nm)	Per Salvo	Type
$Weapon_1$	2	80	1	missile
$Weapon_2$	2	21	1	missile
$Weapon_3$	0.2	4	1	missile
$Weapon_4$	0	1	5	Gun
$Weapon_5$	1	1.8	40	Gun

The air defense weapons are deployed on the seven combat ships in different defensive formation. The types and quantities of weapons equipped on each combat platform of the Ally faction are indicated in Table 2. Zero (0) means that the unit is not equipped with this type of weapon. Figure 4 depicts a typical confrontational setup of the digital battle environment. The survival of middle and large ships is the most critical requirement of a successful defense in this work. In this defense formation, Unit_7 is the center ship, and the task of the entire formation is to intercept as many missiles as possible before the center warship is destroyed.

Table 2. The number of weapons equipped on each warship.

Combat Platform ID	Weapon_1	Weapon_2	Weapon_3	Weapon_4	Weapon_5
Unit_1	48	0	24	20	220
Unit_2	48	0	24	20	220
Unit_3	0	32	0	20	0
Unit_4	0	32	0	20	0
Unit_5	0	32	0	20	0
Unit_6	0	32	0	20	0
Unit_7	0	0	72	20	0

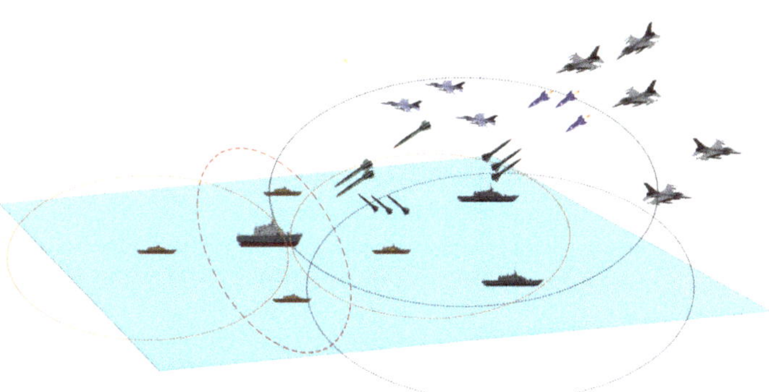

Figure 4. A typical air maritime combat digital battlefield scenario.

5.2. Reinforcement Program

In order for a learning system to improve its performance, reward function design is of great importance. Hence, much attention is given to designing the reward and punishment mechanisms employed in this work. In our case the destruction of Unit_7 is deemed as mission failure. However, the system must be encouraged to search for the best strategy that can enable it to succeed. So, we define two types of wins; Confident Win and Weak Win. The Ally Faction is said to win Confidently if more than two of its units survived in addition to Unit_7 after a confrontation in an episode. On the other hand, a weak win is encountered if at most 2 other units survived in addition to Unit_7. In addition to these specifics, each weapon of the ally and the enemy factions are assigned a cost and reward values respectively. The cost of losing a unit of the ally faction is function of the total number of ammunition on the unit at the time of destruction. However, the cost of losing Unit_7 is always higher in any case. Also, each of the 3 types of enemy units are assigned values which are rewarded to the ally faction when the enemy unit is destroyed. Table 3 shows the costs of weapons consumption. In addition, the interception of a bomber aircraft, fighter jet $T2$ and UAVs $T1$ rewards the Ally Faction with $0.2, 0.15$ and 0.1 respectively. For the weapons deployed by these units of the enemy faction the rewards for intercepting them are $0.25, 0.2$ and 0.15 respectively. Also, the Ally Faction is reward 0 when Unit_7 is destroyed and $+10$ when it survived. A squashed function is used to normalize the final rewards to a range between 0 and 1.

Table 3. Costs of Ally Faction's weapons consumption.

Weapon Type	Cost of Consumption
$Weapon_1$	0.004
$Weapon_2$	0.002
$Weapon_3$	0.001
$Weapon_4$	0.0001
$Weapon_5$	0.0001

5.3. Configuration of Enemy Faction

The Enemy faction consist of multiple hostile air units approaching the Ally faction in formations. A formation consists of a maximum of 8 fighter units and can comprise of different types of units. Each unit has a maximum of 8 ammunition of its weapon type that it can deploy on a defensive unit. However, a unit may run out of fuel and may need to return to based. Hence, not all weapons maybe deployed during its initial engagement. It is assumed that the enemy faction already knows the location of the ally faction, this means they only need to send out attackers. The enemy faction consists of different units comprising bomber aircraft, fighter jet $T2$ and unmanned combat aerial vehicles $T1$. The bombers have the shortest firing range and proximity hence must get closed enough to drop the bombs. In doing so the enemy unit is exposed to the firing range of the ally units. The other two types of units consist can also deploy short to medium range missiles.

5.4. Scenarios Setup

The engagements and battle rounds are design based on three scenarios as follows:

Scenario 1: In this scenario, the defense units know the approach direction of the enemy units. So, the defense units form a formation that ensures that the defended unit is put at the rear of the formation. Also, the attack in this scenario is executed in a single formation deployment. Each of the enemy units can fire only 3 of its ammunition before it must return to base for refueling and there is no second wave of attack.

Scenario 2: In this scenario, a formation is formed around the defended unit which is placed at the center. The enemy units can appear from multiple direction. The attack, in this case, is done on 3 confrontations. In each confrontation, the formations can consist of different units and each unit can fire as many weapons as possible within its maximum allowed capacity. Also, they attack the ally units on different rounds. Each round presents more challenges by increasing the number of enemy units and number of weapons fired.

Scenario 3: In the final scenario, the entire enemy units are divided into 4 groups of 8 each with each group approaching from the north, south, east and west of the ally faction. However, the formation of the ally faction is similar to scenario 2. Also, jamming is enabled in all scenarios except scenario one.

Each of these strategies are employed to train a different model independently from scratch. During training, parallel simulations were run on three different machines simultaneously for almost 2400 training instances (generations). The population of individuals was 20 for each classifier system and maximum number of classifiers in an individual was set to 1000 with Mutation probability = 0.15. The results of the 3 scenarios are shown in Figures 5–7. In each case, the score, win rate and rate of resource utilization is plotted as the training progresses. As can be seen, the win rate for the ally faction increases whiles the total quantity of weapons deployed reduces as training progresses. Except for scenario 3 which obtained a final win rate of around 50%, the other two models of scenario 2 and 3 attained a win rate of about 75%. Similarly, the resource utilization in all scenarios saw a reduction of at least 30% at end of training Figure 8. The comparison of the 3 models with regards to battle costs is presented in Figure 8d.

(**a**) Model training performance score (**b**) Winning rate of model during training

Figure 5. Training performance on scenario 2.

(**a**) Model training performance score (**b**) Winning rate of model during training

Figure 6. Training performance on scenario 1.

(**a**) Model training performance score (**b**) Winning rate of model during training

Figure 7. Training performance on scenario 3.

(**a**) Rate of resource consumption in scenario 1 (**b**) Rate of resource consumption in scenario 2

Figure 8. *Cont.*

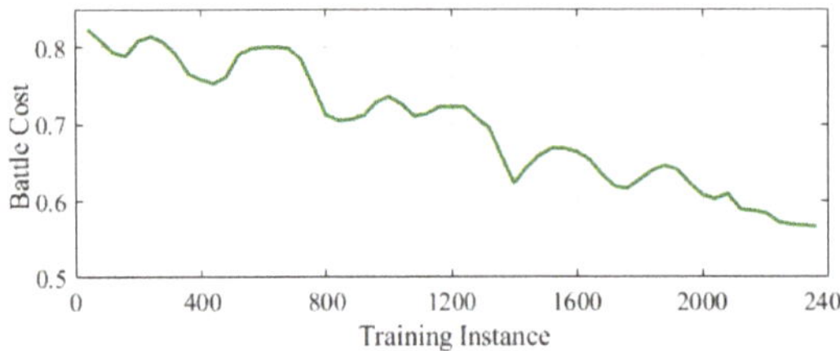

(**c**) Rate of resource consumption in scenario 3

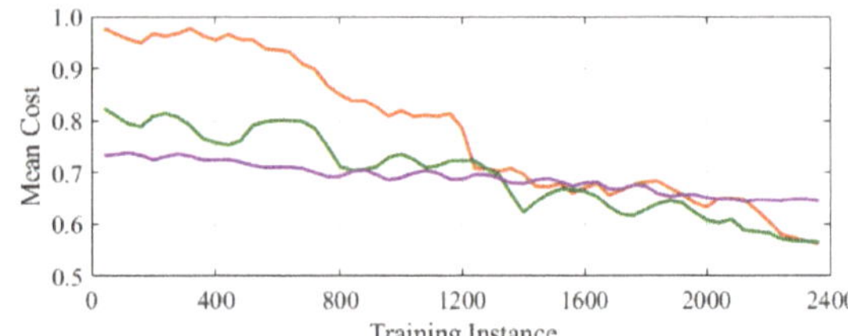

(**d**) Resource utilization of the 3 models during training

Figure 8. Individual and Compared resource utilization of the 3 scenarios models during training.

5.5. Baseline Heuristics

In order to evaluate the performance of the obtained strategies or models of the system, a heuristic assignment strategy was developed. This heuristic model employs multiple rules to govern the decision-making of the system to serve as a credible baseline to benchmark this work.

Hence, in the heuristic baseline,

1. Weapon-target assignment is performed based on priority.
2. To achieve priority-based assignment, the threat value of targets is evaluated based on the assigned targets values, heading and the distance to the Unit7.
3. Based on the computed values, a sorting algorithm is used to sort the targets in ascending order. Targets with low computed values are considered to pose higher threat to the ally faction.
4. After the threat levels are determined and sorted, for each target we select the closest ally unit to attack it with any weapon within range. The number of ammunition to fire is set to a maximum of 4.
5. When intercepting targets, targets identified as weapons are intercept first. This is different from the threat level computation of targets. In other words, targets identified as weapons are assigned first based on their threat level before non-weapons are also assigned based on their threat levels.
6. Also, when a target is within multiple weapon range of a unit, the shortest ranged weapon is utilized.

To evaluate the final models with the baseline, we run the simulations for 30 rounds each. In this case, a round consists of 10 episodes of simulations. The results of the win rate and average costs are presented in Figure 9. From the results, it can be seen that even though model 2 took longer time to make a significant increase in win rate it turns our to perform well in all the scenarios. On the other, model 3 turns to perform better in the other scenarios than it performs in the trained scenario. The three models generalized well to scenarios they were not trained in. Table 4 shows the summary of the percentage wins in the final experiment conducted. Figure 10 shows a successful defense scenario run that employs model 1 of in the first scenario. In this case, the yellow colored units means that those units are not detected by any of the ally units.

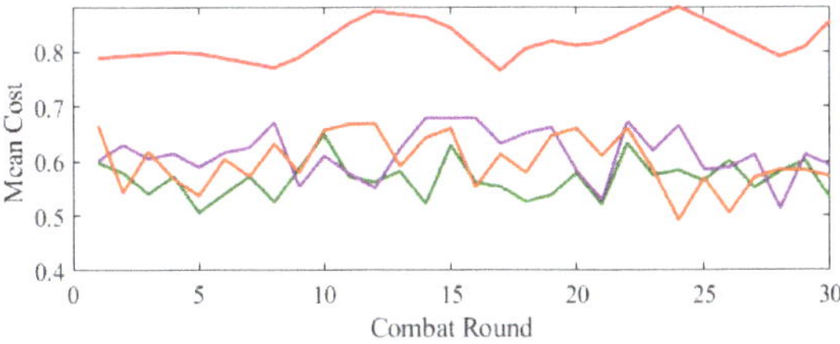

(**a**) Resource consumption for the 3 models and baseline model

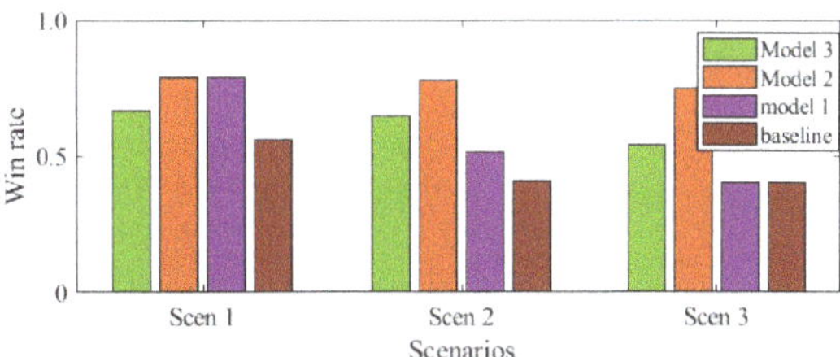

(**b**) Win rate for the 3 models and baseline model

Figure 9. Comparison of performance of the 3 models and baseline model.

Table 4. Detail Performance of trained models.

Scenario/Model	Scenario 1 (% Win)			Scenario 2 (% Win)			Scenario 3 (% Win)		
	Confident	Weak	Total	Confident	Weak	Total	Confident	Weak	Total
Model 1	44	35	79	23.8	28.2	52	20.3	19.7	40
Model 2	40.3	38.8	79.1	44.7	33.4	78.1	39.3	35.8	75.1
Model 3	28.7	37.9	66.7	33.9	30.8	64.7	29.6	24.6	54.2
Baseline	26.6	29.7	56.3	21.5	19.5	41	19.3	21.3	40.6

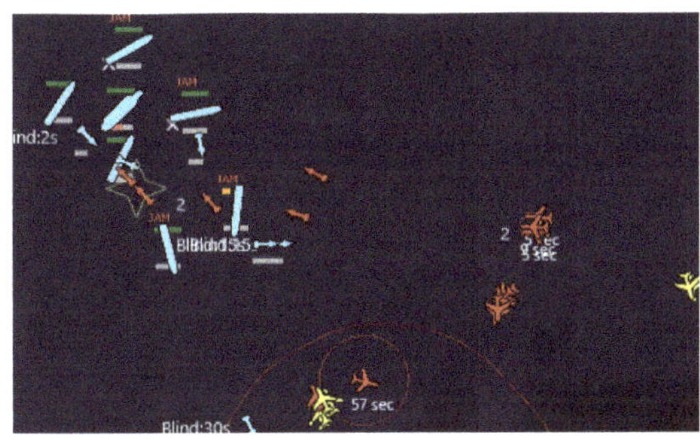

(**a**) Battle state At the beginning of confrontation (t = 10 s)

(**b**) Battle state Mid-way during enemy confrontation (t = 20 s)

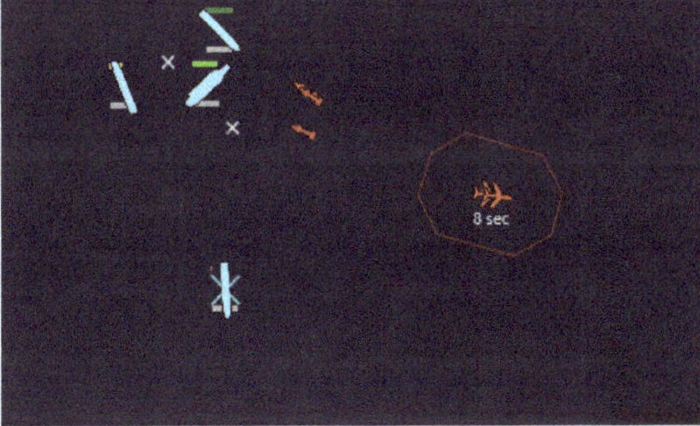

(**c**) Battle state towards the end of confrontation t = 30 s

(**d**) At the end of confrontation (t = 40 s)

Figure 10. A typical confrontation details of model 1 on scenario 1.

5.6. Discussion

From the results presented, the trained models show a more efficient use of resources. This can be realized from the models' winning rate and rate of resource consumption graphs of the respective models. In all cases, in addition to the ally faction winning mores as training progresses, it does so with less resources. The trained models show better weapon distribution and target prioritization strategy compared with the baseline and target prioritization approach. At the initial stages of training, the system will assign weapons to targets that does not pose immediate danger to its survival. This behavior was also peculiar to the baseline heuristic model. However, at the end of the training process targets such as non-weapon units are ignored and only assigned in some instances when no weapon targets were detected. Figure 11 depicts the two situations as indicated. Also, as can be seen in Figure 11b, the model exhibits better cooperative behaviors as the threat level of targets increases. In this case, detected targets are distributed among suitable units for interceptions.

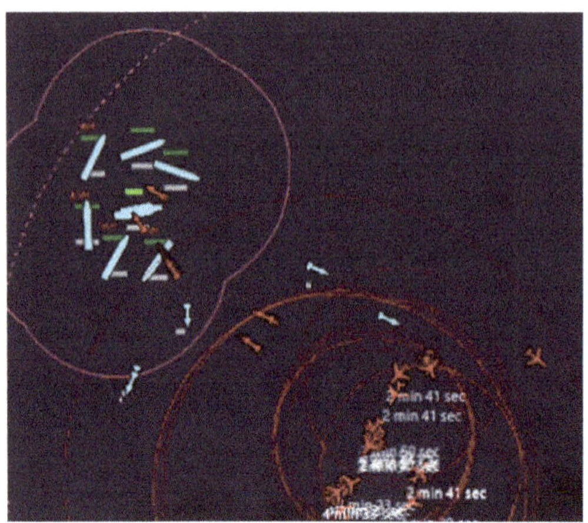

(**a**) Confrontation behavior of model 2 (before)

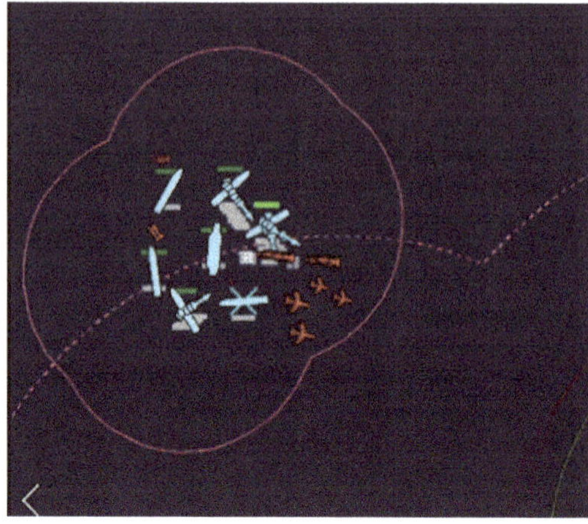

(**b**) Confrontation behavior of model 2 (after)

Figure 11. A typical confrontation details of model 2 before and after training.

In comparison with other approaches such as neural network based models, the system realized by this approach is interpretable and can be refined using prior knowledge. The learning process is facilitate by prior specification and quantization of fuzzy sets for classifier variables and the classifiers filtering introduced by the negative selection mechanism. From the data presented, the performances and winning rates appear to increase more sharply after 400 generations except for model 2. This might be attributed to the complexity level of the scenario model 2 was trained on. Even though model 2 turns out to be the most optimal of the 3. Therefore, even though more complex scenarios might take time for performance to start manifesting, the resulting controllers can achieve more optimal performances.

However, this approach also has its challenges, one of which is the memory consumption to keep track of classifiers and their connections. Also, the computations of classifiers concentrations with respect to other classifiers add additional computational burden on the approach as the size and number of classifiers increase.

6. Conclusions

In this paper, the air defense problem is studied. We explore the application of classifier system and artificial immune models for coordinated self-learning air defense. In particular, this paper presents an approach for multiple unmanned combat systems for coordinated air defense that implements autonomous cooperative threat evaluation and weapon assignment learning approach. We investigate and demonstrate the applicability of a hybrid artificial immune and learning classifiers system for realizing air defense intelligence and presents a hierarchical self-learning approach for multiple unmanned combat systems air defense operations that integrates artificial Immune based algorithms with classifier systems. We further design a mechanism to speed up learning by applying a negative selection mechanism to filter out and condense situations from individual decision units that proposed solutions for the situations in the battlefield using domain knowledge. We evaluate the proposed approach by designing several simulation scenarios. The experimental results of training multiple combat systems demonstrate the learning ability of the approach. Finally the proposed scheme is compared with baseline heuristics by comparing the resource utilization and win rate of the different models obtained and the baseline. Based on the data presented, the most optimal model (model 2) was able to increase its win rate to about 75% and reduce its resource utilization by 30%. The approach is easily scalable since the addition of new units will only require the transfer of knowledge from similar decision units. The ability to transfer and encode prior knowledge facilitates transfer learning and prevent the system from learning from scratch when new units are added to the team. This will speed up the learning process of the system. The model is interpretable by humans compared to 'blackbox' approaches, which is a desirable feature for military applications. The generalization capability of the approach was validated by testing them in the scenarios they were not trained in.

However, more work is still required to improve the performance of the system and approach adopted in this work. For instance, there is a significant level of disparity between the physical world and the simulation environment. In the physical environment, there might not be an immediate confirmation of a successful interception. Hence, the performance in the real battlefield might be degraded to some extend. Also, it will be interesting to consider tuning or learning of the input variables and not just the appropriate consequent as adopted in this approach. In this regard, the learning of the input variables for each classifier system can either be done along with the rule base learning or one after the other. Similarly, further investigation as to how to identify useful knowledge, and condense and merge with other learning agents is needed. We plan to extend this framework to develop a general training system for autonomous decision-making of unmanned combat vehicles for cooperative missions.

Author Contributions: Conceptualization, Supervision and funding acquisition, Y.X.; methodology, software, validation, writing—original draft preparation, S.N.; software and validation, writing—review and editing W.R. All authors have read and agreed to the published version of the manuscript.

Funding: This work was supported in part by the National Natural Science Foundation of China under Grant 61370151 and in part by the National Science and Technology Major Project of China under Grant 2015ZX03003012.

Conflicts of Interest: The funders had no role in the design of the study; in the collection, analyses, or interpretation of data; in the writing of the manuscript, or in the decision to publish the results.

References

1. Zhang, B.; Zhang, M.; Song, Y.; Zhang, L. Combining Evidence Sources in Time Domain With Decision Maker's Preference on Time Sequence. *IEEE Access* **2019**, *7*, 174210–174218. [CrossRef]
2. Zhou, Y.; Tang, Y.; Zhao, X. A Novel Uncertainty Management Approach for Air Combat Situation Assessment Based on Improved Belief Entropy. *Entropy* **2019**, *21*, 495. [CrossRef]
3. Posen, B.R. Foreword: Military doctrine and the management of uncertainty. *J. Strateg. Stud.* **2016**, *39*, 159–173. [CrossRef]
4. Manne, A.S. A Target-Assignment Problem. *Oper. Res.* **1958**, *6*, 346–351. [CrossRef]
5. Karasakal, O. Air defense missile-target allocation models for a naval task group. *Comput. Oper. Res.* **2008**, *35*, 1759–1770. [CrossRef]
6. Bogdanowicz, Z.R. A new efficient algorithm for optimal assignment of smart weapons to targets. *Comput. Math. Appl.* **2009**, *58*, 1965–1969. [CrossRef]
7. Li, X.; Zhou, D.; Pan, Q.; Tang, Y.; Huang, J. Weapon-Target Assignment Problem by Multi-objective Evolutionary Algorithm Based on Decomposition. *Complexity* **2018**, *2018*, 8623051. [CrossRef]
8. Chang, Y.Z.; Li, Z.W.; Kou, Y.X.; Sun, Q.P.; Yang, H.Y.; Zhao, Z.Y. A New Approach to Weapon-Target Assignment in Cooperative Air Combat. *Math. Probl. Eng.* **2017**, *2017*, 2936279. [CrossRef]
9. Naseem, A.; Shah, S.T.H.; Khan, S.A.; Malik, A.W. Decision support system for optimum decision making process in threat evaluation and weapon assignment: Current status, challenges and future directions. *Annu. Rev. Control* **2017**, *43*, 169–187. [CrossRef]

10. Fu, Q.; Fan, C.L.; Song, Y.F.; Guo, X.K. Alpha C2—An Intelligent Air Defense Commander Independent of Human Decision-Making. *IEEE Access* **2020**, 8, 87504–87516. [CrossRef]
11. Li, Y.; Du, W.; Yang, P.; Wu, T.; Zhang, J.; Wu, D.; Perc, M. A Satisficing Conflict Resolution Approach for Multiple UAVs. *IEEE Internet Things J.* **2019**, *6*, 1866–1878. [CrossRef]
12. Pappas, G.J.; Tomlin, C.; Sastry, S.S. Conflict resolution for multi-agent hybrid systems. In Proceedings of the 35th IEEE Conference on Decision and Control, Kobe, Japan, 13 December 1996; Volume 2, pp. 1184–1189. [CrossRef]
13. Sislak, D.; Volf, P.; Pechoucek, M.; Suri, N. Automated Conflict Resolution Utilizing Probability Collectives Optimizer. *IEEE Trans. Syst. Man Cybern. Part C Appl. Rev.* **2011**, *41*, 365–375. [CrossRef]
14. Hunt, J.E.; Cooke, D.E. Learning using an artificial immune system. *J. Netw. Comput. Appl.* **1996**, *19*, 189–212. [CrossRef]
15. Jerne, N. Towards a network theory of the immune system. *Collect. Ann. Inst. Pasteur* **1974**, *125 C*, 373–389.
16. Burnet, F.M. *The Clonal Selection Theory of Acquired Immunity*; Vanderbilt University Press: Nashville, TN, USA, 1959; p. 232. Available online: https://www.biodiversitylibrary.org/bibliography/8281 (accessed on 10 December 2020).
17. Matzinger, P. The danger model: A renewed sense of self. *Science* **2002**, *296*, 301–305. [CrossRef]
18. Kong, X.; Liu, D.; Xiao, J.; Wang, C. A multi-agent optimal bidding strategy in microgrids based on artificial immune system. *Energy* **2019**, *189*, 116154. [CrossRef]
19. Holland, J.H. *Adaptation in Natural and Artificial Systems: An Introductory Analysis with Applications to Biology, Control and Artificial Intelligence*; MIT Press: Cambridge, MA, USA, 1992.
20. Cao, Y.; Ireson, N.; Bull, L.; Miles, R. *Design of a Traffic Junction Controller Using Classifier Systems and Fuzzy Logic*; Springer: Cham, Switzerland, 1999; pp. 342–353.
21. Rochner, F.; Prothmann, H.; Branke, J.; Müller-Schloer, C.; Schmeck, H. An Organic Architecture for Traffic Light Controllers. In Proceedings of the Informatik 2006—Informatik für Menschen, Dresden, Germany, 2–6 October 2006; pp. 120–127.
22. Gershoff, M.; Schulenburg, S. Collective behavior based hierarchical XCS. In Proceedings of the Tenth International Workshop on Learning Classifier Systems (IWLCS 2007), London, UK, 8 July 2007; pp. 2695–2700. [CrossRef]
23. Inoue, H.; Takadama, K.; Shimohara, K. Exploring XCS in multiagent environments. In Proceedings of the Genetic and Evolutionary Computation Conference, GECCO 2005, Washington, DC, USA, 25–26 June 2005; pp. 109–111. [CrossRef]
24. Wang, C.; Chen, H.; Yan, C.; Xiang, X. Reinforcement Learning with an Extended Classifier System in Zero-sum Markov Games. In Proceedings of the 2019 IEEE International Conference on Agents (ICA), Jinan, China, 18–21 October 2019. [CrossRef]
25. Bonarini, A.; Trianni, V. Learning fuzzy classifier systems for multi-agent coordination. Recent Advances in Genetic Fuzzy Systems. *Inf. Sci.* **2001**, *136*, 215–239. [CrossRef]
26. Bacardit, J.; Bernadó-Mansilla, E.; Butz, M. *Learning Classifier Systems: Looking Back and Glimpsing Ahead*; Springer: Berlin/Heidelberg, Germany, 2007; Volume 4998, pp. 1–21.
27. Stalph, P.; Butz, M.; Pedersen, G. *Controlling a Four Degree of Freedom Arm in 3D Using the XCSF Learning Classifier System*; Springer: Cham, Switzerland, 2009; pp. 193–200.
28. Studley, M.; Bull, L. X-TCS: Accuracy-based learning classifier system robotics. In Proceedings of the 2005 IEEE Congress on Evolutionary Computation, Edinburgh, UK, 2–5 September 2005; Volume 3, pp. 2099–2106.
29. Ernest, N. Genetic Fuzzy Trees for Intelligent Control of Unmanned Combat Aerial Vehicles. Ph.D. Thesis, University of Cincinnati, Cincinnati, OH, USA, 2015. [CrossRef]
30. Zhou, K.; Wei, R.; Xu, Z.; Zhang, Q.; Lu, H.; Zhang, G. An Air Combat Decision Learning System Based on a Brain-Like Cognitive Mechanism. *Cogn. Comput.* **2020**, *12*. [CrossRef]
31. Başpınar, B.; Koyuncu, E. Assessment of Aerial Combat Game via Optimization-Based Receding Horizon Control. *IEEE Access* **2020**, *8*, 35853–35863. [CrossRef]
32. Changqiang, H.; Kangsheng, D.; Hanqiao, H.; Shangqin, T.; Zhuoran, Z. Autonomous air combat maneuver decision using Bayesian inference and moving horizon optimization. *J. Syst. Eng. Electron.* **2018**, *29*, 86–97. [CrossRef]
33. Yang, Z.; Zhou, D.; Kong, W.; Piao, H.; Zhang, K.; Zhao, Y. Nondominated Maneuver Strategy Set With Tactical Requirements for a Fighter Against Missiles in a Dogfight. *IEEE Access* **2020**, *8*, 117298–117312. [CrossRef]
34. Ma, X.; Xia, L.; Zhao, Q. Air-Combat Strategy Using Deep Q-Learning. In Proceedings of the 2018 Chinese Automation Congress (CAC), Xi'an, China, 30 November–2 December 2018; pp. 3952–3957. [CrossRef]
35. Han, W.; Zhang, B.; Wang, Q.; Luo, J.; Ran, W.; Xu, Y. A Multi-Agent Based Intelligent Training System for Unmanned Surface Vehicles. *Appl. Sci.* **2019**, *9*, 1089. [CrossRef]
36. Holmes, J.; Lanzi, P.L.; Stolzmann, W.; Wilson, S. Learning classifier systems: New models, successful applications. *Inf. Process. Lett.* **2002**, *82*, 23–30. [CrossRef]
37. Brownlee, J. Learning classifier systems. In *Encyclopedia of Machine Learning and Data Mining*; Springer: Cham, Switzerland, 2017.
38. Stein, A. Learning Classifier Systems: From Principles to Modern Systems. In Proceedings of the GECCO '19, Prague, Czech Republic, 13–17 July 2019; Association for Computing Machinery: New York, NY, USA, 2019; pp. 747–769. [CrossRef]
39. Abbasi, M.; Yaghoobikia, M.; Rafiee, M.; Jolfaei, A.; Khosravi, M.R. Efficient resource management and workload allocation in fog–cloud computing paradigm in IoT using learning classifier systems. *Comput. Commun.* **2020**, *153*, 217–228. [CrossRef]
40. Liang, M.; Palado, G.; Browne, W.N. Identifying Simple Shapes to Classify the Big Picture. In Proceedings of the 2019 International Conference on Image and Vision Computing New Zealand (IVCNZ), Dunedin, New Zealand, 2–4 December 2019; pp. 1–6. [CrossRef]

41. Nordsieck, R.; Heider, M.; Angerer, A.; Hähner, J. Towards Automated Parameter Optimisation of Machinery by Persisting Expert Knowledge. In Proceedings of the ICINCO 2019, Prague, Czech Republic, 29–31 July 2019.
42. Butz, M.; Wilson, S. An Algorithmic Description of XCS. *Soft Comput. Fusion Found. Methodol. Appl.* **2001**, *6*. [CrossRef]
43. De Castro, L.; Timmis, J. *Artificial Immune Systems: A New Computational Intelligence Approach*; Springer: Cham, Switzerland, 2002.
44. De Castro, L.; Von Zuben, F. The Clonal Selection Algorithm with Engineering Applications. *Artif. Immune Syst.* **2001**, *8*. [CrossRef]
45. Aickelin, U.; Cayzer, S. The Danger Theory and Its Application to Artificial Immune Systems. *SSRN Electron. J.* **2008**. [CrossRef]
46. Farmer, J.; Packard, N.H.; Perelson, A.S. The immune system, adaptation, and machine learning. *Phys. D Nonlinear Phenom.* **1986**, *22*, 187–204. [CrossRef]

Article

Assessing a Multi-Objective Genetic Algorithm with a Simulated Environment for Energy-Saving of Air Conditioning Systems with User Preferences

Alejandro Humberto García Ruiz [1], Salvador Ibarra Martínez [1], José Antonio Castán Rocha [1], Jesús David Terán Villanueva [1,*], Julio Laria Menchaca [1], Mayra Guadalupe Treviño Berrones [1], Mirna Patricia Ponce Flores [2,3] and Aurelio Alejandro Santiago Pineda [4]

1 Facultad de Ingeniería, Universidad Autónoma de Tamaulipas, Tampico 89339, Mexico; a2113330227@alumnos.uat.edu.mx (A.H.G.R.); sibarram@docentes.uat.edu.mx (S.I.M.); jacastan@docentes.uat.edu.mx (J.A.C.R.); jlaria@docentes.uat.edu.mx (J.L.M.); mgtrevino@docentes.uat.edu.mx (M.G.T.B.)
2 Graduate Program Division, Tecnológico Nacional de México, Madero 89440, Mexico; D94070293@cdmadero.tecnm.mx
3 Graduate Program Division, Instituto Tecnológico de Ciudad Madero, Cd., Madero 89440, Mexico
4 Information Technology Engineering, Polytechnic University of Altamira, Altamira 89602, Mexico; aurelio.santiago@upalt.edu.mx
* Correspondence: jdteran@docentes.uat.edu.mx

Citation: García Ruiz, A.H.; Ibarra Martínez, S.; Castán Rocha, J.A.; Terán Villanueva, J.D.; Laria Menchaca, J.; Treviño Berrones, M.G.; Ponce Flores, M.P.; Santiago Pineda, A.A. Assessing a Multi-Objective Genetic Algorithm with a Simulated Environment for Energy-Saving of Air Conditioning Systems with User Preferences. *Symmetry* **2021**, *13*, 344. https://doi.org/10.3390/sym13020344

Academic Editor: Jan Awrejcewic

Received: 27 January 2021
Accepted: 16 February 2021
Published: 20 February 2021

Publisher's Note: MDPI stays neutral with regard to jurisdictional claims in published maps and institutional affiliations.

Abstract: Electricity is one of the most important resources for the growth and sustainability of the population. This paper assesses the energy consumption and user satisfaction of a simulated air conditioning system controlled with two different optimization algorithms. The algorithms are a genetic algorithm (GA), implemented from the state of the art, and a non-dominated sorting genetic algorithm II (NSGA II) proposed in this paper; these algorithms control an air conditioning system considering user preferences. It is worth noting that we made several modifications to the objective function's definition to make it more robust. The energy-saving optimization is essential to reduce CO2 emissions and economic costs; on the other hand, it is desirable for the user to feel comfortable, yet it will entail a higher energy consumption. Thus, we integrate user preferences with energy-saving on a single weighted function and a Pareto bi-objective problem to increase user satisfaction and decrease electrical energy consumption. To assess the experimentation, we constructed a simulator by training a backpropagation neural network with real data from a laboratory's air conditioning system. According to the results, we conclude that NSGA II provides better results than the state of the art (GA) regarding user preferences and energy-saving.

Keywords: energy optimization; genetic algorithms; multi-objective optimization; artificial neural network simulator

1. Introduction

Due to constant technological growth, it is now possible to integrate electronic circuits and computer systems to develop, manage, and monitor a wide variety of devices in many of the structures that surround us [1–3]. One of the main advantages of including computational techniques in regular devices is the efficient administration of electrical energy. Researchers have focused their efforts on optimizing these regular devices as well as new research focused on the use of artificial intelligence, neural networks, and fuzzy logic and a study on a new class of optimization problems of constrained interval values [4,5]. This paper implements two evolutionary algorithms, a genetic algorithm (GA) and a non-dominated sorting genetic algorithm (NSGA II), to optimize an air conditioning system's energy consumption. We chose to research air conditioning systems because they consume a large amount of energy regardless of the purpose.

In the literature, it is possible to find some papers focused on ensuring an adequate administration of electrical energy through the so-called home energy management systems which seek to promote renewable energy, lengthen device life, and avoid accidents. In [1], Collota et al. present an example of this system, where a fuzzy controller is used to improve the home energy management scheme; the results showed the system's capability to reduce the maximum load demand for electrical energy. Similarly, Attoue et al. [2] used machine learning techniques, optimization, and data structures to create an energy management system to satisfy the needs of an intelligent building; the results obtained show the proposed mechanisms' effectiveness to improve the performance.

Even though there is a wide variety of papers aiming to save electricity, air conditioners still consume a significant amount of energy. In [6], the authors present a fuzzy controller made in Simulink to lower air conditioning energy while dealing with thermal disturbances. The results demonstrate that the fuzzy controller with triangular membership functions gives the desired performance with an error lower than 1% and saving 25% of energy consumption. Chen, Fu, and Liu [3] proposed a hybrid meta ensemble learning and stacked auto-encoder (SAE), tested with data from the air-conditioning system from a commercial building in Singapore. The proposed meta ensemble learning method showed efficient energy management of the air-conditioning system.

Additionally, Arikiez et al. [7] proposed a heuristic algorithm based on mixed integer linear programming (MILP) to minimize the cost of electrical energy caused by a set of air conditioners. The result reveals that their algorithm can solve a massive problem in a few seconds and gives reasonable suboptimal solutions. Moreover, Ullah and Kim [8] proposed two improved optimization algorithms, a particle swarm optimization and a genetic algorithm, to achieve maximum user comfort in the building environment with minimum energy consumption. Results showed that the proposed optimization algorithms produce better results than the baseline scheme in user comfort and consumed energy.

This paper proposes two metaheuristic algorithms to optimize an air conditioning system while considering the users' preferences, a genetic algorithm (GA) and a non-dominated sorting genetic algorithm (NSGA II). The former intends to replicate the algorithm from [8], and the latter was selected to analyze the performance of a multi-objective optimization algorithm.

The remainder of the paper is as follows: Section 2 addresses the design of the optimization algorithms; Section 3 shows our simulator design; Sections 4 and 5 describe the experimental results and the discussion, respectively. Finally, Section 6 contains the conclusions and future work.

2. Algorithm Design

2.1. Problem Definition

The proposed algorithms aim to tackle two objectives: energy-saving and user satisfaction. These objectives are directly in conflict because the air conditioner system must be turned on for extended periods of time to increase user satisfaction while increasing energy consumption, which conflicts with energy-saving. Such a task seems relatively simple for NSGA II due to its multi-objective nature. Additionally, the genetic algorithm used a weighted objective function to optimize both objectives simultaneously. In this paper, we use the formulation presented in [8] by Ullah and Kim. However, we modify Equations (2)–(4) in two different ways. The first modification adds a cost to maintain the current temperature. The second modification nullifies any energy cost when the laboratory's temperature is lower than the desired temperature. It is important to explain that our climate is warm; therefore, we consider a cooling scheme and turn on the air conditioner system to cool down the room, trying to reach the optimal desired temperature of 22 °C. However, we considered a temperature of 27 °C and below as acceptable yet not no ideal. Furthermore, temperatures below 22 °C are also not desired and would trigger a shutdown of the air conditioner system, nullifying the energy cost and raising the temperature.

For this formulation, expression (1) shows the user-accepted temperature range.

$$UserPreferences = [T_{min}, T_{max}] \tag{1}$$

where T_{min} and T_{max} represent the minimum and maximum temperatures accepted by the user. It is worth noting that for our case, T_{min} is the ideal temperature because we are in a tropical environment and a colder climate is desired. Additionally, the current room temperature is T_c, and the temperature recommended by the optimization algorithms is named $T_o | T_{min} \leq T_o \leq T_{max}$.

With T_o, it is possible to calculate the optimal energy (E_o) needed to change or maintain T_c if it were the same as T_o; see Equation (2). Additionally, if the current temperature is lower than the optimized temperature ($T_c < T_o$), the air conditioner system should be turned off; hence the zero in that case. To calculate this parameter, we need to know the power required per unit change in temperature (P_T).

$$E_o = \begin{cases} P_T + P_T \cdot (T_c - T_o) & if\ T_c \geq T_o \\ 0 & if\ T_c < T_o \end{cases} \tag{2}$$

Similarly, it is possible to calculate the minimum and maximum hypothetical energy consumptions (E_{min}) and (E_{max}), as shown in expressions (3) and (4), which are used later to calculate the energy-saving gain (G_{es}). For these equations, there is a similar condition as Equation (2); where if $T_c < T_{max}$ or $T_c < T_{min}$, it would mean that the current temperature is lower than the maximum or minimum temperature and the air conditioner system will not need to be turned on to reach those temperatures, hence requiring 0 energy.

$$E_{min} = \begin{cases} P_T + P_T \cdot (T_c - T_{max}) & if\ T_c \geq T_{max} \\ 0 & if\ T_c < T_{max} \end{cases} \tag{3}$$

$$E_{max} = \begin{cases} P_T + P_T \cdot (T_c - T_{min}) & if\ T_c \geq T_{min} \\ 0 & if\ T_c < T_{min} \end{cases} \tag{4}$$

Equations (5) and (6) show the user satisfaction gain (G_{us}) and the energy-saving gain (G_{es}) where both gains range from 0 to 1, with 0 representing no gain and 1 represents the most gain. Furthermore, both objectives are in direct conflict because increasing user satisfaction entails an increased use of the air conditioner resulting in higher energy cost.

$$G_{us} = \left(\frac{T_{max} - T_o}{T_{max} - T_{min}} \right)^2 \in [0, 1] \tag{5}$$

$$G_{es} = 1 - \left(\frac{E_o - E_{min}}{E_{max} - E_{min}} \right)^2 \in [0, 1] \tag{6}$$

Additionally, the user relevance factor for each objective is determined by α_{us} for user satisfaction and α_{es} for energy-saving where $\alpha_{us} + \alpha_{es} = 1$. Therefore, the NSGA II would maximize both objectives considering their relevance factor; see Equations (7) and (8).

$$Maximize(\alpha_{us} G_{uc}) \in [0, 1] \tag{7}$$

$$Maximize(\alpha_{es} G_{es}) \in [0, 1] \tag{8}$$

On the other hand, GA uses the weighted function in Equation (9).

$$Maximize(\alpha_{us} G_{uc} + \alpha_{es} G_{es}) \in [0, 1] \tag{9}$$

Figure 1 shows an example of the calculation of user satisfaction gain (G_{us}) and the energy-saving gain (G_{es}) for a given T_o and T_c.

1:	$T_o = 24.9039$ $T_{min} = 22$ $P_T = 5$	$T_c = 30.2000$ $T_{max} = 27$
2:	$E_o = 5 + 5 \cdot (30.2000 - 24.9039)$ $= 31.4805$	
3:	$E_{min} = 5 + 5 \cdot (30.2000 - 27) = 21.00$	$E_{max} = 5 + 5 \cdot (30.2000 - 22) = 46.00$
4:	$G_{us} = \left(\frac{27 - 24.9039}{27 - 22}\right)^2 = 0.1757$	$G_{es} = 1 - \left(\frac{31.4805 - 21.00}{46.00 - 21.00}\right)^2 = 0.8242$

Figure 1. Example of user satisfaction gain (G_{us}) and the energy-saving gain (G_{es}) calculation.

2.2. GA and NSGA II

As we said before, we modify the formal definition of [8], trying to make it more robust by considering new elements; however, we will not be able to compare directly to their results because of the modifications in the definition. Thus, in this paper, we implement their genetic algorithm as explained in [8], and a multi-objective NSGA II as our proposal (both algorithms are available at https://github.com/ahgarciar/Air-Conditioner-System-Controllers (accessed on 19 February 2021)).

We use the same population, selection, crossover, and mutation operators for both algorithms with little modifications for the NSGA II, as explained below.

Population. We use a vector of one hundred solutions for both algorithms, where each solution ($T_{o_i} \in \mathbb{R}$) is a possible temperature between 22 and 27 °C.

Selection. The algorithms use a binary tournament between random individuals of the population to select the parents for the crossover operator. The GA uses a binary tournament, comparing the solutions' objective values, and the winners are selected for crossover. On the other hand, the NSGA II algorithm uses the rank of the Pareto front of each solution to identify the winner, which will be the solution with the lowest level on the Pareto front, meaning that it is less dominated than the other solution. However, suppose both solutions are in the same level of the Pareto front. In that case, we use the crowded comparison operator ($>_n$), which identifies the solution with the largest crowding distance, which will be explained further in this section.

Crossover. As stated before, our GA and NSGA II have a static population of one hundred solutions; then, another one hundred solutions are produced as offspring, using the Simulated Binary Crossover operator [9]. This operator requires two parents ($p1$ and $p2$) to obtain two new individuals (T_{o_i} and $T_{o_{i+1}}$).

Here, we explain the crossover operator for our specific problem. Thus, we will change some of the original notation of this crossover to match our variables.

First, we identify the two parents' maximum and minimum temperature values; see Equations (10) and (11).

$$max = MAX(p1, p2) \tag{10}$$

$$min = MIN(p1, p2) \tag{11}$$

Then, we calculate Equations (12) and (13) to produce $\beta 1$ and $\alpha 1$, which also use a random real number ($u \in [0, 1]$) and a constant $\eta = 20$ in Equation (14) to produce $\beta 1q$. Finally, we produce T_{o_i} as shown in Equation (15).

$$\beta 1 = 1 + \frac{2(min - T_{min})}{max - min} \tag{12}$$

$$\alpha 1 = 2 - (\beta 1)^{-(\eta+1)} \tag{13}$$

$$\beta 1q = \begin{cases} (u \cdot \alpha 1)^{\frac{1}{\eta+1}} & if\ u\ \leq \left(\frac{1}{\alpha 1}\right) \\ \left(\frac{1}{2-u \cdot \alpha 1}\right)^{\frac{1}{\eta+1}} & otherwise \end{cases} \quad (14)$$

$$T_{o_i} = 0.5(max + min - \beta 1q * (max - min)) \quad (15)$$

Then, we carry out a similar process to produce $T_{o_{i+1}}$. However, the difference is that $\beta 2$ is calculated using T_{max} instead of T_{min} as $\beta 1$, see Equations (16)–(19).

$$\beta 2 = 1 + \frac{2(T_{max} - max)}{max - min} \quad (16)$$

$$\alpha 2 = 2 - (\beta 2)^{-(\eta+1)} \quad (17)$$

$$\beta 2q = \begin{cases} (u \cdot \alpha 2)^{\frac{1}{\eta+1}} & if\ u\ \leq \left(\frac{1}{\alpha 2}\right) \\ \left(\frac{1}{2-u \cdot \alpha 2}\right)^{\frac{1}{\eta+1}} & otherwise \end{cases} \quad (18)$$

$$T_{o_{i+1}} = 0.5(max + min - \beta 2q * (max - min)) \quad (19)$$

Figure 2 shows an example of the crossover operator.

1:	$p1 = 25.1457$	$p2 = 25.2698$
2:	$min = 25.1457$	$max = 25.2698$
3:	$\beta 1 = 1 + \frac{2(25.1457 - 22)}{25.2698 - 25.1457} = 51.6962$	$\beta 2 = 1 + \frac{2(27 - 25.2698)}{25.2698 - 25.1457} = 28.8839$
4:	$\alpha 1 = 2 - (51.6962)^{-(20+1)} = 2.00$	$\alpha 2 = 2 - (28.8839)^{-(20+1)} = 2.00$
5:	Considering $u = 0.9504$	
6:	$\beta 1q = \left(\frac{1}{2 - 0.9504 * 2.00}\right)^{\frac{1}{20+1}} = 1.1164$	$\beta 1q = \left(\frac{1}{2 - 0.9504 * 2.00}\right)^{\frac{1}{20+1}} = 1.1164$
7:	$T_{o_i} = 0.5(25.2698 + 25.1457 - 1.1164 * (25.2698 - 25.1457)) = 25.1384$	$T_{o_{i+1}} = 0.5(25.2698 + 25.1457 + 1.1164 * (25.2698 - 25.1457)) = 25.2770$

Figure 2. Example of crossover process.

Mutation. As a mutation operator, we used the polynomial mutation operator [10], which is applied to fifty percent of the solutions generated in the crossover operator.

Here, we modify the temperature produced by the crossover operation depending on the δ value, which ranges from -1 to 1, and also uses a random real number ($u \in [0,1]$) and a constant $\eta = 20$; see Equations (20)–(23).

$$T_{o_i} = T_{o_i} + \delta_q(T_{max} - T_{min}) \quad (20)$$

$$\delta_1 = (T_{o_i} - T_{min})/(T_{max} - T_{min}) \quad (21)$$

$$\delta_2 = (T_{max} - T_{o_i})/(T_{max} - T_{min}) \quad (22)$$

$$\delta_q = \begin{cases} \left[2u + (1.0 - 2.0 * u) * (1.0 - \delta_1)^{\eta+1}\right]^{\frac{1}{(\eta+1)}} - 1 & if\ u \leq\ 0.5 \\ 1 - \left[2(1-u) + 2.0 * (u - 0.5) * (1.0 - \delta_2)^{\eta+1}\right]^{\frac{1}{(\eta+1)}} & otherwise \end{cases} \quad (23)$$

Figure 3 shows an example of the mutation operator.

$T_{o_i} = 25.7195$

$\delta_1 = \dfrac{25.7195 - 22}{(27-22)} = 0.7439 \qquad \delta_2 = \dfrac{27 - 25.7195}{(27-22)} = 0.2561$

Considering $u = 0.3184$

$\delta_q = [2 * 0.3184 + (1.0 - 2.0 * 0.3184) * (1.0 - 0.7439)^{20+1}]^{\frac{1}{(20+1)}} - 1 = -0.0212$

Figure 3. Example of the mutation process.

For this experimentation, we tested several values for constant η, resulting in a value of twenty.

The NSGA II requires two additional procedures:

Fast Non-Dominated Sort. This procedure allocates the solutions of the pool of solutions (P) into Pareto fronts of non-dominated individuals; this allows elitism in selecting the parents for the offspring of NSGA II.

Crowding Distance. A particular solution's crowding distance value is the average distance of its two neighboring solutions [11]. Figure 4 shows the calculation of the crowding distance of the solution i, which is an estimate of the size of the largest cuboid enclosing i without including any other solution. The algorithm used in this implementation is a normalized version of the original algorithm with the minimum (MIN_m) and maximum (MAX_m) values per objective (m) [12]; this algorithm is calculated for each generation for the current population. The algorithm requires a set of non-dominated individuals (I) to calculate their distance.

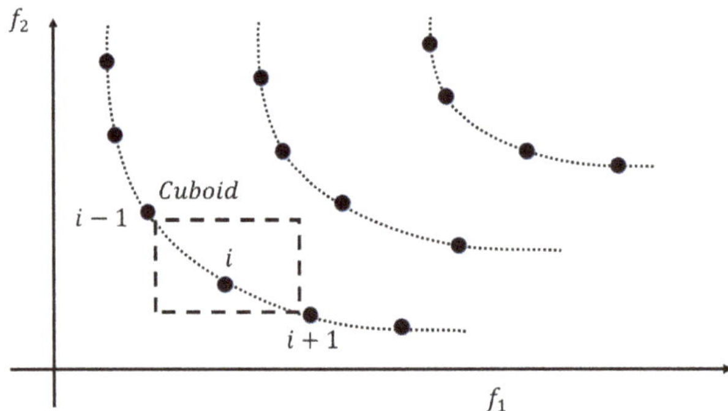

Figure 4. Crowding distance calculation.

3. Simulator

To assess the optimization algorithms, we constructed a neural-network-based simulator with temperature data from a specific room within our laboratory. The data acquired was the internal and external temperature (in Celsius) and the state of the air conditioner system. We recorded this data every five minutes using two temperature sensors (DHT22) with an Arduino Mega controller in the second week of September from Monday to Wednesday. Figure 5 shows a diagram from the data acquisition to the evaluation of the neural-network-based simulator. Regarding the data acquisition, we force fluctuations in the internal temperature (from the external temperature to 24 °C) during different moments of the day, turning on and off the air conditioning system.

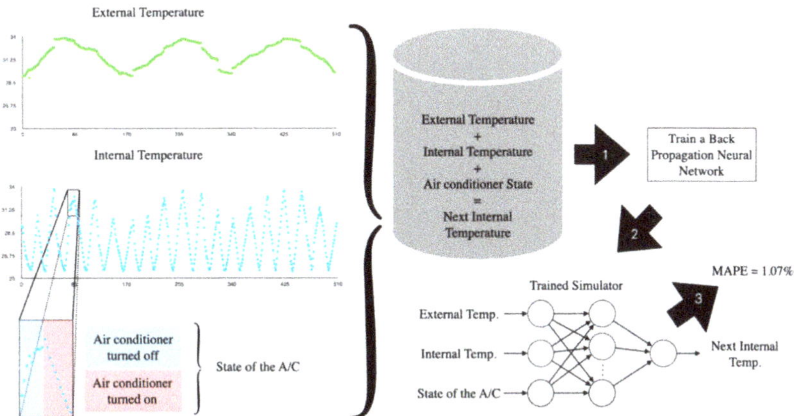

Figure 5. Diagram from data acquisition to the evaluation of the simulator.

Once the data were acquired, we constructed a dataset with the external temperature, internal temperature, air conditioning state, and the next internal temperature taken from the next row. With this dataset, we trained a backpropagation neural network implemented in C# and is available at https://github.com/ahgarciar/Air-Conditioner-System-Controllers (accessed on 19 February 2021). We configured the neural network with three input neurons, ten neurons in a hidden layer, and one output neuron. Additionally, it was tested with the mean average percentage error (MAPE) [10], producing a 1.07% error; this result shows that our simulator has a relatively good performance.

4. Experimental Results

This experimentation aims to assess energy-saving and user satisfaction for five whole days. To do this, we executed both algorithms independently and used the neural network to simulate the temperature changes, given each algorithm's recommendation.

Figure 6 shows the external temperature acquired for five days and the simulated internal temperature obtained by the neuronal-network-based simulator. The simulator requires three parameters to forecast the new internal temperature: current external temperature, current internal temperature, and the state of the air conditioner system; the latest is obtained according to the optimizer recommendation, i.e., if the optimizer recommendation is lower or equal to the internal temperature, the air conditioning system is turned on; otherwise, it is turned off.

To avoid increasing computational costs, we execute the optimization algorithms once every half hour. These experimentations consider three configurations (see Table 1): prioritizing energy-savings (A), a balance between energy-saving and user satisfaction (B), and prioritizing user satisfaction (C). These three configurations constitute the three primary user profiles within our laboratory and, roughly, in any conventional home.

Table 1. Configurations of the user's priorities.

Configuration	α_{us}	α_{es}
A	0.2	0.8
B	0.5	0.5
C	0.8	0.2

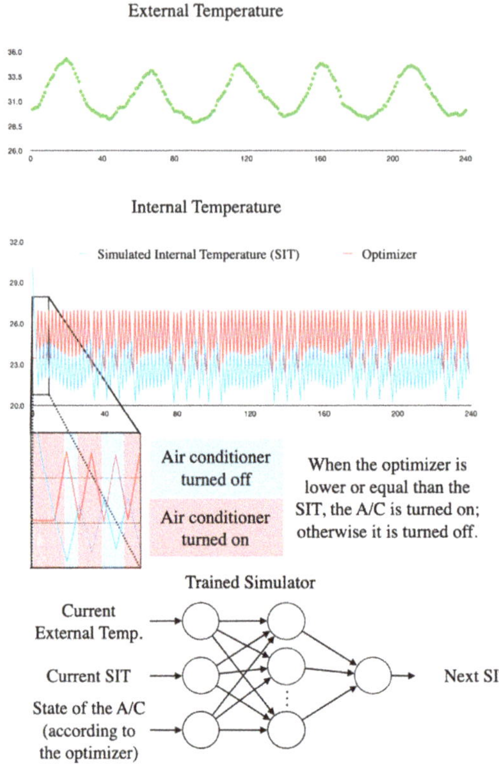

Figure 6. Internal temperature simulation based on the optimizer recommendation.

We configured the algorithms with the values from Table 1; then, the algorithms produce their recommendations for five simulated days; where we use as a recommendation the best solution produced by GA, and a single solution from the first Pareto front for the NSGA II. To keep a fair comparison, we used the nadir point of the Pareto front, which is the solution with the lowest Euclidian distance to a hypothetical optimal point in the space. This hypothetical optimal point is (1, 1) because both objectives intend to maximize their gain, and 1 is the largest possible value for both objectives; see Figure 7.

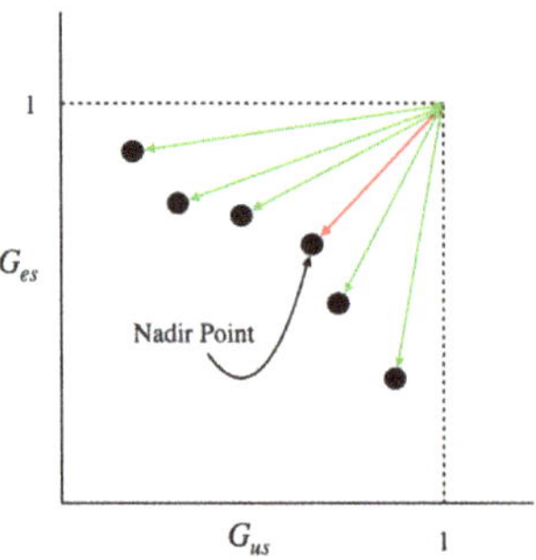

Figure 7. Nadir point calculation.

Figures 8–10 show the temperature recommended by GA and NSGA II for each moment of the day analyzed (every half an hour) for each of the configurations.

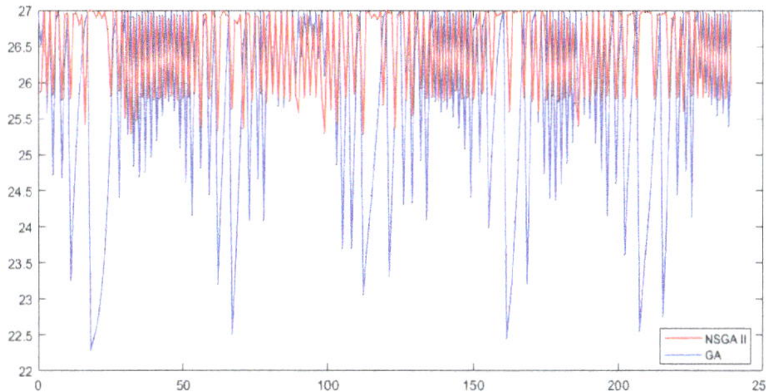

Figure 8. Temperature recommended by genetic algorithms using configuration A.

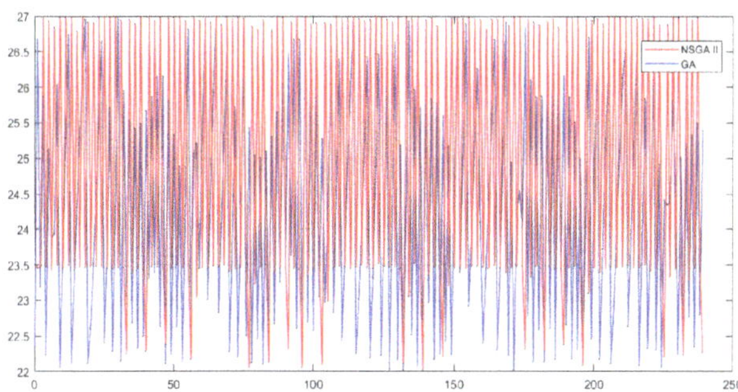

Figure 9. Temperature recommended by genetic algorithms using configuration B.

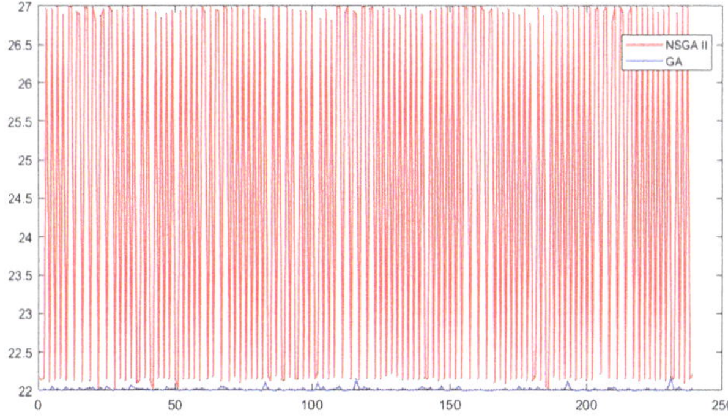

Figure 10. Temperature recommended by genetic algorithms using configuration C.

Here, we can see that for configuration A, both algorithms recommend higher temperatures than the other configurations. However, GA does recommend lower values than NSGA II. For configuration B, GA's recommendations have larger fluctuations than NSGA II, which mainly maintains its recommendations from 27 to 23.5 °C. Finally, for configuration C, GA recommends the lowest temperature, about 22 °C, while NSGA II increases its fluctuation.

Based on each algorithm's recommended temperature, the simulator turns the air conditioning on or off for half an hour to adjust the room temperature.

Figures 11–13 show the room temperature simulation results by applying the algorithms' recommendation for each configuration.

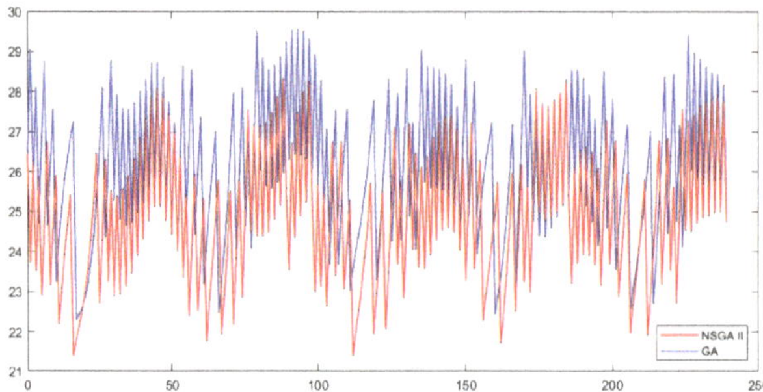

Figure 11. Simulation of temperature according to the recommendation of the genetic algorithms for configuration A.

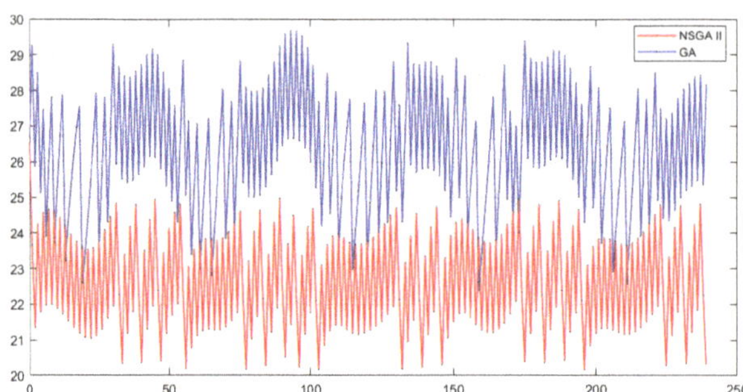

Figure 12. Simulation of temperature according to the recommendation of the genetic algorithms for configuration B.

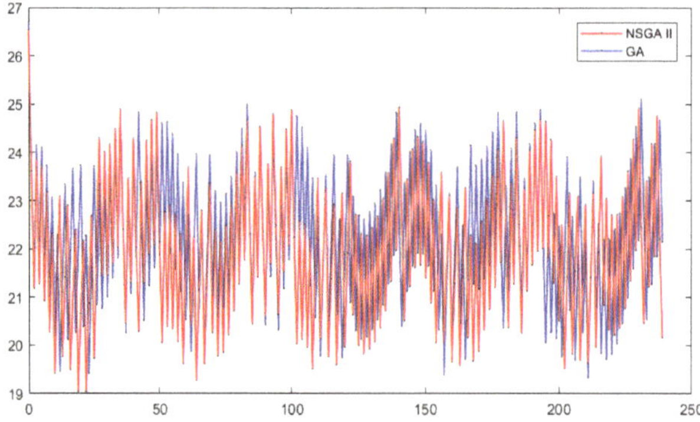

Figure 13. Simulation of temperature according to the recommendation of the genetic algorithms for configuration C.

In Figure 11, configuration A, both algorithms have similar behavior. However, NSGA II produces lower values than GA. Figure 12, configuration B, shows that NSGA II has lower temperature values than GA and both algorithms have considerably different behavior. Finally, Figure 13 shows an almost identical behavior for both algorithms with a slight tendency for the NSGA II to produce lower temperatures than GA.

The satisfaction level increases when the room temperature is closer to the minimum user desired temperature T_{min}. As stated before, we live in a warm environment; thus, we require a cooling scheme for the air conditioning system. Therefore, the algorithm that keeps closer to T_{min} provides greater satisfaction for the user.

Therefore, to measure user satisfaction, we created a satisfaction ratio, defined as the sum of the absolute differences between the current simulated temperature and T_{min} for each moment of the day; this is because values lower than T_{min} are unwanted.

A lower satisfaction ratio implies a higher level of satisfaction. Table 2 presents the satisfaction ratios calculated for each algorithm for all the configurations.

Table 2. Satisfaction ratio for each algorithm.

Configuration	NSGA II	GA
A	698.10343	993.36546
B	316.98898	1058.35429
C	313.84696	313.88824

As we can see, the NSGA II algorithm achieves better performance over the GA in maintaining the room temperature according to the user preferences. For the configuration where the algorithm prioritizes energy-saving (A), the NSGA II outperformed the GA's satisfaction rate by about 295 units. Furthermore, for the second configuration (B), which prioritizes a balance between energy-saving and user satisfaction, the NSGA II outperformed the GA's satisfaction rate by about 741 units. Finally, for the last configuration (C), which prioritizes user satisfaction, both algorithms produced almost the same satisfaction rate.

To measure the energy consumption, we evaluate the maximum and minimum energy consumption by simulating two extreme scenarios for the whole test. The first with the air conditioning system turned off and the second with the air conditioning system turned on (see Figure 14). According to this simulation and during its execution, there was an electrical energy use of 0 and 105 kWh, respectively.

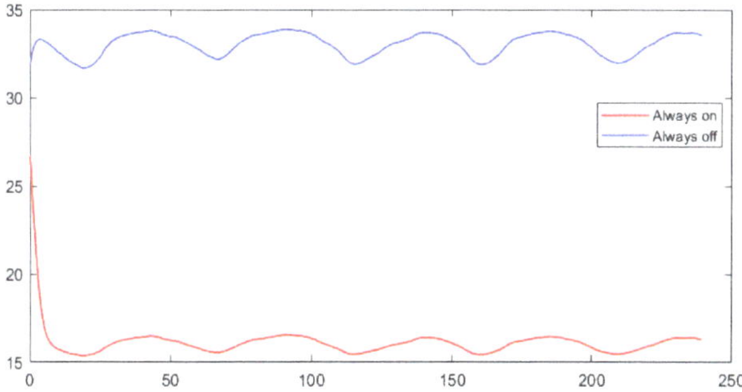

Figure 14. Temperature simulation considering the air conditioning always off and always on.

Table 3 shows calculations of the number of kWh used by all the configurations from Table 1 for both algorithms. Additionally, we estimated the cost of $2.997 Mexican pesos for each kWh consumed. We took this electric cost from the Mexican Federal Electricity Commission (CFE). Therefore, for the previous extreme scenario, where the air conditioning system is turned on for the whole test, the cost produced would be $314.69.

Table 3 shows that NSGA II outperformed GA in the total cost of the electrical energy consumption in two of the three configurations. However, in the non-winning configuration, GA improves by a smaller margin to NSGA II than in the other cases.

Table 3. Cost of maintaining the configurations.

Configuration	NSGA II		GA	
	kWh	Cost	kWh	Cost
A	83.74032	$250.97	86.01629	$257.79
B	94.29362	$282.60	93.56877	$280.43
C	89.98913	$269.70	91.99613	$275.71

5. Discussion

As we can see, NSGA II achieved a better performance than GA regarding user preferences while also outperforming GA in two of three configurations regarding the energy cost.

Although both algorithms adapt to the user preferences and the energy-saving, the NSGA II algorithm provides greater satisfaction. The neural-network-based simulator showed that the room temperature values of the NSGA II were closer to the user ideal temperature for two of three configurations. In this sense, we can conclude that the NSGA II algorithm produced better results than GA to control the air conditioning systems.

It is worth noting that the experimentation results were on a period of five days with a single room in our laboratory. However, by extending the experimentation to a longer time span and more laboratories/rooms, we can easily reach a cost of about $3700 for just two rooms with the air conditioning turned on for one month. On the other hand, we can decrease this cost to about $3000 with the NSGA II configured for energy-saving for the same amount of time. This comparison shows that we can reach an energy-saving of about nineteen percent by using the NSGA II optimizer, highlighting the impact of using an intelligent system on electrical energy consumption.

6. Conclusions

In this paper, we tackle the problem of controlling an air conditioning system using evolutive algorithms to increase energy-saving while considering user satisfaction.

Therefore, we assessed the performance of two genetic algorithms to control an air conditioning system. The first algorithm was an implementation of a genetic algorithm (GA) proposed in [8], while the second was our proposed NSGA II algorithm, which is a multi-objective optimization algorithm. We considered the NSGA II an appropriate alternative because the air conditioning system intends to increase energy-saving while considering the user preference. Thus, both objectives are directly opposed, which is a key identifier of a multi-objective problem, making NSGA II an excellent alternative for solving.

Additionally, we proposed a modification to the definition of the problem, enhancing it to consider new elements, including the cost of maintaining the same temperature and the cancellation of the cost of returning to a warmer temperature, which occurs if the room is colder than the desired temperature.

Furthermore, we designed and implemented a neural-network-based simulator to measure both algorithms' performance in equal circumstances. This simulator showed a relatively low mean average percentage error of 1.07.

Regarding the comparison between GA and NSGA II, the latter showed better energy-savings and user satisfaction than GA. Therefore, our assumption that a multi-objective algorithm would produce better results for controlling an air conditioning system with two objectives proved right.

Finally, we are currently working on expanding the experimentation by adding other services. As future work, we encourage researchers to develop an intelligent system prototype using NSGA II as a decision-making engine, integrating different configurations of user preferences and profiles and further studying the air conditioning system's energy cost.

Author Contributions: Conceptualization, Methodology, Investigation, A.H.G.R., S.I.M., and J.D.T.V.; Software, A.H.G.R., J.D.T.V. and A.A.S.P.; Validation, J.A.C.R. and J.D.T.V.; Formal Analysis, Resources, J.A.C.R., and J.L.M.; Writing—Original Draft Preparation, A.H.G.R.; Writing—Review & Editing, S.I.M., J.D.T.V., and A.A.S.P.; Data Curation, Visualization, Supervision, M.G.T.B., M.P.P.F.; Project Administration, Funding Acquisition, S.I.M. and J.A.C.R. All authors have read and agreed to the published version of the manuscript.

Funding: This work was funded by PNPC of the National Council for Science and Technology (CONACyT) of México, and the Universidad Autónoma de Tamaulipas.

Institutional Review Board Statement: Not applicable.

Informed Consent Statement: Not applicable.

Data Availability Statement: The data presented in this study are available at https://github.com/ahgarciar/Air-Conditioner-System-Controllers (accessed on 19 February 2021).

Acknowledgments: A.H.G.R. would like to thank the CONACyT for the scholarship awarded. The APC was funded by the Universidad Autónoma de Tamaulipas.

Conflicts of Interest: The authors declare no conflict of interest.

References

1. Collotta, M.; Pau, G. A Solution Based on Bluetooth Low Energy for Smart Home Energy Management. *Energies* **2015**, *8*, 11916–11938. [CrossRef]
2. Attoue, N.; Shahrour, I.; Younes, R. Smart Building: Use of the Artificial Neural Network Approach for Indoor Temperature Forecasting. *Energies* **2018**, *11*, 395. [CrossRef]
3. Chen, Y.; Fu, G.; Liu, X. Air-Conditioning Load Forecasting for Prosumer Based on Meta Ensemble Learning. *IEEE Access* **2020**, *8*, 123673–123682. [CrossRef]
4. Treanţă, S. On a Class of Constrained Interval-Valued Optimization Problems Governed by Mechanical Work Cost Functionals. *J. Optim. Theory Appl.* **2021**, 1–12. [CrossRef]
5. Treanţă, S. Efficiency in uncertain variational control problems. *Neural Comput. Appl.* **2020**, 1–14. [CrossRef]
6. Shah, Z.A.; Sindi, H.F.; Ul-Haq, A.; Ali, M.A. Fuzzy Logic-Based Direct Load Control Scheme for Air Conditioning Load to Reduce Energy Consumption. *IEEE Access* **2020**, *8*, 117413–117427. [CrossRef]
7. Arikiez, M.; Grasso, F.; Kowalski, D.; Zito, M. Heuristic algorithm for minimizing the electricity cost of air conditioners on a smart grid. In Proceedings of the 2016 IEEE International Energy Conference (ENERGYCON), Leuven, Belgium, 4–8 April 2016; pp. 1–6.
8. Ullah, I.; Kim, D. An Improved Optimization Function for Maximizing User Comfort with Minimum Energy Consumption in Smart Homes. *Energies* **2017**, *10*, 1818. [CrossRef]
9. Deb, K.; Kumar, A. Real-coded Genetic Algorithms with Simulated Binary Crossover: Studies on Multimodal and Multiobjective Problems. *Complex Syst.* **1995**, *9*, 431–454.
10. Deb, K.; Goyal, M. A combined genetic adaptive search (GeneAS) for engineering design. *Comput. Sci. Inform.* **1996**, *26*, 30–45.
11. Raquel, C.R.; Naval, P.C. An Effective Use of Crowding Distance in Multiobjective Particle Swarm Optimization. In Proceedings of the 7th Annual Conference on Genetic and Evolutionary Computation, Washington, DC, USA, 25–29 June 2005; pp. 257–264.
12. Pedersen, G.K.M.; Goldberg, D.E. Dynamic Uniform Scaling for Multiobjective Genetic Algorithms. *Comput. Vis.* **2004**, *3103*, 11–23.

Article

A Novel Neural Network Training Algorithm for the Identification of Nonlinear Static Systems: Artificial Bee Colony Algorithm Based on Effective Scout Bee Stage

Ebubekir Kaya [1,*] and Ceren Baştemur Kaya [2]

[1] Department of Computer Engineering, Engineering—Architecture Faculty, Nevsehir Haci Bektas Veli University, Nevşehir 50300, Turkey
[2] Department of Computer Technologies, Nevsehir Vocational College, Nevsehir Haci Bektas Veli University, Nevşehir 50300, Turkey; ceren@nevsehir.edu.tr
* Correspondence: ebubekir@nevsehir.edu.tr

Abstract: In this study, a neural network-based approach is proposed for the identification of nonlinear static systems. A variant called ABCES (ABC Based on Effective Scout Bee Stage) is introduced for neural network training. Two important changes are carried out with ABCES. The first is an update of "limit" control parameters. In ABC algorithm, "limit" value is fixed. It is adaptively adjusted according to number of iterations in ABCES. In this way, the efficiency of the scout bee stage is increased. Secondly, a new solution-generating mechanism for the scout bee stage is proposed. In ABC algorithm, new solutions are created randomly. It is aimed at developing previous solutions in the scout bee stage of ABCES. The performance of ABCES is analyzed on two different problem groups. First, its performance is evaluated on 13 numerical benchmark test problems. The results are compared with ABC, GA, PSO and DE. Next, the neural network is trained by ABCES to identify nonlinear static systems. 6 nonlinear static test problems are used. The performance of ABCES in neural network training is compared with ABC, PSO and HS. The results show that ABCES is generally effective in the identification of nonlinear static systems based on neural networks.

Keywords: artificial intelligence; artificial bee colony algorithm; global optimization; neural network; nonlinear static system

Citation: Kaya, E.; Baştemur Kaya, C. A Novel Neural Network Training Algorithm for the Identification of Nonlinear Static Systems: Artificial Bee Colony Algorithm Based on Effective Scout Bee Stage. *Symmetry* **2021**, *13*, 419. https://doi.org/10.3390/sym13030419

Academic Editor: Laszlo T. Koczy and István A. Harmati

Received: 3 February 2021
Accepted: 28 February 2021
Published: 5 March 2021

1. Introduction

One of the most important issues of artificial intelligence is heuristic optimization algorithms. They are used to solve many real-world problems and provide many advantages. Therefore, the number of heuristic optimization algorithms has increased recently. Heuristic optimization algorithms are divided into different classes according to the source of inspiration such as swarm intelligence, bio-inspired, physics and chemistry-based, and other algorithms [1]. ABC algorithm is one of the most popular heuristic algorithms based on swarm intelligence. It is used to solve many problems in different areas.

One of the other important usage areas of ABC algorithm is artificial neural network (ANN) training. ANN produces output values by using input values. It can learn with samples. The learning process continues according to a tolerance value. The information obtained as a result of learning is stored in weights. In this way, using weights can produce suitable results in the face of similar situations. This is a very important advantage of ANN. It is one of the main reasons for choosing ANN within the scope of this study.

In our daily life, we encounter nonlinear systems in all areas. Some exhibit static, others dynamic behavior. In fact, this is why the identification of nonlinear systems is important. When the literature is examined, studies have been carried out on nonlinear dynamic systems generally in system identification. In other words, nonlinear static systems are ignored. Nonlinear dynamic systems are affected by previous times. Time series is one of the most important examples of these systems. Nonlinear static systems are time independent. They are formed as a result of the interaction of independent variables. In fact, many fields such as education, medicine, engineering, finance, business exhibit static behavior. This is the main difference between nonlinear static and dynamic systems. This structural difference affects the complexity level of the systems. This study focuses directly on the identification of nonlinear static systems.

Neuro-fuzzy and ANN are popular methods used in prediction and modeling. Both methods can learn with samples. The learning process makes the methods powerful. This situation allows the generation of an output for input values that they do not know at all. Although both methods have strengths and weaknesses according to problem types, there is no definite inference. Experiences are decisive in this process. In [2–5], ANFIS, one of the neuro-fuzzy models, was trained using ABC algorithm for the identification of nonlinear dynamic/static systems. In these studies, the performance of ANFIS training-based ABC algorithm on system identification was observed. However, the performance of ANN training based on ABC algorithm in the identification of nonlinear static systems is unknown. In [6], ANN was trained using the ABC algorithm for pattern classification. But it is not about system identification. Therefore, ANN training is carried out by using ABC algorithm and its variant to identify nonlinear static systems in this study. The performance of approaches such as ABC algorithm, PSO and HS are analyzed comparatively in the identification of nonlinear static systems for the first time. In this respect, it is a pioneering and innovative study. In fact, it is one of the first studies whose main subject is the identification of nonlinear static systems. Within the scope of this study, only standard ABC algorithm is not used for ANN training. A novel algorithm called ABCES based on ABC algorithm is introduced for ANN training. With ABCES, two important changes are realized in structure of ABC algorithm. The first is adjustment of limit control parameter. Unlike standard ABC algorithm, this control parameter is not fixed. It is determined adaptively. The second is updating of solution generation mechanism belonging to the scout bee stage. In standard ABC algorithm, the solutions for the scout bee stage are generated randomly. In the proposed method, a new solution generation mechanism that uses the global best solution to ensure the continuity of the gains achieved, has been proposed. In this way, the scout bee stage has been made more effective. These changes are an innovative approach to improving the performance of standard ABC algorithm.

2. Related Works

2.1. The Studies on ABC Algorithm

Some studies related to ABC algorithm are presented in this section. Horng [7] suggested a max entropy thresholding (MET) approach based on ABC algorithm for image segmentation. The study compared the results obtained with four different methods: PSO, hybrid cooperative-comprehensive learning-based PSO algorithm (HCOCLPSO), Fast Otsu's method and honey-bee mating optimization (HBMO). Karaboga [8] designed digital infinite impulse response (IIR) filters by using ABC algorithm. Yeh and Hsieh [9] solved reliability redundancy allocation problem by using a variant of ABC algorithm and the results were compared with different methods in the literature. Hemamalini and Simon [10] used ABC algorithm for economic load dispatch problem. Hong [11] proposed a model to predict electric load based on support vector regression (SVR) and ABC algorithm. Şahin [12] used GA and ABC to maximize the thermal performance of a solar air collector. Zaman et al. [13] proposed a method based on ABC algorithm for synthesizing antenna arrays. Deng [14] used ABC algorithm to classify customers in mobile e-commerce environment. Bulut and Tasgetiren [15] developed a variant of ABC algorithm for economic lot scheduling problem. There are many studies related to ABC algorithm out of these [16–18].

Although ABC algorithm's global convergence speed is very good, its different variants have been proposed to increase the speed of local convergence of ABC algorithm. The main purpose here is to improve the performance of ABC algorithm. Bansal et al. [19] proposed adaptive version of ABC algorithm. Here, two important parameters were adjusted according to current fitness values adaptively: step size and "limit" control parameters. Babaeizadeh and Ahmad [20] updated employed, onlooker and scout bee phases in ABC algorithm. Draa and Bouaziz [21] suggested a new ABC algorithm for image contrast enhancement. Karaboga and Gorkemli [22] proposed a variant known as qABC and modified solution generation mechanism belonging to onlooker phase. Gao et al. [23] presented new solution generation mechanisms using more information about the population. Wang [24] made two important updates via generalized opposition-based learning method and local best solution in ABC algorithm. Kıran and Fındık [25] added direction information for each dimension of each food source position to increase the speed of convergence of ABC algorithm. Liang and Lee [26] updated ABC algorithm using different operations and strategies such as elite, solution sharing, instant update, cooperative strategy and population manager strategies. Karaboga and Kaya [4] used arithmetic crossover and adaptive neighborhood radius to improve the performance of ABC algorithm. Different variants of ABC algorithm have been proposed out of these [27–30].

2.2. The Studies on ANN and Neuro-Fuzzy

Due to advantages of ANN, it is seen that it is used successfully in solving many real-world problems [31–35]. Capizzi et al. [36] proposed neural network topology to model surface plasmon polaritons propagation. Sciuto et al. [37] suggested an approach based on a spiking neural network for anaerobic digestion process. Capizzi et al. [38] used a back-propagation neural network (BPNN) for automated oil spill detection by satellite remote sensing. An effective training algorithm should be used to achieve effective results with ANN. Therefore, heuristic algorithms have been used extensively in ANN training recently. ABC algorithm is one of successful heuristic algorithms and it is used in ANN training. Mohmad Hassim and Ghazali [39] suggested an approach for training functional link neural network (FLNN) by using ABC algorithm for time series prediction. They demonstrated that the proposed approach was better than FLNN model based on BP. Zhang et al. [40] suggested a model based on a forward neural network for classifying MR brain image and adjusted with ABC algorithm the parameters of a forward neural network. Ozkan et al. [41] used a model-based neural network and ABC for modeling daily reference evapotranspiration. Chen et al. [42] used an approach based on BPNN and ABC algorithm for prediction of water quality. Karaboga and Ozturk [6] applied ABC algorithm to train FFNNs on pattern classification. The benchmark classification problems were used for performance analysis. Obtained results by using ABC algorithm was compared with the well-known some algorithms. It was reported that training FFNN based on ABC algorithm gave effective results on the related problem. Another usage area of ABC algorithm is ANFIS training. Karaboga and Kaya [2,3] used standard ABC algorithm for adaptive-network-based fuzzy inference system (ANFIS) training for nonlinear dynamic systems identification. In a different study, Karaboga and Kaya [4] proposed a new ANFIS training algorithm called an adaptive and hybrid artificial bee colony algorithm (aABC) to obtain more effective results in the identification of nonlinear dynamic systems. In the next study, Karaboga and Kaya [5] trained ANFIS by using aABC algorithm for nonlinear static systems identification. The performance of aABC algorithm was tested on 5 nonlinear static systems and compared with PSO, GA, HS and ABC algorithm.

3. Materials and Methods

3.1. Standard ABC Algorithm

The ABC algorithm is one of the popular swarm-based heuristic optimization algorithms. It models the food searching behavior of the honey-bees [43]. It includes three different types of bees named employed bees, onlooker bees and scout bees. There are some assumptions in ABC algorithm. Some of these are: half of the colony consists of employed bees. The other half is the onlooker bees. Specifically, the number of employed bees is equal to the number of onlooker bees. The basic steps of ABC algorithm are as follows:

In the initial phase, the positions of food sources are determined randomly by using (1). Here, x_i shows ith solution. i is in range [1, population size]. x_j^{min} is the lower value to be taken by parameter j. Also, x_j^{max} is the upper value.

$$x_{ij} = x_j^{min} + rand(0,1)\left(x_j^{max} - x_j^{min}\right) \tag{1}$$

Every employed bee probabilistically develops a new food source using the solution positions in memory. (2) is used to create a new solution in this process. Here, k is an integer number in range [1, number of employed bees]. Θ_{ij} is a random number in range [−1, 1].

$$v_{ij} = x_{ij} + \Theta_{ij}\left(x_{ij} - x_{kj}\right) \tag{2}$$

If the amount of nectar of the new source is higher than before, information belonging to the previous position is deleted from the memory. At the same time, information belonging to the new food source is written to the memory. Otherwise, the previous position is maintained. After the search process is completed, the employed bees share the food source information with the onlooker bees. A onlooker bee evaluates the information of all bees. It selects a source according to the probability value obtained from (3). As in the employed bee stage, a new solution develops by modifying the current solution. And they control the nectar quantity of the candidate solution. If the nectar amount of the candidate solution is better, information of the previous solution is deleted from the memory.

$$p_i = \frac{fit_i}{\sum_1^{SN} fit_n} \tag{3}$$

"Limit" is one of the important control parameters of ABC algorithm. If a position is not improved up to the limit value, it is assumed that this food source was abandoned. The abandoned food source is replaced by a new food source by scout bee.

3.2. Artificial Bee Colony Algorithm Based on Effective Scout Bee (ABCES)

One of the most important control parameters of ABC algorithm is "limit". Failure counter is the number of failures in producing the solution. When the failure counter reaches "limit" value, a random solution is created instead of the previous solution. This prevents the creation of qualified solutions. In the scout bee phase, instead of creating solutions randomly, it is aimed to be transformed into more qualified individuals. In this study, two major changes have been made in the structure of standard ABC algorithm. The main purpose of these changes is to make the scout bee stage more effective. With this modification, the convergence speed and solution quality of the algorithm are improved.

To make the scout bee stage more efficient, a strategy has been proposed to determine the *limit* control parameter. In the standard ABC algorithm, *limit* value is fixed throughout all iterations. This causes to go the scout bee stage less frequently to produce a new solution. *limit* value is adaptively determined by using (4) to prevent this.

$$limit = 1 + w \times D \times FoodNumber \times ((maxCycle - iter)/maxCycle) \tag{4}$$

Here, the maximum value of *limit* value is 1 + D × *FoodNumber*. maxCycle is the maximum number of iterations, and iter represents the number of current iterations. The limit value is adaptively adjusted according to the number of iterations. Initially, *(maxCycle−iter)/maxCycle* gets the maximum value and *limit* has the greatest value in the same way. In fact, the maximum value of *limit* is adaptively adjusted here. At each iteration, *limit* value changes. *limit* is set according to the value of w. w is a random number in the range [0,1]. It is ensured that the value of the *limit* is within the range [1, $Upper_{limit}$]. $Upper_{limit}$ is found by using (5).

$$Upper_{limit} = 1 + D \times FoodNumber \times ((maxCycle - iter)/maxCycle) \tag{5}$$

The scout bee stage becomes more effective with the change in *limit*. In this case, an effective solution-generating mechanism is needed to obtain more qualified solutions. The main purpose here is to continue with a more effective solution than the previous solution. Therefore, the solution-generating mechanism given in (6) is proposed.

$$v_{ij} = \begin{cases} x_{ij}\gamma + x_j^g(1-\gamma), & r1 < r2 \\ x_{ij} + x_{ij} \times \delta, & \text{other} \end{cases} \tag{6}$$

$$r2 = rand + ((maxCycle - iter)/maxCycle) \tag{7}$$

Here, x^g is the global best solution. $r1$ is a random number in the range [0,1]. $r2$ is determined by (7). γ is the arithmetic crossover rate and is calculated randomly. δ is the step size and is accepted as 0.01. Arithmetic crossover is applied between the current solution and the global best solution. In other words, the quality of the current solution is being improved by approximating the global best solution. The value of $r2$ is randomly generated depending on the number of iterations. Thus, the related solution closes global best solution at first. In this way, the local convergence speed of the algorithm increases. Three different preventions are taken to prevent locally minimal risk. First, the arithmetic crossover rate is randomly selected. Secondly, when not $r1 < r2$, a new solution is produced in the neighborhood of the current solution. This is achieved via the step size (δ). In particular, in high iterations, outside of global best solution, new solutions are produced according to the current solution. Therefore, quality solutions can even be obtained in high iterations. The third is the possibility of updating the current and global best solution in employed and onlooker bee stages. At the same time, it has been ensured that the new solutions are different from the global best solution with three different preventions.

In summary, adaptive adjustment of *limit* value is provided. The effectiveness of the scout bee stage is increased with new *limit* calculation method. A new solution-generating mechanism is proposed for the scout bee stage. In this way, the local convergence speed of the algorithm is increased, and it is provided that better quality solutions are obtained.

3.3. Training Feed Forward Artificial Neural Networks

Artificial neural Networks (ANNs) are one of the artificial intelligence techniques. ANNs consist of the interconnection of artificial neurons. Figure 1 shows the general structure of an artificial neuron. An artificial neuron consists of inputs, weights, bias value, activation and transfer function. In this way, an output is obtained from the inputs of neurons. The output of a neuron is calculated using (8). x is the input value. w are the weight values corresponding to the input. b is bias value. f is the activation function. y corresponds to the output of the artificial neuron.

$$y = f\left(\sum_{i=1}^{m} w_i x_i + b\right) \tag{8}$$

A FFNN consists of 3 layers as input, hidden and output. In FFNN, calculations specified in (8) are performed in each neuron. In this way, each neuron affects the neurons in the next layer. There is no interaction in the same layer. In FFNN, output is obtained corresponding to the input values. This is only possible by creating a model for a related problem. For this, the network needs to be trained. Training the network is the process of determining the weights and bias values. Training algorithms are used for this. One of the learning methods is learning with samples. Training dataset is required for this. It reflects the characteristics of the network and the network learns in the training process. The learning level of the network is related to the error value. Error value refers to the relationship between the real output and predicted output. A low error value is very important for a successful training process. For low error value, an effective training algorithm is required.

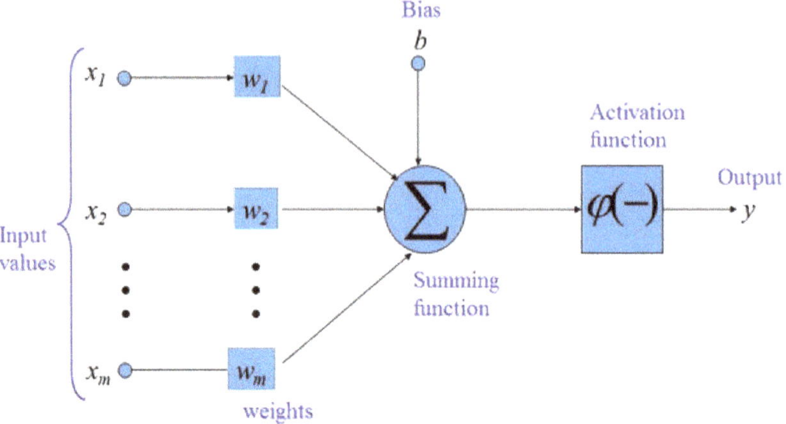

Figure 1. General structure of an artificial neuron.

4. Simulation Results

4.1. Solution of Global Optimization Problems

In applications, 13 numerical test problems are used to analyze the performance of ABCES algorithm. The related problems are given in Table 1. For ABC and ABCES algorithms, population size is taken as 50. The results are obtained for different values of D $\in$ {50,100,150,1000}. The number of evaluations has been used 100,000, 500,000, 1,000,000 values. Each application is run 30 times. Each initial population is determined randomly.

Table 1. Benchmark functions used in experiments.

Function	Formulation
SumSquares	$f(x) = \sum_{i=1}^{n} ix_i^2$
Levy	$f(x) = \sin^2(\pi w_1) + \sum_{i=1}^{n-1}(w_i - 1)^2[1 + 10\sin^2(\pi w_i + 1)] + (w_d - 1)^2 + [1 + \sin^2(2\pi w_d$ $w_i = 1 + \frac{x_i - 1}{4}$
Rosenbrock	$f(x) = \sum_{i=1}^{n-1}[100(x_{i+1} - x_i^2)^2 + (x_i - 1)^2]$
The Sum of Different Powers	$f(x) = \sum_{i=1}^{n}\lvert x_i\rvert^{i+1}$
Zakharov	$f(x) = \sum_{i=1}^{n} x_i^2 + (\sum_{i=1}^{n} 0.5ix_i)^2 + (\sum_{i=1}^{n} 0.5ix_i)^4$
Ackley	$f(x) = -20\exp\left(-0.2\sqrt{\frac{1}{n}\sum_{i=1}^{n} x_i^2}\right) - \exp\left(\frac{1}{n}\sum_{i=1}^{n}\cos(2\pi x_i)\right) + 20 + e$
Step	$f(x) = \sum_{i=1}^{n}(\lfloor x_i + 0.5\rfloor)^2$
Rastrigin	$f(x) = \sum_{i=1}^{n}\left[x_i^2 - 10\cos(2\pi x_i) + 10\right]$
Griewank	$f(x) = \frac{1}{4000}\sum_{i=1}^{n} x_i^2 + \prod_{i=1}^{n}\cos\left(\frac{x_i}{\sqrt{i}}\right) + 1$
Rotated Hyper-Ellipsoid	$f(x) = \sum_{i=1}^{n}\sum_{j=1}^{i} x_j^2$
Dixon–Price	$f(x) = (x_1 - 1)^2 + \sum_{i=2}^{n} i(2x_i^2 - x_{i-1})^2$
Perm	$f(x) = \sum_{k=1}^{n}\left[\sum_{i=1}^{n}(i^k + \beta)\left(\left(\frac{x_i}{i}\right)^k - 1\right)\right]^2$
Sphere	$f(x) = \sum_{i=1}^{n} x_i^2$

The results found with ABC and ABCES algorithms in 100,000 evaluations are given in Table 2. 13 test functions are used, and the results are obtained for D = {50, 100, 150} in each function. In addition to the objective function and standard deviation values are given here. When D is 50, 100 and 150, ABCES algorithm has better results than ABC algorithm in SumSquares, Levy, Sphere, Rosenbrock, The Sum of Different Powers, Zakharov, Ackley, Step, Griewank, Rotated Hyper-Ellipsoid, Dixon–Price and Perm. ABC algorithm is only successful in Rastrigin. Apart from average objective function value, ABCES algorithm is more successful in the standard deviation values. This shows that the results obtained by using ABCES in 100,000 evaluations are more robust. The Wilcoxon signed rank test is used to determine the significance of the results and it is given in Table 3. The evaluation is made according to $p = 0.05$ level. 13 test functions are evaluated in 3 different dimensions (D = 50, 100, 150). Specifically, the significance of 39 results is examined. A significant difference is found in favor of the ABCES algorithm in 34 of these. This result indicates that ABCES algorithm is better in 34 objective function value. There is no significant difference in 4 results. In only one result, there is significant difference indicating that ABC algorithm is more good.

Table 2. Comparison of the results obtained by using ABC and ABCES (D: Dimension, Mean: Mean Values, SD: Standard Deviation, Population Size = 50, Number of Evaluation = 100,000.)

No	Function	Range	D	ABC		ABCES (Proposed)	
				Mean	SD	Mean	SD
1	SumSquares	[−10, 10]	50	9.61×10^{-10}	1.94×10^{-9}	2.25×10^{-14}	2.65×10^{-14}
			100	7.87×10^{-3}	3.96×10^{-2}	4.75×10^{-6}	4.74×10^{-6}
			150	9.90×10^{-1}	2.60	5.80×10^{-4}	4.62×10^{-4}
2	Levy	[−10, 10]	50	1.28×10^{-10}	3.66×10^{-10}	7.69×10^{-11}	1.29×10^{-10}
			100	4.97×10^{-4}	1.26×10^{-3}	1.63×10^{-5}	1.55×10^{-5}
			150	1.57×10^{-2}	3.82×10^{-2}	1.15×10^{-3}	8.41×10^{-4}
3	Sphere	[−100, 100]	50	3.21×10^{-9}	7.92×10^{-9}	1.05×10^{-13}	1.10×10^{-13}
			100	3.53×10^{-2}	6.87×10^{-2}	5.71×10^{-6}	4.65×10^{-6}
			150	7.24×10^{-1}	2.19	9.15×10^{-4}	1.10×10^{-3}
4	Rosenbrock	[−30, 30]	50	4.55	5.63	2.33	2.70
			100	2.10×10^{2}	1.16×10^{2}	1.45×10^{2}	6.80×10^{1}
			150	6.69×10^{2}	4.65×10^{2}	4.63×10^{2}	3.54×10^{2}
5	The Sum of Different Powers	[−1, 1]	50	4.48×10^{-17}	1.92×10^{-17}	2.65×10^{-18}	2.24×10^{-18}
			100	3.34×10^{-11}	1.16×10^{-10}	1.06×10^{-15}	1.50×10^{-15}
			150	1.98×10^{-7}	7.54×10^{-7}	1.83×10^{-11}	4.07×10^{-11}
6	Zakharov	[−10, 10]	50	1.12×10^{3}	9.81×10^{1}	1.35×10^{2}	3.16×10^{1}
			100	2.63×10^{3}	1.32×10^{2}	3.50×10^{2}	5.97×10^{1}
			150	4.31×10^{3}	1.78×10^{2}	5.83×10^{2}	1.28×10^{2}
7	Ackley	[−32, 32]	50	3.06×10^{-5}	4.98×10^{-5}	8.56×10^{-7}	3.71×10^{-7}
			100	2.46×10^{-1}	3.39×10^{-1}	2.70×10^{-2}	1.94×10^{-2}
			150	2.01	4.99×10^{-1}	1.25	3.40×10^{-1}
8	Step	[−100, 100]	50	6.03×10^{-9}	1.82×10^{-8}	2.35×10^{-9}	4.55×10^{-9}
			100	2.80×10^{-2}	7.90×10^{-2}	1.37×10^{-4}	9.29×10^{-5}
			150	8.46×10^{-1}	2.30	4.25×10^{-3}	2.10×10^{-3}
9	Rastrigin	[−5.12, 5.12]	50	7.89×10^{-2}	2.54×10^{-1}	1.89×10^{-1}	4.85×10^{-1}
			100	1.44×10^{1}	4.76	1.67×10^{1}	4.30
			150	5.57×10^{1}	7.85	6.60×10^{1}	7.93
10	Griewank	[−600, 600]	50	2.54×10^{-5}	1.00×10^{-4}	3.86×10^{-11}	9.49×10^{-11}
			100	3.43×10^{-2}	5.47×10^{-2}	2.84×10^{-3}	5.73×10^{-3}
			150	2.65×10^{-1}	3.14×10^{-1}	4.39×10^{-2}	3.95×10^{-2}
11	Rotated Hyper-Ellipsoid	[−65536, 65536]	50	2.03×10^{-4}	8.46×10^{-4}	2.15×10^{-7}	3.50×10^{-7}
			100	4.73×10^{5}	1.89×10^{6}	6.54	5.38
			150	2.81×10^{7}	8.16×10^{7}	2.16×10^{4}	2.54×10^{4}
12	Dixon–Price	[−10, 10]	50	4.71×10^{-2}	1.20×10^{-1}	4.51×10^{-2}	2.53×10^{-2}
			100	2.42×10^{1}	1.43×10^{1}	1.48×10^{1}	8.03
			150	7.48×10^{1}	3.42×10^{1}	5.92×10^{1}	1.68×10^{1}
13	Perm	[−4, 4]	2	2.54×10^{-9}	5.40×10^{-9}	5.50×10^{-12}	1.41×10^{-11}
			4	1.42×10^{-1}	1.01×10^{-1}	9.35×10^{-3}	2.71×10^{-2}
			6	2.01×10^{3}	4.18×10^{3}	1.51×10^{1}	2.20×10^{1}

Table 3. Wilcoxon signed rank test results between standard ABC and ABCES (Population Size = 50, Number of Evaluation = 100,000).

No	Function	Range	D	Statistical Results p-Value	Significance
			50	0.000	+
1	SumSquares	[−10, 10]	100	0.000	+
			150	0.000	+
			50	0.041	+
2	Levy	[-10, 10]	100	0.002	+
			150	0.165	-
			50	0.000	+
3	Sphere	[−100, 100]	100	0.000	+
			150	0.000	+
			50	0.021	+
4	Rosenbrock	[−30, 30]	100	0.019	+
			150	0.019	+
			50	0.000	+
5	The Sum of Different Powers	[−1, 1]	100	0.000	+
			150	0.000	+
			50	0.000	+
6	Zakharov	[−10, 10]	100	0.000	+
			150	0.000	+
			50	0.000	+
7	Ackley	[−32, 32]	100	0.000	+
			150	0.000	+
			50	0.090	-
8	Step	[−100, 100]	100	0.043	+
			150	0.003	+
			50	0.192	-
9	Rastrigin	[−5.12, 5.12]	100	0.131	-
			150	0.000	*
			50	0.000	+
10	Griewank	[−600, 600]	100	0.000	+
			150	0.000	+
			50	0.000	+
11	Rotated Hyper-Ellipsoid	[−65536, 65536]	100	0.000	+
			150	0.000	+
			50	0.006	+
12	Dixon–Price	[−10, 10]	100	0.002	+
			150	0.024	+
			2	0.000	+
13	Perm	[−4, 4]	4	0.000	+
			6	0.000	+

When the results are obtained in 100,000 evaluations are examined, it is seen that fast convergence continues in the problems. Therefore, it is determined that better results can be achieved in high iteration. Thus, the results found in 500,000 evaluations are given in Table 4. In 500,000 evaluations, the quality of the solutions has improved at a high rate according to 100,000 evaluations in Levy, Sphere, The Sum of Different Powers, Zakharov, Ackley, Step, Rastrigin, Griewank and Rotated Hyper-Ellipsoid functions. The objective function values are 10^{-15} and below in all dimensions (D = 50, 100, 150) in SumSquares, Sphere, The Sum of Different Powers, Griewank and Rotated Hyper-Ellipsoid functions. The results obtained for D = 50 are 10^{-15} and below in Levy, Rastrigin and Step functions. They are between 10^{-10} and 10^{-15} for D = 100 and D = 150. Similarly, the objective function value obtained in Ackley function is between 10^{-10} and 10^{-15}. In Perm function, the dimensions affected to the results a lot. Although the objective function value is about 10^{-13} in D = 2, it is about 10^{-2} in D = 4. Along with that, it is about 5 in D = 6. In Rosenbrock function, objective function values between 0.1 and 1 are obtained. In Zakharov function, they are between 51 and 312. This function has the highest objective function value. At the same time, Table 4 compares ABC and ABCDE algorithms. ABC algorithm is only better in Levy, Step and Dixon–Price functions. In other problems, the ABCES algorithm is more successful than the ABC algorithm. Although ABCES has better results in Levy and Step functions in 100,000 evaluations, this situation has changed in favor of the ABC algorithm in 500,000 evaluations. In Rastrigin function, while ABC algorithm is more successful in 100,000 evaluations, ABCES algorithm is better in 500,000 evaluations. The Wilcoxon signed rank test is performed between ABC and ABCES to determine the significance of the results obtained in 500,000 evaluations and it is given Table 5. The analyses are performed according to $p = 0.05$ level. The significance of 39 objective function values is examined. In 25 of them, a significant difference is obtained with ABCES algorithm. This result shows that ABCES algorithm is more successful than with ABC algorithm in these functions. ABC algorithm is only better in 8 of them. These results belong to Levy, Step and Dixon–Price functions which ABC algorithm is effective. In the remaining 6 results, no significant difference is found between ABC and ABCES. Despite ABCES is especially better in Rosenbrock function, it is not significant. Also, as in 100,000 evaluations, the best standard deviation values are generally obtained by ABCES algorithm in 500,000 evaluations.

Table 4. Comparison of the results obtained by using ABC and ABCES (D: Dimension, Mean: Mean Values, SD: Standard Deviation, Population Size = 50, Number of Evaluation = 500,000).

No	Function	Range	D	ABC		ABCES (Proposed)	
				Mean	SD	Mean	SD
			50	8.81×10^{-16}	9.93×10^{-17}	1.63×10^{-16}	3.20×10^{-17}
1	SumSquares	[−10, 10]	100	2.12×10^{-15}	1.96×10^{-16}	4.08×10^{-16}	5.96×10^{-17}
			150	3.72×10^{-15}	3.94×10^{-16}	8.55×10^{-16}	1.32×10^{-16}
			50	8.63×10^{-16}	9.66×10^{-17}	1.71×10^{-15}	8.02×10^{-16}
2	Levy	[−10, 10]	100	2.14×10^{-15}	1.89×10^{-16}	1.03×10^{-13}	1.73×10^{-13}
			150	3.59×10^{-15}	2.05×10^{-16}	6.61×10^{-12}	1.82×10^{-11}
			50	9.10×10^{-16}	9.92×10^{-17}	1.64×10^{-16}	2.95×10^{-17}
3	Sphere	[−100, 100]	100	2.10×10^{-15}	2.19×10^{-16}	3.97×10^{-16}	5.60×10^{-17}
			150	3.88×10^{-15}	3.57×10^{-16}	8.99×10^{-16}	1.42×10^{-16}
			50	2.80×10^{-1}	7.03×10^{-1}	1.61×10^{-1}	4.09×10^{-1}
4	Rosenbrock	[−30, 30]	100	4.52×10^{-1}	7.50×10^{-1}	1.84×10^{-1}	2.39×10^{-1}
			150	1.31	2.35	9.09×10^{-1}	1.91
			50	2.52×10^{-17}	7.33×10^{-18}	9.33×10^{-19}	1.44×10^{-18}
5	The Sum of Different Powers	[−1, 1]	100	4.43×10^{-17}	1.29×10^{-17}	2.69×10^{-18}	4.30×10^{-18}
			150	5.87×10^{-17}	1.85×10^{-17}	3.73×10^{-18}	4.33×10^{-18}
			50	9.24×10^{2}	9.83×10^{1}	5.19×10^{1}	1.57×10^{1}
6	Zakharov	[−10, 10]	100	2.51×10^{3}	1.19×10^{2}	1.88×10^{2}	2.45×10^{1}
			150	4.07×10^{3}	2.04×10^{2}	3.12×10^{2}	5.00×10^{1}
			50	6.59×10^{-14}	5.36×10^{-15}	3.26×10^{-14}	2.89×10^{-15}
7	Ackley	[−32, 32]	100	1.57×10^{-13}	1.03×10^{-14}	7.61×10^{-14}	7.23×10^{-15}
			150	3.34×10^{-10}	4.23×10^{-10}	2.47×10^{-12}	5.58×10^{-13}
			50	8.88×10^{-16}	1.10×10^{-16}	2.00×10^{-15}	1.50×10^{-15}
8	Step	[−100, 100]	100	2.12×10^{-15}	2.25×10^{-16}	6.98×10^{-14}	1.17×10^{-13}
			150	3.76×10^{-15}	2.98×10^{-16}	5.13×10^{-12}	1.18×10^{-11}
			50	0	0	0	0
9	Rastrigin	[−5.12, 5.12]	100	1.14×10^{-13}	7.19×10^{-14}	4.93×10^{-14}	6.35×10^{-14}
			150	2.38×10^{-12}	4.09×10^{-12}	1.65×10^{-12}	6.54×10^{-13}
			50	9.99×10^{-17}	1.66×10^{-16}	0	0
10	Griewank	[−600, 600]	100	4.22×10^{-16}	3.44×10^{-16}	0	0
			150	1.60×10^{-15}	1.05×10^{-15}	2.26×10^{-16}	2.52×10^{-16}
			50	8.83×10^{-16}	1.12×10^{-16}	1.60×10^{-16}	3.63×10^{-17}
11	Rotated Hyper-Ellipsoid	[−65536, 65536]	100	2.13×10^{-15}	2.29×10^{-16}	4.29×10^{-16}	5.87×10^{-17}
			150	1.95×10^{-12}	3.39×10^{-12}	3.24×10^{-15}	6.81×10^{-16}
			50	7.82×10^{-10}	1.37×10^{-9}	3.05×10^{-5}	2.06×10^{-5}
12	Dixon–Price	[−10, 10]	100	2.00×10^{-5}	1.92×10^{-5}	2.03×10^{-3}	1.07×10^{-3}
			150	1.60×10^{-1}	5.82×10^{-1}	2.00×10^{-2}	8.78×10^{-3}
			2	5.97×10^{-12}	2.38×10^{-11}	1.09×10^{-13}	2.58×10^{-13}
13	Perm	[−4, 4]	4	3.39×10^{-2}	2.65×10^{-2}	2.51×10^{-2}	4.81×10^{-2}
			6	2.80×10^{2}	2.83×10^{2}	4.91	8.31

Table 5. Wilcoxon signed rank test results between standard ABC and ABCES (Population Size = 50, Number of Evaluation = 500,000).

No	Function	Range	D	Statistical Results	
				p-Value	Significance
			50	0.000	+
1	SumSquares	$[-10, 10]$	100	0.000	+
			150	0.000	+
			50	0.000	*
2	Levy	$[-10, 10]$	100	0.000	*
			150	0.000	*
			50	0.000	+
3	Sphere	$[-100, 100]$	100	0.000	+
			150	0.000	+
			50	0.271	-
4	Rosenbrock	$[-30, 30]$	100	0.111	-
			150	0.688	-
			50	0.000	+
5	The Sum of Different Powers	$[-1, 1]$	100	0.000	+
			150	0.000	+
			50	0.000	+
6	Zakharov	$[-10, 10]$	100	0.000	+
			150	0.000	+
			50	0.000	+
7	Ackley	$[-32, 32]$	100	0.000	+
			150	0.000	+
			50	0.000	*
8	Step	$[-100, 100]$	100	0.000	*
			150	0.000	*
			50	1.000	-
9	Rastrigin	$[-5.12, 5.12]$	100	0.002	+
			150	0.940	-
			50	0.000	+
10	Griewank	$[-600, 600]$	100	0.000	+
			150	0.000	+
			50	0.000	+
11	Rotated Hyper-Ellipsoid	$[-65536, 65536]$	100	0.000	+
			150	0.000	+
			50	0.000	*
12	Dixon–Price	$[-10, 10]$	100	0.000	*
			150	0.000	+
			2	0.001	+
13	Perm	$[-4, 4]$	4	0.116	-
			6	0.000	+

It is very important the success that optimization algorithms show in high-dimensional problems. Therefore, the results obtained with ABC and ABCES algorithms are given for D = 1000 on SumSquares, Levy, Sphere, Rosenbrock, The Sum of Different Powers, Zakharov, Ackley, Step, Rastrigin, Griewank, Rotated Hyper-Ellipsoid and Dixon–Price functions in Table 6. ABC algorithm is only better in Rastrigin function. ABCES is more successful in all other problems. In particular, in Rotated Hyper-Ellipsoid function, the objective function value is obtained as 6.40×10^8 by ABC algorithm and no effective solution is found. In contrast, it is achieved as 6.25×10^2 by using ABCES. Other than that, while the success rate of ABC algorithm on SumSquares, Sphere and The Sum of Different Powers functions is low, more effective results are obtained with ABCES algorithm. The Wilcoxon signed rank test is used to determine whether the results are significant, and it is given in Table 7. The analyses are performed according to $p = 0.05$ level. The significance status for 12 functions is examined. In 8 of them, a significant difference is found in favor of ABCES. In only one function, a significant difference is obtained with ABC algorithm. No significant difference is found in other functions. In addition, in all functions, the best standard deviation values are achieved by using ABCES. When the results given in Tables 6 and 7 are evaluated, they show that ABCES algorithm is better than ABC algorithm on high-dimensional problems.

Table 6. Comparison of the results obtained by using ABC and ABCES (D: Dimension, Mean: Mean Values, SD: Standard Deviation, Population Size = 50, Number of Evaluation = 1,000,000).

No	Function	Range	D	ABC		ABCES (Proposed)	
				Mean	SD	Mean	SD
1	SumSquares	[−10, 10]	1000	1.33	3.11	2.19×10^{-5}	1.14×10^{-5}
2	Levy	[−10, 10]		2.46×10^{-3}	4.85×10^{-3}	1.35×10^{-4}	1.41×10^{-4}
3	Sphere	[−100, 100]		1.36×10^{-1}	3.13×10^{-1}	6.82×10^{-6}	2.62×10^{-6}
4	Rosenbrock	[−30, 30]		2.34×10^{3}	1.18×10^{3}	1.62×10^{3}	3.11×10^{2}
5	The Sum of Different Powers	[−1, 1]		2.22×10^{-9}	9.43×10^{-9}	3.60×10^{-16}	6.73×10^{-16}
6	Zakharov	[−10, 10]		3.16×10^{4}	4.80×10^{2}	2.38×10^{3}	2.69×10^{2}
7	Ackley	[−32, 32]		6.05×10^{-1}	1.28	5.95×10^{-2}	3.13×10^{-2}
8	Step	[−100, 100]		5.76×10^{-2}	1.30×10^{-1}	4.63×10^{-4}	3.14×10^{-4}
9	Rastrigin	[−5.12, 5.12]		1.86×10^{2}	2.02×10^{2}	2.20×10^{2}	2.16×10^{1}
10	Griewank	[−600, 600]		5.60×10^{-2}	1.13×10^{-1}	4.45×10^{-3}	2.01×10^{-2}
11	Rotated Hyper-Ellipsoid	[−65536, 65536]		6.40×10^{8}	2.97×10^{9}	6.25×10^{2}	2.68×10^{2}
12	Dixon–Price	[−10, 10]		2.23×10^{3}	7.39×10^{2}	1.96×10^{3}	3.67×10^{2}

Table 7. Wilcoxon signed rank test results between standard ABC and ABCES (D: Dimension, Mean: Mean Values, SD: Standard Deviation, Population Size = 50, Number of Evaluation = 1,000,000).

No	Function	Range	D	Statistical Results	
				p-Value	Significance
1	SumSquares	[−10, 10]	1000	0.000	+
2	Levy	[−10, 10]		0.082	-
3	Sphere	[−100, 100]		0.000	+
4	Rosenbrock	[−30, 30]		0.002	+
5	The Sum of Different Powers	[−1, 1]		0.000	+
6	Zakharov	[−10, 10]		0.000	+
7	Ackley	[−32, 32]		0.000	+
8	Step	[−100, 100]		0.339	-
9	Rastrigin	[−5.12, 5.12]		0.000	*
10	Griewank	[−600, 600]		0.000	+
11	Rotated Hyper-Ellipsoid	[−65536, 65536]		0.000	+
12	Dixon–Price	[−10, 10]		0.131	-

Comparison of GA, PSO, DE, ABC and ABCES algorithms is given in Table 8. In the comparison, SumSquares, Sphere, Rosenbrock, Zakharov, Ackley, Step, Rastrigin, Griewank, Dixon–Price and Perm functions are used. Results of GA, PSO, DE and ABC algorithm are taken from [44]. The results are given for population/colony size is 50 and number of evaluations is 500,000. In addition, values below 10^{-12} in [44] are assumed as 0 (zero). For fair comparison, values below 10^{-12} are accepted as 0 (zero) in ABCES algorithm too. When the related table is analyzed, 0 (zero) are obtained with PSO, DE, ABC and ABCES algorithms in SumSquares, Sphere, Step functions. Algorithms other than GA and ABC reach 0 (zero) value in Zakharov function. Also, ABC and ABCES algorithms find 0 (zero) value in Ackley function. The best results for Rastrigin, Griewank and Dixon–Price functions are achieved with ABC and ABCES algorithms. In addition, the best results for Rosenbrock and Perm are obtained by using ABCES Algorithm. These results given in Table 8 show that ABCES algorithm is generally more successful than GA, PSO, DE, and ABC algorithm.

Table 8. Statistical results of 30 runs obtained by GA, PSO, DE, ABC and ABCES [44].

Function	D	Range	Algorithm	Results	
				Mean	Std.
SumSquares	30	[−10, 10]	GA	1.48×10^2	1.24×10^1
			PSO	0	0
			DE	0	0
			ABC	0	0
			ABCES (Proposed)	0	0
Sphere	30	[−100, 100]	GA	1.11×10^3	7.42×10^1
			PSO	0	0
			DE	0	0
			ABC	0	0
			ABCES (Proposed)	0	0
Rosenbrock	30	[−30, 30]	GA	1.96×10^5	3.85×10^4
			PSO	1.52×10^1	2.42×10^1
			DE	1.82×10^1	5.04
			ABC	8.87×10^{-2}	7.74×10^{-2}
			ABCES (Proposed)	5.36×10^{-2}	1.62×10^{-1}
Zakharov	10	[−5, 10]	GA	1.34×10^{-2}	4.53×10^{-3}
			PSO	0	0
			DE	0	0
			ABC	2.48×10^{-4}	1.83×10^{-4}
			ABCES (Proposed)	0	0
Ackley	30	[−32, 32]	GA	1.47×10^1	1.78×10^{-1}
			PSO	4.94×10^{-1}	9.02×10^{-2}
			DE	0	0
			ABC	0	0
			ABCES (Proposed)	0	0
Step	30	[−100, 100]	GA	1.17×10^3	7.66×10^1
			PSO	0	0
			DE	0	0
			ABC	0	0
			ABCES (Proposed)	0	0
Rastrigin	30	[−5.12, 5.12]	GA	5.29×10^1	4.56
			PSO	4.40×10^1	1.17×10^1
			DE	1.17×10^1	2.54
			ABC	0	0
			ABCES (Proposed)	0	0
Griewank	30	[−600, 600]	GA	1.06×10^1	1.16
			PSO	1.74×10^{-2}	2.08×10^{-2}
			DE	1.48×10^{-3}	2.96×10^{-3}
			ABC	0	0
			ABCES (Proposed)	0	0
Dixon–Price	30	[−10, 10]	GA	1.22×10^3	2.66×10^2
			PSO	10^{-8}	1.83×10^{-9}
			DE	10^{-9}	1.83×10^{-10}
			ABC	0	0
			ABCES (Proposed)	0	0
Perm	4	[−4, 4]	GA	3.03×10^{-1}	1.93×10^{-1}
			PSO	3.61×10^{-2}	4.89×10^{-2}
			DE	2.40×10^{-2}	4.60×10^{-2}
			ABC	4.11×10^{-2}	2.31×10^{-2}
			ABCES (Proposed)	6.20×10^{-3}	2.23×10^{-2}

4.2. Training Neural Networks with ABCES Algorithm for the Identification of Nonlinear Static Systems

In this section, the performance of ABCES algorithm is assessed on neural network training for the identification of nonlinear static systems. In the applications, 6 nonlinear static systems ($S_1, S_2, S_3, S_4, S_5, S_6$) given in Table 9 are used. S_1 has one input. S_2 and S_3 consist of two inputs. S_4 and S_5 have three inputs. S_6 has four inputs. Datasets are created using the equations given here. For S_1, S_2 and S_3, y output value is obtained by using the input value(s) in the range of [0, 1]. The dataset contains 100 data for the first 3 systems. 80% of the dataset is used for training process and the rest is used for testing. The input values are in the range of [1,6] for S_4. A dataset consisting of 216 data is created using 6 values for each input. 173 data points of the dataset belong to the training process. The rest are chosen for testing. A dataset with 125 data is created using related equation in S_5. Input values are in the range of [0,1]. For S_6, input values are used in the range of $[-0.25,0.25]$ and a dataset consisting of 125 data is created. In S_5 and S_6, 100 data points are used for the training process. The rest are chosen for testing. According to the dataset index value (i), mod (i, 5) = k operation is applied in all systems. If k = 0, the data is chosen for testing. Otherwise, it is included to dataset of the training process. There are two reasons for applying the mod operation according to 5 value: The first is to choose 80% of the dataset for the training process. In this case, the rest belong to the test dataset. It is ensured that the training dataset covers the whole dataset. This way, a more effective training process is realized. At the same time, the test dataset reflects the whole system. Feed forward neural network (FFNN) is used in this study. Sigmoid function is used for the neurons in the hidden layer and the output layer. Three different network structures are used for each system. 4, 8 and 12 neurons are used in the hidden layer. Training FFNN is realized via ABCES algorithm. Flow chart of FFNN training based on ABCES algorithm for the identification of nonlinear static systems is presented in Figure 2. Before the training, the input and output pairs of the nonlinear static system are normalized in the range of [0,1]. For ABCES algorithm, population size and maximum number of iterations are taken as 20 and 5000, respectively. The number of training and test data used for each system is given in Table 9. MSE (mean squared error) calculated as in (9) is used as error value for training and testing process. Here, n is the number of samples. y_i is real output and $(\bar{y}_i)$ is predicted output. Each application is run 30 times to analyze it statistically. Mean error value (mean) and standard deviation (std) are obtained.

$$MSE = \frac{1}{n}\sum_{i=1}^{n}(y_i - \bar{y}_i)^2 \tag{9}$$

The results obtained with the ABCES algorithm are presented in Table 10. The increase in the number of neurons in the hidden layer in S_1 has increased the solution quality. The best mean error values for training and test are achieved with the 1-12-1 network structure. The number of neurons affects the mean training and test error values in S_2 differently. Although the best mean training error value is found with 2-12-1, the best mean test error value is obtained with 2-8-1. The low number of neurons in S_3 is more effective. The best mean error values for both training and test are achieved with 2-4-1. Close performance is observed in 3-8-1 and 3-12-1 network structures in S_4. Similarly, the best mean training error values for S_5 are found with 3-8-12 and 3-12-1. However, the best mean test error value is obtained by using 3-4-1. All the best results in S_6 are 4-12-1. When all systems are evaluated in general, it is possible to make four basic comments. First, network structure affects performance. Increasing or decreasing the number of neurons exhibits different behaviors depending on the system. Second, there is a difference between training and test errors. This situation can be explained by the selection of the training and test dataset. Third, generally low standard deviation values are obtained. This situation shows the stability of the solutions. Finally, the low error values found indicate that the ABCES algorithm is successful. In Figure 3, the graphs of the output found with ABCES algorithm and the real output are compared. It is seen that effective output graphics are obtained with ABCES algorithm in all systems. In fact, this is an indication that nonlinear static systems are identified with high accuracy.

It is compared with PSO, HS and ABC algorithm to better evaluate the performance of ABCES algorithm. The results are presented in Table 11. In S_1, the best mean training and test error values are found by ABCES algorithm. ABC algorithm is more effective after ABCES algorithm. The same is true for S_2. The best mean training error value in S_3 is found with ABCES. After ABCES, PSO is more effective. Although the best result in the mean test error value is obtained with ABCES, the worst results are found with HS. In S_4 it is clear that ABCES is effective. In S_5, the best mean training error value is found with ABCES, while the best mean test error value is obtained via PSO. The best results in S_6 are obviously found with ABCES. When the results are evaluated in general, ABCES algorithm is more successful in neural network training than others. After ABCES, the performances are listed as ABC algorithm, PSO and HS, respectively.

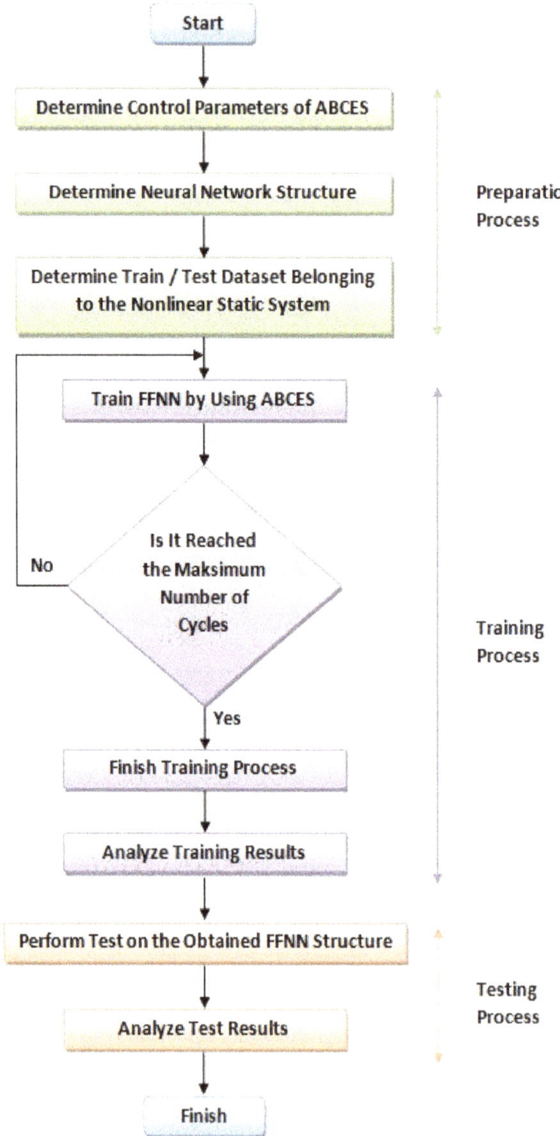

Figure 2. Flowchart for FFNN training based on ABCES algorithm for the identification of nonlinear static systems.

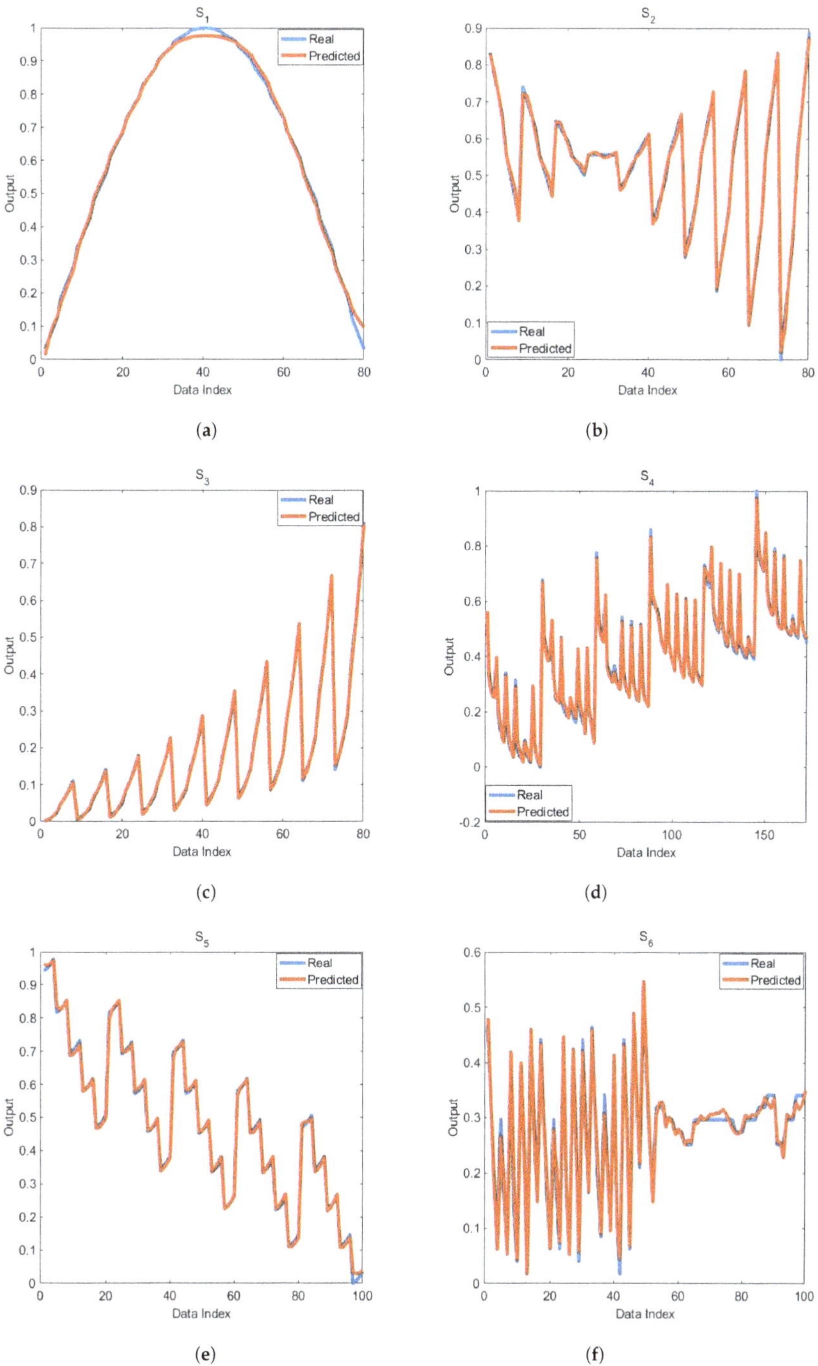

Figure 3. Comparison of real and predicted outputs for (**a**) S_1 (**b**) S_2 (**c**) S_3 (**d**) S_4 (**e**) S_5 (**f**) S_6.

Table 9. Information on nonlinear static systems used.

System	Equation	Inputs	Output	Number of Training / Test Data	Range
S_1	$y = 2\sin(\pi x_1)$	x_1	y	80/20	[0,1]
S_2	$y = 10.391\{(x_1 - 0.4)(x_2 - 0.6) + 0.36\}$	x_1, x_2	y	80/20	[0,1]
S_3	$y = \tanh(x_1 + x_2 - 11)$	x_1, x_2	y	80/20	[0,1]
S_4	$y = 1 + x_1^{0.5} + x_2^{-1} + x_3^{-1.5}$	x_1, x_2, x_3	y	173/43	[1,6]
S_5	$y = (x_1 - 5.5)^2 + (x_2 - 5.5)^2 + x_3^2$	x_1, x_2, x_3	y	100/25	[0,1]
S_6	$y = e^{2x_1 \sin(\pi x_4)} + \sin(x_2 x_3)$	x_1, x_2, x_3, x_4	y	100/25	[−0.25,0.25]

Table 10. Results obtained with ABCES on nonlinear static system identification.

System	Network Structure	Train		Test	
		Mean	Std	Mean	Std
	1-4-1	1.16×10^{-3}	4.86×10^{-4}	2.21×10^{-3}	7.65×10^{-4}
S_1	1-8-1	4.10×10^{-4}	3.53×10^{-4}	1.19×10^{-3}	5.42×10^{-4}
	1-12-1	2.73×10^{-4}	1.64×10^{-4}	1.12×10^{-3}	5.81×10^{-4}
	2-4-1	8.49×10^{-4}	5.09×10^{-4}	5.62×10^{-3}	3.46×10^{-3}
S_2	2-8-1	2.52×10^{-4}	1.03×10^{-4}	2.47×10^{-3}	1.62×10^{-3}
	2-12-1	1.99×10^{-4}	7.06×10^{-5}	2.61×10^{-3}	1.23×10^{-3}
	2-4-1	6.45×10^{-5}	3.64×10^{-5}	2.57×10^{-3}	1.16×10^{-3}
S_3	2-8-1	6.86×10^{-5}	3.74×10^{-5}	2.67×10^{-3}	9.58×10^{-4}
	2-12-1	7.03×10^{-5}	3.33×10^{-5}	3.06×10^{-3}	2.21×10^{-3}
	3-4-1	7.17×10^{-4}	2.13×10^{-4}	1.21×10^{-3}	3.92×10^{-4}
S_4	3-8-1	4.35×10^{-4}	1.87×10^{-4}	9.50×10^{-4}	3.88×10^{-4}
	3-12-1	4.33×10^{-4}	1.67×10^{-4}	1.07×10^{-3}	5.56×10^{-4}
	3-4-1	3.59×10^{-4}	1.74×10^{-4}	3.06×10^{-3}	3.69×10^{-3}
S_5	3-8-1	2.39×10^{-4}	9.59×10^{-5}	6.38×10^{-3}	1.01×10^{-2}
	3-12-1	2.41×10^{-4}	9.06×10^{-5}	6.74×10^{-3}	7.21×10^{-3}
	4-4-1	6.71×10^{-4}	1.82×10^{-4}	9.02×10^{-4}	3.49×10^{-4}
S_6	4-8-1	4.56×10^{-4}	1.46×10^{-4}	8.42×10^{-4}	8.09×10^{-4}
	4-12-1	3.76×10^{-4}	1.19×10^{-4}	6.90×10^{-4}	3.32×10^{-4}

Table 11. Comparison of results found by using GA, PSO, HS, ABC and ABCES for on nonlinear static system identification based on neural network.

System	Network Structure	Train		Test	
		Mean	Std	Mean	Std
	PSO	2.08×10^{-3}	6.86×10^{-4}	3.26×10^{-3}	1.17×10^{-3}
S_1	HS	1.29×10^{-2}	7.38×10^{-3}	1.44×10^{-2}	8.17×10^{-3}
	ABC	5.86×10^{-4}	1.72×10^{-4}	1.22×10^{-3}	4.20×10^{-4}
	ABCES (Proposed)	2.73×10^{-4}	1.64×10^{-4}	1.12×10^{-3}	5.81×10^{-4}
	PSO	1.95×10^{-3}	8.65×10^{-4}	7.56×10^{-3}	3.55×10^{-3}
S_2	HS	2.50×10^{-2}	1.00×10^{-2}	3.94×10^{-2}	1.82×10^{-2}
	ABC	5.53×10^{-4}	2.07×10^{-4}	4.19×10^{-3}	3.30×10^{-3}
	ABCES (Proposed)	1.99×10^{-4}	7.06×10^{-5}	2.61×10^{-3}	1.23×10^{-3}
	PSO	8.17×10^{-5}	3.35×10^{-5}	3.74×10^{-3}	1.29×10^{-3}
S_3	HS	1.21×10^{-3}	7.68×10^{-4}	8.43×10^{-3}	4.73×10^{-3}
	ABC	2.60×10^{-4}	9.02×10^{-5}	3.56×10^{-3}	1.70×10^{-3}
	ABCES (Proposed)	6.45×10^{-5}	3.64×10^{-5}	2.57×10^{-3}	1.16×10^{-3}
	PSO	1.96×10^{-3}	9.17×10^{-4}	2.31×10^{-3}	7.28×10^{-4}
S_4	HS	9.28×10^{-3}	8.00×10^{-3}	8.71×10^{-3}	7.41×10^{-3}
	ABC	1.21×10^{-3}	2.97×10^{-4}	1.84×10^{-3}	4.87×10^{-4}
	ABCES (Proposed)	4.35×10^{-4}	1.87×10^{-4}	9.50×10^{-4}	3.88×10^{-4}
	PSO	4.88×10^{-4}	3.38×10^{-4}	2.16×10^{-3}	4.01×10^{-3}
S_5	HS	2.94×10^{-3}	1.45×10^{-3}	8.58×10^{-3}	5.24×10^{-3}
	ABC	8.62×10^{-4}	2.69×10^{-4}	6.30×10^{-3}	6.43×10^{-3}
	ABCES (Proposed)	3.59×10^{-4}	1.74×10^{-4}	3.06×10^{-3}	3.69×10^{-3}
	PSO	2.72×10^{-3}	1.44×10^{-3}	2.61×10^{-3}	1.33×10^{-3}
S_6	HS	1.58×10^{-2}	7.14×10^{-3}	1.65×10^{-2}	9.56×10^{-3}
	ABC	9.91×10^{-4}	2.63×10^{-4}	1.58×10^{-3}	1.23×10^{-3}
	ABCES (Proposed)	3.76×10^{-4}	1.19×10^{-4}	6.90×10^{-4}	3.32×10^{-4}

In Table 11, it is seen that the solution quality of ABCES algorithm is better than other algorithms. Besides the quality of the solution, the convergence speed is also important.

Therefore, the convergence graphs of PSO, HS, ABC and ABCES on all systems are compared in Figure 4. It is observed that the convergence of ABCES algorithm is more effective on all systems. These graphics show that ABCES algorithm has better convergence speed than other algorithms. After the ABCES algorithm, the best convergence is achieved with the ABC algorithm, except S_3 and S_5. PSO has a more effective convergence than the ABC algorithm on S_3 and S_5.

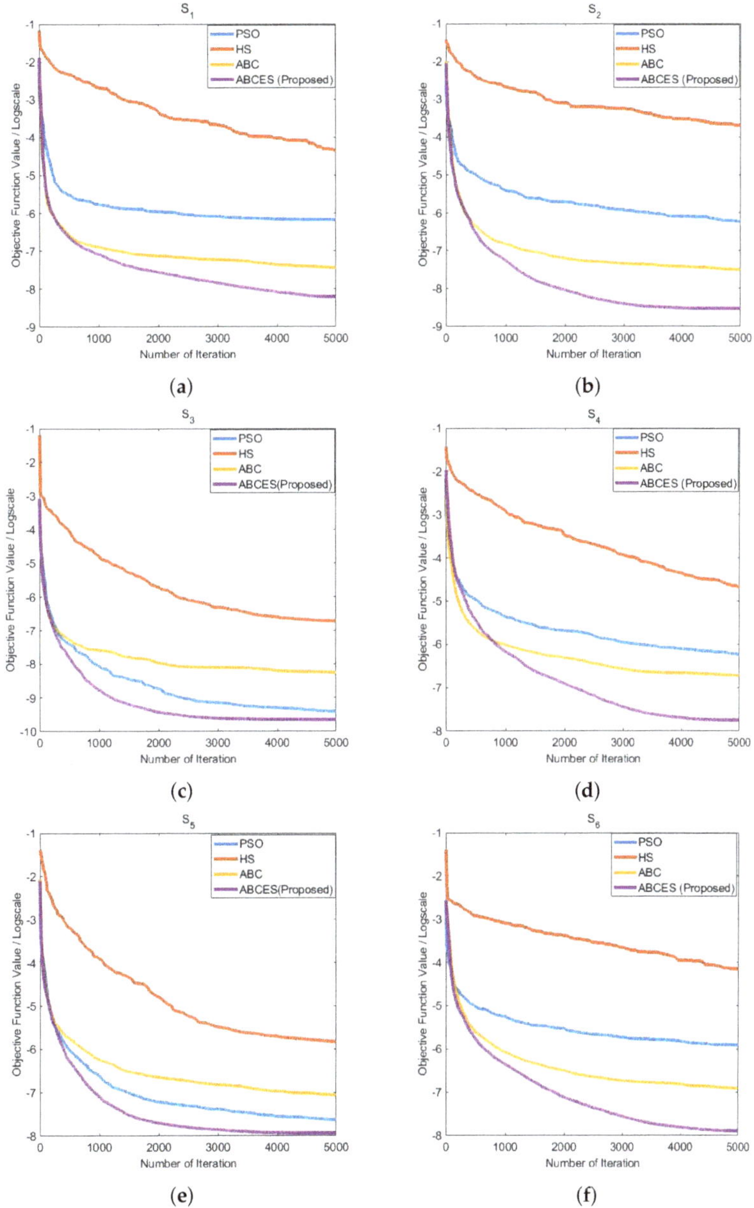

Figure 4. Comparison of convergences of PSO, HS, ABC and ABCES on (**a**) S_1 (**b**) S_2 (**c**) S_3 (**d**) S_4 (**e**) S_5 (**f**) S_6.

5. Discussion

ABCES algorithm generates new solutions by using the information of previous solutions instead of random solution in the scout bee stage. "Limit" value is not fixed and is determined adaptively according to the number of iterations. How these changes affect the performance of ABCES algorithm is examined on two different problem groups: global optimization problems and FFNN training for the identification of nonlinear static systems.

The proposal of a new solution generation mechanism for the scout bee stage has been effective in solving global optimization problems. Many applications are realized in different number of evaluation and different problem dimensions. In these application results, it is observed that ABCES algorithm is generally more effective than ABC algorithm. Especially in high-dimensional problems, the performance of the algorithm has been significantly improved. The occurrence of a clear performance difference between the standard ABC algorithm and ABCES algorithm shows the effect of the scout bee stage and "Limit" control parameter. At the same time, it is seen that ABCES algorithm has more success in general compared to heuristics such as GA, PSO and DE. This is an indication that ABCES algorithm can compete with different heuristic algorithms. ABCES algorithm also finds low standard deviation values parallel to the low error value. This shows that the results are robust.

The identification of nonlinear static systems is one of the difficult problems due to system behavior. The effect of changes on both the scout bee stage and the "limit" control parameter are analyzed on 6 nonlinear static systems. Generally, as the number of neurons in the hidden layer increases, more effective results are obtained. This situation shows that the problem is difficult, and it reveals the necessity of more weight values to explain the relationship. With ABCES algorithm, a performance increase of 50% and above has been achieved in all systems compared to ABC algorithm. The changes in the scout bee stage have increased the convergence speed of ABCES algorithm. ANN training aims to find the closest output to the real output. It is seen from the analyzes that ABCES algorithm is an effective training algorithm in this regard. It is compared with heuristics such as PSO, HS and ABC to better understand the success of ABCES algorithm. The results show that ABCES algorithm is successful in FFNN training.

It is seen that the changes realized on the scout bee stage and limit control parameter with ABCES algorithm positively affect the result. Different solution-generating mechanisms for the scout bee stage can be integrated to further improve the performance of ABCES algorithm. At the same time, different approaches can be put forward to determine "limit" control parameter adaptively.

6. Conclusions

This paper proposes a neural network-based approach for the identification of nonlinear static systems. A new training algorithm called ABCES (ABC Based on Effective Scout Bee Stage) is introduced to achieve effective results in modeling with artificial neural networks. Standard ABC algorithm basically consists of three stages: employed bee, onlooker bee and scout bee. Employed and onlooker bee stages are more efficient than the scout bee stage. When the scout bee stage is reached, it is understood that better new solution is not developed. In this case, a random solution is created in the scout bee stage of standard ABC algorithm. In fact, this means failure to use of information obtained. If this is prevented, a more effective algorithm will be created. For this purpose, ABCES algorithm is proposed to create a more effective scout bee stage. In this algorithm, two important changes are made according to standard ABC algorithm. First, "limit" control parameter is set to adaptive according to the number of iterations. Secondly, a new solution generation mechanism that enables the adjustment of the new position according to the global best solution in the scout bee stage, is proposed. With these changes, an effective ABCES algorithm has been created. The performance of ABCES algorithm is evaluated on two different problem groups. First, the applications are realized on 13 numerical optimization test problems. It is compared with GA, PSO, DE and ABC algorithms. The Wilcoxon signed rank test is applied to determine the significance of the results. The results show that ABCES algorithm is generally more successful than other algorithms in solving numerical optimization problems.

Secondly, FFNN is trained by using ABCES algorithm for the identification nonlinear static systems. Six nonlinear static systems are used in the applications. The effect of different network structures on performance is examined. The performance of ABCES algorithm is compared with PSO, HS and ABC algorithm in terms of solution quality and speed of convergence. The results show that ABCES algorithm is generally more successful than other algorithms in the identification of nonlinear static systems based on neural networks.

In this study, ABCES algorithm is used first time and it is evaluated on global optimization problems and training FFNN. In future studies, it is possible to examine the performance of ABCES algorithm on different types of problems. As a continuation of this study, FFNN training can be performed by using ABCES algorithm to identify nonlinear dynamic systems. Additionally, neuro-fuzzy models can be trained with ABCES algorithm to identify nonlinear dynamic and static systems. Its performance on neuro-fuzzy training can be evaluated. Apart from system identification, ANN and neuro-fuzzy training can be carried out with ABCES for the solution of real-world problems.

Author Contributions: E.K.: Conceptualization, Methodology, Validation, Software, Review and Editing; C.B.K.: Software, Writing, Data curation, Example analysis, Visualization. All authors have read and agreed to the published version of the manuscript.

Funding: This research received no external funding.

Conflicts of Interest: The authors declare no conflict of interest.

References

1. Fister, I., Jr.; Yang, X.-S.; Fister, I.; Brest, J.; Fister, D. A brief review of nature-inspired algorithms for optimization. *arXiv* **2013**, arXiv:1307.4186v1.
2. Karaboga, D.; Kaya, E. Training ANFIS by using the artificial bee colony algorithm. *Turk. J. Electr. Eng. Comput. Sci.* **2017**, *25*, 1669–1679. [CrossRef]
3. Karaboga, D.; Kaya, E. Training ANFIS Using Artificial Bee Colony Algorithm for Nonlinear Dynamic Systems Identification. In Proceedings of the 22nd Signal Processing and Communications Applications Conference (SIU), Trabzon, Turkey, 23–25 April 2014; IEEE: Piscataway, NJ, USA, 2014.
4. Karaboga, D.; Kaya, E. An adaptive and hybrid artificial bee colony algorithm (aABC) for ANFIS training. *Appl. Soft Comput.* **2016**, *49*, 423–436. [CrossRef]
5. Karaboga, D.; Kaya, E. Training ANFIS by using an adaptive and hybrid artificial bee colony algorithm (aABC) for the identification of nonlinear static systems. *Arab. J. Sci. Eng.* **2019**, *44*, 3531–3547. [CrossRef]
6. Karaboga, D.; Ozturk, C. Neural networks training by artificial bee colony algorithm on pattern classification. *Neural Netw. World* **2009**, *19*, 279.
7. Horng, M.-H. Multilevel thresholding selection based on the artificial bee colony algorithm for image segmentation. *Expert Syst. Appl.* **2011**, *38*, 13785–13791. [CrossRef]
8. Karaboga, N. A new design method based on artificial bee colony algorithm for digital IIR filters. *J. Frankl. Inst.* **2009**, *346*, 328–348. [CrossRef]
9. Yeh, W.-C.; Hsieh, T.-J. Solving reliability redundancy allocation problems using an artificial bee colony algorithm. *Comput. Oper. Res.* **2011**, *38*, 1465–1473. [CrossRef]
10. Hemamalini, S.; Simon, S.P. Artificial bee colony algorithm for economic load dispatch problem with non-smooth cost functions. *Electr. Power Components Syst.* **2010**, *38*, 786–803. [CrossRef]
11. Hong, W.-C. Electric load forecasting by seasonal recurrent SVR (support vector regression) with chaotic artificial bee colony algorithm. *Energy* **2011**, *36*, 5568–5578. [CrossRef]
12. Şahin, A.Ş. Optimization of solar air collector using genetic algorithm and artificial bee colony algorithm. *Heat Mass Transf.* **2012**, *48*, 1921–1928. [CrossRef]
13. Zaman, M.A.; Gaffar, M.; Alam, M.M.; Mamun, S.A.; Matin, M.A. Synthesis of antenna arrays using artificial bee colony optimization algorithm. *Int. J. Microw. Opt. Technol.* **2011**, *6*, 234–241.
14. Deng, X. An efficient hybrid artificial bee colony algorithm for customer segmentation in mobile E-commerce. *J. Electron. Commer. Organ. (JECO)* **2013**, *11*, 53–63. [CrossRef]
15. Bulut, O.; Tasgetiren, M.F. An artificial bee colony algorithm for the economic lot scheduling problem. *Int. J. Prod. Res.* **2014**, *52*, 1150–1170. [CrossRef]
16. Bansal, J.C.; Sharma, H.; Jadon, S.S. Artificial bee colony algorithm: A survey. *Int. J. Adv. Intell. Paradig.* **2013**, *5*, 123–159. [CrossRef]

17. Karaboga, D.; Gorkemli, B.; Ozturk, C.; Karaboga, N. A comprehensive survey: Artificial bee colony (ABC) algorithm and applications. *Artif. Intell. Rev.* **2014**, *42*, 21–57. [CrossRef]
18. Ozturk, C.; Hancer, E.; Karaboga, D. Color image quantization: A short review and an application with artificial bee colony algorithm. *Informatica* **2014**, *25*, 485–503. [CrossRef]
19. Bansal, J.C.; Sharma, H.; Arya, K.; Deep, K.; Pant, M. Self-adaptive artificial bee colony. *Optimization* **2014**, *63*, 1513–1532. [CrossRef]
20. Babaeizadeh, S.; Ahmad, R. An efficient artificial bee colony algorithm for constrained optimization problems. *J. Eng. Appl. Sci.* **2014**, *9*, 405–413.
21. Draa, A.; Bouaziz, A. An artificial bee colony algorithm for image contrast enhancement. *Swarm Evol. Comput.* **2014**, *16*, 69–84. [CrossRef]
22. Karaboga, D.; Gorkemli, B. A quick artificial bee colony (qABC) algorithm and its performance on optimization problems. *Appl. Soft Comput.* **2014**, *23*, 227–238. [CrossRef]
23. Gao, W.-F.; Liu, S.-Y.; Huang, L.-L. Enhancing artificial bee colony algorithm using more information-based search equations. *Inf. Sci.* **2014**, *270*, 112–133. [CrossRef]
24. Wang, B. A novel artificial bee colony algorithm based on modified search strategy and generalized opposition-based learning. *J. Intell. Fuzzy Syst.* **2015**, *28*, 1023–1037. [CrossRef]
25. Kıran, M.S.; Fındık, O. A directed artificial bee colony algorithm. *Appl. Soft Comput.* **2015**, *26*, 454–462. [CrossRef]
26. Liang, J.-H.; Lee, C.-H. A modification artificial bee colony algorithm for optimization problems. *Math. Probl. Eng.* **2015**, in press. [CrossRef]
27. Babaeizadeh, S.; Ahmad, R. Enhanced constrained artificial bee colony algorithm for optimization problems. *Int. Arab J. Inf. Technol.* **2017**, *14*, 246–253.
28. Li, G.; Cui, L.; Fu, X.; Wen, Z.; Lu, N.; Lu, J. Artificial bee colony algorithm with gene recombination for numerical function optimization. *Appl. Soft Comput.* **2017**, *52*, 146–159. [CrossRef]
29. Ozturk, C.; Hancer, E.; Karaboga, D. A novel binary artificial bee colony algorithm based on genetic operators. *Inf. Sci.* **2015**, *297*, 154–170. [CrossRef]
30. Yaghoobi, T.; Esmaeili, E. An improved artificial bee colony algorithm for global numerical optimisation. *Int. J. Bio-Inspired Comput.* **2017**, *9*, 251–258. [CrossRef]
31. Dreiseitl, S.; Ohno-Machado, L. Logistic regression and artificial neural network classification models: A methodology review. *J. Biomed. Inform.* **2002**, *35*, 352–359. [CrossRef]
32. Lisboa, P. J.; Taktak, A. F. The use of artificial neural networks in decision support in cancer: A systematic review. *Neural Netw.* **2006**, *19*, 408–415. [CrossRef]
33. Tealab, A. Time series forecasting using artificial neural networks methodologies: A systematic review. *Future Comput. Inform. J.* **2018**, *3*, 334–340. [CrossRef]
34. Dase, R. K.; Pawar, D. D. Application of artificial neural network for stock market predictions: A review of literature. *Int. J. Mach. Intell.* **2010**, *2*, 14–17.
35. Dhillon, A.; Verma, G. K. Convolutional neural network: A review of models, methodologies and applications to object detection. *Prog. Artif. Intell.* **2020**, *9*, 85–112. [CrossRef]
36. Capizzi, G.; Lo Sciuto, G.; Napoli, C.; Tramontana, E. A multithread nested neural network architecture to model surface plasmon polaritons propagation. *Micromachines* **2016**, *7*, 110. [CrossRef]
37. Sciuto, G. L.; Susi, G.; Cammarata, G.; Capizzi, G. A Spiking Neural Network-Based Model for Anaerobic Digestion Process. In Proceedings of the 2016 International Symposium on Power Electronics, Electrical Drives, Automation and Motion (SPEEDAM), Anacapri Capri Island , Italy; IEEE: Piscataway, N.J, USA, 2016.
38. Capizzi, G.; Sciuto, G. L.; Woźniak, M.; Damaševicius, R. A Clustering Based System for Automated Oil Spill Detection by Satellite Remote Sensing. In Proceedings of the International Conference on Artificial Intelligence and Soft Computing, Zakopane, Poland, 12–16 June 2016; Springer, Cham, Switzerland, 2016.
39. Mohmad Hassim, Y.M.; Ghazali, R. An improved functional link neural network learning using artificial bee colony optimisation for time series prediction. *Int. J. Bus. Intell. Data Min.* **2013**, *8*, 307–318.
40. Zhang, Y.; Wu, L.; Wang, S. Magnetic resonance brain image classification by an improved artificial bee colony algorithm. *Prog. Electromagn. Res.* **2011**, *116*, 65–79. [CrossRef]
41. Ozkan, C.; Kisi, O.; Akay, B. Neural networks with artificial bee colony algorithm for modeling daily reference evapotranspiration. *Irrig. Sci.* **2011**, *29*, 431–441. [CrossRef]
42. Chen, S.; Fang, G.; Huang, X.; Zhang, Y. Water quality prediction model of a water diversion project based on the improved artificial bee colony–backpropagation neural network. *Water* **2018**, *10*, 806. [CrossRef]
43. Karaboga, D.; Basturk, B. A powerful and efficient algorithm for numerical function optimization: Artificial bee colony (ABC) algorithm. *J. Glob. Optim.* **2007**, *39*, 459–471. [CrossRef]
44. Karaboga, D.; Akay, B. A comparative study of artificial bee colony algorithm. *Appl. Math. Comput.* **2009**, *214*, 108–132. [CrossRef]

Article

On Present Value Evaluation under the Impact of Behavioural Factors Using Oriented Fuzzy Numbers

Krzysztof Piasecki * and Anna Łyczkowska-Hanćkowiak *

Institute of Economy and Finance, WSB University in Poznań, ul. Powstańców Wielkopolskich 5, 61-895 Poznań, Poland
* Correspondence: krzysztof.piasecki@wsb.poznan.pl (K.P.); anna.lyczkowska-hanckowiak@wsb.poznan.pl (A.Ł.-H.)

Abstract: In general, the present value (PV) concept is ambiguous. Therefore, behavioural factors may influence on the PV evaluation. The main aim of our paper is to propose some method of soft computing PV evaluated under the impact of behavioural factors. The starting point for our discussion is the notion of the Behavioural PV (BPV) defined as an imprecisely real-valued function of distinguished variables which can be evaluated using objective financial knowledge or subjective behavioural premises. In our paper, a BPV is supplemented with a forecast of the asset price closest to changes. Such BPV is called the oriented BPV (O-BPV). We propose to evaluate an O-BPV by oriented fuzzy numbers which are more useful for portfolio analysis than fuzzy numbers. This fact determines the significance of the research described in this article. O-BPV may be applied as input signal for systems supporting invest-making. We consider here six cases of O-BPV: overvalued asset with the prediction of a rise in its price, overvalued asset with the prediction of a fall in its price, undervalued asset with the prediction of a rise in its price, undervalued asset with the prediction of a fall in its price, fully valued asset with the prediction of a rise in its rice and fully valued asset with the prediction of a fall in its rice. All our considerations are illustrated by numerical examples. Presented examples show the way in which we transform superposition of objective market knowledge and subjective investment opinion into simple return rate.

Keywords: behavioural finance; imprecision; oriented fuzzy number; oriented present value; oriented return

Citation: Piasecki, K.; Łyczkowska-Hanćkowiak, A. On Present Value Evaluation under the Impact of Behavioural Factors Using Oriented Fuzzy Numbers. *Symmetry* **2021**, *13*, 468. https://doi.org/10.3390/sym13030468

Academic Editor: Kóczy T. László

Received: 23 February 2021
Accepted: 9 March 2021
Published: 12 March 2021

1. Introduction

The starting point for evaluating any financial asset is its present value (PV), defined as a current equivalent of a cash payable in a fixed moment of the future [1]. It results in the fact that the PV of a future cash flows may be an imprecise value. For this reason, the reliable PV evaluation requires the use of soft computing a commonly accepted model of an imprecise value is a fuzzy number (FN) [2]. The natural consequence of this approach is estimating PV with FNs. Therefore, fuzzy PV may be defined as a discounted fuzzy prediction of a future cash flow value [3].

The concept of using FNs in financial arithmetic comes from Buckley [4]. The Ward's definition [3] was generalized in [5] to the case of imprecisely assessed postponement. Sheen [6] expanded the Ward's definition to the case of fuzzy interest rate. The problems connected to calculating fuzzy PV were considered in [4,7–9]. Huang [10] expanded the Ward's definition to the case of future cash flow described by a fuzzy variable. A more general definition of fuzzy PV was proposed by Tsao [11], who assumed that future cash flow can be treated as a fuzzy probabilistic set. Calzi [12] has formulated an axiomatic definition of fuzzy PV. All those authors depicted PV as a discount of an imprecisely evaluated future cash flow. A different approach was introduced in [13], where the fuzzy PV is determined as imprecise approximation of the current quoted price of an evaluated asset. Some applications of fuzzy PV were considered in [12–18]. FNs are also used in quantitative finance for modelling imprecision of financial data. In most of the papers regarding imprecision in finance, it is assumed a priori that the return rate from a security is a FN [19–29]. Yet, this assumption is connected, in most cases, to uncertain or unclear or incomplete information available to the investor. Then, authors apply mostly possibility theory [30,31] and credibility theory [32]. Kahraman et al. [33] introduced a research in which both cash flows and return rates are evaluated by trapezoidal FNs. Limiting them to the case of trapezoidal FNs stems from the fact that arithmetic for trapezoidal FNs is significantly simpler than arithmetic for any FNs. More information on this topic is presented in competent monographies [34,35].

Ordered FNs are defined by Kosiński et al. [36], who in this way introduced a FN additionally equipped with an orientation. For formal reasons [37], the Kosiński's theory was revised in [38]. If ordered FN is linked to the revised theory, then it is called Oriented FN (OFN) [39–42]. Ordered FNs are applied in decision-making, economics and finance [42–63].

In [1,64], the behavioural PV (BPV) was defined as fair price approximation determined under impact of behavioural factors. Then, BPV is imprecisely estimated by FN. In [65], the BPV is supplemented with a qualitative prediction of the price trend. This subjective prediction is implemented in a BPV model as an orientation of FN. The existence of such forecasts is proved by the observed balance between the supply and demand reported by investors in financial markets. In this way, the BPV was replaced by oriented BPV (O-BPV) described by an ordered FN. This approach makes portfolio analysis difficult [57]. These difficulties arise from the fact that the sum of the FNs may, in fact, not be an ordered FN.

For this reason, in this paper we present a revised approach to O-BPV. Our main goal is to describe O-BPV by means of OFNs. In the future, this approach will facilitate portfolio analysis because of OFNs are more useful for portfolio analysis than FNs. Therefore, we intend to apply O-BPV for management of portfolio risk of imprecision determined by some behavioural factors. Such possibilities of future applications determine the significance of the studies described in this work.

In [65], the O-BPV is determined with use ordered FNs. Moreover, its membership function was described by a logically complicated system of identities. This also caused very difficult application of the proposed O-BPV model. Therefore, we will try to simplify the identities describing the O-BPV membership function.

This paper is organised as follows. Behavioural aspects of PV definition are presented in Section 2. Section 3 contains basic information of OFNs. In Section 4, we explain the notion of oriented fuzzy PV. BPV is generally defined in Section 5. An interval representation of BPV is described in Section 6. Fuzzy representation of BPV is discussed in Section 7. The concept of O-BPV is introduced in Section 8. Some examples of O-BPV applications are presented in Section 9. There, O-BPV is used for determining return rate. Section 10 contains final conclusions.

2. Behavioural Essence of Present Value

Any PV is used for discounting the money value. This is the basic tool of financial arithmetic. The starting point for the financial arithmetic development was the interest theory. Development of the financial arithmetic theory has resulted in axioms formulated by Peccati, who has defined PV as an additive function of the payment value [66]. This theory is developed in recent years [67].

Among other things, it has been proved that any PV fulfils the conditions of Peccati's definition if and only if it is a linear discount [66], meaning that this PV can be represented by the product of the payment value and the discounting factor defined as a nonincreasing function of the payment time. On the basis of the interest theory, assuming constant interest rate, we meet only with exponential discount factors. It has been shown that any compound discount is represented by an exponential discounting factor.

On the other hand, many different kinds of discounting factors were described during the study on a behavioural aspects of dynamic money evaluation. The first mathematical model of behavioural finance was introduced by Ramsey [68], who explained the relationship between the marginal product of capital value, the subjective discount rate and the real interest rate. Samuelson [69] introduced an exponential model of the subjective discount factor. This model was adjusted in [70].

The Samuelson's model [69] assumes that the subjective discount rate is constant over time. This assumption is one of the many criticised problems of the exponential model. The exponential discount model has "anomalies" related to the behavioural effect that subjective discount rate varies in time [71–74]. Dynamically inconsistent time preferences with this effect are described in [75,76].

Imperfections of the exponential discounting were found, causing the creation of new discounting models and approaches. The hyperbolic discounting was introduced by Mazur [77] who generalised some particular function applied in [78,79]. Some kinds of discount hyperbolic models are discussed in [73,80]. In [81], we can find arithmetic discounting. This discounting describes bill discounting method specified by financial law. The hyperbolic discounting was generalised to hyperboloid one in [72]. Quasi-hyperbolic discounting [82] is used by economists hoping to preserve as much of the exponential model as possible. In the discrete time version this discounting method was introduced in [83]. The quasi-hyperbolic discounting can also be used for the case of continuous time [84]. Moreover, as an alternative to the percentage decrement for delayed payments may be used a fixed cost model to exponential discounting [84]. Some authors [82,85–92] consider it to be important to extend the hyperbolic function by the behavioural exponents. Very frequently, those models are only some modifications of hyperbolic or exponential discounting.

Hyperbolic models are also criticised (see, for example, in [75,76,92]). Read [92] proposed his model as such exponential discounting that the discount factor occurs within the given time window. Arguing with Weber–Fechner Law [93], Roelofsma [89] proposed his model as an exponential discounting where the exponent is directly proportional to logarithm of time. Ebert and Prelec [90] defined their Constant Sensitivity discounting factor as the Cobb–Douglas function. The Constant Relative Decreasing Impatience discount factor is defined as the constant relative risk aversion function [94]. In an analogous way, Constant Absolute Decreasing Impatience discount factor is defined as a constant absolute risk aversion function [94]. In [81], a discount factor is defined by means of the assumption that its marginal change follows the Stevens' power law [95,96] which says that "a psychologically effective variable is a power function of its physical cause". Multi-threaded results of studies on the discounting factors were competently discussed in [80]. Moreover, in [97–99] the discount factor is defined for such discounting under continuous capitalisation which is determined by a constant Arrow-Pratt's measure of absolute risk-aversion. The axiomatic theory of discount factors was introduced in [100] and developed in [101]. Rotschedla et al. [102] present such a discounting model that the discounted value is given as a hedonic price dependent on fundamentals. Let us note that PV defined in this way does not meet the axiomatic conditions determined by Peccati for PV.

The variety of discount approaches described above justifies our approach to PV as an ambiguous value.

Let us summarise the PV studies described above. We see that the impact of behavioural premises differentiates the discounting factors used to determine PV. Moreover, Peccati [66] has shown that any PV may be defined by a discounting factor. It means that Peccati's definition of PV depends on behavioural premises which are always subjective. On the other side, financial arithmetic requires an objective PV definition. Therefore, any PV is defined as a current equivalent of a cash flow [1]. There financial flow utility is set by the multicriteria comparison consisting of temporal preference [103] and wealth preferences. This definition of PV is more general than the Peccati's definition. The PV determined by

Rotschedla et al. [102] fulfil the conditions of a generalised PV. Generalised PV is defined without a discounting factor. Therefore, a generalised PV definition is independent of a priori given behavioural phenomena. For this reason, the notion of generalised PV is a valid tool to study the behavioural impact on a discount.

The research domain of behavioural finance is the paradoxes and anomalies in financial markets, which are difficult to explain based on neoclassical economic theory. Behavioural analysis of financial markets points to the psychological aspect of an investment as the reason for this state of affairs. At present, an extensive bibliography is already devoted to the results of those studies. The consequence of this intensive research is aiming to obtain such formal models which explain behavioural mechanisms of the financial market. Here, we can distinguish a few approaches to this topic.

The most typical behavioural finance model is a formal prospect theory [104,105]. In this theory, a subjective transformation of the objective probability is distinguished as a behavioural basis for investment decisions.

Barberis et al. [106] develop the prospect theory. They additionally point out imprecise estimation of PV, as a result of the subjective approach to security valuation.

Daniel et al. [107] show the diversified responses of individual investors to received information as the reason for the disclosure of the market paradoxes. Assumed lack of the strong effectiveness of the financial market is one of characteristics of this theory.

Hong and Stein [108] describe the investment activity, as the game amongst investors applying fundamental analysis and investors applying technical analysis. This interaction of two rational theories produces such market phenomena, which are paradoxes from a point of view of the economic theory. PV behavioural aspects are enclosed here in the choice of a cognitive strategy.

Behavioural experiments [75,109–111] present the regularities of subjective discounting. There it is show that

- the receivables are discounted by a higher discount rate than liabilities and
- smaller amounts are discounted by a lower discount rate than large amounts.

The above behavioural paradoxes may be explained by means of the prospect theory in its final version. Kahneman and Tversky [112] say that any evaluation depends on risk aversion and on loss aversion. The loos aversion implies that the accelerated consumption is less desirable than the delayed one [73,113]. From economical point of view, any receivable may be considered as delated consumption. In analogous way, any liability can be interpreted as accelerated consumption. It explains why we discount liabilities less than receivable. Unfortunately, this explanation is not sufficient for determining such discounting functions which differentiate the receivable discount and liability discount. Behavioural analysis of financial markets points to the psychological aspect of investors' activities as the reason for this state of affairs. The behavioural impact on PV has been confirmed by experiments [81,88,113–115].

In the general case, each of Peccati's models of PV [66] is inconsistent with the results of behavioural experiments which explain the financial practice. The subject of behavioural finance research is the phenomena which are paradoxes, inter alia, from the viewpoint of the interest theory. This intensive research consequence is prospecting for such formal models which explain the observed paradoxes.

In recent years, the concept of financial flow utility has played an important role in the behavioural finance research. This problem is discussed, for example, in [73,80,81,116–121]. PV is defined there as the financial flow utility. As we see, the behavioural approach to PV is similar to the neoclassical one. All PV models mentioned in this paragraph meet the conditions of a generalised PV definition [1]. In this definition, the financial flow utility is defined in a such way that the receivable utility is positive and the liabilities utility is negative. Therefore, generalised PV may be applied for evaluation of behavioural asymmetry between receivable discount and liability discount described in [73,75,109–111,113]. Note that the notion of the negative utility was introduced only by Rabin [122]. All above references prove that many researchers accept the view that the behavioural factors have an impact on evaluating PV.

In [13,64], the BPV was defined as fair price approximation determined under impact of behavioural factors. Such an approach is in line with the approach proposed later by Rotschedla et al. [102]. It causes that any PV estimation is in fact an imprecise number. Because FN is a commonly accepted model of an imprecise number, BPV should be described by an FN.

3. Oriented Fuzzy Numbers—Basic Facts

The symbol $\mathcal{F}(\mathbb{R})$ denotes the family of all fuzzy subsets in the real line $\mathbb{R}$. Any fuzzy subset $\mathcal{A} \in \mathcal{F}(\mathbb{R})$ is described by its membership function $\mu_A \in [0,1]^{\mathbb{R}}$, as the set of ordered pairs

$$\mathcal{A} = \{(x, \mu_A(x)); x \in \mathbb{R}\}. \tag{1}$$

Among other things, this fuzzy subset $\mathcal{A}$ may be characterised by its support closure $[\mathcal{A}]_{0^+}$ given in a following way:

$$[\mathcal{A}]_{0^+} = \lim_{\alpha \to 0^+} \{x \in \mathbb{R} : \mu_A(x) \geq \alpha\}. \tag{2}$$

FN is usually defined as a fuzzy subset of the real line $\mathbb{R}$. The most general definition of FN is given as follows.

Definition 1. *[2] The fuzzy number (FN) is such a fuzzy subset $\mathcal{L} \in \mathcal{F}(\mathbb{R})$ with bounded support closure $[\mathcal{L}]_{0^+}$ that it is represented by its upper semi-continuous membership function $\mu_L \in [0;1]^{\mathbb{R}}$ satisfying the conditions*

$$\exists_{x \in \mathbb{R}}\ \mu_L(x) = 1, \tag{3}$$

$$\forall_{(x,y,z) \in \mathbb{R}^3}\ x \leq y \leq z \Rightarrow \mu_L(y) \geq \min\{\mu_L(x); \mu_L(z)\}. \tag{4}$$

The set of all FN we denote by the symbol $\mathbb{F}$.

Theorem 1. *[123,124] For any FN $\mathcal{L}$ there exists such a non-decreasing sequence $(a,b,c,d) \subset \mathbb{R}$ that $\mathcal{L}(a,b,c,d,L_L,R_L) = \mathcal{L} \in \mathcal{F}(\mathbb{R})$ is determined by its membership function $\mu_L(x|a,b,c,d,L_L, R_L) \in [0,1]^{\mathbb{R}}$ described by the identity*

$$\mu_L(x|a,b,c,d,L_L,R_L) = \begin{cases} 0, & x \notin [a,d], \\ L_L(x), & x \in [a,b[, \\ 1, & x \in [b,c], \\ R_L(x), & x \in]c,d], \end{cases} \tag{5}$$

where the left reference function $L_L \in [0,1[^{[a,b[}$ and the right reference function $R_L \in [0,1[^{]c,d]}$ are upper semi-continuous monotonic ones meeting the condition

$$[\mathcal{L}]_{0^+} = [a,d]. \tag{6}$$

The FN $\mathcal{L}(a,a,a,a,L_L,R_L) = a$ represents the real number $a \in \mathbb{R}$. Therefore, we can say $\mathbb{R} \subset \mathbb{F}$.

Any function $G : \mathbb{R} \supset \mathbb{A} \to \mathbb{R}$ may be considered as an unary operator on real numbers. Using the Zadeh's extension principle, we can extend this operator to the fuzzy case. Let us write this extended operator as the function $\widetilde{G} : \mathbb{F} \supset \mathbb{D} \to \mathbb{F}$ described by the equation

$$\mathcal{K}(e,f,g,h,L_K,R_K) = \widetilde{G}(\mathcal{L}(a,b,c,d,L_L,R_L)), \tag{7}$$

where

- FN $\mathcal{L}(a,b,c,d,L_L,R_L)$ is represented by its membership function

$$\mu_L(x) = \mu_L(x|a,b,c,d,L_L,R_L), \tag{8}$$

- FN $\mathcal{K}(e,f,g,h,L_K,R_K)$ is represented by its membership function

$$\mu_K(x) = \mu_K(x|e,f,g,h,L_K,R_K). \tag{9}$$

In line with the Zadeh extension principle, the unary operator (7) is uniquely defined by its membership function $\mu_K \in [0,1]^{\mathbb{R}}$ given by the identity

$$\mu_K(x) = \min\{\max\{\mu_L(y)\} \ : \ y = G(x)\}. \tag{10}$$

Moreover, the following concepts may be applied for description of the unary operator (7) in detail.

Definition 2. [124] *For any upper semi-continuous non-decreasing function* $L \in [0,\ 1]^{[u,\ v]}$*, its cut-function* $L^\star \in [u,v]^{[0;1]}$ *is determined by the identity*

$$L^\star(\alpha) = \min\{x \in [u,v] : \ L(x) \geq \alpha\}. \tag{11}$$

Definition 3. [124] *For any upper semi-continuous non-increasing function* $R \in [0,\ 1]^{[u,\ v]}$*, its cut-function* $R^\star \in [0,\ 1]^{[u,\ v]}$ *is determined by the identity*

$$R^\star(\alpha) = \max\{x \in [u,v] : \ R(x) \geq \alpha\}. \tag{12}$$

Definition 4. [124] *For any bounded continuous and non-decreasing function* $l \in [l(0),\ l(1)]^{[0,1]}$*, its pseudo-inverse* $l^\triangleleft \in [0,1]^{[l(0),\ l(1)]}$ *is determined by the identity*

$$l^\triangleleft(x) = \max\{\alpha \in [0,1] : l(\alpha) = x\}. \tag{13}$$

Definition 5. [124] *For any bounded continuous and non-increasing function* $r \in [r(0),\ r(1)]^{[0,1]}$*, its pseudo-inverse* $r^\triangleleft \in [0;1]^{[r(1),\ r(0)]}$ *is determined by the identity*

$$r^\triangleleft(x) = \min\{\alpha \in [0,1] : r(\alpha) = x\}. \tag{14}$$

Using the theorems proved by Goetschel and Voxman [124], we can show that if the unary operator $G : \mathbb{R} \supset \mathbb{A} \to \mathbb{R}$ is decreasing, then the unary operator (7) is given by the identity

$$\mathcal{L}(G(d),G(c),G(b),G(a),L_K,R_K) = \widetilde{G}(\mathcal{L}(a,b,c,d,L_L,R_L)), \tag{15}$$

where

$$\forall_{\alpha\in[0,1]} \quad l_K(\alpha) = G(L_L^\star(\alpha)), \tag{16}$$

$$\forall_{\alpha\in[0,1]} \quad r_K(\alpha) = G(R_L^\star(\alpha)), \tag{17}$$

$$\forall_{y\in[G(d),G(c)[} \quad L_K(y) = r_K^\triangleleft(y), \tag{18}$$

$$\forall_{y\in]G(b),G(a)]} \quad R_K(y) = l_K^\triangleleft(y). \tag{19}$$

The notion of ordered FN is introduced by Kosiński et al. [36]. An important disadvantage of Kosiński's theory is that there exist such ordered FNs which are not linked to any membership function [37]. For formal reasons, the Kosiński's theory is revised in [38]. In revised theory, ordered FN is replaced by oriented FN (OFN) defined as follows.

Definition 6. [38] *For any monotonic sequence* $(a,b,c,d) \subset \mathbb{R}$, *OFN* $\overleftrightarrow{\mathcal{L}}(a,b,c,d,S_L,E_L) = \overleftrightarrow{\mathcal{L}}$ *is the pair of orientation* $\overrightarrow{a,d} = (a,d)$ *and FN described by membership function* $\mu_L(\cdot|a,b,c,d,S_L,E_L) \in [0,1]^{\mathbb{R}}$ *given by the identity*

$$\mu_L(x|a,b,c,d,S_L,E_L) = \begin{cases} 0, & x \notin [a,d] \equiv [d,a], \\ S_L(x), & x \in [a,b[\equiv]b,a], \\ 1, & x \in [b,c] \equiv [c,b], \\ E_L(x), & x \in]c,d] \equiv [d,c[, \end{cases} \tag{20}$$

where the starting function $S_L \in [0,1[^{[a,b[}$ *and the ending function* $E_L \in [0,1[^{]c,d]}$ *are upper semi-continuous monotonic ones meeting the condition (6).*

The identity (20) additionally describes such modified notation of intervals which is used in the OFN theory. The notation $\mathcal{I} \equiv \mathcal{K}$ means that "the interval $\mathcal{I}$ may be equivalently replaced by the interval $\mathcal{K}$".

The symbol $\mathbb{K}$ denotes the space of all OFNs. Any OFN describes an imprecise number with additional information about the location of the approximated number. This information is given as an orientation of OFN. If $a < d$, then OFN $\overleftrightarrow{\mathcal{L}}(a,b,c,d,S_L,E_L)$ has the positive orientation $\overrightarrow{a,d}$. If $a > d$, then OFN $\overleftrightarrow{\mathcal{L}}(a,b,c,d,S_L,E_L)$ has the negative orientation $\overrightarrow{a,d}$. If $a = d$, OFN $\overleftrightarrow{\mathcal{L}}(a,a,a,a,S_L,E_L) = a$ describes an unoriented number $a \in \mathbb{R}$.

Kosiński has defined arithmetic operators on ordered FNs in an intuitive way consistent with the results obtained by Goetschel and Voxman [124]. For OFNs, any arithmetic operator is defined alike. In this way, any unary operator $G : \mathbb{R} \supset \mathbb{A} \to \mathbb{R}$ may be extended to OFN case. Using the Kosiński's approach, we define an extended unary operator $\overleftrightarrow{G} : \mathbb{K} \supset \mathbb{H} \to \mathbb{K}$ as follows:

$$\overleftrightarrow{\mathcal{K}}(G(a),G(b),G(c),G(d),S_K,E_K) = \overleftrightarrow{G}(\mathcal{L}(a,b,c,d,S_L,E_L)), \tag{21}$$

where

$$\forall\, \alpha \in [0,1] \quad s_K(\alpha) = G(S_L^\star(\alpha)), \tag{22}$$

$$\forall\, \alpha \in [0,1] \quad e_K(\alpha) = G(E_L^\star(\alpha)), \tag{23}$$

$$\forall\; y \in [G(a),G(b)[\quad S_K(y) = s_K^{\triangleleft}(y), \tag{24}$$

$$\forall\; y \in]G(c),G(d)] \quad E_K(y) = e_K^{\triangleleft}(y). \tag{25}$$

When we compare the dependencies (15–19) and (21–25), then we notice that for the case of decreasing unary operator $G : \mathbb{R} \supset \mathbb{A} \to \mathbb{R}$, its extension to OFNs differs from its extension to FNs. This is an important difference between OFNs and FNs.

In Sections 8 and 9, we restrict our considerations to the case of strictly monotonic starting and ending functions. Then, for any monotonic unary operator $G : \mathbb{R} \supset \mathbb{A} \to \mathbb{R}$, Equation (21) is simpler because then we have

$$\forall\; y \in [G(a),G(b)[\quad S_K(y) = S_L\left(G^{-1}(y)\right), \tag{26}$$

$$\forall\; y \in]G(c),G(d)] \quad E_K(y) = E_L\left(G^{-1}(y)\right). \tag{27}$$

4. Oriented Fuzzy Present Value

The start point for our considerations is a notion of PV defined as a current equivalent. In Section 2, we have shown that PV may be imprecise. This observation agrees with understanding PV as equivalent of future cash flow. The natural consequence of these conclusions is estimating PV with FNs. Therefore, such PV is called a fuzzy one (F-PV).

In Section 1, the evolution of F-PV model is described in detail. In general, fuzzy PV is characterised by a non-decreasing sequence $\left(V_s,\ V_f,\ \check{P},\ V_l,\ V_e\right)$, where

- $\check{P}$ is a quoted price,
- $[V_s, V_e] \subset \mathbb{R}^+$ is an interval of all possible values of PV,
- $\left[V_f, V_l\right] \subset [V_s, V_e]$ is an interval of all prices which do not noticeably differ from a quoted price $\check{P}$.

Then F-PV is estimated by FN

$$\widetilde{PV} = \mathcal{L}\left(V_s,\ V_f,\ V_l,\ V_e\ , S_{PV}, E_{PV}\ \right), \tag{28}$$

where the reference functions $S_{PV} \in [0,1[^{[V_s,V_f[}$ and $E_{PV} \in [0,1[^{]V_l,V_e]}$ are the given ones.

Moreover, the F-PV estimation should be supplemented by a forecast of price closest changes. For example, price closes changes may be predicted with the use of prediction table presented in [125]. In [126], it is shown that OFN application for a portfolio analysis is more useful than the analogous application of FN. For these reasons, an imprecise PV may be evaluated by OFN [42,62]. PV determined in this way is called an oriented one (O-PV). Any O-PV is characterised by a monotonic sequence $\left(V_s,\ V_f,\ \check{P},\ V_l,\ V_e\right)$ and then it is estimated as follows:

$$\overset{\leftrightarrow}{PV} = \overset{\leftrightarrow}{\mathcal{L}}\left(V_s,\ V_f,\ V_l,\ V_e, L_{PV}, R_{PV}\right). \tag{29}$$

If we predict a increase in price, then O-PV has positive orientation. If we predict a fall in price, then O-PV has a negative orientation.

In this paper, O-PV is used for modelling imprecise assessments of PV. After Klir [127], the imprecision is composed of ambiguity and indistinctness. In the considered case, the ambiguity is understood as a lack of clear indication of a one value out of many. An indistinctness is interpreted as a lack of an explicit distinction between values equal to PV and values different from PV.

5. Behavioural Present Value

Let us consider a fixed asset which is the subject of a trade on a financial market. The quoted price $\check{P}$ of this asset may fluctuate over time. Therefore, we can consider a quotation trend. If the demand for this asset is equal to its supply, then the quoted price $\check{P}$ is equal to balanced price P_0. Then, the asset market is balanced, and it meets the market equilibrium condition. In general, the balanced price varies with time. Nevertheless, at any point in time the current value of a balanced price can be determined by the econometric model contained in the Arbitrage Price Theory [128]. In this way, the balanced price P_0 may also be substantively justified by fundamental factors. Of course, the balanced price may be computed in a different way. It is important here that the investor accepts the calculated value P_0 as the balanced price.

Fama [129] introduces a well-known notion of a market informational efficiency. Then, Fama hypothesises that the highly informationally efficient markets do not exist. In 1980, Grossman and Stiglitz [130] show that information efficiency increases with a decrease in the cost of acquiring and analysing information. Since then, we have been observing the intensive development of computerisation, among others in the economic and financial sphere. This development significantly reduces the cost of processing information. Therefore, today we can assume that the considered asset market is highly informationally efficient. This assumption is consistent with a base of the behavioural finance theory [104,105,112].

Using technical analysis, we assume that the quoted price $\check{P}$ converges to the balanced price P_0. If

$$\check{P} < P_0 \tag{30}$$

then the considered asset is undervalued. For the case

$$\check{P} > P_0 \tag{31}$$

the considered asset is overvalued. We call both of these cases financial disequilibrium states. For the case of financial equilibrium

$$\check{P} = P_0 \tag{32}$$

the considered asset is fully valued.

The accrued market knowledge is the unique basis for determining the value of balanced price P_0. This value is a synthetic image of knowledge about the state of financial market. On the highly efficient financial market, each market player determines the same value P_0, which is an objective in this situation. At the same time, all investors observe the identical value of quoted price $\check{P}$. Therefore, this value is an objective in its essence. The knowledge of both of these values is sufficient for rational investment-making. In case (30), the rationale suggests buying the considered asset. This transaction is only possible if a sale offer is proposed. The natural question here is, what were the reasons of the investor selling undervalued asset. Similarly, for case (31) the rationale suggests selling the considered financial instrument. This transaction is only possible if a purchase offer is proposed. This raises the question of what were the reasons for the investor buying the valued asset and what premises direct investors buying this security. We discuss above only balancing the demand and supply reported by the investors. Liquidity traders perform their transactions only when this balance vanishes. This phenomenon is widely observed. The answer to the above two questions can be only one. An explanation is only the influence of behavioural factors.

Financial disequilibrium is described by the alternative of inequalities (30) or (31). If a highly efficient financial market is under the financial disequilibrium, then the market equilibrium is maintained due to irrational premises. It causes that every transaction is concluded under the influence of irrational premises. These premises may be behavioural in nature. Therefore, consideration of behavioural factors helps to clarify the paradox of the paradox of maintaining market equilibrium in the conditions of financial disequilibrium in highly efficient market.

The finance theory suggests that PV should be equal to the quoted price $\check{P}$. On the other hand, a balanced price P_0 may influence the PV deviation from the observed quoted price. This deviation depends on the investor receptivity to behavioural factors.

Therefore, PV evaluation is determined, inter alia, under the influence of behavioural factors. Each behavioural evaluation is subjective. In Section 2, we have shown that subjective assessment of PV is ambiguous. Each of the valuation alternatives is called a possible PV (PPV). The behavioural PV (BPV) is defined as a convex set of all PPV. According to the classical finance theory, the quoted price $\check{P}$ is also PPV. For this reason, BPV is a strong generalisation of PV model proposed by Rotschedla et al. [102]. The BPV dependence on subjective financial factors means that each investor designates their own version of BPV. Thus, we will conduct all further considerations for the established single investor.

6. Interval Representation of Behavioural Present Value

Any convex crisp subset of real numbers is an interval. For this reason, the subject of our first considerations is BPV represented by an interval.

We begin our considerations on BPV by discussing case (32) of financial equilibrium. This assumption causes that any PPV is approximation of the quoted price $\check{P}$. The considered PPV domain always depends on a specific investor's susceptibility to the influence of behavioural factors. Therefore, each investor subjectively distinguishes the following values:

- P_{min} the minimal PPV expected under the financial equilibrium condition (32),

- P_{max} the maximal PPV expected under the financial equilibrium condition (32).

In general, the investor may not be aware of these values. However, when preparing any investment decision support system, we can ask the investor about these values. Therefore, we assume the values P_{min} and P_{max} are known to us.

In the considered case, each investor must take into account the possibility of changes in quotations. Then, the range of PPV variability fulfils the condition

$$0 < P_{min} < P_0 < P_{max}. \tag{33}$$

Numerical interval $[P_{min}, P_{max}]$ is the BPV image determined for the case of the financial equilibrium.

We lead further considerations on BPV for any quoted price $\check{P}$. Then BPV should be dependent on deviation

$$\Delta P = \check{P} - P_0. \tag{34}$$

of the quoted price from the balanced price. For each investor, we determine the following values:

- V_{min} the minimal PPV expected for the quoted price $\check{P}$,
- V_{max} the maximal PPV expected for the quoted price $\check{P}$.

In line with Barberis et al. [106], we assume that both values are dependent on the sentiment index $\alpha \in]0,1]$. We consider sentiment index as an individual investor's characteristic. The value $\alpha \in]0,1]$ describes the degree of the influence of cognitive conservatism phenomenon [131]. This phenomenon is a frequent topic of discussion in behavioural finance. The value $1-\alpha \in [0;1[$ informs us about the intensity of the impact of deviation ΔP on the investor's beliefs.

The investor determines the minimal PPV V_{min} as the weighted average of the assumed minimal PPV P_{min} and its current correction $P_{min} + \Delta P$. The weight of the minimal PPV P_{min} is equal to the value α of the investor's sentiment index. In determining the minimal PPV V_{min}, the investor must consider that the minimal PPV is not greater than the quoted price $\check{P}$. We get

$$\begin{aligned} V_{min} &= \min\{(1-\alpha)\cdot(P_{min} + \Delta P) + \alpha\cdot P_{min}, \check{P}\} = \min\{P_{min} + (1-\alpha)\cdot\Delta P, \check{P}\} \\ &= \min\{P_{min} + (1-\alpha)\cdot(\check{P} - P_0), \check{P}\}. \end{aligned} \tag{35}$$

The investor determines the maximal PPV V_{max}, as the weighted average of the assumed maximal PPV P_{max} and its current correction $P_{max} + \Delta P$. The weight of the maximal PPV P_{max} is equal to the value α of the investor's sentiment index. In determining the maximal PPV V_{max}, the investor must consider that the maximal PPV is not less than the quoted price $\check{P}$. We get

$$\begin{aligned} V_{max} &= \max\{(1-\alpha)\cdot(P_{max} + \Delta P) + \alpha\cdot P_{max}, \check{P}\} = \max\{P_{max} + (1-\alpha)\cdot\Delta P, \check{P}\} \\ &= \max\{P_{max} + (1-\alpha)\cdot(\check{P} - P_0), \check{P}\}. \end{aligned} \tag{36}$$

We note that in case

$$\check{P} \le \frac{P_{min} - P_0}{\alpha} + P_0 \tag{37}$$

the minimal PPV V_{min} is equal to the quoted price $\check{P}$. Then, the BPV model excludes a possibility of decline in quotation because of the considered asset is uniquely identified as undervalued. Furthermore, when

$$\check{P} \ge \frac{P_{max} - P_0}{\alpha} + P_0 \tag{38}$$

then maximal PPV V_{max} is equal to the quoted price $\check{P}$. Then, the BPV model excludes a possibility of rise in quotation because of the considered asset is uniquely identified as overvalued.

We see that only in the case of large deviations ΔP, rationale is the only reason to invest. Range of behavioural reasons' impact is described by the inequalities

$$\frac{P_{min} - P_0}{\alpha} + P_0 < \check{P} < \frac{P_{max} - P_0}{\alpha} + P_0. \tag{39}$$

Finally, for each investor we can determine a PPV variability range

$$\overline{BPV}(\check{P}) = [V_{min}, V_{max}] = \begin{cases} [\check{P},\ P_{max} + (1-\alpha)\cdot\Delta P] & for\ (37) \\ [P_{min} + (1-\alpha)\cdot\Delta P,\ P_{max} + (1-\alpha)\cdot\Delta P] & for\ (39) \\ [P_{min} + (1-\alpha)\cdot\Delta P,\ \check{P}] & for\ (38) \end{cases}. \tag{40}$$

Analogous results were obtained in [63–65]. In this way, we have set the image of market information impact on the investor's beliefs. Determined above range $[V_{min}, V_{max}]$ is an interval representation of BPV (I-BPV) depending on the variables below:

- $\check{P}$ a quoted price,
- P_0 a balanced price,
- P_{min} the minimal PPV expected under financial equilibrium condition (32),
- P_{max} the maximal PPV expected under financial equilibrium condition (32),
- α a sentiment index.

The assumed PPV range $[P_{min}, P_{max}]$ and a value of sentiment index $\alpha \in \,]0;1]$ are dependent on the investor's receptivity to influence of behavioural factors. Thus, each investor is characterised by different values of these variables. In Section 5, we pointed out that the quoted price $\check{P}$ and the balanced price P_0 are objective in nature. Any BPV model considered in this paper will be characterised by the vector $(\check{P}, P_0, P_{min}, P_{max}, \alpha)$. Finally, we note that for a given vector $(\check{P}, P_0, P_{min}, P_{max}, \alpha)$, the vector $(\check{P}, P_0, V_{min}, V_{max})$ is uniquely defined. Therefore, we keep our further discussion for a given value of the second parameters' vector.

Example 1. *For considered asset $\mathcal{Y}$ its substantially justified balanced price is $P_0 = 30\$$. A financial analyst assumes that under financial equilibrium condition (32)*

- minimal PPV is $P_{min} = 10\$$,
- maximal PPV is $P_{max} = 60\$$.

An investor observes a quoted price $\check{P} = 60\$$. We have $\Delta P = 30\$$. Therefore, we see that the asset $\mathcal{Y}$ is overvalued. Because of investor's cognitive conservatism characterised by a sentiment index $\alpha = \frac{1}{3}$, using (40) we obtain I-BPV given as the interval $\overline{BPV}(60) = [30, 80] = [V_{min}, V_{max}]$.

7. Fuzzy Representation of Behavioural Present Value

Let a fixed parameter vector $\boldsymbol{v} = (\check{P}, P_0, V_{min}, V_{max})$ be given. In the considered case, the interval $\overline{BPV}(\check{P}) = [V_{min}, V_{max}]$ of PPV variability is determined explicitly. A relative distance between any $x \in \mathbb{R}$ and the quoted price $\check{P}$ is determined by function $\beta(\cdot|\boldsymbol{v}) \in [0,1]^{\mathbb{R}}$ given as follows:

$$\beta(x|\boldsymbol{v}) = \begin{cases} 1, & x \notin [V_{min}, V_{max}], \\ \frac{x-\check{P}}{\check{P}-V_{min}}, & x \in [V_{min}, \check{P}[, \\ 0, & x \in [\check{P}, \check{P}], \\ \frac{x-\check{P}}{V_{max}-\check{P}}, & x \in \,]\check{P},\ V_{max}]. \end{cases} \tag{41}$$

Then, the degree of a similarity [132] to the quoted price $\check{P}$ is defined as a function $\gamma(\cdot|\boldsymbol{v}) \in [0,1]^{\mathbb{R}}$ given by the identity

$$\gamma(x|\boldsymbol{v}) = 1 - \beta(x|\boldsymbol{v}). \tag{42}$$

Using any I-BPV model, we treat all PPV as equally acceptable. Nevertheless, we can suppose that the investor accepts more PPVs that are closer to the quoted price. This implies that individual PPVs differ in their degrees of acceptance. We see that I-BPV insufficiently describes the behavioural effects complexity. This means that it necessary to create a BPV model describing the variability of an individual PPV acceptance, leading directly to defining BPV as a fuzzy subset $\widetilde{BPV}(\boldsymbol{v}) \in \mathcal{F}(\mathbb{R})$ of all PPV. In this way we determine fuzzy BPV (F-BPV). F-BPV is described by its membership function $\mu_{BPV}(\cdot|\boldsymbol{v}) \in [0,1]^{\mathbb{R}}$ assigning an acceptance degree to each PPV. It means that the value $\mu_{BPV}(x|\boldsymbol{v})$ is a truth value of the sentence "the number $x \in \mathbb{R}$ is accepted as PPV".

The similarity degree $\gamma(\cdot|\boldsymbol{v})$ will be considered as the first reference point for determining the F-BPV membership function $\mu_{BPV}(\cdot|\boldsymbol{v})$. As the second reference point for determining this membership function, we take a rational forecast of next change in the quotation. This forecast is based on the following facts:

- if the disequilibrium condition (30) is met, then rationale excludes the decrease in a quotation;
- if the disequilibrium condition (31) is met, then rationale excludes the increase in a quotation; and
- if the equilibrium condition (32) is met, then rationale cannot exclude any future quotation.

Thus, the rational forecast of next quotation change is described by a rationality degree $\Theta(\cdot|\boldsymbol{v}) \in \{0,1\}^{\mathbb{R}}$ given as follows:

$$\Theta(x|\boldsymbol{v}) = \begin{cases} 0, & (x-\check{P})\cdot\delta P > 0, \\ 1, & (x-\check{P})\cdot\delta P \leq 0, \end{cases} \tag{43}$$

where

$$\delta P = \frac{\check{P} - P_0}{\check{P}}. \tag{44}$$

For any quoted price $\check{P}$, an investor assesses the acceptance degree as a weighted average of rationality degree and the similarity degree. The weights are appointed regarding to the assumption that the influence of the rationality degree increases with an increase in a relative distance $|\delta P|$ between the quoted and balanced price and with the increase in a similarity degree $\gamma(\cdot|\boldsymbol{v})$. Therefore, without the generality loss we can assume that the weight of a rationality degree is directly proportional to the product $\gamma(x|\boldsymbol{v})\cdot|\delta P|$. Then, the acceptance degree is given as follows:

$$\mu_{BPV}(x|\boldsymbol{v}) = \frac{1}{1+\gamma(x|\boldsymbol{v})\cdot|\delta P|}\cdot\gamma(x|\boldsymbol{v}) + \frac{\gamma(x|\boldsymbol{v})\cdot|\delta P|}{1+\gamma(x|\boldsymbol{v})\cdot|\delta P|}\cdot\Theta(x|\boldsymbol{v}) = \frac{\gamma(x|\boldsymbol{v})\cdot(1+|\delta P|\cdot\Theta(\beta|\boldsymbol{v}))}{1+\gamma(x|\boldsymbol{v})\cdot|\delta P|}. \tag{45}$$

It implies that the membership function $\mu_{BPV}(\cdot|\boldsymbol{v})$ is determined by the identity

$$\mu_{BPV}(x|\boldsymbol{v}) = \begin{cases} 0, & x \notin [V_{min}, V_{max}], \\ h(x|\boldsymbol{v}). & x \in [V_{min}, \check{P}[, \\ 1, & x \in [\check{P}, \check{P}], \\ k(x|\boldsymbol{v}), & x \in]\check{P}, V_{max}], \end{cases} \tag{46}$$

where the reference functions $h(\cdot|\boldsymbol{v}) \in [0,1[^{[V_{min},\check{P}[}$ and $k(\cdot|\boldsymbol{v}) \in [0,1[^{]\check{P},V_{max}]}$ are defined by the identities

$$h(x|\boldsymbol{v}) = \begin{cases} \frac{(x-V_{min})(1+\delta P)}{\check{P}-V_{min}+(x-V_{min})\cdot\delta P}, & \delta P > 0, \\ \frac{x-V_{min}}{\check{P}-V_{min}-(x-V_{min})\cdot\delta P}, & \delta P \leq 0, \end{cases} \tag{47}$$

$$k(x|\boldsymbol{v}) = \begin{cases} \frac{V_{max}-x}{V_{max}-\check{P}+(V_{max}-x)\cdot\delta P}, & \delta P > 0, \\ \frac{(V_{max}-x)(1-\delta P)}{V_{max}-\check{P}-(V_{max}-x)\cdot\delta P}, & \delta P \leq 0. \end{cases} \tag{48}$$

We observe that both reference functions are strictly monotonic. This fact is very important for future considerations in Section 9.

In line with Theorem 1, F-BPV is FN

$$\widetilde{BPV}(\boldsymbol{v}) = \mathcal{L}(V_{min}, \check{P}, \check{P}, V_{max}, h(\cdot|\boldsymbol{v}), k(\cdot|\boldsymbol{v})) \tag{49}$$

which approximates the quoted price $\check{P}$. F-BPV is described by its membership function μ_{BPV} determined separately for undervalued assets fulfilling the condition (30), fully valued assets fulfilling the condition (32), and overvalued assets fulfilling the condition (31). Figure 1a–c shows a graph of these membership functions.

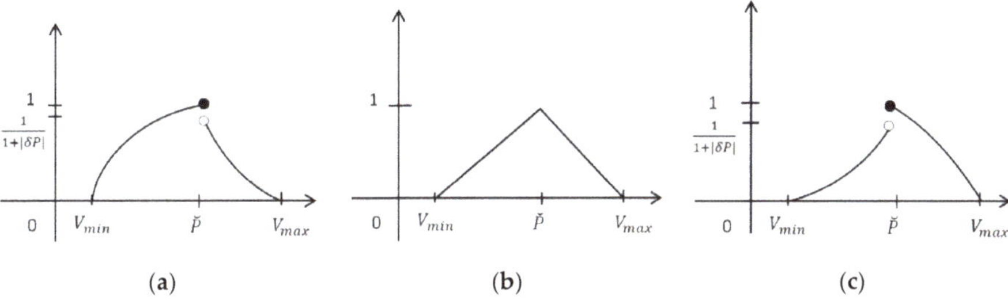

Figure 1. A graphs of membership function of F-BPV (**a**) for overvalued assets, (**b**) for fully valued assets and (**c**) for undervalued assets.

Example 2. *In Example 1, we have evaluated the asset $\mathcal{Y}$ by means of I-BPV. This asset is represented by the parameters vector $\boldsymbol{v} = (60\$, 30\$, 30\$, 80\$)$. Now, using (28–30) we evaluate an asset $\mathcal{Y}$ by F-BPV equal to FN*

$$\widetilde{BPV}(\boldsymbol{v}) = \mathcal{L}(30, 60, 60, 80, h(\cdot|\boldsymbol{v}), k(\cdot|\boldsymbol{v})), \tag{50}$$

where

$$h(x|\boldsymbol{v}) = \frac{4x-120}{x+60} \text{for } x \in [30, 60[, \tag{51}$$

$$k(x|\boldsymbol{v}) = \frac{3x-240}{x-140} \text{ for } x \in]60, 80]. \tag{52}$$

In line with (48), the membership function $\mu_{BPV}(\cdot|\boldsymbol{v}) \in [0,1]^{\mathbb{R}}$ F-BPV is given in following way:

$$\mu_{BPV}(x|\boldsymbol{v}) = \begin{cases} 0, & x \notin [30, 80], \\ \frac{4x-120}{x+60}, & x \in [30, 60[, \\ 1, & x \in [60, 60], \\ \frac{3x-240}{x-140}, & x \in]60, 80]. \end{cases} \tag{53}$$

We see that F-BPV is fuzzy extension of PV model proposed by Rotschedla et al. [102].

8. Behavioural Present Value Represented by Oriented Fuzzy Numbers

Let us give the fixed parameter vector $\boldsymbol{v} = (\check{P}, P_0, V_{min}, V_{max})$ representing evaluated asset. In the previous chapter, we have considered its F-BPV represented by FN (46). The behavioural nature of investors is discussed in [133]. Among other things, this discussion shows that investors are also guided by their subjective predictions of quoted price closest changes. If we take into account these predictions, then we substitute F-BPV by oriented BPV (O-BPV) given as OFN

$$\overleftrightarrow{BPV} = \overleftrightarrow{\mathcal{L}}(V_s, \check{P}, \check{P}, V_e, S_{BPV}, E_{BPV}), \tag{54}$$

where

$$[V_s, V_e] \in \{[V_{min}, V_{max}], [V_{max}, V_{min}]\} \tag{55}$$

is the interval of all PPV,

$$S_{BPV}(x) = \begin{cases} h(x|\boldsymbol{v}) & V_s < V_e, \\ k(x|\boldsymbol{v}) & V_s > V_e, \end{cases} \tag{56}$$

$$E_{BPV}(x) = \begin{cases} k(x|\boldsymbol{v}) & V_s < V_e, \\ h(x|\boldsymbol{v}) & V_s > V_e. \end{cases} \tag{57}$$

The membership function $\mu_{BPV}(\cdot|\boldsymbol{v})$ of OFN $\overleftrightarrow{BPV}(\boldsymbol{v})$ is given by the identity (46).

Positive O-BPV orientation predicts a rise in assets price. Then, O-BPV is given by the formula

$$\overleftrightarrow{BPV}(\boldsymbol{v}) = \overleftrightarrow{\mathcal{L}}(V_{min}, \check{P}, \check{P}, V_{max}, h(\cdot|\boldsymbol{v}), k(\cdot|\boldsymbol{v})). \tag{58}$$

In this way, we obtain three cases of O-BPV predicting a rise in asset price: for overvalued assets, for fully valued assets and for undervalued assets. The membership functions of these O-BPV kinds are presented in Figure 2a–c.

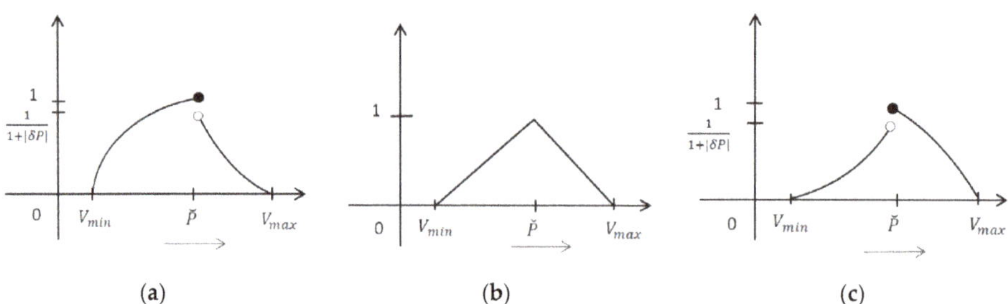

Figure 2. A graph of membership function of O-BPV predicting rise in price (**a**) for overvalued assets, (**b**) for fully valued assets and (**c**) for undervalued assets.

Negative O-BPV orientation predicts a fall in asset price. Then, O-BPV is given by the formula

$$\overleftrightarrow{BPV}(\boldsymbol{v}) = \overleftrightarrow{\mathcal{L}}(V_{max}, \check{P}, \check{P}, V_{min}, k(\cdot|\boldsymbol{v}), h(\cdot|\boldsymbol{v})). \tag{59}$$

In this way, we obtain three cases of O-BPV predicting fall in asset price: for overvalued assets, for fully valued assets and for undervalued assets. The membership functions of these OBPV kinds are presented in Figure 3a–c.

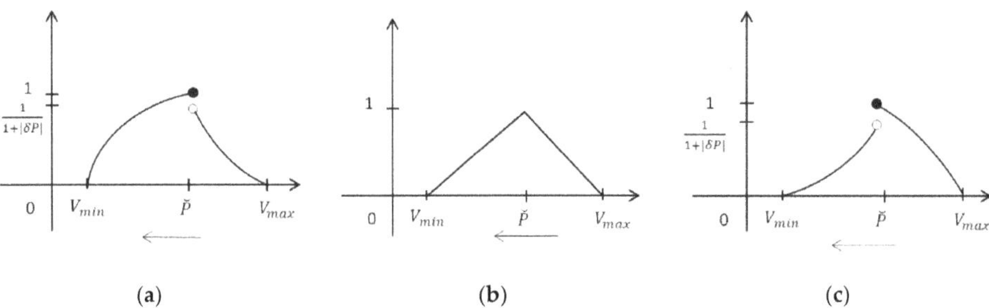

Figure 3. A graph of membership function of O-BPV predicting fall in price (**a**) for overvalued assets, (**b**) for fully valued assets and (**c**) for undervalued assets.

Example 3. *Among other things, in Example 1 we show that asset $\mathcal{Y}$ is overvalued. In Example 2, we evaluate asset $\mathcal{Y}$ by means of F-BPV (50).*

Andrew and Helen are two people whose subjective forecasts of the change of future prices differ. Contrary to the recommendations of the economic theory, Andrew believes that $\mathcal{Y}$ quotations will increase in the near future. Therefore, he evaluates the asset $\mathcal{Y}$ by O-BPV:

$$\overleftrightarrow{BPV}(\boldsymbol{v}) = \overleftrightarrow{\mathcal{L}}(30, 60,\ 60, 80,\ h(\cdot|\boldsymbol{v}), k(\cdot|\boldsymbol{v})), \tag{60}$$

where the functions $h(\cdot|\boldsymbol{v})$ and $k(\cdot|\boldsymbol{v})$ are given, respectively, by (51) and (52). The O-BPV (60) is positively oriented. Its membership function is determined by (53).

In line with the economic theory, Helen is sure that the $\mathcal{Y}$ quotations will decrease in the near future. Therefore, she evaluates the asset $\mathcal{Y}$ by O-BPV:

$$\overleftrightarrow{BPV}(\boldsymbol{v}) = \overleftrightarrow{\mathcal{L}}(80, 60,\ 60, 30,\ k(\cdot|\boldsymbol{v}), h(\cdot|\boldsymbol{v})). \tag{61}$$

The O-BPV (61) is negatively oriented. Its membership function is given by (53). Moreover, this membership function may be equivalently determined as follows:

$$\mu_{BPV}(x|\boldsymbol{v}) = \begin{cases} 0, & x \notin [80,\ 30], \\ \frac{3x-240}{x-140}, & x \in [80, 60[, \\ 1, & x \in [60, 60], \\ \frac{4x-120}{x+60}, & x \in]60, 30]. \end{cases} \tag{62}$$

Let us note that membership functions (53) and (62) have the same graphs.

Each membership function of F-BPV or O-BPV is represented by a graph called the shortly BPV graph. The main objective of the presentation in Figures 1–3 is to show the similarities between BPV graphs dedicated to different kinds of assets related to the same vector $(P_0, P_{min}, P_{max}, \alpha)$. We see that these graphs are similar. In particular, for considered case, we have here

- all overvalued assets have identical BPV graphs,
- all fully valued assets have identical BPV graphs,
- all undervalued assets have identical BPV graphs.

Therefore, we can conclude that a BPV graph and a BPV orientation are independent characteristics of BPV.

If we change vector $(P_0, P_{min}, P_{max}, \alpha)$, then obtained BPV graphs will only differ from one another in volatility range and its intensity of convexity.

9. Oriented Expected Return Determined by Behavioural Present Value

In this section, we apply O-BPV to determining the return rate similarly as Japanese candles were used in [42]. We will consider the asset represented parameters vector $\boldsymbol{v} = \left(\check{P}, P_0, V_{min}, V_{max}\right)$.

For given due date $t > 0$, the considered asset is characterised by following values:

- Predicted FV V_t,
- Evaluated PV V_0.

The benefits from owning this asset are characterised with use of a simple return rate r_t determined as follows:

$$r_t = \frac{V_t - V_0}{V_0} = \frac{V_t}{V_0} - 1. \tag{63}$$

In [42], it is justified in detail that FV is a random variable $\widetilde{V}_t : \Omega \to \mathbb{R}^+$ where the symbol Ω denotes a space of all elementary states ω of the financial market. In a conventional approach to a return rate estimation, an asset PV is equal to quoted price $\check{P}$. Then, the return rate is a random variable given in the following way:

$$r_t(\omega) = \frac{\widetilde{V}_t(\omega) - \check{P}}{\check{P}}. \tag{64}$$

We define any risk as a possibility of negative effects of taken actions. Uncertainty risk results from the lack of knowledge about the future conditions of the activities undertaken. In a financial analysis, an uncertainty risk is usually described by the probability distribution of return rate (64). The expected value $\bar{r}$ of this distribution is called expected return rate. We can assume that expected return rate $\bar{r}$ exists. The mentioned probability distribution can always be described by its cumulative distribution function $F_r(\cdot|\bar{r}) : \mathbb{R} \to [0,1]$. From (64), we immediately get

$$\widetilde{V}_t(\omega) = \check{P} \cdot (1 + r_t(\omega)). \tag{65}$$

If we take together (63) and (65), then we obtain the following formula describing the return rate:

$$r_t = r_t(V_0, \omega) = \frac{\check{P} \cdot (1 + r_t(\omega))}{V_0} - 1. \tag{66}$$

It implies that the expected return rate is given by formula

$$\mathcal{R}(V_0) = \int_{-\infty}^{+\infty} \frac{\check{P} \cdot (1+y)}{V_0} - 1 dF_r(y|\bar{r}) = \frac{\check{P} \cdot (1+\bar{r})}{V_0} - 1. \tag{67}$$

In this manner, we determine the expected return rate $\mathcal{R} : \mathbb{R}^+ \to \mathbb{R}$ as a unary operator transforming PV. If PV is imprecisely estimated by O-PV, then using the Kosinski's approach, we define the expected return rate by an extension $\overleftrightarrow{\mathcal{R}} : \mathbb{K} \to \mathbb{K}$ of a unary operator (67).

We consider now the case of O-PV equal to O-BPV $\overleftrightarrow{BPV}(\boldsymbol{v})$ given by (54). Its starting and ending functions are strictly monotonic. Therefore, the identities (21), (26), (27), and (67) imply that the expected return rate is given by an equation

$$\overleftrightarrow{\mathcal{L}}\left(\frac{\check{P} \cdot (1+\bar{r})}{V_s} - 1, \bar{r}, \bar{r}, \frac{\check{P} \cdot (1+\bar{r})}{V_e} - 1, S_R, E_R\right) = \overleftrightarrow{\mathcal{R}}\left(\overleftrightarrow{\mathcal{L}}(V_s, \check{P}, \check{P}, V_e, S_{BPV}, E_{BPV})\right), \tag{68}$$

where

$$\forall_{r \in \left[\frac{\check{P} \cdot (1+\bar{r})}{V_s} - 1, \bar{r}\right[} : \quad S_R(r) = S_{BPV}\left(\mathcal{R}^{-1}(r)\right) = S_{BPV}\left(\check{P} \cdot \frac{1+\bar{r}}{1+r}\right), \tag{69}$$

$$\forall_{r \in \left]\bar{r}, \frac{\check{P} \cdot (1+\bar{r})}{V_e} - 1\right]} : \quad E_R(r) = E_{BPV}\left(\mathcal{R}^{-1}(r)\right) = E_{BPV}\left(\check{P} \cdot \frac{1+\bar{r}}{1+r}\right), \tag{70}$$

where the interval $[V_s, V_e]$ is determined by (55). If we compare (69) and (70) with (56) and (57), then we get

$$S_R(r) = \begin{cases} h\left(\check{P}\cdot\frac{1+\bar{r}}{1+r}\middle|\boldsymbol{v}\right), & V_s < V_e, \\ k\left(\check{P}\cdot\frac{1+\bar{r}}{1+r}\middle|\boldsymbol{v}\right), & V_s > V_e, \end{cases} \tag{71}$$

$$E_R(r) = \begin{cases} k\left(\check{P}\cdot\frac{1+\bar{r}}{1+r}\middle|\boldsymbol{v}\right), & V_s < V_e, \\ h\left(\check{P}\cdot\frac{1+\bar{r}}{1+r}\middle|\boldsymbol{v}\right), & V_s > V_e. \end{cases} \tag{72}$$

The identities (54) and (68) show that O-BPV and the expected return rate determined by are it appositively oriented. Therefore, we can say the following.

- If O-BPV describes a subjective belief about rise in quotations, then we can anticipate a decline in the expected return rate.
- If O-BPV describes a subjective belief about fall in quotations, then we can anticipate an upturn in the expected return rate.

In finance, both of above facts are well known. This observation proves that the extension of F-PV model to the case of O-PV model is an appropriate direction for the development of fuzzy finance theory.

Example 4. *The asset $\mathcal{Y}$ is overvalued. Despite this, Andrew believes that $\mathcal{Y}$ quotations will increase in the near future. Therefore, he evaluates the asset $\mathcal{Y}$ by positively oriented O-BPV (60). On the other hand, the $\mathcal{Y}$ quotations are characterised by an expected quarterly return rate $\bar{r} = 0.02$. If Andrew determines the expected return rate $\overset{\leftrightarrow}{\mathcal{R}}$ with use O-BPV, then he gets*

$$\overset{\leftrightarrow}{\mathcal{R}} = \overset{\leftrightarrow}{\mathcal{L}}(1.04,\ 0.02,\ 0.02, -0.235, S_R,\ E_R) \tag{73}$$

where

$$S_R(r) = \frac{124.8 - 120\cdot r}{121.2 + 60\cdot r}, \tag{74}$$

$$E_R(r) = \frac{56.4 + 240\cdot r}{201.2 + 140\cdot r}. \tag{75}$$

We see that expected return is negatively oriented. Moreover, Andrew shows that membership function of the expected return rate $\overset{\leftrightarrow}{\mathcal{R}}$ is given as follows

$$\rho(r|\boldsymbol{v}) = \mu_{BPV}\left(\frac{61.2}{1+r}\middle|\boldsymbol{v}\right) = \begin{cases} 0, & r \notin [1.04,\ -0.235], \\ \frac{124.8-120\cdot r}{121.2+60\cdot r}, & r \in [1.04,\ 0.02[, \\ 1, & r \in [0.02,\ 0.02], \\ \frac{56.4+240\cdot r}{201.2+140\cdot r}, & r \in]0.02,\ -0.235]. \end{cases} \tag{76}$$

10. Conclusions

Apart from the theory of interest, any PV is an ambiguous value determined under the influence of, among others, behavioural premises. This view was fully substantiated by the literature study presented in Section 2. This sufficiently proves the need to use soft computing techniques for PV evaluation.

For this reason, in Section 5, BPV is generally defined as the set of all real numbers equal to possible PVs. It is obvious that BPV is an imprecise number. In this paper, we discuss BPV approximation given as following kinds of imprecision numbers:

- I-BPV approximated by interval numbers in Section 6,
- F-BPV approximated by FNs in Section 7,
- O-BPV approximated by OFNs in Section 8.

Interval numbers are definitely a poorer form of information than FNs. For this reason, we should always replace I-BPV with F-BPV. This replacement does not require any additional data. In [126], it is shown that oriented PV application for a portfolio analysis is more useful than the analogous application of fuzzy PV. This makes the use of O-BPV more preferred than the use of F-BPV.

Each of the proposed BPV models is determined by a specific membership function. We are of the opinion that each of the above-mentioned models can be described by means of different membership functions. The search for new membership function proposals may be very fruitful direction for further research.

In [134,135], it is shown that BPV may be valued by intuitionistic FNs [136]. We believe that other types of imprecise numbers can also be used as BPV models. Looking for such opportunities is an interesting direction for further research. However, we must remember that each proposed modelling method for BPV should be carefully justified by serious financial or behavioural reasons. Proposing new BPV models, researchers should also remember about the results contained in [137].

Section 9 describes in detail an application of O-BPV for determining return rate. This result facilitates the use of O-BPV for an analysis of assets with PV estimated by OFN. It is expedient to further develop the fuzzy finance theory based on OFN. On the other hand, the O-BPV may be applied in such algorithms based on financial technical analysis which support invest-making. For example, here we can apply the advice-making algorithms described in [61,62]. Moreover, we can use O-BPV as input signal to fuzzy or neuro-fuzzy systems explained in detail in [138]. The assessment of the suitability of BPV to support decisions is an interesting direction for further research.

Author Contributions: Conceptualisation, K.P.; methodology, K.P. and A.Ł.-H.; validation, A.Ł.-H.; formal analysis, K.P.; investigation, K.P. and A.Ł.-H.; writing—original draft preparation, K.P.; writing—review and editing, K.P. and A.Ł.-H.; visualisation, A.Ł.-H. Both authors have read and agreed to the published version of the manuscript.

Funding: This research received no external funding.

Institutional Review Board Statement: Non applicsable.

Informed Consent Statement: Non applicable.

Data Availability Statement: Non applicable.

Acknowledgments: Authors are very grateful to the anonymous reviewers for their insightful and constructive comments and suggestions.

Conflicts of Interest: The authors declare no conflict of interest.

References

1. Piasecki, K. Basis of Financial Arithmetic from the Viewpoint of the Utility Theory. *Oper. Res. Decis.* **2012**, *22*, 37–53.
2. Dubois, D.; Prade, H. Operations on Fuzzy Numbers. *Int. J. Syst. Sci.* **1978**, *9*, 613–629. [CrossRef]
3. Ward, T.L. Discounted Fuzzy Cash Flow Analysis. *Fall Ind. Eng. Conf. Proc.* **1985**, *476*, 481.
4. Buckley, J.J. The Fuzzy Mathematics of Finance. *Fuzzy Sets Syst.* **1987**, *21*, 257–273. [CrossRef]
5. Greenhut, J.G.; Norman, G.; Temponi, C.T. Towards a Fuzzy Theory of Oligopolistic Competition. In Proceedings of the 3rd International Symposium on Uncertainty Modeling and Analysis and Annual Conference of the North American Fuzzy Information Processing Society, College Park, MD, USA, 17–20 September 1995; IEEE: College Park, MD, USA; pp. 286–291.
6. Sheen, J.N. Fuzzy Financial Profitability Analyses of Demand Side Management Alternatives from Participant Perspective. *Inf. Sci.* **2005**, *169*, 329–364. [CrossRef]
7. Gutierez, I. Fuzzy Numbers and Net Present Value. *Scand. J. Mgmt.* **1989**, *5*, 149–159. [CrossRef]
8. Kuchta, D. Fuzzy Capital Budgeting. *Fuzzy Sets Syst.* **2000**, *111*, 367–385. [CrossRef]
9. Lesage, C. Discounted Cash-flows Analysis. An Interactive Fuzzy Arithmetic Approach. *Eur. J. Econ. Soc. Syst.* **2001**, *15*, 49–68. [CrossRef]
10. Huang, X. Two New Models for Portfolio Selection with Stochastic Returns Taking Fuzzy Information. *Eur. J. Oper. Res.* **2007**, *180*, 396–405. [CrossRef]
11. Tsao, C.T. Assessing the Probabilistic Fuzzy Net Present Value for a Capital, Investment Choice Using Fuzzy Arithmetic. *J. Chin. Ins. Ind. Eng.* **2005**, *22*, 106–118. [CrossRef]
12. Calzi, M.L. Towards a General Setting for the Fuzzy Mathematics of Finance. *Fuzzy Sets Syst.* **1990**, *35*, 265–280. [CrossRef]
13. Piasecki, K. Behavioural Present Value. *SSRN Electron. J.* **2011**, *1*, 1–17. [CrossRef]

14. Baerecke, T.; Bouchon-Meunier, B.; Detyniecki, M. Fuzzy Present Value. In Proceedings of the 2011 IEEE Computational Intelligence for Financial Engineering and Economics, Paris, France, 11–15 April 2011; IEEE Symposium: Paris, France, 2011; pp. 1–6.
15. Biswas, P.; Pramanik, S. Fuzzy Approach to Replacement Problem with Value of Money Changes with Time. *Int. J. Comput. Appl.* **2011**, *30*, 28–33. [CrossRef]
16. Nosratpour, M.; Nazeri, A.; Meftahi, H. Fuzzy Net Present Value for Engineering Analysis. *Manag. Sci. Lett.* **2012**, *2*, 2153–2158. [CrossRef]
17. Piasecki, K. Effectiveness of Securities with Fuzzy Probabilistic Return. *Oper. Res. Decis.* **2011**, *21*, 65–78.
18. Piasecki, K. On Imprecise Investment Recommendations. *Stud. Log. Gramm. Rhetor.* **2014**, *37*, 179–194. [CrossRef]
19. Li, X.; Quin, Z.; Kar, S. Mean-Variance-Skewness Model for Portfolio Selection with Fuzzy Returns. *Eur. J. Oper. Res.* **2010**, *202*, 239–247. [CrossRef]
20. Quin, Z.; Wen, M.; Gu, C. Mean-absolute Deviation Portfolio Selection Model with Fuzzy Returns. *Iran. J. Fuzzy Syst.* **2011**, *8*, 61–75.
21. Tsaur, R. Fuzzy Portfolio Model with Different Investor Risk Attitudes. *Eur. J. Oper. Res.* **2013**, *227*, 385–390. [CrossRef]
22. Tanaka, H.; Guo, P.; Turksen, B. Portfolio Selection Based on Fuzzy Probabilities and Possibility Distributions. *Fuzzy Sets Syst.* **2000**, *111*, 387–397. [CrossRef]
23. Duan, L.; Stahlecker, P. A Portfolio Selection Model Using Fuzzy Returns. *Fuzzy Optim. Decis. Mak.* **2011**, *10*, 167–191. [CrossRef]
24. Guo, H.; Sun, B.; Karimi, H.R.; Ge, Y.; Jin, W. Fuzzy Investment Portfolio Selection Models Based on Interval Analysis Approach. *Math. Probl. Eng.* **2012**. [CrossRef]
25. Gupta, P.; Mittal, G.; Mehlawat, M.K. Multiobjective Expected Value Model for Portfolio Selection in Fuzzy Environment. *Optim. Lett.* **2013**, *7*, 1765–1791. [CrossRef]
26. Gupta, P.; Mittal, G.; Saxena, A. Asset Portfolio Optimization Using Fuzzy Mathematical Programming. *Inf. Sci.* **2008**, *178*, 1734–1755. [CrossRef]
27. Huang, X.; Shen, W. Multi-period Mean-variance Model with Transaction Cost for Fuzzy Portfolio Selection. In Proceedings of the Seventh International Conference on Fuzzy Systems and Knowledge Discovery, Yantai, Shandong, China, 10–12 August 2010. [CrossRef]
28. Liu, Y.; Sun, L. Comparative Research of Portfolio Model Using Fuzzy Theory. In Proceedings of the Fourth International Conference on Fuzzy Systems and Knowledge Discovery, Haikou, China, 24–27 August 2007. [CrossRef]
29. Liu, Y.J.; Zhang, W.G. Fuzzy Portfolio Optimization Model Under Real Constraints. *Insur. Math. Econ.* **2013**, *53*, 704–711. [CrossRef]
30. Wu, X.L.; Liu, Y.K. Optimizing Fuzzy Portfolio Selection Problems by Parametric Quadratic Programming. *Fuzzy Optim. Decis. Mak.* **2012**, *11*, 411–449. [CrossRef]
31. Wang, S.; Zhu, S. On Fuzzy Portfolio Selection Problems. *Fuzzy Optim. Decis. Mak.* **2002**, *1*, 361–377. [CrossRef]
32. Mehlawat, M.K. Credibilistic Mean-entropy Models for Multi-period Portfolio Selection with Multi-choice Aspiration Levels. *Inf. Sci.* **2016**, *345*, 9–26. [CrossRef]
33. Kahraman, C.; Ruan, D.; Tolga, E. Capital Budgeting Techniques Using Discounted Fuzzy Versus Probabilistic Cash Flows. *Inf. Sci.* **2002**, *142*, 57–76. [CrossRef]
34. Fang, Y.; Lai, K.K.; Wang, S. Fuzzy Portfolio Optimization. Theory and Methods. In *Lecture Notes in Economics and Mathematical Systems*; Springer: Berlin, Germany, 2008; Volume 609.
35. Gupta, P.; Mehlawat, M.K.; Inuiguchi, M.; Chandra, S. Fuzzy Portfolio Optimization. Advances in Hybrid Multi-criteria Methodologies. In *Studies in Fuzziness and Soft Computing*; Springer: Berlin, Germany, 2014; Volume 316. [CrossRef]
36. Kosiński, W.; Prokopowicz, P.; Ślęzak, D. Fuzzy Numbers with Algebraic Operations: Algorithmic Approach. In Proceedings of the Eight International Conference on Information, Logistics & Supply Chain–ILS 2020, Austin, TX, USA, 22–24 April 2020; Klopotek, M., Wierzchoń, S.T., Michalewicz, M., Eds.; Physica Verlag: Sopot, Poland; Springer: Berlin/Heidelberg, Germany, 2002; pp. 311–320.
37. Kosiński, W. On Fuzzy Number Calculus. *Int. J. Appl. Math. Comput. Sci.* **2006**, *16*, 51–57.
38. Piasecki, K. Revision of the Kosiński's Theory of Ordered Fuzzy Numbers. *Axioms* **2018**, *7*, 16. [CrossRef]
39. Prokopowicz, P. The Directed Inference for the Kosinski's Fuzzy Number Model. In Proceedings of the Second International Afro-European Conference for Industrial Advancement, Villejuif, France, 9–11 September 2015.
40. Prokopowicz, P.; Pedrycz, W. The Directed Compatibility Between Ordered Fuzzy Numbers—A Base Tool for a Direction Sensitive Fuzzy Information Processing. In *Advances in Inteligent Systems and Computing*; Abraham, A., Wegrzyn-Wolska, K., Hassanien, A.E., Snasel, V., Alimi, A.M., Eds.; Springer: Cham, Switzerland, 2015; Volume 127, pp. 493–505. [CrossRef]
41. Prokopowicz, P.; Pedrycz, W. The Directed Compatibility between Ordered Fuzzy Numbers–A Base Tool for a Direction Sensitive Fuzzy Information Processing. In Proceedings of the Artificial Intelligence and Soft Computing ICAISC 2015, Lecture Notes in Computer Science. Zakopane, Poland, 14–18 June 2015; Rutkowski, L., Korytkowski, M., Scherer, R., Tadeusiewicz, R., Zadeh, L., Zurada, J., Eds.; Springer: Cham, Switzeland, 2015; Volume 9119. [CrossRef]
42. Piasecki, K. Relation "Greater Than or Equal to" Between Ordered Fuzzy Numbers. *Appl. Syst. Innov.* **2019**, *2*, 26. [CrossRef]
43. Piasecki, K.; Łyczkowska-Hanćkowiak, A. Representation of Japanese Candlesticks by Oriented Fuzzy Numbers. *Economics* **2020**, *8*, 1. [CrossRef]

44. Prokopowicz, P.; Czerniak, J.; Mikołajewski, D.; Apiecionek, Ł.; Ślęzak, D. Analysis of Temporospatial Gait Parameters. In *Theory and Applications of Ordered Fuzzy Number*; Springer: Berlin, Germany, 2017; Volume 356, pp. 289–302.
45. Kacprzak, D.; Kosiński, W.; Kosiński, W.K. Financial Stock Data and Ordered Fuzzy Numbers. In Proceedings of the Artificial Intelligence and Soft Computing. ICAISC 2013, Zakopane, Poland, 9–13 June 2013; Lecture Notes in Computer Science. Springer: Berlin/Heidelberg, Germany, 2013; Volume 7894, pp. 259–270.
46. Kacprzak, D.; Kosiński, W. Optimizing Firm Inventory Costs as a Fuzzy Problem. *Stud. Log. Gramm. Rhetor.* **2014**, *37*, 89–105. [CrossRef]
47. Chwastyk, A.; Pisz, I. OFN Capital Budgeting Under Uncertainty and Risk. In *Theory and Applications of Ordered Fuzzy Number*; Prokopowicz, P., Czerniak, J., Mikołajewski, D., Apiecionek, Ł., Ślęzak, D., Eds.; Springer: Berlin, Germany, 2017; Volume 356, pp. 157–170.
48. Czerniak, J.M. OFN Ant Method Based on TSP Ant Colony Optimization. In *Theory and Applications of Ordered Fuzzy Number*; Prokopowicz, P., Czerniak, J., Mikołajewski, D., Apiecionek, Ł., Ślęzak, D., Eds.; Springer: Berlin, Germany, 2017; Volume 356, pp. 207–222.
49. Ewald, D.; Czerniak, J.M.; Paprzycki, M. A New OFN Bee Method as an Example of Fuzzy Observance Applied for ABC Optimization. In *Theory and Applications of Ordered Fuzzy Number*; Prokopowicz, P., Czerniak, J., Mikołajewski, D., Apiecionek, Ł., Ślęzak, D., Eds.; Springer: Berlin, Germany, 2017; Volume 356, pp. 223–238.
50. Kacprzak, D. Input-Output Model Based on Ordered Fuzzy Numbers. In *Theory and Applications of Ordered Fuzzy Number*; Prokopowicz, P., Czerniak, J., Mikołajewski, D., Apiecionek, Ł., Ślęzak, D., Eds.; Springer: Berlin, Germany, 2017; Volume 356, pp. 171–182.
51. Kacprzak, D. Objective Weights Based on Ordered Fuzzy Numbers for Fuzzy Multiple Criteria Decision-making Methods. *Entropy* **2017**, *19*, 373. [CrossRef]
52. Marszałek, A.; Burczyński, T. Ordered fuzzy candlesticks. In *Theory and Applications of Ordered Fuzzy Number*; Prokopowicz, P., Czerniak, J., Mikołajewski, D., Apiecionek, Ł., Ślęzak, D., Eds.; Springer: Berlin, Germany, 2017; Volume 356, pp. 183–194. [CrossRef]
53. Piasecki, K. Expected Return Rate Determined as Oriented Fuzzy Number. In Proceedings of the 35th International Conference Mathematical Methods in Economics Conference, Hradec Králové, Czech Republic, 13–15 September 2017; pp. 561–565.
54. Roszkowska, E.; Kacprzak, D. The Fuzzy SAW and Fuzzy TOPSIS Procedures Based on Ordered Fuzzy Numbers. *Inf. Sci.* **2016**, *369*, 564–584. [CrossRef]
55. Rudnik, K.; Kacprzak, D. Fuzzy TOPSIS Method with Ordered Fuzzy Numbers for Flow Control in a Manufacturing System. *Appl. Soft Comput.* **2017**, *52*, 1020–1041. [CrossRef]
56. Zarzycki, H.; Czerniak, J.M.; Dobrosielski, W.T. Detecting Nasdaq Composite Index Trends with OFNs. In *Theory and Applications of Ordered Fuzzy Number*; Prokopowicz, P., Czerniak, J., Mikołajewski, D., Apiecionek, Ł., Ślęzak, D., Eds.; Springer: Berlin, Germany, 2017; Volume 356, pp. 195–206.
57. Łyczkowska-Hanćkowiak, A.; Piasecki, K. Two-assets Portfolio with Trapezoidal Oriented Fuzzy Present values. In Proceedings of the 36th International Conference Mathematical Methods in Economics Conference, Jindřichův Hradec, Czech Republic, 12–14 September 2018; pp. 306–311.
58. Łyczkowska-Hanćkowiak, A.; Piasecki, K. Present Value of Portfolio of Assets with Present Values Determined by Trapezoidal Ordered Fuzzy Numbers. *Oper. Res. Decis.* **2018**, *28*, 41–56.
59. Piasecki, K.; Roszkowska, E. On Application of Ordered Fuzzy Numbers in Ranking Linguistically Evaluated Negotiation Offers. *Adv. Fuzzy Syst.* **2018**. [CrossRef]
60. Piasecki, K.; Roszkowska, E.; Łyczkowska-Hanćkowiak, A. Simple Additive Weighting Method Equipped with Fuzzy Ranking of Evaluated Alternatives. *Symmetry* **2019**, *11*, 482. [CrossRef]
61. Piasecki, K.; Roszkowska, E.; Łyczkowska-Hanćkowiak, A. Impact of the Orientation of the Ordered Fuzzy Assessment on the Simple Additive Weighted Method. *Symmetry* **2019**, *11*, 1104. [CrossRef]
62. Łyczkowska-Hanćkowiak, A. Sharpe's Ratio for Oriented Fuzzy Discount Factor. *Mathematics* **2019**, *7*, 272. [CrossRef]
63. Łyczkowska-Hanćkowiak, A. On Application Oriented Fuzzy Numbers for Imprecise Investment Recommendations. *Symmetry* **2020**, *12*, 1672. [CrossRef]
64. Piasecki, K.; Siwek, J. Behavioural Present Value Defined as Fuzzy Number a New Approach. *Folia Oeconomica Stetin.* **2015**, *15*, 27–41. [CrossRef]
65. Łyczkowska-Hanćkowiak, A. Behavioural Present Value Determined by Ordered Fuzzy Number. *SSRN Electron. J.* **2017**. [CrossRef]
66. Peccati, L. Su di Una Caratterizzazione del Principio del Criterio Dell'attualizzazione. In *Studium Parmense*; Università di Parma: Parma, Italy, 1972.
67. Janssen, J.; Manca, R.; Volpe di Prignano, E. *Mathematical Finance. Deterministic and Stochastic Models*; John Wiley & Sons: London, UK, 2009.
68. Ramsey, F.P.A. Mathematical Theory of Saving. *Econ. J.* **1928**, *38*, 543–559. [CrossRef]
69. Samuelson, P.A. A Note on Measurement of Utility. *Rev. Econ. Stud.* **1937**, *4*, 155–161. [CrossRef]
70. Koopmans, T.C.; Diamond, P.A.; Williamson, R.E. Stationary Utility and Time Perspective. *Econom. J. Econom. Soc.* **1964**, *32*, 82–100. [CrossRef]

71. Strotz, R.H. Myopia and Inconsistency in Dynamic Utility Maximization. *Rev. Econ. Stud.* **1955**, *23*. [CrossRef]
72. Loewenstein, G.; Prelec, D. Anomalies in Intertemporal Choice: Evidence and Interpretation. *Q. J. Econ.* **1992**, *107*, 573–597. [CrossRef]
73. Frederick, S.; Loewenstein, G.; O'Donoghue, T. Time Discounting and Time Preference: A Critical Review. *J. Econ. Lit.* **2002**, *40*, 351–401. [CrossRef]
74. Streich, P.; Levy, J.S. Time Horizons, Discounting, and Intertemporal Choice. *J. Confl. Resolut.* **2007**, *51*, 199–226. [CrossRef]
75. Thaler, R.H. Some Empirical Evidence on Dynamic Inconsistency. *Econ. Lett.* **1981**, *8*. [CrossRef]
76. Herrnstein, R.J. Rational Choice Theory: Necessary but not Sufficient. *Am. Psychol.* **1990**, *45*. [CrossRef]
77. Mazur, J.E. An Adjusting Procedure for Studying Delayed Reinforcement. In *The Effect of Delay and of Intervening Events on Reinforcement Value. Quantitative Analysis of Behavior*; Commons, M.L., Mazur, J.E., Nevin, J.A., Rachlin, H., Eds.; Lawrence Erlbaum Associates, Inc.: Mahwah, NJ, USA, 1987; pp. 55–73.
78. Ainslie, G. Specious Reward: A Behavioral Theory of Impulsiveness and Impulse Control. *Psychol. Bull.* **1975**, *82*. [CrossRef]
79. Herrnstein, R.J. Self-control as Response Strength. In *Quantification of Steady-state Operant Behavior*; Elsevier/North Holland Biomedical Press: Amsterdam, The Netherlands, 1981.
80. Doyle, J.R. Survey of Time Preference, Delay Discounting Model. *Judgm. Decis. Mak.* **2013**, *8*, 116–135. [CrossRef]
81. Killeen, P.R. An Additive-utility Model of Delay Discounting. *Psychol. Rev.* **2009**, *116*, 602–619. [CrossRef] [PubMed]
82. Rachlin, H. Notes on Discounting. *J. Exp. Anal. Behav.* **2006**, *85*, 425–435. [CrossRef]
83. Laibson, D. Golden Eggs and Hyperbolic Discounting. *Q. J. Econ.* **1997**, *112*, 443–477. [CrossRef]
84. Benhabib, J.; Bisin, A.; Schotter, A. Present Bias, Quasi-hyperbolic Discounting, and Fixed Costs. *Games Econ. Behav.* **2010**, *69*, 205–223. [CrossRef]
85. Commons, M.L. How Reinforcement Density is Discriminated and Scaled. In *Quantitative Analyses of Behavior. Discriminative Properties of Reinforcement Schedules*; Commons, M.L., Nevin, J.A., Eds.; Ballinger: Cambridge, MA, USA, 1981; Volume 1.
86. Commons, M.L.; Woodford, M.; Ducheny, J.R. The Relationship Between Perceived Density of Reinforcement in a Schedule Sample Audits Reinforcing Value. In *Quantitative Analysis of Behavior. Matching and Maximizing Accounts*; Commons, M.L., Nevin, J.A., Eds.; Ballinger: Cambridge, MA, USA, 1982; Volume 2.
87. Davison, M.C. Preference for Mixed-interval Versus Fixed-interval Schedules. *J. Exp. Anal. Behav.* **1969**, *12*, 247–252. [CrossRef]
88. Du, W.; Green, L.; Myerson, J. Cross-cultural Comparisons of Discounting Delayed and Probabilistic Rewards. *Psychol. Rec.* **2002**, *52*, 479–492. [CrossRef]
89. Roelofsma, P.H.M.P. Modeling Intertemporal Choices: An Anomaly Approach. *Acta Psychol.* **1996**, *93*, 5–22. [CrossRef]
90. Ebert, J.E.; Prelec, D. The Fragility of Time: Time-insensitivity and Valuation of the Near and Future. *Manag. Sci.* **2007**, *53*, 1423–1438. [CrossRef]
91. Myerson, J.; Green, L. Discounting of Delayed Rewards: Models of Individual Choice. *J. Exp. Anal. Behav.* **1995**, *64*, 263–276. [CrossRef] [PubMed]
92. Read, D. Is time Discounting Hyperbolic or Subadditive? *J. Risk Uncertain.* **2001**, *23*, 5–32. [CrossRef]
93. Masin, S.C.; Zudini, V.; Antonelli, M. Early Alternative Derivations of Fechner's Law. *J. Hist. Behav. Sci.* **2009**, *45*, 56–65. [CrossRef] [PubMed]
94. Bleichrodt, H.; Rohde, K.I.M.; Wakker, P.P. Non-hyperbolic Time Inconsistency. *Games Econ. Behav.* **2009**, *66*, 27–38. [CrossRef]
95. Stevens, S.S. On the Psychophysical Law. *Psychol. Rev.* **1957**, *64*, 153–181. [CrossRef]
96. Stevens, S.S. To Honor Fechner and Repeal His Law. *Science* **1961**, *133*, 80–86. [CrossRef] [PubMed]
97. Piasecki, K. Discounting Under Impact of Temporal Risk Aversion—A Case of Discrete Time. *Res. Pap. Wrocław Univ. Econ.* **2015**, *381*, 289–298. [CrossRef]
98. Piasecki, K. Discounting Under Impact of Temporal Risk Aversion—A Case of Continuous Time. In *Economic Development and Management of Regions*; Jedlička, P., Ed.; Gaudeamus, University of Hradec Kralove: Hradec Králové, Chezh Republic, 2015; Volume 5.
99. Piasecki, K. Discounting Under Impact of Risk Aversion. *SSRN Electron. J.* **2015**. [CrossRef]
100. Ok, E.A.; Masatlioglu, Y. A Theory of (Relative) Discounting. *J. Econ. Theory* **2007**, *137*, 214–245. [CrossRef]
101. Dubra, J. A Theory of Time Preferences Over Risky Outcomes. *J. Math. Econ.* **2009**, *45*, 576–588. [CrossRef]
102. Rotschedl, J.; Kaderabkova, B.; Čermáková, K. Parametric Discounting Model of Utility. *Procedia Econ. Financ.* **2015**, *30*, 730–741. [CrossRef]
103. Von Mises, L. *The Ultimate Foundation of Economic Science. An Essay on Method*; D. Van Nostrand Company, Inc.: Princeton, NJ, USA, 1962.
104. Tversky, A.; Kahneman, D. Availability: A Heuristic for Judging Frequency and Probability. *Cogn. Psychol.* **1973**, *5*, 207–232. [CrossRef]
105. Tversky, A.; Kahneman, D. Judgment Under Uncertainty: Heuristic and Biases. *Science* **1974**, *185*, 1124–1131. [CrossRef]
106. Barberis, N.; Shleifer, A.; Vishny, R. A Model of Investor Sentiment. *J. Financ. Econ.* **1998**, *49*, 307–343. [CrossRef]
107. Daniel, K.; Hirshleifer, D.; Subrahmanyam, A. Overconfidence, Arbitrage and Equilibrium Asset Pricing. *J. Financ.* **2001**, *56*, 921–965. [CrossRef]
108. Hong, H.; Stein, J. A Unified Theory of Under Reaction, Momentum Trading and Over Reaction in Asset Market. *J. Financ.* **1999**, *54*, 2143–2184. [CrossRef]

109. Ainslie, G.; Haendel, V. The Motives of the Will. In *Etiologic Aspects of Alcohol and Drug Abuse, Springfield*; Gottheil, E., Druley, K.A., Skoloda, T.E., Waxman, H., Eds.; Charles, C. Thomas: Springfield, IL, USA, 1983; pp. 119–140.
110. Kirby, K.N. Bidding on the Future: Evidence Against Normative Discounting of Delayed Rewards. *J. Exp. Psychol. Gen.* **1997**, *126*, 54–70. [CrossRef]
111. Kirby, K.N.; Marakovič, N.N. Modeling Myopic Decision: Evidence for Hyperbolic Delay-Discounting with Subjects and Amounts. *Organ. Behav. Hum. Decis. Process.* **1995**, *64*, 22–30. [CrossRef]
112. Kahneman, D.; Tversky, A. Prospect Theory: An Analysis of Decision under Risk. *Econometica* **1979**, *47*, 263–292. [CrossRef]
113. Loewenstein, G. Frames of Mind in Intertemporal Choice. *Manag. Sci.* **1988**, *34*, 200–214. [CrossRef]
114. Shelley, M.K. Outcome Signs, Question Frames and Discount Rates. *Manag. Sci.* **1993**, *39*, 806–815. [CrossRef]
115. Kirby, K.N.; Santiesteban, M. Concave Utility, Transaction Costs and Risk in Measuring Discounting of Delayed Rewards. *J. Exp. Psychol. Learn. Mem. Cogn.* **2003**, *29*, 66–79. [CrossRef]
116. Fishburn, P.; Edwards, W. Discount Neutral Utility Models for Denumerable Time Streams. *Theory Decis.* **1997**, *43*, 139–166. [CrossRef]
117. Dacey, R.; Zielonka, P. A Detailed Prospect Theory Explanation of the Disposition Effect. *J. Behav. Financ.* **2008**, *9*, 43–50. [CrossRef]
118. Zauberman, G.; Kyu Kim, B.; Malkoc, S.; Bettman, J.R. Discounting Time and Time Discounting: Subjective Time Perception and Intertemporal Preferences. *J. Mark. Res.* **2009**, *46*, 543–556. [CrossRef]
119. Kim, B.K.; Zauberman, G. Perception of Anticipatory Time in Temporal Discounting. *J. Neurosci. Psychol. Econ.* **2009**, *2*, 91–101. [CrossRef]
120. Epper, T.; Fehr-Duda, H.; Bruhin, A. Uncertainty Breeds Decreasing Impatience: The Role of Risk Preferences in Time Discounting. *SSRN Electron. J.* **2009**, *412*. [CrossRef]
121. Kontek, K. Decision Utility Theory: Back to von Neumann, Morgenstern, and Markowitz. *SSRN Electron. J.* **2009**. [CrossRef]
122. Rabin, M. Incorporating Fairness into Game Theory and Economics. *Am. Econ. Rev.* **1993**, *83*, 1281–1302.
123. Delgado, M.; Vila, M.A.; Voxman, W. On a Canonical Representation of Fuzzy Numbers. *Fuzzy Sets Syst.* **1998**, *93*, 125–135. [CrossRef]
124. Goetschel, R.; Voxman, W. Elementary Fuzzy Calculus. *Fuzzy Sets Syst.* **1986**, *18*, 31–43. [CrossRef]
125. Piasecki, K.; Stasiak, M.D. The Forex Trading System for Speculation with Constant Magnitude of Unit Return. *Mathematics* **2019**, *7*, 623. [CrossRef]
126. Piasecki, K.; Łyczkowska-Hanćkowiak, A. Oriented Fuzzy Numbers vs. Fuzzy Numbers. *Mathematics* **2021**, *9*, 523. [CrossRef]
127. Klir, G.J. Developments in Uncertainty-based Information. *Adv. Comput.* **1993**, *36*, 255–332. [CrossRef]
128. Ross, S.A. The Arbitrage Theory of Capital Asset Pricing. *J. Econ. Theory* **1976**, *13*, 341–360. [CrossRef]
129. Fama, E.F. Efficient Capital Markets: A Review of Theory and Empirical Work. *J. Financ.* **1970**, *25*, 383–417. [CrossRef]
130. Grossman, S.J.; Stiglitz, J.E. On the Impossibility of Informationally Efficient Markets. *Am. Econ. Rev.* **1980**, *70*, 393–408.
131. Edwards, W. Conservatism in Human Information Processing. In *Formal Representation of Human Judgment*; Klienmutz, B., Ed.; Wiley: New York, NY, USA, 1968; pp. 17–52.
132. Tversky, A. Features of Similarity. *Psychol. Rev.* **1977**, *84*, 327–352. [CrossRef]
133. Akerlof, G.A.; Shiller, R.I. *Animal Spirits: How Human Psychology Drives the Economy, and Why It Matters for Global Capitalism*; Princeton University Press: Princeton, NJ, USA, 2009.
134. Piasecki, K. On Return Rate Estimated by Intuitionistic Fuzzy Probabilistic Set. In *Mathematical Methods in Economics MME 2015*; Martincik, D., Ircingova, J., Janecek, P., Eds.; Faculty of Economics, University of West Bohemian: Plzen, Poland, 2015; pp. 641–646.
135. Piasecki, K. The Intuitionistic Fuzzy Investment Recommendations. In Proceedings of the Mathematical Methods in Economics MME 2016 Conference Proceedings, Liberec, Czech Republic, 6–9 September 2016; Kocourek, A., Vavroušek, M., Eds.; Technical University of Liberec: Liberec, Czech Republic, 2016; pp. 681–686.
136. Atanassov, K.T. Intuitionistic Fuzzy Sets. *Fuzzy Sets Syst.* **1986**, *20*, 87–96. [CrossRef]
137. Bustince, H.; Barrenechea, E.; Pagola, M.; Fernandez, J.; Xu, Z.; Bedregal, B.; Montero, J.; Hagras, H.; Herrera, F.; De Baets, B. A Historical Account of Types of Fuzzy Sets and Their Relationships. *IEEE Trans. Fuzzy Syst.* **2016**, *24*, 179–194. [CrossRef]
138. Versaci, M.; Calcagno, S.; Cacciola, M.; Morabito, F.; Palamara, I.; Pellicanò, D. Standard Soft Computing Techniques for Characterization of Defects in Nondestructive Evaluation. In *Ultrasonic Nondestructive Evaluation Systems*; Burrascano, P., Callegari, S., Montisci, A., Ricci, M., Versaci, M., Eds.; Springer: Cham, Switzeland, 2015.

MDPI

Article

An Enhanced Spectral Clustering Algorithm with S-Distance

Krishna Kumar Sharma [1,2], Ayan Seal [1,3,*], Enrique Herrera-Viedma [4,5] and Ondrej Krejcar [3,6]

1 Department of Computer Science & Engineering, PDPM Indian Institute of Information Technology, Design and Manufacturing, Jabalpur 482005, India; krisshna.sharma@gmail.com or krishnakumarsharma@iiitdmj.ac.in
2 Department of Computer Science & Informatics, University of Kota, Kota, Rajasthan 324022, India
3 Center for Basic and Applied Science, Faculty of Informatics and Management, University of Hradec Králové, Hradec 50003 Králové, Czech Republic; ondrej.krejcar@uhk.cz
4 Andalusian Research Institute in Data Science and Computational Intelligence (DaSCI), University of Granada, 18071 Granada, Spain; viedma@decsai.ugr.es
5 Faculty of Engineering, King Abdulaziz University, Jeddah 21589, Saudi Arabia
6 Malaysia-Japan International Institute of Technology (MJIIT), Universiti Teknologi Malaysia, Kuala Lumpur 54100, Malaysia
* Correspondence: ayan@iiitdmj.ac.in

Abstract: Calculating and monitoring customer churn metrics is important for companies to retain customers and earn more profit in business. In this study, a churn prediction framework is developed by modified spectral clustering (SC). However, the similarity measure plays an imperative role in clustering for predicting churn with better accuracy by analyzing industrial data. The linear Euclidean distance in the traditional SC is replaced by the non-linear S-distance (*Sd*). The *Sd* is deduced from the concept of S-divergence (*SD*). Several characteristics of *Sd* are discussed in this work. Assays are conducted to endorse the proposed clustering algorithm on four synthetics, eight UCI, two industrial databases and one telecommunications database related to customer churn. Three existing clustering algorithms—*k*-means, density-based spatial clustering of applications with noise and conventional SC—are also implemented on the above-mentioned 15 databases. The empirical outcomes show that the proposed clustering algorithm beats three existing clustering algorithms in terms of its Jaccard index, f-score, recall, precision and accuracy. Finally, we also test the significance of the clustering results by the Wilcoxon's signed-rank test, Wilcoxon's rank-sum test,and sign tests. The relative study shows that the outcomes of the proposed algorithm are interesting, especially in the case of clusters of arbitrary shape.

Keywords: S-divergence; S-distance; spectral clustering

Citation: Kumar Sharma, K.; Seal, A.; Herrera-Viedma, E.; Krejcar, O. An Enhanced Spectral Clustering Algorithm with S-Distance. *Symmetry* **2021**, *13*, 596. https://doi.org/10.3390/sym13040596

Academic Editors: Kóczy T. László and István A. Harmati

Received: 4 March 2021
Accepted: 25 March 2021
Published: 2 April 2021

1. Introduction

Advancements in information technology have given rise to digital innovation in the service industry; e.g., the e-commerce industry, banking, telecom, airline industry, etc. [1]. Customers now have easy access to enormous amounts of data for their desired service or consumables. This in turn has generated a scenario in which companies are finding it a very difficult task to retain their existing customer base. Companies have thus become more cautious to increase customer acquisition and to control customer churn. Consumers switching from one firm to another for a specified period negatively impacts the economy of the company. Thus, customer acquisition and churn management have become key factors for the service sector. Several methods exist to effectively increase customer acquisition and to manage customer churn, such as improving customer acquisition by nurturing effective relationship with customers, identifying the customers who are likely to leave and giving proactive solutions to the causes of their dissatisfaction, improving sales approaches and improving marketing strategies and customer services. Technology is also responsible for the reframing of marketing to increase customer loyalty through the examination of stored information and customer metrics. It also allows customer relations to be connected with business demand [2]. However, the problem of identifying the best set of clients who can subscribe to a product or service is considered NP-hard [3].

It is important to utilize and allocate resources effectively and efficiently by distinguishing high-value customers. It is also imperative for industrial enterprises to customize marketing strategies in such a way that they can achieve an edge over their competitors. There is a need to use an unsupervised machine learning clustering algorithm in order to group customers according to some similarity or common trend, especially when the customer database is growing constantly, when the average transaction size increases or when the frequency of transactions per customer increases. In other words, a clustering algorithm helps to analyze the different needs of different groups of customers or to customize marketing strategies for an organization to acquire customers and to manage customer churn. These problems can be handled using analytical methods, which use the concepts of statistics, machine learning and data mining [4]. In [5], data mining by evolutionary learning using the genetic algorithm was presented for churn predictions for telecom subscriber data. Machine learning algorithms—for instance, decision tree (DT) and neural networks (NN)—have been exploited to predict customer churn by considering billing information, demographics, call detail, contract/service status, records and service change logs [6]. An approach to recognizing potential bank customers who may react to a promotional offer in direct marketing based on customer historical data using support vector machine (SVM) was presented in [7]. Churn prediction in online games using records of players' login was addressed using the k-nearest neighbors (KNN) algorithm in [8,9]. Various machine learning algorithms such as logistic regression, DT, NN and SVM were adopted and compared to anticipate the victory of telemarketing calls for selling bank long-term investments in [10]. SVM [11] and KNN [12] were also used to predict potential online buyers based on browser session data and hypertext transfer protocol level information. In [13], deciding the active demand reduction potential of wet appliances was considered and solved using the expectation-maximization clustering algorithm. Hierarchical and fuzzy k-means clustering were compared in order to improve business models in demand response programs [14]. In [15], density and grid-based (DGB), density-based spatial clustering of applications with noise (DBSCAN), fast search and find of density peaks (FSFDP) and other clustering algorithms were exploited for DNA microarray industrial data, finding DGB is more suitable for clustering databases with arbitrary shapes than the traditional approaches. E-customer behavior characterization was done by utilizing Web server log data using association rules in [16].

It can be observed from the literature that almost all the conventional unsupervised machine learning algorithms have been exploited in industrial applications, especially in churn prediction, by analyzing the behaviors of customers. However, the performance of an unsupervised clustering algorithm relies on data/features, similarity/distance measure, objective functions, initial cluster centers and the clustering algorithm itself. The similarity measure plays an important role in disclosing hidden patterns and understanding the massive industrial data properly. A substantial amount of research work has been done for the study of clustering using various linear distance measures such as Euclidean, Manhattan, Pearson correlation, Eisen cosine correlation, Spearman correlation, Kendall correlation, Bit-Vector, Hamming, the Jaccard Index and the Dice Index, but this has drawn little attention, especially in terms of introducing non-linearity into similarity measures for data clustering [17,18]. Surprisingly few of these approaches do not abide by the triangle inequality property [19]. The aim of investigating non-linearity in clustering algorithms is to identify a more accurate boundary between two groups. The Bregman divergence was considered as a measure of similarity and merged with the traditional k-means to increase its efficacy in [19]. Currently, a few studies on various divergence-based similarity measures in clustering are underway [20,21]. In this work, the spectral clustering (SC) algorithm is adopted and modified using the non-linear S-distance (Sd), which is obtained from the S-divergence (SD). Some characteristics of Sd are also discussed in this study. The proposed

SC algorithm is implemented on four toy databases, eight real-world UCI databases, two service industrial databases and one telecommunications database related to customer churn. The proposed SC algorithm is compared with conventional SC algorithms; i.e., the SC algorithm with linear Euclidean distance (*Ed*) [22], *k*-means [22] and DBSCAN [15]. All the achieved outcomes show that the proposed clustering algorithm performs better than the three existing approaches.

The rest of the article is structured as follows: *Sd* and its properties are presented in Section 2. The graph Laplacian and its characteristics are shown in Section 3. The modified SC algorithm and its proof of convergence are addressed in Section 4. Section 5 presents empirical outcomes and discussion. Section 6 concludes the work.

2. S-Distance and Its Properties

In d-dimensional Euclidean space $\Re^d_+$, $\mathbf{p}$ and $\mathbf{q}$ are two points [23]. Equation (1) is employed to compute the *Sd*.

Definition 1. $dist_s : \Re^d_+ \times \Re^d_+ \rightarrow \Re_+ \cup \{0\}$ as

$$dist_s^2(\mathbf{p},\mathbf{q}) = \sum_{l=1}^{d}[\log((p_l+q_l)/2) - (\log(p_l)+\log(q_l))/2] \tag{1}$$

Let f be an injective function stated as $f : \Re^d_+ \rightarrow M_d$ such that f($\mathbf{p}$) = diag$((p_1, p_2, ..., p_d))$, where M_d represents the positive definite matrices of size $d \times d$. Thus, the *Sd* is well-stated. The *Sd* is obtained from the idea of SD, which is denoted arithmetically by Equation (2).

$$dist_{sd}{}^2(P,Q) = \log\left(\left|\frac{P+Q}{2}\right|\right) - \frac{\log(|P|)+\log(|Q|)}{2}, \tag{2}$$

where $|.|$ is a determinant of a matrix and $dist_s(\mathbf{p},\mathbf{q}) = dist_{sd}(f(\mathbf{p}), f(\mathbf{q}))$. At the moment, we ensure that *Sd* meets all the characteristics for becoming a metric. The characteristics are given below:

Proposition 1. *Non-negativity:* $dist_s(\boldsymbol{p},\boldsymbol{q}) \geq 0$

Proof. The modified form of Equation (1) is presented below:
$dist_s^2(\mathbf{p},\mathbf{q}) = \Sigma_{l=1}^d[\log((p_l+q_l)/2) + \log((p_lq_l)^{\frac{-1}{2}})]$

$$\implies \Sigma_{l=1}^d\left[\log\left(\frac{(p_l+q_l)}{2\sqrt{p_lq_l}}\right)\right] = \Sigma_{l=1}^d\left[\log\left(\frac{1}{2}\left(\sqrt{\frac{p_l}{q_l}}+\sqrt{\frac{q_l}{p_l}}\right)\right)\right] \geq 0$$

$\therefore dist_s(\mathbf{p},\mathbf{q}) \geq 0$ □

Proposition 2. *Equality:* $dist_s(\boldsymbol{p},\boldsymbol{q}) = 0$ *iff* $\boldsymbol{p} = \boldsymbol{q}$.

Proof. Proposition 2 can be written as $dist_s^2(\mathbf{p},\mathbf{q}) = \Sigma_{l=1}^d\left[\log\left(\frac{1}{2}\left(\sqrt{\frac{p_l}{q_l}}+\sqrt{\frac{q_l}{p_l}}\right)\right)\right]$ Now, if $\mathbf{p}$ and $\mathbf{q}$ are the same then $\mathbf{q}$ can be substituted by $\mathbf{p}$ in the above Equation and the adjusted Equation is $dist_s^2(\mathbf{p},\mathbf{q}) = \Sigma_{l=1}^d\left[\log\left(\frac{1}{2}\left(\sqrt{\frac{p_l}{p_l}}+\sqrt{\frac{p_l}{p_l}}\right)\right)\right] \implies d[\log(1)] = 0$

$\therefore dist_s(\mathbf{p},\mathbf{q}) = 0$ iff $p = q$. □

Proposition 3. *Symmetry:* $dist_s^2(\boldsymbol{p},\boldsymbol{q}) = dist_s^2(\boldsymbol{q},\boldsymbol{p})$

Proof. The *Sd* amid $\mathbf{p}$ and $\mathbf{q}$ is expressed as given below:
$dist_s^2(\mathbf{p},\mathbf{q}) = \Sigma_{l=1}^d\left[\log\left(\frac{1}{2}\left(\sqrt{\frac{p_l}{q_l}}+\sqrt{\frac{q_l}{p_l}}\right)\right)\right]$ [as already noted in Proposition 1] $= dist_s^2(\mathbf{q},\mathbf{p})$

$\therefore dist_s(\mathbf{p},\mathbf{q}) = dist_s(\mathbf{q},\mathbf{p})$. Thisimplies that the *Sd* also abides by the symmetric metric property. □

Proposition 4. *Triangle Inequality: In d-dimensional Euclidean space* $\Re^d_+$, $\boldsymbol{p}$, $\boldsymbol{q}$ *and* $\boldsymbol{o}$ *are any three points. Then, this proposition states that the sum of any two sides—namely,* $dist_s(\boldsymbol{p},\boldsymbol{o})$ *and* $dist_s(\boldsymbol{o},\boldsymbol{q})$*—of a triangle is equal to or exceeds the length of the third side* $dist_s(\boldsymbol{p},\boldsymbol{q})$*. Mathematically,* $dist_s(\boldsymbol{p},\boldsymbol{q}) \leq dist_s(\boldsymbol{p},\boldsymbol{o}) + dist_s(\boldsymbol{o},\boldsymbol{q})$.

Proof. The following can be written by utilizing propositions 1 and 2:
$dist_s(\mathbf{p},\mathbf{q}) \geq 0$, $dist_s(\mathbf{p},\mathbf{o}) \geq 0$, and $dist_s(\mathbf{o},\mathbf{q}) \geq 0$
$\therefore dist_s(\mathbf{p},\mathbf{q}) \leq dist_s(\mathbf{p},\mathbf{o}) + dist_s(\mathbf{o},\mathbf{q})$.
□

Thus, *Sd* is a metric. At this time, some of the characteristics of *Sd* are presented below:

Theorem 1. *Sd is not a Bregman divergence.*

Proof. This may be demonstrated by refutation. At the beginning, we can assume that the *Sd* is a Bregman divergence. This implies that $dist_s(\mathbf{p},\mathbf{q})$ is rigorously convex in $\mathbf{p}$. It will be necessary to demonstrate that $dist_s(\mathbf{p},\mathbf{q})$ is not convex in $\mathbf{p}$.

Take the double derivative for both sides of the following Equation with regard to p_l. Then,

$$\frac{\partial dist_s^2}{\partial p_l} = \frac{1}{p_l+q_l} - \frac{1}{2p_l}$$

If $l \neq r$, then $\frac{\partial^2 dist_s^2}{\partial p_r \partial p_l} = 0$;

otherwise, $\frac{\partial^2 dist_s^2}{\partial p_l^2} = \frac{-1}{(p_l+q_l)^2} + \frac{1}{(2p_l)^2}$ We get $\frac{\partial^2 dist_s^2}{\partial p_l^2} < 0$ only when $q_l < (\sqrt{2}-1)p_l$ $\forall l \in \{1,\ldots,d\}$, we get $\frac{\partial^2 dist_s^2}{\partial p_l^2} \leq 0$ for $l \in \{1,\ldots,d\}$. Thus, a Hassian matrix that has negative diagonal entries would be attained. So, we have verified that the *Sd* is not a Bregman divergence. □

Theorem 2. *The* $a \circ \boldsymbol{p}$ *is employed to denote the Hadamard product amid a and* $\boldsymbol{p}$*. Then, this can be written* $dist_s^2(\boldsymbol{a}\circ\boldsymbol{p},\boldsymbol{a}\circ\boldsymbol{q}) = dist_s^2(\boldsymbol{p},\boldsymbol{q})$ *for* $\boldsymbol{a} \in \Re^d_+$.

Proof. This can be written as $\mathbf{a}\circ\mathbf{p} = (a_1p_1,\ldots,a_dp_d)$ as per Hadamard product. Thus,
$dist_s^2(a_lp_l,aq_l) = \log((a_lp_l+a_lq_l)/2) - 0.5(\log(a_lp_l)+\log(a_lq_l)) = \log((a_l(p_l+q_l))/2) - 0.5(\log(p_l)+\log(q_l)+2\log(a_l))$
$= \log((p_l+q_l)/2) - 0.5(\log(p_l)+\log(q_l))$
$= dist_s^2(\mathbf{p},\mathbf{q}) \implies \Sigma_{l=1}^d dist_s^2(a_lp_l,a_lq_l) = \Sigma_{l=1}^d dist_s^2(p_l,q_l)$
$\therefore dist_s^2(\mathbf{a}\circ\mathbf{p},\mathbf{a}\circ\mathbf{q}) = dist_s^2(\mathbf{p},\mathbf{q})$ □

Theorem 3. *Sd is not an f-divergence.*

Proof. If q_l is substituted by p_lv_l, where $v_l \in \Re^d_+$ in Equation (1), then

$$dist_s^2(\mathbf{p},\mathbf{q}) = \Sigma_{l=1}^d[\log((p_l+p_lv_l)/2) - (\log(p_l)+\log(p_l)v_l)/2]$$
$$\implies dist_s^2(\mathbf{p},\mathbf{q}) = \Sigma_{l=1}^d[\log((1+v_l)/2) - (\log(v_l))/2]$$
$$\implies \Sigma_{l=1}^d U(v_l) = \Sigma_{l=1}^d U\left(\frac{q_l}{p_l}\right)$$

Thus, $dist_s^2(\mathbf{p},\mathbf{q})$ cannot be denoted as $\Sigma_{l=1}^d p_l diag\left(\frac{q_l}{p_l}\right)$.

$\therefore dist_s()$ is not an f-divergence. □

Remark 1. *Figure 1 displays the line of the norm-balls of the Sd and Ed around the point* $(5000, 5000)$ *in* $\Re^d_+$, *where* $d = 2$. *One can observe from Fig. 1 that the lines of Sd and Ed look like distorted triangles and concentric circles, respectively. Further, the contour plots of Sd approach each other as we get close to the origin. When two points get close to the origin, then Sd would be high. On the contrary, the Sd of two points would be low when they are far from the origin. In contrast, the Ed of two points would be the same in both cases. Thus, Sd works well if the larger clusters are far from the origin and the smaller clusters are nearer to the origin.*

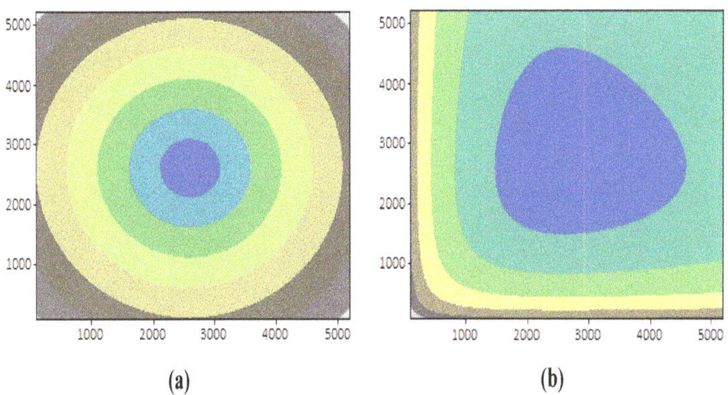

Figure 1. Contour plot of the norm balls for (**a**) *Ed* and (**b**) S-distance (*Sd*).

Remark 2. *Sd abides by the principle of Burbea–Rao divergence in* $\Re^d_+$ *with condition* $f(\boldsymbol{p}) = -\sum_{l=1}^{d} \log(p_l)$. *Thus,* $f(\boldsymbol{p})$ *is convex in* $\Re^d_+$.

3. Graph Laplacian and Its Properties

Consider a database $D = \{\mathbf{p}_1, \ldots, \mathbf{p}_n\}$ with n number of points in d-dimensional Euclidean space, where $\mathbf{p}_i \in \Re^d_+$ expresses the ith point. $W = (\aleph, \Psi, A)$ is another representation of the same database, where $\aleph$ and Ψ is the set of points and edges of these points, respectively. The A is used to express an affinity matrix or a symmetric weighted matrix of the graph W. In order to build a W, we consider the local neighborhood relationships of these points. Some approaches are available in the literature to construct affinity matrices [24]. Despite that, we have utilized a symmetry-favored KNN to increase the modeling of a graph and reduce outliers and the effect of noise. The graph W may be expressed by the underlying manifold characteristics of the data space [25,26]. In SC, the proper selection of the pairwise similarity measure is crucial [24,26]. Equation (3) is employed to produce an asymmetry-weighted matrix $\Pi \in \Re^{n \times n}$ connected to W.

$$\Pi_{i,j} = \begin{cases} \exp\left(-\frac{dist_s^2(\mathbf{p}_i, \mathbf{p}_j)}{dist_s(\mathbf{p}_i, {}^{\mathbf{k}}\mathbf{p}_i) dist_s(\mathbf{p}_j, {}^{\mathbf{k}}\mathbf{p}_j)}\right), & \text{if } \mathbf{p}_j \in dist_{\mathbf{k}}(\mathbf{p}_i) \\ 0, & \text{otherwise} \end{cases}, \tag{3}$$

where $dist_s()$ is the Sd between two data points $\mathbf{p}_i$ and $\mathbf{p}_j$, ${}^{\mathbf{k}}\mathbf{p}_i$ represents the $\mathbf{k}^{th}$ NN of $\mathbf{p}_i \in \aleph$ and $dist_{\mathbf{k}}()$ is the set of KNN of $\mathbf{p}_i$.

The weighted symmetric matrix of graph W is achieved by utilizing Π using Equation (4).

$$A_{i,j} = \begin{cases} 1, & \text{if } \mathbf{p}_j \in dist_{\mathbf{k}}(\mathbf{p}_i) \text{ and } \mathbf{p}_i \in dist_{\mathbf{k}}(\mathbf{p}_j) \\ \Pi_{i,j}, & \text{if } \mathbf{p}_j \in dist_{\mathbf{k}}(\mathbf{p}_i) \text{ and } \mathbf{p}_i \notin dist_{\mathbf{k}}(\mathbf{p}_j) \\ \Pi_{j,i}, & \text{otherwise} \end{cases} \tag{4}$$

Equation (4) is adopted to build the symmetry-favored KNN graph of W. Figure 2 shows a pictorial representation of the difference between a symmetry-favored KNN graph and a KNN graph. The weights of symmetric edges of W are higher than the asymmetric edges because the points associated with symmetric edges belong to the same sub-manifold.

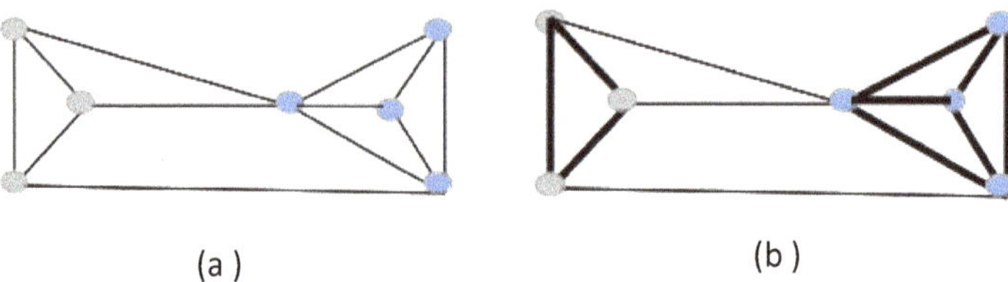

Figure 2. (**a**) 3NN and (**b**) symmetry-favored 3NN (higher edge weights are denoted by bold edges).

The degree matrix $\zeta = \begin{Bmatrix} \sigma_1 & \dots & \dots & \dots \\ \dots & \sigma_2 & \dots & \dots \\ \vdots & \vdots & \sigma_i & \vdots \\ \dots & \dots & \dots & \sigma_n \end{Bmatrix}$, where σ_i is determined using Equation (5)

$$\sigma_i = \sum_{j=1}^{n} \Pi_{i,j} \tag{5}$$

The essential components of an SC are graph Laplacian matrices, which are of two types: normalized and unnormalized [27]. Equation (6) is employed to estimate the unnormalized graph Laplacian matrix.

$$W_{un} = \zeta - A \tag{6}$$

In contrast, Equation (7) is exploited to calculate the normalized graph Laplacian matrix.

$$W_{no} = \zeta^{-\frac{1}{2}} W_{un} \zeta^{\frac{1}{2}} = I - \zeta^{-\frac{1}{2}} A \zeta^{\frac{1}{2}}, \tag{7}$$

where I is an identity matrix. The $\mu_0, \dots, \mu_{n-1}$ and $\tau_0, \dots, \tau_{n-1}$ are the eigenvalues and eigenvectors of W_{no}, respectively. Proposition 5 presents a discussion of the properties of W_{no}.

Proposition 5. *Three properties of W_{no} are given below:*

1. *We have*

$$g^T W_{no} g = \frac{1}{2} \sum_{i,j=1}^{n} a_{i,j} \left(\frac{g_i}{\sqrt{\sigma_i}} - \frac{g_j}{\sqrt{\sigma_j}}\right)^2 \tag{8}$$

   for every $g \in \Re^n$.
2. *W_{no} is symmetric and positive semidefinite.*
3. *W_{no} consists of n non-negative and real-valued eigenvalues $0 = \mu_0 \leq \dots \leq \mu_{n-1}$, where n is the number of points in D.*

4. Proposed Spectral Clustering Algorithm and Analysis

In SC, a graph partitioning problem is approximated in a manner so that low weights are assigned to edges, which are between clusters. This means that the association between clusters is low or clusters are not similar. On the other hand, high edge weights are assigned when clusters are similar. In [28], a similarity graph with an adjacency matrix

A is partitioned by solving the mincut problem. This consists of the selection of partition $B_1, \ldots, B_k$, which minimizes Equation (9).

$$cut(B_1, \ldots, B_k) = \sum_{i=1}^{c} cut(B_i, \overline{B_i}), \tag{9}$$

Here, $\overline{B_i}$ is the complement of B_i, where B_i is a disjoint subset of $\aleph$ points. In reality, the mincut problem does not give us satisfactory partitions. So, this problem can be solved using a normalized cut, Ncut, which is defined by Equation (10).

$$Ncut(B_1, \ldots, B_k) = \sum_{i=1}^{k} \frac{cut(B_i, \overline{B_i})}{vol(B_i)}, \tag{10}$$

The Ncut problem can be relaxed and helps to derive the normalized SC [24]. Equation (11) is used to represent the cluster indicator vector, $g_j = (g_{1,j}, \ldots, g_{n,j})'$.

$$g_{i,j} = \begin{cases} \frac{1}{\sqrt{vol(B_j)}}, & \text{if } \mathbf{p}_i \in B_j \\ 0, & otherwise \end{cases}, \tag{11}$$

where $1 \le i \le n, 1 \le j \le k$ and a matrix G can be constructed as $G = (g_{i,j})_{1 \le i \le n, 1 \le j \le k}$ and $G'G = I$ with $g_i{}'\zeta g_i = 1$ and $g_i{}' W_{no} g_i = 2\frac{cut(B_i, \overline{B_i})}{vol(B_i)}$. So, Equation (12) is utilized to denote the minimization of Ncut.

$$\min_{B_1, \ldots, B_k} Tr(G' W_{no} G), \text{ subject to } G'\zeta G = I \tag{12}$$

where Tr is the trace of a matrix. After relaxing the discreteness condition and replacing $V = \zeta^{\frac{1}{2}} G$, the relaxed problem is as shown in Equation (13):

$$\min_{V \in \Re^{n \times k}} Tr(V' \zeta^{\frac{-1}{2}} W_{no} \zeta^{\frac{-1}{2}} V) \text{ subject to } V'V = I \tag{13}$$

Equation (13) consists of a matrix V that contains the first k eigenvectors of W_{no} as columns. Let $V = \{v_1, \ldots, v_n\}$ be a given set of vectors in $\Re^k_+$. Equation (13) can be further simplified as Equation (14).

$$\min_{V \in \Re^{n \times c}} Tr(V' W_{no} V) \text{ subject to } V'V = I \tag{14}$$

Equation (13) is the trace minimization problem that is solved by a matrix V, containing the first k eigenvectors of W_{no} in columns. We want to assign $v_i \in V$ to any mutually exclusive class such that $2 \le k \le n$. A mathematical way to design this problem as follows:

$$\chi: \text{ minimize } h(Q, C) = \sum_{i=1}^{n} \sum_{j=1}^{k} a_{ij} dist_s^2(v_i, \mathbf{c}_j) \text{ subject to}$$
$$\sum_{j=1}^{k} a_{ij} = 1 \text{ where } a_{ij} \in \{1, 0\} \text{ and } C = \{\mathbf{c}_1, \ldots, \mathbf{c}_k\}, \mathbf{c}_j \in \Re^k_+$$
$$\forall i \in \{1, \ldots, n\}, \forall j \in \{1, \ldots, k\} \tag{15}$$

The solution to the above problem χ uses k-means with Sd, which converges to a local optimal solution of χ in finite iterations [29]. Algorithm 1 shows the modified SC.

Algorithm 1 The proposed SC algorithm.

Input: $\aleph$, **k**, k ▷ $\aleph$—a set of points, nearest neighbors for affinity matrix and number of clusters
Output: Cluster labels of all the points

1. Compute the KNN graph and the weight matrix A using (3)–(4)
2. To get the normalized graph Laplacian W_{no} by (5)–(7) as $W_{no} = \zeta^{-\frac{1}{2}} W_{un} \zeta^{\frac{1}{2}} = I - \zeta^{-\frac{1}{2}} A \zeta^{\frac{1}{2}}$
3. Calculate the k smallest eigenvalues $\{\mu_i\}_{i=1,\ldots,c}$ and their corresponding eigenvectors $\{\sigma_i\}_{i=1,\ldots,k}$ using the affinity matrix W_{no} in (7) and form a matrix $\mathrm{Y} \in \Re^{n\times k}$
4. Convert Y matrix to $\Gamma \in \Re^{n\times k}$ by normalizing the Y such that the rows to have unit length (i.e. $\Gamma_{ij} = \frac{\mathrm{Y}_{ij}}{(\sum_l \mathrm{Y}_{il}^2)^{\frac{1}{2}}}$)
5. Cluster data points $\Gamma_{i=1,\ldots,n} \in \Re^k$ in to k clusters via k-means clustering with either Ed or Sd
6. At the end, allot each point p_i to cluster j if and only if i^{th} row of matrix Γ was allotted to cluster j

5. Experimental Results and Discussion

A laptop Intel(R) Core(TM) i7-2620M CPU@2.70GHz and 4-GB RAM running on Windows 10 with a 64-bit Python 3.6.5 compiler was used for this study. Every aspect of the work was done in the Spyder 3.2.8 Python development environment.

5.1. Database Description

In total, 15 databases of three classes are considered in this work to compare the performance of the proposed clustering algorithm with three existing approaches.

5.1.1. Synthetic Databases

Four synthetic/toy databases were considered. In varied distributed database (DB1), data points are distributed with varied variances. Four concentric circles are present in noisy four-circles (DB2), where each circle represents a class. The blob database (DB3) consists of isotropic Gaussian blobs with three classes. The data point distribution is anisotropic in nature for the anisotropically distributed database (DB4). Table 1 presents the title of the toy databases, the number of sample points of these databases, the number of facets in each point and the number of clusters. The data distribution in the two-dimensional Euclidean space of each of these four synthetic databases is shown in Figure 3. The x-axis and y-axis of each plotted distribution denote feature 1 and feature 2, respectively, as two features are present in each of the toy databases.

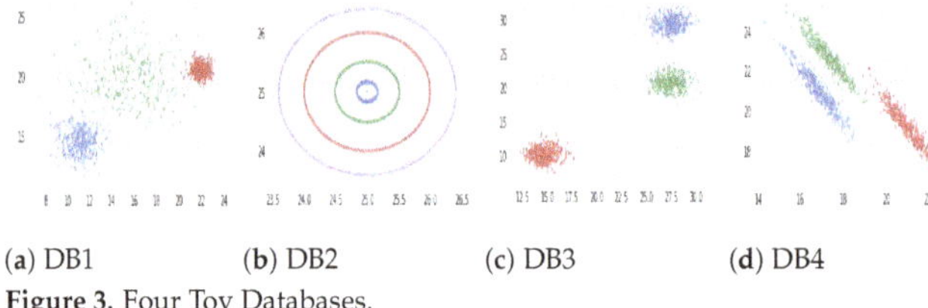

(**a**) DB1 (**b**) DB2 (**c**) DB3 (**d**) DB4

Figure 3. Four Toy Databases.

Table 1. Databases.

S.No.	Category	Databases	Samples	Features	Clusters
1	Synthetic	Varied Distributed Data (DB1)	1500	2	3
2	Synthetic	Noisy 4-Circles (DB2)	1500	2	4
3	Synthetic	Blobs (DB3)	1500	2	3
4	Synthetic	Anisotropically Distributed Data (DB4)	1500	2	3
5	UCI	Avila database (DB5)	10430	10	12
6	UCI	Breast Cancer database (DB6)	569	30	2
7	UCI	Digits database (DB7)	1797	64	10
8	UCI	Iris database (DB8)	150	4	3
9	UCI	Letter Recognition database (DB9)	20000	16	26
10	UCI	Poker Hand database (DB10)	25010	10	10
11	UCI	Shuttle database (DB11)	43500	9	7
12	UCI	Wine database (DB12)	178	13	3
13	Industrial	Banking marketing database (DB13)	45211	12	2
14	Industrial	Online Shoppers' Purchasing Intention (DB14)	12330	18	2
15	Industrial	Telecommunication customer churn database (DB15)	7044	21	2

5.1.2. UCI and Industrial Databases

Eight popular realistic databases—Digits, Iris, Breast cancer, Wine, Avila, Letter, Poker and Shuttle—were adopted from the UCI repository [30,31]. A brief portrayal of these UCI databases is given in Table 1. On the other hand, two industrial databases—namely, Bank telemarketing [10] and Purchasing intention [32]—are considered in this work. A database related to telecommunication customer churn was adopted from the Kaggle repository to study the customer data for retaining and maximizing benefit by devising suitable business plans. Brief details of these databases are given in Table 1. Outliers and data reconciliation are not handled separately in this work. However, normalization was carried out before applying the proposed algorithm to model the data correctly. As mentioned in section 2, *Sd* is defined in d-dimensional Euclidean space $\Re^d_+$; thus, raw data were normalized to obtain a positive scale by shifting data with the absolute of the most negative value such that the most negative value would be the minimum positive non-zero value and all other data points would be positive.

5.2. *Evaluation Indices*

Accuracy is one of the most adopted validation indices. It denotes the ratio of correct outcomes that a machine learning algorithm has attained. The higher the accuracy obtained by an algorithm, the better and more useful that algorithm is. However, this may mislead researchers due to the accuracy paradox. Accuracy adopted along with other indices; for instance, the Jaccard index, f-score, recall, and precision [33–35]. Interested readers are referred to [36] to learn about the various validation indices in depth. Non-parametric statistical hypothesis tests, called the Wilcoxon's signed-rank test, Wilcoxon's rank-sum test and sign test, were conducted as well at the 5% significance level to determine whether two dependent samples were chosen from the data [37,38].

5.3. Results and Discussion

In this study, the proposed SC—i.e., an SC with *Sd* (SC-S)—is compared with the conventional SC (SC-E) [22], *k*-means [22] and DBSCAN [15] on the basis of 14 databases. As we know, an affinity matrix helps to represent data points graphically and the affinity matrix depends on a symmetric favored KNN. So, in the first experiment, two methods—SC-S and SC-E—were executed on four synthetic databases only and the performances were judged based on five validation indices; namely, the Jaccard index, f-score, recall, precision and accuracy. A significant amount of time has been devoted by the research community to deciding the best value of **k** for KNN. Still, this is an open problem. So, the value of *k* is determined based on empirical results in this work. Initially, 10 is considered as the value of **k**. Later on, this reaches 30 with a step size of 5. The achieved Jaccard index, f-score, recall, precision and accuracy using SC-S and SC-E are shown in Figures 4–8, respectively. It is observed from Figures 4–8 that the SC-S always outperforms the SC-E for the five evaluation metrics. Moreover, KNN was stable when the value of *k* was 20, which is used for the rest of the work [25].

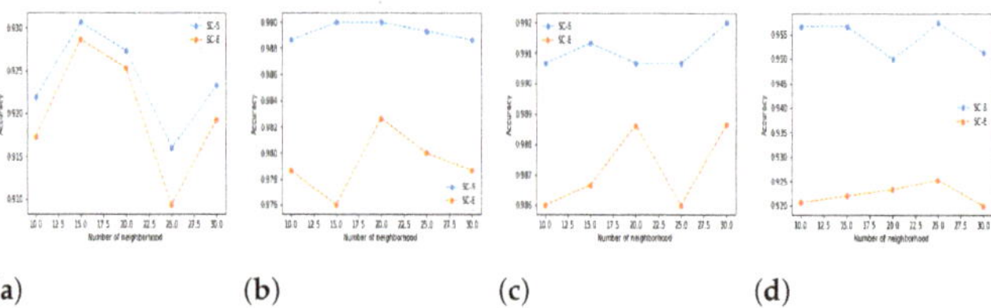

(**a**) (**b**) (**c**) (**d**)

Figure 4. Comparative analysis using the accuracy index on various toy databases using SC-E and SC-S in the case of varying neighborhoods for the construction of an affinity matrix. (**a**) DB1 (**b**) DB2 (**c**) DB3 and (**d**) DB4.

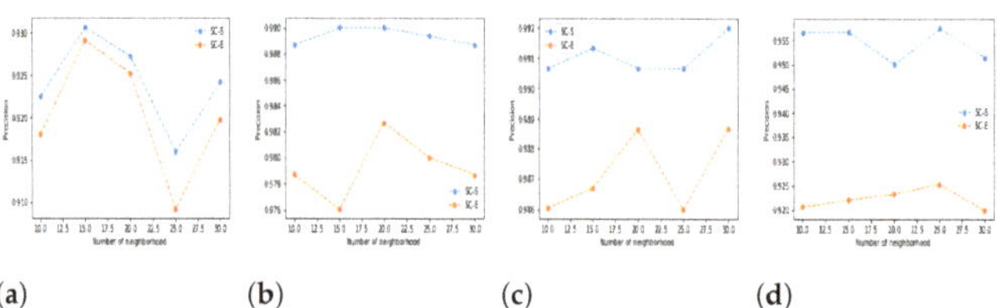

(**a**) (**b**) (**c**) (**d**)

Figure 5. Comparative analysis using the precision index on various toy databases using SC-E and SC-S in the case of varying neighborhoods for the construction of an affinity matrix. (**a**) DB1 (**b**) DB2 (**c**) DB3 and (**d**) DB4.

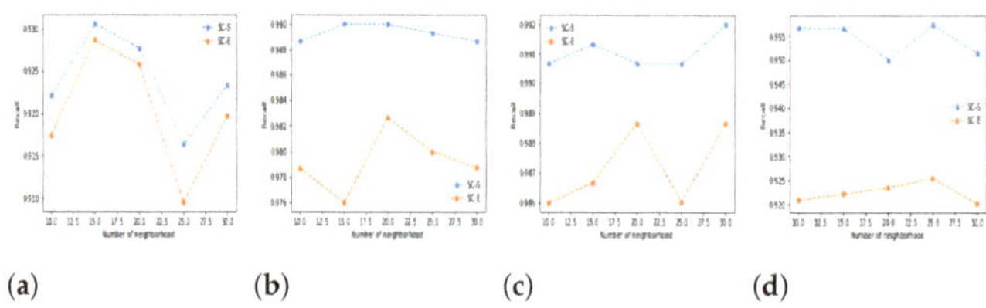

(**a**) (**b**) (**c**) (**d**)

Figure 6. Comparative analysis using the recall index on various toy databases using SC-E and SC-S in the case of varying neighborhoods for the construction of an affinity matrix. (**a**) DB1 (**b**) DB2 (**c**) DB3 and (**d**) DB4.

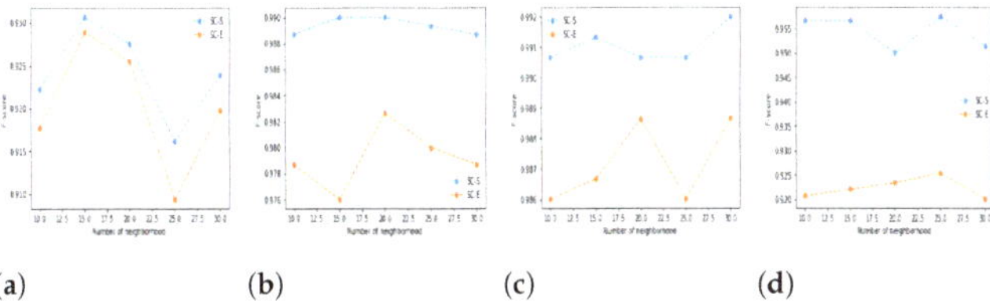

(**a**) (**b**) (**c**) (**d**)

Figure 7. Comparative analysis using the fscore index on various toy databases using SC-E and SC-S in the case of varying neighborhoods for the construction of an affinity matrix. (**a**) DB1 (**b**) DB2 (**c**) DB3 and (**d**) DB4.

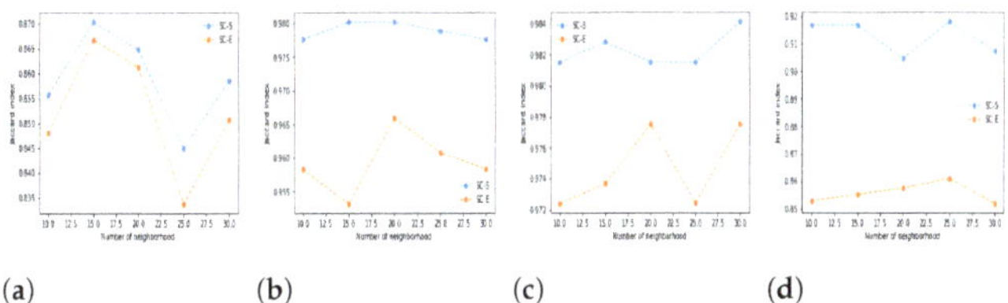

(**a**) (**b**) (**c**) (**d**)

Figure 8. Comparative analysis using the Jaccard index on various toy databases using SC-E and SC-S in the case of varying neighborhoods for the construction of an affinity matrix. (**a**) DB1 (**b**) DB2 (**c**) DB3 and (**d**) DB4.

In the second experiment, the SC-S was compared with SC-E, *k*-means and DBSCAN on 14 databases. Figures 9–12 show the data distribution of each toy database separately after applying four clustering algorithms. Here, different colors are used to denote different clusters. The number of colors depends on the number of clusters in each database. However, the colors are assigned in each database randomly. So, no color is used to fix a particular cluster. It is clear from Figures 9–12 that the *k*-means performs worst compared to the rest of the three methods, but it is difficult to comment on these three methods with regard only to Figures 9–12. In Fig. 10, the result of *k*-means shows that *k*-means works better in the case of spherical data only. While the other methods perform better compared to *k*-means, more information is required to say more about the four clustering algorithms. Figure 13 shows the obtained Jaccard index, f-score, recall, precision and accuracy using the four clustering algorithms. Here, two parameters—the radius (Eps) and a minimum number of points (MinPts)—are required to execute DBSCAN. The values of Eps and MinPts are 0.5 and 3, respectively [39]. Figure 13 illustrates that the proposed clustering algorithm SC-S is the best among the four used clustering algorithms in terms of the five evaluation metrics.

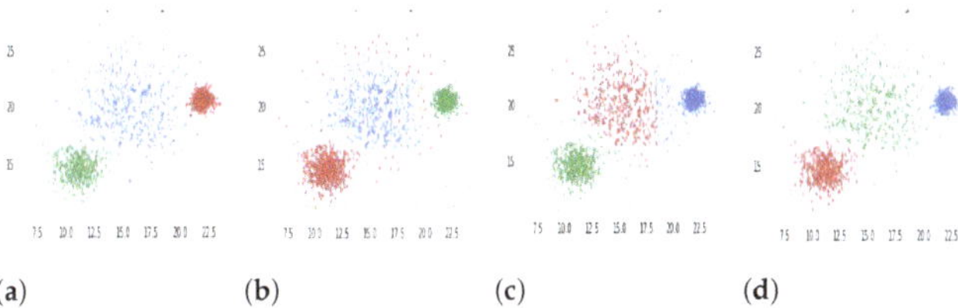

(**a**) (**b**) (**c**) (**d**)

Figure 9. Result of clustering algorithm on DB1 using four methods: (**a**) SC-E (**b**) DBSCAN, (**c**) *k*-means clustering and (**d**) SC-S.

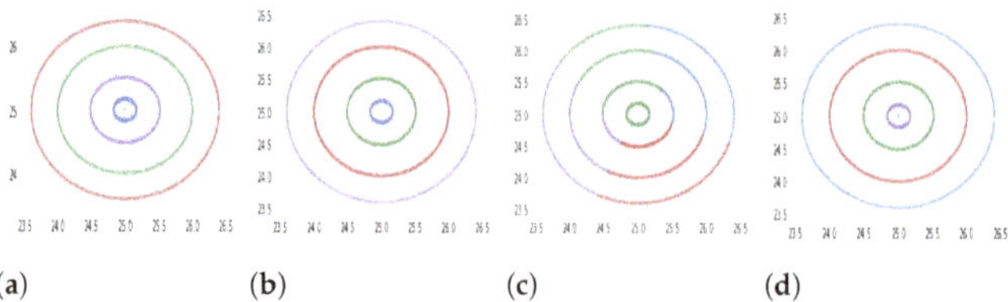

(**a**) (**b**) (**c**) (**d**)

Figure 10. Result of Ccustering algorithm on DB2 using four methods: (**a**) SC-E (**b**) DBSCAN, (**c**) *k*-means clustering and (**d**) SC-S.

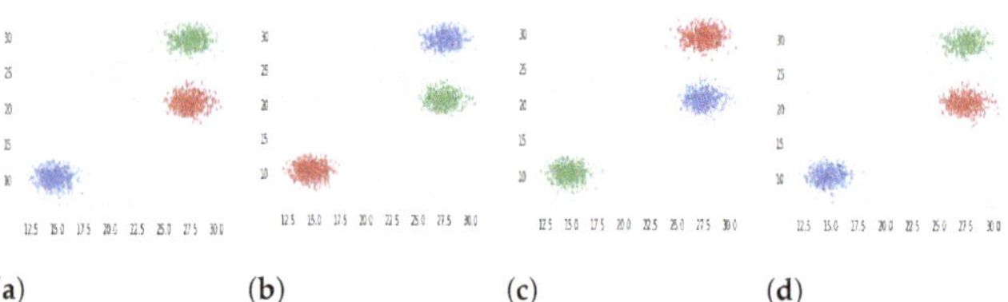

(**a**) (**b**) (**c**) (**d**)

Figure 11. Result of clustering algorithm on DB3 using four methods: (**a**) SC-E (**b**) DBSCAN, (**c**) *k*-means clustering and (**d**) SC-S.

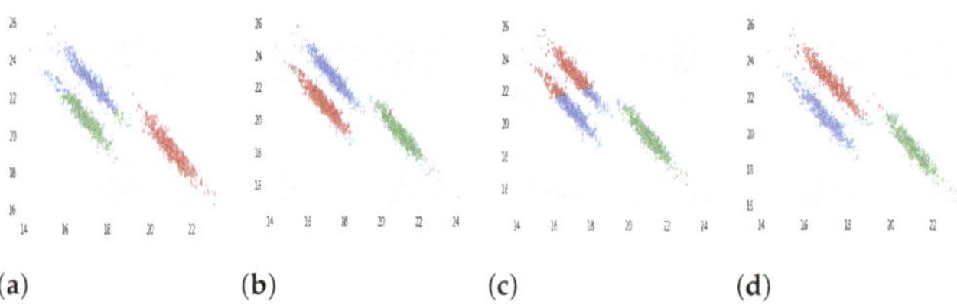

(**a**) (**b**) (**c**) (**d**)

Figure 12. Result of clustering algorithm on DB4 sets using four methods: (**a**) SC-E (**b**) DBSCAN, (**c**) *k*-means clustering and (**d**) SC-S.

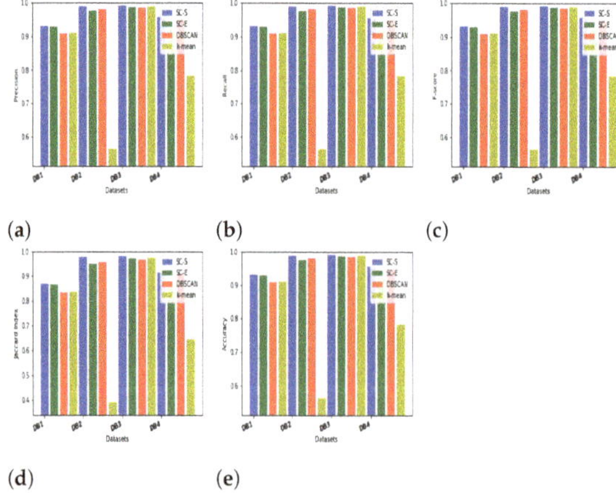

Figure 13. Comparative analysis of SC-E, SC-S, DBSCAN and *k*-means clustering on toy data sets using various validation indices: (**a**) precision, (**b**) recall, (**c**) f-score, (**d**) Jaccard index and (**e**) accuracy.

The proposed method along with three existing approaches was executed on eight UCI databases as discussed in Figure 14. In addition, two industrial databases and one telecommunication database were used for customer churn analysis, and the achieved results are displayed in Figure 15. Figure 15 shows the accuracy and TP rates obtained for the test cases with regard to the prediction horizon, which is calculated as the number of tasks performed by the users before leaving the commercial web site. As shown in Figure 15, it is clear that the SC-S has the highest accuracy compared to other existing approaches. Long-term deposits are favored by banks to maintain funds with minimal interest. Thus, this long-term deposit policy is better at generating higher successful sales, even if it requires some effort in communicating with customers. Under these circumstances, the proposed model SC-S shows higher accuracy as compared to the other existing approaches, as discussed in Figure 15. In this type of database, human agents have less probability to convert any call into successful calls. The telecommunication database is clustered into two clusters; namely, stable customers and churning customers. The objective is to predict customer behavior in the future based on these features. This analysis using clustering can help enterprises to develop efficient marketing strategies to select valuable customers and those that are necessary to retain, while customers that are going to churn can be contacted with appropriate retention measures to maximize profits. Further, enterprises can perform a deep analysis of the stable customers and can target more similar customers to increase their market space.

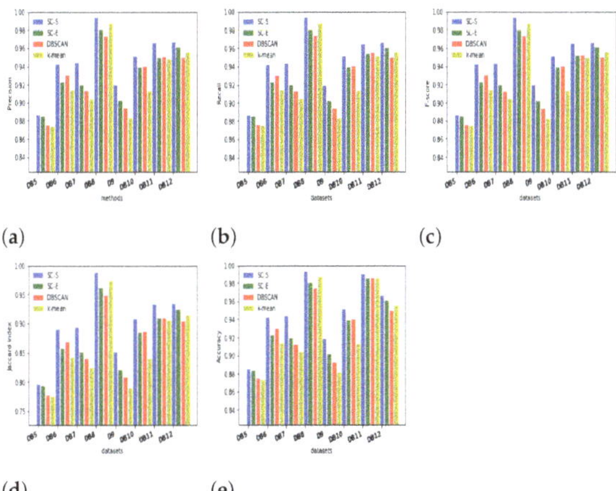

(a) (b) (c)

(d) (e)

Figure 14. Comparative analysis of SC-E, SC-S, DBSCAN and *k*-means clustering on UCI data sets using various validation indices: (**a**) precision, (**b**) recall, (**c**) f-score, (**d**) Jaccard index and (**e**) accuracy.

In this experiment, the non-parametric significance test of the SC-S was compared with other methods: SC-E, *k*-means and DBSCAN. First, a pairwise comparison of the SC-C with SC-E was performed and is labeled as "M1". Second, a pairwise comparison was done with *k*-means and marked as "M2". Finally, SC-S was compared with DBSCAN and denoted as "M3". This pairwise experiment was performed for three indices; namely, the Wilcoxon's signed-rank test, Wilcoxon's rank-sum test and sign tests [40]. These pairwise tests are the easiest ways to test statistics that can be conducted within the framework of an empirical study by a researcher. These non-parametric tests are also executed by considering the *p*-values (in Table 2) that are obtained based on accuracy only. The results of Table 2 allow us to refute the null hypothesis at a 5% level of significance. So, SC-S is statistically superior to the three existing approaches. Some insignificant *p*-values higher than 0.05 are also reported in Table 2.

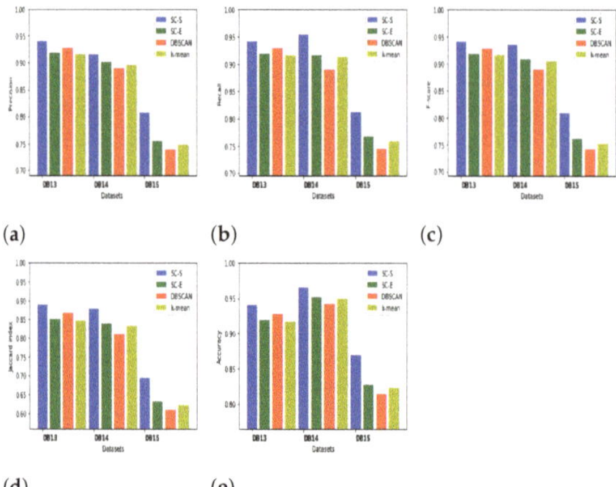

Figure 15. Comparative analysis of SC-E, SC-S, DBSCAN and *k*-means clustering on industry data sets using various validation indices: (**a**) precision, (**b**) recall, (**c**) f-score, (**d**) Jaccard index and (**e**) accuracy.

Table 2. Significance test using the accuracy validation index.

	Rank-Sum			**Sign-Rank**			**Sign Test**		
Databases	**M1**	**M2**	**M3**	**M1**	**M2**	**M3**	**M1**	**M2**	**M3**
DB1	0.04937	0.00016	0.00195	0.0047	0.00492	0.00492	0.00195	0.04937	0.0015
DB2	0.00016	0.00016	0.00016	0.0047	0.00448	0.00448	0.00195	0.00195	0.00195
DB3	0.00195	0.00016	0.00016	0.0047	0.00492	0.00492	0.00195	0.00016	0.00195
DB4	0.28992	0.00016	0.0015	0.0047	0.00492	0.00492	0.00195	0.00195	0.0015
DB5	0.00016	0.00289	0.00016	0.00492	0.00492	0.00492	0.00195	0.0015	0.00195
DB6	0.00016	0.00016	0.00016	0.00492	0.00492	0.00492	0.00195	0.00195	0.00195
DB7	0.00016	0.00016	0.00016	0.00492	0.00492	0.0047	0.0015	0.00016	0.00195
DB8	0.00289	0.00016	0.00016	0.00492	0.00492	0.00492	0.00195	0.00195	0.00492
DB9	0.00016	0.00492	0.0015	0.00448	0.00448	0.0047	0.00195	0.00195	0.00195
DB10	0.00016	0.00016	0.00016	0.00492	0.00492	0.00492	0.0015	0.00195	0.00195
DB11	0.00016	0.00492	0.00016	0.00492	0.00492	0.00492	0.00195	0.00195	0.00492
DB12	0.00016	0.00016	0.0015	0.00492	0.00492	0.00492	0.00195	0.00195	0.00195
DB13	0.0015	0.00016	0.00016	0.00407	0.0047	0.00492	0.00195	0.00195	0.00195
DB14	0.00016	0.00016	0.00016	0.00492	0.00492	0.00492	0.00195	0.00195	0.00492
DB15	0.0047	0.04937	0.00195	0.00492	0.00448	0.00492	0.00289	0.00195	0.0047

6. Conclusions

In this work, an enhanced SC based on *Sd* is proposed to predict customer churn with better accuracy by analyzing industrial data. The traditional KNN is replaced by a symmetric-favored KNN in the proposed algorithm in order to increase the efficacy of clustering. Extensive experiments are performed on four synthetics, eight UCI, two industrial databases and one telecommunication database for customer churn analysis, validating the proposed algorithm by the comparison with three existing clustering algorithms; namely, SC-E, *k*-means and DBSCAN. All the outcomes show that the proposed algorithm performs better than the three existing approaches in terms of five validation metrics: the Jaccard index, f-score, recall, precision and accuracy. The statistical significance of the SC-S is also measured by considering Wilcoxon's signed-rank test, Wilcoxon's rank-sum test and sign tests. This study can be extended to large databases by optimizing the step of eigenvalue computation using either the Hadoop architecture or parallel computation. The real-world databases consist of categorical as well as numerical attributes. This study proves that the

SC-S works well on databases with numerical attributes only. However, the SC-S cannot work on databases with categorical attributes, which deserves further study.

Author Contributions: Conceptualization, K.K.S., A.S., E.H.-V., and O.K.; methodology, K.K.S., and A.S.; software, K.K.S. and A.S.; validation, E.H.-V. and O.K.; formal analysis, K.K.S., and A.S.; investigation, A.S., E.H.-V. and O.K.; resources, A.S. and O.K.; data curation, K.K.S. and A.S.; writing—original draft preparation, K.K.S. and A.S.; writing—review and editing, K.K.S., A.S., E.H.-V. and O.K.; visualization, K.K.S., A.S. and E.H.-V.; supervision, A.S., E.H.-V. and O.K.; project administration, A.S., E.H.-V. and O.K. All authors have read and agreed to the published version of the manuscript.

Funding: This research received no external funding.

Institutional Review Board Statement: Not applicable.

Informed Consent Statement: Not applicable.

Data Availability Statement: Not applicable.

Acknowledgments: This work is partially supported by the project "Prediction of diseases through computer assisted diagnosis system using images captured by minimally-invasive and non-invasive modalities", Computer Science and Engineering, PDPM Indian Institute of Information Technology, Design and Manufacturing, Jabalpur India (under ID: SPARCMHRD-231). This work is also partially supported by the project "Smart Solutions in Ubiquitous Computing Environments", Grant Agency of Excellence, University of Hradec Kralove, Faculty of Informatics and Management, Czech Republic (under ID: UHK-FIM-GE-2204/2021); project at Universiti Teknologi Malaysia (UTM) under Research University Grant Vot-20H04, Malaysia Research University Network (MRUN) Vot 4L876 and the Fundamental Research Grant Scheme (FRGS) Vot5F073 supported by the Ministry of Education Malaysia for the completion of the research.

Conflicts of Interest: The authors declare no conflict of interest.

References

1. Mohan, K.; Seal, A.; Krejcar, O.; Yazidi, A. Facial Expression Recognition Using Local Gravitational Force Descriptor-Based Deep Convolution Neural Networks. *IEEE Trans. Instrum. Meas.* **2020**, *70*, 1–12. [CrossRef]
2. Rust, R.T.; Moorman, C.; Bhalla, G. Rethinking marketing. *Harv. Bus. Rev.* **2010**, *88*, 94–101.
3. Nobibon, F.T.; Leus, R.; Spieksma, F.C. Optimization models for targeted offers in direct marketing: Exact and heuristic algorithms. *Eur. J. Oper. Res.* **2011**, *210*, 670–683. [CrossRef]
4. Abbasi, A.A.; Younis, M. A survey on clustering algorithms for wireless sensor networks. *Comput. Commun.* **2007**, *30*, 2826–2841. [CrossRef]
5. Au, W.H.; Chan, K.C.; Yao, X. A novel evolutionary data mining algorithm with applications to churn prediction. *IEEE Trans. Evol. Comput.* **2003**, *7*, 532–545.
6. Hung, S.Y.; Yen, D.C.; Wang, H.Y. Applying data mining to telecom churn management. *Expert Syst. Appl.* **2006**, *31*, 515–524. [CrossRef]
7. Hossein Javaheri, S. Response Modeling in Direct Marketing: A Data Mining Based Approach for Target Selection. 2008. Available online: https://www.researchgate.net/publication/292282619_Response_modeling_in_direct_marketing_A_data_mining_based_approach_for_target_selection (accessed on 25 March 2021).
8. Castro, E.G.; Tsuzuki, M.S. Churn prediction in online games using players' login records: A frequency analysis approach. *IEEE Trans. Comput. Intell. Games* **2015**, *7*, 255–265. [CrossRef]
9. Sharma, K.K.; Seal, A. Clustering analysis using an adaptive fused distance. *Eng. Appl. Artif. Intell.* **2020**, *96*, 103928. [CrossRef]
10. Moro, S.; Cortez, P.; Rita, P. A data-driven approach to predict the success of bank telemarketing. *Decis. Support Syst.* **2014**, *62*, 22–31. [CrossRef]
11. Suchacka, G.; Skolimowska-Kulig, M.; Potempa, A. Classification Of E-Customer Sessions Based On Support Vector Machine. *ECMS* **2015**, *15*, 594–600.
12. Suchacka, G.; Skolimowska-Kulig, M.; Potempa, A. A k-Nearest Neighbors method for classifying user sessions in e-commerce scenario. *J. Telecommun. Inf. Technol.* **2015**. Available online: https://yadda.icm.edu.pl/baztech/element/bwmeta1.element.baztech-40e29335-8f5f-4d8c-aa93-8c13a90d1b2d (accessed on 25 March 2021).
13. Labeeuw, W.; Stragier, J.; Deconinck, G. Potential of active demand reduction with residential wet appliances: A case study for Belgium. *IEEE Trans. Smart Grid* **2015**, *6*, 315–323. [CrossRef]
14. Faria, P.; Spínola, J.; Vale, Z. Aggregation and remuneration of electricity consumers and producers for the definition of demand-response programs. *IEEE Trans. Ind. Inform.* **2016**, *12*, 952–961. [CrossRef]
15. Wu, B.; Wilamowski, B.M. A fast density and grid based clustering method for data with arbitrary shapes and noise. *IEEE Trans. Ind. Inform.* **2017**, *13*, 1620–1628. [CrossRef]

16. Suchacka, G.; Chodak, G. Using association rules to assess purchase probability in online stores. *Inf. Syst. Bus. Manag.* **2017**, *15*, 751–780. [CrossRef]
17. Bottou, L.; Bengio, Y. *Convergence Properties of the K-Means Algorithms*; Advances in Neural Information Processing Systems: Vancouver, BC, Canada, 1995; pp. 585–592.
18. Sharma, K.K.; Seal, A. Spectral embedded generalized mean based k-nearest neighbors clustering with s-distance. *Expert Syst. Appl.* **2020**, *169*, 114326. [CrossRef]
19. Banerjee, A.; Merugu, S.; Dhillon, I.S.; Ghosh, J. Clustering with Bregman divergences. *J. Mach. Learn. Res.* **2005**, *6*, 1705–1749.
20. Nock, R.; Nielsen, F.; Amari, S.I. On conformal divergences and their population minimizers. *IEEE Trans. Inf. Theory* **2016**, *62*. [CrossRef]
21. Notsu, A.; Komori, O.; Eguchi, S. Spontaneous clustering via minimum gamma-divergence. *Neural Comput.* **2014**, *26*, 421–448. [CrossRef] [PubMed]
22. Chang, E.C.; Huang, S.C.; Wu, H.H. Using K-means method and spectral clustering technique in an outfitter's value analysis. *Qual. Quant.* **2010**, *44*, 807–815. [CrossRef]
23. Sra, S. Positive definite matrices and the S-divergence. *Proc. Am. Math. Soc.* **2016**, *144*, 2787–2797. [CrossRef]
24. Von Luxburg, U. A tutorial on spectral clustering. *Stat. Comput.* **2007**, *17*, 395–416. [CrossRef]
25. Jiao, L.; Shang, F.; Wang, F.; Liu, Y. Fast semi-supervised clustering with enhanced spectral embedding. *Pattern Recognit.* **2012**, *45*, 4358–4369. [CrossRef]
26. Kim, T.H.; Lee, K.M.; Lee, S.U. Learning full pairwise affinities for spectral segmentation. *IEEE Trans. Pattern Anal. Mach. Intell.* **2013**, *35*, 1690–1703. [PubMed]
27. Chen, W.; Feng, G. Spectral clustering: a semi-supervised approach. *Neurocomputing* **2012**, *77*, 229–242. [CrossRef]
28. Shi, J.; Malik, J. Normalized cuts and image segmentation. *IEEE Trans. Pattern Anal. Mach. Intell.* **2000**, *22*, 888–905.
29. Selim, S.Z.; Ismail, M.A. K-means-type algorithms: A generalized convergence theorem and characterization of local optimality. *IEEE Trans. Pattern Anal. Mach. Intell.* **1984**, *1*, 81–87. [CrossRef] [PubMed]
30. De Stefano, C.; Maniaci, M.; Fontanella, F.; di Freca, A.S. Reliable writer identification in medieval manuscripts through page layout features: The "Avila" Bible case. *Eng. Appl. Artif. Intell.* **2018**, *72*, 99–110. [CrossRef]
31. Dheeru, D.; Karra Taniskidou, E. *UCI Machine Learning Repository*; UCI: Irvine, CA, USA, 2017.
32. Sakar, C.O.; Polat, S.O.; Katircioglu, M.; Kastro, Y. Real-time prediction of online shoppers' purchasing intention using multilayer perceptron and LSTM recurrent neural networks. *Neural Comput. Appl.* **2019**, *31*, 6893–6908. [CrossRef]
33. Sharma, K.K.; Seal, A. Modeling uncertain data using Monte Carlo integration method for clustering. *Expert Syst. Appl.* **2019**, *137*, 100–116. [CrossRef]
34. Seal, A.; Karlekar, A.; Krejcar, O.; Gonzalo-Martin, C. Fuzzy c-means clustering using Jeffreys-divergence based similarity measure. *Appl. Soft Comput.* **2020**, *88*, 106016. [CrossRef]
35. Sharma, K.K.; Seal, A. Multi-view spectral clustering for uncertain objects. *Inf. Sci.* **2021**, *547*, 723–745. [CrossRef]
36. Fawcett, T. An introduction to ROC analysis. *Pattern Recognit. Lett.* **2006**, *27*, 861–874. [CrossRef]
37. Karlekar, A.; Seal, A.; Krejcar, O.; Gonzalo-Martin, C. Fuzzy k-means using non-linear s-distance. *IEEE Access* **2019**, *7*, 55121–55131. [CrossRef]
38. Sharma, K.K.; Seal, A. Outlier-robust multi-view clustering for uncertain data. *Knowl. Based Syst.* **2021**, *211*, 106567. [CrossRef]
39. Kriegel, H.P.; Pfeifle, M. Density-based clustering of uncertain data. In Proceedings of the Eleventh ACM SIGKDD International Conference on Knowledge Discovery in Data Mining, Chicago, IL, USA, 21–24 August 2005; pp. 672–677.
40. Richardson, A.M. Nonparametric Statistics: A Step-by-Step Approach. *Int. Stat. Rev.* **2015**, *83*, 163–164. [CrossRef]

Short Biography of Authors

Krishna Kumar Sharma is currently pursuing PhD with the Computer Science and Engineering Department, PDPM Indian Institute of Information Technology, Design and Manufacturing Jabalpur, Madhya Pradesh, India and has received the M.Tech.(Information Technology) degree from IIIT Allahabad, Uttar Pradesh, India, in 2011. He is also currently an Assistant Professor with the Computer Science and Informatics Department, University of Kota, Kota, Rajasthan, India. His current research interest includes pattern recognition.

Ayan Seal received the PhD degree in engineering from Jadavpur University, West Bengal, India, in 2014. He is currently an Assistant Professor with the Computer Science and Engineering Department, PDPM Indian Institute of Information Technology, Design and Manufacturing Jabalpur, Madhya Pradesh, India. He is also associated with the Center for Basic and Applied Science, Faculty of Informatics and Management, University of Hradec Králové, Hradec Králové, Czech Republic. He has visited the Universidad Politecnica de Madrid, Spain as a visiting research scholar. He is the recipient of several awards including Sir Visvesvaraya Young Faculty Research Fellowship from Media Lab Asia, Ministry of Electronics and Information Technology, Government of India. He has authored or co-authored of several journals, conferences and book chapters in the area of computer vision and machine learning. His current research interests include image processing and pattern recognition.

Enrique Herrera-Viedma received the M.Sc. and Ph.D. degrees in computer science from the University of Granada, Granada, Spain, in 1993 and 1996, respectively. He is a Professor of computer science and the Vice-President for Research and Knowledge Transfer with University of Granada, Granada, Spain. His h-index is 81 with more than 23 000 citations received in Web of Science and 97 in Google Scholar with more than 37000 cites received. He has been identified as one of the world's most influential researchers by the Shanghai Center and Thomson Reuters/Clarivate Analytics in both computer science and engineering in the years 2014, 2015, 2016, 2017, 2018, 2019 and 2020. His current research interests include group decision making, consensus models, linguistic modeling, aggregation of information, information retrieval, bibliometric, digital libraries, web quality evaluation, recommender systems, and social media. Dr. Herrera-Viedma is Vice President for Publications in System Man & Cybernetic Society and an Associate Editor in several journals such as IEEE Transactions on Fuzzy Systems, IEEE Transactions on Systems, Man, and Cybernetics: Systems, IEEE Transactions on Intelligent Transport System, Information Sciences, Applied Soft Computing, Soft Computing, Fuzzy Optimization and Decision Making, and Knowledge-Based Systems.

Ondrej Krejcar is a full professor in systems engineering and informatics at the University of Hradec Kralove, Czech Republic. In 2008 he received his Ph.D. title in technical cybernetics at Technical University of Ostrava, Czech Republic. He is currently a vice-rector for science and creative activities of the University of Hradec Kralove from June 2020. At present, he is also a director of the Center for Basic and Applied Research at the University of Hradec Kralove. In years 2016-2020 he was vice-dean for science and research at Faculty of Informatics and Management, UHK. His h-index is 19, with more than 1250 citations received in the Web of Science. In 2018, he was the 14th top peer reviewer in Multidisciplinary in the World according to Publons and a Top Reviewer in the Global Peer Review Awards 2019 by Publons. Currently, he is on the editorial board of the MDPI Sensors IF journal (Q1/Q2 at JCR), and several other ESCI indexed journals. He is a Vice-leader and Management Committee member at WG4 at project COST CA17136, since 2018. He has also been a Management Committee member substitute at project COST CA16226 since 2017. Since 2019, he has been Chairman of the Program Committee of the KAPPA Program, Technological Agency of the Czech Republic as a regulator of the EEA/Norwegian Financial Mechanism in the Czech Republic (2019–2024). Since 2020, he has been Chairman of the Panel 1 (Computer, Physical and Chemical Sciences) of the ZETA Program, Technological Agency of the Czech Republic. Since 2014 until 2019, he has been Deputy Chairman of the Panel 7 (Processing Industry, Robotics, and Electrical Engineering) of the Epsilon Program, Technological Agency of the Czech Republic. At the University of Hradec Kralove, he is a guarantee of the doctoral study program in Applied Informatics, where he is focusing on lecturing on Smart Approaches to the Development of Information Systems and Applications in Ubiquitous Computing Environments. His research interests include Control Systems, Smart Sensors, Ubiquitous Computing, Manufacturing, Wireless Technology, Portable Devices, biomedicine, image segmentation and recognition, biometrics, technical cybernetics, and ubiquitous computing. His second area of interest is in Biomedicine (image analysis), as well as Biotelemetric System Architecture (portable device architecture, wireless biosensors), development of applications for mobile devices with use of remote or embedded biomedical sensors.

Article

Extractive Summarization Based on Dynamic Memory Network

Ping Li * and Jiong Yu

College of Information Science and Engineering, Xinjiang University, Urumqi 830046, China; yujiong@xju.edu.cn

Abstract: We present an extractive summarization model based on the Bert and dynamic memory network. The model based on Bert uses the transformer to extract text features and uses the pre-trained model to construct the sentence embeddings. The model based on Bert labels the sentences automatically without using any hand-crafted features and the datasets are symmetry labeled. We also present a dynamic memory network method for extractive summarization. Experiments are conducted on several summarization benchmark datasets. Our model shows comparable performance compared with other extractive summarization methods.

Keywords: text summarization; recurrent neural network; embedding; dynamic memory network

Citation: Li, P.; Yu, J. Extractive Summarization Based on Dynamic Memory Network. *Symmetry* **2021**, *13*, 600. https://doi.org/10.3390/sym13040600

Academic Editors: Kóczy T. László and István A. Harmati

Received: 6 March 2021
Accepted: 1 April 2021
Published: 3 April 2021

Publisher's Note: MDPI stays neutral with regard to jurisdictional claims in published maps and institutional affiliations.

1. Introduction

Summarization is an import problem of natural language understanding and information retrieval. The aim of the summarization is to condense the input text and remain the core meaning of the input text. The methods of the summarization are classified into two categories: extractive summarization method and abstractive summarization method. These two methods are symmetry important. The extractive summarization method selects the salient content from the documents while the abstractive summarization method paraphrases the content of the document. The earlier research mainly concentrated on extractive summarization method and the recent research focuses on neural extractive summarization and neural abstractive summarization. In this paper, we only pay attention to the extractive summarization method.

The early work of the extractive summarization method which was done by Edmundson [1] scores the sentences by considering the title words, clue words, and the sentence positions. Lin [2] uses some regulations to find the topic sentences and trains a model to predict the topic sentences based on the positions.

As the development of deep learning, researchers mainly concentrate on using the neural network method to resolve the extractive summarization problem. In particular, the development of the neural network language model [3] and the text representation methods [4] make the natural language processing take off. Cao [5] applies the neural network to extractive query-focused summarization which is a task of information retrieval. In their model, they employ the CNN(Convolutional Neural Network) to project the sentences of the document and the query to latent space. To get the document representation, they use the weighted-sum pooling over the sentence embedding. Lastly, they rank and select the sentences of document after comparing the similarity between the sentence embedding and document embedding.

Because of the success of the RNN (Recurrent Neural Network) in machine translation [6], Rush [7] first employs the RNN based on an attention mechanism for abstractive summarization.

Nallapati [8] uses the sequence model based on RNN to extract summarization of a single document, which is the problem we are focusing on. In their model, they see the extractive summarization task as a binary classification task and use the RNN model as a sentence classifier. Recently, Zhou [9] integrates the MMR (Maximal Marginal Relevance) selection strategy proposed by Carbonell and Goldstein [10] into the scoring model. In their model, they employ the BiGRU (Bidirectional Gated Recurrent Unit) [4] as their encoder to get the sentence representation and document representation and they construct their labeled training data by maximizing the ROUGE-2 F1 score [2] . These neural extractive summarization methods mentioned above are all using the RNN as their encoder and the labeled data construction method is computationally expensive. Narayan [11] employs the CNNs [12] as the sentence encoder and employs the RNN as the document encoder. Because of the strong ability to extract text features, we will use the transformer [13] as the encoder just like Bert [14] does.

Our main contributions are as follows:

1. We propose an extractive summarization model that achieves the comparable result against other baselines.
2. We propose a simple and effective sentence label method used in the extractive summarization problem.
3. We incorporate the positional encoding to a dynamic memory network.
4. We propose to use a dynamic memory network method for extractive summarization.

2. Materials and Methods

2.1. Problem Formulation

A document is represented as D and S represents the sentence of the document. D is defined as Equation (1):

$$D = \{S_1, S_2, \ldots, S_n\} \tag{1}$$

The n in Equation (1) is represented as the max number of the document sentences. The target of the extractive summarization task is to extract r sentences of the document ($r < n$) and the r sentences maintain the core information of the document. We make the extractive summarization task as a binary classification task just like Cheng [15] and Nallapati [8] do. Given a sentence s_i $(0 \le i \le n)$, which consists of tokens represented as w_j $(0 \le j \le m)$, we predict the probability of $P(Y|S)$ $(Y \in \{0,1\})$. The last hidden output of the Bert is denoted as h_k $(0 \le k \le m)$ and the output of the sentence encoder is denoted as $s_i^{'}$, which is the embedding representation of sentence i. The target sentence in training set is denoted as t_l $(0 \le l \le p)$ and p is the length of the target summaries. The embedding of the t_l is denoted as $t_l^{'}$. The b_l is denoted as the sentence labeled with 1.

2.2. Extractive Summarization Based on Bert

2.2.1. Sentence Encoder

Differentiated with Narayan [9,11], we don't use the hierarchical encoder and only use the sentence encoder. The architecture of the sentence encoder is shown in Figure 1. Because of the strong ability to extract the text features, we use the Bert to get the word embeddings of the sentences. We refer you to the detailed description of the pre-training Bert model in the paper written by Jacob [14]. Using the Bi-Transformer and the masked language model makes the Bert better than other pre-training models.

The input of the sentence encoder is the one-hot representations of the words in a sentence and the model parameters of pre-training are loaded in the sentence encoder model. Given sentence I, through the pre-training Bert model, we will get a list of embeddings. The embedding of sentence I is calculated as Figure 2. We just use this sentence encoder in the process of building a labeled training set and the process of the prediction and extractor will use the fine-tuned Bert directly:

$$S_i^{'} = (h_1 + h_2 + \cdots + h_m)/m \tag{2}$$

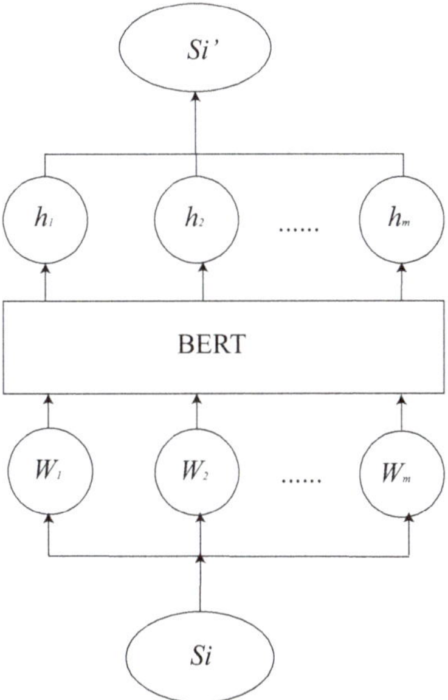

Figure 1. Overview of the sentence encoder based on Bert.

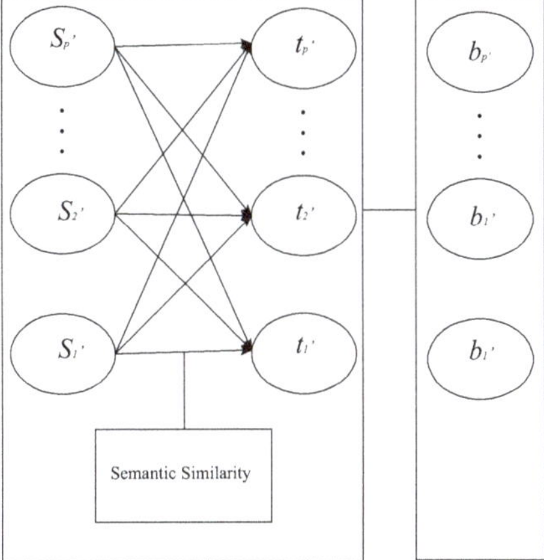

Figure 2. Overview of the labeling process.

2.2.2. Label Training Set

We conduct our experiments on the CNN/Dailymail dataset [16] and several other datasets. We will train a binary classifier on these datasets. However, there is no labeled training dataset. In order to construct this labeled dataset, we need to label the sentence of the document with 0 and 1. The label 0 shows that this sentence should not be included in the summaries, and label 1 shows that this sentence should include the summaries. Differentiated with Zhou [9,11], we don't use the rouge method to build the supervised datasets, and we will use the semantic similarity method to label the training set. The overview of the labeling process is shown in Figure 2.

Given the $s_i^{'}$ and the $t_l^{'}$, the equations of calculating the semantic similarity are shown as Equations (3) and (4). For every sentence in the target summaries, we find the maximum similarity sentence from the source document. These selected sentences are labeled with 1 and the others are labeled with 0:

$$similarity = \cos(s_i{'}, t_l{'}) \tag{3}$$

$$\cos(s_i{'}, t_l{'}) = \frac{s_i{'} \cdot t_l{'}}{\|s_i{'}\| \times \|t_l{'}\|} \tag{4}$$

2.2.3. Sentence Extractor

We directly use the fine-tuned Bert model to train our binary classifier which will predict the probability of the sentence to be extracted. The classify model that loaded the pre-training parameters reads a single sentence, and the classified probability is denoted as Equation (5). In Equation (5), the C is the final hidden state of the input representation in the Bert model, and the W are the new parameters to be learned. The loss function is a cross entropy loss function that is denoted as Equation (5). In the test phase, we predict the probability of the label when we input the test dataset. We assume that the probability of label 1, which is greater than the probability of label 0, should be the extractive summary. The max sequence length of the input sequence is set to 128:

$$P(Y|S) = Softmax(CW^{\top}) \tag{5}$$

$$Loss = -logP(Y|S) \tag{6}$$

2.3. Extractive Summarization Based on the Dynamic Memory Network

Kumar et al. first introduce the DMN (dynamic memory network) for the QA (question answering) problem [17]. Caiming et al. [18] propose several improvements over the base dynamic memory network. The DMN consists of input module, question module, episodic memory module, and answer module. The episodic memory module will produce the focusing parts of the inputs through the input module and question module. The answer module will generate the answers based on the outputs of the memory module. Because the similarity of the QA problem and the summarization problem, we employ the DMN as our base model to classify the sentences.

Dynamic Memory Network

In order to extract the summarizations from the document, we propose a dynamic memory network that is composed of input module, memory module, summarization module, and linear module. What we are focusing on is to memorize the salient content of the sentence and then classify the facts. The modules of our model are showed in Figure 3.

Input Module: the inputs of our model are the all sentences of our document sets. We need to encode the sentences by using an encoder. The encoder of [17] is a GRU (gated recurrent network) [4,19,20]. The input of the GRU is the word embeddings and then the encoder calculates the hidden states of the words, which are concatenated as the sentence representations. The hidden state of the token i can be defined as $h_i = GRU(x_i, h_{i-1})$, where x_i is the embedding of token i and h_{i-1} is the hidden state of previous token. The GRU is defined as:

$$z_t = \sigma\left(W^{(z)} x_t + U^{(z)} h_{t-1} + b^{(z)}\right) \tag{7}$$

$$r_t = \sigma\left(W^{(r)} x_t + U^{(r)} h_{t-1} + b^{(r)}\right) \tag{8}$$

$$\tilde{h}_t = \tanh\left(W x_t + r_t \circ U h_{t-1} + b^{(h)}\right) \tag{9}$$

$$h_t = z_t \circ h_{t-1} + (1 - z_t) \circ \tilde{h}_t \tag{10}$$

where z_t is the update gate and r_t is the reset gate, the W and U are the weights, and the b is the bias. The σ and the $tanh$ are the activation functions.

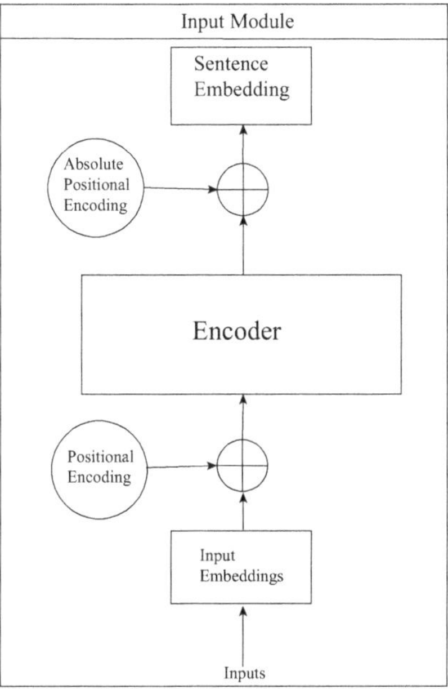

Figure 3. Input module.

The input of our model is one sentence, and we encode one sentence by using a bi-GRU. Thus, we can use not only the information of previous words but also use the information of the afterwards words. When we encode the sentences, we also consider the positions of the words and the positions of the sentences. Differentiated with [17], we combine the word positional encoder and sentence positional encoder with the sentence encoder. The word positional encoder [13] is defined as:

$$WPE_{(wordPos,2i)} = \sin\left(wordPos/10000^{2i/d}\right) \tag{11}$$

$$WPE_{(wordPos,2i+1)} = \cos\left(wordPos/10000^{2i/d}\right) \tag{12}$$

where the $wordPos$ is the position of the word, i is the dimension, and d is the dimension of the word embedding. From previous research, we can find that the position of the sentence is very important when extracting the sentences from the document. We define the sentence positional encoder as:

$$SPE_{(i)} = p_i \tag{13}$$

where the p_i is the absolute position of the sentence. In our experiments, we also consider using the Bi-RNN [21] and the transformer [13] as the encoder. In the BiRNN model, the forward RNN encodes the input sequence as

$$\overrightarrow{f} = \left(\overrightarrow{h}_1, \cdots, \overrightarrow{h}_{T_x}\right) \tag{14}$$

and the backward RNN encodes the input sequence as

$$\overleftarrow{f} = \left(\overleftarrow{h}_1, \cdots, \overleftarrow{h}_{T_x}\right). \tag{15}$$

The final representation of word j is denoted as

$$\vec{h}_j = \left[\overrightarrow{h}_j^{\top}; \overleftarrow{h}_j^{\top}\right]^{\top}. \tag{16}$$

Through the $\vec{h}_j$, we will get the sentence embedding.

Summarization Module: in the problem of question and answer, the dynamic memory model should encode the question by using the question module. However, in our model, there is no question. In order to memory the salient content in the source sentence, we need to encode the salient content, which is the abstracts of the document in the training phase, while, in the testing phase, it is the first three sentences of the document. We call this encoder module as the summarization module. The summarization module is very similar to the input module. The difference between the summarization module and input module is that we use the RNN as encoder when we encode the content .

Memory Module: the memory module mainly gets the similarity representation of the sentence to be classified and the salient content of the document. The input of the memory module is the sentence representation that is the output of the input module and the summarization representation, which is the output of the summarization module. In order to get the similarity representation, we employ the same mechanism as [17] to compute the representation of similarity. In our module, we only compute three episodes. In order to capture the similarities of the input module and summarization module, we define the similarity content as

$$z(I,S) = Concat(I * S, S * memory, |S - memory|, |I - S|,) \tag{17}$$

where I is the output of the input module, and S is the output of the summarization module. The memory is initialized with I. The gate function in our model is defined as

$$g = \sigma\left(W^{(2)} \tanh\left(W^{(1)} z(I,S) + b^{(1)}\right) + b^{(2)}\right). \tag{18}$$

With the gate function g and the similarity content representation z, we compute the episode as:

$$h_t = g_t RNN\left(\vec{I}, h_{t-1}\right) + (1 - g_t) h_{t-1}. \tag{19}$$

The memory is updated as

$$memory = GRU(h_t). \tag{20}$$

Linear Module: in the linear module, we need to map the similarity representation to a two-dimensional output by using a linear layer and then get a probability by using a softmax layer. Through this probability, we can train our model with the labeled training set:

$$out_i = L\left(RNN(h_t, \vec{S})\right) \tag{21}$$

$$P_i = softmax(out_i) \tag{22}$$

3. Experimental Setup

We will present our experimental setup for assessing the performance of our model which we call ESBOB and SDMN in this section. We will discuss the datasets used for training and evaluation. The implementation details and the evaluation method are described for comparison.

3.1. Datasets

We train and evaluate our model on the non-anonymized CNN and DailyMail datasets [16], which are developed for the question-answering system. We split the dataset for 287, 226 training pairs, 13,368 validation pairs, and 11,490 testing pairs followed by [22–24]. Because the CNN/Dailymail datasets don't include the reference extractive summarization, we will use the abstractive summarizations as the reference summarizations.

The second dataset we use is the WikiHow developed by Koupaee and Wang [25] . This dataset is a diverse dataset that is extracted from an online knowledge base. The dataset has 168,128 training pairs and 6000 testing pairs

The third dataset we use is the XSum developed by Narayan [26] . This dataset is a one-sentence summary dataset. The summaries in this dataset are written professionally.

3.2. Baselines

In order to compare, we choose some approaches as our baselines. These baselines are listed below:

(1) Leading three sentences (Lead-3). This method constructs the summary by extracting the first three sentences of the document. We have given our lead-3 result and the Lead3 result of [23] .
(2) Cheng and Lapata [15] . Cheng and Lapata propose an extractive model which consist of the hierarchical document encoder and an attention-based extractor.
(3) SummaRuNNer [8] . This model is based on a recurrent neural network.
(4) REFRESH [11] . Narayan makes the extractive summarization task as a sentence ranking task and optimizes the rouge metrics through the reinforcement learning object.
(5) NEUSUM [27] . This model makes the sentence scoring and sentence selection to an end-to-end neural network framework.
(6) BANDITSUM [28] . In the field of text summarization, many methods have been proposed with reinforcement learning. In this model, a policy gradient reinforcement learning algorithm is used to train to select the summarization sentences.

3.3. Implementation Details

The pre-trained Bert we use in the process of labeling, training, and testing is the uncased Bert-base model. In the pre-trained model, there are 12 transformer blocks. The model employs 768 hidden sizes and 12 self-attention heads. The optimizer used in our model is Adam optimizer [29] with initial learning rate 0.001. We use a batch size of 128 and an epoch of 3 at training time. We train our sentence classifier on 4 Tesla K80 GPU. At test time, we extract the first three sentences as the baseline because of the LEAD3 is a commonly used baseline. We also extract the sentences of the training set as the reference summarizations based on the semantic similarity and the abstractive summarizations.

When we train our dynamic memory network, we also use the Adam optimizer [29] with an initial learning rate of 0.001. The batch size we use is 32, and we train our model in two epochs. The training set we use is the data labeled by Bert. The summarizations we use in the summarization module are the first three sentences of the document. We train an extractive system based on the dynamic memory network on 1 Tesla K80 GPU.

3.4. Evaluation

The F1 ROUGE value [30] is used to evaluate our summarization model. We will report the f scores of ROUGE-1, ROUGE-2, and ROUGE-L. We will compare our model against the lead-3 baseline which just selects the first three sentences in the document as the summary. On the CNN/Daily mail dataset, the result of the approach put forward by Cheng and Lapata [15] is reported. We compare our model against the approach called BanditSum [28] and the approach called Refresh [11], which trains the extractive summarization with reinforcement learning. We also compare our model against the approach called SummaRuNNer [8], which treats the extractive summarization as a binary classify task. The result of approach called Neusum [9] is also reported.

4. Results

4.1. Results on CNN/Daily Mail

Table 1 shows the experiment results by using the dynamic memory network called SDMN. We can find that the SDMN is effective in processing the summarization problem. Incorporation the pre-training method into the SDMN model, the experiment result performs very well. There are three encoders that we have used in our experiments. The Bi-LSTM (Bi directional Long Short Term Memory) encoder with the dynamic memory network shows the best result. Thus, in some extent, the Bi-LSTM has not been replaced by transformers.

Table 1. ROUGE evaluation (%) on the CNN/DailyMail test set using dynamic memory network (SDMN) with different encoders.

Method	ROUGE-1	ROUGE-2	ROUGE-L
SDMN + BiGRU + pe	36.69	15.53	33.14
SDMN + BiLSTM + pe	40.24	17.53	36.49
SDMN + trans + pe	40.2	17.5	36.48

Table 2 shows the experiment results using automatic metrics. We report the lead3 result which is supplied by [23], and the second method result is supplied by [11]. From the table, we can find that the neusum [9] achieves the state-of-the-art result. The method trained by reinforcement learning depends on the rouge score when labeling the dataset and training the model, which leads to the hard improvement on the experiment result. The semantic experiment is conducted by us, which extracts the summarizations by semantic similarity. However, this semantic similarity approach can not be applied into the inference phase. From the result, we can find out that the training set built on semantic similarity is effective to a certain extent. In order to improve the experiment result, we need to find some more effective method to label the training set or use some methods to extract the summarizations without labeling the training set such as reinforcement learning method. Our method does not beat the method neusum and method banditsum. However, our method named SDMNTransPe achieves a comparable result as other methods. From this, we can find that our method is effective.

Table 2. ROUGE evaluation (%) on the CNN/DailyMail test set. Models marked with * are trained and evaluated on the anonymized dataset, and so are not comparable to our results.

Method	ROUGE-1	ROUGE-2	ROUGE-L
LEAD3 (ours)	39.89	17.24	36.12
LEAD3 (See et al.)	40.3	17.7	36.6
SemanticSim	50.54	27.63	46.77
Cheng and Lapata	35.5	14.7	32.2
SummaRuNNer *	39.6	16.2	35.3
REFRESH	40.0	18.2	36.6
NEUSUM	41.59	19.01	37.98
BANDITSUM	41.5	18.7	37.6
SDMNTransPe	**40.2**	**17.5**	**36.48**

4.2. Results on WikiHow and XSum

WikiHow is a summarization dataset that has short summaries. The XSum dataset has symmetry long summaries. We will evaluate the short summarization dataset to find the result by using the extractive summarization method. In Table 3, the first section contains the lead-1 result and the second section contains the groundtruth result. From Table 3, we can reconfirm that the first line in the document contains the important information.

The last section in Table 3 shows the result of our extractive summarization method. The last result is improved on the lead-1 result and is not comparable to oracle results. Thus, we can use the extractive summarization to improve the result on short summarization datasets.

The summary in the WikiHow contains one sentence. The average number of the sentences in the XSum summary is 8.4 and the average number of the sentences in the CNN/DailyMail dataset is 4.8. The XSum dataset is the long summaries dataset. Our model is effective in processing short summaries dataset and one sentence summary dataset. In order to prove that our model is effective on the long summaries dataset, we conduct the experiments on the XSum dataset. When we train our model, we select the one sentence, two sentences, and all sentences from the summaries as the candidate summary. From Table 4, we can find that the more sentences we select, the better performance the extractive summarization model gets.

Table 3. ROUGE evaluation (%) on the WikiHow test set. Lead indicates selecting the first sentence as the summary.

Method	ROUGE-1	ROUGE-2	ROUGE-L
Lead	24.97	5.83	23.24
oracle	35.59	12.98	32.68
SDMN	30.23	7.58	27.34

Table 4. ROUGE evaluation (%) on the XSum test set. Num indicates the number of sentences we choose to form a candidate summary. All indicates the sentences the model selected from the candidate summary.

Method	ROUGE-1	ROUGE-2	ROUGE-L
SDMN (1)	21.15	4.11	15.23
SDMN (2)	22.82	4.21	16.54
SDMN (all)	23.51	4.35	17.43

5. Analysis

Our analysis is driven by the following two questions:

(1) Where are the salient sentences in the document?
(2) Why is our method effective compared with other methods?

In order to find the discipline about the positions of the salient sentences in the document. We choose the CNN/DM, WikiHow, and XSum datasets. There are relations between the extractive summarization method and the abstractive summarization method. The abstractive summarization model is to find the salient sentences or salient words in the document and then paraphrase the selected contents. What the extractive summarization model does is just what the abstractive method needs. The discipline of the positions of the salient sentences in the document is very important to the abstractive summarization problem and the extractive summarization problem.

Figures 4–6 show the salient sentences' positions inspected by the Bert in the three datasets. We can find that the XSum dataset contains one sentence summary, the CNN/Dailymail dataset contains the medium summary sentences, and the WikiHow dataset contains the large summary sentences. The salient sentences are not always in the first line of the document. The salient sentences are always located in the whole article and the two salient sentences have nearly equal distance. Table 5 shows the title, first sentence and the labeled sentence in the XSum dataset. We can find that the sentence labeled by the Bert Model is more adherent to the title than the first sentence of the document. When the authors are editing the article, they always highlight the gist in some paragraphs. Thus, when we write the summarization automatically, we should not just concentrate on the first lines of the document but on the whole paper.

The document we are processing is always the long text. In order to select the salient sentences from the document, we need to classify the sentences of the document. Relatively speaking, the sentences of the document are short text. The transformer encoder is good at processing the long text with the attention mechanism while the LSTM encoder is good at processing the short text. That makes our model get equal performance when we choose the transformer and the LSTM as the encoder. The model based on the reinforcement learning depends on the rouge score when they label the dataset and train their model while we label our dataset using the semantic matching method which makes our method comparable with other methods. Our method makes full use of the reference summaries when training our model by using the dynamic memory network. The dynamic memory network mechanism can help our model find the most important sentence in the document. Our method is slightly weaker than the methods based on the reinforcement learning.

In the encoder phase, we employ the GRU, LSTM, and transformer as our encoder. The difference between our model and someone else's model is that the sentences labeled for the classification model are identified by pre-training model Bert. While other models only use the sentences' information, we incorporate the sentence position information into the encoder model. Our model makes full use of the referenced summarizations. First, we use the referenced summarizations and the Bert model to spot the salient sentences in the document. Second, we use the dynamic memory network to compute the similarity representations of the document sentences and the reference summarizations for the classification model. The similarity feature is computed dynamically, which is good for the classification model. The characteristics of our model described above make our model comparable with other extractive models.

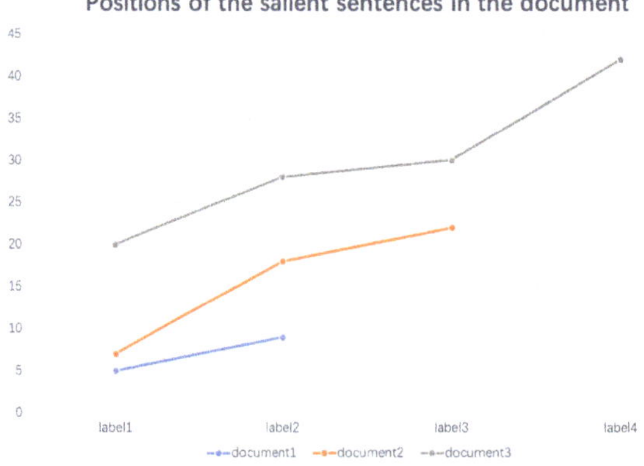

Figure 4. Positions of the salient sentences in the document of the cnndm dataset.

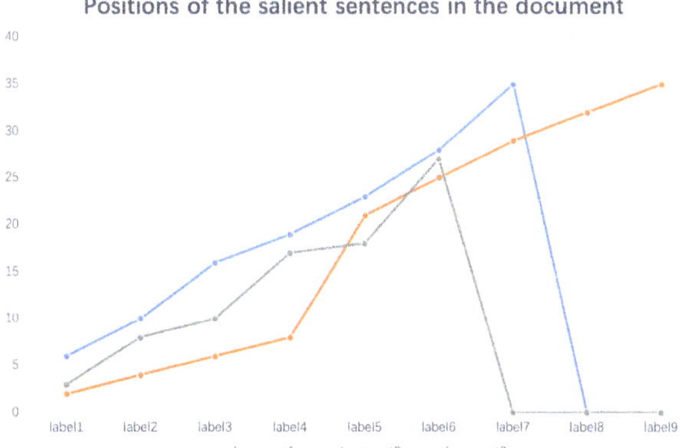

Figure 5. Positions of the salient sentences in the document of the wikihow dataset.

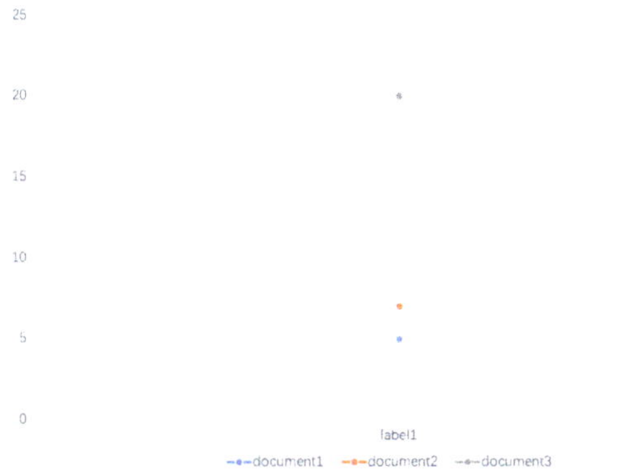

Figure 6. Positions of the salient sentences in the document of the xsum dataset.

Table 5. The sentences in the xsum document.

Doc	Title I First, Sentence I Labeled Sentence
doc1	Spend £3.3 m fund on Wales-based stars, says Gareth Davies
doc2	Alliance Party east Belfast alert was a hoax, PSNI say
doc3	UK energy policy 'deters investors'
doc1	New Welsh Rugby Union chairman Gareth Davies believes a joint £3.3 m WRU-regions fund should be used to retain home-based talent such as Liam Williams, not …
doc2	A suspicious package left outside an Alliance Party office in east Belfast has been declared a hoax
doc3	The UK's international reputation for a strong and well-balanced energy policy has taken another knock
doc1	3 m should be spent on ensuring current Wales-based stars remain there
doc2	Condemning the latest hoax, Alliance MLA Chris Lyttle said: "It is a serious incident for the local area, it causes serious disruption, it puts people's lives at risk, … "
doc3	A spokesman for her department, commenting on the WEC report, said: "We've made record investments in renewables and are committed to lower-carbon … "

6. Related Work

Recently, with the springing up of the deep learning method, all kinds of neural network methods are applied in extractive summarization [5,9,11,15,28].

Cheng [15] sees the task of sentence extractive summarization and word extractive summarization as a binary classifier task. In the sentence encoder, they employ a hierarchical encoder which uses a single-layer CNN for obtaining the sentence-level representations and use a recurrent neural network for obtaining document representations. The sentence extractor in [15] is an MLP(Multi-layer Perceptron) with an attention mechanism. When building the training dataset, Cheng uses the rule-based method including taking into account the position of the sentence in the document. Cao [5] presents a query-focused extractive summarization system which is used in information retrieval. This extractive summarization system consists of the CNN Layer, Pooling Layer, and Ranking Layer. The sentence ranking process only compares the semantic similarity between the sentence embedding and the document embedding. Nallapati [8] presents an RNN based sequence model for extractive summarization of documents. In their work, they also treat extractive summarization as a sequence classification problem. They use a two-layer bi-GRU for obtaining the word-level representation and the sentence-level representation. The sentence extractor uses a logistic layer. They build the labeled training dataset by maximizing the rouge score with respect to gold summaries. Zhou [9] obtains the sentence representations by using a hierarchical encoder which consists of BiGRU. They couple the sentence scoring step and sentence selection step. The sentence extractor consists of a BiGRU and a MLP(Multi-Layer Perceptron). Narayan [11] conceptualizes extractive summarization as a sentence ranking task and proposes a novel training algorithm with a reinforcement learning objective [31] which optimizes the rouge metric. They use a convolutional sentence encoder and a LSTM document encoder. The sentence extractor consists of LSTM cells and a softmax layer. When they rank the sentences, they train their model in a reinforcement learning framework. This reinforcement learning avoids labeling the training set and makes the model better at discriminating among sentences. Dong [28] treats the extractive summarization as a contextual bandit problem. They also use a policy gradient reinforcement learning algorithm to select sentences and maximize rouge score. The difference between [11] and [28] is the action space. In [11], they approximate the action space while Ref. [28] uses the true action space.

There are some extractive summarization methods that used the pretrained model [32–35].

7. Conclusions

In this work, we put forward an extractive summarization model that is based on Bert and dynamic memory network. In our model, we use a simple semantic matching method to label the training set and train our model using the pre-trained Bert model. A strong ability to extract the text features makes the model effective. Experimental results show that the model based on Bert and dynamic memory network achieves the comparable result against other extractive systems on the datasets. The dynamic memory network with the bi-LSTM encoder we use for the extractive summarization problem achieves good results. In the future, we will incorporate this extractive summarization method to the abstractive method.

Author Contributions: Conceptualization, P.L.; methodology, J.Y.; software, P.L.; validation, J.Y.; formal analysis, J.Y.; data curation, P.L.; writing—original draft preparation, P.L.; writing—review and editing, J.Y.; visualization, P.L.; supervision, J.Y.; project administration, J.Y.; funding acquisition, J.Y. All authors have read and agreed to the published version of the manuscript.

Funding: This research was funded by the National Natural Science Foundation of China (No. 61862060, No. 61562086, No. 61562087).

Institutional Review Board Statement: Not applicable.

Informed Consent Statement: Not applicable.

Data Availability Statement: Not applicable.

Acknowledgments: We would like to thank the anonymous reviewers for their valuable feedback.

Conflicts of Interest: The authors declare no conflict of interest.

References

1. Edmundson, H.P. New Methods in Automatic Extracting. *J. ACM* **1969**, *16*, 264–285. [CrossRef]
2. Lin, C.Y.; Hovy, E. Identifying Topics by Position. In Proceedings of the Fifth Conference on Applied Natural Language Processing, Washington, DC, USA, 31 March–3 April 1997; Association for Computational Linguistics: Stroudsburg, PA, USA, 1997; pp. 283–290. [CrossRef]
3. Bengio, Y.; Ducharme, R.; Vincent, P.; Janvin, C. A Neural Probabilistic Language Model. *J. Mach. Learn. Res.* **2003**, *3*, 1137–1155.
4. Cho, K.; van Merrienboer, B.; Gulcehre, C.; Bahdanau, D.; Bougares, F.; Schwenk, H.; Bengio, Y. Learning Phrase Representations using RNN Encoder–Decoder for Statistical Machine Translation. In Proceedings of the 2014 Conference on Empirical Methods in Natural Language Processing (EMNLP), Doha, Qatar, 25–29 October 2014; pp. 1724–1734.
5. Cao, Z.; Li, W.; Li, S.; Wei, F.; Li, Y. AttSum: Joint Learning of Focusing and Summarization with Neural Attention. In Proceedings of the COLING 2016, the 26th International Conference on Computational Linguistics: Technical Papers, Osaka, Japan, 11–16 December 2016; Association for Computational Linguistics:Stroudsburg, PA, USA, 2016; pp. 547–556.
6. Bahdanau, D.; Cho, K.; Bengio, Y. Neural Machine Translation by Jointly Learning to Align and Translate. *arXiv* **2014**, arXiv:1409.0473.
7. Rush, A.M.; Chopra, S.; Weston, J. A Neural Attention Model for Abstractive Sentence Summarization. In Proceedings of the 2015 Conference on Empirical Methods in Natural Language Processing, Lisbon, Portugal, 17–21 September 2015. [CrossRef]
8. Nallapati, R.; Zhai, F.; Zhou, B. SummaRuNNer: A Recurrent Neural Network Based Sequence Model for Extractive Summarization of Documents. In Proceedings of the Thirtieth AAAI Conference on Artificial Intelligence, Phoenix, AZ, USA, 12–17 February 2016; pp. 3075–3081.
9. Zhou, Q.; Yang, N.; Wei, F.; Huang, S.; Zhou, M.; Zhao, T. Neural Document Summarization by Jointly Learning to Score and Select Sentences. In Proceedings of the 56th Annual Meeting of the Association for Computational Linguistics (Volume 1: Long Papers), Melbourne, Australia, 15–20 July 2018; Association for Computational Linguistics: Stroudsburg, PA, USA,2018; pp. 654–663.
10. Carbonell, J.; Goldstein, J. The Use of MMR, Diversity-based Reranking for Reordering Documents and Producing Summaries. In Proceedings of the 21st Annual International ACM SIGIR Conference on Research and Development in Information Retrieval, Melbourne, Australia, 24–28 August 1998; ACM: New York, NY, USA, 1998; pp. 335–336. [CrossRef]
11. Narayan, S.; Cohen, S.B.; Lapata, M. Ranking Sentences for Extractive Summarization with Reinforcement Learning. In Proceedings of the 2018 Conference of the North American Chapter of the Association for Computational Linguistics: Human Language Technologies, Volume 1 (Long Papers), New Orleans, Louisiana, 1–6 June2018; Association for Computational Linguistics: Stroudsburg, PA, USA, 2018; pp. 1747–1759. [CrossRef]
12. ColloBert, R.; Weston, J.; Bottou, L.; Karlen, M.; Kavukcuoglu, K.; Kuksa, P. Natural Language Processing (Almost) from Scratch. *J. Mach. Learn. Res.* **2011**, *12*, 2493–2537.
13. Vaswani, A.; Shazeer, N.; Parmar, N.; Uszkoreit, J.; Jones, L.; Gomez, A.N.; Kaiser, L.; Polosukhin, I. Attention is All you Need. In Proceedings of the Neural Information Processing Systems, Long Beach, CA, USA, 4–9 December 2017; pp. 5998–6008.
14. Devlin, J.; Chang, M.; Lee, K.; Toutanova, K. BERT: Pre-training of Deep Bidirectional Transformers for Language Understanding. *arXiv* **2018**, arXiv:1810.04805.
15. Cheng, J.; Lapata, M. Neural Summarization by Extracting Sentences and Words. In Proceedings of the 54th Annual Meeting of the Association for Computational Linguistics (Volume 1: Long Papers), Berlin, Germany, 7–12 August 2016; Association for Computational Linguistics: Stroudsburg, PA, USA, 2016; pp. 484–494. [CrossRef]
16. Hermann, K.M.; Kocisky, T.; Grefenstette, E.; Espeholt, L.; Kay, W.; Suleyman, M.; Blunsom, P. Teaching Machines to Read and Comprehend. In *Advances in Neural Information Processing Systems 28*; Cortes, C., Lawrence, N.D., Lee, D.D., Sugiyama, M., Garnett, R., Eds.; Association for Computational Linguistics: Stroudsburg, PA, USA, 2015; pp. 1693–1701.
17. Kumar, A.; Irsoy, O.; Ondruska, P.; Iyyer, M.; Bradbury, J.; Gulrajani, I.; Zhong, V.; Paulus, R.; Socher, R. Ask me anything: Dynamic memory networks for natural language processing. In Proceedings of the 33rd International Conference on Machine Learning, New York, USA, 19–24 June 2016; pp. 1378–1387.
18. Xiong, C.; Merity, S.; Socher, R. Dynamic memory networks for visual and textual question answering. In Proceedings of the 33rd International Conference on Machine Learning, New York, USA, 19–24 June 2016; pp. 2397–2406.
19. Chung, J.; Gulcehre, C.; Cho, K.; Bengio, Y. Empirical Evaluation of Gated Recurrent Neural Networks on Sequence Modeling. *arXiv* **2014**, arXiv:1412.3555.
20. Cho, K.; van Merrienboer, B.; Bahdanau, D.; Bengio, Y. On the Properties of Neural Machine Translation: Encoder–Decoder Approaches. In Proceedings of the SSST-8, Eighth Workshop on Syntax, Semantics and Structure in Statistical Translation, Doha, Qatar, 25 October 2014; pp. 103–111.
21. Schuster, M.; Paliwal, K.K. Bidirectional recurrent neural networks. *IEEE Trans. Signal Process.* **1997**, *45*, 2673–2681. [CrossRef]
22. Nallapati, R.; Zhou, B.; dos Santos, C.; Gulcehre, C.; Xiang, B. Abstractive Text Summarization using Sequence-to-sequence RNNs and Beyond. In Proceedings of the 20th SIGNLL Conference on Computational Natural Language Learning, Berlin, Germany, 7–12 August 2016; Association for Computational Linguistics: Stroudsburg, PA, USA,2016; pp. 280–290. [CrossRef]

23. See, A.; Liu, P.J.; Manning, C.D. Get To The Point: Summarization with Pointer-Generator Networks. In Proceedings of the 55th Annual Meeting of the Association for Computational Linguistics (Volume 1: Long Papers), Vancouver, Canada, July 30–August 4 2017; Association for Computational Linguistics: Stroudsburg, PA, USA, 2017; pp. 1073–1083. [CrossRef]
24. Gehrmann, S.; Deng, Y.; Rush, A. Bottom-Up Abstractive Summarization. In Proceedings of the 2018 Conference on Empirical Methods in Natural Language Processing, Brussels, Belgium, 31 October–4 November 2018; Association for Computational Linguistics: Stroudsburg, PA, USA, 2018; pp. 4098–4109.
25. Koupaee, M.; Wang, W.Y. WikiHow: A Large Scale Text Summarization Dataset. *arXiv* **2018**, arXiv:1810.09305.
26. Narayan, S.; Cohen, S.B.; Lapata, M. Don't Give Me the Details, Just the Summary! Topic-Aware Convolutional Neural Networks for Extreme Summarization. In Proceedings of the 2018 Conference on Empirical Methods in Natural Language Processing, Brussels, Belgium, 31 October–4 November 2018; pp. 1797–1807.
27. Zhou, Q.; Yang, N.; Wei, F.; Huang, S.; Zhou, M.; Zhao, T. Neural Document Summarization by Jointly Learning to Score and Select Sentences. *arXiv* **2018**, arXiv:1807.02305.
28. Dong, Y.; Shen, Y.; Crawford, E.; van Hoof, H.; Cheung, J.C.K. BanditSum: Extractive Summarization as a Contextual Bandit. In Proceedings of the EMNLP 2018: 2018 Conference on Empirical Methods in Natural Language Processing, Brussels, Belgium, 31 October–4 November 2018; pp. 3739–3748.
29. Kingma, D.P.; Ba, J. Adam: A Method for Stochastic Optimization. *arXiv* **2014**, arXiv:1412.6980.
30. Lin, C.Y. ROUGE: A Package for Automatic Evaluation of Summaries. In Proceedings of the Text Summarization Branches Out, Barcelona, Spain, 25–26 July 2004.
31. Montague, P. Reinforcement Learning: An Introduction, by Sutton, R.S. and Barto, A.G. *Trends Cogn. Sci.* **1999**, 3, 360. [CrossRef]
32. Zhang, X.; Wei, F.; Zhou, M. HIBERT: Document Level Pre-training of Hierarchical Bidirectional Transformers for Document Summarization. In Proceedings of the 57th Annual Meeting of the Association for Computational Linguistics, Brussels, Belgium, 31 October–4 November 2018; Association for Computational Linguistics: Florence, Italy, 2019; pp. 5059–5069. [CrossRef]
33. Zhong, M.; Liu, P.; Wang, D.; Qiu, X.; Huang, X. Searching for Effective Neural Extractive Summarization: What Works and What's Next,. In Proceedings of the 57th Annual Meeting of the Association for Computational Linguistics, Florence, Italy, 28 July–2 August 2019; Association for Computational Linguistics: Florence, Italy, 2019; pp. 1049–1058. [CrossRef]
34. Bae, S.; Kim, T.; Kim, J.; Lee, S.G. Summary Level Training of Sentence Rewriting for Abstractive Summarization. In Proceedings of the 2nd Workshop on New Frontiers in Summarization, Hong Kong, China, 4 November 2019; Association for Computational Linguistics: Hong Kong, China, 2019; pp. 10–20. [CrossRef]
35. Liu, Y.; Lapata, M. Text Summarization with Pretrained Encoders. In Proceedings of the 2019 Conference on Empirical Methods in Natural Language Processing and the 9th International Joint Conference on Natural Language Processing (EMNLP-IJCNLP), Hong Kong, China, 3–7 November 2019; Association for Computational Linguistics: Hong Kong, China, 2019; pp. 3730–3740. [CrossRef]

symmetry

MDPI

Article

Dynamics of Fuzzy-Rough Cognitive Networks

István Á. Harmati

Department of Mathematics and Computational Sciences, Széchenyi István University, 9026 Győr, Hungary; harmati@sze.hu

Abstract: Fuzzy-rough cognitive networks (FRCNs) are interpretable recurrent neural networks, primarily designed for solving classification problems. Their structure is simple and transparent, while the performance is comparable to the well-known black-box classifiers. Although there are many applications on fuzzy cognitive maps and recently for FRCNS, only a very limited number of studies discuss the theoretical issues of these models. In this paper, we examine the behaviour of FRCNs viewing them as discrete dynamical systems. It will be shown that their mathematical properties highly depend on the size of the network, i.e., there are structural differences between the long-term behaviour of FRCN models of different size, which may influence the performance of these modelling tools.

Keywords: fuzzy-rough cognitive network; fuzzy cognitive map; granular computing; fuzzy-rough sets; stability; convergence

Citation: Harmati, I.Á. Dynamics of Fuzzy-Rough Cognitive Networks. *Symmetry* **2021**, *13*, 881. https://doi.org/10.3390/sym13050881

Academic Editor: José Carlos R. Alcantud

Received: 1 April 2021
Accepted: 11 May 2021
Published: 15 May 2021

1. Introduction

Artificial Intelligence (AI) models and methods are parts of our lives. However, most of the AI techniques are blackboxes in the sense, that they do not explain how and why they arrived at a specific conclusion. Explainable AI try to overcome this situation, developing models with interpretable semantics and transparency [1,2]. Fuzzy Cognitive Maps (FCMs) are one of the earliest EAI models, introduced by B. Kosko [3]. FCMs are recurrent neural networks employing weighted causal relation between the model's concepts. Due to their modelling ability and interpretability, these models have a wide range of application [4,5].

Although FCMs have an enormous number of applications, only a few studies are devoted to the analytical and not empirical discussion of their behaviour. Boutalis et al. [6] examined the existence and uniqueness of fixed points of FCMs. Lee and Kwon studied the stability of FCMs using Lyapunov method [7]. Knight et al. [8] analyzed FCMs with linear and sigmoid transfer functions. In [9], the authors generalized the findings of [6] to FCMs with arbitrary sigmoid function. All of these studies arrived to the conclusion (although in a different form) that when the parameter of the sigmoid threshold function is small enough, then the FCM converges to a unique fixed point, regardless of the initial activation values.

The hybridisation of rough set theory and fuzzy set theory [10,11] provide a not only promising, but fruitful combination of different methods of handling and modelling uncertainty [12,13].

The application areas encompass a wide variety of sciences, so only a few of them are mentioned here, without the need for completeness. Fuzzy rough sets and fuzzy rough neural networks have been applied in feature selection problems [14], evolutionary fuzzy rough neural networks have been developed for stock prediction [15]. Fuzzy rough set models are used in multi-criteria decision-making in [16]. Classification tasks have been solved by fuzzy rough granular neural networks in [17]. The combination of unsupervised convolutional neural networks and fuzzy-rough C-mean was used effectively for clustering of large-scale image dataset in [18]. The environmental impact of a renewable energy system was estimated using fuzzy rough sets in [19]. The interval-valued fuzzy-rough based Delphi method was applied for evaluating the siting criteria of offshore wind farms in [20]. Another current direction is the fusion of neutrosophic theory and rough set theory [21]. An example of its application is the emission-based prioritization of bridge maintenance projects [22]. Nevertheless, in the aspect of this paper, fuzzy rough granular networks [23] are the most exciting applications of the synergy of fuzzy and rough theories.

Granular Computing [24] uses information granules, such as classes, clusters, subsets etc., just like humans do. Granular Neural Networks (GNNs) [25] make a synergy between the celebrated neural networks and granular computing [26]. Rough Cognitive Networks (RCN) are GNNs, introduced by Nápoles et al. [27], combining the abstract semantic of the three-way decision model with the neural reasoning mechanism of Fuzzy Cognitive Maps for addressing numerical decision-making problems. The information space is discretized (granulated) by using Rough Set Theory [28,29], which has many other interesting applications [30–33]. According to simulation results, RNN was capable to outperform standard classifiers. On the other hand, learning the similarity threshold parameter had significant computational cost.

Rough Cognitive Ensembles (RCEs) was proposed to overcome this computational burden [34]. It employs a collection of Rough Cognitive Networks as base classifiers, each operating at a different granularity level. This allows suppressing the requirement of learning a similarity threshold. Nevertheless, this model is still very sensitive to the similarity threshold upon which the rough information granules are built.

Fuzzy-Rough Cognitive Maps (FRCNs) has been introduced by Nápoles et al. [35]. The main feature of FRCNs is that the crisp information granules are replaced with fuzzy-rough granules. Based on simulation results, FRCNs show performance comparable to the best blackbox classifiers.

Vanloffelt et al. [36] studied the contributions of building blocks to the FRCNs' performance via empirical simulations with several different network topologies. They concluded that the connections between positive neurons might not necessary to maintain the performance of FRCNs. The theoretical study by Concepción et al. [37] discussed the contribution of negative and boundary neurons. Moreover, they arrived at the conclusion that negative neurons have no impact on the decision, and the ranking between positive neuron remains invariant during the whole reasoning process.

Besides the results presented in [37], this paper was motivated by the fact that only a few studies are discussing the behaviour of cognitive networks from the strict mathematical point of view. Nevertheless, such studies may provide us with information about what we can or cannot achieve with these models. Analyzing the behaviour and contribution of the building blocks unveils the exact role of components of the complex structure: which part is crucial and which one is unnecessary etc. In this paper, we do not develop, neither implement another new fuzzy-rough model. Instead, we analyze the behaviour of FRCNs, which are comparable in performance to the best black-box classifiers. Because of their proven competitiveness [35], there is no need for further model verification and validation.

In the current paper, the dynamical behaviour of fuzzy-rough cognitive networks is examined. The main contributions are the following: first, we show that stable positive neurons have at most two different activation values for any initial activation vectors. Then we show that a certain point with equal coordinates (called *trivial fixed point*) is always a fixed point, nevertheless, not always a fixed point attractor. Furthermore, a condition for the existence of a unique, globally attractive fixed point is also stated. Complete analysis of the dynamics of positive neurons for two and three decision classes is provided. Finally, we show that for a higher number of classes, the occurrence of limit cycles is a necessity and the vast majority of initial activation values lead to oscillation. The rest of the paper is organized as follows. In Section 2, we recall the construction of fuzzy-rough cognitive maps and overview the existing results about their behaviour. In Section 3, a summary of the mathematical background necessary for further investigation of the dynamics of FRCNs is provided, including contraction mapping and elements of bifurcation theory. Section 4 presents general result about the dynamics of positive neurons, condition for unique fixed point attractor and refinement of some findings of [37]. Section 5 introduces size-specific results for FRCNs, providing a complete description for the case of two and three decision classes and pointing out that over a specific size, oscillation behaviour will be naturally present. Section 6 discusses the relation of the behaviour of positive neurons to the final decision class of FRCNs. The paper ends with a short conclusion in Section 7.

2. Fuzzy-Rough Cognitive Networks

In this section, we shortly summarize the basic notions of fuzzy-rough cognitive networks and the findings about their dynamical properties reported in the literature. It is based on the works [35–37].

2.1. Construction of FRCNs

Building up FRCNs includes the following three steps: information space granulation, network construction and finally, network exploitation.

Information space granulation means dividing available information into granules. Let $\mathcal{U}$ denote the universe of discourse and $X = \{X_1, \ldots, X_N\}$, $X \subset \mathcal{U}$. Then $X_c \subset \mathcal{U}$ is the subset containing all objects assigned (labelled) to decision class D_c. The membership degree of $x \in \mathcal{U}$ to X_c is computed in a binary way:

$$\mu_{X_c}(x) = \begin{cases} 1 & , x \in X_c \\ 0 & , x \notin X_c \end{cases}. \tag{1}$$

Membership function $\mu_P(y,x)$ is the next component, using similiraty degree between two instances x and y:

$$\mu_P(y,x) = \mu_{X_c}(x)\varphi(x,y) = \mu_{X_c}(x)(1 - \delta(x,y)), \tag{2}$$

where $\mu_P \colon U \times U \to [0,1]$ is the membership degree of y to X_c, given that x belongs to X_c (in this sense, it is a conditional membership degree). It is composed by the previously defined binary membership function $\mu_{X_c}(x)$ and the similarity degree $\varphi(x,y)$. The later is based on a normalized distance measure $\delta(x,y)$. Membership functions for the lower and upper approximation for any fuzzy set X_c, respectively:

$$\mu_{P_*(X_c)}(x) = \min\left\{\mu_{X_c}(x), \inf_{y \in \mathcal{U}} \mathcal{I}(\mu_P(y,x), \mu_{X_c}(x))\right\}, \tag{3}$$

$$\mu_{P^*(X_c)}(x) = \max\left\{\mu_{X_c}(x), \sup_{y \in \mathcal{U}} \mathcal{T}(\mu_P(y,x), \mu_{X_c}(x))\right\}. \tag{4}$$

Here $\mathcal{I}$ denotes an implication function and $\mathcal{T}$ denotes a conjunction function. As a next step, we calculate the membership functions associated to the positive, negative and boundary regions:

$$\mu_{POS(X_c)}(x) = \mu_{P_*(X_c)}(x), \tag{5}$$

$$\mu_{NEG(X_c)}(x) = 1 - \mu_{P^*(X_c)}(x), \tag{6}$$

$$\mu_{BND(X_c)}(x) = \mu_{P^*(X_c)}(x) - \mu_{P_*(X_c)}(x). \tag{7}$$

After determining the membership functions of the decision classes, we construct a network using four type of neurons:

- $\mathcal{D} = \{D_1, \ldots, D_N\}$ is the set of decision neurons,
- $\mathcal{P} = \{P_1, \ldots, P_N\}$ is the set of positive neurons,
- $\mathcal{N} = \{N_1, \ldots, N_N\}$ is the set of negative neurons,
- $\mathcal{B} = \{B_1, \ldots, B_N\}$ is the set of boundary neurons.

The recurrent neural network have N output neurons (decision neurons). The number of input neuron is between $2N$ and $3N$, depending on the number of non-empty boundary regions. The 'wiring' between the neurons is based on the following steps (see Algorithm 1). Positive, negative and boundary neurons influence themselves with intensity 1. Each positive neuron influence the decision neuron related to it with intensity 1, moreover these neurons act on the other positive and decision neurons with intensity -1. Finally, if two decision classes share non-empty boundary regions, then each boundary neuron influences both decision neurons with intensity 0.5. As it is clear from this setting, the weights of a FRCN are not determined by learning methods, they are based on the semantic relations between the information granules.

Algorithm 1: The construction procedure of the fuzzy-rough cognitive network.

```
for each subset X_c do
    Add a neuron P_c as the c-th positive region
    Add a neuron N_c as the c-th negative region
    Add a neuron B_c as the c-th boundary region
    Add a neuron D_c as the c-th decision class
end
for each neuron C_i do
    if C_i ≠ D_i then
        w_ii = 1.0
    end
    for each neuron C_j do
        if C_i = P_c then
            if C_j = D_c then
                w_ij = 1.0
            end
            else if C_j = D_{v≠c} then
                w_ij = −1.0
            end
            else if C_j = P_{v≠c} then
                w_ij = −1.0
            end
        end
        if C_i = N_c and C_j = D_c then
            w_ij = −1.0
        end
        if C_i = B_c and C_j = D_v and min_{x∈U}{μ_BND(X_c)(x), μ_BND(X_v)(x)} > 0 then
            w_ij = 0.5
        end
    end
end
```

The computation of initial activation values ($A_i^{(0)}$s) of the neurons is based in the similiraty degree between the new object y and all $x \in \mathcal{U}$, and the membership degree of every x to positive, negative and boundary regions. The initial activation value of decision neurons is zero.

$$\text{Positive neurons } (P_i): \quad A_i^{(0)} = \frac{\sum_{x \in U} \mathcal{T}(\varphi(x,y), \mu_{POS(X_c)}(x))}{\sum_{x \in U} \mu_{POS(X_c)}(x)} \tag{8}$$

$$\text{Negative neurons } (N_i): \quad A_i^{(0)} = \frac{\sum_{x \in U} \mathcal{T}(\varphi(x,y), \mu_{NEG(X_c)}(x))}{\sum_{x \in U} \mu_{NEG(X_c)}(x)} \tag{9}$$

$$\text{Boundary neurons } (B_i): \quad A_i^{(0)} = \frac{\sum_{x \in U} \mathcal{T}(\varphi(x,y), \mu_{BND(X_c)}(x))}{\sum_{x \in U} \mu_{BND(X_c)}(x)} \tag{10}$$

After determining the initial values, the next step is network exploitation, by the iteration rule

$$A_i^{(t+1)} = f\left(\sum_{j=1}^{N} w_{ij} A_j^{(t)}\right). \tag{11}$$

or in matrix vector form $A^{(t+1)} = f(WA^{(t)})$, where f is understood coordinate-wise. Here w_{ij} is the weight of the connection from neuron C_j to neuron C_i. Function $f(x) = \frac{1}{1+e^{-\lambda x}}$ keeps the values in the $(0, 1)$ range. For high λ values, $f(x)$ tends to behave like the step function, while for very small λ values, the activation values converge to a unique fixed point attractor, regardless of the initial activation values, producing the same output for every input values. The $\lambda = 5$ choice is widely used in FCM applications, FRCNs employ this parameter value, too. The iteration stops when

- activation values converge to a fixed point attractor (practically it means that the consecutive values do not show reasonable change), or
- a predefined maximal number of iteration is reached.

Finally, the label of the decision neuron with highest activation value is assigned the classified object. Figure 1 shows the FRCN's structure for a binary classification problem.

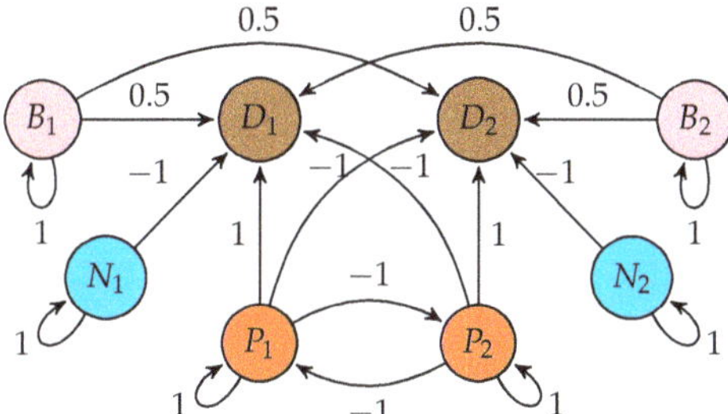

Figure 1. The structure of a fuzzy-rough cognitive network for binary classification.

2.2. Preliminary Results on the Dynamics of FRCNs

In this subsection, we shortly summarize the main contributions of [37] regarding the dynamics of fuzzy-rough cognitive networks. For detailed explanations and proofs, see [37].

- Negative and boundary neurons always converge to a unique fixed value. This value depends on parameter λ, but the convergence and uniqueness are independent of λ.
- The ranking between positive neurons remains invariant during the recurrent reasoning. As we will see in Section 5, although this statement is absolutely true in the strict mathematical sense, from the practical point if view, in some cases it has very limited applicability.
- In an FRCN with N decision classes, there will always be $N-1$ positive neurons with activation values less than or equal to $1/2$ (after at least one iteration step).
- In an FRCN there will be at most one neuron with activation value higher or equal to $1/2$ (after at least one iteration step).

Consider the updating rule $A_i^{(t+1)} = f\left(\sum_{j=1}^{N} w_{ij} A_j^{(t)}\right)$ (see Equation (11)), where f is the sigmoid function with parameter λ. Assuming that positive neurons reach stable states, for every i it holds, that $P_i^{(t+1)} = P_i^{(t)} = f\left(\sum_{j=1}^{N} w_{ij} P_j^{(t)}\right)$. Recall that each positive neuron is influenced by the other positive neurons with weight -1, while influences itself with weight 1 (i.e., if $i = j$, then $w_{ij} = 1$, otherwise $w_{ij} = -1$). Consequently, we have

$$P_i^{(t)} = P_i^{(t+1)} = \frac{1}{1 + e^{-\lambda\left(P_i^{(t)} - \sum_{j=1, j \neq i}^{N} P_j^{(t)}\right)}} = \frac{1}{1 + e^{-\lambda\left(2P_i^{(t)} - \sum_{j=1}^{N} P_j^{(t)}\right)}}. \tag{12}$$

Now let us further investigate the equation $P_i^{(t)} = \frac{1}{1 + e^{-\lambda\left(2P_i^{(t)} - \sum_{j=1}^{N} P_j^{(t)}\right)}}$. Expressing $\sum_{j=1}^{N} P_j^{(t)}$ we get the following equation:

$$\sum_{j=1}^{N} P_j^{(t)} = \frac{1}{\lambda} \ln \frac{1 - P_i^{(t)}}{P_i^{(t)}} + 2P_i^{(t)}, \tag{13}$$

on the left hand side we find the sum of all activation values, which depends on the value of $P_i^{(t)}$ (recall that $P_i^{(t)}$ is the stabilized activation value of any positive neuron). The sum of the activation values is a function of $P_i^{(t)}$, of course. Define the following real function:

$$s(x) = \frac{1}{\lambda} \ln \frac{1 - x}{x} + 2x. \tag{14}$$

From the behaviour of $s(x)$, the authors derived some properties of the positive neurons:

- If $\lambda \leq 2$, then $s(x)$ is monotone decreasing, thus a specific value can be produced by a single input value x. It means that if $\lambda \leq 2$ and the positive neurons are stable (converge to a fixed point), then they have the same activation value.
- If $\lambda > 2$, then $s(x)$ produces the same value for at most three different input values in $(0, 1)$. It means that if the vector of positive neurons converges to a fixed point, then this vector has at most three different coordinate values.

In Sections 4 and 5, we refine these statements and introduce some additional results regarding the dynamics of FRCNs.

3. Mathematical Background

The updating rule of FRCNs suggest to handle them as discrete dynamical systems. In this section, we briefly summarize the most important notions and methods used in the forthcoming sections.

3.1. Contraction Mapping

In the network exploitation phase, we apply the iteration rule Equation (11) again and again until the activation values stabilize (or the number of iterations reaches the predefined maximum). If the activation values stabilize (i.e., arrive an equilibrium state), then the difference between the outputs of two consecutive iteration steps will be smaller and smaller. In other words, the iteration contracts a subset of the state space (or the whole state space) into a single point. The following definition provides a strict mathematical description of this property.

Definition 1 (see [38], p. 220)**.** *Let (X, d) be a metric space, with metric d. If φ maps X into X and if there is a number $c < 1$ such that*

$$d(\varphi(x), \varphi(y)) \leq c d(x, y) \tag{15}$$

for all $x, y \in X$, then φ is said to be a contraction *of X into X.*

If the iteration reaches an equilibrium state, then this state will not change by applying the updating rule. Mathematically speaking, it is a fixed point of the mapping generating the iteration. The following famous theorem establishes connection between the contraction mapping and a unique fixed point.

Theorem 1 (Contraction mapping theorem or Banach's fixed point theorem, see [38], pp. 220–221)**.** *If X is a complete metric space, and if φ is a contraction of X into X, then there exists one and only one $x \in X$ such that $\varphi(x) = x$.*

The proof to this theorem is constructive (see [38], p. 221) and offers a straightforward way to find this unique fixed point. We only have to pick an arbitrary element of X and apply mapping φ again and again. The limit will be the unique fixed point of φ: $x_n = \varphi(x_{n-1})$, then $\lim_{n\to\infty} x_n = x^*$ and $\varphi(x^*) = x^*$.

An arbitrary fixed point x^* is said to be *asymptotically stable* if starting the iteration *close enough* to x^*, the limit will be x^*. Moreover, the fixed point x^* is said to be *globally asymptotically stable* if starting the iteration from *any* element of the state space, the limit will be x^*. Based on Theorem 1 and its constructive proof, it is clear that the unique fixed point of a contraction is globally asymptotically stable.

3.2. Elements of Bifurcation Theory

The dynamics of a discrete dynamical system may change, if its parameters are varied. A qualitative change of the dynamical behaviour, e.g., a transition from a unique stable fixed point to multiple fixed points or oscillation, is called bifurcation and the corresponding critical parameter is called bifurcation point.

The detailed description of the dynamics of FRCNs requires the application of the elements of bifurctaion theory. Here only the most important notions are listed. For more details, see [39,40].

Consider a discrete-time dynamical system depending on only one parameter λ ($G\colon \mathbb{R}^n \to \mathbb{R}^n$):

$$x_{k+1} = G(x_k, \lambda), \qquad x_k \in \mathbb{R}^n, \quad \lambda \in \mathbb{R}, \tag{16}$$

where function G is smooth with respect to x and λ. Let's assume that x_0 is a fixed point of the mapping with parameter λ_0. The local stability (resistance agianst small perturbations) of the fixed point depends on the eigenvalues of the Jacobian evaluated at x_0 ($J_G(x_0)$). If the Jacobian has no eigenvalues on the unit circle, x_0 is said to be *hyperbolic*. Hyperbolic fixed points can be categorized according to the eigenvalues of $J_G(x_0)$:

- If all of the eigenvalues lie inside the unit circle (i.e., the absolute value of the eigenvalues is less than one), then it is an asymptotically stable fixed point. In other words, this fixed point attracts the space in every direction in its (sometimes very small) neighborhood. Consequently, its basin of attraction is an n dimensional subset of then n dimensional space.
- If there are some eigenvalues (at least one) with absolute value greater than one, and there are some (at least one) eigenvalues with absolute value less than one, then it is a saddle point. It means that this fixed point may attract points of the space in some, but not every direction in its neighborhood. Consequently, the dimension of its basin of attraction is less than n.
- If all of the eigenvalues has absolute value greater than one, then it is an unstable fixed point (repeller fixed point).

In the FCM theory and similary in FRCN terminology, stable fixed point is a fixed point, that can be a limit of the iteration. In this sense, stable and saddle fixed points are considered as stable in the FCM (FRCN) sense. Nevertheless, their dynamical behavior could be very different. We will see in Section 5 that different type of fixed points have significantly different size of basin of attractions. In other words, some fixed points are less important than others.

The simplest ways the hyperbolicity can be violeted are the following:

- A simple positive eigenvalue crosses the unit circle at $\alpha = 1$. The bifurcation associated to this scenario is called saddle-node bifurcation. It is a birth of new fixed points.
- A simple negative eigenvalue crosses the unit circle at $\alpha = -1$. This causes period-doubling bifurcation, the birth of limit cycles.
- A pair of complex conjugate eigenvalues approaches the unit circle at $\alpha_{1,2} = e^{i\phi}$, $\phi \in (0, \pi)$. This is the so-called Neimark-Sacker bifurcation, causes the occurence of a closed invariant curve.

It will be shown in Section 5, that depending on the size of the FRCN, different types of bifurcations determine the main dynamics of the system.

4. General Results on the Dynamics of Positive Neurons

In this section, we introduce some general results about the dynamics of positive neurons. Size specific results are presented in Section 5.

With start with the refinement of a result from [37]. Further investigation of the function $s(x)$ (see Equation (14)) provides more information about the possible fixed points of positive neurons (see Figure 2). It has been shown that there are at most one positive neuron with activation value higher than $1/2$. If $s(x) > 1$, then a specific value may be produced by at most three different values of x. But two of these values are higher than $1/2$, thus only one of them get a role in the activation vector. It means that one coordinate of the activation vector is greater than $1/2$ and the remaining ones are less than $1/2$ and they have equal values.

Consider now the case, when $s(x) < 1$. Observe that the graph of $s(x)$ is symmetrical about the point $(1/2, 1)$ (it can be easily verified analytically). Let us choose a specific value $\gamma = s(x)$, such that the horizontal line $y = \gamma$ has three intersection points with the graph. Denote the first coordinates of theses point by x_1, x_2, x_3 ($x_1 < x_2 < x_3$). Using the symmetry of the function, we can conclude that $x_1 > 1 - x_3$. Consequently, $x_1 + x_3 > 1$. Since the sum of the activation values is less than 1 ($s(x) < 1$), it follows that there can be at most two different values (x_1 and x_2) in the activation vector for any given $s(x)$.

Summarizing this short argument, if the activation vector of positive neurons converges to a fixed point, then it may have at most *two* different coordinate values.

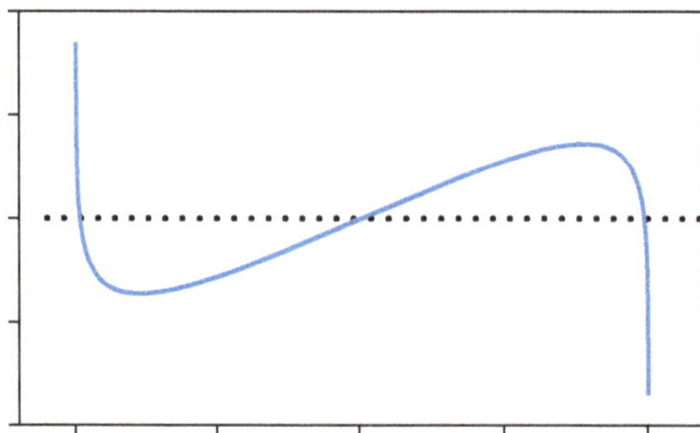

Figure 2. The graph of $s(x) = \frac{1}{\lambda} \ln \frac{1-x}{x} + 2x$, with $\lambda = 5$. Observe that the curve is symmetrical about the point $(1/2, 1)$.

The iteration rule for the updating of the neurons' activation values has the form $A^{(t+1)} = f\left(WA^{(t)}\right)$, where the sigmoid is applied coordinate-wise and W is the connection matrix of the network. Based on the construction of the network (see Algorithm 1), it has the following block form:

$$W = \begin{array}{c} \text{neuron sets} \\ \mathcal{D} \\ \mathcal{B} \\ \mathcal{N} \\ \mathcal{P} \end{array} \begin{array}{c} \begin{array}{cccc} \mathcal{D} & \mathcal{B} & \mathcal{N} & \mathcal{P} \end{array} \\ \begin{bmatrix} O & W_B & -I & W_D \\ O & I & O & O \\ O & O & I & O \\ O & O & O & W_P \end{bmatrix} \end{array}, \tag{17}$$

where O, I denote zero matrix and identity matrix, respectively. W_B describes the connections from boundary to decision neurons (it contains 0 s and 0.5 s, their positions depend on non-empty boundary regions), W_P describes the connections between positive neurons (if $i = j$, then $w_{ij} = 1$, else $w_{ij} = -1$), $W_D = W_P$ contains the connections from positive neurons to decision neurons.

Because of the upper-diagonal block structure, instead of dealing with the whole matrix, we can use the blocks. It has been proved in [37] that activation values of the negative and boundary neurons converge to the same unique value, which depends on λ, but independent of the initial activation values. Positive neurons influence themselves, each other and decision neurons, but do not receive input from other set of neurons. Their activation values are propagated to the decision neurons. In a long run, when neurons reach a stable state (or the iteration is stopped by achieving the maximal number of iterations), the propagated value is their stable (or final) state. In the following, we examine the long-term behaviour of positive neurons.

Lemma 1. *For every $\lambda > 0$ and every number of decision classes N, there always exists a fixed point of the positive neurons, whose coordinates are the same. Nevertheless, this fixed point is not always a fixed point attractor.*

Proof. Consider the fixed point equation for every $1 \leq j \leq N$:

$$P = f(W_P P) \tag{18}$$

$$\begin{bmatrix} P_1 \\ \vdots \\ P_N \end{bmatrix} = f\left(W_P \begin{bmatrix} P_1 \\ \vdots \\ P_N \end{bmatrix} \right) \tag{19}$$

In coordinate-wise form:

$$P_j = \frac{1}{1 + e^{-\lambda\left(P_j - \sum\limits_{i=1, j\neq j} P_i\right)}} = \frac{1}{1 + e^{-\lambda\left(2P_j - \sum\limits_{i=1} P_i\right)}} \tag{20}$$

If $P_j = x$ for every $1 \leq j \leq N$, then it simplifies to the following equation:

$$x = \frac{1}{1 + e^{-\lambda(2-N)x}} = \frac{1}{1 + e^{\lambda(N-2)x}} \tag{21}$$

We show that there always exists a unique solution to this equation. Let us introduce the function

$$g(x) = x - \frac{1}{1 + e^{\lambda(N-2)x}}. \tag{22}$$

Function $g(x)$ is continuous and differentiable, moreover $g(0) = -0.5 < 0$ and $g(1) = 1 - \frac{1}{1+e^{\lambda(N-2)}} > 0$, thus it has at least one zero in $(0, 1)$. According to Rolle's theorem, between two zeros of a differentiable function its derivative has a zero. The derivative is

$$g'(x) = 1 + \frac{\lambda(N-2)e^{\lambda(N-2)x}}{(1 + e^{\lambda(N-2)x})^2}, \tag{23}$$

which is always positive. It means that there is exactly one zero of $g(x)$ in $(0,1)$. Consequently, we have shown that for any given $\lambda > 0$ and N, there is exactly one fixed point of the positive neurons with equal coordinates. There may be other fixed points, but their coordinates are not all the same. □

The following lemma plays a crucial role in the proof of Theorem 2 and in the examination of the Jacobian of the iteration mapping.

Lemma 2. *Let W_P be an $N \times N$ matrix with the following entries:*

$$w_{ij} = \begin{cases} 1 & \text{if } i = j \\ -1 & \text{if } i \neq j \end{cases} \tag{24}$$

Then the eigenvalues of W_P are $2-N$ (with multiplicity one) and 2 (with multiplicity $N-1$).

Proof. Basic linear algebra. □

Theorem 2. *Consider a fuzzy cognitive map (recurrent neural network) with sigmoid transfer function ($f(x) = 1/(1+e^{-\lambda x})$) and with weight matrix W_p whose entries are*

$$w_{ij} = \begin{cases} 1 & \text{if } i = j \\ -1 & \text{if } i \neq j \end{cases} \tag{25}$$

If

$$\lambda < \frac{4}{\max\{2, |2-N|\}},$$

then it has exactly one fixed point. Moreover, this fixed point is a global attractor, i.e., the iteration starting from any initial activation vector ends at this point.

Proof. We are going to show that if the condition in theorem is fulfilled, then the mapping $P \to f(W_P P)$ is contraction, thus according to Banach's theorem, it has exactly one fixed point and this fixed point is globally asymptotically stable, i.e., iterations starting from any initial vectors arrive to this fixed point. Let us choose two different initial vectors, P and P'. Then

$$\|f(W_P P) - f(W_P P')\|_2 = \left(\sum_{i=1}^{N} \left[f\left(\sum_{j=1}^{N} w_{ij} p_j \right) - f\left(\sum_{j=1}^{N} w_{ij} p'_j \right) \right]^2 \right)^{1/2} \tag{26}$$

$$\leq \left(\sum_{i=1}^{N} \left[\frac{\lambda}{4} \left(\sum_{j=1}^{N} w_{ij} p_j - \sum_{j=1}^{N} w_{ij} p'_j \right) \right]^2 \right)^{1/2} = \frac{\lambda}{4} \left(\sum_{i=1}^{N} \left[\sum_{j=1}^{N} w_{ij} (p_j - p'_j) \right]^2 \right)^{1/2} \tag{27}$$

$$= \frac{\lambda}{4} \|W_p(P-P')\|_2 = \frac{\lambda}{4} \frac{\|W_p(P-P')\|_2}{\|P-P'\|_2} \|P-P'\|_2 \tag{28}$$

$$\leq \frac{\lambda}{4} \|W_p\|_2 \|P-P'\|_2 \tag{29}$$

Here the first inequality comes from the fact that the derivative of the sigmoid function $f(x)$ is less than or equals $\lambda/4$ and $f(x)$ is Lipschitzian, while the second inequality comes from the definition of the induced matrix norm. Since W_p is a real, symmetric matrix its spectral norm ($\| * \|_2$) equals the maximal absolute values of its eigenvalues. By Lemma 2, $\|W_p\|_2 = \max\{2, |2-N|\}$. According to the definition of contraction (Equation (15)), if the coefficient of $\|P-P'\|_2$ is less than one, then the mapping is a contraction and by

Theorem 1 it has exactly one fixed point and this fixed point is globally asymptotically stable. The inequality in the Theorem comes by a simple rearranging:

$$\frac{\lambda}{4}\|W_p\|_2 = \frac{\lambda}{4}\max\{2, |2-N|\} < 1 \iff \lambda < \frac{4}{\max\{2, |2-N|\}}. \tag{30}$$

□

The immediate corollary of Theorem 2 and Lemma 1 is if there is a unique globally attracting fixed point, then its coordinates are equal. We will refer to fixed point with equal coordinates as *trivial fixed point*. The whole complex behaviour of positive neurons (and in such a way, fuzzy-rough cognitive networks) evolves from this trivial fixed point via bifurcations (see the flowchart Figure 3 for the way to the first bifurcation). In Section 5, we show that different size of FRCNs (different number of decision classes N) may show significantly different qualitative behaviour.

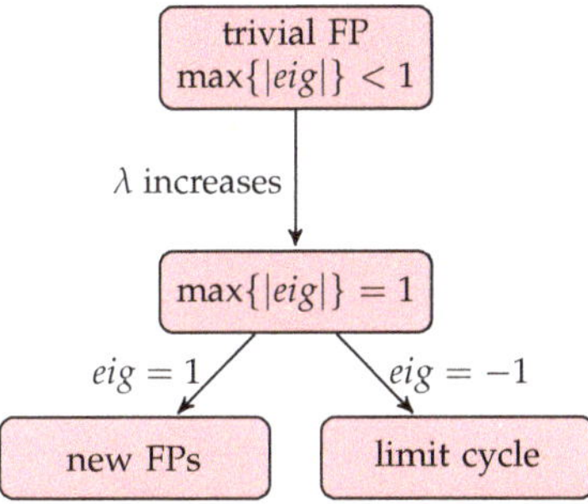

Figure 3. Flowchart of the way to first bifurcation. The trivial fixed point loses its global stability when the eigenvalues of the Jacobian (evaluated at the trivial fixed point) reach ± 1 and the first bifurcation occurs. As parameter λ increases, the absolute values of the eigenvalues increase. Due to the symmetry of the weight matrix, we do not have to count with complex eigenvalues.

5. Dynamics of Positive Neurons

First we provide the Jacobian at the trivial fixed point. In general (except the case $N = 2$), this fixed point is a function of λ and N. Let us denote the coordinates of the trivial fixed point by p^*. Then the (i,j) entry of the Jacobian of the mapping $f(W_pP)$ at this point, using the fact that $f(W_pP^*) = p^*$ and for the sigmoid function $f' = \lambda f(1-f)$:

$$\left.\frac{\partial f}{\partial p_i}\right|_{P=P^*} = \lambda f(W_pP^*)(1 - f(W_pP^*))w_{ij} = \lambda p^*(1-p^*)w_{ij} \tag{31}$$

The whoole Jacobian matrix evaluated at the trivial fixed point is the following:

$$J_{p^*} = \lambda p^*(1-p^*)W_p. \tag{32}$$

Its eigenvalues are $\lambda p^*(1-p^*)$ times the eigenvalues of W_p: $(2-N)\lambda p^*(1-p^*)$ and $2\lambda p^*(1-p^*)$. As the value of λ increases, at a certain point the absolute value of the eigenvalue with the highest modulus reaches one, the trivial fixed point loses its global stability and a bifurcation occurs. The type of this bifurcation has great effect on the further evolution and dynamics of the system. Based on the eigenvalues of W_P, we see that Neimark-Sacker bifurcation does not occur here, but saddle-node and perido-doubling bifurcations do play an important role.

5.1. $N = 2$

Consider first the case when we have only two decision classes. The relations between the positive neurons can be seen in Figure 1. The weight matrix describing the connections is the following (subscript P refers to positive):

$$W_P = \begin{bmatrix} 1 & -1 \\ -1 & 1 \end{bmatrix} \tag{33}$$

Easy to check the the point $[0.5, 0.5]^T$ is always a fixed point of the mapping $f(W_p P) = P$, since

$$\begin{bmatrix} 1 & -1 \\ -1 & 1 \end{bmatrix} \begin{bmatrix} 0.5 \\ 0.5 \end{bmatrix} = \begin{bmatrix} 0 \\ 0 \end{bmatrix} \tag{34}$$

and $f(0) = 0.5$. According to Theorem 2, if $\lambda < 2$, then it is the only fixed point, moreover it is globally asymptotically stable, i.e., strating from any initial activation vector, the iteration will converge to this fixed point. The Jacobian of the mapping at this fixed point is

$$\lambda \cdot 0.5(1 - 0.5))\begin{bmatrix} 1 & -1 \\ -1 & 1 \end{bmatrix} = \frac{\lambda}{4}\begin{bmatrix} 1 & -1 \\ -1 & 1 \end{bmatrix} \tag{35}$$

and its eigenvalues are 0 and $\lambda/2$. When the eigenvalue $\lambda/2 = 1$ ($\lambda = 2$), a bifurcation occurs, giving birth to two new fixed points. In the following, we are going to show that for every $\lambda > 2$, there are exactly three fixed points, moreover these fixed points have the following coordinates: $[0.5, 0.5]^T$, $[x^*, 1 - x^*]$ and $[1 - x^*, x^*]$, where x^* is a fixed point of a one dimensional mapping described below.

Let us assume that $(x_1, x_2)^T$ is a fixed point of the mapping, then

$$\begin{bmatrix} x_1 \\ x_2 \end{bmatrix} = f\left(\begin{bmatrix} 1 & -1 \\ -1 & 1 \end{bmatrix}\begin{bmatrix} x_1 \\ x_2 \end{bmatrix}\right) = \begin{pmatrix} f(x_1 - x_2) \\ f(x_2 - x_1) \end{pmatrix} \tag{36}$$

Since f is the sigmoid function, we have that $f(-x) = 1 - f(x)$, consequently

$$x_1 = f(x_1 - x_2), \tag{37}$$

$$x_2 = f(x_2 - x_1) = f(-(x_1 - x_2)) = 1 - f(x_1 - x_2) = 1 - x_1. \tag{38}$$

So for a fixed point the coordinates are $(x_1, 1 - x_1)$. The first equation leads to the following fixed point equation:

$$x_1 = f(2x_1 - 1). \tag{39}$$

It means that the FPs of the positive neurons can be determined by solving Equation (39). From the graphical point of view, easy to see that if $\lambda \le 2$, then it has exactly one solution ($x_1 = 0.5$), but if $\lambda > 2$, then there are three different solutions: 0.5, x^* and $1 - x^*$ (see Figure 4).

From the analytical viewpoint, we have to solve the equation

$$x = \frac{1}{1 + e^{-\lambda(2x-1)}}. \tag{40}$$

Applying the inverse of $f(x)$ and rearranginig the terms:

$$\frac{1}{\lambda}\ln\frac{1-x}{x} + 2x = 1 \tag{41}$$

As it was pointed out in [37], if $\lambda > 2$, then the left hand side has local minimum at $\frac{1}{2} - \sqrt{\frac{\lambda-2}{4\lambda}}$ less than one, and local maximum at $\frac{1}{2} + \sqrt{\frac{\lambda-2}{4\lambda}}$ greater than one. If $\lambda \le 2$, then the function is strictly monotone decreasing. Using continuity of the function, we conclude that there are exactly three solutions for every $\lambda > 2$ and there is a unique solution if $\lambda \leq 2$ (see Figure 5).

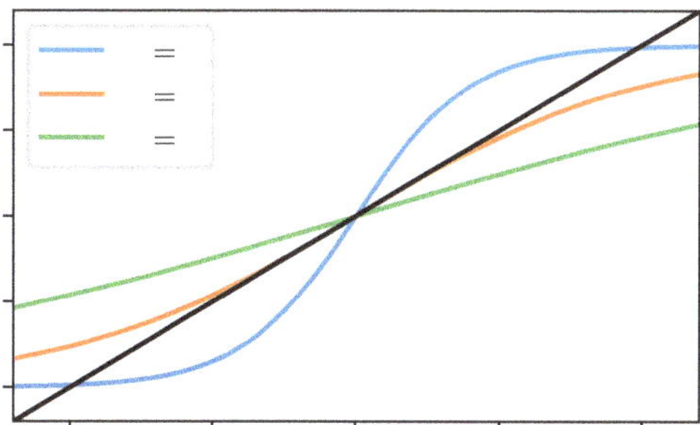

Figure 4. The graph of the function $f(2x-1)$. It has exactly one fixed point if $\lambda \leq 2$ and three fixed points for $\lambda > 2$. One of them is unstable, referring to the trivial fixed point of positive neurons, the other two fixed points are stable.

For $\lambda = 5$, the fixed points are $(0.5,\ 0.5)$, $(0.0072,\ 0.9928)$ and $(0.9928,\ 0.0072)$.

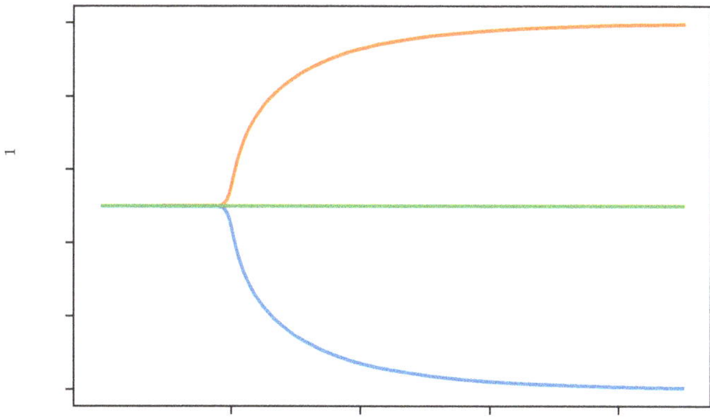

Figure 5. Possible values of the positive neurons in function of parameter λ, for $N = 2$ decision classes. Green line denotes the value of the trivial fixed point, while orange and blue denote the high and low value of the other two fixed points. Observe the pitchfork bifurcation at $\lambda = 2$.

Let us examine the basins of attraction for the three different fixed points, i.e., $\lambda > 2$ and the fixed points are $[0.5, 0.5]^T$, $[x^*, 1-x^*]$ and $[1-x^*, x^*]$, with $x^* > 1/2$. Consider a point $(x_1, x_2)^T$ as initial activation vector (see Figure 6).

- If $x_1 = x_2$, then the iteration leads to the fixed point $(0.5, 0.5)$, since

$$f\left(\begin{bmatrix} 1 & -1 \\ -1 & 1 \end{bmatrix}\begin{bmatrix} x_1 \\ x_2 \end{bmatrix}\right) = \begin{bmatrix} f(x_1-x_2) \\ f(x_2-x_1) \end{bmatrix} = \begin{bmatrix} f(0) \\ f(0) \end{bmatrix} = \begin{bmatrix} 0.5 \\ 0.5 \end{bmatrix}. \tag{42}$$

- If $x_1 > x_2$, then $f(x_1-x_2) > f(0) > f(x_2-x_1)$, so this ordering remains invariant during the iteration process. Moreover, after the first iteration step it reduces to a one dimensional iteration with initial value $x = f(x_1-x_2) > 1/2$ and updating equation $x = f(2x-1)$. If $x > 1/2$, then fixed point x^* attracts this one-dimensional iteration. Consequently, the original two-dimensional itertaion converges to $(x^*, 1-x^*)^T$.

- Similarly, if $x_1 < x_2$, then the iteration ends in $(1 - x^*, x^*)^T$.

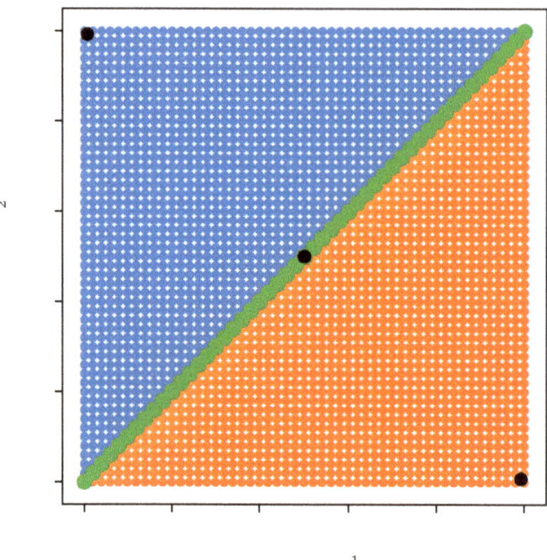

Figure 6. Basin of attraction of the fixed points for $N = 2$ decision classes and $\lambda = 5$ in the (P_1, P_2) plane. Fixed point attractors are denoted by large black dots.

The size of the basin of attraction can be considered as the number of its points. In the strict mathematical sense it is infinity, of course. On the other hand, the basin of fixed point $(0.5, 0.5)$ is a one dimensional object (line segment), while the basins of $(x^*, 1 - x^*)$ and $(1 - x^*, x^*)$ are two-dimensional sets (triangles), so they are 'much more bigger' sets.

In applications, we always work with sometimes large, but finite numbers of points, based on the required and available precision. Let us define the level of granularity as the length of subintervals, when we divide the unit interval into n equal parts. Then the division points are $0, 1/n, 2/n, \ldots, 1$, so we have $n + 1$ points. The basin of fixed point $(0.5, 0.5)$ contains $n + 1$ points, while the basins of the two other fixed points have $n(n+1)/2, n(n+1)/2$ points. By increasing the number of division points, the proportion of the basins tend to zero and $1/2$, as it was expected.

$$\lim_{n \to \infty} \frac{n+1}{(n+1)^2} = 0 \tag{43}$$

$$\lim_{n \to \infty} \frac{n(n+1)/2}{(n+1)^2} = \frac{1}{2} \tag{44}$$

In a certain sense, it means that fixed points $(x^*, 1 - x^*)^T$ and $(1 - x^*, x^*)^T$ are more important than fixed point $(0.5, 0.5)^T$, since much more initial activation values lead to these points.

5.2. $N = 3$

The structure of the connections between positive neurons can be seen in Figure 7. In this case, the eigenvalues of W_P are -1 (with multiplicity one) and 2 (with multiplicity two). The fixed point with equal coordinates loses its global asymptotic stability when the absolute value of its larger eigenvalue equals one. Since the positive eigenvalue has the higher absolute value, this bifurcation results in new fixed points. Nevertheless, this eigenvalue has multiplicity two, so it is not a simple bifurcation, i.e., not only a pair of new fixed points arise, but a couple of new FPs. The trivial fixed point becomes a saddle point, i.e., it attracts points in a certain direction, but repells them in other directions. If we further increase the value of the parameter λ, then the absolute value of the negative eigenvalue reaches one and the trivial fixed point suffers a bifurcation again. Since the eigenvalue is -1, this is a period-doubling bifurcation, giving birth to a two-period limit cycle.

We show that there are three type of fixed points:

- The trivial fixed point with equal coordinates (FP_0);
- Fixed points with one *high* and two *low* values (FP_1);
- Fixed points with one *low* and two *medium* coordinates (FP_2).

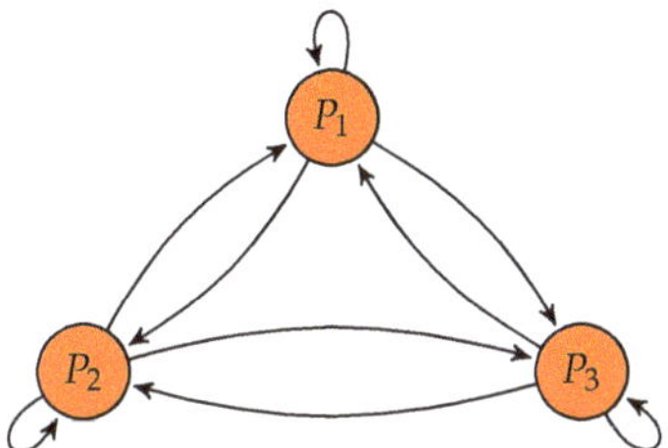

Figure 7. Topology of the connections between the positive neurons in case of $N = 3$ decision classes. Self connections have weight 1, other connections have weight -1.

The existence of FP_0 is clear, as it was shown by Lemma 1. As it was pointed out in Section 4, the non-trivial fixed points have two different coordinate values. Let us denote these values by x, x and y. Then the fixed point equation is

$$\begin{bmatrix} x \\ x \\ y \end{bmatrix} = f\left(\begin{bmatrix} 1 & -1 & -1 \\ -1 & 1 & -1 \\ -1 & -1 & 1 \end{bmatrix}\begin{bmatrix} x \\ x \\ y \end{bmatrix}\right) = \begin{pmatrix} f(-y) \\ f(-y) \\ f(y-2x) \end{pmatrix}, \tag{45}$$

which simplifies to the following system of equations:

$$x = f(-y) \tag{46}$$

$$y = f(y-2x) \tag{47}$$

By substituting x, we have

$$y = f(y-2x) = f(y-2f(-y)) = \frac{1}{1+e^{-\lambda(y-2f(-y))}} = \frac{1}{1+e^{-\lambda\left(y-2\frac{1}{1+e^{\lambda y}}\right)}} \tag{48}$$

It is again a fixed point equation, whose number of solutions depends on the value of parameter λ (see Figure 8):

- if $\lambda \leq 2.2857$ (rounded), then there is exactly one solution, it refers to the trivial fixed point (FP_0);
- if $\lambda > 2.2857$, then there are three different solutions, one refers to FP_0, one to FP_1s and one to FP_2.

Using the values of y, we can determine the values of x. Furthermore, if y is *high*, then based on the equation $x = f(-y) = 1 - f(y)$, we may conclude that x is *low*. Similarly, if y is *low*, then x is *medium*. Finally, there are seven fixed points:

- the fixed point with equal coordinates (FP_0);
- three fixed points with one *high* and two *low* values (FP_1);
- three fixed points with one *low* and *two* medium values (FP_2s).

For $\lambda = 5$, these fixed points are (rounded to four decimals)

- $(0.2355, 0.2355, 0.2355)$ (FP_0),
- $(0.9926, 0.0069, 0.0069)$ and its permutations (FP_1),
- $(0.4904, 0.4904, 0.0076)$ and its permutations (FP_2).

Initial activation values lead to these fixed points according to the following:

- if $P_1^0 = P_2^{(0)} = P_3^{(0)}$, then the iteration converges to FP_0;
- if $P_1^0 > P_2^{(0)} \geq P_3^{(0)}$, then the iteration converges to FP_1;
- if $P_1^0 = P_2^{(0)} > P_3^{(0)}$, then the iteration converges to FP_2.

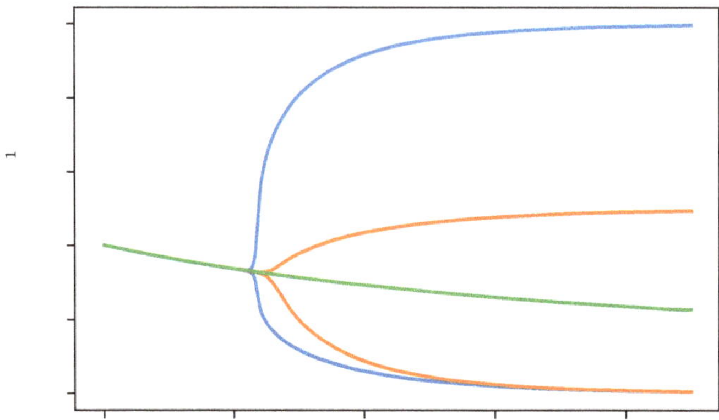

Figure 8. Possible values of positive neurons in function of parameter λ, for $N = 3$ decision classes.

Ranking is preserved between positive neurons, in the sense that if $P_x^{(0)} \geq P_y^{(0)}$, then $P_x^* \geq P_y^*$. Since the number of possible outcomes is very limited (only three cases without permutations), it means that some differences in the initial activation values will magnified, for example if the initial activation vector is $(0.3, 0.2, 0.1)$, then the iteration converges to $(0.9926, 0.0069, 0.0069)$. On the other hand, some large differences will be hidden: initial activation vector $(1, 0.95, 0)$ leads again to $(0.9926, 0.0069, 0.0069)$.

Limit cycle occurs, when the negative eigenvalue of the Jacobian computed at the trivial fixed point reaches -1 (at about $\lambda = 5.8695$). Similarly to the trivial fixed point, the elements of the limity cycle have equal coordinates. Let us denote these points by (x_1, x_1, x_1) and (x_2, x_2, x_2). The members of a two-period limit cycle are fixed points of the double iterated function.

$$\begin{bmatrix} x_1 \\ x_1 \\ x_1 \end{bmatrix} = f\left(W_P \begin{bmatrix} x_2 \\ x_2 \\ x_2 \end{bmatrix}\right) \quad \text{and} \quad \begin{bmatrix} x_2 \\ x_2 \\ x_2 \end{bmatrix} = f\left(W_P \begin{bmatrix} x_1 \\ x_1 \\ x_1 \end{bmatrix}\right) \tag{49}$$

In coordinate-wise form this provides the following system of equations:

$$x_1 = f(-x_2) \tag{50}$$

$$x_2 = f(-x_1) \tag{51}$$

From which we have

$$x_2 = f(-x_1) = f(-f(-x_2)) = \frac{1}{1 + e^{\lambda f(-x_2)}} = \frac{1}{1 + e^{\lambda \frac{1}{1 + e^{\lambda x_2}}}} \tag{52}$$

If $\lambda \leq 5.8695$, then it has a unique solution. It refers to the trivial fixed point, since this point is a fixed point of the double-iterated function, too. If $\lambda > 5.8695$, then there are two other solutions, *low* and *medium*, these are the coordinates of the two-period limit cycle. For example, for $\lambda = 7$, these points are 0.0563 and 0.4027.

For a general case, basins of attraction of a dynamical systems are difficult to determine and sometimes it is analytically not feasible task [41,42], enough to mention the famous graph of Newton's method's basin ofattractions [43]. We examined the basin of attraction of the fixed points by putting an equally spaced grid on the set of possible initial values of positive neurons P_1, P_2 and P_3, and applied the grid points as initial activation values. Table 1 shows the sizes of the basins of attraction for different granurality. Results are visualized in Figures 9 and 10.

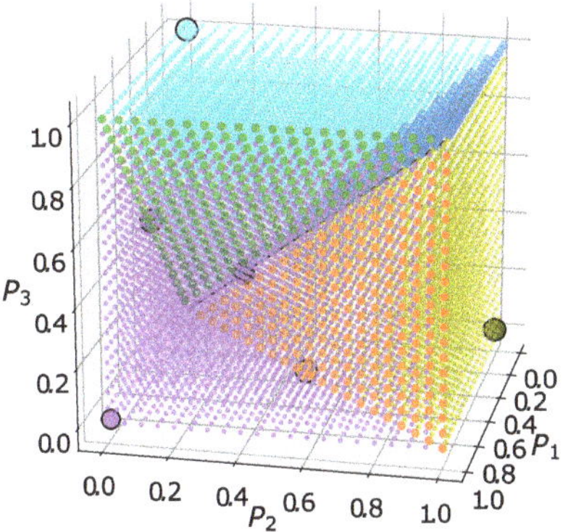

Figure 9. Basins of attraction of all the fixed points of the positive neurons for $\lambda = 5$, $N = 3$ in the (P_1, P_2, P_3) space. The fixed points are (0.2355, 0.2355, 0.2355), (0.4904, 0.4904, 0.0076) and its permutation, (0.9926, 0.0069, 0.0069) and its permutations. The fixed points are denoted by large dots.

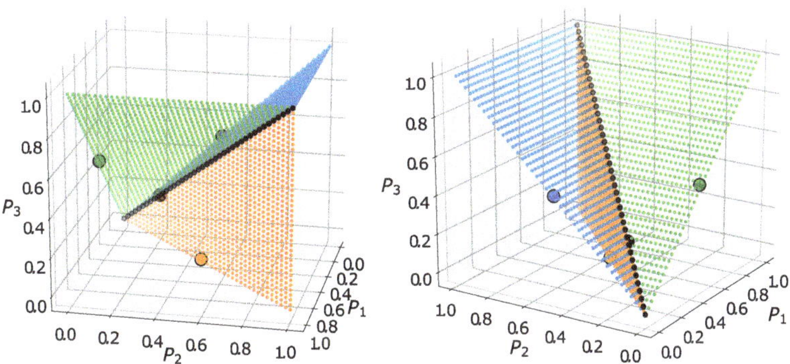

Figure 10. Basin of attraction of trivial fixed point (0.2355, 0.2355, 0.2355) and fixed points with two medium and one low values (0.4904, 0.4904, 0.0076) and its permutation in the (P_1, P_2, P_3) space. The fixed points are denoted by large dots.

Table 1. Number of points in basin of attraction in percentage of the total number of points for $N = 3$ decision classes, $\lambda = 5$. FP_0 refers to fixed point with equal coordinates, FP_1 refers to fixed points with one high and two low values, while F_2 refers to fixed points with two medium and one low values.

Number of Classes	Granurality	FP_0	FP_1	FP_2	LC	Total Number of Points
$N = 3$	0.5	11.11	55.56	33.33	0	3^3
	0.25	4.00	72.00	24.00	0	5^3
	0.2	2.78	76.39	20.83	0	6^3
	0.1	0.83	86.78	12.39	0	11^3
	0.05	0.23	92.97	6.80	0	21^3
	0.01	0.01	98.52	1.47	0	101^3

5.3. $N \geq 4$

If the FRCN has $N = 4$ decision classes, then the eigenvalues of the Jacobian at the trivial fixed point have the same magnitude, but with different sign. So positive and negative eigenvalues reaches one (in absolute value) at the same value of parameter λ (see Figure 11), causing the appearance of new fixed points and limit cycle simultaneuosly. The trivial fixed point is no longer an attractor, but fixed points with pattern one high, three low and two medium, two low values do exists.

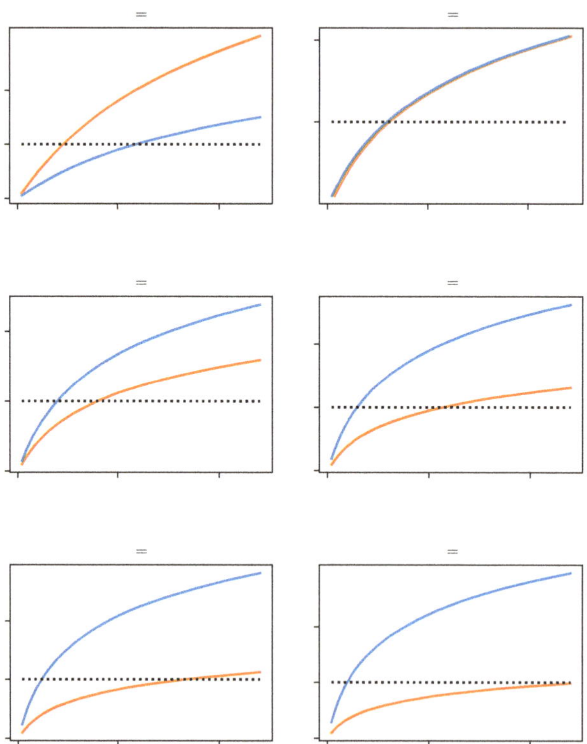

Figure 11. Eigenvalues of Jacobian evaluated at the trivial fixed point. Orange: positive eigenvalue, blue: absolute value of the negative eigenvalue. Curves are slightly shifted for visibility in $N = 4$.

If $N \geq 5$, then the absolute value of the negative eigenvalue of the Jacobian evaluated at the trivial fixed point is higher then the positive value, consequently first a period-doubling bifurcation occurs and the trivial fixed point loses its attractiveness. We should note that occurence of fixed points of type FP_1 and FP_2 is not linked to the other (positive) eigenvalue, since they occur earlier, for a smaller value of λ.

For the general case (N decision classes), there exist two types of fixed points with the following patterns: one high, $N-1$ low values and two medium, $N-2$ low values. The fixed point equations for the one high (x_1), $N-1$ low (x_2) pattern

$$x_1 = f(x_1 - (N-1)x_2) \tag{53}$$
$$x_2 = f(-x_1 - (N-3)x_2), \tag{54}$$

which leads to the following one-dimensional fixed point problem:

$$x_1 = (N-1)f\left(\frac{4-2N}{N-1} + \frac{N-3}{N-1}f^{-1}(x_1)\right) + f^{-1}(x_1) \tag{55}$$

The pattern two medium (x_1), $N-2$ low (x_2) values leads to the following equations:

$$x_1 = f(-(N-2)x_2) \tag{56}$$
$$x_2 = f(-2x_1 - (N-4)x_2), \tag{57}$$

from which we get the a one-dimensional fixed point problem:

$$x_2 = -\frac{1}{N-4}f^{-1}(x_2) - \frac{2}{N-4}f(-(N-2)x_2) \tag{58}$$

Nevertheless, these fixed points are less important for multiple decision classes.

Finally, we provide a geometrical reasoning of the structure of fixed points. Consider two fixed points of type FP_1, i.e., they have one high and $N-1$ low coordinates. Their basins of attractions are separated by a set, whose points do belong to none of them, but lie on an $N-1$ dimensional hyperplane 'between' them. Without loss of generality, we may assume that one fixed point is $P_1^* = (\alpha, \beta, \ldots, \beta)$ and the other one is $P_2^* = (\beta, \alpha, \beta, \ldots, \beta)$. Because of symmetry, the hyperplane is perpendicular the line connecting P_1^* and P_2^*, i.e., its normal vector is paralel to $\overrightarrow{P_1^* P_2^*} = (\beta - \alpha, \alpha - \beta, 0, \ldots, 0)$. Additionally, the hyperplane crosses the line at the middle point of P_1^* and P_2^*, which has coordinates $\left(\frac{\alpha+\beta}{2}, \frac{\alpha+\beta}{2}, \beta, \ldots, \beta\right)$. Consequently, the equation of the separating hyperplane is $x_1 = x_2$. The separating set is a subset of this plane with the additional constrain $x_i < x_1$, for every $i \neq 1, 2$. Consequently, a fixed point of type FP_2 has two medium coordinates with equal values ($x_1 = x_2$), and $N-2$ equal, but low coordinates. Since there are N fixed points of type FP_1, there are $\binom{N}{2}$ fixed points of type FP_2.

Simulation results show, that for $N \geq 4$, limit cycles tend to steal the show. Limit cycle oscillates between two activation vectors with equal coordinates (the equality of the coordinates is an immediate consequence of symmetry). Let us denote these points by $(x_1, \ldots, x_1)$ and $(x_2, \ldots, x_2)$. The members of a two-period limit cycle are fixed points of the double iterated function.

$$\begin{bmatrix} x_1 \\ \vdots \\ x_1 \end{bmatrix} = f\left(W_P \begin{bmatrix} x_2 \\ \vdots \\ x_2 \end{bmatrix}\right) \quad \text{and} \quad \begin{bmatrix} x_2 \\ \vdots \\ x_2 \end{bmatrix} = f\left(W_P \begin{bmatrix} x_1 \\ \vdots \\ x_1 \end{bmatrix}\right) \tag{59}$$

In coordinate-wise form this gives the following system of equations:

$$x_1 = f(-(N-2)x_2) \tag{60}$$

$$x_2 = f(-(N-2)x_1) \tag{61}$$

From which we have

$$x_2 = f(-(N-2)x_1) = f(-(N-2)f(-(N-2)x_2)) = \frac{1}{1+e^{\lambda(N-2)\frac{1}{1+e^{\lambda(N-2)x_2}}}}. \tag{62}$$

For λ values generally applied in fuzzy cognitive maps and for $\lambda = 5$ used in FRCNs, this fixed point equation has three solutions: one refers to the trivial fixed point, which is no longer a fixed point attractor, the other two (*low* and *medium*) are coordinates of the elements of the limit cycle (see Figure 12).

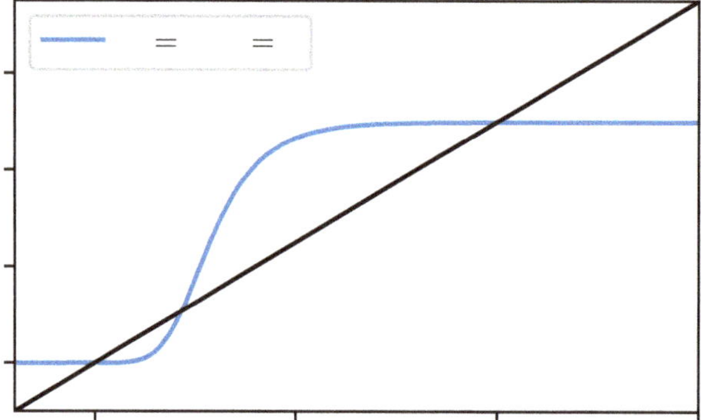

Figure 12. Fixed point of the double-iterated function (Equation (62)), for $N = 6$ and $\lambda = 5$. At the middle fixed point, the value of the derivative is greater than one, so it is a repelling fixed point. The two other fixed points, with low and medium values are stable.

Simulation results show that by increasing the number of decision classes, more and more initial values arrive to a limit cycle. We put equally spaced N dimensional grid on the set $[0,1]^N$ with stepsize 0.5, 0.25, 0.2 and 0.1, then applied the gridpoints as initial activation values of the positive neurons. Since any particular real-life dataset finally turns into initial activation values for the FRCN model, it means that these gridpoints can be viewed as representations of possible datasets, up to the predefined precision (i.e., step size). Iteration stopped when convergence or limit cycle was detected or the predefined maximum numbers of steps reached. As we can observe in Table 2, most of the initial values finally arrive to a limit cycle.

Table 2. Number of points in basin of attraction in percentage of the total number of points, for different number of classes (N) and level of granularity, $\lambda = 5$. FP_1 refers to fixed points with one high and $N-1$ low values, while F_2 refers to fixed points with two medium and $N-2$ low values, LC stands for limit cycle.

Number of Classes	Granurality	FP_0	FP_1	FP_2	LC	Total Number of Points
$N = 4$	0.5	0	39.51	14.81	45.68	3^4
	0.25	0	48.64	9.60	41.76	5^4
	0.2	0	45.68	6.02	48.30	6^4
	0.1	0	48.77	3.36	47.87	11^4
	0.05	0	50.34	1.41	48.25	21^4
$N = 5$	0.5	0	24.69	8.23	67.08	3^5
	0.25	0	27.52	0.96	71.52	5^5
	0.2	0	23.21	2.06	74.73	6^5
	0.1	0	17.34	0.50	82.16	11^5
	0.05	0	16.24	0.14	83.62	21^5
$N = 6$	0.5	0	13.99	4.12	81.89	3^6
	0.25	0	7.22	0.29	92.49	5^6
	0.2	0	7.52	0.13	92.35	6^6
	0.1	0	4.75	0.04	95.21	11^6
$N = 7$	0.5	0	7.36	1.92	90.72	3^7
	0.25	0	2.72	0.08	97.20	5^7
	0.2	0	1.90	0.03	98.07	6^7
	0.1	0	1.13	0.00	98.87	11^7
$N = 8$	0.5	0	3.66	0.00	96.34	3^8
	0.25	0	0.95	0.00	99.05	5^8
	0.2	0	0.60	0.00	99.40	6^8
	0.1	0	0.03	0.00	99.97	11^8

6. Relation to Decision

It has been proved previously, that values of negative and boundary neurons converge to the same value (it is ≈ 0.9930 for $\lambda = 5$) and the dynamical behaviour of positive neurons was analyzed in the preceding sections. Now we examine their effect on the decision neurons and such a way, on the final decision.

Decision neurons have only input values, they do not influence each other, neither other types of neurons. As a consequence, the sigmoid transfer function $f(x)$ only transform their values into the $(0,1)$ interval, but does not change the order of the values (with respect to the ordering relation '$\leq$'), since $f(x)$ is strictly monotone increasing. Before analyzing the effects of the results of the previous sections, we briefly summarize the conclusion of [37]: *assuming that the activation values of positive neurons reach a stable state*, they concluded that negative neurons have no influence on FRCNs' performance, but the ranking of positive neurons' activation values and the number of boundary neurons connected to each decision neuron have high impact. Based on the previous sections, below we ad some more insights to this result.

If positive neurons reach a stable state (fixed point), then this stable state have either the pattern one *high* and $N-1$ *low values* (FP_1) or two *medium* and $N-2$ *low* values (FP_2), the trivial fixed point with equal coordinates (FP_0) plays a rule only for 2 and 3 decision classes. These values are unique and completely determined by the parameter λ and the number of decision classes N. It means that the number of possible final states is very limited. This fact was mentioned in the case of $N = 3$ decision classes, but valid for every $N \geq 3$ cases. Namely, small differences between the initial activation vales could be magnified by the exploitation phase. Almost equal initial activation values with proper maximum lead to the pattern of one *high* and $N - 1$ *low values*, resulting in something like a winner-takes-all rule. Although the runner-up has only a little smaller initial value, after reaching the stable state, it needs the same number of boundary connections to overcome the winner, as the one with very low initial value needs.

If the maximal number of iterations is reached without convergence (i.e., the activation value vector oscillates in a limit cycle), then the iteration is stopped and the last activation vector is taken. It has either $(low, \ldots, low)$ or $(medium, \ldots, medium)$ pattern with equal coordinates. In this case, the positive neurons have absolutely no effect on the final decision. The classification goes to the neuron with the highest number of boundary connections, regardless of the small or large differences between the initial activation values of the positive neurons.

7. Conclusions and Future Work

The behaviour of fuzzy-rough cognitive networks was studied applying the theory of discrete dynamical systems and their bifurcations. The dynamics of negative and positive neurons was fully discussed in lthe iterature, so we focused on the behaviour of positive neurons. It was pointed out, that the number of fixed points is very limited and their coordinate values follow a specific pattern (FP_0, FP_1, FP_2). Additionally, it was proved that when the number of decision classes is greater than three, the limit cycles unavoidably occur, causing the recurrent reasoning inconclusive. Simulations show that proportion of initial activation values leading to limit cycles increases with the number of decision classes, and the waste number of scenarios lead to oscillation. In this case, the decision relies totally on the number of boundary neuron connected to each decision neurons, regardless of the initial activation value of positive neurons.

The method applied in the paper may be followed in the analysis of other FCM-like models. As we have seen, if the parameter of the sigmoid threshold function is small enough, then an FCM has one and only one fixed point, which is globally asymptotically stable. If we increase the value of the parameter, then a fixed-point bifurcation occurs, causing an entirely different dynamical behaviour. If the weight matrix has a nice structure, for example in the case of positive neurons, then there is a chance to find the unique fixed point in a simple form or as a limit of a lower-dimensional iteration and determine the parameter value at the bifurcation point. Similarly, based on the eigenvalues of the Jacobian evaluated at this fixed point we can determine the type of bifurcation. Nevertheless, general FCMs have no well-structured weight matrices, since weights are usually determined by human experts or learning methods. It means some limitations on the generalization of the method applied. Theoretically, we can find unique fixed points and the bifurcation point, but this task is much more difficult for a general weight matrix.

Another exciting and important research direction is the possible generalization of the results to the extensions of fuzzy cognitive maps. Some well-known extensions are fuzzy grey cognitive maps (FGCMs) [44], interval-valued fuzzy cognitive maps (IVFCMs) [45], intuitionistic fuzzy cognitive maps (IFCMs) [46,47], temporal IFCMs [48], the combination of fuzzy grey and intuitionistic FCMs [49], interval-valued intuitionistic fuzzy cognitive maps (IVIFCMs) [50]. This future work probable requires deep mathematical inspection on interval-valued dynamical systems and may lead to several new theoretical and practical results on interval-valued cognitive networks, as well.

Funding: This research was supported by National Research, Development and Innovation Office (NKFIH) K124055.

Institutional Review Board Statement: Not applicable.

Informed Consent Statement: Not applicable.

Data Availability Statement: No new data were created or analyzed in this study. Data sharing is not applicable to this article.

Conflicts of Interest: The author declares no conflict of interest.

References

1. Samek, W.; Müller, K.R. Towards explainable artificial intelligence. In *Explainable AI: Interpreting, Explaining and Visualizing Deep Learning*; Springer: Berlin, Germany, 2019; pp. 5–22.
2. Došilović, F.K.; Brčić, M.; Hlupić, N. Explainable artificial intelligence: A survey. In Proceedings of the 2018 41st International Convention on Information and Communication Technology, Electronics and Microelectronics (MIPRO), Opatija, Croatia, 21–25 May 2018; pp. 0210–0215.
3. Kosko, B. Fuzzy cognitive maps. *Int. J. Man Mach. Stud.* **1986**, *24*, 65–75. [CrossRef]
4. Papageorgiou, E.I.; Salmeron, J.L. A review of fuzzy cognitive maps research during the last decade. *IEEE Trans. Fuzzy Syst.* **2012**, *21*, 66–79. [CrossRef]
5. Szwed, P. Classification and feature transformation with Fuzzy Cognitive Maps. *Appl. Soft Comput.* **2021**, 107271. [CrossRef]
6. Boutalis, Y.; Kottas, T.L.; Christodoulou, M. Adaptive Estimation of Fuzzy Cognitive Maps With Proven Stability and Parameter Convergence. *IEEE Trans. Fuzzy Syst.* **2009**, *17*, 874–889. [CrossRef]
7. Lee, I.K.; Kwon, S.H. Design of sigmoid activation functions for fuzzy cognitive maps via Lyapunov stability analysis. *IEICE Trans. Inf. Syst.* **2010**, *93*, 2883–2886. [CrossRef]
8. Knight, C.J.; Lloyd, D.J.; Penn, A.S. Linear and sigmoidal fuzzy cognitive maps: An analysis of fixed points. *Appl. Soft Comput.* **2014**, *15*, 193–202. [CrossRef]
9. Harmati, I.Á.; Hatwágner, M.F.; Kóczy, L.T. On the existence and uniqueness of fixed points of fuzzy cognitive maps. In *International Conference on Information Processing and Management of Uncertainty in Knowledge-Based Systems*; Springer: Berlin, Germany, 2018; pp. 490–500.
10. Zadeh, L.A. Fuzzy sets as a basis for a theory of possibility. *Fuzzy Sets Syst.* **1978**, *1*, 3–28. [CrossRef]
11. Zadeh, L.A. Fuzzy sets and information granularity. *Adv. Fuzzy Set Theory Appl.* **1979**, *11*, 3–18.
12. Lingras, P.; Jensen, R. Survey of rough and fuzzy hybridization. In Proceedings of the 2007 IEEE International Fuzzy Systems Conference, London, UK, 23–26 July 2007; pp. 1–6.
13. Dubois, D.; Prade, H. Rough fuzzy sets and fuzzy rough sets. *Int. J. Gen. Syst.* **1990**, *17*, 191–209. [CrossRef]
14. Ji, W.; Pang, Y.; Jia, X.; Wang, Z.; Hou, F.; Song, B.; Liu, M.; Wang, R. Fuzzy rough sets and fuzzy rough neural networks for feature selection: A review. *Wiley Interdiscip. Rev. Data Min. Knowl. Discov.* **2021**, *11*, e1402. [CrossRef]
15. Cao, B.; Zhao, J.; Lv, Z.; Gu, Y.; Yang, P.; Halgamuge, S.K. Multiobjective evolution of fuzzy rough neural network via distributed parallelism for stock prediction. *IEEE Trans. Fuzzy Syst.* **2020**, *28*, 939–952. [CrossRef]
16. Zhang, K.; Zhan, J.; Wu, W.Z. Novel fuzzy rough set models and corresponding applications to multi-criteria decision-making. *Fuzzy Sets Syst.* **2020**, *383*, 92–126. [CrossRef]
17. Pal, S.K.; Ray, S.S.; Ganivada, A. Classification using fuzzy rough granular neural networks. In *Granular Neural Networks, Pattern Recognition and Bioinformatics*; Springer: Berlin, Germany, 2017; pp. 39–76.
18. Riaz, S.; Arshad, A.; Jiao, L. Fuzzy rough C-mean based unsupervised CNN clustering for large-scale image data. *Appl. Sci.* **2018**, *8*, 1869. [CrossRef]
19. Li, C.; Wang, N.; Zhang, H.; Liu, Q.; Chai, Y.; Shen, X.; Yang, Z.; Yang, Y. Environmental impact evaluation of distributed renewable energy system based on life cycle assessment and fuzzy rough sets. *Energies* **2019**, *12*, 4214. [CrossRef]
20. Deveci, M.; Özcan, E.; John, R.; Covrig, C.F.; Pamucar, D. A study on offshore wind farm siting criteria using a novel interval-valued fuzzy-rough based Delphi method. *J. Environ. Manag.* **2020**, *270*, 110916. [CrossRef]
21. Zhang, C.; Li, D.; Kang, X.; Song, D.; Sangaiah, A.K.; Broumi, S. Neutrosophic fusion of rough set theory: An overview. *Comput. Ind.* **2020**, *115*, 103117. [CrossRef]
22. Gokasar, I.; Deveci, M.; Kalan, O. CO_2 Emission based prioritization of bridge maintenance projects using neutrosophic fuzzy sets based decision making approach. *Res. Transp. Econ.* **2021**, 101029. [CrossRef]
23. Ganivada, A.; Dutta, S.; Pal, S.K. Fuzzy rough granular neural networks, fuzzy granules, and classification. *Theor. Comput. Sci.* **2011**, *412*, 5834–5853. [CrossRef]
24. Pedrycz, W. *Granular Computing: An Emerging Paradigm*; Springer Science & Business Media: Berlin, Germany, 2001; Volume 70.
25. Pedrycz, W.; Vukovich, G. Granular neural networks. *Neurocomputing* **2001**, *36*, 205–224. [CrossRef]
26. Falcon, R.; Nápoles, G.; Bello, R.; Vanhoof, K. Granular cognitive maps: A review. *Granul. Comput.* **2019**, *4*, 451–467. [CrossRef]
27. Nápoles, G.; Grau, I.; Papageorgiou, E.; Bello, R.; Vanhoof, K. Rough cognitive networks. *Knowl.-Based Syst.* **2016**, *91*, 46–61. [CrossRef]
28. Pawlak, Z. Rough sets. *Int. J. Comput. Inf. Sci.* **1982**, *11*, 341–356. [CrossRef]
29. Pawlak, Z. Vagueness and uncertainty: A rough set perspective. *Comput. Intell.* **1995**, *11*, 227–232. [CrossRef]
30. Pawlak, Z. Rough set theory and its applications. *J. Telecommun. Inf. Technol.* **2002**, 7–10.
31. Skowron, A.; Dutta, S. Rough sets: past, present, and future. *Nat. Comput.* **2018**, *17*, 855–876. [CrossRef] [PubMed]
32. Li, X.; Luo, C. An intelligent stock trading decision support system based on rough cognitive reasoning. *Expert Syst. Appl.* **2020**, *160*, 113763. [CrossRef]
33. Zhang, H.Y.; Yang, S.Y. Three-way group decisions with interval-valued decision-theoretic rough sets based on aggregating inclusion measures. *Int. J. Approx. Reason.* **2019**, *110*, 31–45. [CrossRef]
34. Nápoles, G.; Falcon, R.; Papageorgiou, E.; Bello, R.; Vanhoof, K. Rough cognitive ensembles. *Int. J. Approx. Reason.* **2017**, *85*, 79–96. [CrossRef]

35. Nápoles, G.; Mosquera, C.; Falcon, R.; Grau, I.; Bello, R.; Vanhoof, K. Fuzzy-rough cognitive networks. *Neural Netw.* **2018**, *97*, 19–27. [CrossRef]
36. Vanloffelt, M.; Nápoles, G.; Vanhoof, K. Fuzzy-Rough Cognitive Networks: Building Blocks and Their Contribution to Performance. In Proceedings of the 2019 18th IEEE International Conference On Machine Learning And Applications (ICMLA), Boca Raton, FL, USA, 16–19 December 2019; pp. 922–928.
37. Concepción, L.; Nápoles, G.; Grau, I.; Pedrycz, W. Fuzzy-Rough Cognitive Networks: Theoretical Analysis and Simpler Models. *IEEE Trans. Cybern.* **2020**. [CrossRef]
38. Rudin, W. *Principles of Mathematical Analysis*, 3rd ed.; McGraw-Hill: New York, NY, USA, 1976.
39. Kuznetsov, Y.A. *Elements of Applied Bifurcation Theory*; Springer Science & Business Media: Berlin, Germany, 2013; Volume 112.
40. Wiggins, S. *Introduction to Applied Nonlinear Dynamical Systems and Chaos*; Springer Science & Business Media: Berlin, Germany, 2003; Volume 2.
41. Schultz, P.; Menck, P.J.; Heitzig, J.; Kurths, J. Potentials and limits to basin stability estimation. *New J. Phys.* **2017**, *19*, 023005. [CrossRef]
42. Giesl, P. On the determination of the basin of attraction of discrete dynamical systems. *J. Differ. Equ. Appl.* **2007**, *13*, 523–546. [CrossRef]
43. Susanto, H.; Karjanto, N. Newton's method's basins of attraction revisited. *Appl. Math. Comput.* **2009**, *215*, 1084–1090. [CrossRef]
44. Salmeron, J.L. Modelling grey uncertainty with fuzzy grey cognitive maps. *Expert Syst. Appl.* **2010**, *37*, 7581–7588. [CrossRef]
45. Hajek, P.; Prochazka, O. Interval-valued fuzzy cognitive maps for supporting business decisions. In Proceedings of the 2016 IEEE International Conference on Fuzzy Systems (FUZZ-IEEE), Vancouver, BC, Canada, 24–29 July 2016; pp. 531–536.
46. Papageorgiou, E.I.; Iakovidis, D.K. Intuitionistic fuzzy cognitive maps. *IEEE Trans. Fuzzy Syst.* **2012**, *21*, 342–354. [CrossRef]
47. Hadjistoykov, P.; Atanassov, K. Remark on intuitionistic fuzzy cognitive maps. *Notes Intuit. Fuzzy Sets* **2013**, *19*, 1–6.
48. Hadjistoykov, P.P.; Atanassov, K.T. On temporal intuitionistic fuzzy cognitive maps. *C. R. L'Acad. Bulg. Des Sci. Sci. Math. Nat.* **2014**, *67*, 1233–1240.
49. Hajek, P.; Froelich, W.; Prochazka, O. Intuitionistic fuzzy grey cognitive maps for forecasting interval-valued time series. *Neurocomputing* **2020**, *400*, 173–185. [CrossRef]
50. Hajek, P.; Prochazka, O. Interval-valued intuitionistic fuzzy cognitive maps for supplier selection. In *International Conference on Intelligent Decision Technologies*; Springer: Cham, Switzerland, 2017; pp. 207–217.

symmetry

MDPI

Article

A Hybrid Discrete Bacterial Memetic Algorithm with Simulated Annealing for Optimization of the Flow Shop Scheduling Problem

Anita Agárdi [1], Károly Nehéz [1], Olivér Hornyák [1] and László T. Kóczy [2,3,*]

[1] Institute of Information Science, University of Miskolc, 3515 Miskolc, Hungary; agardianita@iit.uni-miskolc.hu (A.A.); aitnehez@uni-miskolc.hu (K.N.); oliver.hornyak@uni-miskolc.hu (O.H.)
[2] Department of Information Technology, Széchenyi István University, 9026 Győr, Hungary
[3] Department of Telecommunication and Media Informatics, Budapest University of Technology and Economics, 1111 Budapest, Hungary
* Correspondence: koczy@tmit.bme.hu or koczy@tmit.bme.hul.com

Abstract: This paper deals with the flow shop scheduling problem. To find the optimal solution is an NP-hard problem. The paper reviews some algorithms from the literature and applies a benchmark dataset to evaluate their efficiency. In this research work, the discrete bacterial memetic evolutionary algorithm (DBMEA) as a global searcher was investigated. The proposed algorithm improves the local search by applying the simulated annealing algorithm (SA). This paper presents the experimental results of solving the no-idle flow shop scheduling problem. To compare the proposed algorithm with other researchers' work, a benchmark problem set was used. The calculated makespan times were compared against the best-known solutions in the literature. The proposed hybrid algorithm has provided better results than methods using genetic algorithm variants, thus it is a major improvement for the memetic algorithm family solving production scheduling problems.

Keywords: discrete bacterial memetic evolutionary algorithm; simulated annealing; flow shop scheduling problem

Citation: Agárdi, A.; Nehéz, K.; Hornyák, O.; Kóczy, L.T. A Hybrid Discrete Bacterial Memetic Algorithm with Simulated Annealing for Optimization of the Flow Shop Scheduling Problem. *Symmetry* **2021**, *13*, 1131. https://doi.org/10.3390/sym13071131

Academic Editor: José Carlos R. Alcantud

Received: 19 April 2021
Accepted: 11 June 2021
Published: 24 June 2021

1. Introduction

This paper investigates whether a new meta-heuristic proposed by the authors, an improved and modified version of the discrete bacterial memetic evolutionary algorithm, is capable of solving the flow shop scheduling problem (FSSP) in an efficient way, possibly in a more efficient way than other approaches proposed by other authors. FSSP was first published in 1954 by Johnson [1]. Although there exist exact solution algorithms, they are not feasible for the large-sized scheduling problem, as FSSP is an NP-hard problem. Many researchers have addressed the problem since its introduction. The overview of the proposed solution is as follows.

Wei et al. [2] introduced a hybrid genetic simulated annealing algorithm, which combines the individual steps and operators of the simulated annealing and the genetic algorithm (HSGA). In their solution, the genetic algorithm (GA) finds a new optimal solution, and the simulated annealing attempts to improve that solution. To compare their solution, the widely used Taillard data set [3] was used, which is a reference benchmark for FSSP. The Taillard data set contains benchmark data between 20 and 500 jobs and between 5 and 20 machines for the flow shop scheduling problem. The following state-of-the-art algorithms were compared in [2]: memetic algorithm, iterated greedy algorithm with a referenced insertion scheme, hybrid genetic algorithm (i.e., GA with improved local search method that searches in a larger neighborhood), improved iterated greedy algorithm (using Tabu mechanism to escape from local minima), and discrete self-organizing migrating algorithm. Their hybrid genetic simulated annealing algorithm proved to be the best for the test data set.

Another hybrid genetic algorithm, which combined two local search methods with GA, was introduced by Tseng et al. [4]. Their hybrid genetic algorithm is compared with the primitive genetic algorithm, genetic algorithm+ insertion search, and genetic algorithm+ insertion search with cut-and-repair. Their algorithm found better results than the reference benchmark problems.

The results for the flow shop problem were also compared with the following algorithms: Johnson's algorithm, Nawaz–Enscore–Ham (NEH) heuristic, iterated local search algorithm, and iterated greedy algorithm. Based on the paper of Belabid et al. [5], the NEH heuristic gives better results than the others.

The flow shop scheduling problem has been solved with a relatively new algorithm called the flower pollination algorithm by Qu et al [6]. Compared with other heuristic algorithms, such as Tabu-based reconstruction strategy (TMIIG), discrete particle swarm optimization algorithm (DPSO), improved iterated greedy algorithm (IIGA), effective hybrid particle swarm optimization (HPSO), hybrid differential evolution approach (HDE), and genetic algorithm (GA), the flower pollination approach was the most efficient one.

An invasive weed optimization (IWO) algorithm was introduced in [7] for FSSP. The authors also used the Taillard benchmark to test the efficiency of their algorithm. The algorithm was compared with Nawaz–Enscore–Ham (NEH) algorithm. Their solution obtained better makespan than the NEH algorithm for every instance of 12 different scale benchmarks. It has proved to be better in terms of both final accuracy and convergence. The reason is that the global exploration in [7] based on normal distribution is better than the other algorithms.

Simulated annealing is another efficient optimization algorithm; there are several articles on the topic that solve the flow shop scheduling problem with this algorithm, for example, Ogbu and Smith [8], Lin et al. [9], and Aurich et al. [10].

The following section formulates the FSPP problem itself, and then an overview is given on the state-of-the-art of approximate solving algorithms. Next, the proposed new memetic algorithm is presented, which may be considered as a further development and improvement of the discrete bacterial memetic evolutionary algorithm, an approach that has already been successfully applied to the solution of other discrete NP-hard problems. The authors executed some benchmark-based tests and compared the results with the algorithms discussed in the overview of the state-of-the-art. The table of the results with comparisons and explanations and, finally, some concluding notes are presented in the last section.

2. Formulation of the Flow Shop Scheduling Problem

In the case of the flow shop scheduling problem [2], n jobs and m machines are given with the following constraints: (1) The jobs have the same processing route. (2) Each job must be run on all machines exactly once. (3) All jobs and machines must be ready to work at time zero. (3) All jobs have m processing steps. (4) Neither machines nor jobs have priority. (5) A single machine can do a single job at a time. (6) If a job is started on a machine, no process can interfere. During processing, all jobs must follow the first in–first out (FIFO) rule. The mathematical model of the flow shop scheduling problem can be written in the following way:

$J = \{1, 2, \ldots, n\}$: Set of jobs
$M = \{1, 2, \ldots, m\}$: Set of machines
$p_{i,j}$: processing time of job i on machine j
$S_{i,j}$ starting time when job i is processed on machine j
$C_{i,j}$ finishing time when job i is processed on machine j
$\pi = \{\pi_1, \pi_2, \ldots, \pi_n\}$ the job sequence
$C_{max}(\pi)$: the makespan of a job sequence

The objective function is the minimization of the makespan. Makespan is the overall length of finishing the job sequence. The objective function can be written in the following way:

$$(C_{max}(\pi)) \rightarrow \min \tag{1}$$

where the following conditions must be met:

(1) A job can only be processed by one machine at a time:

$$\text{If } X^{t}_{i,j} > 0 \text{ then } p_{i,k} = 0 \; \forall k \in [1, 2, \ldots, n], k \neq i \tag{2}$$

(2) A machine can only process a single job at a time:

$$\text{If } p_{i,j} > 0 \text{ then } p_{k,j} = 0 \; \forall k \in [1, 2, \ldots, m], k \neq j \tag{3}$$

(3) The next job cannot start until the current job is completed on the given machine:

$$C_{i,j} \leq S_{i+1,j} \tag{4}$$

(4) If machine j + 1 is not ready, the job will delay at machine j until machine j + 1 will be free:

$$C_{i,j} \geq C_{i-1,j+1} \tag{5}$$

(5) For any job i, the completion time is the starting and processing time on machine j:

$$C_{i,j} = S_{i,j} + p_{i,j} \tag{6}$$

(6) The makespan of the schedule is the time when the last job finishes on the last machine:

$$C_{max}(\pi) = C_{\pi n,m} \tag{7}$$

Benchmark Data Sets

To compare the efficiency of flow shop scheduling problems, researchers use benchmark data sets. In this paper, the Taillard data set will be used. It consists of 120 benchmark instances. It also provides the best-known upper bounds for the makespan criterion. There are other benchmark data sets provided by Carlier (8 instances) [11], Heller (2 instances) [12], and Revees (21 instances) [13]. These are smaller problems that are straightforward to solve by simple algorithms.

There are two significant indices for measuring the performance of an algorithm. The first one is the solution quality, which can be represented by relative percentage deviation (RPD) over the best-known upper bound. The second important indicator is running time (t_r), which is counted until the cycle taken to reach the last improvement. In this paper we had no investigation into running time.

3. The Family of Bacterial Evolutionary Memetic Algorithms

Bacterial evolutionary algorithm (BEA) is an evolutionary computing algorithm, which was inspired by microbial evolution [14].

BEA is inspired by the interesting process of bacterial recombination. The algorithm uses two operators, bacterial mutation and gene transfer. The first step is to generate an initial population. Then, those two genetic type operations are employed to create new individuals and evaluate them by a fitness function. These operations are repeated until the stop condition.

As the first implementation, the bacterial evolutionary algorithm was only used for finding the optimal parameters of a fuzzy rule-based system. Over the years, it turned out to be efficiently applicable to many other optimization tasks, e.g., interactive nurse scheduling optimization problem [15], automatic data clustering [16], and three-dimensional bin packing problem [17].

According to the early definitions, memetic algorithms are modified genetic algorithms that use an additional local search operator; see Moscato et al. [18]. The idea of combining the BEA with local search came first when an attempt was taken to improve the approximation and optimization capability of the approach when estimating the parameters of fuzzy rule bases. As the latter may be interpreted as black boxes generating input–output functions, the scope of potential benchmark problems was extended to several additional areas, beyond the examples the original Nawa and Furuhashi paper had discussed. Mechanical, chemical, and electrical engineering problems were tested along with transcendental mathematical functions, while the local search applied was a second-order gradient-based method (the Levenberg–Marquardt algorithm). The results turned out to be better than the original ones, even better than any other approach applied in the literature for optimizing the parameters of trapezoidal fuzzy membership functions in fuzzy rule-based "function generators [19]. Later, first-order gradient methods were also tested as local search algorithms, and the results were promising [20].

In the next step, the idea of bacterial memetic evolutionary algorithm was tested on discrete, permutation-based problems, where, as a matter of course, the local search applied was also a discrete process. The first proposed operator family was the n-opt local search, and after some simulations, it was narrowed to the subsequent application of the be 2-opt or 3-opt operators, as when applying $n \geq 4$, the overhead time proved to be too large, thus the efficiency of the algorithm was decreased. The 2-opt operator was first applied to the traveling salesman problem [21], where two edges are exchanged in the graph. 3-opt [22] is similar to the 2-opt operator; the only difference is that, here, three edges are exchanged in one step.

The pseudo-code of the DBMEA is given in Algorithm 1 [23,24]. The algorithm has five input parameters, these are as follows: N_{ind}, N_{clones}, N_{inf}, I_{seg}, *and* I_{trans}; N_{ind} is the number of individuals in the population, N_{clones} is the number of clones in the bacterial mutation, N_{inf} is the number of infections in the gene transfer, I_{seg} is the number of segments in the bacterial mutation, and I_{trans} is the length of the gene transferred part. Step 1 generates an initial population. Step 2 is the application of the bacterial mutation. The third step is the local search (also called as the memetic step). Local search tries to improve on a particular solution (by producing neighbor solutions) until it finds a better neighbor solution. The algorithm stops when it can no longer find a better individual in the neighborhood, i.e., it finds a local optimum. Then, the gene transfer operation is performed. The algorithm repeats steps 2–4 until the termination condition triggers. Then, it returns the best solution.

The following notation is used:

x_1: the actual gene transferred solution,
x_{best}: the best solution found so far,
f: fitness function,
N_{ind}: the number of individuals.

3.1. Discrete Bacterial Memetic Evolutionary Algorithm

Algorithm 1 Discrete Bacterial Memetic Evolutionary algorithm

BEGIN PROCEDURE DBMEA (N_{ind}, N_{clones}, N_{inf}, I_{seg}, I_{trans})
Step 1: Generate initial population P.
WHILE (termination criteria is not met) DO
Step 2: Bacterial mutation (Population, N_{clones} I_{seg}*)*
Step 3: local search
Step 4: x_1 = *Gene transfer (Population,* N_{inf}, I_{trans}*)*
IF $f(x_1) < f(xbest)$ *THEN DO*
Step 5: $x_{best} = x_1$
END IF
END WHILE
RETURN x_{best}
END PROCEDURE

3.2. Bacterial Mutation

The bacterial mutation [23] operates throughout the population by performing special mutation operations on each individual (see Algorithm 2). As input parameters, the initial population, the number of clones (N_{clones}), and the length of the segment (I_{seg}) are passed. Steps 1–6 are performed on each element of the population. A certain number of clones (N_{clones}) are made from each bacterium; see Figure 1. The original bacterium is broken down into segments. As the algorithm shows, it happens with high probability for coherent segments, and with low probability for loose segments. During both the coherent and loose segment operations, we go through each segment. We select a non-mutated segment. First, the elements of the segment are inverted to form the first clone. Then, we randomly change the elements of the segment to create other clones. This way, we generate a total of N_{clones} clones. At the end of the mutation, the best clone takes the place of the original bacterium.

Algorithm 2 Bacterial Mutation

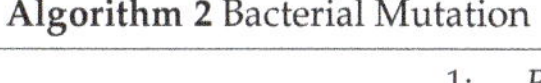

BEGIN PROCEDURE Bacterial mutation (Population, N_{clones}, I_{seg})
FOR i to size(Population) DO
Step 1. Create a random number between 0 and 1
Step 2. Get the ith element of the population: p = Population(i)
Step 3. create N_{clones} clones of p and a random number r [0..1]
IF ($r \leq$ COHERENT_LOOSE_RATE)
Step 4. cut p into coherent segments with I_{seg} length
ELSE
Step 5. cut p into loose segments with I_{seg} length
END IF
Step 6. replace Population(i) with the best set of the clones and p
END FOR
RETURN Population
END PROCEDURE

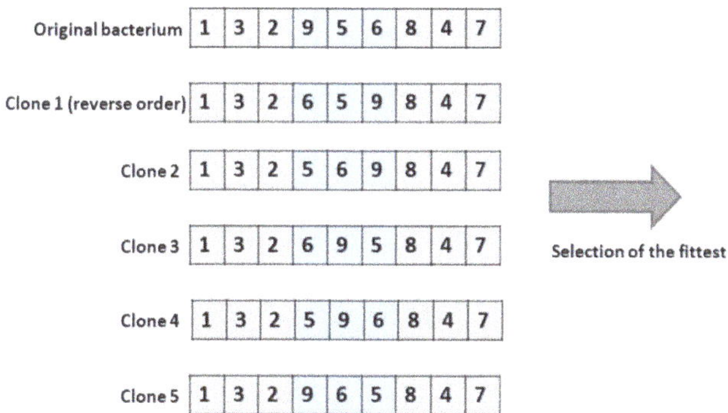

Figure 1. Bacterial mutation.

In the case of a coherent segment, the segments are arranged one after the other (Figure 2). In the case of a loose segment, the elements of the segments are not adjacent (Figure 3).

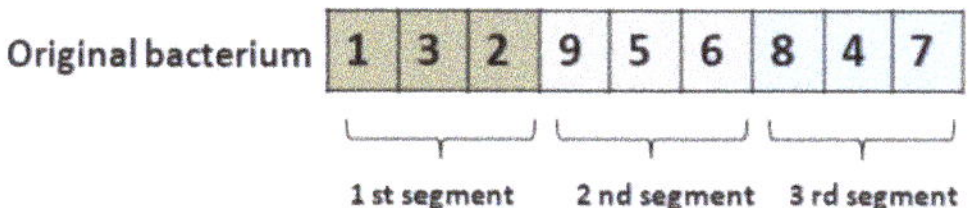

Figure 2. The coherent segment mutation.

Figure 3. The loose segment mutation.

3.3. Gene Transfer

The gene transfer [23] operates on the whole population. At first, the elements of the population are ranked based on the fitness values. Then, the population is divided into two parts, a superior and an inferior part, according to the fitness values. As the next step, the gene transfer operator is executed N_{inf} times with a randomly selected element from the superior and one from the inferior part. During the gene transfer operation, a randomly selected segment with length I_{trans} is transferred from the superior bacterium to the inferior bacterium, so that there are no duplicates in the thus established new bacterium. This process is illustrated in Figure 4. Algorithm 3 presents the process.

Algorithm 3 Gene transfer

BEGIN PROCEDURE Gene transfer (Population, N_{inf}, I_{trans})
Step 1. sort the Population according to the fitness values
Step 2. divide the population into superior and inferior parts based on the fitness values
FOR i to N_{inf} DO
Step 3. selecting a random bacterium from the superior part (p_{source})
Step 4. selecting a random bacterium from the inferior part ($p_{destination}$)
Step 5. selecting a random segment from p_{source} with I_{trans} length
Step 6. copying the segment into $p_{destination}$ bacterium in a random position
Step 7. eliminating the duplicates in $p_{destination}$
END FOR
RETURN Population
END PROCEDURE

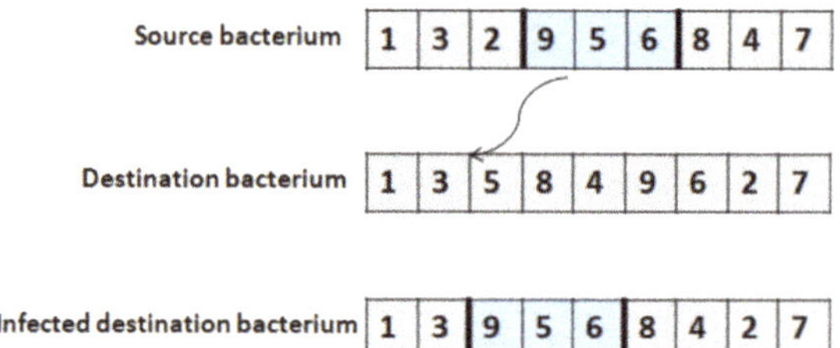

Figure 4. The gene transfer operator.

4. The Simulated Annealing Algorithm

Simulated annealing (SA) [25] operates on a single solution rather than on a whole population. SA first produces a random new neighboring solution. If this neighbor is better than the current solution, it accepts it deterministically as new current solution. If it is not better, the algorithm still may decide probabilistically whether to keep the current one, or replace it with the new neighboring one. The input parameter of the algorithm is the "temperature" (T), which determines the probability of accepting worse solutions in the algorithm. The temperature continuously decreases; it is determined by the temperature control parameter (α). Algorithm 4 illustrates the SA algorithm.

Algorithm 4 Simulated Annealing

```
SIMULATED ANNEALING
BEGIN PROCEDURE Bacterial mutation (T,α,L)
WHILE termination condition is not met DO
WHILE L processing length is not reached DO
Step 1. Create the neighbor (SN) of the current solution (SC)
Step 2. Calculate ΔE as follows: ΔE = E(SN) − E(SC)
IF ΔE > 0 THEN
Step 3. SC = SN
ELSE IF P(E(SN), E(SC), T) > rand[0, 1]) THEN DO
Step 4. SC = SN
END WHILE
Step 5. Reduce temperature (T)
END WHILE
END PROCEDURE
```

5. A Novel Algorithm: A Hybrid Discrete Bacterial Memetic Evolutionary Algorithm with Simulated Annealing

We could see in the DBMEA pseudo-code that the algorithm uses discrete local search. The originally proposed algorithm applies the 2-opt and 3-opt methods, which operate on individual elements of the population. To improve the efficiency of the local search in the solution of the particular problem on hand, we propose now to use simulated annealing for local search instead. This latter accepts a worse new solution with some probability, and this way allows getting out of small local optimum areas. In addition, we have introduced a mortality rate (N_{mort}). It practically means that certain elements of the population are to be dropped and replaced by randomly generated new individuals. The pseudo-code of our algorithm is shown in Algorithm 5, where N_{ind} is the number of individuals; N_{clones} is the number of clones; N_{inf} means the number of times the gene transfer operator is executed; I_{seg} is the length of the segment; I_{trans} is the length of the segment, which is transferred from the source to the destination bacterium; T is the temperature, which is decreasing along the iterations by α; and L is the iteration control parameter.

Algorithm 5 The Discrete Bacterial Memetic Evolutionary Algorithm with Simulated Annealing

```
BEGIN PROCEDURE DBMEA_SA (N_ind, N_clones, N_inf, I_seg, I_trans, T, α, L, N_mort)
Step 1: Generate initial population P.
WHILE (termination criteria is not met) DO
Step 2: Bacterial mutation (P, N_clones, I_seg)
FOR i IN Population DO
Step 3: Simulated annealing (P(i), T, α, L)
END FOR
Step 4: Gene transfer (P, N_inf, I_trans)
Step 5: Sort the population, based on the fitness values
Step 6: Generate new elements in the population in place of the worst N_mort elements
Step 7: Store the best solution
END WHILE
RETURN best solution
END PROCEDURE
```

6. Experimental Results

The proposed hybrid algorithm was implemented for a personal computer with 8th generation Intel i7 CPU and 16 GB memory. The Typescript programming language was used, as it allowed the ability to quickly implement the algorithm variants in a portable way. The authors ran the tests on a Windows 10 operating system. Full source codes are available at [26]. To compare the results of the new method with other ones known from the literature, the Taillard benchmark dataset [3] was used. There are 10 individuals for each benchmark data type in the set. Table 1 contains the makespan values calculated by the following algorithms.

Table 1. Experimental results compared with other up-to-date approaches.

Instance	n × m	Lower Bound	Best Known	DBMEA + SA	IWO [7]	HGSA [2]	HGA [4]	HMM-PFA [6]
Ta001	20 × 5	1232	1278	1283	1389	1324	1449	1486
Ta002	20 × 5	1290	1359	1360	-	1442	1460	1528
Ta003	20 × 5	1073	1081	1081	-	1098	1386	1460
Ta004	20 × 5	1268	1293	1293	-	1469	1521	1588
Ta005	20 × 5	1198	1235	1235	-	1291	1403	1449
Ta011	20 × 10	1448	1582	1587	2207	1713	1955	2044
Ta012	20 × 10	1479	1659	1681	-	1718	2123	2166
Ta013	20 × 10	1407	1496	1510	-	1555	1912	1940
Ta014	20 × 10	1308	1377	1384	-	1516	1782	1811
Ta015	20 × 10	1325	1419	1420	-	1573	1933	1933
Ta021	20 × 20	1911	2297	2308	3226	2331	2912	2973
Ta022	20 × 20	1711	2099	2120	-	2280	2780	2852
Ta023	20 × 20	1844	2326	2349	-	2480	2922	3013
Ta024	20 × 20	1810	2223	2223	-	2362	2967	3001
Ta025	20 × 20	1899	2291	2316	-	2507	2953	3003
Ta031	50 × 5	2712	2724	2724	3020	2731	3127	3160
Ta032	50 × 5	2808	2834	2848	-	2934	3438	3432
Ta033	50 × 5	2596	2621	2634	-	2638	3182	3210
Ta034	50 × 5	2740	2751	2776	-	2785	3289	3338
Ta035	50 × 5	2837	2863	2864	-	2864	3315	3356
Ta041	50 × 10	2907	2991	3059	3465	3198	4251	4274
Ta042	50 × 10	2821	2867	2933	-	3020	4139	4177
Ta043	50 × 10	2801	2839	2931	-	3055	4083	4099
Ta044	50 × 10	2968	3063	3077	-	3124	4480	4399
Ta045	50 × 10	2908	2976	3041	-	3129	4316	4322
Ta051	50 × 20	3480	3850	3957	5475	4105	6138	6129
Ta052	50 × 20	3424	3704	3823	-	3992	5721	5725
Ta053	50 × 20	3351	3640	3760	-	3900	5847	5862
Ta054	50 × 20	3336	3720	3823	-	3921	5781	5788
Ta055	50 × 20	3313	3610	3737	-	4020	5891	5886
Ta061	100 × 5	5437	5493	5495	5839	5536	6492	6361
Ta062	100 × 5	5208	5268	5290	-	5302	6353	6212
Ta063	100 × 5	5130	5175	5213	-	5221	6148	6104
Ta064	100 × 5	4963	5014	5023	-	5044	6080	5999
Ta065	100 × 5	5195	5250	5265	-	5358	6254	6179
Ta071	100 × 10	5759	5770	5825	6815	5964	8115	8055
Ta072	100 × 10	5345	5349	5414	-	5596	7986	7853
Ta073	100 × 10	5623	5676	5727	-	5796	8057	8016
Ta074	100 × 10	5732	5781	5892	-	5928	8327	8328
Ta075	100 × 10	5431	5467	5567	-	5748	7991	7936
Ta081	100 × 20	5851	6202	6407	9405	6395	10,745	10,675
Ta082	100 × 20	6099	6183	6334	-	6433	10,655	10,562
Ta083	100 × 20	6099	6271	6480	-	6689	10,672	10,587
Ta084	100 × 20	6072	6269	6409	-	6419	10,630	10,588
Ta085	100 × 20	6009	6314	6518	-	6536	10,548	10,506
Ta091	200 × 10	10,816	10,862	11,002	11,783	11,120	15,739	15,225
Ta092	200 × 10	10,422	10,480	10,627	-	10,658	15,534	14,990
Ta093	200 × 10	10,886	10,922	11,088	-	11,224	15,755	15,257
Ta094	200 × 10	10,794	10,889	11,004	-	11,075	15,842	15,103
Ta095	200 × 10	10,437	10,524	10,666	-	10,793	15,692	15,088
Ta101	200 × 20	10,979	11,195	11,483	15,217	11,642	20,148	19,531

Table 1. *Cont.*

Instance	n × m	Lower Bound	Best Known	DBMEA + SA	IWO [7]	HGSA [2]	HGA [4]	HMM-PFA [6]
Ta102	200 × 20	10,947	11,203	11,535	-	11,683	20,539	19,942
Ta103	200 × 20	11,150	11,281	11,603	-	11,930	20,511	19,759
Ta104	200 × 20	11,127	11,275	11,634	-	11,791	20,461	19,759
Ta105	200 × 20	11,132	11,259	11,549	-	11,728	20,339	19,697
Ta111	500 × 20	25,922	26,059	26,652	30,730	26,859	49,095	46,121
Ta112	500 × 20	26,353	26,520	27,115	-	27,220	49,461	46,627
Ta113	500 × 20	26,320	26,371	n/a	-	27,511	48,777	46,013
Ta114	500 × 20	26,424	26,456	26,974	-	26,912	49,283	46,396
Ta115	500 × 20	26,181	26,334	n/a	-	26,930	48,950	46,251

- DBMEA + SA: Discrete Bacterial Memetic Algorithm + Simulated Annealing;
- IWO: Invasive Weed Optimization [7];
- HGSA Hybrid Genetic Simulated Annealing [2];
- HGA: Hybrid Genetic Algorithm [4];
- HMM-PFA: Hormone Modulation Mechanism Flower Pollination Algorithm [6].

In the test we carried out, twelve different problem sizes were selected; these can be found in the column "$n \times m$: job and machine numbers", namely, 20 × 5, 20 × 10, 20 × 20, 50 × 5, 50 × 10, 50 × 20, 100 × 5, 100 × 10, 100 × 20, 200 × 5, 200 × 10, and 200 × 20. The instance names run from Ta001 to Ta120. Table 1 shows five instances for each problem set, while the full table with comparisons of the results applying the five approaches mentioned above is published also in [26].

For the evaluation of the obtained optima, it is worthwhile to compare both our own results and the ones obtained by other authors. There is an estimation method for an absolute theoretic lower bound, which is proposed in [3] and can be calculated as follows: let b_i the minimum amount of time before machine starts working and a_i is the minimum time until it remains inactive after the end of the operation and let T_i be its total processing time:

$$b_i = \min_j \left(\sum_{k=1}^{i-1} p_{kj} \right) \quad (8)$$

$$a_i = \min_j \left(\sum_{k=i+1}^{m} p_{kj} \right) \quad (9)$$

$$T_i = \sum_{j=1}^{n} p_{ij} \quad (10)$$

Let C_{max} denote the optimal makespan time; it must be greater or equal to the maximum between the minimum of time required by the machines and the minimum of time required each job. This value is called "lower bound", and Table 1 displays this theoretical minimum in the third column:

$$Lower\ bound = max\left\{ \max_i (b_i + a_i + T_i), \max_j \left(\sum_{i=1}^{m} p_{ij} \right) \right\} \leq C_{max} \quad (11)$$

Our hybrid DBMEA + SA algorithm ran within a reasonably short time, even though no direct measurements were done. In the case of some large instances, however, the running time exceeded the limitation of the available computer resources. Those cases are indicated in Table 1 by n/a entries. Our algorithm always found a better or equal result compared with all other approaches in the literature. It found the best-known solution in 9 cases out of 120. In a further 56 cases, the deviation from the optimal solution was less than 1%; in the remaining 52 cases, the difference was between 1% and 3%. In three cases, the running time exceeded the set limit. Where the algorithm did not find the best-known solution, it got very close to it. Compared with all other algorithms published by other authors, as mentioned above, the proposed new algorithm provided much better results in all cases.

7. Conclusions

There is an interesting symmetry–asymmetry issue when solving complex problems, setting up models for complex systems, and developing algorithms for search and optimization in them. In this paper, the highly complex and mathematically intractable flow shop scheduling problem is in one pan of the scale, while in the other, the new modified discrete bacterial memetic evolutionary algorithm (DBMA) is found. By proper weighing of the costs, namely, the error in the accuracy of the optimization in one pan and the need for resources, especially, the running time of the optimization meta-heuristics in the other one must be brought to equilibrium, this way generating a symmetry in the solution. The exact position of the symmetrical (balanced) solution can, however, be calibrated by the designer of the solution, thus it may fit the application context of the concrete problem, considering the available resources and the expected quality of the quasi-optimum found. Thus, the asymmetric role played by problem to solve and model/algorithm for solution must be balanced and, that way, the whole problem–solution complex must be brought in a symmetrical configuration.

In our approach, there is, however, another aspect of the symmetry–asymmetry concept present. Memetic algorithms consist of two essential components, the "outer" shell that is an evolutionary or population based global searcher, and the "inner" core that is a local searcher, whether traditional gradient based, or exhaustive search type, respectively; or, as in the novel algorithm proposed in this paper, another meta-heuristic method. The two components must also form a symmetric combination in the above sense: the two must be in proper balance of resource intensity and need. Many results have shown that too much local search will slow down the whole optimization procedure, while too little (compared to the outer global search) may lead to ever randomly wandering attempts to approach the optimum, where even the most efficient local search can only produce a local optimum. We trust that, in this novel algorithm, a very efficient and well balanced, let us say, symmetric enough, solution for the combination of the two components of the memetic algorithm was found.

In our paper, the DBMA was very successfully applied for the approximate solution of other, similarly NP-hard discrete problems. Namely, the original DBMA with n-opt type local search method was developed for TSP problems. We found, however, that this local search provided relatively poor results for solving the flow shop scheduling problem. In the proposed new and improved algorithm, we have replaced the local search by the simulated annealing algorithm, a method that has been applied with some success itself for solving similar tasks. We found that this hybrid DBMEA and SA algorithm became unambiguously more efficient, compared with all other population-based metaheuristic approaches proposed by other researchers. The authors calculated the make span times for a known benchmark data set and compared the results with the algorithms in other papers as well as with the best known solutions (it should be clearly stated that the optimal solution cannot be calculated owing to the size of the problem, so the best known published makespan times were used as the basis of the comparison). The proposed new algorithm indeed over-performed all the state-of-the-art algorithms. The calculated makespan times were very close to the best known solutions, while the computing time still remained reasonable, even on a standard personal computer. So, the proposed algorithm has the capability to so far most efficiently solve large-scale FSPP problems.

Author Contributions: Conceptualization, L.T.K. and K.N.; methodology, K.N. and A.A.; software, K.N. and A.A.; validation, O.H.; formal analysis, O.H. and A.A.; investigation, A.A. and K.N.; resources, A.A.; data curation, A.A. and K.N.; writing—original draft preparation, A.A. and O.H.; writing—review and editing, A.A., O.H., L.T.K. and K.N.; visualization, A.A.; supervision, L.T.K. and K.N.; project administration, O.H.; funding acquisition, L.T.K. All authors have read and agreed to the published version of the manuscript.

Funding: This research is supported by the National Office of Research, Development, and Innovation grant NKFIH K124055.

Institutional Review Board Statement: Not applicable.

Informed Consent Statement: Not applicable.

Data Availability Statement: The data presented in this study are openly available in GitHub at https://doi.org/10.5281/zenodo.4635757 (accessed on 17 April 2021).

Conflicts of Interest: The authors declare no conflict of interest.

References

1. Johnson, S.M. Optimal two-and three-stage production schedules with setup times included. *Nav. Res. Logist. Q.* **1954**, *1*, 61–68. [CrossRef]
2. Wei, H.; Li, S.; Jiang, H.; Hu, J.; Hu, J. Hybrid genetic simulated annealing algorithm for improved flow shop scheduling with makespan criterion. *Appl. Sci.* **2018**, *8*, 2621. [CrossRef]
3. Taillard, E. Benchmarks for basic scheduling problems. *EJOR* **1993**, *64*, 278–285. [CrossRef]
4. Tseng, L.Y.; Lin, Y.T. A hybrid genetic algorithm for no-wait flowshop scheduling problem. *Int. J. Prod. Econ.* **2010**, *128*, 144–152. [CrossRef]
5. Belabid, J.; Aqil, S.; Allali, K. Solving Permutation Flow Shop Scheduling Problem with Sequence-Independent Setup Time. *J. Appl. Math.* **2020**, *2020*, 7132469. [CrossRef]
6. Qu, C.; Fu, Y.; Yi, Z.; Tan, J. Solutions to no-wait flow shop scheduling problem using the flower pollination algorithm based on the hormone modulation mechanism. *Complexity* **2018**, *2018*, 1973604. [CrossRef]
7. Zhou, Y.; Chen, H.; Zhou, G. Invasive weed optimization algorithm for optimization no-idle flow shop scheduling problem. *Neurocomputing* **2014**, *137*, 285–292. [CrossRef]
8. Ogbu, F.A.; Smith, D.K. The application of the simulated annealing algorithm to the solution of the n/m/Cmax flowshop problem. *Comput. Oper. Res.* **1990**, *17*, 243–253. [CrossRef]
9. Lin, S.W.; Cheng, C.Y.; Pourhejazy, P.; Ying, K.C. Multi-temperature simulated annealing for optimizing mixed-blocking permutation flowshop scheduling problems. *Expert Syst. Appl.* **2020**, *165*, 113837. [CrossRef]
10. Aurich, P.; Nahhas, A.; Reggelin, T.; Tolujew, J. Simulation-based optimization for solving a hybrid flow shop scheduling problem. In Proceedings of the 2016 Winter Simulation Conference (WSC), Arlington, VA, USA, 11–14 December 2016; pp. 2809–2819.
11. Carlier, J. Ordonnancements a contraintes disjonctives. *RAIRORecherche Oper.* **1978**, *12*, 333–351. [CrossRef]
12. Heller, J. Some numerical experiments for an M×J flow shop and itsdecision-theoretical aspects. *Oper. Res.* **1960**, *8*, 178–184. [CrossRef]
13. Reeves, C. A genetic algorithm for flowshop sequencing. *Comput. Oper. Res.* **1995**, *22*, 5–13. [CrossRef]
14. Nawa, N.E.; Furuhashi, T. Fuzzy system parameters discovery by bacterial evolutionary algorithm. *IEEE Trans. Fuzzy Syst.* **1999**, *7*, 608–616. [CrossRef]
15. Inoue, T.; Furuhashi, T.; Maeda, H.; Takaba, M. A study on interactive nurse scheduling support system using bacterial evolutionary algorithm engine. *Trans. Inst. Elect. Eng. Jpn.* **2002**, *122*, 1803–1811.
16. Das, S.; Chowdhury, A.; Abraham, A. A bacterial evolutionary algorithm for automatic data clustering. In Proceedings of the IEEE Congress on Evolutionary Computation 2009 (CEC '09), Trondheim, Norway, 18–21 May 2009; pp. 2403–2410.
17. Hoos, H.H.; Stutzle, T. *Stochastic Local Search: Foundations and Applications*; Morgan Kaufmann: San Francisco, CA, USA, 2005.
18. Moscato, P.; Mathieson, L. Memetic Algorithms for Business Analytics and Data Science: A Brief Survey. *Bus. Consum. Anal. New Ideas* **2019**, 545–608. [CrossRef]
19. Gong, G.; Deng, Q.; Chiong, R.; Gong, X.; Huang, H. An effective memetic algorithm for multi-objective job-shop scheduling. *Knowl. Based Syst.* **2019**, *182*, 104840. [CrossRef]
20. Botzheim, J.; Cabrita, C.; Kóczy, L.T.; Ruano, A.E. Fuzzy rule extraction by bacterial memetic algorithms. *Int. J. Intell. Syst.* **2009**, *24*, 312–339. [CrossRef]
21. Muyldermans, L.; Beullens, P.; Cattrysse, D.; Van Oudheusden, D. Exploring variants of 2-opt and 3-opt for the general routing problem. *Oper. Res.* **2005**, *53*, 982–995. [CrossRef]
22. Balazs, K.; Koczy, L.T. Hierarchical-interpolative fuzzy system construction by genetic and bacterial memetic programming approaches. *Int. J. Uncertain. Fuzziness Knowl. Based Syst.* **2012**, *20*, 105–131. [CrossRef]

23. Kóczy, L.T.; Földesi, P.; Tüű-Szabó, B. An effective discrete bacterial memetic evolutionary algorithm for the traveling salesman problem. *Int. J. Intell. Syst.* **2017**, *32*, 862–876. [CrossRef]
24. Tüű-Szabó, B.; Földesi, P.; Kóczy, L.T. An Efficient Evolutionary Metaheuristic for the Traveling Repairman (Minimum Latency) Problem. *Int. J. Comput. Intell. Syst.* **2020**, *13*, 781–793. [CrossRef]
25. Dai, M.; Tang, D.; Giret, A.; Salido, M.A.; Li, W.D. Energy-efficient scheduling for a flexible flow shop using an improved genetic-simulated annealing algorithm. *Robot. Comput. Integr. Manuf.* **2013**, *29*, 418–429. [CrossRef]
26. Agárdi, A.; Nehéz, K. Flow Shop Scheduling Problem Optimization with Discrete Bacterial Memetic Evolutionary Algorithm and Simulated Annealing. 2021. Available online: https://github.com/anitaagardi/production-optimization-DBMEA (accessed on 25 March 2021).

Article

Flattening Layer Pruning in Convolutional Neural Networks

Ernest Jeczmionek [1,†] and Piotr A. Kowalski [2,3,*,†]

1 AGH Doctoral School, AGH University of Science and Technology, al. Mickiewicza 30, 30-059 Cracow, Poland; ejec@agh.edu.pl
2 Faculty of Physics and Applied Computer Science, AGH University of Science and Technology, al. Mickiewicza 30, 30-059 Cracow, Poland
3 Systems Research Institute, Polish Academy of Sciences, ul. Newelska 6, 01-447 Warsaw, Poland
* Correspondence: pkowal@agh.edu.pl
† These authors contributed equally to this work.

Abstract: The rapid growth of performance in the field of neural networks has also increased their sizes. Pruning methods are getting more and more attention in order to overcome the problem of non-impactful parameters and overgrowth of neurons. In this article, the application of Global Sensitivity Analysis (GSA) methods demonstrates the impact of input variables on the model's output variables. GSA gives the ability to mark out the least meaningful arguments and build reduction algorithms on these. Using several popular datasets, the study shows how different levels of pruning correlate to network accuracy and how levels of reduction negligibly impact accuracy. In doing so, pre- and post-reduction sizes of neural networks are compared. This paper shows how Sobol and FAST methods with common norms can largely decrease the size of a network, while keeping accuracy relatively high. On the basis of the obtained results, it is possible to create a thesis about the asymmetry between the elements removed from the network topology and the quality of the neural network.

Keywords: global sensitivity analysis; Sobol procedure; fast algorithm; convolutional neural network; structure reduction; pruning; quality

Citation: Jeczmionek, E.; Kowalski, P.A. Flattening Layer Pruning in Convolutional Neural Networks. *Symmetry* **2021**, *13*, 1147. https://doi.org/10.3390/sym13071147

Academic Editors: Kóczy T. László and István A. Harmati

Received: 12 May 2021
Accepted: 9 June 2021
Published: 27 June 2021

1. Introduction

Since the beginning of the twenty-first century, computational intelligence (CI) [1], in the guise of artificial intelligence and machine learning, has been experiencing great strides in its development, both practically [2] and theoretically [3]. The term embraces fuzzy logic, genetic and evolutionary algorithms, swarming intelligence, rough sets and artificial neural networks. Artificial neural networks (ANNs) are computational systems developed on the basis of the biological structure of the brains of living organisms, and they consist of objects representing neurons and the connections between them. In ANNs, there are layers made of neurons. The first (input layer) is used to enter data into the network, and the last (output layer) returns the generated results [4], while a number of hidden layers exists in between. As the number of layers increases, the network processing time automatically increases. Therefore, the optimization of the structure of the neural network is an important issue.

ANNs play a very important and often unheralded role in the modern world. Calculations made with their help can be found, among others, in forecasting air pollution [5] or when conversing via cell phone [6]. A popular and recent development is the use of neural networks in natural language processing, in particular, for text generation [7], automatic text translation [8], text analysis [9], spam message detection [10] and spoken text recording [11]. Due to their versatility and the possibility of modeling non-linear processes, ANNs are used in the automotive industry (navigation systems, autopilot), telecommunications and robotics [12].

Convolutional neural networks (CNNs) were introduced in 1980–1982, under the term "neocognitron" [13], but their rather dynamic development began to be felt only from around 2010. Since then, many ready-made CNN neural networks have been created. Indeed, it can be said that this type of neural network is the basic structure in the class of deep neural networks [14].

CNNs require the use of high-powered computers, as they are, in reality, often oversized for purpose. To prevent the exaggerated growth of size, pruning methods are needed and have been subjected to research. There are two categories of pruning based on purpose: pruning for performance [15] or pruning for size [16]. The method shown in this article can be applied in order to minimize the number of reduction cycles and the number of neurons in each reduction cycle.

Sensitivity analysis methods were designed to assess the impact of a model's input on its output. Two major subgroups are Local Sensitivity Analysis (LSA) [17] and Global Sensitivity Analysis (GSA) [18]. The first method measures sensitivity by varying only one input parameter, while GSA changes all inputs simultaneously.

The main focus of this article is to propose a reduction layer based on sensitivity analysis that is combined with a flattening layer. The presence of a large number of convoluted feature matrices and the substantial size of first fully connected layer generates enormous numbers of permutation. In some cases, these weights are responsible for 90% of the total parameters in the network. This research concentrates on minimizing unnecessary connections between the convolutional layer and the fully connected network.

In a large number of algorithms, the removal of certain elements from their structure determines the simplification of such procedures. Unfortunately, such symmetry reduces the quality of the results obtained. In the proposed algorithm for reducing the structure of the neural network, asymmetry between the reduction in the topological structure of the neural network and the obtained results is observed. The above is largely realized by way of the use of sensitivity analysis precursors, thanks to which the weakest links of the tested CNN flattening layer are determined.

Decomposition and pruning are common techniques applied to compress the architectures of neural networks. Tucker decomposition is a well-known Low-Rank factorization method to decompose both convolutional and fully connected layers [19]. Another popular method is tensor rank decomposition. This is based on the superdiagonal core tensor of Tucker decomposition [20,21]. Filter pruning is a natural approach in CNN compression. Moreover, a team of Nvidia researchers has presented a kernel pruning algorithm based on a minimization of the Taylor series expansion of the error [16]. The attention mechanism introduced by a Google research team [22] has also been applied to prune CNN with regard to the classification problem [23]. Both decomposition and pruning methods can be combined for better network compression [24].

This article is divided into the following sections. Section 2 describes the concept of a convolutional neural network, and it includes a discussion on what is sensitivity analysis and what methods are utilized in the research. Section 3 presents the algorithm used for a proposed reduction method. Section 4 consists of a description of the datasets and an analysis of the results obtained from applying variants of the reduction algorithm. The last section, Section 5, summarizes the results and and provides the obtained conclusions.

2. Methods

2.1. Convolutional Neural Network

The convolutional neural network (CNN) is a modern architecture of a neural network used in anomaly detection [25] and natural language processing (such as in sentence modeling) [26], as well as in classification [27].

The major area of CNN application is in computer vision, including object detection [2], image classification [28] and segmentation [29]. CNNs that have been around for a few years, such as GoogLeNet, present human-like accuracy of classification [30], and performance of these networks has been under continuous improvement. In this paper, attention is focused upon the classification problem. CNN, mostly applied on problems concerning images, is built upon the input image undergoing a series of convolutional operations. The mathematical formula for convolution operation is presented in Equation (1).

$$\begin{bmatrix} A_{1,1} & \cdots & A_{1,N} \\ \vdots & \ddots & \vdots \\ A_{M,1} & \cdots & A_{M,N} \end{bmatrix} * \begin{bmatrix} B_{1,1} & \cdots & B_{1,N} \\ \vdots & \ddots & \vdots \\ B_{M,1} & \cdots & B_{M,N} \end{bmatrix} = \sum_{i=0}^{M-1} \sum_{j=0}^{N-1} A_{(M-i),(N-j)} B_{(i+1),(j+1)} \quad (1)$$

This CNN includes many kernels (also called 'filters'). The task of the kernel is to learn feature extraction. If many convolutional layers are stacked, the first layers are responsible for deriving so-called 'high level features', such as the image's basic outlines or curves. The following layers, through the application of matrices of convoluted features, extract more and more detailed characteristics. Generally, multiple convolutions of an image would generate large matrices, extending computation time and memory usage. To resolve the issue, a 'pooling layer' was introduced. Its first purpose is to reduce the size of a matrix by applying the functions of a pooling kernel. The most common of these are averaging and maximizing. The second task of the pooling layer is to reduce non-dominant properties by leaving only the most important feature, hence reducing image 'noise'. Convolutional and pooling layers are put together in many combinations. A popular approach to CNN modeling is to form a stack of one or two convolutional layers, followed by a pooling layer.

The two previously described layers are responsible for learning an image's features. They result in a vector of low-level convoluted features matrices. To assign an image to a category, a classification network is required. Before this can be done, the output of convolutional and pooling layers has to be flattened to a single vector. A fully Connected Network (FCN) is a feed-forward network, the purpose of which is to learn the likelihood of membership to a category. However, having an output that is a vector of the number of features assigned to each individual class is not desirable. To simplify the results and make these more understandable, a softmax function is incorporated. Softmax, acting in the form of an activation function in the last layer, returns only one value assigned to a category, with the highest number of features corresponding to that category being grouped together.

In this article, two CNNs were used, Figure 1 for 2D datasets and Figure 2 for 1D datasets. For faster convergence, a dropout layer was added to each CNN. This layer zeroes weights with set probability. Hence, all networks end with a FCL of a size corresponding to the number of categories in the dataset, and they use ReLU as an activation function. In both the 1D CNN and the 2D CNN, categorical cross-entropy is considered a loss function. Their treatment differs in that the Adam optimizer was set in the 1D CNN and a Stochastic Gradient Descent was established in case of 2D CNN. The 1D datasets of the 1D CNN are composed of double 64 3×3 kernels, followed by a dropout layer with probability of 50%. Subsequently, a max pooling layer of size 2×2 and consisting of 100 neurons in a fully connected layer is attached. 2D CNN is more complicated and is structured with a double sequence of two 32 3×3 filters, followed by a 2×2 max pooling layer and the 25% dropout layer. The 2D CNN ends with a 512 neuron FCL and a 50% dropout layer. Table 2 lists the number of neurons in the first layer of the FCN, the total number of parameters in the CNN, the number of frozen non-trainable parameters in their convolutional layers and, finally, the number of trainable parameters in FCN.

```
Model: "sequential"
_________________________________________________________________
Layer (type)                 Output Shape              Param #
=================================================================
conv2d_4 (Conv2D)            (None, 26, 26, 32)        320
_________________________________________________________________
conv2d_5 (Conv2D)            (None, 24, 24, 32)        9248
_________________________________________________________________
max_pooling2d_2 (MaxPooling2 (None, 12, 12, 32)        0
_________________________________________________________________
dropout_3 (Dropout)          (None, 12, 12, 32)        0
_________________________________________________________________
conv2d_6 (Conv2D)            (None, 10, 10, 64)        18496
_________________________________________________________________
conv2d_7 (Conv2D)            (None, 8, 8, 64)          36928
_________________________________________________________________
max_pooling2d_3 (MaxPooling2 (None, 4, 4, 64)          0
_________________________________________________________________
dropout_4 (Dropout)          (None, 4, 4, 64)          0
_________________________________________________________________
flatten_1 (Flatten)          (None, 1024)              0
_________________________________________________________________
input_reduction (InputReduct (None, 1024)              0
_________________________________________________________________
dense (Dense)                (None, 512)               524800
_________________________________________________________________
dropout (Dropout)            (None, 512)               0
_________________________________________________________________
dense_1 (Dense)              (None, 10)                5130
=================================================================
Total params: 594,922
Trainable params: 529,930
Non-trainable params: 64,992
```

Figure 1. CNN model architecture dedicated for 2D image input.

```
Model: "sequential"
_________________________________________________________________
Layer (type)                 Output Shape              Param #
=================================================================
conv1d (Conv1D)              (None, 14, 64)            256
_________________________________________________________________
conv1d_1 (Conv1D)            (None, 12, 64)            12352
_________________________________________________________________
dropout (Dropout)            (None, 12, 64)            0
_________________________________________________________________
max_pooling1d (MaxPooling1D) (None, 6, 64)             0
_________________________________________________________________
flatten (Flatten)            (None, 384)               0
_________________________________________________________________
input_reduction (InputReduct (None, 384)               0
_________________________________________________________________
dense (Dense)                (None, 100)               38500
_________________________________________________________________
dense_1 (Dense)              (None, 7)                 707
=================================================================
Total params: 51,815
Trainable params: 39,207
Non-trainable params: 12,608
```

Figure 2. CNN model architecture dedicated for 1D series input.

2.2. Global Sensitivity Analysis

Sensitivity analysis (SA) consists of a group of methods used for finding how the uncertainty in the model output can be assigned to the uncertainty of the model input [31]; hence, they are used to discover the connection between uncertainty of the model input and output [32]. Local Sensitivity Analysis (LSA) and Global Sensitivity Analysis (GSA) are subgroups of SA. The LSA approach alters one input parameter at a time with all others remaining constant [17], while the GSA approach modifies all input parameters concurrently. The most common approaches for evaluating the impact on the models' output are regression methods, screening algorithms [33] and variance-based methods. The variance-based procedures used in this article are Sobol [34], Fourier Amplitude Sensitivity Test (FAST) [35] and extended Fourier Amplitude Sensitivity Test (eFAST) [36,37].

The difference between FAST and eFAST methods is that the former calculates only the first-order sensitivity, while the latter also calculates total order sensitivity. For simplicity, both these algorithms are heretofore called FAST. In this article, both first and total order sensitivities are used.

Sobol's method is based on decomposition of output variance into a sum of input variances. It measures the impact of each individual input and the permutations between them on the output. It is achieved by calculating first-order, second-order, higher-order and total-order sensitivity indices. To calculate the indices, a Monte Carlo integration is applied.

The FAST method is derived from the time series Fourier decomposition in signal theory. The original FAST method provided only first-order indices, but extension of the method generates higher-order sensitivities. To compute the indices pattern, a search based on sinusoidal functions is applied.

3. Pruning Algorithm in Flattening Layer

This section describes the algorithm (Algorithm 1) used to prune the input of FCN. Global sensitivity analysis has not been previously applied to compress CNNs, and pruning could be its natural application. The aim of the algorithm is to provide a flattening layer pruning algorithm. This approach only reduces the weights between the convolutional and fully connected layers. The proposed procedure can be easily stacked with other pruning and decomposition compression procedures. In fact, stacking a few different procedures can lead to better compression [24]. The presented algorithm was applied to a simple CNN to validate its utility. As mentioned before, developers should not solely rely on this method as a standalone solution. GSA methods are still under research, the intent being to create an algorithm to compress all the layers of a CNN.

First the CNN has to be created and trained on the chosen dataset. The algorithm will then execute R sensitivity calculations, and each time it will prune D parameters. For each reduction cycle, pretrained CNN has to be loaded. Weights of convolutional layers must also be kept unchanged, the only part of the network subjected to training should be the FCN. What is more, all convolutional parameters have to be frozen. The next step is to join pretrained CNN and freshly initialized FCN with a reduction layer. This reduction layer is a flattening layer that has the ability to filter neurons. The large number of outputs from the convolutional layers, allied with the large number of input neurons of the FCN, results in permutations that unnecessarily consume resources. The task of the reduction layer is to prune non-impactful connections inputs. This leads to a reduction in the size of the network. The subsequent step is to calculate the sensitivity by applying one of the previously mentioned methods and aggregate it with the chosen norm. The inputs are then sorted by their sensitivity, and the least impactful ones are removed by the reduction layer. The process is repeated until reaching a given number of reduction cycles. The algorithm was implemented in Python and uses the Keras library. Computations were executed via Google's Colaboratory service.

Algorithm 1: FCN input reduction

```
Result: CNN with reduction layer
load dataset
pretrain CNN
for number of reductions R do
    load pretrained CNN and freeze convolutional layers
    join convoluitional layers with reduction layer and FCN
    retrain the network
    calculate sensitivity
    aggragte sensitivity using one of norms
    sort FCN inputs by sensitivity
    for number of pruned parameters D do
        prune the input with the smallest sensitivity from reduction layer
    end
end
```

4. Results

As previously mentioned, CNNs consist of two main parts. The first embodies the convolutional and pooling layers that produce the extracted features. The second incorporates classification layers, mostly FCN. Between these two parts, a custom reduction layer is proposed. It disables the least significant parameters coming from the flatten layer. In each dataset, reduction is applied R times, with D parameters pruned each time. The convolutional part of the network is pretrained, all its parameters are non-trainable, and the FCN alone is trained from scratch each time to adjust to the reduced input. The focus of the research is to sustain or improve test accuracy. To do this, two methods with two sensitivities—Sobol (first order), Sobol total (order), FAST (first order) and FAST total (order)—and three norms—Euclidean, absolute value and maximum—are compared. In each training cycle, when sensitivity matrices are calculated for all outputs, a selected norm is applied to aggregate output sensitivity matrices. FCN input weights are then sorted by aggregated sensitivities of output. Weights with the least sensitive values are pruned, and the procedure is repeated.

4.1. Data Sets

This research is based on a total of four classification datasets (Table 1). Two of these are vectors of features, while the others contain images. The credit card fraud dataset is a Kaggle dataset [38] detecting frauds on the basis of 28 parameters. With regard to the set, confidentiality of financial data has forced application of PCA transformation. In the original dataset only 492 cases out of 284,807 transactions are marked as frauds. A new dataset had, therefore, to be created to overcome the unbalance of data. It contains 984 transactions, of which 50% are fraud cases. The dataset was randomly divided into 80% for the training set and 20% for the test set.

The Beans dataset is a 2020 UCI dataset [39] classifying dry bean grains into seven species, registered with a high-resolution camera. It contains 13,611, 16-element vectors of features. Each dataset record is composed of 12 dimension parameters and 4 shape forms. In this case, the dataset was also divided randomly into training and test sets in a ratio of 80%/20%.

MNIST and FASHION MNIST are image classification datasets made easily available through the Keras library [40]. Both are composed of 28×28 grayscale images with 60,000 elements in the training sets and 10,000 elements in the test sets. The differences between these datasets are the types of images found. MNIST is made up of images of digits, while FASHION MNIST consists of images of clothes and accessories. MNIST is a dataset created from the US National Institute of Standards and Technology's Special Database 1 and Special Database 3. It gathers handwritten digits from around 250 writers, where writers for training and test sets were disjoined. FASHION MNIST is an attempt to replace the MNIST dataset. The authors of FASHION MNIST criticized MNIST in that it is too easy to achieve high accuracy, is overused and does not represent modern computer vision tasks.

Table 1. Datasets comparison.

Dataset	Data Type	Train/Test Set Size	Labels
MNIST	28 × 28 greyscale images	60 k/10 k	10
FAHION MNIST	28 × 28 greyscale images	60 k/10 k	10
Beans	16 parameters	10,888/2723	7
Credit Card Fraud	28 parameters after PCA	786/198	2

4.2. MNIST and FASHION MNIST

MNIST and FASHION MNIST are similar datasets with different types of images. Results and conclusions for both datasets are similar and will be placed in a joined section. Figures 3 and 4 present the detailed results that are further discussed in this section. For these two datasets, the test accuracy was found to be larger than the train accuracy. This was not expected and was probably caused by high dropout percentage, where, through probability, neurons were not considered for training. In the case of the CNN used for these datasets, its flatten layer output size was 1024, and a total number of 20 reduction cycles was performed. In each cycle, 50 least impactful parameters were pruned. The original total number of parameters was 594,922, of which 529,930 were considered trainable. At the last step, the number of trainable parameters was decreased to 43,530. This is an over ten times reduction. The cost for so high a reduction in size is a decrease of 10% for test accuracy. A gentle decline in accuracy is observed until around 600 pruned neurons, and further reduction resulted in a more radical decrease. All methods and norms presented very competitive results.

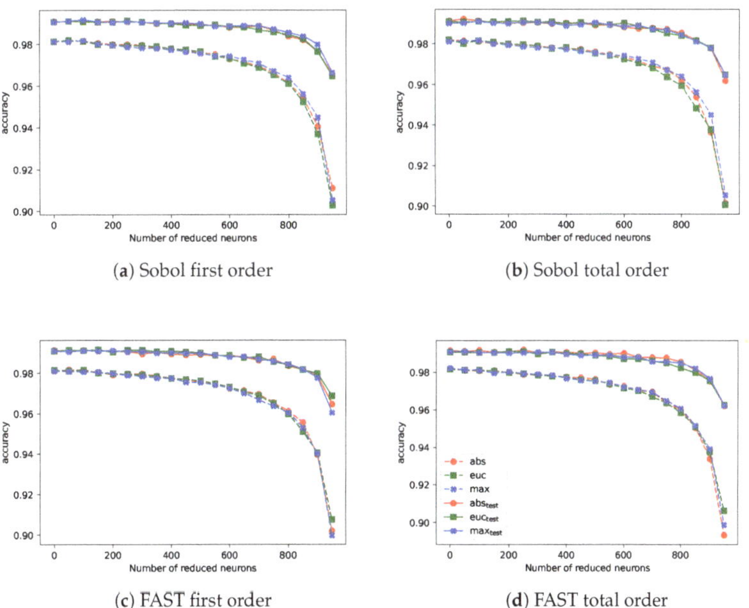

(**a**) Sobol first order (**b**) Sobol total order

(**c**) FAST first order (**d**) FAST total order

Figure 3. FCN reduction for the MNIST dataset by methods and norms on train and test sets.

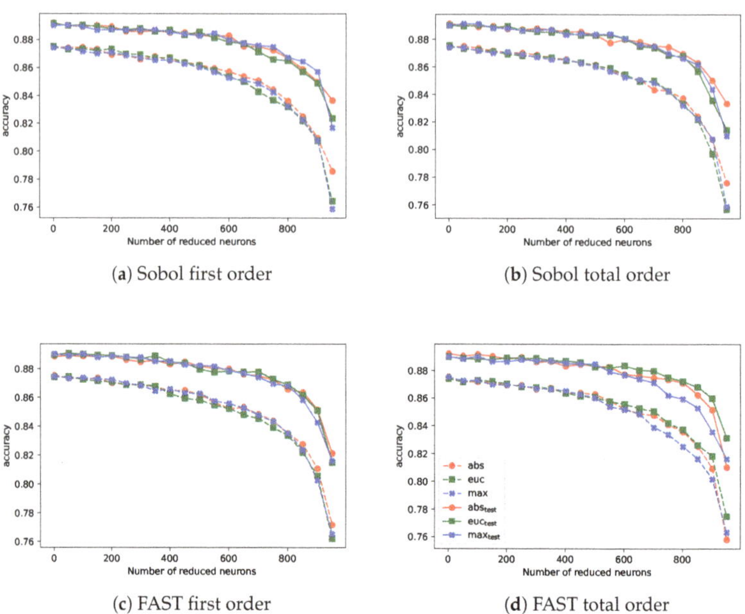

(**a**) Sobol first order (**b**) Sobol total order

(**c**) FAST first order (**d**) FAST total order

Figure 4. FCN reduction for the Fashion MNIST dataset training and test sets, by methods and norms.

4.3. Credit Card Fraud

The credit card fraud data are a post-PCA transformation. This suggests that there is no large potential for further pruning. In our experiment, the reduction was performed 17 times, each time reducing 30 parameters. The original FCN inputs had 768 parameters and were reduced to 288 parameters at the last run. This led to a drop of trainable parameters from 77,102 to 29,102, while the total number of parameters dropped from 89,710 to 41,710. We found that the in the PCA transformation, reduction had little influence on accuracy. For both methods, Sobol and FAST, the test accuracy did not change or was improved at some point in each scenario. As seen in Figure 5, the Sobol scenario test accuracy results are more flat, while in the case of FAST, they slightly decrease. In most of the cases, the maximum cost for reducing the size of the network by half is up to 2 percentage points of test accuracy drop.

4.4. Beans

This set was reduced 15 times with 20 pruned parameters at each run. The original number of FCN input parameters was 384 and was reduced to 104. This resulted in a decrease in total network parameters from 51,815 to 23,815, while FCN parameter accuracy fell to 50%. In the case of the first-order Sobol method, abs and euc norms test accuracy reached a peak training accuracy. Similar results are observed when the Sobol total method was applied. Here, test accuracy, when max norm was applied, reached the level of training accuracy after a cycle of 100 reductions. In contrast, Abs test accuracy approached training accuracy at the end of a reduction after a high drop. Test accuracy reached or exceeded train accuracy in all cases when FAST and FAST total methods were applied. As seen in Figure 6, abs and euc norms test accuracy reached training accuracy after just 50 neurons were reduced, with accuracy maintained to the end of the run. Max norm was found to outperform other norms, but only when a large number of FCN inputs were reduced. Finally, abs norms preserved the highest test accuracy. This was only a few percentage points lower than the training accuracy of the original non-reduced network. What is more, Euc norms presented similar results for both FAST and FAST total scenarios, while abs norm significantly reduced test accuracy for FAST total, when large reductions are taken into account. For FAST total, max norm had the highest test accuracy, outperforming training accuracy in the last reduction runs.

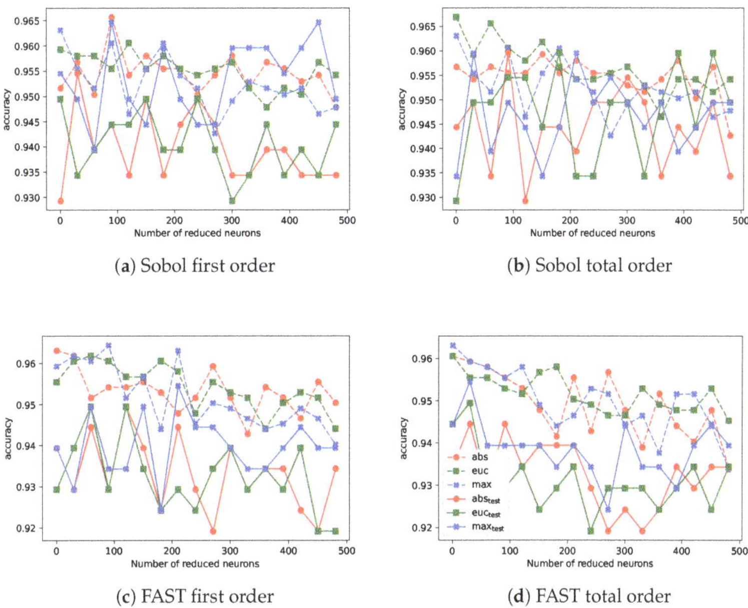

(**a**) Sobol first order (**b**) Sobol total order

(**c**) FAST first order (**d**) FAST total order

Figure 5. FCN reduction for the Credit Card Fraud dataset training and test set, by methods and norms.

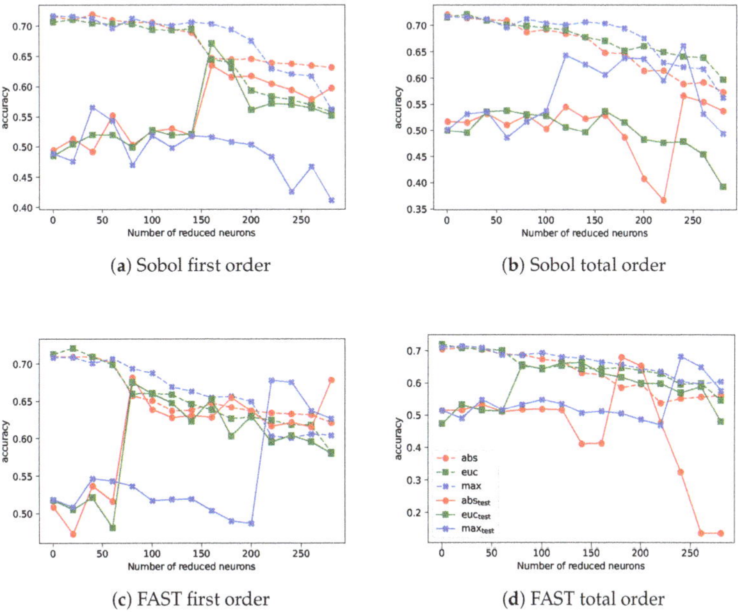

(**a**) Sobol first order (**b**) Sobol total order

(**c**) FAST first order (**d**) FAST total order

Figure 6. FCN reduction for the Bean dataset training and test set, by methods and norms.

4.5. Discussion of the Results

Sobol and Fashion MNIST sets sustained stable, low-accuracy drops until half the parameters were reduced. Subsequently, the accuracy drop reached 10% with 90% reduction in FCL neurons. The fraud dataset showed that reduction of non-meaningful data does not have to impact the accuracy. In that case, accuracy fluctuated around a constant value. The Beans dataset is the most surprising. Here, test accuracy largely exceeded train accuracy. We also noted that GSA was proven not only to be able to reduce the impact, but also to keep high accuracy and, indeed, to improve test accuracy. Table 3 presents post-reduction numbers of input FCN parameters, total number of parameters and number of parameters for CNN and FCN, similarly to Table 2.

On comparing Tables 2 and 3, in extreme cases, it can be seen that the size of the network was reduced from 54% for the Credit Card Fraud and Bean datasets to up to 82% in the case of MNIST and Fashion MNIST datasets.

Table 2. CNN parameters pre-reduction.

		Before Reduction		
Datasets	**Input Size**	**Total**	**Non-Trainable**	**Trainable**
MINST/Fashion MNIST	1024	594,922	64,992	529,930
Credit Card Fraud	768	89,710	12,608	77,102
Beans	384	51,815	12,608	39,207

Table 3. CNN parameters post-reduction.

		Final Maximal Reduction		
Datasets	**Input Size**	**Total**	**Non-Trainable**	**Trainable**
MINST/Fashion MNIST	74	108,522	64,992	43,530
Credit Card Fraud	288	41,710	12,608	29,102
Beans	104	23,815	12,608	11,207

The proposed procedure was expected to prune flattening layer connections and minimize accuracy loss with pretrained convolutional layers. This was observed in the case of MNIST and FASHION MNIST datasets. Utilization of the Credit card fraud dataset presented surprising results. Besides a general training accuracy drop with deleted connections, some norm functions were able to improve the accuracy of the applied test sets. The same, yet more clearer, phenomenon was observed in the results of Beans classification. In all cases, besides the Sobol total order indices method, a Euclidean norm function boosted the test set accuracy to the level of the training set accuracy. In this scenario, the presented GSA-based pruning algorithm not only decreased network size, but also vastly improved test set accuracy of the network. The algorithm pruned only the flattening layer. In further work, GSA methods are going to be applied to other layers of the CNN to create a pruning method for the whole structure. We hope to create a pruning method that is able not only to compress the structure, but also to increase its performance.

5. Conclusions

This article applied the GSA methods of Sobol and FAST to reduce the number of FCN input neurons in CNNs. Originally, the full number of connections between matrices of convoluted features and FCN led to a large number of training parameters. When we applied the described reduction algorithms that are based on GSA methods and three norms, we were able to cut-down the number of unnecessary parameters while keeping near to the original accuracy levels. For some datasets, the proposed pruning provided accuracy levels close to the original solution.

The reduction in the structure of the internal neural network has a very positive effect on several aspects. The first is faster computation time. This applies both to the current time related to the recovery mode and also to learning time, as neural networks often need to be retrained when new data come in, and the greater volume of data, the more time is needed in doing so. The second aspect discussed here is the issue of neuronal structure overfitting. In practice, various types of treatments are often used to address the problem of overfitting, such as data reorganization, drop-out procedure and the use of a special penalty function during training. However, the method proposed in the text of the article solves the problem naturally because removing redundant neurons implies the existence of fewer points of freedom and, therefore, the likelihood of overfitting being minimized.

For datasets other than that originally used, accuracy was decreased, but it was disproportionately less when compared to the size of the reduction. As the algorithm is able to significantly reduce the size of the network with a cost of small performance drop, it can, therefore, enable the use of previously overlarge networks on devices with less memory, such as mobile systems. The smaller number of parameters also directly relates to improvement in the network's prediction time.

The proposed procedure can be applied as either a supervised or unsupervised algorithm. In the first case, it is necessary to leave the validation sample on the basis of which the quality of the network computation can be checked in each iteration of neuron removal. Of course, this is related to more computation time. In the case of treating the reduction procedure as unsupervised, we can assume in advance the number of neurons we want to remove.

The above proposal of the neural network reduction algorithm also addresses research related to the analysis of the significance of individual components of CNN. This is especially true because, in most cases, neural networks are treated as black-box models, their internal elements being not subject to analysis. What is more, during the neural network system synthesis, one does not try to understand the meaning of individual elements but bases the application exclusively on empiric performance.

Further research plans will be related to understanding and assessing the use of reduction methods inside CNN neural networks, in particular, in the layers of the fully connected part. This action will be aimed at reducing the topological structure of the neural network and, as a consequence, slashing computation time and enhancing the efficiency of neural computations.

Author Contributions: Conceptualization, P.A.K. and E.J.; software, E.J.; validation, P.A.K.; formal analysis, P.A.K. and E.J.; investigation, E.J.; resources, E.J.; data curation, E.J.; Writing—Original draft preparation, P.A.K. and E.J.; Writing—Review and editing, P.A.K.; visualization, E.J.; supervision, P.A.K.; project administration, P.A.K. All authors have read and agreed to the published version of the manuscript.

Funding: This research is partially financed by Grants for Statutory Activity from Faculty of Physics and Applied Computer Science of the AGH University of Science and Technology in Cracow.

Institutional Review Board Statement: Not applicable.

Informed Consent Statement: Not applicable.

Data Availability Statement: MNIST/ Fashion MNIST https://keras.io/api/datasets/ (accessed on 23 January 2021); Credit Card Fraud https://www.kaggle.com/mlg-ulb/creditcardfraud (accessed on 23 January 2021) ; Beans https://archive.ics.uci.edu/ml/datasets/Dry+Bean+Dataset (accessed on 23 January 2021).

Acknowledgments: This article is based upon work from COST Action 17124 DigForAsp, supported by COST (European Cooperation in Science and Technology), www.cost.eu and is partially financed by Grants for Statutory Activity from Faculty of Physics and Applied Computer Science of the AGH University of Science and Technology in Cracow. The authors would also like to thank the anonymous referees for their careful reading of the paper and their contribution of useful suggestions that helped to improve this article.

Conflicts of Interest: The authors declare no conflict of interest.

Abbreviations

The following abbreviations are used in this manuscript:

CI	Computational Intelligence
ANN	Artificial Neural Networks
CNN	Convolutional Neural Network
FCN	Fully Connected Network
FCL	Fully Connected Layer
SA	Sensitivity Analysis
GSA	Global Sensitivity Analysis
LSA	Local Sensitivity Analysis
ReLU	Rectified Linear Unit

References

1. Rutkowski, L. Computational. In *Intelligence Methods and, Techniques;* Springer: Berlin/Heidelberg, Germany, 2008.
2. Dhillon, A.; Verma, G.K. Convolutional neural network: A review of models, methodologies and applications to object detection. *Prog. Artif. Intell.* **2020**, *9*, 85–112. [CrossRef]
3. De Wilde, P. *Neural Network Models: Theory and Projects;* Springer Science & Business Media: London, UK, 2013.
4. Haykin, S. Neural Networks: A comprehensive foundation. *Neural Netw.* **2004**, *2*, 41.
5. Kowalski, P.A.; Sapała, K.; Warchałowski, W. PM10 forecasting through applying convolution neural network techniques. *Air Pollut. Stud.* **2020**, *3*, 31–43 . [CrossRef]
6. Balcerek, J.; Pawłowski, P.; Dąbrowski, A. Classification of emergency phone conversations with artificial neural network. In Proceedings of the 2017 Signal Processing: Algorithms, Architectures, Arrangements, and Applications (SPA), Poznan, Poland, 20–22 September 2017; pp. 343–348.
7. Kiddon, C.; Zettlemoyer, L.; Choi, Y. Globally coherent text generation with neural checklist models. In Proceedings of the 2016 Conference on Empirical Methods in Natural Language Processing, Austin, TX, USA, 1–5 November 2016, pp. 329–339.
8. Zhang, J.; Zong, C. Deep Neural Networks in Machine Translation: An Overview. *IEEE Intell. Syst.* **2015**, *30*, 16–25. [CrossRef]
9. Roberts, C.W. *Text Analysis for the Social Sciences: Methods for Drawing Statistical Inferences from Texts and Transcripts;* Routledge: New York, NY, USA, 2020.
10. Wu, C.H. Behavior-based spam detection using a hybrid method of rule-based techniques and neural networks. *Expert Syst. Appl.* **2009**, *36*, 4321–4330. [CrossRef]
11. Gazeau, V.; Varol, C. Automatic spoken language recognition with neural networks. *Int. J. Inf. Technol. Comput. Sci.* **2018**, *10*, 11–17. [CrossRef]
12. Pierson, H.; Gashler, M. Deep Learning in Robotics: A Review of Recent Research. *Adv. Robot.* **2017**, *31*, 821–835. [CrossRef]
13. Fukushima, K.; Miyake, S. Neocognitron: A self-organizing neural network model for a mechanism of visual pattern recognition. In *Competition and Cooperation in Neural Nets;* Springer: Berlin/Heidelberg, Germany, 1982; pp. 267–285.
14. Bengio, Y.; Goodfellow, I.; Courville, A. *Deep learning;* MIT Press: Cambridge, MA, USA, 2017; Volume 1.
15. Li, H.; Kadav, A.; Durdanovic, I.; Samet, H.; Graf, H. Pruning Filters for Efficient ConvNets. *arXiv* **2016**, arXiv:1608.08710.
16. Molchanov, P.; Tyree, S.; Karras, T.; Aila, T.; Kautz, J. Pruning Convolutional Neural Networks for Resource Efficient Transfer Learning. *arXiv* **2016**, arXiv:1611.06440.
17. Kowalski, P.A.; Kusy, M. Sensitivity analysis for probabilistic neural network structure reduction. *IEEE Trans. Neural Netw. Learn. Syst.* **2017**, *29*, 1919–1932. [CrossRef] [PubMed]
18. Fock, E. Global sensitivity analysis approach for input selection and system identification purposes—A new framework for feedforward neural networks. *IEEE Trans. Neural Netw. Learn. Syst.* **2013**, *25*, 1484–1495. [CrossRef]
19. Kim, Y.D.; Park, E.; Yoo, S.; Choi, T.; Yang, L.; Shin, D. Compression of Deep Convolutional Neural Networks for Fast and Low Power Mobile Applications. *arXiv* **2016**, arXiv:1511.06530.
20. Harshman, R.A. Foundations of the PARAFAC procedure: Models and conditions for an "explanatory" multi-modal factor analysis. *UCLA Work. Pap. Phon.* **1970**, *16*, 1–84.
21. Zhou, M.; Liu, Y.; Long, Z.; Chen, L.; Zhu, C. Tensor rank learning in CP decomposition via convolutional neural network . *Signal Process. Image Commun.* **2019**, *73*, 12–21. [CrossRef]
22. Vaswani, A.; Shazeer, N.; Parmar, N.; Uszkoreit, J.; Jones, L.; Gomez, A.; Kaiser, L.; Polosukhin, I. Attention Is All You Need. *arXiv* **2017**, arXiv:1706.03762.
23. Zhang, S.; Wu, G.; Gu, J.; Han, J. Pruning Convolutional Neural Networks with an Attention Mechanism for Remote Sensing Image Classification. *Electronics* **2020**, *9*, 1209. [CrossRef]
24. Goyal, S.; Roy Choudhury, A.; Sharma, V. Compression of Deep Neural Networks by Combining Pruning and Low Rank Decomposition. In Proceedings of the 2019 IEEE International Parallel and Distributed Processing Symposium Workshops (IPDPSW), Rio de Janeiro, Brazil, 20–24 May 2019; pp. 952–958. [CrossRef]

25. Ren, H.; Xu, B.; Wang, Y.; Yi, C.; Huang, C.; Kou, X.; Xing, T.; Yang, M.; Tong, J.; Zhang, Q. Time-Series Anomaly Detection Service at Microsoft. In Proceedings of the 25th ACM SIGKDD International Conference on Knowledge Discovery & Data Mining; Association for Computing Machinery, New York, NY, USA, 4–8 August 2019; pp. 3009–3017. [CrossRef]
26. Kalchbrenner, N.; Grefenstette, E.; Blunsom, P. A Convolutional Neural Network for Modelling Sentences. In Proceedings of the 52nd Annual Meeting of the Association for Computational Linguistics, ACL 2014, Baltimore, MD, USA, 22–27 June 2014; Volume 1. [CrossRef]
27. Kim, Y. Convolutional Neural Networks for Sentence Classification. In Proceedings of the 2014 Conference on Empirical Methods in Natural Language Processing, Doha, Qatar, 25–29 October 2014. [CrossRef]
28. Vedaldi, A.; Zisserman, A. Deep Inside Convolutional Networks: Visualising Image Classification Models and Saliency Maps. *arXiv* **2013**, arXiv:1312.6034.
29. Renz, K.; Stache, N.; Albanie, S.; Varol, G. Sign language segmentation with temporal convolutional networks. In Proceedings of the ICASSP 2021–2021 IEEE International Conference on Acoustics, Speech and Signal Processing (ICASSP), Toronto, ON, Canada, 6–11 June 2021.
30. Szegedy, C.; Liu, W.; Jia, Y.; Sermanet, P.; Reed, S.; Anguelov, D.; Erhan, D.; Vanhoucke, V.; Rabinovich, A. Going deeper with convolutions. In Proceedings of the IEEE Conference on Computer Vision and Pattern Recognition, Boston, MA, USA, 7–12 June 2015; pp. 1–9. [CrossRef]
31. Saltelli, A.; Tarantola, S.; Campolongo, F.; Ratto, M. *Sensitivity Analysis in Practice: A Guide to Assessing Scientific Models*; Wiley: New York, NY, USA, 2004. [CrossRef]
32. Saltelli, A.; Ratto, M.; Andres, T.; Campolongo, F.; Cariboni, J.; Gatelli, D.; Saisana, M.; Tarantola, S. *Global Sensitivity Analysis: The Primer*; John Wiley & Sons: Hoboken, NJ, USA, 2008; Volume 304. [CrossRef]
33. Morris, M.D. Factorial Sampling Plans for Preliminary Computational Experiments. *Technometrics* **1991**, *33*, 161–174. [CrossRef]
34. Saltelli, A.; Annoni, P.; Azzini, I.; Campolongo, F.; Ratto, M.; Tarantola, S. Variance based sensitivity analysis of model output. Design and estimator for the total sensitivity index. *Comput. Phys. Commun.* **2010**, *181*, 259–270. [CrossRef]
35. Saltelli, A.; Bolado, R. An alternative way to compute Fourier amplitude sensitivity test (FAST). *Comput. Stat. Data Anal.* **1998**, *26*, 445–460. [CrossRef]
36. Saltelli, A.; Tarantola, S.; Chan, K. A Quantitative Model-Independent Method for Global Sensitivity Analysis of Model Output. *Technometrics* **2012**, *41*. [CrossRef]
37. Lauret, P.; Fock, E.; Mara, T. A Node Pruning Algorithm Based on a Fourier Amplitude Sensitivity Test Method. *IEEE Trans. Neural Netw.* **2006**, *17*, 273–293. [CrossRef]
38. Kaggle Datasets. Available online: https://www.kaggle.com/ (accessed on 23 January 2021).
39. Dua, D.; Graff, C. UCI Machine Learning Repository. 2017. Available online: http://archive.ics.uci.edu/ml/index.php (accessed on 23 January 2021).
40. Keras Datasets. Available online: https://keras.io/api/datasets/ (accessed on 23 January 2021).

Article

A New Feature Selection Method Based on a Self-Variant Genetic Algorithm Applied to Android Malware Detection

Le Wang [1], Yuelin Gao [2,*], Shanshan Gao [1] and Xin Yong [1]

[1] School of Computer Science and Engineering, Northern Minzu University, Yinchuan 750021, China; sdnywll@163.com (L.W.); 13008030953@163.com (S.G.); yongxin_Azure@163.com (X.Y.)
[2] Ningxia Province Key Laboratory of Intelligent Information and Data Processing, North Minzu University, Yinchuan 750021, China
* Correspondence: gaoyuelin@263.net; Tel.: +86-139-9510-0900

Abstract: In solving classification problems in the field of machine learning and pattern recognition, the pre-processing of data is particularly important. The processing of high-dimensional feature datasets increases the time and space complexity of computer processing and reduces the accuracy of classification models. Hence, the proposal of a good feature selection method is essential. This paper presents a new algorithm for solving feature selection, retaining the selection and mutation operators from traditional genetic algorithms. On the one hand, the global search capability of the algorithm is ensured by changing the population size, on the other hand, finding the optimal mutation probability for solving the feature selection problem based on different population sizes. During the iteration of the algorithm, the population size does not change, no matter how many transformations are made, and is the same as the initialized population size; this spatial invariance is physically defined as symmetry. The proposed method is compared with other algorithms and validated on different datasets. The experimental results show good performance of the algorithm, in addition to which we apply the algorithm to a practical Android software classification problem and the results also show the superiority of the algorithm.

Keywords: feature selection; machine learning; asexual; genetic algorithm; android malicious application detection

Citation: Wang, L.; Gao, Y.; Gao, S.; Yong, X. A New Feature Selection Method Based on a Self-Variant Genetic Algorithm Applied to Android Malware Detection. *Symmetry* **2021**, *13*, 1290. https://doi.org/10.3390/sym13071290

Academic Editors: Kóczy T. László and István A. Harmati

Received: 10 June 2021
Accepted: 14 July 2021
Published: 18 July 2021

Publisher's Note: MDPI stays neutral with regard to jurisdictional claims in published maps and institutional affiliations.

1. Introduction

Data classification is one of the tasks of data mining in the field of machine learning and in the framework of pattern recognition [1]; the quality of the data has a significant impact on the performance of these data mining methods. When training machine learning models, irrelevant, redundant, and noisy data have an enormous impact on the time and the spatial complexity of the machine and can also affect the algorithm's performance. Therefore, pre-processing techniques for data are necessary [2]. In machine learning classification tasks, the dataset's size determines the number of features in the dataset, but not all features are helpful for training classifier models, and high-dimensional features can instead lead to dimensionality disasters. Data dimensionality reduction methods include feature extraction(FE), where features are transformed into a smaller dimension, and feature selection(FS) [3], where features are selected from the complete set of features to build a subset of features without transformation [4]. The method chosen in this paper is feature selection, the aim of which is to identify the most distinct subset of features in the whole feature set and thus provide a suitable recognition rate for a particular classifier [5].

Traditional feature selection methods can be divided into three main categories: filter, wrapper and embedded algorithms. In filter algorithm, the feature selection phase is carried out independently of the training learner phase. The method uses traditional information theory, chi-square tests, mutual information and correlation coefficients to make a rough selection of features. For example, the Mutual Information based Feature

Selection method [6,7] and the Conditional Mutual Information Maximization [8] etc. These methods propose to pre-process the feature set and filter out the least relevant features to reduce the feature set's dimensionality, but due to the complexity of the formulae and the high time, complexity can only be used for smaller data sets and pre-classification. Wrapper algorithm differ from filter algorithm in that the performance results of the learner determine the selection of a subset of features. Embedded algorithm complete feature selection and learner training in the same optimization process. Wrapper algorithms select features mostly in conjunction with machine learning classifiers and intelligent algorithms. In this paper, the meta-heuristic algorithm decides whether a feature in the full set of features is added to the feature subset, and the performance of the algorithm determines the number of features selected, so the use of the algorithm to determine feature selection belongs to the wrapper algorithm.

Assuming that there are n features in the feature set, the search space for feature selection is 2^n . Feature selection being an NP-hard problem, traversing it to get all possible solutions is impossible in some cases [9]. Meta-heuristics have the advantages of high efficiency, superiority, and robustness compared to the traditional greedy algorithm hill-climbing algorithm and the ability to obtain a solution or several near-optimal solutions within a sufficient space and time scale, and are therefore increasingly used by researchers to solve complex optimization problems. As part of the meta-heuristic algorithm, the evolutionary algorithm is inspired by the phenomenon of biological evolution in nature. Dr. Holland first proposed the genetic algorithm in 1975 [10], which follows Darwin's 'survival of the fittest, natural selection' law of evolution, whereby populations are renewed by three leading evolutionary operators: selection, crossover, and mutation. However, in nature, species reproduce not only sexually in pairs but also in a few species that have only one parent and do not require gametes. The brood itself does not combine with sex cells to produce offspring such as lower multicellular animals, unicellular plants, algae, ferns, fungi, and bacteria. These species are characterized by a small number of species and a single community that can reproduce in a short period. However, this is both an advantage and a disadvantage; when there is a sudden change in the environment, the community's organisms die off in large numbers, indicating that the populations produced by asexual reproduction are not well adapted. In 2009, J Cantó et al. [11] proposed an asexual genetic algorithm for solving complex mathematical function maximization problems with two variables and optimizing the parameters of the chi-square test in model fitting. The algorithm does not require crossover operators to generate offspring, and offspring are renewed like the way bacterial cells are divided by a single parent randomly selecting different points within a narrow domain, and both parents are always retained if they are more suitable than the offspring. Experimental results showed that this codeless asexual genetic approach is more efficient than traditional genetic algorithms in solving continuous optimization problems, and it is also computationally cheap and requires fewer generations to reach a global solution. In 2010, Alireza Farasat et al. [12] built mathematical models based on the budding mechanism of asexual reproduction to solve optimization problems and decision problems. The experimental results proved the convergence of the algorithm. They verified that the asexual reproduction algorithm has excellent advantages in solving real-time decision problems by exploring the search space without limiting the convergence time and has superior performance compared to these swarm intelligence algorithms of PSO. In 2013, Anabela Simões et al. [13] proposed an asexual permutation genetic algorithm inspired by the DNA sequence structure discovered by Barbara McClintock in the 1950s. Unlike the simple permutation of a sexual mechanism, the asexual permutation genetic algorithm has single parents, while transposons and insertion points are made on a single individual. The genetic algorithm is compared to genetic algorithms with single, multiple, and random crossover operators and finds that it always finds better optima than other sexual reproduction algorithms for both large and small populations. In 2015, Mehrdad Amirghasemi et al. [14] proposed an effective asexual genetic algorithm to solve JSP problems, which combined asexual genetics, an elite pool, and tabu search, with the biased mutation to increase the diversity of the search space and update to the elite pool used to balance exploration and exploitation. The results also demonstrated the effectiveness and efficiency of using this asexual genetic algorithm in solving JSP problems.

The advantages of asexual genetic algorithms in solving optimization problems in different domains have been reviewed above. The genetic algorithm itself, a discrete coding approach, is naturally well equipped to solve feature selection problems, and there is a dearth of research on single asexual genetic algorithms for FS problems, so the following analysis is given in this paper to demonstrate that the algorithm performs equally well in solving FS problems. The main contributions of this paper are concluded as follows:

(1) The effect of population size on the genetic algorithm was verified.
(2) There is no crossover operator in the asexual genetic algorithm, so this paper verifies the effect of different mutation probabilities on the algorithm foe feature selection.
(3) The performance of the improved genetic algorithm is demonstrated in Android malicious application detection.
(4) The improved genetic algorithm is implemented for feature selection.

The rest of this paper is organized as follows. Section 2 introduces background and method. Section 3 proposes the algorithm of this paper. Section 4 applies the algorithm to the Android malware detection problem. Section 5 demonstrates the effectiveness of the algorithm proposed in this paper through experiments. Section 6 summarizes the article.

2. Background and Method

2.1. Feature Selection

The feature selection problem differs from traditional optimization problems. It is identified as a discrete binary problem where the search space is an n-dimensional lattice space of Boolean type, and the solution to feature selection is to display and update at each corner of the hypercube [15].

$$X_i = (x_{i1}, x_{i2}, \ldots.x_{iD}),\ x_{ij} \in \{0,1\} \tag{1}$$

where $x_{ij} = 1$ represents the j-th feature is selected into the i-th feature subset x_i, whereas $x_{ij} = 0$ means this feature is not selected.

Thus, the feature selection problem can be formulated as the following optimization problem:

$$\begin{cases} max\ f(X) \\ s.t.X = (x_1, x_2, \ldots.x_D), x_i \in \{0,1\} \\ 1 \leqslant \{X_i\} \leqslant D \end{cases} \tag{2}$$

where $\{X\}$ represents the set of selected feature subsets, i.e., a subset of features. $f(X)$ denotes the fitness of the selected feature subset X, which is the accuracy of the classification.

2.2. Genetic Algorithm

The purpose of feature selection is to select some of the most compelling features from the original features in order to reduce the dimensionality of the dataset. A subset of the selected features will result in higher classification accuracy [9]. As a class of optimization problems, many researchers have applied evolutionary algorithms and swarm intelligence optimization algorithms to this. The particle swarm optimization algorithm is a collaborative group-based search algorithm that simulates the foraging behavior of a flock of birds, using the individual extremum *pbest* and the group extremum *gbest* to find the algorithmic optimum (solution) [16,17]. The artificial bee colony algorithm is an optimization algorithm based on the honey bee colony's honey harvesting behavior [18]. The grey wolf optimization algorithm, inspired by the predatory behavior of grey wolf packs, finds optimal solutions through collaboration between wolf packs [19,20]. The pigeon flocking algorithm simulates pigeon homing behavior with minimal adjustment parameters and is easy to implement [21]. Genetic algorithm, one of the most classical evolutionary algorithms, is a learning method inspired by biology. It is a random search and optimization algorithm. Genetic algorithms balance the exploration and exploitation of algorithms through three major evolutionary operators: selection, crossover and mutation, preventing premature maturation of the algorithm and thus finding the optimal solution. It has now been widely used in various fields, such as circuit wiring problems, task scheduling, and machine learning classification tasks. For example, in 1998 Jing-Wein Wang et al. [22] proposed the use of a genetic algorithm as an evaluation function for feature selection, which largely improved the performance of the selected subset of features. The algorithm searches a huge candidate object space and finds the best performing objects according to the fitness function. Individuals with good fitness will be retained during the iteration. Flawed individuals are eliminated or selectively mutated to enter the next iteration. As the iterative process increases, the initial population is updated until the termination conditions are satisfied, or a certain threshold is reached to obtain the final individuals. The flow of the above algorithm to solve the FS problem is shown in Figure 1.

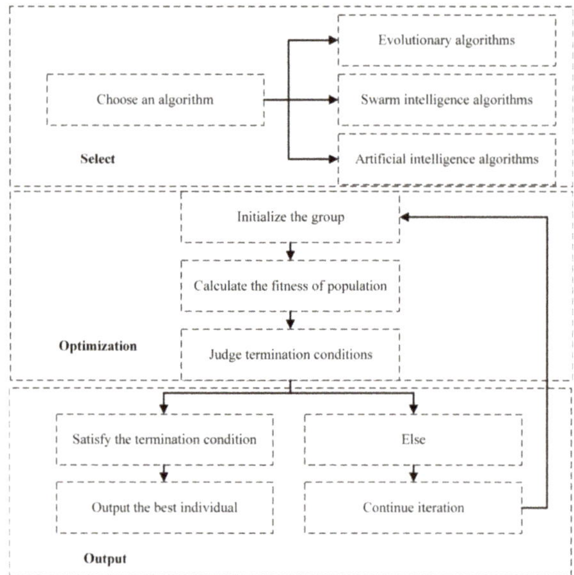

Figure 1. Intelligence algorithm optimization model.

The standard genetic algorithm simulates the evolution process of natural organisms, and the selection operator embodies the environmental selection process of "natural selection, the survival of the fittest" in the evolution of organisms. The crossover and mutation operators play a key role in population renewal during the iterative process. The crossover operator simulates the mating process of individuals in nature, thus increasing the population diversity and theoretically improving the global search capability of the algorithm; unlike the crossover operator, the mutation operator simulates the genetic mutation of individuals in the population by mutating a gene position of an individual (chromosome) from the individual itself, thus improving the local search capability of the algorithm.

2.3. Random Forest Algorithm

The random forest algorithm is part of a large branch of machine learning currently known as ensemble learning. As the name suggests, the algorithm is derived from the decision tree algorithm, where a number of weak classifiers make predictions and then the final strong classifier gives the result. The more correlated any two trees in the forest are, the greater the error rate, while the stronger the classification ability of each tree, the lower the error rate of the whole forest, i.e., the random forest integrates the advantages of each tree's classification result. The classifiers before the fourth section all use the random forest algorithm.

3. The Proposed Self-Variant Genetic Algorithm (SV-GA)

3.1. Theoretical Basis

In general, the feature selection problem is encoded in binary, and even if the feature values correspond to real values, they are mapped to binary space when solving the problem.To theoretically demonstrate the redundancy of the crossover operator, we first simulated the process of chromosome change in the algorithm.

A population $M(t)$ with n chromosomes is randomly generated, and each chromosome is composed of a string p of size N. Each position i in the string represents the locus of each chromosome, $p\_i = 1$ means that the feature is selected, $p\_i = 0$ means that the feature is not selected, that is, the current gene position is assigned a value of 1, and the number of 1 in the chromosome gene position represents the number of features carried by the current chromosome. The chromosome initialization code is shown in Figure 2.

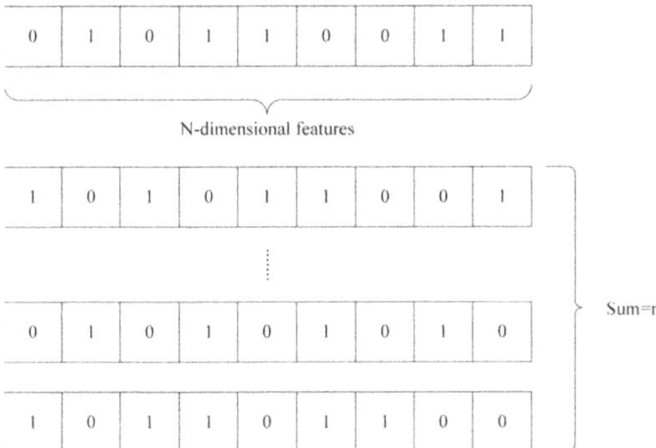

Figure 2. Feature encoding method.

To better understand the advantages of asexual genetic algorithms in feature selection, this paper proposes to analyze its principles by fictionalizing different individuals.

Individual 1 with random 0–1 sequence, as shown in Figure 3:

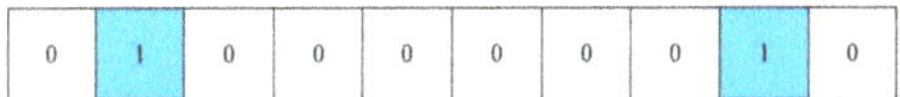

Figure 3. Individual 1.

Individual 2 with random 0–1 sequence, as shown in Figure 4:

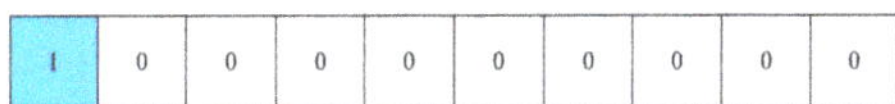

Figure 4. Individual 2.

Individual 3 with random 0–1 sequence, as shown in Figure 5:

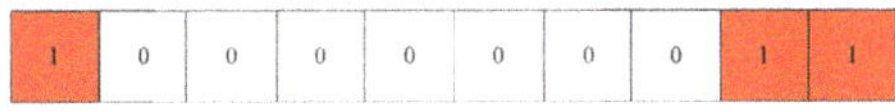

Figure 5. Individual 3.

In a standard genetic algorithm, two different or identical chromosomes can be crossed by a single point to obtain a new and different chromosome from the parent, that is, randomly find a cut, exchange its head or tail to obtain a new individual, and reproduce the process biological reproduction. The single-point crossing process of the above two benign individuals is as follows:

The single point of intersection of individual 1 and individual 2 is shown in Figure 6:

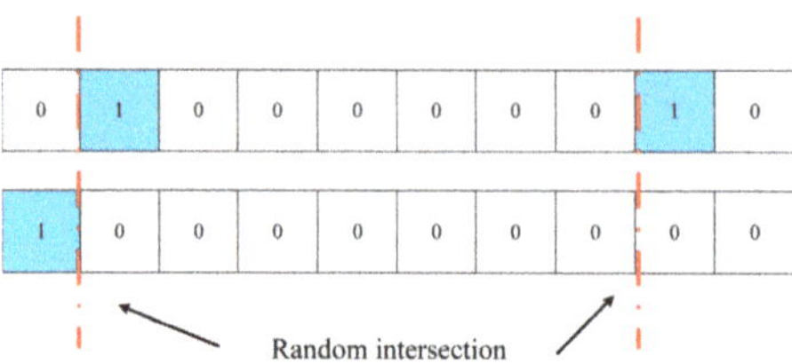

Figure 6. Schematic diagram of single point crossing.

Get new offspring individuals 3, 4, as shown in Figure 7:

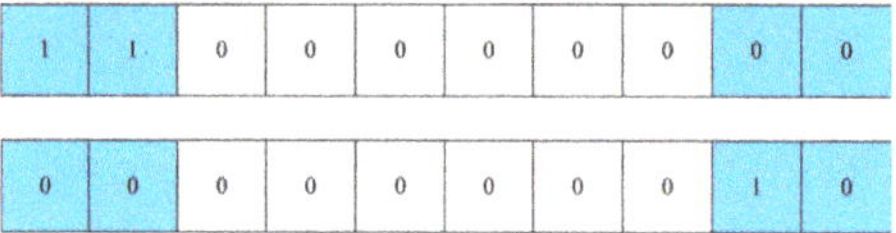

Figure 7. Crossingto produce new individuals.

The Individual 2 mutates to obtain the Individual 3 situation, as shown in Figure 8:

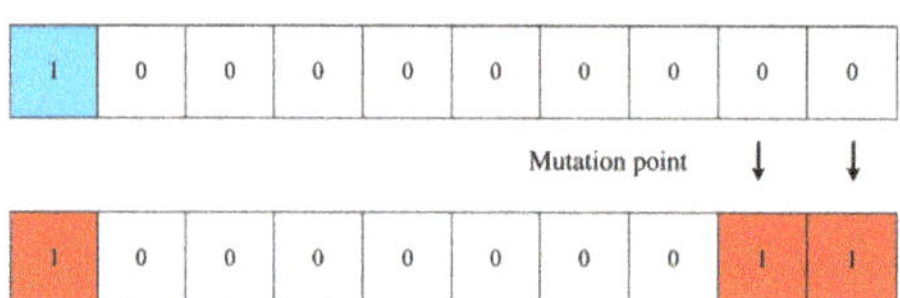

Figure 8. Schematic representation of the changes in individuals after mutation.

From the above crossover variation process, it is clear that the paternal chromosome crossover may only change the value of one gene locus, with no change in the rest of the gene locus other than the crossover point, and the same is true for two-point and multi-point crossovers, where only the value of the gene near the crossover point changes.

3.2. Fitness Function

In machine learning classification tasks, the accuracy of the classification is usually used as a fitness function. Still, for feature selection problems, the fitness function is determined not only by considering the classification accuracy but also by taking into account the number of feature subsets. When different algorithms have the same accuracy, it is better to choose the algorithm with fewer features. The fitness function is defined as follows:

$$Fitness(X_i) = f(L_i), f(L_i) = \frac{correctly\ predicted\ samples}{total\ number\ of\ samples} \tag{3}$$

$$Fitness = \alpha f(L_i) + \beta \frac{n}{N} \tag{4}$$

where X_i and L_i are the i-th individual and corresponding feature subset, $f(L_i)$ is the accuracy of the random forest classifier, n is the number of features selected, and N is the number of features in the dataset, where α usually takes 0.99 and β takes 0.01 as know in [23].

3.3. The Algorithm Flow of the SV-GA

According to the above description, we can find that in solving the feature selection problem, what affects the accuracy of the classifier is the number of selected feature bits in the individual, i.e., the number of sign bits of 1. Then the corresponding ones in the genetic algorithm are the gene bits of each chromosome. When the chromosome undergoes mutation, each gene bit can be operated by changing 0 to 1 and 1 to 0. At this point we can roughly assume that a single mutation operation can satisfy the needs of the features when performing classification. The steps of its application in feature selection are as follows:

Step1: Initialize the population N, t is the number of current iterations, and the total number of features is P.

Step2: Calculate the fitness value of individuals in the current population, and obtain the fitness value $F(n)$ of each chromosome after n calculations, that is, obtain the fitness value of each individual with different characteristics.

Step3: Selection. In the SV-GA algorithm, the selection method is tournament selection, and the number of individuals selected each time is 3, and the tournament selection method is replacement sampling. Three individuals are randomly selected from the population to calculate their fitness, and the better individual directly enters the next generation.

Step4: Mutation. According to a certain mutation probability, a mutation operation is performed on each chromosome in the population, and the mutated individuals differ from those after the selection operation and carry different characteristics, and these individuals with different characteristics constitute a new population.

Step5: Algorithm termination conditions. If it satisfies the artificially set number of iterations, the algorithm terminates, set the number of iterations then output the feature subset selected by the algorithm and skip to *Step6* at the end of the iteration. If not, then execute Step2.

Step6: Output results. The individual with the highest fitness value is output and the feature with a gene position of 1.

3.4. Computational Complexity Analysis

In SV-GA, the factors affecting the time complexity of the algorithm are not only the population size N, but also the number of iterations t. The feature selection problem, as a class of optimization problems, aims to improve the classification accuracy by reducing the number of features for training, and the dimensionality D of the features in the sample set also affects the efficiency of the algorithm in the process of algorithm optimization. In summary, the time complexity of the algorithm can be summarized as $O(t * N * D)$, and the number of individuals in the population and the feature dimension are the main factors affecting the computational complexity.

3.5. Numerical Analysis

The SV-GA changes the iterative process of the traditional genetic algorithm, and increases the diversity of individuals in the population by changing the population size. The impact of mutation operations on the performance of the algorithm should not be underestimated. In this section we first verified the effect of different mutation probabilities on the algorithm, comparing the performance of GA and SV-GA, and then we verified the effect of different population sizes on the algorithm. All results are mean results from 10 independent runs with 100 iterations.

3.5.1. Datasets

To verify the effectiveness and applicability of the algorithms, this paper uses SV-GA to test on different UCI standard binary classification test datasets [24]. Table 1 provides the name of the dataset, the number of samples contained in the dataset, the total number of features, and the number of features selected by different algorithms.

Table 1. Dataset Properties.

No.	Dataset	Number of Samples	Number of Features
D1	SPECT Heart	267	22
D2	CMC	962	9
D3	Sonar	207	60
D4	Credit6000	6000	65
D5	Heart-statlog	270	13
D6	Spambase	4600	57

3.5.2. Results with Different Mutation Rate

In the meta-heuristic algorithm, the search operator will deal with the overall selection pressure, convergence problem, randomization, and diversity, which are all dedicated to exploration and exploitation [25]. Based on the characteristics of the feature selection problem, we retain the selection and mutation operators in the SV-GA algorithm. On the one hand, we change the number of initial population sizes to increase the diversity of the population, i.e., to improve the global search capability of the algorithm, and on the other hand, we experimentally verify the effect of the mutation rate on the accuracy while keeping the population size constant. This subsection is to verify the effect of mutation probabilities on the performance of SV-GA.

To ensure that the variables are unique, the population size for the results of the iterative curve shown in Figure 9 is set to 30 (pop = 30). The results on the six datasets show that the algorithm performs poorly when the mutation rate is set to 0.01, 0.02 and 0.6. The results on all six datasets are the worst when the rate is 0.01. It is not difficult to analyze that the performance of the genetic algorithm with the crossover operator omitted depends on the magnitude of the variance, and when the variance is set to 0.01, the algorithm is close to 0 variance, i.e., the features involved in the classification are initialized features, the structure of the population does not change significantly and the algorithm falls into a local optimum solution. Overall, on both the SPECT heart and the Spambase, the algorithm performed optimally when the probability of variation of the algorithm was set to 0.2, when the population was best adapted.

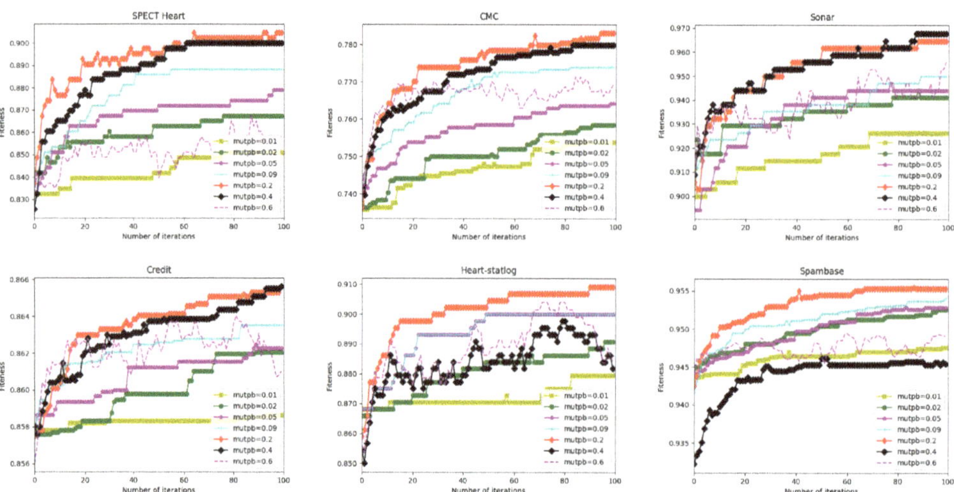

Figure 9. The evolutionary curves of different mutation rate for datasets.

3.5.3. Comparison between GA and SV-GA

Table 2 shows the parameter settings for GA and SV-GA when comparing experiments on different UCI datasets.

Table 2. Parameter settings.

Algorithms	Parameter	Value
	population	30
GA	Selection	Tournament Tournize = 3
	Crossover	One-point Cxpb = 0.5
	Simple Mutate	Mupb = 0.2
	population	30, 50, 100
SV-GA	Selection	Tournament Tournize = 3
	Simple Mutate	Mupb = 0.2

The iteration curves shown in Figure 10 show a comparison of the algorithms for GA and SV-GA. The algorithms were both iterated 100 times, with a population size of 30 for both GA and SV-GA and a mutation probability of 0.2, with a crossover probability of 0.5 for the former. From the experimental results, it is easy to see that the fitness value of SV-GA is increasing with the number of iterations. The effect is more evident on the SPECT Heart, CMC, Sonar, Credit6000, and Spambase datasets, while in Heart-statlog, the fitness value of the algorithm is unstable if not as good as GA. Still, the algorithm later convergence is faster than the former. It is easy to see from the graphs of the iterations of the two algorithms on different datasets that the asexual genetic algorithm has an advantage in solving the feature selection problem.

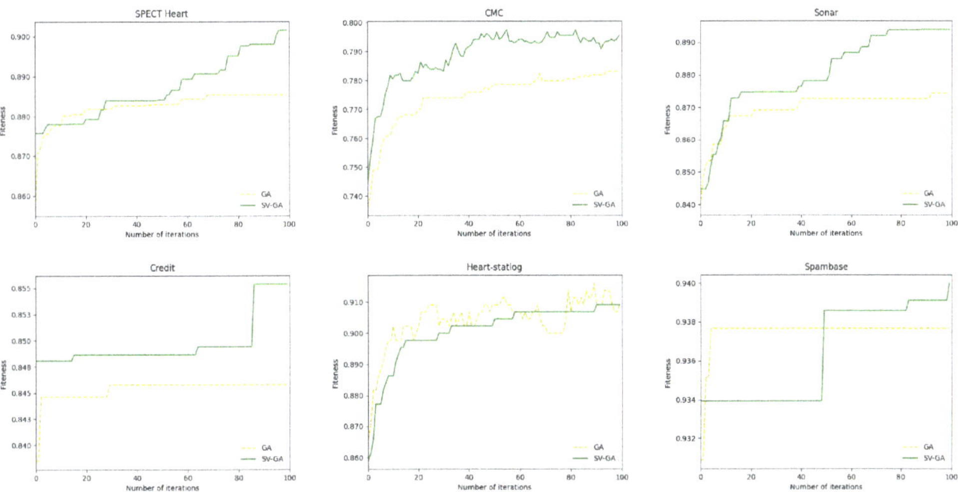

Figure 10. The evolutionary curves of GA and SV-GA.

3.5.4. Symmetry Theory and Algorithms

Symmetry is defined in both mathematics and physics. Mathematically, symmetry is defined as graphical symmetry and numerical palindromes; physically, symmetry is defined as invariance after some operation, both in time and space. In the asexual genetic algorithm proposed in this paper for solving the feature selection problem, the population size does not change due to a change in strategy. For example, if the number of individuals in the initialized population is 30, then as the number of iterations increases, the number of individuals remains at 30. The optimal solution is the best value (maximum or minimum) among the 30 individuals when solving the optimization problem. Therefore, in the following subsection, we verify the algorithm results for population sizes of 30, 50, and 100, respectively.

3.5.5. Results of the Population Size

A comparison of the fitness values of the standard GA with those of the SV-GA algorithm for different population sizes is shown in Table 3, where the SV-GA algorithm selects the operator for tournament selection and the mutation probability is set to the best mutation probability of 0.2 as experimentally demonstrated above. Analysis of the values in this table shows that the algorithm with a population size of 100 performs better than the algorithm with population sizes of 30 and 50 while keeping the variation probability constant.

Table 3. Mean and standard deviation results of acc on fitness and its competitors for six Datesets.

	GA		SV-GA(30)		SV-GA(50)		SV-GA(100)	
	Mean	**(Std)**	**Mean**	**(Std)**	**Mean**	**(Std)**	**Mean**	**(Std)**
D1	0.865	(0.006)	0.882	(0.003)	0.887	(0.008)	**0.897**	(0.006)
D2	0.752	(0.008)	0.747	(0.004)	0.748	(0.006)	**0.755**	(0.007)
D3	0.868	(0.015)	0.871	(0.012)	0.882	(0.013)	**0.887**	(0.013)
D4	0.846	(0.002)	0.849	(0.001)	0.849	(0.002)	**0.850**	(0.002)
D5	0.891	(0.018)	0.894	(0.020)	0.898	(0.016)	**0.904**	(0.012)
D6	0.938	(0.002)	0.939	(0.001)	0.940	(0.002)	**0.943**	(0.002)

The reason for this is that the local search capability of the algorithm is stable when the variation probability is constant, while the more individuals within the population, the greater the diversity of the population, i.e., the greater the global search capability of the algorithm. By varying the population size and variation probability, the SV-GA balances the exploitation and exploration capability of the algorithm, thus facilitating the algorithm to find the global optimal solution.

4. Android Malicious Application Detection Based on SV-GA Algorithm

Feature selection is an important step in processing classification tasks and in the pre-processing phase of data mining, with the aim of improving the accuracy of the classifier. In order to verify the performance benefits of the proposed algorithm, we apply it to the Android malicious application detection problem and propose a framework process to solve the problem.

4.1. Android Malicious Application Detection

With the continuous advancement of time and the increase in the number of third-party application markets, many researchers have explored the detection methods of Android malicious applications from a software perspective to achieve the purpose of protecting system stability. In this paper, we focus on solving the security problem at the application level of Android by compiling the corresponding feature sets based on the source code information obtained by the decompiler tool and feeding them into the classifier model for training to classify benign and malicious applications. Due to the high dimensionality of the acquired feature attributes, it inevitably causes a dimensional disaster or increases the time and space complexity of model training, which affects the efficiency of Android malicious application detection. It is vital to choose a suitable feature selection method to reduce the number of features while improving the accuracy of the classification model.

Malicious application detection methods are frequently updated and changed with the deepening of research. The main research trends are in the following two aspects: one is based on the improvement of a single machine learning classification algorithm; the other is based on the study of feature selection methods. Machine learning methods are currently the most widely used technical means in the artificial field. The classification methods in supervised learning play a driving role in the detection of Android malicious applications (such as the KNN algorithm, Naive Bayes algorithm, Logistic Regression, and Decision Tree algorithm). The disadvantage is that the training model speed needs to be improved, and it can only make a simple judgment of the malicious software that has already appeared on the market and cannot realize the detection of unknown types of applications. Feizollah A, Nor B, Salleh R et al. [26] evaluated the performance of K-means and Mini batch K-means clustering algorithms in Android malware detection and analyzed the network traffic of benign and malware on two algorithms. The result showed that the overall performance of the Mini batch K-means clustering algorithm was better than the K-means algorithm. Nath, Hiran V et al. [27] applied the classification algorithm in machine learning to features such as n-gram model and byte sequence extracted from malware. The algorithm included classification methods such as decision trees and boosted decision trees. The results proved the machine learning classification algorithm could realize the simple classification of malware. However, only using machine learning algorithms to judge the quality of the application is slightly thin. Rajesh Kumar et al. [28] proposed a method based on the combination of probabilistic statistical analysis and machine learning algorithms to reduce the dimensionality of features and achieved classification between known and unknown benign and malicious software.

Android malicious application detection based on feature selection method is divided into static analysis and dynamic analysis methods [29]. Static analysis involves obtaining the source code of an Android application by decompiling software without running the application, analyzing it to extract relevant syntactic and semantic information, permission information in configuration files, intent, the corresponding API calls, etc., coding and mapping its code integration into vector space, and combining it with machine learning classification in order to achieve malicious application classification. In contrast, dynamic analysis is similar to the black-box testing of software. The source code structure is not taken into account, and only relevant features are obtained during the installation or use of the application, such as network traffic analysis, application power consumption, user behavioral features. Dynamic analysis has the advantage of a large feature selection space and a wide range of input classifiers. For example, Zarni Aung et al. [30] extracted single permission as a feature for training, designed and implemented a framework based on machine learning technology classify malware and benign software. Shanshan Wang et al. [31] conducted an in-depth study on the behavior of network traffic generated by the application during use, mapped the mobile terminal traffic information flow to the server-side, analyzed the network traffic characteristics, and combined the C4.5 algorithm to complete the detection of malicious applications. Du W, Yin H et al. [32] proposed a method to describe Android malware that relied on API calls and package-level information in bytecode and determined the category of unknown application software based on known Android malicious applications. Compared with the classifier based on permission features, the KNN classifier's accuracy was as high as 99%. Daniel Arp et al. proposed a lightweight detection framework: Drebin [33]. This method extensively collected application characteristics (permissions, hardware combinations, etc.) obtained from static analysis and mapped them to the joint vector space, using traditional PCA to reduce dimensional method selection features. The biggest advantage of this framework is the ability to identify malicious applications on smartphones directly. Wang W, Gao Z, Zhao M et al. [34] proposed an Android malicious application detection model: DroidEnsemble. The classification features in the model analyze static string features such as permissions in each application code pattern and include structural feature s such as control flow graphs and data flow graphs, such as function call graphs. The results of classifying these two types of features show that the model's detection accuracy is greatly improved, while the false alarm rate is also reduced. The approach based on feature selection has certain advantages, but it can lead to the high dimensionality of the feature combinations, leading to the high complexity of the algorithm's training process in space and time and affects the accuracy of the machine learning and data mining methods. Therefore, the selection of features with good detection performance is key to the method.

4.2. Construction of Feature Sets

Feature selection aims to reduce the number of features used for classification while maintaining classification accuracy [35]. Based on the dynamic and static analysis mentioned above, we have chosen the static analysis method to classify the software. One is that the static analysis method is not only simpler but also less harmful to mobile devices. When using a mobile device, the application will request a permission to respond during installation and the system will simultaneously check whether the permission is invoked. Permissions become one of the indispensable static features for detecting malicious Android applications [36].

4.3. APK Pre-Processing

The Android application package (APK) of the third-party application market is not presented in source code but is similar to the packaged file format (zip). In order to obtain the information in the package, it is necessary to use a decompiler tool to realize the pre-processing of APK decompression. The tool used in this article, Apktool [37], is a lightweight decompilation tool, a closed binary Android application tool, which can decode resources and applications into the most primitive state of java source code, and automatically realize file structure processing. It can be used locally and supports multiple platform analysis applications. As shown in Figure 11 is the file resource list obtained by decompiling a real Android application using this apktool tool.

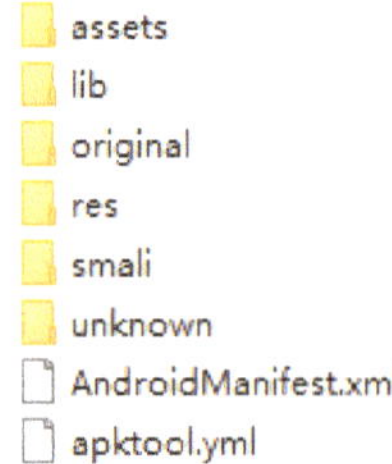

Figure 11. Example of file resource list.

After decompiling the APK, it can see the permission information in the total configuration file Androidmanifest.xml. Figure 12 shows a partial list of permissions for a test APK. P is the total set of possible permissions that some applications in the android platform can request. Moreover, each android application is represented in the framework of the required permission set. Therefore, assuming that the size of P is N, each application is represented by a binary string p of size N, where each position i of the string represents the i-th permission in a set of possible permissions, so that $p\_iin0, 1$. If the application does not need permission i, then $p\_i = 0$; if the application requires permission i, then $p\_i = 1$.

```
<uses-permission android:name="android.permission.READ_PHONE_STATE"/>
<uses-permission android:name="android.permission.GET_ACCOUNTS"/>
<uses-permission android:name="android.permission.WAKE_LOCK"/>
<uses-permission android:name="android.permission.ACCESS_COARSE_LOCATION"/>
<uses-permission android:name="android.permission.ACCESS_FINE_LOCATION"/>
<uses-permission android:name="android.permission.MOUNT_UNMOUNT_FILESYSTEMS"/>
<uses-permission android:name="android.permission.WRITE_EXTERNAL_STORAGE"/>
<uses-permission android:name="android.permission.RECORD_AUDIO"/>
<uses-permission android:name="android.permission.VIBRATE"/>
```

Figure 12. Example of file resource list.

4.4. Coding

When classifying Android malware, permission features are binary coded in such a way that if total permission sets of Android applications are defined as P, then the permissions applied by each software are $N(N <= P)$, coded one by one, and if a feature in permission set P appears, then a permission bit in N is flagged as 1, otherwise it is 0. An example of matching permissions for a single application is shown below :

0,0,0,0,0,0,0,0,0,0,1,0,1,0,0,0,0,0,0,0,0,0,0,0,0,0,1,0,0,0,0,1,0,0,0,1,0,0,1,0,0,0,0,0,1,0,0,1, 0,0,0,0,1,0,1,1,0,0,0,1,1,0,0,0,0,0,1,0,1,0,1,0,0,0,0,0,1,0,1,1,0,0,1,0,1

There are 88 permission bits in this example. The exact number of settings is described in the numerical experiments section. When each benign application is developed, the permissions applied are very few (compared to the official permission set). The number of "1" s in the gene position of each chromosome is relatively small. The number of permissions requested by malicious applications is greater than or equal to that of benign applications.

4.5. System Framework

This paper proposes a system model based on the permissions requested by Android applications, including three modules: decompilation, feature selection, and application classification. The block diagram of Android malicious application detection is shown in Figure 13. The proposed model consists of the following modules:

1. Decompilation module. Use the decompilation tool "Apktool" [37] to decompress each application package file to obtain the total configuration file Androidmanifest.xml of the application and obtain the permission list for each APK. All the extracted permissions are used as the original feature set.
2. Feature selection module. This module optimizes the original data set and selects the feature subset that dramatically impacts the classification effect. The original subset is randomly selected as the initial iterative group, then iterated through the improved genetic algorithm to screen the group, and finally, get the set of permission features that optimize the classification effect.
3. Classification module. The feature data set extracted by the decompiler module is selected by an asexual genetic algorithm to obtain a feature subset, which is fed into a machine learning classifier for training, and its classification accuracy judges the feature selection method to achieve the classification of Android software.

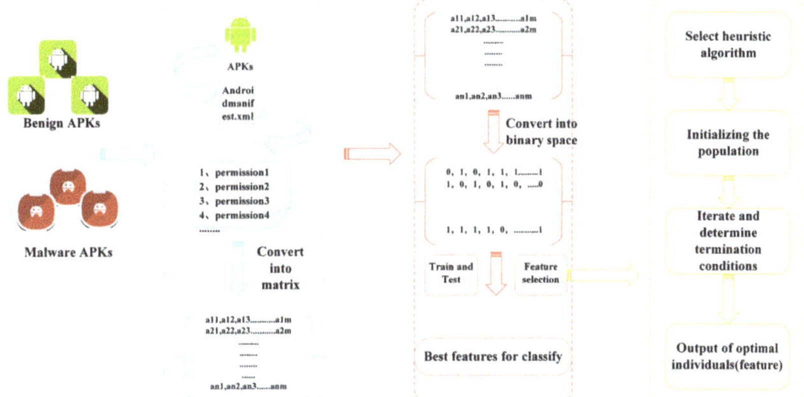

Figure 13. System model diagram.

5. Experimental Research

In this subsection, the feasibility of the asexual genetic algorithm for binary feature selection is further demonstrated with experimental data using different data sets and comparisons between different feature selection methods.

5.1. Numerical Verification

There are 1788 benign samples and 932 malicious samples in the experimental data of this paper. Among them, 532 malicious samples are from the Canadian Institute of Cyber Security [38], the samples contain typical Android platform malware, such as malicious ransomware applications, threatening SMS applications, and advertising applications; there are also 400 malicious samples from the data set of Dr. Wang's repository (http://infosec.bjtu.edu.cn/wangwei/?page_id=85 (accessed on 17 July 2020)). Total 188 benign Android applications, mainly from Google play and Xiaomi App Market. After security testing, these apps are released to the app market and are therefore deemed benign and safe. Besides, there are 1600 benign samples from Dr. Wang's database.

5.1.1. Permission Description

The number of permissions matched by the samples from Canadian Institute, Google play market, and Xiaomi app market is the official 144 permissions, while the number of permissions given by Dr. Wang's dataset is the most frequently selected 88 permissions after filtering, so the number of permissions for unified features in this paper is set to 88. The 88-feature datasets are input to logistic regression (LR), random forest (RF), Gaussian Naive Bayes (GNB) and K-nearest neighbor (KNN) classifiers for training and testing, respectively.

5.1.2. Classifier Model Evaluation

The primary objective when classifying Android applications is to flag malicious applications from the list, as only malicious software poses a risk to the system and user security, and not all software can be correctly classified when the classifier is trained, so we need an confusion matrix to describe the different types of errors and measure the severity of the errors separately. In the confusion matrix, as shown in Table 4, each column represents the instances in a predicted class, and each row represents the instances in an actual class. A specific table layout allows visualization of the performance of the classifier [17]. In this article, the positive class represented malicious applications, and the negative class represented benign applications.

Table 4. Classification results of the four classifiers under 88 permission features.

	Malware Class	**Benign Class**
Malware prediction	TP	FP
Benign prediction	FN	TN

Let *TP* (True Positive) be the number of malicious applications correctly predicted as Android malware. *FP* (False Positive) be the number of benign applications incorrectly predicted as Android malware. *FN* (False Negative) is the number of malicious applications that are incorrectly detected as benign applications, and *TN* (True Negative) be the number of benign applications correctly detected as benign applications. The following is the evaluation standard formula of the classifier:

Sample Accuracy (*ACC*) represents the percentage of the overall data set that is correctly classified. The higher the *ACC* value, the better the classification effect. It is defined as follows:

$$ACC = \frac{TP + TN}{TP + TN + FP + FN} \quad (5)$$

True positive rate (*TPR*) represents the probability that a positive sample will be correctly predicted as positive. The higher the value, the more effective the classifier is. It is defined as follows:

$$TPR = \frac{TP}{TP + FN} \quad (6)$$

False Positive Rate (FPR) represents the probability that a negative sample is falsely predicted to be positive. The higher the value, the worse the effect of the classifier. It is defined as follows:

$$FPR = \frac{FP}{FP + TN} \tag{7}$$

Precision (P) represents the probability that a positive sample is predicted to be positive, and is defined as follows:

$$P = \frac{TP}{TP + FP} \tag{8}$$

Recall is the probability that a positive sample is predicted to be positive, and the equation is the same as the true rates. It is defined as follows:

$$Recall = \frac{TP}{TP + FN} \tag{9}$$

F1-score ($F - S$) This indicator considers both precision rate and recall rate, so that both reach the highest at the same time, defined as follows:

$$F - S = \frac{2 \cdot P \cdot Recall}{P + Recall} \tag{10}$$

In order to find a classifier that makes the best classification performance among many machine learning classification models, this paper uses unselected (unpreprocessed) datasets of features input to different classifiers, relying on the goodness of the above classification metrics to determine the classification model applied in the following comparison experiments. The experimental results are shown in Table 5.

Table 5. Classification results of the four classifiers under 88 permission features.

No.	Classifier	ACC	TPR	FPR	$F - S$
1	LR	0.901	0.833	0.097	0.372
2	RF	0.867	0.958	0.136	0.337
3	GNB	0.955	0.958	0.136	0.337
4	KNN	0.933	0.208	0.018	0.247

The purpose of feature selection is to improve the classification accuracy while reducing the number of features for training the classification model; therefore, high precision and high true rate become the indicators for judging the good and bad pairs of classification models. The experimental results show that the GNB has the highest ACC and TPR and the model's best overall performance when classifying the dataset without feature selection.

5.2. Discrete and Continuous Algorithms

In a feature selection optimization problem, the search space is a hypercube, and all solutions lie only at values of 0 or 1 [20]. Genetic algorithms do not require changes to them in solving FS problems due to their natural discrete encoding. In contrast, in swarm intelligence algorithms such as GWO, WOA, and ACO, where the search range of the algorithm is a continuous space, many researchers have proposed a binary form of the algorithm in order to enable optimization of discrete problems. Kennedy proposed a binary version of PSO in 1997 [39], and much work has been done around this version. The algorithm uses a sigmoid function to map a vector in continuous space to a two-dimensional space, and the mapping equation is shown below.

$$S\left(X_{ij}^{t}\right) = \frac{1}{1 + e^{-X_{ij}^{t}}} \tag{11}$$

$$X_{ij}^{t+1} = \begin{cases} 1 & if\ S\left(X_{ij}^t\right) > \sigma \\ 0 & otherwise \end{cases} \tag{12}$$

$$X_{ij}^{t+1} = \begin{cases} \left(X_{ij}^t\right)^{-1} & if\ T\left(X_{ij}^t\right) > \sigma \\ X_{ij}^t & otherwise \end{cases} \tag{13}$$

In addition to this, four S-shaped and V-shaped mapping functions were introduced by Shahrzad Saremi et al. [40] to transform the continuous space. The equation for the change in X_{ij}^{t+1} when using the S-shaped mapping function is Equation (12) and for the V-shaped mapping function is Equation (13). The mapping functions are shown in the Table 6. Any of the continuous optimization algorithms can achieve binary conversion by means of S-shaped and V-shaped map functions.

Table 6. S-shaped and V-shaped mapping functions.

S-Shaped		**V-Shaped**	
S1	$T(x) = \frac{1}{1+e^{-2x}}$	V1	$T(x) = \left\lvert \frac{\sqrt{2}}{\pi} \int_{-0}^{(\sqrt{\pi}/2)x} e^{t^2} dt \right\rvert$
S2	$T(x) = \frac{1}{1+e^{-x}}$	V2	$T(x) = \lvert tanh(x) \rvert$
S3	$T(x) = \frac{1}{1+e^{-(x/2)}}$	V3	$T(x) = \left\lvert (x)/\sqrt{1+x^2} \right\rvert$
S4	$T(x) = \frac{1}{1+e^{-(x/3)}}$	V4	$T(x) = \left\lvert \frac{2}{\pi} arctan(\frac{\pi}{2}x) \right\rvert$

5.3. Comparison Algorithm and Their Parameter Setting

The purpose of feature selection is to select a subset of feasible features by eliminating irrelevant, redundant, or noisy features [41]. In order to evaluate the proposed algorithm, a number of existing feature selection algorithms were selected, such as the grey wolf optimization algorithm(GWO) [42], the whale optimization algorithm(WOA) [43] and the ant colony algorithm(ACO) [44]. The methods used in evolutionary algorithms to solve the feature selection problem are as follows:

- Traditional genetic algorithms [10];
- A new binary version of the grey wolf optimization algorithm [23];
- S-type binary whale optimization algorithm [45];
- Ant colony algorithm for binary encoding [46]

In order to be fair, the proposed algorithm is compared with these algorithms under the same parameter settings. The population size for all algorithms is 30, except for the population size for SV-GA, the number of runs is equal to be 10, the iterations is equal to 100. In [46], this paper draws on the parameter settings of α and β in the ant colony algorithm tested in that paper to get the best results, so that $\alpha = 1$ and $\beta = 3$, for comparison with the algorithm in this paper. To verify the performance of the proposed algorithm, we divide each dataset into a training set and a test set, e.g., k-fold cross-validation, dividing the sample set into k subsamples of equal size, selecting k−1 subsamples as the training set and all the remaining subsamples as the test set. Among the many supervised classifiers, we choose the better-performing GNB.

5.4. Experimental Results and Analysis

5.4.1. Fitness Performance

In the simulation experiments section, we compare the SV-GA algorithm with population sizes of 30, 50, and 100 with the GA, GWO, ACO, and WOA algorithms. All experimental data are mean results of independent runs.

Table 7 shows the statistical results using Equation (4) as the fitness function. To show that our proposed algorithm's performance is significantly better than that of the comparison algorithms, we used a non-parametric statistical test: Wilcoxon rank-sum test with a significance level of $\alpha = 0.05$. The null hypothesis is that the proposed algorithm is not significantly different from the comparison algorithm, and the alternative hypothesis is that the proposed algorithm is significantly different from the comparison algorithm. We use the symbols $+, =, -$ to indicate that the proposed algorithm's performance is significantly better, no significant difference, and significantly inferior to the corresponding comparison algorithm.

Table 7. Comparison of Fitness results on different algorithms for different datasets.

Datasets	GA	SV-GA(30)	SV-GA(50)	SV-GA(100)	GWO	ACO	WOA
	Mean(std)	**Mean(std)**	**Mean(std)**	**Mean(std)**	**Mean(std)**	**Mean(std)**	**Mean(std)**
Android Dataset	0.922(0.004)+	0.927(0.003)+	0.930(0.003)+	0.936(0.002)	**0.947**(0.005)−	0.935(0.004)+	0.925(0.004)+
SPECT Heart	0.865(0.006)+	0.882(0.003)+	0.887(0.008)+	**0.897**(0.006)	0.801(0.033)+	0.830(0.029)+	0.856(0.012)=
CMC	0.752(0.008)=	0.747(0.004)+	0.748(0.006)=	**0.755**(0.007)	0.647(0.028)+	0.644(0.048)+	0.733(0.009)+
Sonar	0.868(0.015)+	0.871(0.012)+	0.882(0.013)=	**0.887**(0.013)	0.755(0.011)+	0.750(0.045)+	0.878(0.020)+
Credit6000	0.846(0.002)+	0.849(0.001)=	0.849(0.002)=	**0.850**(0.002)	0.842(0.004)+	0.774(0.061)+	0.841(0.002)+
Heart-statlog	0.891(0.018)=	0.894(0.020)=	0.898(0.016)=	**0.904**(0.012)	0.709(0.058)+	0.725(0.125)+	0.841(0.029)+
Spambase	0.938(0.002)+	0.939(0.001)+	0.940(0.002)+	**0.943**(0.002)	0.926(0.012)+	0.833(0.047)+	0.921(0.003)+
+/−/=	5/0/2	5/0/2	3/0/4		6/1/0	7/0/0	6/0/1

As can be seen from the table, the fitness value of SV-GA with an initial population size of 100 on the Android, SPECT Heart, CMC, Sonar, Credit6000, Heart-statlog, and Spambase data sets are all higher than those of traditional GA and SV-GA of population size of 30 and 50, the fitness values are 93.6%, 89.7%, 75.5%, 88.7%, 85.0%, 90.4%, and 94.3%. When compared with other algorithms, WOA has the best performance on the Android dataset. In addition, the fitness value of SV-GA with a population size of 100 far exceeds other algorithms. The results of the statistical tests showed that on the Android dataset, SV-GA(100) results were significantly more significant than the other algorithms on the Android dataset, except for the non-significant results compared with the GWO algorithm. On the SPECT Heart dataset, WOA was not significantly different, and the results were significant on the rest of the datasets. On the CMC, Sonar, Credit6000, Heart-statlog, and Spambase datasets, the results were significant compared to the GWO, ACO, and WOA algorithms. Overall, SV-GA(100) ranked first in terms of performance.

5.4.2. Classification and Selected Features

Table 8 shows the results of using Equation (3) as the fitness function. Unlike Equation (4), Equation (3) does not take into account the number of features, and the results of the classifier are used directly as the fitness function. From the experimental results in Table 8, the performance of the genetic algorithm without the crossover operator is worse than that of the traditional genetic algorithm on the SPECT Heart, CMC, Credit6000, and Heart-statlog datasets. The reason for this result is, on the one hand, because of the small number of samples in the dataset and the relatively small number of features in the examples. When no crossover operation is performed, the population's diversity is not guaranteed. The algorithm tends to fall into premature maturity, making it difficult to find the global optimal solution. The results in Table 9 show that on the seven datasets of Android, SPECT Heart, CMC, Sonar, Credit6000, Heart-statlog, and Spambase, the SV-GA with a population size of 100 selected the smallest proportion of features in more than half of the datasets, which is consistent with the goal of optimizing the feature selection problem, i.e., is the accuracy of the classifier is not high, and the number of features selected is low. Still, the final results obtained for both in proportion to specific parameters indicate that the algorithm performance is good. It can be demonstrated that the algorithm has some advantage in solving the FS problem. The graphical results of Tables 8 and 9 are shown in Figure 14a,b. The red bars are the algorithms compared with other algorithms in all experiments.

Table 8. Comparison of ACC results on different algorithms for different datasets.

Datasets	GA	SV-GA(30)	SV-GA(50)	SV-GA(100)	GWO	ACO	WOA
	Mean(std)	**Mean(std)**	**Mean(std)**	**Mean(std)**	**Mean(std)**	**Mean(std)**	**Mean(std)**
Android Dataset	0.917(0.004)	0.923(0.002)	0.941(0.004)	0.940(0.004)	**0.947**(0.005)	0.935(0.004)	0.925(0.004)
SPECT Heart	**0.925**(0.004)	0.811(0.006)	0.817(0.010)	0.860(0.008)	0.801(0.033)	0.830(0.029)	0.856(0.012)
CMC	**0.900**(0.072)	0.762(0.009)	0.821(0.103)	0.806(0.064)	0.647(0.028)	0.644(0.048)	0.733(0.009)
Sonar	0.884(0.014)	**0.886**(0.012)	0.871(0.012)	0.869(0.009)	0.755(0.011)	0.750(0.045)	0.878(0.020)
Credit6000	**0.855**(0.001)	0.853(0.002)	0.854(0.002)	0.849(0.001)	0.842(0.004)	0.774(0.061)	0.841(0.002)
Heart-statlog	**0.958**(0.027)	0.824(0.010)	0.892(0.011)	0.862(0.013)	0.709(0.058)	0.725(0.125)	0.841(0.029)
Spambase	0.936(0.003)	0.950(0.002)	0.947(0.002)	**0.951**(0.002)	0.926(0.012)	0.833(0.047)	0.921(0.003)

Table 9. Comparison of feature number on different algorithms for different datasets.

Datasets	GA	SV-GA(30)	SV-GA(50)	SV-GA(100)	GWO	ACO	WOA
	SF(%)	SF(%)	SF(%)	SF(%)	SF(%)	SF(%)	SF(%)
Android Datase	46.5(52.84)	47.9(54.43)	46.8(53.18)	43.307(49.213)	43.800(49.773)	**42.5**(47.852)	64.400(73.182)
SPECT Heart	11.5(52.27)	10.5(47.73)	12.2(54.45)	**9.9**(45)	11.8(53.636)	17.4(79.090)	8.1(36.818)
CMC	6(66.67)	5.6(62.22)	3.9(43.33)	**3.769**(41.880)	3.9(43.333)	5.5(61.111)	5(55.556)
Sonar	29.6(49.33)	32.6(54.33)	31.3(52.17)	29.6(49.33)	27(45)	31.7(47.692)	**22.4**(37.333)
Credit6000	31.9(49.08)	30.7(47.23)	34.5(53.08)	31.7(48.77)	32.7(50.308)	**31**(47.692)	6.1(9.384)
Heart-statlog	7.368(56.680)	9(69.23)	7.6(58.46)	6.227(47.902)	7(53.846)	8(61.538)	**4.9**(37.692)
Spambase	32.1(56.32)	34.2(60)	32.3(56.67)	**27.176**(47.678)	27.6(48.421)	35.9(62.982)	40.2(70.526)

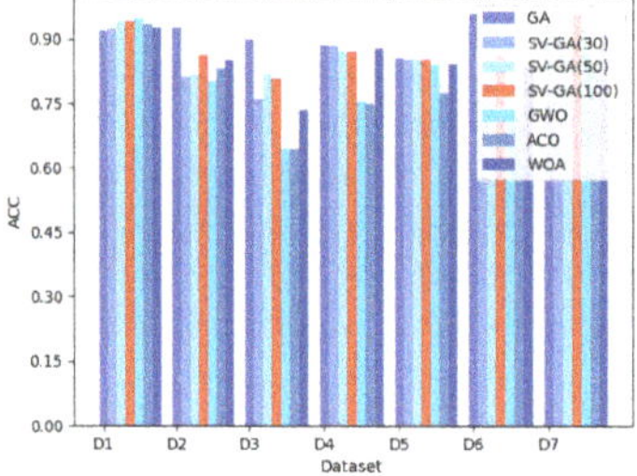

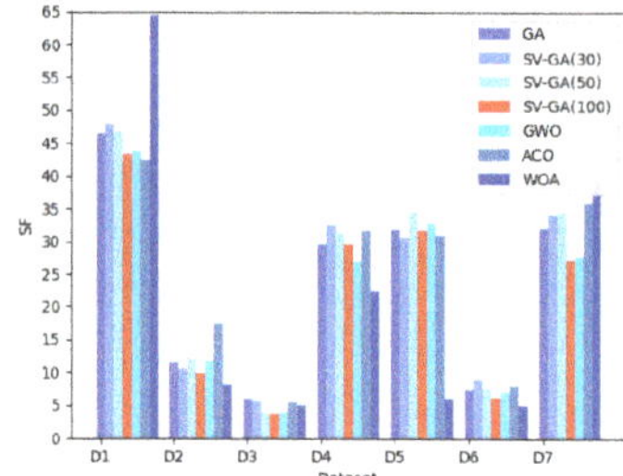

(**a**) Comparison of accuracy of different datasets (**b**) Comparison of the number of features in different datasets

Figure 14. Comparison between different methods and different datasets on ACC and SF.

5.4.3. Running Time

In this paper, the SV-GA is influenced not only by the initial population size but also by the variation rate. Table 10 shows the mean results for ten independent runs of the different algorithms. It is easy to see from the results that the GA times are much higher than the SV-GA for arbitrary populations. The reason for this result is that the crossover operator increases the algorithm's time complexity. When compared to the other algorithms, GWO takes the least time, which may depend on the co-evolutionary strategy of the algorithm itself to speed up finding the optimal solution. Still, the overall SV-GA time does not pull away from this algorithm by a large margin, and to some extent, there is no significant gap. When solving the feature selection problem, the time complexity depends heavily on the number of samples and the number of features. For example, Credit6000 has the highest number of instances of any of the seven datasets, so no matter how well the algorithm performs, the time on that dataset is bound to exceed that on any of the remaining datasets.

Table 10. Comparison of running-time on different algorithms for different datasets.

Datasets	GA	SV-GA(30)	SV-GA(50)	SV-GA(100)	GWO	ACO	WOA
Android Dataset	102.739	10.195	9.254	9.174	**8.595**	10.265	198.043
SPECT Heart	57.345	4.582	4.559	5.405	3.229	**1.360**	14.501
CMC	70.033	5.403	6.032	7.025	**3.278**	11.812	39.057
Sonar	99.635	4.938	5.573	6.459	4.189	**3.865**	11.678
Credit6000	70.528	33.297	35.857	52.305	**15.421**	217.895	391.099
Heart-statlog	59.316	4.770	5.267	6.453	**3.134**	4.455	11.919
Spamabse	247.696	19.128	23.238	22.507	**10.612**	91.235	445.082

5.5. Discussions

It is necessary to understand the drawbacks of each proposed stochastic algorithm. The algorithm proposed in this paper is designed to solve the specific optimization problem of feature selection, and the crossover operator redundancy is only for this class of optimization problems; the method may not be effective when solving continuous optimization problems. Secondly, the FS problem is optimized for a dataset, and the number of samples and the dimensionality of the data features are the most critical factors affecting the running time of the algorithm, so how to apply the algorithm to high-dimensional instances without increasing the computational complexity is a significant task for future research.

6. Conclusions

By summarizing the characteristics of the asexual genetic algorithm and analyzing the feature selection problem, we use the asexual genetic algorithm for the first time to solve the feature selection problem. The article demonstrates for the first time the advantages of the asexual genetic algorithm for solving this type of problem from a theoretical point of view, followed by the development and exploration ability of the algorithm by changing the population size and mutation rate to balance the algorithm to get the best variation rate for solving this type of problem, and finally the algorithm is used to solve the actual problem of Android Malware Application Detection, the results demonstrate that the algorithm is a feasible alternative in solving the feature selection problem.

Author Contributions: Conceptualization, L.W. and Y.G.; methodology, L.W.; formal analysis, S.G.; review and editing, X.Y. All authors have read and agreed to the published version of the manuscript.

Funding: This research was supported by the National Natural Science Foundation of China under Grant (11961001), the Construction Project of first-class subjects in Ningxia higher Education (NXYLXK2017B09), the major proprietary funded project of North Minzu University (ZDZX201901), and postgraduate Innovation Project Funding of Northern University for Nationalities (YCX20087).

Institutional Review Board Statement: Article does not involve human research.

Informed Consent Statement: Article does not involve human research.

Data Availability Statement: The study did not report any data.

Acknowledgments: In this section you can acknowledge any support given which is not covered by the author contribution or funding sections. This may include administrative and technical support, or donations in kind (e.g., materials used for experiments).

Conflicts of Interest: The authors declare no conflict of interest.

References

1. Jain, A.K. Statistical pattern recognition: A review. *IEEE Trans. Pattern Anal. Mach. Intell.* **2002**, *27*, 4–37. [CrossRef]
2. Jesus, J.; Canuto, A.; Araujo, D. An exploratory analysis of data noisy scenarios in a Pareto-front based dynamic feature selection method. *Appl. Soft Comput.* **2020**, *100*, 106951. [CrossRef]
3. Nguyen, B.H.; Xue, B.; Zhang, M. A survey on swarm intelligence approaches to feature selection in data mining. *Swarm Evol. Comput.* **2020**, *54*, 100663. [CrossRef]

4. Ray, P.; Reddy, S.S.; Banerjee, T. Various dimension reduction techniques for high dimensional data analysis: A review. *Artif. Intell. Rev.* **2021**, *54*, 3473–3515. [CrossRef]
5. Rodrigues, D.; Pereira, L.A.M.; Almeida, T.N.S.; Papa, J.P.; Yang, X.S. BCS: A Binary Cuckoo Search algorithm for feature selection. *Proc. IEEE Int. Symp. Circuits Syst.* **2013**. [CrossRef]
6. Battiti, R. Using mutual information for selecting features in supervised neural net learning. *Neural Netw. IEEE Trans.* **1994**. [CrossRef] [PubMed]
7. Razniewski, S.; Strzelecki, M. Evaluation of texture features based on mutual information. *Ispa Int. Symp. Image Signal Process. Anal.* **2005**. [CrossRef]
8. Fleuret, F. Fast Binary Feature Selection with Conditional Mutual Information. *J. Mach. Learn. Res.* **2004**, *5*, 1531–1555.
9. Dash, M.; Liu, H. Feature selection for classification. *Intell. Data Anal.* **1997**, *1*, 131–156. [CrossRef]
10. Holland, J. Adaptation in natural and artificial systems: An introductory analysis with application to biology. *Control Artif. Intell.* **1975**. [CrossRef]
11. Cantó, J.; Curiel, S.; Martínez-Gómez, E. A simple algorithm for optimization and model fitting: AGA (asexual genetic algorithm). *Astron Astrophys.* **2009**, *501*, 1259–1268. [CrossRef]
12. Farasat, A.; Menhaj, M.B.; Mansouri, T.; Moghadam, M.R. ARO: A new model-free optimization algorithm inspired from asexual reproduction. *Appl. Soft Comput.* **2010**, *10*, 1284–1292. [CrossRef]
13. Simoes, A.; Costa, E. Using genetic algorithms with sexual or asexual transposition: a comparative study. *Proc. CEC00* **2000**, *10*, 1196–1203.
14. Amirghasemi, M.; Zamani, R. An effective asexual genetic algorithm for solving the job shop scheduling problem. *Comput. Ind. Eng.* **2015**, *83*, 123–138. [CrossRef]
15. Salesi, S.; Cosma, G. A novel extended binary cuckoo search algorithm for feature selection. In Proceedings of the International Conference on Knowledge Engineering and Applications, London, UK, 21–23 October 2017; pp. 6–12.
16. Ab Razak, M.F.; Anuar, N.B.; Othman, F.; Firdaus, A.; Afifi, F.; Salleh, R. Bio-inspired for Features Optimization and Malware Detection. *Arab. J. Sci. Eng.* **2018**, *43*, 6963–6979. [CrossRef]
17. Zhang, Y.; Wang, S.; Phillips, P.; Ji, G. Binary PSO with mutation operator for feature selection using decision tree applied to spam detection. *Knowl. Based Syst.* **2014**, *64*, 22–31. [CrossRef]
18. Palanisamy, S.; Kanmani, S. Artificial Bee Colony Approach for Optimizing Feature Selection. *IJCSI* **2012**, *9*, 432–438.
19. Sreedharan, N.P.; Ganesan, B.; Raveendran, R.; Sarala, P.; Dennis, B. Grey Wolf Optimization-based Feature Selection and Classification for Facial Emotion Recognition. *IET Biom.* **2018**, *7*. [CrossRef]
20. Hu, P.; Pan, J.S.; Chu, S.C. Improved Binary Grey Wolf Optimizer and Its application for feature selection. *Knowl. Based Syst.* **2020**, *195*, 105746. [CrossRef]
21. Pan, J.S.; Tian, A.Q.; Chu, S.C.; Li, J.B. Improved binary pigeon-inspired optimization and its application for feature selection. *Appl. Intell.* **2021**. [CrossRef]
22. Wang, J.W.; Chen, C.H.; Pan, J.S. Genetic Feature Selection for Texture Classification Using 2-D Non-Separable Wavelet Bases. *IEICE Trans. Fundam. Electron. Commun. Comput. Sci.* **1998**, *E81A*, 1635–1644.
23. Emary, E.; Zawbaa, H.M.; Hassanien, A.E. Binary grey wolf optimization approaches for feature selection. *Neurocomputing* **2016**, *172*, 371–381. [CrossRef]
24. Dua, D.; Graff, C. *UCI Machine Learning Repository*; University of California, School of Information and Computer Science: Irvine, CA, USA, 2019. Available online: http://archive.ics.uci.edu/ml (accessed on 17 July 2020).
25. Tein, L.H.; Ramli, R. Recent advancements of nurse scheduling models and a potential path. In Proceedings of the ICMSA 2010, Grand Seasons Hotel, Kuala Lumpur, Malaysia, 3–4 November 2010; pp. 395–409.
26. Feizollah, A.; Nor, B.; Salleh, R.; Amalina, F. Comparative study of k-means and mini batch k-means clustering algorithms in android malware detection using network traffic analysis. In Proceedings of the ISBAST 2014, Kuala Lumpur, Malaysia, 26–27 August 2014. [CrossRef]
27. Nath, H.V.; Mehtre, B.M. Static Malware Analysis Using Machine Learning Methods. In Proceedings of the SNDS-2014, Trivandrum, India, 13–14 March 2014; pp. 440–450.
28. Xiaosong, Z.; Khan, R.U.; Kumar, J.; Ahad, I.; Kumar, R. Effective and Explainable Detection of Android Malware Based on Machine Learning Algorithms. In Proceedings of the ICCAI 2018, Chengdu, China, 12–14 March 2018; pp. 35–40.
29. Zhao, X.; Fang, J.; Wang, X. Android malware detection based on permissions. In Proceedings of the ICICT 2014, Nanjing, China, 2 October 2014. [CrossRef]
30. Aung, Z.; Zaw, W. Permission-Based Android Malware Detection. *IJSTR* **2013**, *2*, 228–234.
31. Wang, S.; Chen, Z.; Yan, Q.; Yang, B.; Peng, L.; Jia, Z. A mobile malware detection method using behavior features in network traffic. *J. Netw. Comput. Appl.* **2019**, *133*, 15–25. [CrossRef]
32. Aafer, Y.; Du, W.; Yin, H. DroidAPIMiner: Mining API-Level Features for Robust Malware Detection in Android. *Springer Int. Publ.* **2013**, *127*, 86–103.
33. Arp, D.; Spreitzenbarth, M.; Hubner, M.; Gascon, H.; Rieck, K.; Siemens, C.E. DREBIN: Effective and Explainable Detection of Android Malware in Your Pocket. In Proceedings of the NDSS, San Diego, CA, USA, August 2014. [CrossRef]
34. Wang, W.; Gao, Z.; Zhao, M.; Li, Y.; Liu, J.; Zhang, X. DroidEnsemble: Detecting Android Malicious Applications with Ensemble of String and Structural Static Features. *IEEE Access* **2018**, *6*, 31798–31807. [CrossRef]

35. Raymer, M.L.; Punch, W.F.; Goodman, E.D.; Kuhn Leslie, A.; Jain, A.K. Dimensionality reduction using genetic algorithms. *IEEE Trans. Evol. Comput.* **2000**, *4*, 164–171. [CrossRef]
36. Bhattacharya, A.; Goswami, R.T.; Mukherjee, K. A feature selection technique based on rough set and improvised PSO algorithm (PSORS-FS) for permission based detection of Android malwares. *Int. J. Mach. Learn. Cybern.* **2018**, *10*, 1893–1907. [CrossRef]
37. *Apktool*. May 2015. [Online]. Available online: https://ibotpeaches.github.io/Apktool/ (accessed on 17 July 2020).
38. Taheri, L.; Kadir, A.F.; Lashkari, A.H. Extensible Android Malware Detection and Family Classification Using Network-Flows and API-Calls. In Proceedings of the ICCST 2019, Cairo, Egypt, 1–3 October 2019. [CrossRef]
39. Kennedy, J.; Eberhart, R.C. A discrete binary version of the particle swarm algorithm. In Proceedings of the 1997 IEEE International Conference on Systems, Man, and Cybernetics. Computational Cybernetics and Simulation, Orlando, FL, USA, 12–15 October 1997. [CrossRef]
40. Saremi, S.; Mirjalili, S.; Lewis, A. *How Important Is a Transfer Function in Discrete Heuristic Algorithms*; Springer: Berlin/Heidelberg, Germany, 2014; pp. 625–640.
41. Hilda, G.T.; Rajalaxmi, R.R. Effective feature selection for supervised learning using genetic algorithm. In Proceedings of the ICECS, Coimbatore, India, 26–27 February 2015; pp. 909–914.
42. Mirjalili, S.; Mirjalili, S.M.; Lewis, A. Grey Wolf Optimizer. *Adv. Eng. Softw.* **2014**, *69*, 46–61. [CrossRef]
43. Mirjalili, S.; Lewis, A. The Whale Optimization Algorithm. *Adv. Eng. Softw.* **2016**, *95*, 51–67. [CrossRef]
44. Colorni, A. Distributed optimization by ant colonies. *Proc. ECAL* **1991**, *142*, 134–142.
45. Hussien, A.G.; Hassanien, A.E.; Houssein, E.H.; Bhattacharyya, S.; Amin, M. S-shaped Binary Whale Optimization Algorithm for Feature Selection. *Recent Trends Signal Image Process.* **2019**, *727*, 79–87.
46. Wan, Y.; Wang, M.; Ye, Z.; Lai, X. A Feature Selection Method Based on Modified Binary Coded Ant Colony Optimization Algorithm. *Appl. Soft Comput.* **2016**, *49*, 248–258. [CrossRef]

MDPI

Article

An Improved Bees Algorithm for Training Deep Recurrent Networks for Sentiment Classification

Sultan Zeybek [1,*], Duc Truong Pham [2], Ebubekir Koç [3] and Aydın Seçer [4]

1 Department of Computer Engineering, Fatih Sultan Mehmet Vakif University, Istanbul 34445, Turkey
2 Department of Mechanical Engineering, University of Birmingham, Birmingham B15 2TT, UK; d.t.pham@bham.ac.uk
3 Department of Biomedical Engineering, Fatih Sultan Mehmet Vakif University, Istanbul 34445, Turkey; ekoc@fsm.edu.tr
4 Department of Mathematical Engineering, Yildiz Technical University, Istanbul 34220, Turkey; asecer@yildiz.edu.tr
* Correspondence: szeybek@fsm.edu.tr

Abstract: Recurrent neural networks (RNNs) are powerful tools for learning information from temporal sequences. Designing an optimum deep RNN is difficult due to configuration and training issues, such as vanishing and exploding gradients. In this paper, a novel metaheuristic optimisation approach is proposed for training deep RNNs for the sentiment classification task. The approach employs an enhanced Ternary Bees Algorithm (BA-3+), which operates for large dataset classification problems by considering only three individual solutions in each iteration. BA-3+ combines the collaborative search of three bees to find the optimal set of trainable parameters of the proposed deep recurrent learning architecture. Local learning with exploitative search utilises the greedy selection strategy. Stochastic gradient descent (SGD) learning with singular value decomposition (SVD) aims to handle vanishing and exploding gradients of the decision parameters with the stabilisation strategy of SVD. Global learning with explorative search achieves faster convergence without getting trapped at local optima to find the optimal set of trainable parameters of the proposed deep recurrent learning architecture. BA-3+ has been tested on the sentiment classification task to classify symmetric and asymmetric distribution of the datasets from different domains, including Twitter, product reviews, and movie reviews. Comparative results have been obtained for advanced deep language models and Differential Evolution (DE) and Particle Swarm Optimization (PSO) algorithms. BA-3+ converged to the global minimum faster than the DE and PSO algorithms, and it outperformed the SGD, DE, and PSO algorithms for the Turkish and English datasets. The accuracy value and F1 measure have improved at least with a 30–40% improvement than the standard SGD algorithm for all classification datasets. Accuracy rates in the RNN model trained with BA-3+ ranged from 80% to 90%, while the RNN trained with SGD was able to achieve between 50% and 60% for most datasets. The performance of the RNN model with BA-3+ has as good as for Tree-LSTMs and Recursive Neural Tensor Networks (RNTNs) language models, which achieved accuracy results of up to 90% for some datasets. The improved accuracy and convergence results show that BA-3+ is an efficient, stable algorithm for the complex classification task, and it can handle the vanishing and exploding gradients problem of deep RNNs.

Keywords: bees algorithm; training deep neural networks; metaheuristics; opinion mining; recurrent neural networks; sentiment classification; natural language processing

Citation: Zeybek, S.; Pham, D. T.; Koç, E.; Seçer, A. An Improved Bees Algorithm for Training Deep Recurrent Networks for Sentiment Classification. *Symmetry* **2021**, *13*, 1347. https://doi.org/10.3390/sym13081347

Academic Editors: Kóczy T. László and István A. Harmati

Received: 8 July 2021
Accepted: 22 July 2021
Published: 26 July 2021

1. Introduction

Deep recurrent neural networks (RNNs) are powerful deep learning models with the ability to learn from the large sets of sequential data that characterise many tasks such as natural language processing [1], time series prediction [2], machine translation [3] and image captioning [4]. Deep RNNs have self-looped connected deep layers, which can retain information from the past and make it possible to learn arbitrarily long time sequences. However, despite their theoretical power, they have well-known computational issues such as training difficulties due to vanishing and exploding gradients [5], the need for implementation in hardware and memory limitations [6]. Besides, designing a deep learning model to perform a particular task could be very time-consuming as it involves many optimisation steps such as selecting a proper network architecture, finding the optimum hyperparameters of the selected architecture, and choosing the correct training algorithm for the model. Training a deep RNN is making it learn higher-level nonlinear features from large amounts of sequential data, which is typically a nonconvex optimisation problem [6]. This problem can be formulated as the minimisation of nonlinear loss functions with multiple local optima and saddle points. From the perspective of optimisation, even convex optimisation problems have many challenges. Additional difficulties therefore arise in training deep neural networks because of the nonconvex nature of the problem. For example, Stochastic Gradient Descent (SGD), which is a commonly used training algorithm, could easily get trapped at local minima or saddle points, and it cannot guarantee convergence to the global optimum because of the nonlinear transformations in each hidden layer. Moreover, the gradient of nonlinear activation functions cannot be computed backward through the network layers without vanishing or exploding over many training time steps, which causes the loss of direction in parameter updating to reach a feasible solution [7].

To date, researchers have mainly focused on two alternative pathways to deal with long-term dependencies. The first pathway is to devise new network architectures such as Long Short-Term Memory (LSTM) models [8], Gated Recurrent Units (GRU) [9] and Temporal Restricted Boltzmann Machines (TRBM) [10]. Although these architectures have proved successful in many applications, they are more complex to implement and require long implementation and computation times, in addition to specialised software and powerful hardware. The second pathway is to develop search methods and optimisation algorithms specifically to handle the vanishing and exploding gradient problem. Recently, two popular methods, gradient clipping and gradient scaling, were proposed to avoid the gradient explosion issue. Gradient clipping [5] which employs a shrinking strategy when the gradient becomes too large, is used to avoid remembering only recent training steps. Shrinking has also been employed by second-order optimisation algorithms, but these have been replaced by simple SGD as a fair and practical technique because of the computational cost of Hessian matrices in second-order optimisation [11].

The learning performance of deep learning models does not depend only on improving the training algorithm. The initial design parameters also play a key role in the ability to find global optima without becoming trapped at local stationary points. For example, the initial weights of a deep network can significantly affect training performance and good solutions often cannot be reached with gradient-based training algorithms because of the nonlinearity and "butterfly-effects" of the iterative updating procedure [5]. Generally, design parameters are adjusted manually, and the designer has to evaluate the model performance repeatedly to determine the best objective functions, learning rates, or training algorithm for their task. Besides, even when the optimal model could be designed, additional regularisation strategies such as dropout [12] are required to handle the overfitting problem of a deep model. It is well-known that these procedures are very time-consuming, and new strategies are needed to develop practical solutions.

Numerical methods and exact algorithms cannot handle the nonconvexity of the objective functions of deep RNNs, which are unable to capture curvature information, causing the optimisation process to be trapped at local solutions. Nature-inspired metaheuristic algorithms have been developed to handle nonlinear, multi-constraint and multi-modal optimisation problems. They have proved to be robust and efficient optimisation tools that can avoid the issue of local optima. They can adapt to problem conditions like the nature of the search space (i.e., continuous or discrete), decision parameters, varying constraints and other challenges encountered in the training and designing of RNN models. Previous research into the optimisation of deep learning models has focused on three main areas, namely, hyperparameter optimisation, neural architecture or topology optimisation, and weight optimisation. These studies have been conducted for specified tasks with the numbers of hidden and recurrent neurons limited to a maximum of five, and new practical approaches are needed to be useful for deeper RNN models [13].

This paper proposes using an enhanced Ternary Bees Algorithm (BA-3+) to obtain the optimum weights of a deep RNN model for sentiment classification. Existing population-based optimisation algorithms need to operate with large populations and, as a result, are generally slow. The Bees Algorithm [14] is a population-based algorithm that has been successfully employed to solve many complex real-world optimisation problems including continuous [15] and combinatorial [16] optimisation problems. It is able to find both local and global optima without needing to calculate the gradient of the objective function. The Ternary Bees Algorithm (BA-3) first described in [17] is an improvement on other population-based algorithms that employs a population of just three individual solutions. The BA-3+ algorithm presented in this paper is an enhanced version of BA-3, that also uses only three individual solutions, the global-best solution, the worst solution and an in-between solution. BA-3+ combines the exploration power of the basic Bees Algorithm to escape from local optima and the greedy exploitation drive of new local search operators to improve solutions. The new local search operators comprise one for neighbourhood search using Stochastic Gradient Descent (SGD) and one for search control employing Singular-Value Decomposition (SVD). SGD is a greedy operator for reaching a local optimum quickly. SVD is adopted to stabilise the trainable parameters of the model and overcome the problem of vanishing and exploding gradients of the selected weights when SGD is applied to derive the in-between solution. The aim is to use the strengths of gradient-based backpropagation training as the most commonly used RNN training method, but without its limitations like local optimum traps and vanishing and exploding gradients through long time dependencies. As the proposed algorithm uses only three individual bees, it is very fast, being able to find the global optimum within polynomially-bounded computation times [17]. Experiments with the sentiment classification of English and Turkish movie reviews and Twitter tweets show that the Ternary BA performs well, providing faster and more accurate results compared to previous studies.

The rest of the paper is structured as follows. Section 2 briefly reviews methods to handle vanishing and exploding gradients (VEG) problem of the deep RNNs. Section 3 presents detailed information about deep RNNs and the difficulties with training them. Section 4 details the proposed algorithm and its local search operators, and describes its configuration for training deep RNNs for sentiment classification. Section 5 provides information about the datasets used, the hyperparameters of the model, and the experimental results obtained. Section 6 concludes the paper.

2. Related Work

This section reviews the approaches that have been used to handle the vanishing and exploding gradient (VEG) problem in deep RNN training.

The first way to handle the VEG problem is to use newer types of RNN architectures such as Long-Short-Term Memory (LSTM) [8], Gated-Recurrent Units (GRUs) [9] and Echo-State-Networks (ESNs) [18]. These architectures can model sequences and they produced good results for many applications [19]. However, they have issues such as limited non-linearity learning abilities [20], training times that can sometimes be many days or even months, and are still not completely free from the same gradient problem. Some metaheuristic approaches have been implemented to handle these issues of the advanced deep recurrent networks. Yang et al. proposed an improved whale optimization algorithm (IWOA) to predict the carbon prices, hybrid model, incorporating modified ensemble empirical mode decomposition (MEEMD) and LSTM [21]. Peng et al. proposed a fruit fly optimization algorithm (FOA) to find optimal hyper-parameter of the LSTM network to solve time series problems [22]. ElSaid et al. have also proposed employing the ACO algorithm to evolve the LSTM network structure [23]. Rashid et al. proposed to use Harmony Search (HS), Gray Wolf Optimizer (GWO), Sine Cosine (SCA), and Ant Lion Optimization algorithms (ALOA) algorithms to train LSTMs for classification and analysis of real and medical time series data sets [24]. Besides these, hyperparameter optimisation and initial parameter tuning are also needed to improve their performances [25]. For example, an improved version of the sine cosine optimization algorithm (SCOA) was used to identify the optimal hyperparameters of LSTM [26]. Similarly, Bouktif et al. proposed to use GA and PSO algorithms to find optimum hyperparameters of the LSTM-RNN model for electric load forecasting [27]. In a similar way to the using of new architectures, some researchers have proposed to use new activation functions. [28] had been proposed to use Rectified linear unit (ReLU) function instead of hyperbolic tangent or sigmoid functions. Similarly, Glorot et al. has proposed Deep Sparse Rectifier Neural Networks (SRNNs), that helps optimizing weights during training with rectifier units [29]. However, these approaches have limitations as well. For example, since the ReLU function is positive definite, it causes a bias shift effect and behaves like a bias term for the next layer of the model [30]. Hence, it decreases the learning capacity of the model [31].

The second way to handle the VEG problem is the stabilisation of the updated recurrent weights [32]. Gradient clipping [5] is a well-known heuristic approach to rescale gradients. It controls parameter updating by using a given threshold and prevents unexpected falls to zero or rises to infinity before operating the gradient-descent learning rule. L_1 and L_2 regularisation has also been applied to the recurrent weights to prevent overfitting. They are used as a penalty term during training mainly to bring weights closer to zero [6]. Initialisation methods have also been employed to limit the values of the updated parameters by using the identity or orthogonal shared matrix. Le et al. showed that combining the proper initialisation with rectified linear units can handle the VEG problems of RNNs [33]. Xu et al. proposed a hybrid deep learning model by combining RNNs and CNNs, which is used Rectified Linear Units(ReLUs) and initialised with the identity matrix [34]. Vorontsov et al. used Singular Value Decomposition (SVD) to find the orthogonal matrices of the weight matrix, They proposed to update the parameters at each iteration by using geodesic GD and Cayley methods [35]. Similarly, [36] proposed to use the SVD operator to stabilise gradients of deep neural network, which has been proposed as a Spectral-RNN. However, these methods require the computation of inverse matrices and the unitary initial matrices cannot be held after many training iterations, and the same issues arise again.

In addition to the aforementioned approaches, Hessian-free (HF) optimisation methods and novel training algorithms have also been proposed to model the curvature of the nonlinear functions of deep RNN models using random initialisation. Martens et al. has proposed to train RNNs by using hessian-free optimisation [20]. They have been inspired by the second-order derivative method and Newton optimisation method, which is also called a truncated Newton or the pseudo-Newton method [37]. Nevertheless, besides their sophisticated nature, they do not have enough generalisation ability to learn and need additional damping among hidden layers when they have been applied to large-scale architectures [11]. To date, Kag et al. proposed a novel forward propagation algorithm, (FPTT) to handle the VEG problem, which has been outperformed by the BPTT for many tasks including language modeling [38]. Some gradient independent methods have been developed to address the training difficulties of the Depp RNNs. One of the first best-known heuristic approaches is the simulated annealing method that performs a random neighborhood search to find the optimal weights of the system [7]. In addition to their advantages, the simulated annealing training period may be very long, hence Bengio et al. have recommended improving alternative practical training algorithms [7].

To date, metaheuristic algorithms have been successfully applied to solve many nonlinear optimisation problems with their good initialisation strategies and local search abilities that bring crucial advantages to handle local optima issues such as getting trapped at local optima [39]. Although the number of studies for optimisation of deep architectures is less than that for conventional architectures [40], some studies have been carried out to improve the optimisation performances for the specified and generalized tasks using the intelligent nature of population-based algorithms [41]. The studies are mainly focused on hybridisation approaches, which are used to evolve deep architectures and to optimise the hyperparameters of the deep learning models. They can focuses on various application field such as time-series forecasting [42], classification problem [43], prediction problem [44] and design problem [45].

Studies for evolving network topology with metaheuristics aim to design optimum network architecture. For example, Dessel et al. have proposed to evolve a deep RNN network by using ACO [46]. They present a strategy to design an optimal RNN model with five hidden and five recurrent layers to predict aviation flights. Similarly, Juang et al. proposed a hybrid training algorithm combining GA and PSO for evolving RNN architecture. They called the algorithm HGAPSO and applied the algorithm for RNN design to a temporal sequence production problem [47]. The NeuroEvolution of Augmenting Topologies (NEAT) approach has been developed based on the GA for the optimisation of neural model architectures [48]. Desell et al. have used Ant Colony Optimisation (ACO) to design a deep RNN architecture with five hidden and five recurrent units for predicting flight data [46]. Similarly, Ororbia et al. have implemented Evolutionary eXploration of Augmenting Memory Models (EXAMM) and different versions of it such as GRU, LSTM, MGU and, UGRNN to evolve RNNs [49]. Wang et al. proposed an evolutionary recurrent neural network algorithm for the proxy of image captioning task [50]. A Random Error Sampling-based Neuroevolution (RESN) has been proposed as an evolutionary algorithm to evolve RNN architecture for prediction task [51]. Mo et al. proposed an EA for topology optimisation of the hybrid LSTM-CNN network for remaining useful life prediction [52].

Studies to find the optimum weights and to handle VEG problem of the RNN network focused on hybrid training algorithms. Kang et al. proposed a hybrid training algorithm to get rid of the local optima and saddle points by using the PSO and backpropagation algorithm. They got an improvement on convergence and accuracy results of four different datasets [53]. Ge et al. have presented the modified particle Swarm Optimisation (MPSO) [54] algorithm for training dynamic Recurrent Elman Networks [55]. The proposed method aims to find the initial network structure and initial parameters to learn the optimal value of the network weights for controlling Ultrasonic Motors. Xiao et al. have proposed a hybrid training algorithm with PSO and backpropagation (BP) for Impedance Identification [56]. The RNN architecture has been trained based on finding the minimum MSE and the largest gradient. Zhang et al. have also proposed hybrid PSO and Evolutionary Algorithm (EA) to train RNN for solar radiation prediction [57]. Likewise, Cai et al. have used hybrid PSO-EA for time series prediction with RNN [58]. A real-coded (continuous) Genetic Algorithm (GA) has been employed for training RNN by updating weight parameters using random real-valued chromosomes [59]. Nawi et al. proposed a Cuckoo Search (CS) algorithm for training Elman Recurrent Networks combined with backpropagation for data classification compared to the Artificial Bee Colony and conventional backpropagation algorithm [60]. A recurrent NARX neural network has been trained by a Genetic Algorithm (GA) to improve the state of charge (SOC) of lithium batteries [61].

Although the proposed hybrid approaches can train or optimize the topology of the RNNs, those networks do not have so many hidden layers that they can be considered as deep architectures, since the number of hidden layers of most studies is not as high as deep learning architectures. For example, Bas et al. proposed RNN models that have two to five hidden layers for forecasting using the PSO algorithm. The performance of the proposed algorithm was compared to the LSTM and Pi-Sigma NN architectures, which are trained by using gradient-based algorithms that performed similar [62]. There have only been limited studies into optimizing the architecture of a deep RNN [46] or deep LSTM [23] models. The authors of [46] have used ACO to convert fully connected RNNs into less complex Elman ANNs.

In addition to these studies, some examples of metaheuristic approach focusing on network training in recent years. A neural network training algorithm was proposed by

Kaya et al. namely ABCES (Artificial Bee Colony Algorithm Based on Effective Scout Bee Stage) [63]. They proposed to use ABCES to train a feedforward neural network model to detect the nonlinearity of given static systems, including 13 different numerical optimisation problems. Shettigar et al. proposed an ANN model for surface quality detection, which is trained by using traditional backpropagation algorithm, GA, ABC algorithms compared to the RNN architecture. RNN and BP-NN algorithms performed comparable, and ABC-NN and RNN models gave better results compared to the others [64]. A BeeM-NN algorithm has been proposed as a bee mutation optimizer for training the RNN model for cloud computing application [65]. A Parallel Memetic Algorithm (PMA) has been proposed to train RNNs for the energy efficiency problem by Ruiz et al. [66]. Hu et al. implement a hybrid grey wolf optimizer (GWO) and PSO to determine the endometrial carcinoma disease with Elman RNN. on the [67]. Tian et al. have also proposed a metaheuristics recommendation system for training deep RNNs to optimise real-world optimization problems, such as the aerodynamic design of turbine engines and automated trading [68]. Roy et al. have proposed an Ant-Lion optimizer for training RNN to find energy scheduling in micro grid-connected system [69]. A data-driven deep learning model has been proposed by Aziz et al. by using 10 different classification datasets [70]. Elman RNN and NN models have been trained by PSO that improved the classification accuracy. Similarly, Hassib et al. proposed a data-driven classification framework using Whale Optimization Algorithm (WOA) for feature selection and training the Bidirectional Recurrent Neural Network [71]. A Global Guided Artificial Bee Colony (GGABC) algorithm proposed for Recurrent Neural Network training by for breast cancer prediction dataset [72]. Kumar et al. proposed hybrid flower pollination and PSO algorithm for training LSTM-RNN model to predict Intra-day stock market [73].

According to the review study, recently over two hundred studies have been made focusing on evolutionary swarm intelligence and deep learning models for topology optimization, hyper-parameter optimization, and training parameter optimisation [74]. Table 1 reports some of the selected works related to deep RNNs. Even though there are many studies focusing on training artificial neural networks, most of the proposed metaheuristics for training recurrent deep learning models do not comprise many deep hidden layers that could cause the VEG problem. The existing proposed methods have worked over big population numbers, and they are not trying to optimize the deep RNN architectures. The proposed BA-3+ algorithm has only three bees as a population number, and as we proved in Section 5, the total training time is much lower than the Differential Evolution (DE) and Particle Swarm Optimization (PSO). In addition, the choice of the BA-3+ algorithm is motivated by the No Free Lunch theorem [75], as it released there is no universally efficient algorithm for all kinds of problems. Hence we can say if algorithm A could perform better than algorithm B in some class of problems and datasets, algorithm B could perform better than algorithm A in some other class of problems and datasets. Hence in this study, we focus on exploring the advantages of the BA-3+ approach for improving the deep recurrent learning abilities as a solution for the VEG problem, which has been discussed in detail in the next section.

Table 1. Selected related work to handle the VEG problem of the deep RNNs.

Study/Year	Proposed Solution	Algorithm/Function
[62] 2021	Metaheuristic Training	PSO
[70] 2021	Metaheuristic Training	PSO
[63] 2021	Metaheuristic Training	ABC
[38] 2021	New Gradient-based Training	FPTT
[65] 2021	Metaheuristic Training	BeeM-NN
[22] 2021	Hyperparameter optimization	FOA
[52] 2021	Topology Optimisation	EA
[73] 2021	Hybrid Metaheuristics Training	FP & PSO
[76] 2021	Metaheuristic Training	BSO
[64] 2020	Hybrid Metaheuristics Training	GA & ABC
[71] 2020	Metaheuristic Training	WOA
[21] 2020	Metaheuristic Training	IWOA
[26] 2020	Hyperparameter optimization	SCOA
[27] 2020	Hyperparameter optimization	GA & PSO
[77] 2020	Metaheuristic Training	ABSA
[51] 2020	Topology Optimisation	RESN
[72] 2019	Hybrid Metaheuristic Training	GGABC
[49] 2019	Topology Optimisation	EXAMM
[61] 2019	Metaheuristic Training	GA
[69] 2019	Hybrid Metaheuristic Training	Ant-Lion
[67] 2019	Hybrid Metaheuristic Training	GWO & PSO
[66] 2019	Metaheuristic Training	PMA
[36] 2018	Gradient Stabilisation	SVD
[23] 2018	Topology Optimisation	ACO
[24] 2018	Metaheuristic Training	GWO, ALOA, SCA, HS
[44] 2018	Metaheuristic Training	PSO
[53] 2017	Hybrid Metaheuristic Training	PSO & BP
[45] 2017	Metaheuristic Training	DE
[34] 2016	Initialisation	Training tricks for RNNs
[33] 2015	A New Deep Architecture	RNN& CNN
[46] 2015	Topology Optimisation	ACO
[60] 2015	Metaheuristic Training	CS
[57] 2013	Hybrid Metaheuristic Training	PSO & EA
[29] 2011	A New Deep Architecture	SRNNs
[20] 2011	Hessian-free methods	Hessian-free optimisation
[28] 2010	A New Activation Function	ReLU
[56] 2007	Hybrid Metaheuristic Training	PSO & BP
[58] 2007	Hybrid Metaheuristic Training	PSO & EA
[55] 2007	Hybrid Metaheuristic Training	MPSO & BP
[18] 2004	A New Deep Architecture	Echo-State-Networks (ESNs)
[47] 2004	Hybrid Metaheuristic Training	GA & PSO
[48] 2002	Topology Optimisation	GA (NEAT)
[59] 2001	Metaheuristic Training	GA

3. Recurrent Neural Networks and Problem Preliminaries

3.1. Problems with Training Deep Recurrent Neural Networks

In this section, a deep recurrent neural network (RNN) model is described based on the standard RNN model for the sentiment classification task for both the English and Turkish languages. A many-to-one deep RNN model is presented and formulated for a clear understanding of the training difficulties of the proposed model.

3.2. Model Description and Problem Preliminaries

Let $x(i) = (x_1, x_2, \ldots, x_{t-1}, x_t)$ is a sequential feature vector of the sequence of words, i.e., n-dimensional word embedding or word vector for each observation in the dataset. The proposed deep RNN model has been constructed using the sequences of the following recurrence formula and defined below for each time step t with two commonly used nonlinear activation functions, tanh and sigmoid:

$$h_t = tanh(W_{hh}\, h_{t-1} + W_{xh}\, x_t + b_h) \tag{1}$$

$$y = sigmoid(W_{hy}\, h_t + b_y) \tag{2}$$

The hidden state of the model h_t passes the information from the previous time step h_{t-1}, and uses it to classify the given observation $x^{(i)}$. The sigmoid function is used to predict the sentiment class of $x^{(i)}$, i.e., each review or tweet from the dataset. The objective of the RNN model is to maximise the correct estimation by using each pair of $(y^{(i)},\ t^{(i)})$ or to minimise the BCE error between predicted $y^{(i)}$ and target data $t^{(i)}$.

The training algorithm searches for the optimal values of the learnable parameters $W_{xh},\ W_{hh},\ W_{hy},\ b_h,\ b_y$. Most of the training algorithms, such as SGD are based on the gradient descent learning rule by backpropagation through time (BPTT) [78] using the following update rule for each time step t as follows:

$$\theta^{t+1} = \theta^t - \eta\ \nabla_\theta\ L(f(x_i, \theta),\ t_i)) \tag{3}$$

Here η is the learning rate, that is one of the most important hyperparameters of the deep learning models. Once the loss function is calculated at the feedforward step, the proposed partial derivatives $\frac{\partial L}{\partial W_{hy}}, \frac{\partial L}{\partial b_y}, \frac{\partial L}{\partial W_{hh}}, \frac{\partial L}{\partial W_{xh}}, \frac{\partial L}{\partial b_x}$ represented as ∇_θ in the (3) above need to be calculated for updating the set of trainable parameters of the system.

In this study, we focused on binary sentiment classification datasets for the English and Turkish languages. Turkish classification datasets include movie reviews [79], multi-domain product reviews [79] and Twitter dataset [80] with a raw text format. English dataset includes the huge English IMDB movie reviews dataset [81], small movie reviews dataset [82] and Yelp dataset including reviews about businesses, check-in, photos, and tips of the users [83]. Table 2 presents the detailed information about the datasets. Here, the Yelp dataset is asymmetric distributed and the others are symmetric distributed datasets.

Table 2. Turkish and English datasets.

Dataset	Size
TR Books [79]	700 P, 700 N
TR DVD [79]	700 P, 700 N
TR Electronics [79]	700 P, 700 N
TR Kitchen Appliances [79]	700 P, 700 N
Turkish Movie Reviews [79]	5331 P, 5331 N
Turkish Twitter Dataset [80]	12,490 P, 12,490 N
English IMDB Movie Reviews [81]	12,500 P, 12,500 N
English Movie Reviews [82]	1000 P, 1000 N
English Yelp dataset [83]	3337 P, 749 N

During the backpropagation, since the same weight parameters are shared at each time step, the recursive nature of the training process causes the *vanishing and exploding gradients problem* which leads to a huge loss of information across the deep hidden layers. From the perspective of dynamical systems, when the model keeps its state in the same stable state for a long time, the same issue occurs and the updating information about the system is lost [7].

3.3. Vanishing and Exploding Gradients (VEG) Problem

This section contains a mathematical proof of the vanishing and exploding gradients (VEG) problem encountered when training deep RNNs.

The objective of the RNN model is to maximise the correct estimation by using each pair of $(y^{(i)},\ t^{(i)})$ or to minimise the BCE error between predicted $y^{(i)}$ and target data $t^{(i)}$ as follows:

$$
\begin{aligned}
L(t,y) &= \sum_{i=1}^{N} L_t(y_i,t_i) \\
\frac{\partial L}{\partial W_{hh}} &= \sum_{t=1}^{N_{out}} \frac{\partial L_t}{\partial W_{hh}} \\
&= \sum_{t=1}^{N_{out}} \frac{\partial L_t}{\partial y_t}\frac{\partial y_t}{\partial h_t}\frac{\partial h_t}{\partial W_{hh}} \\
\frac{\partial h_t}{\partial W_{hh}} &= tanh_t'(W_{hh}h_{t-1} + W_{xh}x_t + b_h)[h_{t-1} + \ W_{hh}\frac{\partial h_{t-1}}{\partial W_{hh}}]
\end{aligned}
\tag{4}
$$

h_t and h_{t-1} are both a function of W_{hh}, so the product rule of the derivative should be used for every time step as follows:

$$
\frac{\partial h_{t-1}}{\partial W_{hh}} = tanh_{t-1}'(W_{hh}h_{t-2} + W_{xh}x_{t-1})[h_{t-2} + \ W_{hh}\frac{\partial h_{t-2}}{\partial W_{hh}}]
\tag{5}
$$

$$
\frac{\partial h_{t-2}}{\partial W_{hh}} = tanh_{t-2}'(W_{hh}h_{t-3} + W_{xh}x_{t-2})[h_{t-3} + \ W_{hh}\frac{\partial h_{t-3}}{\partial W_{hh}}]
\tag{6}
$$

The rightmost term of the should be expanded until t = 1 to calculate $\frac{\partial h_1}{\partial W_{hh}}$, and the following backpropagated sequence is found if $tanh_t'(W_{hh}h_{t-2} + W_{xh}x_{t-1} + b_h)$ is represented as $tanh_t'$:

$$
\begin{aligned}
\frac{\partial h_t}{\partial W_{hh}} &= tanh_t'\left[h_{t-1} + \ W_{hh}tanh_{t-1}'\left[h_{t-2} + \dots \ + W_{hh}\frac{\partial h_1}{\partial W_{hh}}\right]\right] \\
&= tanh_t'\ h_{t-1} + tanh_t'\ W_{hh}tanh_{t-1}'h_{t-2} + \ \dots
\end{aligned}
\tag{7}
$$

Here the partial derivatives of h_t are calculated with respect to the previous time step h_k. Therefore, the total loss can backpropagated according to W_{hh} as follows:

$$
\frac{\partial L}{\partial W_{hh}} = \sum_{k=1}^{t} \frac{\partial L_t}{\partial y_t}\frac{\partial y_t}{\partial h_t}\frac{\partial h_t}{\partial h_k}\frac{\partial h_k}{\partial W_{hh}}
\tag{8}
$$

Here $\frac{\partial h_t}{\partial h_k} = \prod_{s=k+1}^{t} \frac{\partial h_s}{\partial h_{s-1}}$ for any time state s of the system, and each $\frac{\partial h_s}{\partial h_{s-1}}$ is the Jacobian matrix for h $\in \mathbb{R}^{D_n}$ defined as follows:

$$
\begin{aligned}
\frac{\partial h_s}{\partial h_{s-1}} &= \left[\frac{\partial h_s}{\partial h_{s-1,1}} \cdots \frac{\partial h_s}{\partial h_{s-1,D_n}}\right] \\
&= \begin{bmatrix} \frac{\partial h_{s,1}}{\partial h_{s-1,1}} & \cdots & \frac{\partial h_{s,1}}{\partial h_{s-1,D_n}} \\ \vdots & \ddots & \vdots \\ \frac{\partial h_{s,D_n}}{\partial h_{s-1,1}} & \cdots & \frac{\partial h_{s,D_n}}{\partial h_{s-1,D_n}} \end{bmatrix}
\end{aligned}
\tag{9}
$$

Equation (8) becomes the following:

$$
\frac{\partial L}{\partial W_{hh}} = \sum_{t=1}^{T}\sum_{k=1}^{t} \frac{\partial L_t}{\partial y_t}\frac{\partial y_t}{\partial h_t}\left(\prod_{s=k+1}^{t} \frac{\partial h_s}{\partial h_{s-1}}\right)\frac{\partial h_k}{\partial W_{hh}}
\tag{10}
$$

Since $h_t = tanh\ (W_{hh}h_{t-1} + W_{xh}x_t + b_h)$, the term $\prod_{s=k+1}^{t} \frac{\partial h_s}{\partial h_{s-1}}$ gives the diagonal matrix from (9) and (10), that can be seen from the following equation:

$$\prod_{s=k+1}^{t} \frac{\partial h_s}{\partial h_{s-1}} = \prod_{s=k+1}^{t} W_{hh}^T diag(tanh'(W_{hh}h_{s-1})\ ) \tag{11}$$

Let a and b be both the upper bounds, or largest singular values of the matrices W_{hh}^T and $diag(tanh'(W_{hh}h_{s-1}))$, respectively. The 2-norms of these matrices are bounded by ab defined as follows for any time state s of the system:

$$\left\| \frac{\partial h_s}{\partial h_{s-1}} \right\| = \left\| W_{hh}^T \right\| \left\| diag\left(tanh'(W_{hh}h_{s-1})\right) \right\| \leq ab \tag{12}$$

$$\left\| \frac{\partial h_t}{\partial h_k} \right\| = \left\| \prod_{s=k+1}^{t} \frac{\partial h_s}{\partial h_{s-1}} \right\| \leq (ab)^{t-k} \tag{13}$$

During training, the same weight matrix W_{hh} is used across the layers, so as $t \to \infty$ the term $(a.b)^{t-k}$ vanishes at the very small value such as $(ab)^{t-k} \to 0$ or explodes to the extremely large value such as $(a.b)^{t-k} \to \infty$.

It has been shown that [5,7], if the absolute value of the largest eigenvalue of the W_{hh}^T is smaller than $\frac{1}{b}$, then the gradients vanish as follows:

$$\begin{aligned}
\forall\ s, \left\| \frac{\partial h_s}{\partial h_{s-1}} \right\| &\leq \left\| W_{hh}^T \right\| \left\| diag\left(tanh'(W_{hh}h_{s-1})\right) \right\| \\
&\leq ab < \frac{1}{b}.b < 1 \\
\exists\ \gamma, \left\| \frac{\partial h_s}{\partial h_{s-1}} \right\| &\leq \gamma < 1 \\
\left\| \frac{\partial h_t}{\partial h_k} \right\| &= \prod_{s=k+1}^{t} \frac{\partial h_s}{\partial h_{s-1}} \leq (\gamma)^{t-k}
\end{aligned} \tag{14}$$

As $t \to \infty$, it is clear that $\lim_{t\to\infty} \prod_{s=k+1}^{t} \frac{\partial h_s}{\partial h_{s-1}} = 0$. Similarly, when the largest eigenvalue of the W_{hh}^T is bigger than $1/b$, gradients explode and $\lim_{t\to\infty} \prod_{s=k+1}^{t} \frac{\partial h_s}{\partial h_{s-1}} = \infty$.

4. An Enhanced Ternary Bees Algorithm (BA-3+) to Handle VEG Problem of Training Deep RNN

In this section, a population-based search algorithm for training deep RNNs is presented. The learnable parameters $\theta = (W_{xh},\ W_{hh},\ W_{hy}\ , b_h,\ b_y)$ are the same as in SGD, which is defined as a candidate solution in BA-3+, and try to minimise the binary cross-entropy loss function $L(y,\ t)$ for each pair of the sequential input $(x_1,\ x_2,\ \ldots, x_{t-1},\ x_t)$, the desired-targeted output t, and the predicted value y.

Gradient-based learning algorithms are particularly sensitive to the initial value of the weights and noise variance of the dataset in non-convex optimisation. Hence, the difficulty of the training deep RNN model depends on not only keeping the information through long-term time but also initial values of the parameters. Most initialisation methods are generally based on the random initialisation [84] or researchers choose to initiate the weights as an identity matrix or close to the identity conventionally [85]. Therefore, finding optimum initial parameters for a specified model and exploring which parameters should be updated and learned are still remains an open difficult optimisation task, due to the lack of the exact knowledge about the which properties of these parameters are kept or learned, under which conditions [6].

As mentioned above, this work uses an enhanced Ternary Bees Algorithm (BA-3+) for training deep RNNs. BA-3+ combines exploitative local search with explorative global search [17]. Improvements to the training of deep RNN models with BA-3+ have been made in three key areas: finding promising candidate solutions and initialising the model with good initial weights and biases, improving local search strategies to enhance good solutions by neighbourhood search, particularly to overcome the vanishing and exploding gradients problem, and performing exploration to find new potential solutions with global search.

4.1. Representation of Bees for Deep RNN Model

The Bees Algorithm was developed by Pham et al. with inspiration from clever foraging behaviours of the honey bees in nature [14]. In the proposed method the bee represents a sequential deep RNN model, which is modelled for the binary sentiment classification task. As can be seen in Figure 1, every bee (Sequential model) instance has the input layer, hidden deep RNN layers, and the output layer. The proposed model has the learnable parameters $\theta = (W_{xh}, W_{hh}, W_{hy}, b_h, b_y)$, and aims to classify the sequential input data $(x_1, x_2, \ldots, x_{t-1}, x_t)$ to its targeted class t.

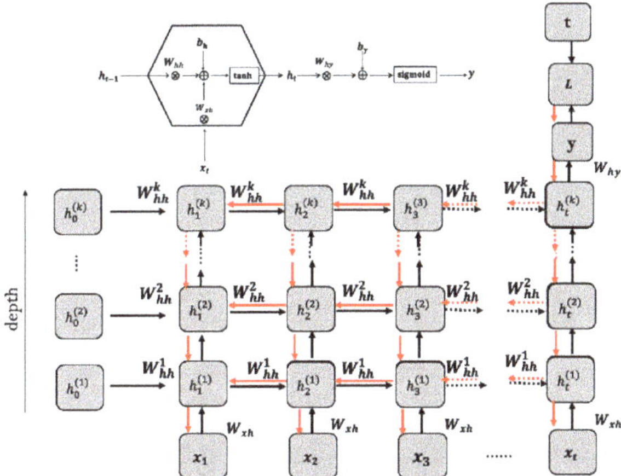

Figure 1. A Deep RNN architecture representing a bee in the proposed algorithm. Black lines are the forward pass of RNN cell at time t (unfolded version at upper) and red lines representing the error backpropagation through long-term dependencies.

Based on the training procedure of the RNN model, each "bee model" has its own forward propagation action to calculate the initial solutions, local search procedure by gradient descent training with singular value decomposition (SVD), and global search actions to find the optimal parameters $\theta = (W_{xh}, W_{hh}, W_{hy}, b_h, b_y)$ via the binary cross-entropy loss function (fitness function) $L_{BCE}(f(x^{(i)}, \theta), y^{(i)})$ as defined Section 3.3.

BA-3+ does not require a large population, which is a drawback with other population-based methods. BA-3+ employs only three individual bees for each training time. Each iteration begins with these three initial solutions as a forward pass of the model and continues with specified search strategies including exploitative local search, stochastic gradient descent (SGD) stabilised by Singular Value Decomposition (SVD), and explorative global search.

As with the basic Bees Algorithm [14], the initial candidate solutions are sorted. The maximum fitness value is selected as the best RNN bee for the local exploitative search. The worst fitness value (third bee) is selected for global search to avoid getting trapped at local optima, and the remaining RNN, i.e., the middle RNN bee is selected for stochastic gradient-descent learning with the stabilisation strategy of SVD operator to update weights and biases without vanishing and exploding gradients.

Figure 2 represents the flowchart of the proposed algorithm. $(x^{(train)}, t^{(train)})$ is the training sample from the dataset, that $x^{(i)} = (x_1, x_2, \ldots, x_{t-1}, x_t)$ is defined as an n-dimensional sequential input and $t^{(i)} \in \{0, 1\}$ is its targeted sentiment class. Parameters of the deep RNN model trained by using BA-3+ are shown in Table 3. Three initial solutions are calculated with the initial trainable parameters θ (see Table 4) and sorted according to the loss function. The elite (best) RNN bee performs the local search operator, the middle RNN bee performs stochastic gradient-descent with SGD operator, and the third RNN bee performs global search. The optimisation continues with a new population of bees until the stopping criteria met; in other words, until the loss value is converged to zero.

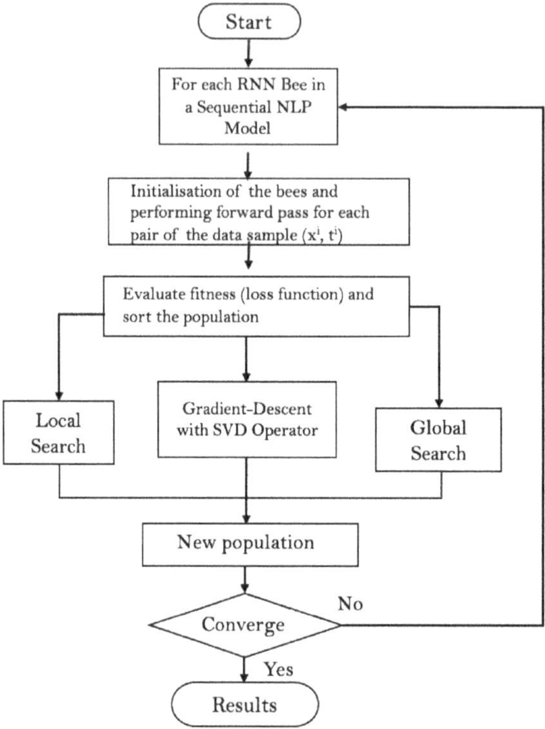

Figure 2. Flowchart of the proposed enhanced Ternary Bees Algorithm (BA-3+).

Table 3. Parameters of the deep RNN model trained by using BA-3+.

Parameters	Information
$x^{(i)} = (x_1, x_2, \ldots, x_{t-1}, x_t)$	ith temporal sequential input from the input training set
$t^{(i)}$	ith targeted class of the output training set
$y^{(i)}$	ith predicted class of the $x^{(i)}$
W_{xh}	Weight matrix from input layer to hidden layer
W_{hh}	Weight matrix from hidden layer to hidden layer
W_{hy}	Weight matrix from hidden layer to output layer
b_h, b_y	Biases for the hidden layer and the output layer
nScout	Number of scout bees for initialisation
ngh	Neighbourhood size for the local search
nhidden	Hidden layer size
η	Learning rate for SGD

Table 4. Learnable (trainable) parameters.

Parameters	Dimension
W_{xh}	$\mathbb{R}^{number\ of\ hidden\ layers\ \times\ dimension\ of\ each\ word}$
W_{hh}	$\mathbb{R}^{number\ of\ hidden\ layers\ \times\ number\ of\ hidden\ layers}$
W_{hy}	$\mathbb{R}^{\lvert V\rvert \times\ number\ of\ hidden\ layers}$
b_h	$\mathbb{R}^{number\ of\ hidden\ layers}$
b_y	$\mathbb{R}^{number\ of\ output\ layers}$
$\lvert V\rvert$:	*number of words in the vocabulary*

4.2. Local Search Operator

The local search procedure in the basic Bees Algorithm includes improving a promising solution within the neighbourhood of the selected solution parameters. In Algorithms 1 and 2, ngh represents the initial size of the neighbourhood for the local search. The neighbourhood begins as a large area and it is reduced by using a shrinking method [86] at each iteration according to the formula $ngh(t+1) = \alpha\ ngh(t)$. Here, α is usually a number between 0 and 1. The neighbourhood matrix is generated with the same dimension of each weight matrix of the learnable parameters $\theta = (W_{xh}, W_{hh}, W_{hy}, b_h, b_y)$, and then is aded to the original weight matrix to obtain the updated weights. The pseudo-code to generate neighbourhood weights is given in Algorithm 2. The updated local weights are used for the local search of BA-3+ that can be seen in Algorithm 4.

Algorithm 1: Pseudo-code to generate neighbourhood weights.

```
Input: weight matrix : w, ngh
Output: updated ngh weights
Function generateNghWeight(w, ngh):
    for each w_e ∈ w do
        ngh ← random number ∈ [−ngh, ngh]
        w_ngh ← w_e + ngh
    return w_ngh
```

Algorithm 2: Pseudo-code of the local search operator of BA-3+.

```
Input: Bee, ngh : neighbourhood radius
Output: Local Bee with Updated Parameters
Function LocalSearch(Bee, ngh):
    for each w ∈ θ = (W_xh, W_hh, W_hy, b_h, b_y) do
        dim_w ← dimension(w)
        w_ngh ← generateNghWeight(w, ngh)
        Bee.w ← w_ngh
    return Bee
```

4.3. Enhanced Local Search by SGD and Singular Value Decomposition (SVD) Operator

As analysed in Sections 3.2 and 3.3 due to the sharing the same hidden matrix W_{hh} across the deep hidden layers and multiplying it again and again at every time step of the BPTT algorithm, the eigenvalues of the Jacobian matrix exponentially grow or vanish after t time steps. To handle this issue, it has been proposed to use a singular value decomposition of the hidden layer matrix to stabilise the eigenvalues of the updated matrix in the enhanced local search of BA-3+. As an example, assume that the eigenvalues of the W_{hh} are represented $\lambda_1, \lambda_2, \ldots, \lambda_n$ The singular values of W_{hh} can be founded by using the positive eigenvalues of the matrix $W_{hh}{W_{hh}}^T$, for every $\lambda_i \geq 0 \in \lambda_1, \lambda_2, \ldots, \lambda_n$, and $S_i = \sqrt{\lambda_i}$ if W_{hh} is positive semi-definite square matrix [87]. Since the learnable parameters of the RNN can also be rectangular matrices, it is needed to find singular values of an arbitrary matrix A.

It is well-known that every arbitrary real matrix can be represented by the product of three matrices as $A = USV^T$, which is called singular value decomposition (SVD) of matrix A, which is used to find the singular values [88]. Figure 3 represents the SVD of an $n \times m$ dimensional matrix. Here, S is the $r \times r$ dimensional diagonal matrix $S_{n \times n} = diag[S_1, S_2, \ldots, \ldots, S_n]$ that each S_i represents the singular values of the matrix A, and U and V contain the corresponding singular vectors where U and V are orthogonal matrices with the $n \times r$ and $r \times m$ dimensions, respectively.

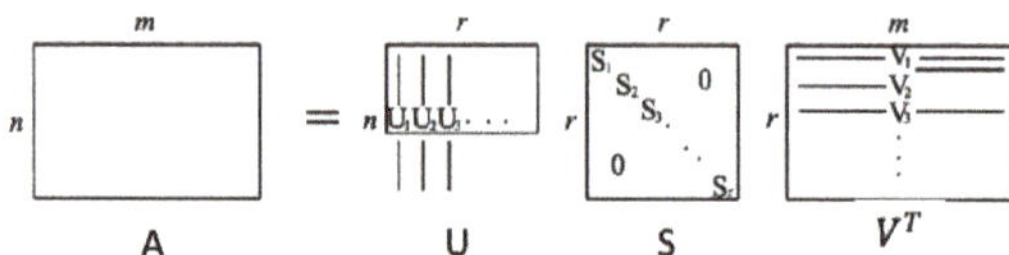

Figure 3. Singular Value Decomposition (SVD) of matrix A.

After updating each parameter of the $\theta = (W_{xh}, W_{hh}, W_{hy}, b_h, b_y)$ by SGD rule, the SVD operator has used to control the eigenvalues of each parameter. The method aims to keep the singular values of the updated matrix close to 1 for gradient stabilisation. To this end, the SVD decomposition of the updated matrix is performed to find the singular values, and then every singular vector is controlled to be close to the unit vector. As given in Algorithm 3, the singular values of the updated weight matrix are restricted to the interval $[1/(1+ngh), 1+ngh]$ to avoid updating in the wrong direction. Here, *ngh* is the initial neighbourhood size, which is chosen between $(0, 1)$. As a result, W_{hh} can be updated over time without vanishing or exploding gradients.

Algorithm 3: Pseudo-code of SGD with the SVD operator.

Input: Learning rate:η, *Bee*, ngh : *neighbourhood radius*
Output: *Bee with Updated Parameters*

Function SGDSVD(η, θ):
  for *each* $w \in \theta = (W_{xh}, W_{hh}, W_{hy}, b_h, b_y)$ **do**
    // Update θ by using gradient descent rule
    $\theta^{t+1} = \theta^t - \eta \nabla_\theta L(f(x_i, \theta), y_i))$
    Calculate SVD for each w :
    U, S, V^T = SVD(w)
    for *each* $s_i \in S$ **do**
      if $s_i \geq (1 + ngh)$ **then**
        $s_i = 1 + ngh$
      else if $s_i \leq 1/(1 + ngh)$ **then**
        $s_i = 1/(1 + ngh)$
    $Bee.w \leftarrow w_{SVD}$
  return *Bee*

4.4. Global Search Operator

Besides the enhanced local search procedures, the proposed algorithm also includes a global search operator that combines random sampling chances which is also a good strategy for escaping local optimum points of the solution space. The third bee in a colony is used for the random exploration for potential new solutions of the search space. If the updated random weights gave a better solution for the loss function, then the third bee is updated with new global searched weights. This procedure gives the advantage to escape getting trapped at local optima, which results in converging to the global optimum faster during the training process. Algorithm 4 shows the pseudo-code of the proposed enhanced Ternary Bees Algorithm (BA-3+). The source code of the proposed algorithm is given at Appendix A.

Algorithm 4: Pseudo-code of the enhanced Ternary Bees Algorithm (BA-3+) for training deep RNN model.

```
Input: nScout, learning rate:η, ngh: neighbourhood
radius, dataset
Function BA-3+(nScout,η, ngh, dataset):
    Start
    inputs ← CreateInputs(dataset)
    targets ← labels(y)
    items ← convert dataset to list of // x=sentences and t=targets
    for each (x^(i), t^(i)) ∈ items do
        Initialize population with ternary RNN_Bee
        while stopping criterion not met do
            Evaluate fitness of the population
            y, Loss ← FORWARD(RNN_Bee, x^(i))
            Sort population according to loss values
            local_bee ← LOCALSEARCH(best_Bee, ngh)
            Evaluate fitness of local_Bee
            if local_Bee better than best_Bee then
                // Update First Bee
                best_Bee = local_Bee
            SGDSVD_Bee ← SGDSVD(second_Bee, ngh)
            Evaluate fitness of SGDSVD_Bee
            if SGDSVD_Bee better than second_Bee then
                // Update Second Bee
                second_Bee = SGDSVD_Bee
            global_Bee ← GLOBALSEARCH(third_Bee)
            Evaluate fitness of global_Bee
            if global_Bee better than third_Bee then
                // Update Third Bee
                third_Bee = global_Bee
            Evaluate fitness of the new population
            Sort population according to loss values
            Best Model = Best Bee
    return Best Model (Best Bee)
```

In this study, sentiment analysis is considered as a binary classification problem. The F1-score or F1 measure was used as a statistical measure for the analysis of the binary classification problem in addition to the accuracy measure. F1 measure is calculated as follows:

$$F_1 = \frac{2\ Precision\ Recall}{Precision + Recall} \tag{15}$$

$$Precision = \frac{TP}{(TP + FP)} \tag{16}$$

$$Recall = \frac{TP}{(TP + FN)} \tag{17}$$

$$Accuracy = \frac{TP + TN}{(TP + TN + FP + FN)} \tag{18}$$

Precision is the total count of true positives divided by the total number of positive results. The recall is the total number of true positive results divided by the number of all samples that should have been classified as positive. F1 measure can also be defined as a harmonic mean of the precision and recall value. The next section reports the details of the proposed algorithm's experimental setup and performance results and benchmarks.

5. Results

5.1. Experimental Setup

The proposed algorithms were implemented using the Tensorflow library with Keras Sequential model in Python on the macOS Catalina on MacBook Pro, 3.1 GHz quad-core Intel Core i5 hardware. The proposed BA-3+ algorithm was run with batch size 1 for each $(x^{(i)} = (x_1, x_2, \ldots, x_{t-1}, x_t), t^{(i)})$ pair of datasets. Each dataset was divided into a training set (%80 of the dataset) and a validation set (%20 of the dataset) by using 5-fold cross-validation. The training was performed with BA-3+ and SGD according to BCE loss value over 100 independent runs, each involving 100 epochs. The BA-3+ training procedure was online learning and happened incrementally over each iteration, which means the learnable parameters were updated after each forward and backward propagation of each training sample [89]. Figure 4 represents the flowchart of the proposed classification model. The model implemented with Python Programming language with on Google Colab IDE [90] by using various tools and libraries, including TensorFlow [91], Keras [92], SciPy [93], NumPy [94], Pandas [95], and Matplotlib [96].

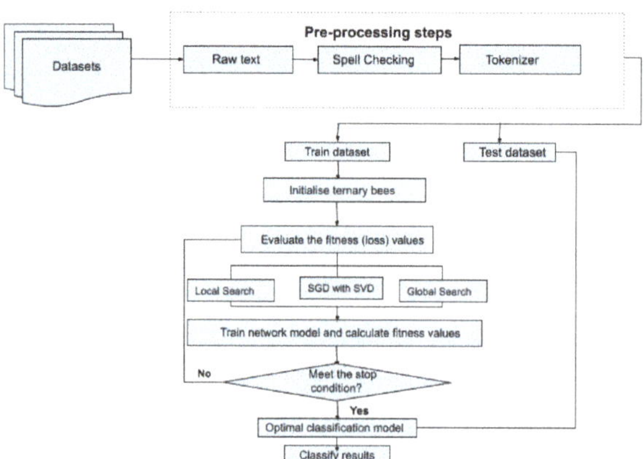

Figure 4. Flowchart of the proposed classification model.

5.2. Parameter Tuning

Table 5 reports the parameters of BA-3+. The number of scout bees represents the number of sequential RNN networks. Each training sample of the input data $x^{(i)} = (x_1, x_2, \ldots, x_{t-1}, x_t)$ is padded to the maximum sequence length after pre-processing steps of the given dataset. Nine widely-used sentiment classification datasets in Turkish and English from three different domains (movie reviews, multi-domain product reviews, and Twitter reviews) were adopted to verify the proposed algorithm.

Table 5. Parameter setting.

Parameter	Value
$x^{(i)} = (x_1, x_2, \ldots, x_{t-1}, x_t)$	max length = 200
nScout	3
nEiteSiteBee	1
nSelectedSiteBee	1
ngh	0.5
nhidden	32
epoch	100
independent runs (solution numbers)	100
batch size	1
upper limit of max singular value	$1 + ngh$
lower limit of min singular value	$1/(1 + ngh)$
η (learning rate)	0.01

For each dataset, the sequential model was implemented with the same hyperparameters for the sake of fair comparison. Every sequential model was constructed with the input layer, hidden RNN layers, and the output layer. The embedding layer was used as the input layer, which converts the indexes of the sequential input to the fixed size dense vectors as an input of the model. The vocabulary size of each dataset was used as the input dimension of the embedding layer. The random uniform function was used as a weight initialiser for both the embedding layer and hidden layers.

As a critical hyper-parameter, the neighbourhood size (ngh) of the BA-3+ is set to 0.5. The learning rate (η) of the SGD is set to 0.01. The neighbourhood size is used for both the local search and the stabilisation of the largest and the smallest eigenvalue of the updated parameters. The number of hidden layers is set to 32 for each model. Each element of the learnable (trainable) parameter of the $\theta = (W_{xh}, W_{hh}, W_{hy}, b_h, b_y)$, the dimension of the weight matrices is changed according to hidden layer numbers, and the corresponding dimensions of the weight matrix can be seen in Table 4, in Section 4.2.

5.3. Results and Discussion

Table 6 reports the best and average accuracy, loss, and F1 measures obtained by both the BA-3+ and SGD algorithms. The accuracy and loss values of the training and validation datasets are given as percentages. The average values were calculated after 100 independent training runs. Figure 5 shows the loss values for each dataset after one independent run which contains 100 training epochs. According to experimental results, the proposed BA-3+ algorithm guarantees fast convergence to the optimum value and the lowest error rate for each dataset. As can be clearly seen in Figures 6 and 7, BA-3+ obtained the best solution within the first 20 iterations for each dataset. Besides, the gap between training and validation performance is small, which means BA-3+ prevents overfitting.

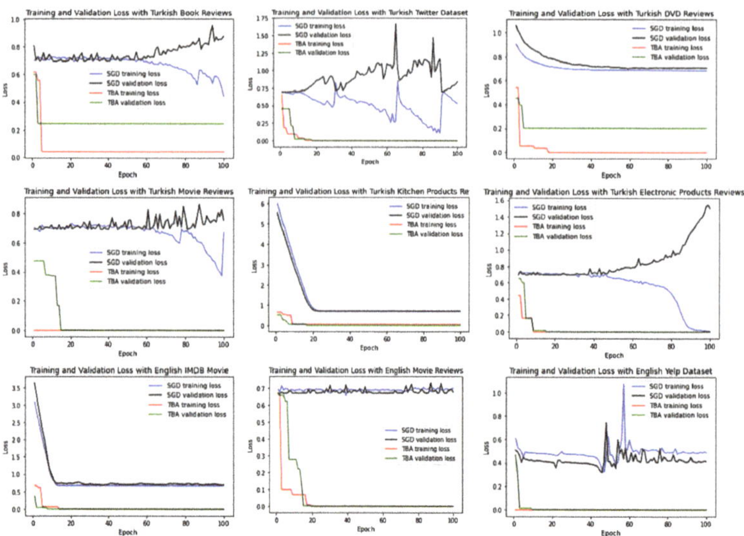

Figure 5. Comparison of the loss values of BA-3+ (TBA) and SGD for one independent run.

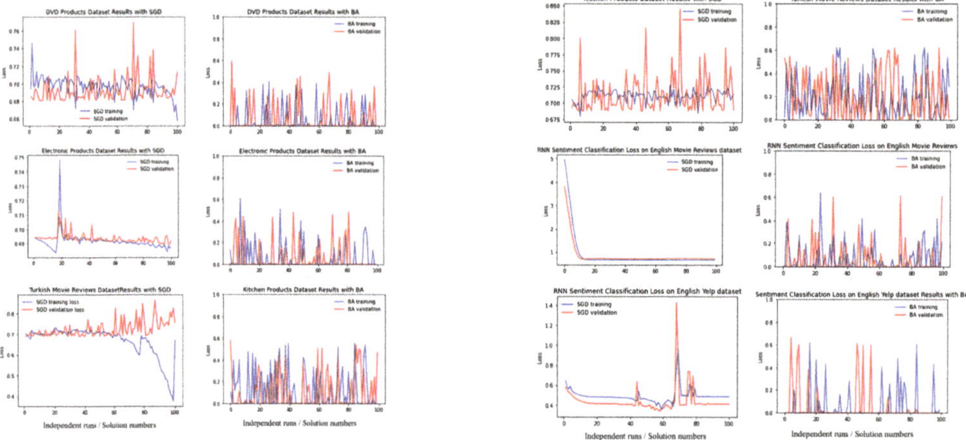

Figure 6. Distributions of the loss values of the training and validation dataset for 100 independent runs.

Figures 6 and 7 represent the distributions of the loss values for both BA-3+ and SGD over 100 independent runs. Table 6 reports the best and average accuracy and loss values of the training datasets and validation datasets. BA-3+ performs better compared to traditional SGD. Results showed that BA-3+ can be used as an effective learning method for the sentiment classification task. The performances of BA-3+ were better both in terms of the best and average accuracy compared to SGD. Similarly, the best and average values of BA-3+ were lower than for SGD for each dataset.

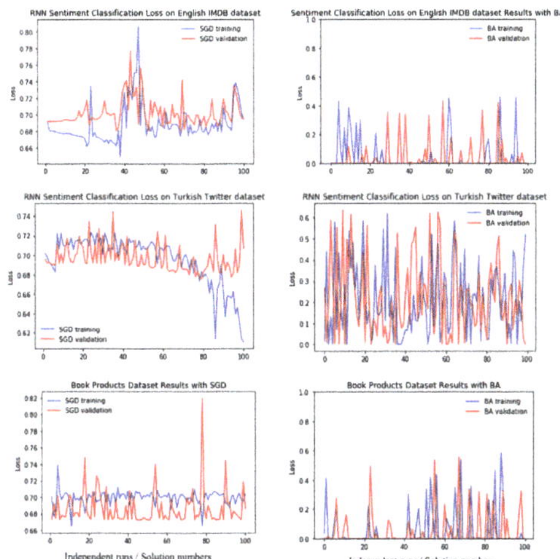

Figure 7. Distributions of the loss values of the training and validation dataset for 100 independent runs.

Table 6. Comparison of the results of 100 independent experiments with 100 epochs.

		Accuracy Results				Loss Results				F1 Score	
Datasets	**Alg.**	$Train_{Best}$	Val_{Best}	$Train_{Avg}$	Val_{Avg}	$Train_{Best}$	Val_{Best}	$Train_{Avg}$	Val_{Avg}	$Train_{Avg}$	Val_{Avg}
TR Book	BA-3+	0.99	0.99	0.801	0.882	0.00	0.00	0.0769	0.117	0.81	0.78
	SGD	0.73	0.64	0.527	0.51	0.686	0.688	0.695	0.697	0.68	0.67
TR DVD	BA-3+	0.99	0.99	0.923	0.91	0.00	0.00	0.076	0.089	0.80	0.78
	SGD	0.52	0.39	0.519	0.389	0.686	0.701	0.708	0.748	0.70	0.70
TR Elect.	BA-3+	0.99	0.99	0.915	0.916	0.00	0.00	0.084	0.083	0.83	0.81
	SGD	0.99	0.58	0.615	0.549	0.092	0.684	0.640	0.849	0.75	0.73
TR Kitchen	BA-3+	0.99	0.99	0.81	0.810	0.00	0.00	0.189	0.191	0.81	0.75
	SGD	0.52	0.54	0.507	0.525	0.692	0.689	0.640	0.813	0.70	0.67
EN IMDB	BA-3+	0.99	0.99	0.911	0.90	0.00	0.00	0.032	0.006	0.77	0.75
	SGD	0.58	0.47	0.579	0.469	0.68	0.693	0.81	0.870	0.71	0.70
TR Twitter	BA-3+	0.99	0.99	0.914	0.830	0.00	0.00	0.017	0.029	0.76	0.73
	SGD	0.99	0.55	0.747	0.456	0.112	0.691	0.51	0.909	0.59	0.57
TR Movie	BA-3+	0.99	0.99	0.93	0.855	0.00	0.00	0.00	0.05	0.80	0.77
	SGD	0.9	0.58	0.58	0.494	0.376	0.689	0.668	0.725	0.66	0.63
EN Movie	BA-3+	0.99	0.99	0.896	0.87	0.00	0.00	0.024	0.058	0.81	0.80
	SGD	0.640	0.610	0.585	0.601	0.655	0.668	0.69	0.678	0.77	0.75
EN Yelp	BA-3+	0.99	0.99	0.87	0.86	0.00	0.00	0.030	0.02	0.75	0.70
	SGD	0.87	0.809	0.79	0.854	0.325	0.318	0.490	0.424	0.61	0.57

BA-3+ has been compared with Differential Evolution (DE) and Particle Swarm Optimization (PSO) algorithms. For the sake of comparison, PSO has been evaluated with three particles as we propose to employ only three scout bees over 100 epochs. However, DE has been employed with 100 generations since it gave too low accuracy when it employees with only three generations. Table 7 reports the time consumption of the SGD, BA-3+, DE, and PSO algorithms in second. As can be seen in Table 8, BA-3+ has outperformed the DE and PSO, and its computation time of the BA-3+ is lower from DE and PSO. Although BA-3+ time consumption was longer than standard SGD, the accuracy value and F1 measure have improved for all classification datasets, at least with a 30–40% improvement. Furthermore, BA-3+ was more stable and converged faster than the DE and PSO algorithms since it employs only three individual bees as a population. As is expected SGD model has the lowest computation time, but the accuracy result is also lower compared to all other metaheuristic training algorithms.

Table 7. Total training time (sec) of the BA-3+, DE, PSO and SGD algorithms.

Datasets	SGD Total Time	BA3+ Total Time	DE Total Time	PSO Total Time
TR Book	237.36	393.139	871.608	544.914
TR DVD	233.92	407.187	860.439	520.700
TR Elect.	233.15	411.297	1317.971	512.482
TR Kitchen	235.73	417.765	855.125	513.929
TR Movie	1021.38	1394.631	3827.576	3969.091
TR Twitter	4334.98	8340.9	16,178.52	8978.6
EN IMDB	12,100.00	22,347.5	31,724.76	25,674.5
EN Movie	1293.718	2206.8	3361.95	2399.71
EN Yelp	4691.78	12,112.8	21,157.6	13,894.9

Table 8. Comparison of BA-3+ performance with DE and PSO and SGD.

Datasets	SGD Acc.	BA3+ Acc.	DE Acc.	PSO Acc.
TR Book	0.527	0.801	0.789	0.690
TR DVD	0.519	0.923	0.808	0.678
TR Elect.	0.58	0.915	0.786	0.696
TR Kitchen	0.507	0.81	0.794	0.686
TR Movie	0.58	0.93	0.887	0.759
TR Twitter	0.747	0.914	0.74	0.709
EN IMDB	0.579	0.91	0.778	0.701
EN Movie	0.585	0.896	0.78	0.69
EN Yelp	0.79	0.87	0.71	0.68

Since we aim to improve the learning capacity of RNN as good as advanced deep learning language models such as LSTMs, we compared the proposed training algorithm with the advanced neural language models, including chain-structured and tree-structured language models. For this purpose, LSTM has been modeled with the same hidden layer number and training optimizer. Tree-LSTM has been modeled with similar hyperparameters to the RNTN model, which operates over MS-TR treebank [97]. The results of the Recursive Neural Tensor Network (RNTN) model have been taken from [97]. Table 9 reports the hyperparameters of the advanced deep language models and Table 10 reports comparisons of accuracy results of advanced recurrent and recursive language models for Turkish binary classification datasets over test dataset. According to the experiments, it is observed that the RNN model combined with the BA-3+ training algorithm performed close or as good as advanced recurrent and recursive deep language models and gave comparable results.

Table 9. Parameter setting for LSTM and Tree-LSTM models.

Parameter	Value
$x^{(i)} = (x_1, x_2, \ldots, x_{t-1}, x_t)$	max length = 50
embedding dimension	300
nhidden	32
epoch	100
batch size	32
optimizer	SGD
learning rate	0.01

Table 10. Comparisons of average accuracy results of advanced recurrent and recursive language models for Turkish binary classification datasets.

Models	Book	Electr.	DVD	Kitchen	Movie	Twitter
LSTM	0.823	0.725	0.751	0.75	0.835	0.895
Tree-LSTM	0.883	0.853	0.85	0.82	0.88	0.89
RNTN	0.86	0.866	0.824	0.798	0.88	0.835
RNN BA3+	0.88	0.801	0.866	0.818	0.854	0.838
RNN SGD	0.808	0.75	0.734	0.704	0.721	0.744

As it can be clearly seen from Figure 8, RNN-SGD performed well for only one dataset. Accordingly, we can say that the performance of the systems can be better when the models are trained with huge datasets such as the Twitter classification dataset (see Table 2). However, in all other cases, we found that the BA algorithm performed well and yielding results just as well as advanced language models. The experimental results demonstrate that BA-3+ gives us a chance to get rid of the disadvantages of the SGD algorithm and can handle the VEG problem. Although BA-3+ time consumption is longer than SGD, the accuracy value and F1 measure are higher for all classification datasets. Furthermore, since BA-3+ employs only three scout bees, the total training time is shorter than the DE and PSO algorithms and it outperformed the DE and PSO (see Tables 7 and 8). BA-3+ and PSO runtimes are similar, but BA-3+ gave better results in terms of accuracy. Even though the DE algorithm was tested for 100 generations, it could not reach the BA-3+ performance.

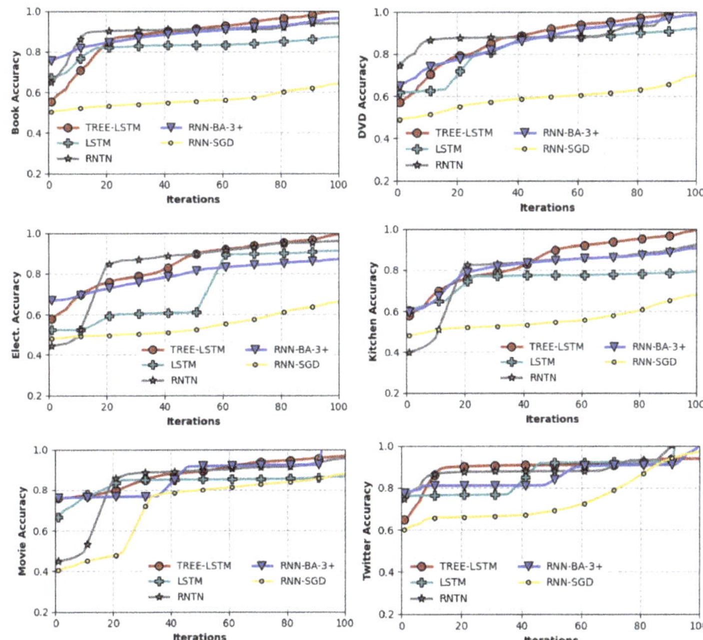

Figure 8. Comparison of BA-3+ performance with advanced models and RNN model trained with SGD for some datasets.

The success of BA-3+ depends on its hybrid metaheuristic nature. BA-3+ evaluates only three candidate models, each having the same deep RNN architecture with different learnable parameters. The training process starts with these three candidate models (number of scout bees), which dynamically search for the optimum values of the learnable parameters, and continues selecting the best model for local search to find better solutions.

This is the feature of BA-3+ that brings good initial parameters for the iterative learning process. After each iteration of the proposed algorithm, if more optimum values of the learnable parameters are found, they will be updated incrementally. This feature of BA-3+ yields faster convergence. Additionally, as BA-3+ combines local SGD training with the SVD, it can exploit the learning ability of SGD while controlling vanishing and exploding gradients, making BA-3+ a hybrid meta-learning algorithm combining gradient-free and gradient-based optimisation. Finally, BA-3+ has a random exploration operator, namely, a global search operator, which enables the exploration of new promising solutions while preventing the optimisation process from being trapped at local stationary points. Besides these advantages, critical hyperparameters, such as learning rate and neighbourhood size, were empirically selected for BA-3+ as for SGD.

In the future, the proposed algorithm will be compared to the recent metaheuristic methods, such as advanced backtracking search optimisation algorithm (ABSA) [77], which has proved its performance superiority compared to the GA, DE, and backtracking search optimization algorithm (BSA) for global optimisation [76]. Since we recommend using only three individual bees as population size, we think it would be unfair to compare BA-3+ with all other population-based metaheuristic algorithms using higher population sizes without any SGD-SVD local search. Therefore, the previous population-based metaheuristics should be improved with hybrid approaches to work well with lower population sizes in the future. Furthermore, the hyperparameter tuning study will be done since the deep learning model and metaheuristic algorithms are sensible to critical hyperparameters. The proposed algorithm will also be tested for fine-grained classification problems with more than two classes. Although the accuracy rate of the proposed algorithm is high, strategies should be found to ensure faster convergence in terms of time. Learning large datasets may also require the parallel running of the proposed algorithm and more powerful hardware resources.

6. Conclusions

This paper has described the use of the Ternary Bees Algorithm (BA-3+) as a training algorithm for finding the optimal set of parameters of a sequential deep RNN language model for the sentiment classification task. BA-3+ has been modeled as a sequential model and tested on nine different datasets including Turkish and English text. The model conducts an exploitative local search with the best bee and an explorative global search with the worst bee. The in-between bee is used for improved Stochastic Gradient Descent (SGD) learning with Singular Value Decomposition (SVD) to stabilise the updated model parameters after the application of SGD. This strategy is adopted to prevent the vanishing and exploding gradients problem of SGD training. The proposed BA-3+ algorithm guarantees fast convergence as it combines local search, global search, and SGD learning with SVD, and it is faster than other iterative, metaheuristic algorithms, as it employs only three candidate solutions in each training step. BA-3+ has been tested on the sentiment classification task with different datasets, and comparative results were obtained for chain-structured and tree-structured deep language models, Differential Evolution (DE), and Particle Swarm Optimization (PSO) algorithms. BA-3+ converged to the global minimum faster compared to the DE and PSO algorithms and it outperformed the SGD, DE, and PSO algorithms both for the Turkish and English datasets. According to the experimental results, BA-3+ performed better compared to the standard SGD training algorithm and RNN combined with BA-3+ performs as good as for advanced deep neural language models. The small differences between the training and validation results for the nine datasets have confirmed the efficiency of the proposed algorithm, with BA-3+ indeed offering better generalisation and faster convergence than the SGD algorithm.

Author Contributions: Conceptualization, S.Z., D.T.P.; methodology, S.Z.; validation, S.Z., D.T.P., E.K., A.S.; formal analysis, S.Z.; investigation, S.Z.; writing—original draft preparation, S.Z., D.T.P.; writing—review and editing, S.Z., D.T.P., E.K., A.S.; visualization, S.Z.; supervision, D.T.P., E.K., A.S.; funding acquisition, S.Z. All authors have read and agreed to the published version of the manuscript.

Funding: This research was funded by The Scientific and Technological Research Council of Turkey (TUBITAK), 2214-A International Research Fellowship Programme, Grant No. 1059B141800193.

Institutional Review Board Statement: Not applicable.

Informed Consent Statement: Not applicable.

Acknowledgments: This research was supported by The Scientific and Technological Research Council of Turkey (TUBITAK), (Grant No: 1059B14180019) and Fatih Sultan Mehmet Vakif University (FSMVU).

Conflicts of Interest: The authors declare no conflict of interest. The funders had no role in the design of the study; in the collection, analyses, or interpretation of data; in the writing of the manuscript, or in the decision to publish the results.

Appendix A

The source code of the proposed method can be found at https://tinyurl.com/4hh8msy2 (accessed on 24 July 2021).

References

1. Mikolov, T.; Karafiát, M.; Burget, L.; Jan, C.; Khudanpur, S. Recurrent neural network based language model. In Proceedings of the 11th Annual Conference of the International Speech Communication Association, Interspeech 2010, Makuhari, Japan, 26–30 September 2010; pp. 1045–1048.
2. Che, Z.; Purushotham, S.; Cho, K.; Sontag, D.; Liu, Y. Recurrent Neural Networks for Multivariate Time Series with Missing Values. *Sci. Rep.* **2018**, *8*. [CrossRef]
3. Kalchbrenner, N.; Blunsom, P. Recurrent Continuous Translation Models. In Proceedings of the 2013 Conference on Empirical Methods in Natural Language Processing, Association for Computational Linguistics, Seattle, WA, USA, 18–21 October 2013; pp. 1700–1709.
4. You, Q.; Jin, H.; Wang, Z.; Fang, C.; Luo, J. Image captioning with semantic attention. In Proceedings of the IEEE Computer Society Conference on Computer Vision and Pattern Recognition, Las Vegas, NV, USA, 27–30 June 2016. [CrossRef]
5. Pascanu, R.; Mikolov, T.; Bengio, Y. On the Difficulty of Training Recurrent Neural Networks. In Proceedings of the 30th International Conference on International Conference on Machine Learning-Volume 28, JMLR.org (ICML'13), Atlanta, GA, USA, 16–21 June 2013; pp. III-1310–III-1318.
6. Lecun, Y.; Bengio, Y.; Hinton, G. *Deep Learning*; MIT Press: Cambridge, MA, USA, 2015. [CrossRef]
7. Bengio, Y.; Simard, P.; Frasconi, P. Learning Long-Term Dependencies with Gradient Descent is Difficult. *IEEE Trans. Neural Netw.* **1994**. [CrossRef] [PubMed]
8. Hochreiter, S.; Schmidhuber, J. Long Short-Term Memory. *Neural Comput.* **1997**. [CrossRef] [PubMed]
9. Chung, J.; Gulcehre, C.; Cho, K.; Bengio, Y. Gated Feedback Recurrent Neural Networks. In Proceedings of the 32nd International Conference on Machine Learning-Volume 37. JMLR.org (ICML'15), Lille, France, 6–11 July 2015; pp. 2067–2075.
10. Sutskever, I.; Hinton, G. Learning multilevel distributed representations for high-dimensional sequences. *J. Mach. Learn. Res.* **2007**, 2, 548–555.
11. Sutskever, I. Training Recurrent Neural Networks. Ph.D. Thesis, University of Toronto, Toronto, ON, Canada, 2013.
12. Srivastava, N.; Hinton, G.; Krizhevsky, A.; Sutskever, I.; Salakhutdinov, R. Dropout: A simple way to prevent neural networks from overfitting. *J. Mach. Learn. Res.* **2014**, *15*, 1929–1958.
13. Darwish, A.; Hassanien, A.E.; Das, S. A survey of swarm and evolutionary computing approaches for deep learning. *Artif. Intell. Rev.* **2020**. [CrossRef]
14. Pham, D.T.; Ghanbarzadeh, A.; Koç, E.; Otri, S.; Rahim, S.; Zaidi, M. The Bees Algorithm-A Novel Tool for Complex Optimisation Problems. In Proceedings of the Intelligent Production Machines and Systems-2nd I*PROMS Virtual International Conference, Online, 3–14 July 2006. [CrossRef]
15. Pham, D.T.; Koç, E.; Ghanbarzadeh, A. *Optimization of the Weights of Multi-Layered Perceptions Using the Bees Algorithm.* In Proceedings of the International Symposium on Intelligent Manufacturing Systems, Online, 3–14 July 2006; Volume 5, pp. 38–46.
16. Ismail, A.H.; Hartono, N.; Zeybek, S.; Pham, D.T. Using the Bees Algorithm to solve combinatorial optimisation problems for TSPLIB. In Proceedings of the IOP Conference Series: Materials Science and Engineering, Batu, Indonesia, 17–19 March 2020; Volume 847. [CrossRef]
17. Laili, Y.; Tao, F.; Pham, D.T.; Wang, Y.; Zhang, L. Robotic disassembly re-planning using a two-pointer detection strategy and a super-fast bees algorithm. *Robot. Comput. Integr. Manuf.* **2019**. [CrossRef]
18. Jaeger, H.; Haas, H. Harnessing Nonlinearity: Predicting Chaotic Systems and Saving Energy in Wireless Communication. *Science* **2004**. [CrossRef]
19. Greff, K.; Srivastava, R.K.; Koutnik, J.; Steunebrink, B.R.; Schmidhuber, J. LSTM: A Search Space Odyssey. *IEEE Trans. Neural Netw. Learn. Syst.* **2017**. [CrossRef]

20. Martens, J.; Sutskever, I. Learning recurrent neural networks with Hessian-free optimization. In Proceedings of the 28th International Conference on Machine Learning (ICML 2011), Bellevue, WA, USA, 28 June–2 July 2011.
21. Yang, S.; Chen, D.; Li, S.; Wang, W. Carbon price forecasting based on modified ensemble empirical mode decomposition and long short-term memory optimized by improved whale optimization algorithm. *Sci. Total Environ.* **2020**, *716*, 137117. [CrossRef] [PubMed]
22. Peng, L.; Zhu, Q.; Lv, S.X.; Wang, L. Effective long short-term memory with fruit fly optimization algorithm for time series forecasting. *Soft Comput.* **2020**, *24*, 15059–15079. [CrossRef]
23. ElSaid, A.E.R.; El Jamiy, F.; Higgins, J.; Wild, B.; Desell, T. Optimizing long short-term memory recurrent neural networks using ant colony optimization to predict turbine engine vibration. *Appl. Soft Comput. J.* **2018**. [CrossRef]
24. Rashid, T.A.; Fattah, P.; Awla, D.K. Using Accuracy Measure for Improving the Training of LSTM with Metaheuristic Algorithms. *Procedia Comput. Sci.* **2018**, *140*, 324–333. [CrossRef]
25. Srivastava, R.K.; Greff, K.; Schmidhuber, J. Training Very Deep Networks. In Proceedings of the 28th International Conference on Neural Information Processing Systems-Volume 2 (NIPS'15), Montreal, QC, Canada, 7–12 December 2015; MIT Press: Cambridge, MA, USA, 2015; pp. 2377–2385.
26. Somu, N.; M R, G.R.; Ramamritham, K. A hybrid model for building energy consumption forecasting using long short term memory networks. *Appl. Energy* **2020**, *261*, 114131. [CrossRef]
27. Bouktif, S.; Fiaz, A.; Ouni, A.; Serhani, M.A. Multi-Sequence LSTM-RNN Deep Learning and Metaheuristics for Electric Load Forecasting. *Energies* **2020**, *13*, 391. [CrossRef]
28. Nair, V.; Hinton, G.E. Rectified linear units improve Restricted Boltzmann machines. In Proceedings of the ICML 2010-Proceedings, 27th International Conference on Machine Learning, Haifa, Israel, 21–24 June 2010.
29. Glorot, X.; Bordes, A.; Bengio, Y. Deep sparse rectifier neural networks. *J. Mach. Learn. Res.* **2011**, *15*, 315–323.
30. Glorot, X.; Bengio, Y. Understanding the difficulty of training deep feedforward neural networks. *J. Mach. Learn. Res.* **2010**, *9*, 249–256.
31. Clevert, D.A.; Unterthiner, T.; Hochreiter, S. Fast and accurate deep network learning by exponential linear units (ELUs). In Proceedings of the 4th International Conference on Learning Representations, ICLR 2016-Conference Track Proceedings, San Juan, PR, USA, 2–4 May 2016.
32. Bengio, Y.; Boulanger-Lewandowski, N.; Pascanu, R. Advances in optimizing recurrent networks. In Proceedings of the 2013 IEEE International Conference on Acoustics, Speech and Signal Processing, Vancouver, BC, Canada, 26–31 May 2013; pp. 8624–8628. [CrossRef]
33. Le, Q.V.; Jaitly, N.; Hinton, G.E. A Simple Way to Initialize Recurrent Networks of Rectified Linear Units. *arXiv* **2015**, arXiv:1504.00941.
34. Xu, X.; Ge, H.; Li, S. An improvement on recurrent neural network by combining convolution neural network and a simple initialization of the weights. In Proceedings of the 2016 the IEEE International Conference of Online Analysis and Computing Science (ICOACS 2016), Chongqing, China, 28–29 May 2016. [CrossRef]
35. Vorontsov, E.; Trabelsi, C.; Kadoury, S.; Pal, C. On orthogonality and learning recurrent networks with long term dependencies. In Proceedings of the 34th International Conference on Machine Learning (ICML 2017), Sydney, Australia, 6–11 August 2017.
36. Zhang, J.; Lei, Q.; Dhillon, I. Stabilizing Gradients for Deep Neural Networks via Efficient SVD Parameterization. In Proceedings of the 35th International Conference on Machine Learning (PMLR), Stockholm, Sweden, 10–15 July 2018; Dy, J., Krause, A., Eds.; Volume 80, pp. 5806–5814.
37. Becker, S.; le Cun, Y. Improving the Convergence of Back-Propagation Learning with Second Order Methods. In Proceedings of the 1988 Connectionist Models Summer School, San Francisco, CA, USA, 17–26 June 1989; pp. 29–37.
38. Kag, A.; Saligrama, V. Training Recurrent Neural Networks via Forward Propagation Through Time. In Proceedings of the 38th International Conference on Machine Learning (PMLR), Virtual, Online, 18–24 July 2021; Volume 139, pp. 5189–5200.
39. Yang, X.S. *Nature-Inspired Metaheuristic Algorithms*; Luniver Press: London, UK, 2008; ISBN 1905986106.
40. Chiroma, H.; Gital, A.Y.; Rana, N.; Abdulhamid, S.M.; Muhammad, A.N.; Umar, A.Y.; Abubakar, A.I. Nature Inspired Meta-heuristic Algorithms for Deep Learning: Recent Progress and Novel Perspective. *Adv. Intell. Syst. Comput.* **2020**, *943*, 59–70. [CrossRef]
41. Chong, H.Y.; Yap, H.J.; Tan, S.C.; Yap, K.S.; Wong, S.Y. Advances of metaheuristic algorithms in training neural networks for industrial applications. *Soft Comput.* **2021**. [CrossRef]
42. Ouaarab, A.; Osaba, E.; Bouziane, M.; Bencharef, O. Review of Swarm Intelligence for Improving Time Series Forecasting. In *Applied Optimization and Swarm Intelligence*; Osaba, E., Yang, X.S., Eds.; Springer: Singapore, 2021; pp. 61–79. [CrossRef]
43. Muñoz-Ordóñez, J.; Cobos, C.; Mendoza, M.; Herrera-Viedma, E.; Herrera, F.; Tabik, S. Framework for the Training of Deep Neural Networks in TensorFlow Using Metaheuristics. In *Intelligent Data Engineering and Automated Learning–IDEAL 2018*; Yin, H., Camacho, D., Novais, P., Tallón-Ballesteros, A.J., Eds.; Springer International Publishing: Cham, Switzerland, 2018; pp. 801–811.
44. Ibrahim, A.M.; El-Amary, N.H. Particle Swarm Optimization trained recurrent neural network for voltage instability prediction. *J. Electr. Syst. Inf. Technol.* **2018**, *5*, 216–228. [CrossRef]
45. Duchanoy, C.A.; Moreno-Armendáriz, M.A.; Urbina, L.; Cruz-Villar, C.A.; Calvo, H.; Rubio, J.D.J. A novel recurrent neural network soft sensor via a differential evolution training algorithm for the tire contact patch. *Neurocomputing* **2017**, *235*, 71–82. [CrossRef]

46. Desell, T.; Clachar, S.; Higgins, J.; Wild, B. Evolving deep recurrent neural networks using ant colony optimization. In *European Conference on Evolutionary Computation in Combinatorial Optimization;* Lecture Notes in Computer Science (Including Subseries Lecture Notes in Artificial Intelligence and Lecture Notes in Bioinformatics); Springer International Publishing: Cham, Switzerland, 2015; Volume 9026, pp. 86–98. [CrossRef]
47. Juang, C.F. A Hybrid of Genetic Algorithm and Particle Swarm Optimization for Recurrent Network Design. *IEEE Trans. Syst. Man Cybern. Part Cybern.* **2004**. [CrossRef]
48. Stanley, K.O.; Miikkulainen, R. Evolving neural networks through augmenting topologies. *Evol. Comput.* **2002**. [CrossRef]
49. Ororbia, A.; ElSaid, A.E.R.; Desell, T. Investigating recurrent neural network memory structures using neuro-evolution. In Proceedings of the GECCO 2019-Proceedings of the 2019 Genetic and Evolutionary Computation Conference, Prague, Czech Republic, 13–17 July 2019; pp. 446–455. [CrossRef]
50. Wang, H.; Wang, H.; Xu, K. Evolutionary recurrent neural network for image captioning. *Neurocomputing* **2020**, *401*, 249–256. [CrossRef]
51. Camero, A.; Toutouh, J.; Alba, E. Random error sampling-based recurrent neural network architecture optimization. *Eng. Appl. Artif. Intell.* **2020**, *96*, 103946. [CrossRef]
52. Mo, H.; Custode, L.L.; Iacca, G. Evolutionary neural architecture search for remaining useful life prediction. *Appl. Soft Comput.* **2021**, *108*, 107474. [CrossRef]
53. Kang, Q.; Liao, W.K.; Agrawal, A.; Choudhary, A. A hybrid training algorithm for recurrent neural network using particle swarm optimization-based preprocessing and temporal error aggregation. In Proceedings of the IEEE International Conference on Data Mining Workshops, ICDMW, New Orleans, LA, USA, 18–21 November 2017. [CrossRef]
54. Eberhart, R.; Kennedy, J. New optimizer using particle swarm theory. In Proceedings of the International Symposium on Micro Machine and Human Science, Nagoya, Japan, 4–6 October 1995. [CrossRef]
55. Ge, H.W.; Liang, Y.C.; Marchese, M. A modified particle swarm optimization-based dynamic recurrent neural network for identifying and controlling nonlinear systems. *Comput. Struct.* **2007**. [CrossRef]
56. Xiao, P.; Venayagamoorthy, G.K.; Corzine, K.A. Combined training of recurrent neural networks with particle swarm optimization and backpropagation algorithms for impedance identification. In Proceedings of the 2007 IEEE Swarm Intelligence Symposium (SIS 2007), Honolulu, HI, USA, 1–5 April 2007. [CrossRef]
57. Zhang, N.; Behera, P.K.; Williams, C. Solar radiation prediction based on particle swarm optimization and evolutionary algorithm using recurrent neural networks. In Proceedings of the SysCon 2013-7th Annual IEEE International Systems Conference, Proceedings, Orlando, FL, USA, 15–18 April 2013. [CrossRef]
58. Cai, X.; Zhang, N.; Venayagamoorthy, G.K.; Wunsch, D.C. Time series prediction with recurrent neural networks trained by a hybrid PSO-EA algorithm. *Neurocomputing* **2007**. [CrossRef]
59. Blanco, A.; Delgado, M.; Pegalajar, M.C. A real-coded genetic algorithm for training recurrent neural networks. *Neural Netw.* **2001**. [CrossRef]
60. Nawi, N.M.; Khan, A.; Rehman, M.Z.; Chiroma, H.; Herawan, T. Weight Optimization in Recurrent Neural Networks with Hybrid Metaheuristic Cuckoo Search Techniques for Data Classification. *Math. Probl. Eng.* **2015**, *2015*, 868375. [CrossRef]
61. Chuangxin, G.; Gen, Y.; Chengzhi, Z.; Xueping, W.; Xiu, C. SoC estimation for lithium-ion battery using recurrent NARX neural network and genetic algorithm. In Proceedings of the IOP Conference Series: Materials Science and Engineering, Hangzhou, China, 28–31 March 2019. [CrossRef]
62. Bas, E.; Egrioglu, E.; Kolemen, E. Training simple recurrent deep artificial neural network for forecasting using particle swarm optimization. *Granul. Comput.* **2021**. [CrossRef]
63. Kaya, E.; Baştemur Kaya, C. A Novel Neural Network Training Algorithm for the Identification of Nonlinear Static Systems: Artificial Bee Colony Algorithm Based on Effective Scout Bee Stage. *Symmetry* **2021**, *13*, 419. [CrossRef]
64. Shettigar, A.K.; Patel, G.C.M.; Chate, G.R.; Vundavilli, P.R.; Parappagoudar, M.B. Artificial bee colony, genetic, back propagation and recurrent neural networks for developing intelligent system of turning process. *SN Appl. Sci.* **2020**, *2*, 660. [CrossRef]
65. Shishira, S.R.; Kandasamy, A. BeeM-NN: An efficient workload optimization using Bee Mutation Neural Network in federated cloud environment. *J. Ambient Intell. Humaniz. Comput.* **2021**, *12*, 3151–3167. [CrossRef]
66. Ruiz, L.G.B.; Capel, M.I.; Pegalajar, M.C. Parallel memetic algorithm for training recurrent neural networks for the energy efficiency problem. *Appl. Soft Comput.* **2019**, *76*, 356–368. [CrossRef]
67. Hu, H.; Wang, H.; Bai, Y.; Liu, M. Determination of endometrial carcinoma with gene expression based on optimized Elman neural network. *Appl. Math. Comput.* **2019**, *341*, 204–214. [CrossRef]
68. Tian, Y.; Peng, S.; Zhang, X.; Rodemann, T.; Tan, K.C.; Jin, Y. A Recommender System for Metaheuristic Algorithms for Continuous Optimization Based on Deep Recurrent Neural Networks. *IEEE Trans. Artif. Intell.* **2020**, *1*, 5–18. [CrossRef]
69. Roy, K.; Mandal, K.K.; Mandal, A.C. Ant-Lion Optimizer algorithm and recurrent neural network for energy management of micro grid connected system. *Energy* **2019**, *167*, 402–416. [CrossRef]
70. Ab Aziz, M.F.; Mostafa, S.A.; Mohd Foozy, C.F.; Mohammed, M.A.; Elhoseny, M.; Abualkishik, A.Z. Integrating Elman recurrent neural network with particle swarm optimization algorithms for an improved hybrid training of multidisciplinary datasets. *Expert Syst. Appl.* **2021**, *183*, 115441. [CrossRef]
71. Hassib, E.M.; El-Desouky, A.I.; Labib, L.M.; El-kenawy, E.S.M. WOA + BRNN: An imbalanced big data classification framework using Whale optimization and deep neural network. *Soft Comput.* **2020**, *24*, 5573–5592. [CrossRef]

72. Shah, H.; Chiroma, H.; Herawan, T.; Ghazali, R.; Tairan, N. An Efficient Bio-inspired Bees Colony for Breast Cancer Prediction. In *Proceedings of the International Conference on Data Engineering 2015 (DaEng-2015)*; Abawajy, J.H., Othman, M., Ghazali, R., Deris, M.M., Mahdin, H., Herawan, T., Eds.; Springer: Singapore, 2019; pp. 597–608.
73. Kumar, K.; Haider, M.T.U. Enhanced Prediction of Intra-day Stock Market Using Metaheuristic Optimization on RNN–LSTM Network. *New Gener. Comput.* **2021**, *39*, 231–272. [CrossRef]
74. Martinez, A.D.; Del Ser, J.; Villar-Rodriguez, E.; Osaba, E.; Poyatos, J.; Tabik, S.; Molina, D.; Herrera, F. Lights and shadows in Evolutionary Deep Learning: Taxonomy, critical methodological analysis, cases of study, learned lessons, recommendations and challenges. *Inf. Fusion* **2021**, *67*, 161–194. [CrossRef]
75. Wolpert, D.H.; Macready, W.G. No Free Lunch Theorems for Optimization. *Trans. Evol. Comp* **1997**, *1*, 67–82. [CrossRef]
76. Zhang, Y. Backtracking search algorithm with specular reflection learning for global optimization. *Knowl. Based Syst.* **2021**, *212*, 106546. [CrossRef]
77. Wang, L.; Peng, L.; Wang, S.; Liu, S. Advanced backtracking search optimization algorithm for a new joint replenishment problem under trade credit with grouping constraint. *Appl. Soft Comput.* **2020**, *86*, 105953. [CrossRef]
78. Rumelhart, D.E.; Hinton, G.E.; Williams, R.J. Learning Internal Representations by Error Propagation. In *Readings in Cognitive Science: A Perspective from Psychology and Artificial Intelligence*; Elsevier: Amsterdam, The Netherlands, 2013. [CrossRef]
79. Demirtas, E.; Pechenizkiy, M. Cross-lingual polarity detection with machine translation. In Proceedings of the 2nd International Workshop on Issues of Sentiment Discovery and Opinion Mining, WISDOM 2013-Held in Conjunction with SIGKDD 2013, Chicago, IL, USA, 11 August 2013. [CrossRef]
80. Hayran, A.; Sert, M. Kelime Gömme ve Füzyon Tekniklerine Dayali Mikroblog Verileri Üzerinde Duygu Analizi. In Proceedings of the 2017 25th Signal Processing and Communications Applications Conference (SIU 2017), Antalya, Turkey, 15–18 May 2017. [CrossRef]
81. Maas, A.L.; Daly, R.E.; Pham, P.T.; Huang, D.; Ng, A.Y.; Potts, C. Learning Word Vectors for Sentiment Analysis. In *Proceedings of the 49th Annual Meeting of the Association for Computational Linguistics: Human Language Technologies*; Association for Computational Linguistics: Portland, OR, USA, 2011; pp. 142–150.
82. Pang, B.; Lee, L.; Vaithyanathan, S. Thumbs up? Sentiment Classification using Machine Learning Techniques. *arXiv* **2002**, arXiv:cs/0205070.
83. Asghar, N. Yelp Dataset Challenge: Review Rating Prediction. *arXiv* **2016**, arXiv:1605.05362.
84. Erhan, D.; Manzagol, P.A.; Bengio, Y.; Bengio, S.; Vincent, P. The Difficulty of Training Deep Architectures and the Effect of Unsupervised Pre-Training. In Proceedings of the Twelth International Conference on Artificial Intelligence and Statistics, Clearwater, FL, USA, 16–19 April 2009; van Dyk, D., Welling, M., Eds.; PMLR, Hilton Clearwater Beach Resort: Clearwater Beach, FL, USA, 2009; Volume 5, pp. 153–160.
85. Talathi, S.S.; Vartak, A. Improving Performance of Recurrent Neural Network with Relu Nonlinearity. *arXiv* **2015**, arXiv:1511.03771.
86. Pham, D.T.; Castellani, M. A comparative study of the Bees Algorithm as a tool for function optimisation. *Cogent Eng.* **2015**. [CrossRef]
87. Burden, R.L.; Faires, J.D. *Numerical Analysis*, 9th ed.; Brooks/Cole, Cengage Learning: Boston, MA, USA, 2011. [CrossRef]
88. Trefethen, L.N.; Bau, D. *Numerical Linear Algebra*; SIAM: Philadelphia, PA, USA, 1997. [CrossRef]
89. Bottou, L. On-line Learning and Stochastic Approximations. In *On-Line Learning in Neural Networks*; Saad, D., Ed.; Publications of the Newton Institute, Cambridge University Press: Cambridge, UK, 1999; pp. 9–42. [CrossRef]
90. Bisong, E. Google Colaboratory. In *Building Machine Learning and Deep Learning Models on Google Cloud Platform: A Comprehensive Guide for Beginners*; Apress: Berkeley, CA, USA, 2019; pp. 59–64. [CrossRef]
91. Abadi, M.; Agarwal, A.; Barham, P.; Brevdo, E.; Chen, Z.; Citro, C.; Corrado, G.S.; Davis, A.; Dean, J.; Devin, M.; et al. TensorFlow: Large-Scale Machine Learning on Heterogeneous Systems. 2015. Available online: https://www.tensorflow.org (accessed on 24 July 2021).
92. Chollet, F. Keras. 2015. Available online: https://github.com/fchollet/keras (accessed on 24 July 2021).
93. Virtanen, P.; Gommers, R.; Oliphant, T.E.; Haberland, M.; Reddy, T.; Cournapeau, D.; Burovski, E.; Peterson, P.; Weckesser, W.; Bright, J.; et al. SciPy 1.0: Fundamental Algorithms for Scientific Computing in Python. *Nat. Methods* **2020**, *17*, 261–272. [CrossRef]
94. Harris, C.R.; Millman, K.J.; van der Walt, S.J.; Gommers, R.; Virtanen, P.; Cournapeau, D.; Wieser, E.; Taylor, J.; Berg, S.; Smith, N.J.; et al. Array programming with NumPy. *Nature* **2020**, *585*, 357–362. [CrossRef] [PubMed]
95. McKinney, W. Data Structures for Statistical Computing in Python. In Proceedings of the 9th Python in Science Conference, Austin, TX, USA, 28 June–3 July 2010; pp. 51–56.
96. Hunter, J.D. Matplotlib: A 2D graphics environment. *Comput. Sci. Eng.* **2007**, *9*, 90–95. [CrossRef]
97. Zeybek, S.; Koc, E.; Seçer, A. MS-TR: A Morphologically Enriched Sentiment Treebank and Recursive Deep Models for Compositional Semantics in Turkish. *Cogent Eng.* **2021**. [CrossRef]

Article

StyleGANs and Transfer Learning for Generating Synthetic Images in Industrial Applications

Harold Achicanoy [1,2,*,†], Deisy Chaves [2,3,*,†] and Maria Trujillo [2,†]

1 Alliance of Bioversity International and CIAT, Km 17 Recta Cali-Palmira, Palmira 763537, Colombia
2 School of Computer and Systems Engineering, Universidad del Valle, Cali 760001, Colombia; maria.trujillo@correounivalle.edu.co
3 Department of Electrical, Systems and Automation, Universidad de León, 24007 León, Spain
* Correspondence: h.a.achicanoy@cgiar.org (H.A.); deisy.chaves@unileon.es (D.C.)
† These authors contributed equally to this work.

Abstract: Deep learning applications on computer vision involve the use of large-volume and representative data to obtain state-of-the-art results due to the massive number of parameters to optimise in deep models. However, data are limited with asymmetric distributions in industrial applications due to rare cases, legal restrictions, and high image-acquisition costs. Data augmentation based on deep learning generative adversarial networks, such as StyleGAN, has arisen as a way to create training data with symmetric distributions that may improve the generalisation capability of built models. StyleGAN generates highly realistic images in a variety of domains as a data augmentation strategy but requires a large amount of data to build image generators. Thus, transfer learning in conjunction with generative models are used to build models with small datasets. However, there are no reports on the impact of pre-trained generative models, using transfer learning. In this paper, we evaluate a StyleGAN generative model with transfer learning on different application domains—training with paintings, portraits, Pokémon, bedrooms, and cats—to generate target images with different levels of content variability: bean seeds (low variability), faces of subjects between 5 and 19 years old (medium variability), and charcoal (high variability). We used the first version of StyleGAN due to the large number of publicly available pre-trained models. The Fréchet Inception Distance was used for evaluating the quality of synthetic images. We found that StyleGAN with transfer learning produced good quality images, being an alternative for generating realistic synthetic images in the evaluated domains.

Keywords: data augmentation; fine-tuning; generative models; StyleGAN; transfer learning

Citation: Achicanoy, H.; Chaves, D.; Trujillo, M. StyleGANs and Transfer Learning for Generating Synthetic Images in Industrial Applications. *Symmetry* **2021**, *13*, 1497. https://doi.org/10.3390/sym13081497

Academic Editor: José Carlos R. Alcantud

Received: 27 June 2021
Accepted: 6 August 2021
Published: 16 August 2021

1. Introduction

Deep learning methods, a subset of machine learning techniques, have achieved outstanding results on challenging computer vision problems, such as image classification, object detection, face recognition, and motion recognition, among others [1]. However, the use of deep learning requires a large volume of representative annotated data to learn general models that achieve accurate results [2]; data are still scarce with asymmetric distributions, i.e., disproportionate number of examples between classes, in most applications related to healthcare, security and industry, due to legal/ethical restrictions, unusual patterns/cases, and image annotation costs [3–5].

As an alternative, image data augmentation has emerged to create training data with symmetric distributions by increasing data and reducing overfitting in deep learning models [6]. Data augmentation has been used through simple transformations, such as rotations, mirroring, and noise addition [4]. However, simple transformations produce a reduced number of valid data which usually are highly correlated and produce overfit models with poor generalisation capacity.

Generative Adversarial Networks (GANs) [7] have emerged as an alternative to create synthetic images by learning the probability distribution from data and generating images with high diversity and low correlation that can be used to build deep learning models [4,5,8–13]. GANs are used in medical applications, such as CT image segmentation [4,14] and disease/injure detection [12,13,15–17]. Nevertheless, since GANs are deep-learning-based models, they also require a significant amount of data and computational time to be trained from scratch. This drawback limits the use of GANs in generating images in applications where data are scarce, such as security and industry. A way to cope with this disadvantage is the use of transfer-learning techniques, which allow building new models from pre-trained ones in other applications or source domains with an abundance of training data by transferring the main features and reducing the training time [2].

Transfer learning has been widely used to address image classification [18–20] and segmentation [21,22] problems. However, the effect of StyleGAN-transfer learning on generating image quality is poorly reported. Wang et al. [23] evaluated the transferability of features, using different source and target domains to build generative models applying transfer learning, with some limitations, such as the generation of low-resolution images and the lack of evaluation of the impact of content variability in target domains.

In this paper, we evaluate a data augmentation strategy based on transfer learning and StyleGAN [24]. We use the first version of StyleGAN due to the large number of publicly available pre-trained models. In particular, we evaluate the capability of StyleGAN and transfer learning to generate synthetic images (data augmentation) considering variability levels on content. Thus, we assess quantitatively and visually the quality of the generated images, using three target domains with fine-tuned StyleGANs from five pre-trained models—source domains. The evaluated target domains correspond to three image sets derived from industrial processes with different levels of content variability, shown in Figure 1: bean seeds (low variability), faces of people aged between 5 and 19 years (medium variability), and chars obtained during coal combustion (high variability). The assessed source domains, to transfer features and build generative models, correspond to five pre-trained StyleGANs with images of paintings, portraits, Pokémon, bedrooms, and cats. Distinct from the commonly used transfer learning strategy, which consists of using related source and target domains, the evaluation is focused on source and target domains that are completely different. Obtaining the results shown, StyleGAN with transfer learning is suitable for the generation of high-resolution images in industrial applications due to having a good generalisation capability regarding content variability of the target image.

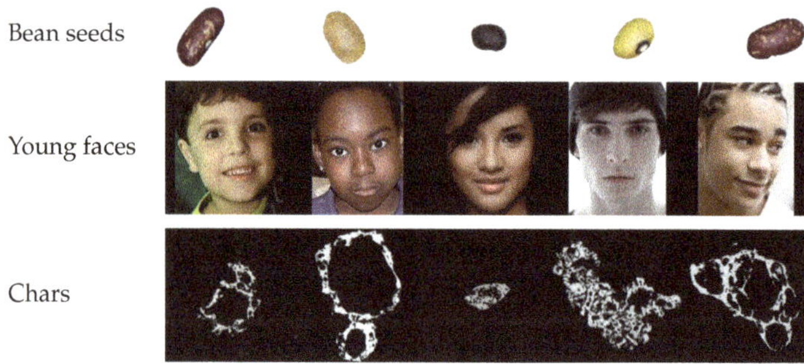

Figure 1. Illustration of target domains for image generation.

The rest of the paper is structured as follows: Section 2 presents the theoretical background on StyleGAN, GANs assessment, and transfer learning. Section 3 summarises the relevant related works. Section 4 details the data augmentation strategy used as an evaluation pipeline. Section 5 describes the performed experiments and results. Section 6 presents the discussion on the obtained results, focusing on the effect of pre-trained models for synthetic image generation; Section 7 depicts the conclusions and future research lines.

2. Theoretical Background

2.1. StyleGAN

StyleGAN [24] combines the architecture of Progressive Growing GAN [25] with the style transfer principles [26] in Figure 2. StyleGAN's architecture addresses some limitations of the GAN models, such as stability during training and lack of control over the images generated.

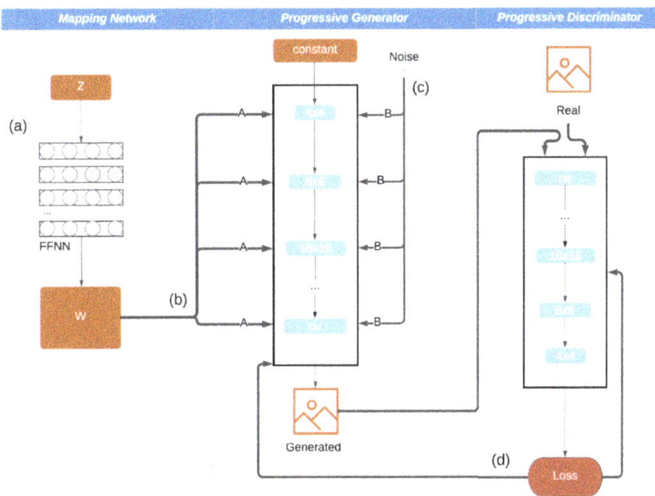

Figure 2. The StyleGAN architecture [27] is composed of three neural networks. (**a**) Mapping network, which converts a random into a style signal. (**b**,**c**) Progressive generator network, which receives the style signal (A) and random noise (B), and produces images progressively. (**d**) The progressive discriminator network, which compares real and generated images to update all the weights for the three networks, improving their performance.

In a traditional GAN model, a generative network receives as input a random vector Z, or latent vector, to generate a new image. In contrast, in the StyleGAN architecture, a latent vector Z (512-dimensional) feeds an 8-layer neural network, called a mapping network, that transforms the latent vector into an intermediate space W (512-dimensional), which defines the style of the image to be generated; see Figure 2a.

An image style defined in the intermediate space W is transferred to the progressive generative network (Figure 2b), where the technique AdaIN (Adaptive Instance Normalization) [26] transforms the latent vector W into two scalars (scale and bias) that control the style of the image generated at each resolution level. In addition to the style guide provided by AdaIN, the progressive generator network has as an input a constant argument. This constant corresponds to an array of $4 \times 4 \times 512$ dimensions, i.e., an image of 4×4 pixels with 512 channels, which is learned during network training and contains a kind of sketch with the general characteristics of the training set images. Furthermore, StyleGAN has noise sources injected at each resolution level to introduce slight variations in the generated images (Figure 2c). The improvements in the StyleGAN generative network allow optimising the quality of the generated synthetic images. Finally, the back propagation algorithm is applied to adjust the weights of the three networks, improving the quality of the images generated during the following iterations (Figure 2d).

2.2. GANs Model Evaluation

Evaluating GAN architectures is particularly hard because there is not a consensus on a unique metric that assesses the quality and diversity of generated images [4,23].

However, a widely used metric in the literature is the Fréchet Inception Distance (FID) [28], defined as follows:

$$FID = \|\mu_r - \mu_g\|^2 + Tr\left(\Sigma_r + \Sigma_g - 2(\Sigma_r \Sigma_g)^{(\frac{1}{2})}\right), \quad (1)$$

where μ_r, μ_g are mean vectors and Σ_r, Σ_g are variance–covariance matrices. Equation (1) is a distance measurement between real and generated images. $X_r \sim N(\mu_r, \Sigma_r)$ and $X_g \sim N(\mu_g, \Sigma_g)$ are multidimensional normal distributions of 2048 dimensions of real and generated images, respectively, which are extracted from the third layer of polling of the Inception-v3 network [29].

The closer the distributions, the lower the metric value. FID values close to zero correspond to a larger similarity between real and generated images, resulting in the differences between the distributions.

2.3. Transfer Learning

Transfer learning models involve applying the knowledge learned in a source domain—where a large amount of training data are available—to a target domain that has a reduced amount of data, as illustrated in Figure 3. The objective is to transfer most characteristics obtained from a source domain into a target domain in order to reduce training time and use deep learning models with limited data [30].

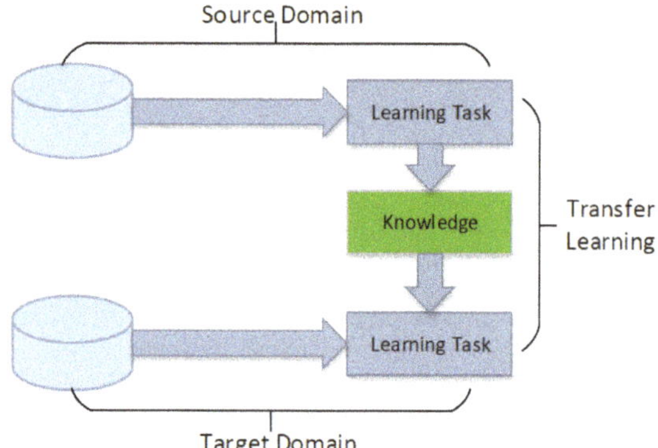

Figure 3. Transfer learning scheme [2]. The learning task to be accomplished in the source domain is provided by a vast number of training data. The target domain usually has a limited number of training data.

The transfer learning is defined as follows: given a source domain D_S, a source learning task T_S, a target domain D_T, and a target learning task T_T, the transfer learning aims to improve the learning of the target predictive function $f_T(.)$ in D_T using the knowledge learned in D_S and T_S, where $D_S \neq D_T$ or $T_S \neq T_T$. A domain D is defined by two components: a feature space X and a marginal probability distribution $P(X)$, $D = \{X, P(X)\}$. Similarly, a learning task T consists of two components: a labels space Y and a target predictive function $f(.)$, $T = \{Y, f(.)\}$.

In unsupervised models, such as GANs, the Y labels space does not exist, and the learning task objective is to estimate the generative distribution of data. For this purpose, a particular transfer learning technique, known as fine tuning, is used. Fine tuning is a transfer learning technique in which a model that has been trained for a specific task is used to perform a new task, with similar characteristics to the first task, by adjusting model parameters. This process means that a model is not built from scratch, but rather takes advantage of the characteristics learned from the original task.

3. Related Works

GANs are capable of image generation in two categories: low-resolution [4,8,12,13,15,16,23] and high-resolution [14,17,31–41]. A summary of these approaches is presented in Table 1. The first group comprises mostly studies between 2017 and 2018. A predominance of medical image augmentation focus on cancer detection [4,15], cerebral diseases [16] and COVID-19 [12,13] is observed. The image generation is motivated by the high cost of medical images acquisition that translates into a limited number of samples to train deep learning models. Publications include studies that evaluated the effectiveness of GANs data augmentation, using computer vision benchmarks datasets [8,23]. Particularly, the work of Wang et al. [23] corresponds to the only attempt to use transfer learning with generative models. This study concluded that it is possible to apply transfer learning in a Wasserstein GAN model [42], using source and target domains with low-resolution images.

The second group includes studies from 2019 to date, driven by new GAN architectures, such as StyleGAN [24], generating images with high-resolution. There are three application areas: agriculture [31,32], medical [14,17,36,37], and electrical domains [35]. Particularly, Fetty et al. [17] presented a complete analysis of StyleGAN models trained from scratch for data augmentation of pelvic malignancies images.

Transfer learning on generative models for limited data has been the subject of study for the last three years [33,34,38–41], focusing on evaluating the impact of freezing the lower generator layers [33,34], the lower discriminator layers [39], and both the generator and discriminator lower layers [40], using mainly general purposes datasets of indoors (e.g., LSUN, Bedroons) and faces (e.g., CelebHQ, FFHQ, CelebA). The results show a reduction in the overfitting derived from the knowledge transfer and training time. However, transfer learning in conjunction with generative models has not been evaluated with a focus on the capability to generate synthetic images with high-resolution, considering the variability levels of content.

We aim to fill this literature gap using the proposed pipeline to evaluate the transfer capability of the knowledge obtained from certain source domains to target domains with different levels of content variability, such as bean seed images (with simple shape, texture, and colour), young faces and chars images (with more complex visual features). These target domains correspond to real industrial applications.

Table 1. Summary of GANs approaches for image generation.

Application Description				Model Description			
Year	Context	Images Number	Images Quality	Generative Model	Transfer Learning	Training Time	Evaluation Metric
2017 [8]	Data augmentation in alphabets, handwritten characters, and faces	Alphabets: 1200, Characters: 3400, Faces: 1802	Low-resolution	Conditional GAN	Yes	Not available	Not available
2018 [4]	Data augmentation using computed tomography images	Different variations	Low-resolution	Progressive Growing GAN	No	~36 GPU hours	Not available
2018 [15]	Data augmentation using GANs for liver injuries classification	Computed tomography images: 182	Low-resolution	Deep Convolutional GAN (DCGAN)	No	Not available	Not available
2018 [16]	Medical image data augmentation using GANs	Alzheimer's Disease Neuroimaging Initiative: 3416	Low-resolution	Pix2Pix GAN	No	Not available	Not available
2018 [23]	Transfer learning in GANs for data augmentation	Varies from 1000 to 100,000	Low-resolution	Wasserstein GAN with Gradient Penality (WGAN-GP)	Yes	Not available	FID minimum: 7.16, FID maximum: 122.46
2020 [12]	Transfer learning in GANs for COVID-19 detection on chest X-ray images	Normal cases: 79, COVID-19: 69, Pneumonia bacterial: 79, Pneumonia virus: 79	Low-resolution	Shallow GAN	Yes	Not available	Not available
2020 [13]	COVID-19 Screening on chest X-ray images	Normal cases: 1341, Pneumonia: 1345	Low-resolution	Deep Convolutional GAN (DCGAN)	No	GPU Titan RTX; 100 epochs	Not available
2019 [31]	Plant diseases detection	Plant leafs: 79,265	High-resolution	StyleGAN	No	Not available	Not available
2019 [32]	Data augmentation on plant leaf diseases	Plant leaf disease: 54,305	High-resolution	DCGAN & WGAN	No	1000 epochs	Not available
2019 [14]	Data augmentation using GANs to improve CT segmentation	Pancreas CT: 10,681	High-resolution	CycleGAN	No	3 M iterations	Qualitative evaluation
2019 [33]	Image generation from small datasets via batch statistics adaptation	Face, Anime, and Flowers: 251–10,000	High-resolution	SNGAN, BigGAN	Yes	3000, 6000–10,000 iterations	FID: 84–130

Table 1. *Cont.*

Application Description				Model Description			
Year	**Context**	**Images Number**	**Images Quality**	**Generative Model**	**Transfer Learning**	**Training Time**	**Evaluation Metric**
2019 [34]	Mind2Mind: transfer learning for GANs	MNIST, KMNIST, and CelebHQ: 30,000–60,000	High-resolution	MindGAN	Yes	Not available	FID: 19.21
2020 [35]	Data augmentation using GANs for electrical insulator anomaly detection	Individual insulators: 3861	High-resolution	BGAN, AC-GAN, PGAN, StyleGAN, BPGAN	No	Not available	Not available
2020 [17]	Data augmentation using StyleGAN for pelvic malignancies images	17,542	High-resolution	StyleGAN	No	One GPU month	FID: 12.3
2020 [36]	Data augmentation for Magnetic Resonance Brain Images	50,000	High-resolution	StyleGAN and Variational autoencoders	No	Not available	Not available
2020 [38]	MineGAN: effective knowledge transfer from GANs to target domains with few images	MNIST, CelebA, and LSUN: 1000	High-resolution	Progressive GAN, SNGAN, and BigGAN	Yes	200 iterations	FID: 40–160
2020 [39]	Freeze the discriminator: a simple baseline for fine-tuning GANs	Animal face, Flowers: 1000	High-resolution	StyleGAN, SNGAN	Yes	50,000 iterations	FID: 24–80
2020 [40]	On leveraging pretrained GANs for generation with limited data	CelebA, Flowers, Cars, and Cathedral: 1000	High-resolution	GP-GAN	Yes	60,000 iterations	FID: 10–80
2021 [37]	Data augmentation for Cardiac Magnetic Resonance	6000	High-resolution	Conditional GAN-based method	Yes	Not available	Not available
2021 [41]	MineGAN++: mining generative models for efficient knowledge transfer to limited data domains	MNIST, FFHQ, Anime, Bedroom, and Tower: 1000	High-resolution	BigGAN, Progressive GAN, and StyleGAN	Yes	200 iterations	FID: 40–100

4. Evaluation Synthetic Images Generation Pipeline

We proposed a five-step pipeline based on the fine tuning of StyleGAN pre-trained models from five source domains, as is shown in Figure 4—paintings, portraits, Pokémon, bedrooms, and cats—in order to generate synthetic images in three target domains: bean seeds, young faces, and chars. Although, there is a new version of StyleGAN, called StyleGAN2 [43], we selected the first version of StyleGAN, due to the large number of publicly available pre-trained StyleGAN models. Moreover, in practice, training a StyleGAN model from scratch requires a huge number of images, computational resources (preferably with multiple GPUs) and processing time.

In the proposed image generation pipeline, first, we select the images target domain of bean seeds, young faces, or chars, as input. Second, images from the target domain are pre-processed to improve the transferability of features by adjusting image resolution. Third, pre-trained StyleGAN models are fine tuned with pre-processed images. Fourth, the FID evaluation metric is used for selecting the best source domain. Fifth, the synthetic images for the input target domain are generated with the best source domain.

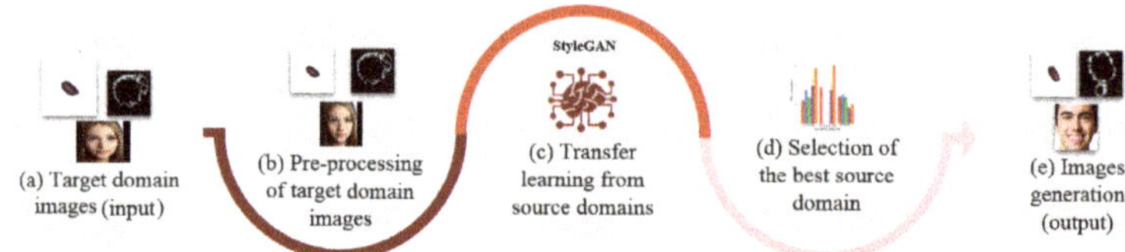

Figure 4. Evaluation pipeline: (**a**) Input images from a target domain. (**b**) Pre-processing of target domain images to improve the transferability of the features. (**c**) Transfer learning using StyleGAN pre-trained models from source domains. (**d**) Selection of the best source domain using the FID metric. (**e**) Generation of synthetic images for the target domain.

4.1. Input Target Domain Images

We used images from three application domains—bean seeds, young faces and chars—with different variability levels of content and potential industrial applications; see Figure 1.

Bean seeds: The dataset [44] has 1500 seed images from 16 bean varieties. Bean seed images are classified as low content variability since shape, colour, and texture characteristics are homogeneous for the analysed seed varieties, corresponding to oval shapes with limited range of red, cream, black, and white colours. In addition, the acquired images share the same background colour. Synthetic images of bean seeds are valuable in developing evaluation tools of genetic breeding trials. These tools are used to preserve the genetic pedigree of seeds over time, accomplish market quality requirements, and increase production levels [45].

Young faces: The images set consists of 3000 images randomly selected from publicly available datasets with reference to age estimation problems: IMDB-WIKI [46], APPA-Real [47], AgeDB [48]. Images correspond to individuals aged between 5 and 19 years. This range of ages was selected because it presents the lowest frequencies in the considered facial datasets. In addition, young faces are crucial in cybersecurity applications, such as access control, the detection of Child Sexual Exploitation Material or the identification of victims of child abuse [49–51]. Young facial images are considered to be of medium variability content since faces have a similar shape structure; the StyleGAN model was originally designed for faces generation.

Chars: The dataset contains 2928 segmented char particle images from coals of high, medium, and low reactivity. Char images are considered to be of high variability content due to the complex particle shapes and lack of colours. Synthetic images of char particles are useful to train models to estimate the combustion parameters in power generation plants [52].

Table 2 contains a summary of the target domains, including the source, number of images, number of classes, and content variability type.

Table 2. Description of target domain datasets.

Target Domain	# of Images	# of Classes	Content Variability
Bean seeds [44]	1500	16	Low
Young faces [46–48]	3000	14	Medium
Chars [52]	2928	3	High

4.2. Pre-Processing Target Domain Images

Input images are pre-processed to improve the transfer of features from the source domain images into the target domain. The pre-processing consists of equally adjusting the images resolution to the resolution of the source domain. In short, operations of resizing, up-scaling, or down-scaling are applied, depending on the difference between the target and source images' dimensions.

4.3. Transfer Learning from Source Domains

We used five pre-trained StyleGAN models—paintings, portraits, Pokémon, bedrooms, and cats—shown in Figure 5. The selection of pre-trained models was based on the (i) public availability of models, and (ii) diversity of images used to build models. Table 3 presents relevant information about pre-trained models: the images source, required images resolution, and the number of iterations used for training.

Figure 5. Illustration of generated images using the training source domain models.

Table 3. Description of pre-trained StyleGAN models (source domains).

Source Domain	Image Resolution	Number of Iterations
Paintings [53]	512 × 512	8040
Portraits [54]	512 × 512	11,125
Pokemon [55]	512 × 512	7961
Bedrooms [56]	256 × 256	7000
Cats [56]	256 × 256	7000

Transfer learning is performed by fine tuning the pre-trained StyleGAN models (source domains) with images of the target domain to build new image generators. During the fine tuning of StyleGAN models, the learning rate is set to 0.001 and the number of minibatch repetitions is set to 1 based on those reported in [57]. The selected values for the learning rate and minibatch repetition increase the stability and speed during training.

4.4. Selection of the Best Source Domain

Once the StyleGAN models are fine tuned, the best source domain is selected based on the best FID metric value to generate images of a target domain. In particular, the StyleGAN model with the lowest FID value has a better performance since the model generates images with a distribution similar to the target domain.

4.5. Synthetic Images Generation (Output)

The StyleGAN with the best performance is used to generate as many images as needed for the target domain (data augmentation).

5. Experimental Evaluation

We assess the transfer learning capability of pre-trained models from five source domains—paintings, portraits, Pokémon, bedrooms, and cats—to build StyleGAN models for generating images of unrelated target domains with different levels of content variability: bean seeds, young faces, and chars. Pre-processed images from target domains are used to fine tune pre-trained StyleGAN models (source domains) over 1000 iterations, using the hyperparameters described in Section 4.3. We run the experiments on a GNU/Linux machine with a GPU Nvidia TITAN Xp 11GB, Cuda 10.1, and CuNDD 7. The source code is available at https://github.com/haachicanoy/stylegan_augmentation_tl (accessed on 28 July 2021).

Table 4 presents the FID values obtained for the fine-tuned StyleGAN models, and Table 5 illustrates the generated images by the target domain.

Table 4. FID values for StyleGAN models built for target domains by fine tuning pre-trained models from different source domains. The best source domain to generate images of a target domain by transfer learning is highlighted in bold. Lower FID values indicate better performance.

Source Target	Bean Seeds	Young Faces	Chars
Paintings	**23.26**	27.77	38.13
Portraits	35.04	30.11	—
Pokémon	27.06	27.56	—
Bedrooms	39.31	**16.98**	**34.81**
Cats	57.92	20.48	61.52

Table 5. Generated images for target domain using fine tuned StyleGAN models.

Source Target	Bean Seeds	Young Faces	Chars
Original image			
Paintings			
Portraits			—
Pokémon			—
Bedrooms			
Cats			

The results show that StyleGAN models are able to generate bean seed images (low content variability) through transfer learning with excellent performance. FID values range between 23.26 and 57.92, corresponding to the source domains of paintings and cats, respectively. Hence, the best source domain to generate bean seed images is paintings. It indicates that the colour pattern in these images is more similar to beans in comparison to the other source domains.

Regarding the generation of young face images (medium content variability), the bedrooms source domain achieves the best results (FID of 16.98). Bedrooms are one of the most complete source domains, standing out for colour and shape features that are efficiently transferred to generate facial images. It is essential to highlight that the five source domains yield FID values lower than 30.11. This performance may be related to the fact that the StyleGAN architecture was specifically developed for generating synthetic face images.

During the fine tuning of models to generate char images, we observed that the source domains of Pokémon and portraits do not converge, leading the training to fail. This is presumably because of the high content variability of chars with complex shapes, varying sizes, and changes in colour intensities that make it difficult to adapt the features from these two source domains. The remaining source domains (paintings, bedrooms, and cats) have features that can be successfully transferred to the generation of char images. In particular, bedrooms achieve the best performance (FID of 34.81).

In most of the cases, fine tuned models generate images of bean seeds, young faces and chars with visual characteristics that are similar to the original ones (see Table 5). However, in some cases, the generated images have visual defects; see Figure 6. Defects on bean seed images comprise bean shape deformation, stains, and changes in colour intensities. Defects on young facial images correspond to colour spots and alterations in the hair, skin, and smile. Defects on chars images include blurring and undefined shapes. Therefore, the generated images have to be filtered to remove images with defects before using them in any application, e.g., the training of an image classifier. A sample of 1000 synthetic images of target domains generated from the best source domains is available at https://doi.org/10.7910/DVN/HHSJY8 (accessed on 23 July 2021).

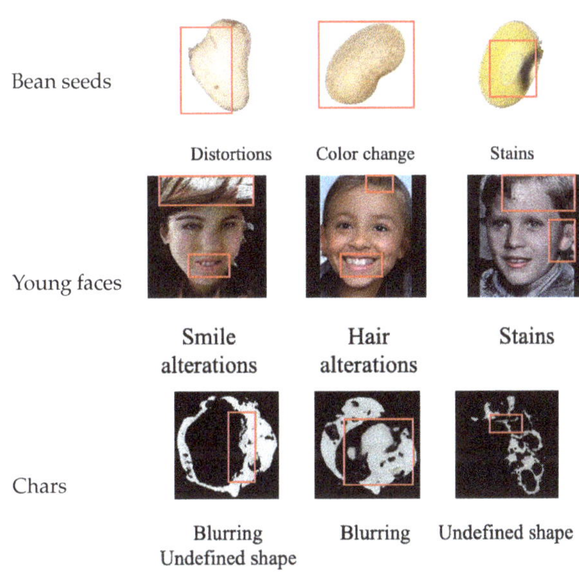

Figure 6. Defects in generated images by target domain.

6. Effect of Pre-Trained Models on Synthetic Images Generation

Figure 7 shows the evolution of the FID metric across the iterations of fine tune StyleGAN models for evaluating target domains (bean seeds, chars and young faces). The results show that, in most cases, the trained models for a target domain—regardless of the source domain—tend to stabilise and converge to a constant FID value, after a certain number of iterations. This occurs for all cases, except for target domains with high content variability (chars) where some source domains do not converge—portraits and Pokémon. Hence, we conclude that transfer learning and generative models, such as StyleGAN, can be successfully used to build generators of images with low and medium content variability, such as seeds and faces. However, the generation of synthetic images with high content variability is limited by the characteristics of source domains. In particular, the best FID values are obtained for the source domains of paintings (bean seeds) and bedrooms (chars and young faces).

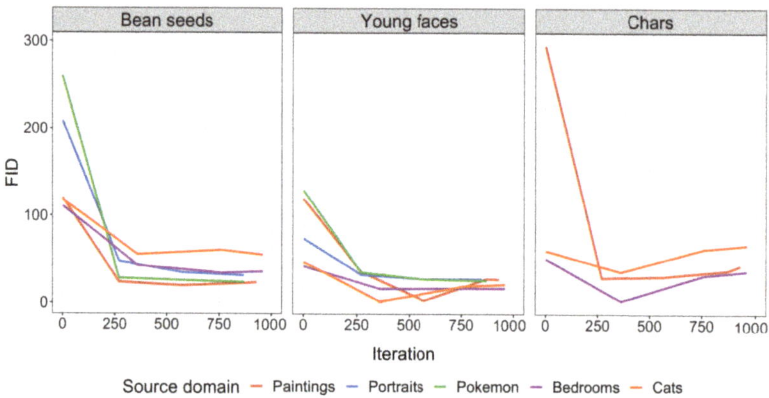

Figure 7. FID metric vs. StyleGAN training iterations for the evaluated source domains grouped by target domain.

We also analysed the loss score values of the generator and discriminator networks during training, shown in Figure 8. Similar to the observed for the FID values, the loss scores of the source domains exhibit a steady behaviour, except for the cats' domain, indicating an instability around 500 iterations.

Furthermore, the use of transfer learning reduces significantly the number of images (up to 1500 for beans) and iterations (up to 1000 or 2 days) required to build StyleGAN models, in comparison to models trained from scratch (70,000 images and 14 days) [24].

Regarding the quality of the synthetic images, Figure 9 presents a bar graph of the FID values by source domains—paintings, portraits, Pokémon, bedrooms, and cats—grouped by the target domain: bean seeds, chars, and young faces. The bar highs correspond to median values of the FID obtained during the fine tuning of StyleGAN models by a source domain, while the black line on the bars represents dispersion of the median FID values. The length of the line denotes the range of the dispersion. Larger lines indicate that the images generated from a source domain differ significantly from the target domain.

The source domains with the best performance—lower median FID value and dispersion—are bedrooms, for the young faces and chars, and paintings, for the bean seeds. Despite bedrooms yielding the lowest FID value in the target domain of chars, this source domain is the one with the largest dispersion, indicating a possible generation of char images with defects.

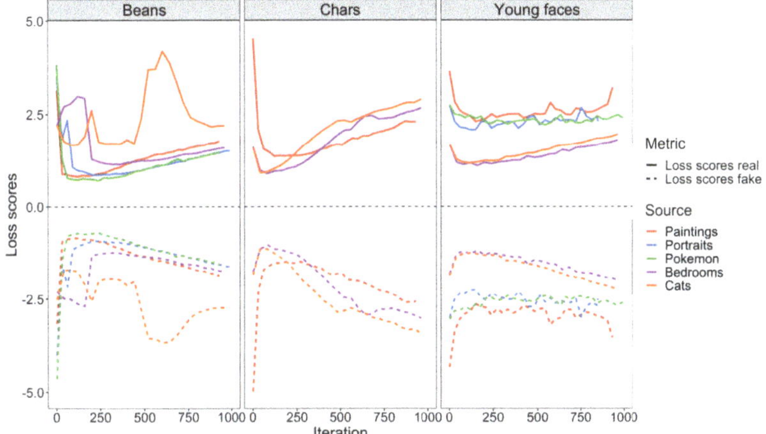

Figure 8. Loss scores vs. StyleGAN training iterations for the evaluated source domains grouped by target domain. Solid and dotted lines correspond to discriminator and generator network scores, respectively; following the source domains colours.

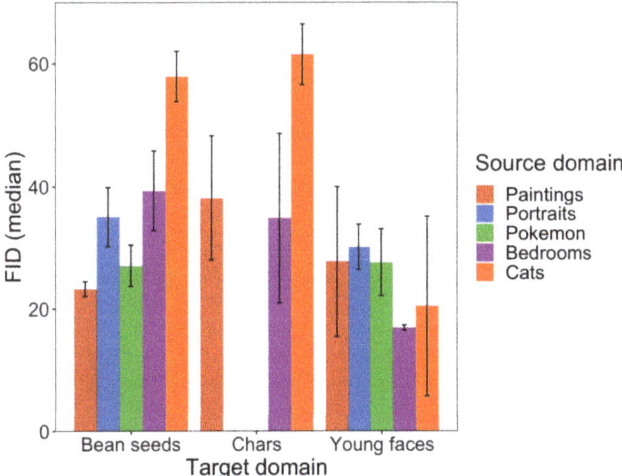

Figure 9. FID (median) for target domains and their source domains.

7. Conclusions

StyleGAN with transfer learning is a strategy for generating synthetic images with a limit number of images from the target domain. We evaluated the application of StyleGAN with transfer learning on generating high-resolution images by a pipeline based on the fine tuning of StyleGAN models. The evaluation was conducted using three target domains from industrial applications with different content variability (bean seeds, chars, and young faces) and five source domains from general applications (paintings, portraits, Pokémon, bedrooms, and cats) to perform transfer learning.

The experimental evaluation confirmed the potential of StyleGAN with transfer learning for generating synthetic images for industrial applications. The proposed pipeline performed better with target domains with low and medium content variability in terms of colour and shape, such as bean seeds and young faces. Moreover, the time and number of images required to build the models were reduced in all cases, which validates the use of StyleGAN with transfer learning for generating synthetic images.

As future work, strategies to optimise the fine tuning hyper-parameters will be evaluated to improve the performance of image generators with high content variability and reduce the defects in synthetic images. General purpose datasets will be assessed, such as FFHQ and LSUN.

Author Contributions: Conceptualisation, H.A., D.C. and M.T.; methodology, H.A., D.C. and M.T.; software, H.A.; validation, H.A. and D.C.; formal analysis, H.A., D.C. and M.T.; investigation, H.A., D.C. and M.T.; resources, H.A., D.C. and M.T; data curation, H.A.; writing—original draft preparation, H.A.; writing—review and editing, D.C. and M.T.; visualisation, H.A. and D.C.; supervision, D.C. and M.T.; project administration, D.C. and M.T.; funding acquisition, H.A., D.C. and M.T. All authors have read and agreed to the published version of the manuscript.

Funding: This research was funded with support from the European Commission under the 4NSEEK project with Grant Agreement 821966. This publication reflects the views only of the author; the European Commission cannot be held responsible for any use that may be made of the information contained therein.

Data Availability Statement: Generated data can be found at https://doi.org/10.7910/DVN/HHSJY8 (accessed on 28 July 2021).

Acknowledgments: This work was supported by the framework agreement between the Universidad de León and INCIBE (Spanish National Cybersecurity Institute) under Addendum 01. Additionally, the authors thank Alliance Bioversity-CIAT for providing the bean seed images.

Conflicts of Interest: The authors declare no conflict of interest. The funders had no role in the design of the study.

References

1. Dargan, S.; Kumar, M.; Ayyagari, M.R.; Kumar, G. A Survey of Deep Learning and Its Applications: A New Paradigm to Machine Learning. *Arch. Comput. Methods Eng.* **2019**, *27*, 1071–1092. [CrossRef]
2. Tan, C.; Sun, F.; Kong, T.; Zhang, W.; Yang, C.; Liu, C. A Survey on Deep Transfer Learning. *arXiv* **2018**, arXiv:1808.01974.
3. Haixiang, G.; Yijing, L.; Shang, J.; Mingyun, G.; Yuanyue, H.; Bing, G. Learning from class-imbalanced data: Review of methods and applications. *Expert Syst. Appl.* **2017**, *73*, 220–239. [CrossRef]
4. Bowles, C.; Chen, L.; Guerrero, R.; Bentley, P.; Gunn, R.; Hammers, A.; Dickie, D.A.; Hernández, M.V.; Wardlaw, J.; Rueckert, D. GAN augmentation: Augmenting training data using generative adversarial networks. *arXiv* **2018**, arXiv:1810.10863.
5. Tanaka, F.H.K.d.S.; Aranha, C. Data Augmentation Using GANs. *arXiv* **2019**, arXiv:1904.09135.
6. Shorten, C.; Khoshgoftaar, T.M. A survey on Image Data Augmentation for Deep Learning. *J. Big Data* **2019**, *6*, 60. [CrossRef]
7. Goodfellow, I.; Pouget-Abadie, J.; Mirza, M.; Xu, B.; Warde-Farley, D.; Ozair, S.; Courville, A.; Bengio, Y. Generative Adversarial Nets. In *Advances in Neural Information Processing Systems 27*; Ghahramani, Z., Welling, M., Cortes, C., Lawrence, N.D., Weinberger, K.Q., Eds.; Curran Associates, Inc.: New York, NY, USA, 2014; pp. 2672–2680.
8. Antoniou, A.; Storkey, A.; Edwards, H. Data augmentation generative adversarial networks. *arXiv* **2017**, arXiv:1711.04340.
9. Garcia Torres, D. Generation of Synthetic Data with Generative Adversarial Networks. Ph.D. Thesis, School of Electrical Engineering and Computer Science, KTH Royal Institute of Technology, Karlskrona, Sweden, 2018.
10. Zeid Baker, M. Generation of Synthetic Images with Generative Adversarial Networks. Master's Thesis, Department of Computer Science and Engineering, Blekinge Institute of Technology, Karlskrona, Sweden, 2018.
11. Ma, Y.; Liu, K.; Guan, Z.; Xu, X.; Qian, X.; Bao, H. Background Augmentation Generative Adversarial Networks (BAGANs): Effective Data Generation Based on GAN-Augmented 3D Synthesizing. *Symmetry* **2018**, *10*, 734. [CrossRef]
12. Loey, M.; Smarandache, F.; Khalifa, N.E.M. Within the Lack of Chest COVID-19 X-ray Dataset: A Novel Detection Model Based on GAN and Deep Transfer Learning. *Symmetry* **2020**, *12*, 651. [CrossRef]
13. Zulkifley, M.A.; Abdani, S.R.; Zulkifley, N.H. COVID-19 Screening Using a Lightweight Convolutional Neural Network with Generative Adversarial Network Data Augmentation. *Symmetry* **2020**, *12*, 1530. [CrossRef]
14. Sandfort, V.; Yan, K.; Pickhardt, P.J.; Summers, R.M. Data augmentation using generative adversarial networks (CycleGAN) to improve generalizability in CT segmentation tasks. *Sci. Rep.* **2019**, *9*, 16884. [CrossRef]
15. Frid-Adar, M.; Diamant, I.; Klang, E.; Amitai, M.; Goldberger, J.; Greenspan, H. GAN-based Synthetic Medical Image Augmentation for increased CNN Performance in Liver Lesion Classification. *Neurocomputing* **2018**, *321*, 321–331. [CrossRef]
16. Shin, H.C.; Tenenholtz, N.A.; Rogers, J.K.; Schwarz, C.G.; Senjem, M.L.; Gunter, J.L.; Andriole, K.; Michalski, M. Medical Image Synthesis for Data Augmentation and Anonymization using Generative Adversarial Networks. *arXiv* **2018**, arXiv:1807.10225.
17. Fetty, L.; Bylund, M.; Kuess, P.; Heilemann, G.; Nyholm, T.; Georg, D.; Löfstedt, T. Latent Space Manipulation for High-Resolution Medical Image Synthesis via the StyleGAN. *Z. Med. Phys.* **2020**, *30*, 305–314. [CrossRef]

18. Coulibaly, S.; Kamsu-Foguem, B.; Kamissoko, D.; Traore, D. Deep neural networks with transfer learning in millet crop images. *Comput. Ind.* **2019**, *108*, 115–120. [CrossRef]
19. Abu Mallouh, A.; Qawaqneh, Z.; Barkana, B.D. Utilizing CNNs and transfer learning of pre-trained models for age range classification from unconstrained face images. *Image Vis. Comput.* **2019**, *88*, 41–51. [CrossRef]
20. Liu, X.; Wang, C.; Bai, J.; Liao, G. Fine-tuning Pre-trained Convolutional Neural Networks for Gastric Precancerous Disease Classification on Magnification Narrow-band Imaging Images. *Neurocomputing* **2020**, *392*, 253–267. [CrossRef]
21. Ferguson, M.; Ak, R.; Lee, Y.T.T.; Law, K.H. Detection and Segmentation of Manufacturing Defects with Convolutional Neural Networks and Transfer Learning. *Smart Sustain. Manuf. Syst.* **2018**, *2*, 137–164. [CrossRef]
22. Abdalla, A.; Cen, H.; Wan, L.; Rashid, R.; Weng, H.; Zhou, W.; He, Y. Fine-tuning convolutional neural network with transfer learning for semantic segmentation of ground-level oilseed rape images in a field with high weed pressure. *Comput. Electron. Agric.* **2019**, *167*, 105091. [CrossRef]
23. Wang, Y.; Wu, C.; Herranz, L.; van de Weijer, J.; Gonzalez-Garcia, A.; Raducanu, B. Transferring GANs: Generating images from limited data. *arXiv* **2018**, arXiv:1805.01677.
24. Karras, T.; Laine, S.; Aila, T. A style-based generator architecture for generative adversarial networks. In Proceedings of the IEEE Conference on Computer Vision and Pattern Recognition, Long Beach, CA, USA, 16–20 June 2019; pp. 4401–4410.
25. Karras, T.; Aila, T.; Laine, S.; Lehtinen, J. Progressive Growing of GANs for Improved Quality, Stability, and Variation. *arXiv* **2018**, arXiv:1710.10196.
26. Huang, X.; Belongie, S. Arbitrary Style Transfer in Real-time with Adaptive Instance Normalization. *arXiv* **2017**, arXiv:1703.06868.
27. Oeldorf, C. Conditional Implementation for NVIDIA's StyleGAN Architecture. 2019. Available online: https://github.com/cedricoeldorf/ConditionalStyleGAN (accessed on 13 November 2019).
28. Heusel, M.; Ramsauer, H.; Unterthiner, T.; Nessler, B.; Hochreiter, S. GANs Trained by a Two Time-Scale Update Rule Converge to a Local Nash Equilibrium. In *Advances in Neural Information Processing Systems 30*; Guyon, I., Luxburg, U.V., Bengio, S., Wallach, H., Fergus, R., Vishwanathan, S., Garnett, R., Eds.; Curran Associates, Inc.: New York, NY, USA, 2017; pp. 6626–6637.
29. Szegedy, C.; Vanhoucke, V.; Ioffe, S.; Shlens, J.; Wojna, Z. Rethinking the Inception Architecture for Computer Vision. *arXiv* **2015**, arXiv:1512.00567.
30. Pan, S.J.; Yang, Q. A Survey on Transfer Learning. *IEEE Trans. Knowl. Data Eng.* **2010**, *22*, 1345–1359. [CrossRef]
31. Arsenovic, M.; Karanovic, M.; Sladojevic, S.; Anderla, A.; Stefanovic, D. Solving Current Limitations of Deep Learning Based Approaches for Plant Disease Detection. *Symmetry* **2019**, *11*, 939. [CrossRef]
32. Arun Pandian, J.; Geetharamani, G.; Annette, B. Data Augmentation on Plant Leaf Disease Image Dataset Using Image Manipulation and Deep Learning Techniques. In Proceedings of the 2019 IEEE 9th International Conference on Advanced Computing (IACC), Tiruchirappalli, India, 13–14 December 2019; pp. 199–204. [CrossRef]
33. Noguchi, A.; Harada, T. Image generation from small datasets via batch statistics adaptation. In Proceedings of the IEEE/CVF International Conference on Computer Vision, Seoul, Korea, 27–28 October 2019; pp. 2750–2758.
34. Fregier, Y.; Gouray, J.B. Mind2Mind: transfer learning for GANs. In Proceedings of the International Conference on Geometric Science of Information. Paris, France, 21–23 July 2021; Springer: 2021; pp. 851–859.
35. Luo, L.; Hsu, W.; Wang, S. Data Augmentation Using Generative Adversarial Networks for Electrical Insulator Anomaly Detection. In Proceedings of the 2020 2nd International Conference on Management Science and Industrial Engineering, Osaka, Japan, 7–9 April 2020; Association for Computing Machinery: Osaka, Japan, 2020; pp. 231–236. [CrossRef]
36. Hirte, A.U.; Platscher, M.; Joyce, T.; Heit, J.J.; Tranvinh, E.; Federau, C. Diffusion-Weighted Magnetic Resonance Brain Images Generation with Generative Adversarial Networks and Variational Autoencoders: A Comparison Study. *arXiv* **2020**, arXiv:2006.13944.
37. Xia, Y.; Ravikumar, N.; Greenwood, J.P.; Neubauer, S.; Petersen, S.E.; Frangi, A.F. Super-Resolution of Cardiac MR Cine Imaging using Conditional GANs and Unsupervised Transfer Learning. *Med. Image Anal.* **2021**, *71*, 102037. [CrossRef] [PubMed]
38. Wang, Y.; Gonzalez-Garcia, A.; Berga, D.; Herranz, L.; Khan, F.S.; Weijer, J.V.D. Minegan: effective knowledge transfer from gans to target domains with few images. In Proceedings of the IEEE/CVF Conference on Computer Vision and Pattern Recognition, Seattle, WA, USA, 14–19 June 2020; pp. 9332–9341.
39. Mo, S.; Cho, M.; Shin, J. Freeze the discriminator: A simple baseline for fine-tuning gans. *arXiv* **2020**, arXiv:2002.10964.
40. Zhao, M.; Cong, Y.; Carin, L. On leveraging pretrained GANs for generation with limited data. In Proceedings of the 37th International Conference on Machine Learning, Virtual Event, 13–18 July 2020; pp. 11340–11351.
41. Wang, Y.; Gonzalez-Garcia, A.; Wu, C.; Herranz, L.; Khan, F.S.; Jui, S.; van de Weijer, J. MineGAN++: Mining Generative Models for Efficient Knowledge Transfer to Limited Data Domains. *arXiv* **2021**, arXiv:2104.13742.
42. Arjovsky, M.; Chintala, S.; Bottou, L. Wasserstein Generative Adversarial Networks. In Proceedings of the 34th International Conference on Machine Learning, Sydney, Australia, 6–11 August 2017; Precup, D., Teh, Y.W., Eds.; International Convention Centre: Sydney, Australia, 2017; Volume 70, pp. 214–223.
43. Karras, T.; Laine, S.; Aittala, M.; Hellsten, J.; Lehtinen, J.; Aila, T. Analyzing and Improving the Image Quality of StyleGAN. In Proceedings of the IEEE/CVF Conference on Computer Vision and Pattern Recognition, Seattle, WA, USA, 14–19 June 2020.
44. Diaz, H. Bean seeds images calibration dataset. Alliance of Bioversity International and CIAT. Unpublished raw data, 2019.
45. Beebe, S. Breeding in the Tropics. In *Plant Breeding Reviews*; John Wiley & Sons, Ltd.: New York, NY, USA, 2012; Chapter 5, pp. 357–426.

46. Rothe, R.; Timofte, R.; Gool, L.V. DEX: Deep EXpectation of apparent age from a single image. In Proceedings of the IEEE International Conference on Computer Vision Workshops (ICCVW), Santiago, Chile, 7–13 December 2015; pp. 10–15.
47. Agustsson, E.; Timofte, R.; Escalera, S.; Baró, X.; Guyon, I.; Rothe, R. Apparent and real age estimation in still images with deep residual regressors on APPA-REAL database. In Proceedings of the FG 2017—12th IEEE International Conference on Automatic Face and Gesture Recognition, Washington, DC, USA, 30 May–3 June 2017; pp. 1–12.
48. Moschoglou, S.; Papaioannou, A.; Sagonas, C.; Deng, J.; Kotsia, I.; Zafeiriou, S. AgeDB: The First Manually Collected, In-the-Wild Age Database. In Proceedings of the 2017 IEEE Conference on Computer Vision and Pattern Recognition Workshops (CVPRW), Honolulu, HI, USA, 21–26 July 2017; pp. 1997–2005.
49. Ye, L.; Li, B.; Mohammed, N.; Wang, Y.; Liang, J. Privacy-Preserving Age Estimation for Content Rating. In Proceedings of the 2018 IEEE 20th International Workshop on Multimedia Signal Processing (MMSP), Vancouver, BC, Canada, 29–31 August 2018; pp. 1–6.
50. Anda, F.; Lillis, D.; Kanta, A.; Becker, B.A.; Bou-Harb, E.; Le-Khac, N.A.; Scanlon, M. Improving Borderline Adulthood Facial Age Estimation through Ensemble Learning. In Proceedings of the 14th International Conference on Availability, Reliability and Security, Canterbury, UK, 26–29 August 2019; Association for Computing Machinery: New York, NY, USA, 2019; pp. 1–8. [CrossRef]
51. Chaves, D.; Fidalgo, E.; Alegre, E.; Jáñez-Martino, F.; Biswas, R. Improving Age Estimation in Minors and Young Adults with Occluded Faces to Fight Against Child Sexual Exploitation. In Proceedings of the 15th International Joint Conference on Computer Vision, Imaging and Computer Graphics Theory and Applications—Volume 5: VISAPP, Valletta, Malta, 27–19 February 2020; INSTICC; SciTePress: Setúbal, Portugal, 2020; pp. 721–729. [CrossRef]
52. Chaves, D.; Fernández-Robles, L.; Bernal, J.; Alegre, E.; Trujillo, M. Automatic characterisation of chars from the combustion of pulverised coals using machine vision. *Powder Technol.* **2018**, *338*, 110–118. [CrossRef]
53. StyleGAN Trained on Paintings (512 × 512). Available online: https://colab.research.google.com/drive/1cFKK0CBnev2BF8z9BOHxePk7E-f7TtUi (accessed on 9 March 2020).
54. StyleGAN-Art. Available online: https://github.com/ak9250/stylegan-art (accessed on 15 March 2020).
55. StyleGAN-Pokemon. Available online: https://www.kaggle.com/ahsenk/stylegan-pokemon (accessed on 15 March 2020).
56. StyleGAN—Official TensorFlow Implementation. Available online: https://github.com/NVlabs/stylegan (accessed on 19 December 2020).
57. Gwern. Making Anime Faces With StyleGAN. 2019. Available online: https://www.gwern.net/Faces (accessed on 9 March 2020).

 MDPI

Article

The Use of Trapezoidal Oriented Fuzzy Numbers in Portfolio Analysis

Anna Łyczkowska-Hanćkowiak

Institute of Economy and Finance, WSB University in Poznan, ul. Powstańców Wielkopolskich 5, 61-895 Poznań, Poland; anna.lyczkowska-hanckowiak@wsb.poznan.pl

Abstract: Oriented fuzzy numbers are a convenient tool to manage an investment portfolio as they enable the inclusion of uncertain and imprecise information about the financial market in a portfolio analysis. This kind of portfolio analysis is based on the discount factor. Thanks to this fact, this analysis is simpler than a portfolio analysis based on the return rate. The present value is imprecise due to the fact that it is modelled with the use of oriented fuzzy numbers. In such a case, the expected discount factor is also an oriented fuzzy number. The main objective of this paper is to conduct a portfolio analysis consisting of the instruments with the present value estimated as a trapezoidal oriented fuzzy number. We consider the portfolio elements as being positively and negatively oriented. We test their discount factor. Due to the fact that adding oriented fuzzy numbers is not associative, a weighted sum of positively oriented discount factors and a weighted sum of negatively oriented factors is calculated and consequently a portfolio discount factor is obtained as a weighted addition of both sums. Also, the imprecision risk of the obtained investment portfolio is estimated using measures of energy and entropy. All theoretical considerations are illustrated by an empirical case study.

Keywords: oriented fuzzy number; imprecision; present value; discount factor; portfolio; finance

Citation: Łyczkowska-Hanćkowiak, A. The Use of Trapezoidal Oriented Fuzzy Numbers in Portfolio Analysis. *Symmetry* **2021**, *13*, 1722. https://doi.org/10.3390/sym13091722

Academic Editors: Kóczy T. László and István A. Harmati

Received: 20 August 2021
Accepted: 14 September 2021
Published: 17 September 2021

1. Introduction

Imprecision is one of the characteristic features of information on financial markets. According to the uncertainty theory, any unknown future state of affairs is uncertain [1,2]. Uncertainty arises from our ignorance of the future state of affairs. This means that it can be modelled with a certain probability, as long as we are able to indicate a specific time at which the considered state of affairs will be known to the observer [3–9]. This was initially conceived by Kolmogorov [5,6] and is called "Kolmogorov's postulate". Imprecision is perceived as a combination of ambiguity and indistinctness of information [10]. Ambiguity is understood as a lack of a clear recommendation between one alternative among various others. Indistinctness is interpreted as a lack of explicit distinction between recommended and unrecommended alternatives. Modelling the imprecision by its membership function of the fuzzy set is a commonly used method [11]. In this paper, we describe imprecision using oriented fuzzy numbers. In portfolio analyses, we use uncertain and imprecise information about the financial market.

A security is an authorization to receive future financial income that is payable to a specified maturity. The value of this income is described as an anticipated future value of capital. The present value (PV) is defined as a current equivalent of a payment available at a stated time in the future [12]. It is assumed that the present value of future cash flows is an approximated value and therefore it can be modelled by fuzzy numbers. In this paper, PV is modelled by oriented fuzzy numbers (OFN) defined in [13]. In recent years, OFNs have been increasingly used to describe and analyze economic [14,15], financial [16–25] and decision-making [26–33] problems. The application of OFNs in financial analysis may minimize imprecision risk, which was presented, for instance, in [32]. The family of all oriented fuzzy numbers has a symmetry axis that is equal to the family R of all real numbers, which was discussed in detail in [32,34].

The base of any security assessment is the return rate, generally defined as an arbitrary non-increasing function of present value and a non-decreasing function of future value. An arbitrary, finite set of securities will be called a financial portfolio. After Markowitz, we also assume simple return rates, which means that the current value is positive and the future value is a random variable with a normal distribution [35]. It allows for some of the parameters, considered in existing portfolio theory, such as return rate or present value [36–38] or probability distribution parameters [39], to be fuzzy. Fuzzy systems are increasingly used in portfolio analysis [40–45]. A detailed evolution of research was presented in [17,32].

In financial arithmetic, fuzzy numbers appeared in 1987 due to Buckley [46]. The definition, which was proposed by Ward [47], was generalized to the case of imprecisely assessed postponement [48], fuzzy nominal interest rate [49], future cash flow described by a fuzzy variable [50], and to the case when future cash flow can be treated as a fuzzy probabilistic set [51]. The current fuzzy value was axiomatically defined by Calzi [52], while in Piasecki [53] the fuzzy PV was estimated based on the current quoted price of a financial asset. More detailed elaboration on that topic can be found in [17,32], among others.

A portfolio with a trapezoidal fuzzy PV was investigated, for instance, in [45]. In financial portfolio management, utilizing oriented fuzzy numbers is more useful than utilizing fuzzy numbers. We use trapezoidal oriented fuzzy numbers in the portfolio analysis because performing summation operations on them is much easier than on oriented fuzzy numbers. Oriented fuzzy numbers can be approximated by trapezoidal fuzzy numbers, as shown in [54]. Operations on oriented fuzzy numbers are much more computationally complicated, as can be seen in [17]. In [21], Piasecki and Łyczkowska-Hanćkowiak showed a universal method of representing Japanese candlesticks with oriented fuzzy numbers. In our paper, we investigate portfolios with PVs described by oriented fuzzy numbers. Moreover, a portfolio analysis based on a fuzzy discount factor is simpler than a portfolio analysis based on a return rate. For this reason, here we use a discount factor evaluated by oriented fuzzy numbers as defined in [25]. This paper presents an original method of using the imprecision expected discount factor in the portfolio analysis. We also show that portfolio diversification may lower uncertainty risk and imprecision risk. The addition of oriented fuzzy numbers is not associative. Therefore, we calculate the weighted sum of positively oriented discount factors and the sum of negatively oriented discount factors separately. Then, the portfolio discount factor is obtained by the weighted addition of these sums. Such a procedure for determining the discount factor of the portfolio is justified by economic premises. The criterion of the maximization of the expected return rate is replaced by the criterion of the minimalization of the expected discount factor. In decision analysis, a convenient tool for measuring the indistinctness risk is the entropy measure. In each of the considered cases, a return rate is a function of future value, which is uncertain by its nature. This uncertainty results from the ignorance of the future state of matters. This ignorance causes the investor to be unsure of future gains or losses. An increase in uncertainty may increase the uncertainty risk, which is the risk of making an incorrect financial decision. An increase in ambiguity means an increase in the amount of highly recommended alternative information about the state of affairs. This increase in the ambiguity of imprecise expected discount factors indicates that the number of alternative decisions that can be selected will be greater. This implies an increased ambiguity risk, i.e., selecting an incorrect assessment from among the recommended ones. This can cause a decision that will result in less than the maximum profit. Ambiguity and uncertainty risk have the same description of the dangers caused by risks taken and the effects of decisions made but they differ in their causes. For this reason, both types of risk should be assessed by different methods. We estimate the ambiguity risk of encumbering the imprecise expected discount factor with the energy measure. An increase in the indistinctness of the imprecise expected discount factor suggests that the differences between the recommended and unrecommended decision alternatives are harder to differentiate. This suggests that the indistinctness risk may increase. This means that we can choose an option that is not recommended. We estimate the indistinctness risk of the imprecise expected discount factor using the entropy measure. The risk of imprecision is the combination of the ambiguity and indistinctness risk. To assess the ambiguity and indistinctness risk, we use the energy measure and entropy measure, respectively. The main objective of this paper is to prove that it is possible to effectively manage a financial asset portfolio while accounting for both imprecision and uncertainty of information about the market. By using OFNs in the analysis, we minimize the imprecision risk.

The paper is organized as follows. Section 2 outlines trapezoidal fuzzy numbers and their basic properties as a theoretical background for our future considerations. Section 3 contains the evaluation of imprecision for oriented fuzzy numbers. In Section 4, we introduce the notion of imprecisely estimated PVs, which are evaluated using trapezoidal oriented fuzzy numbers. We use the obtained model in Section 5, where the expected discount factor is researched. In Section 6, we compare the original method for determining multi-asset portfolios with trapezoidal oriented fuzzy PVs. Our research is illustrated by a case study. This case study sufficiently explains the proposed method of portfolio evaluation. Finally, Section 7 concludes the article and summarizes the main findings.

2. Trapezoidal Oriented Fuzzy Numbers—Basic Facts

The family of all fuzzy subsets in the real line $\mathbb{R}$ we denote by $\mathcal{F}(\mathbb{R})$. Any fuzzy subset $\mathcal{A} \in \mathcal{F}(\mathbb{R})$ is described by its membership function $\mu_A \in [0,1]^{\mathbb{R}}$ as a set of ordered pairs:

$$\mathcal{A} = \{(x, \mu_A(x)); x \in \mathbb{R}\}. \tag{1}$$

Fuzzy subset $\mathcal{A}$ may be characterized by its support closure $[\mathcal{A}]_{0^+}$, given in the following way:

$$[\mathcal{A}]_{0^+} = \lim_{\alpha \to 0^+} \{x \in \mathbb{R} : \mu_A(x) \geq \alpha\}. \tag{2}$$

The core $\mathcal{C}ore(A)$ is defined by the formula:

$$\mathcal{C}ore(A) = \{x \in \mathbb{X} : \mu_A(x) = 1\}. \tag{3}$$

A fuzzy number (FN) is usually defined as a fuzzy subset of the real line $\mathbb{R}$. The most general definition of FN is given as follows:

Definition 1 [55]. *The fuzzy number is such a fuzzy subset $\mathcal{L} \in \mathcal{F}(\mathbb{R})$ with a bounded support closure $[\mathcal{L}]_{0^+}$ that it is represented by its upper semi-continuous membership function $\mu_L \in [0;1]^{\mathbb{R}}$, satisfying the conditions:*

$$\exists_{x \in \mathbb{R}}\ \mu_L(x) = 1, \tag{4}$$

$$\forall_{(x,y,z) \in \mathbb{R}^3}\ x \leq y \leq z \implies \mu_L(y) \geq \min\{\mu_L(x); \mu_L(z)\}. \tag{5}$$

The set of all FNs we denote by the symbol $\mathbb{F}$. Any FN may be represented in the following way:

Theorem 1 [56,57]. *For any FN $\mathcal{L}$ there exists such a non-decreasing sequence $(a,b,c,d) \subset \mathbb{R}$ that $\mathcal{L}(a,b,c,d,L_L,R_L) = \mathcal{L} \in \mathcal{F}(\mathbb{R})$ is determined by its membership function $\mu_L(\cdot\ |a,b,c,d,L_L,R_L) \in [0,1]^{\mathbb{R}}$ described by the identity:*

$$\mu_L(x|a,b,c,d,L_L,R_L) = \begin{cases} 0,\ x \notin [a,d], \\ L_L(x),\ x \in [a,b], \\ 1,\ x \in [b,c], \\ R_L(x),\ x \in [c,d], \end{cases} \tag{6}$$

where the left reference function $L_L \in [0,1]^{[a,b]}$ *and the right reference function* $R_L \in [0,1]^{[c,d]}$ *are upper semi-continuous monotonic ones meeting the condition:*

$$[\mathscr{L}]_{0^+} = [a, d]. \tag{7}$$

The FN $\mathscr{L}(a, a, a, a, L_L, R_L) = [\![a]\!]$ represents the real number $a \in \mathbb{R}$, i.e., $\mathbb{R} \subset \mathbb{F}$.

The notion of ordered FNs is introduced by Kosiński et al. in [58]. For formal reasons, Kosiński's theory is revised in [13]. In the revised theory, the notion of ordered FNs is narrowed down to the notion of oriented FNs (OFN), defined as follows:

Definition 2 [13]. *For any monotonic sequence* $(a, b, c, d) \subset \mathbb{R}$*, an oriented fuzzy number* $\overleftrightarrow{\mathscr{L}}(a, b, c, d, S_L, E_L) = \overleftrightarrow{\mathscr{L}}$ *is the pair of orientation* $\overrightarrow{a, d} = (a, d)$ *and FN* $\mathscr{L} \in \mathbb{F}$ *is described by membership of the function* $\mu_L(\cdot|a, b, c, d, S_L, E_L) \in [0,1]^{\mathbb{R}}$*, given by the identity:*

$$\mu_L(x|a, b, c, d, S_L, E_L) = \begin{cases} 0, \ x \notin [a, d] \equiv [d, a], \\ S_L(x), \ x \in [a, b] \equiv [b, a], \\ 1, \ x \in [b, c] \equiv [c, b], \\ E_L(x), \ x \in [c, d] \equiv [d, c], \end{cases} \tag{8}$$

where the starting function $S_L \in [0,1]^{[a,b]}$ *and the ending function* $E_L \in [0,1]^{[c,d]}$ *are upper semi-continuous monotonic ones meeting condition (7).*

In Equation (8), we use modified interval notation, often used in the theory of oriented fuzzy numbers. The notation $\mathscr{J} \equiv \mathscr{K}$ means that "the interval $\mathscr{J}$ may be equivalently replaced by the interval $\mathscr{K}$". The relationships between FNs, ordered FNs, and OFNs are discussed in detail in [34].

The symbol $\mathbb{K}$ denotes the space of all OFNs. If $a < d$, then OFN $\overleftrightarrow{\mathscr{L}}(a, b, c, d, S_L, E_L)$ has a positive orientation $\overrightarrow{a, d}$. It informs us about the possibility of an increase in the approximated number. If $a > d$, then OFN $\overleftrightarrow{\mathscr{L}}(a, b, c, d, S_L, E_L)$ has the negative orientation $\overrightarrow{a, d}$. It informs us about the possibility of a decrease in the approximated number. If $a = d$, then OFN $\overleftrightarrow{\mathscr{L}}(a, a, a, a, S_L, E_L) = [\![a]\!]$ describes the real number $a \in \mathbb{R}$.

Trapezoidal fuzzy numbers (TrOFNs) are a special case of OFNs.

Definition 3 [13]. *For any monotonic sequence* $(a, b, c, d) \subset \mathbb{R}$*, TrOFN* $\overleftrightarrow{Tr}(a, b, c, d) = \overleftrightarrow{\mathscr{T}}$ *is OFN* $\overleftrightarrow{\mathscr{T}} \in \mathbb{K}$*, determined explicitly by its membership functions* $\mu_T \in [0,1]^{\mathbb{R}}$ *as follows:*

$$\mu_T(x) = \mu_{Tr}(x|a, b, c, d) = \begin{cases} 0, \ x \notin [a, d] \equiv [d, a], \\ \frac{x-a}{b-a}, \ x \in [a, \ b] \equiv [a, b], \\ 1, \ x \in [b, c] \equiv [c, b], \\ \frac{x-d}{c-d}, \ x \in [c, d] \equiv [c, d]. \end{cases} \tag{9}$$

The space of all trapezoidal oriented fuzzy numbers, TrOFNs, we denote as $\mathbb{K}_{Tr}$. By $\mathbb{K}_{Tr}^{+}$ we denote the space of all TrOFNs that have a positive orientation and by $\mathbb{K}_{Tr}^{-}$ we denote the space of all TrOFNs that have a negative orientation. A crisp number $a \in \mathbb{R}$, which is unoriented, is denoted as the following trapezoidal oriented fuzzy number $\overleftrightarrow{Tr}(a, a, a, a) = [\![a]\!]$.

Let symbol $*$ denote any arithmetic operation defined in $\mathbb{R}$ and symbol $\boxed{*}$ denote an extension of an arithmetic operation $*$ to $\mathbb{K}$ [13]. Kosiński has defined arithmetic operators on ordered FNs in an intuitive way, but the addition of a dot product extended to the space $\mathbb{K}$ has a very high level of formal complexity [34]. Therefore, in many applications researchers limit their calculations to arithmetic operations determined in the space $\mathbb{K}_{Tr}$.

In line with Kosinski's approach, we can extend the basic arithmetic operators in the case of $\mathbb{K}_{Tr}$ in such way that for any pair $\left(\overleftrightarrow{Tr}(a,b,c,d), \overleftrightarrow{Tr}(p-a,q-b,r-c,s-d)\right) \in \mathbb{K}^2_{Tr}$ and $\beta \in \mathbb{R}$, arithmetic operations of extended sum $\boxplus$ and dot product $\boxdot$ are defined as follows [34]:

$$\begin{aligned} &\overleftrightarrow{Tr}(a,b,\ c,d) \boxplus \overleftrightarrow{Tr}(p-a,q-b,\ r-c,s-d) = \\ &= \begin{cases} \overleftrightarrow{Tr}(\min\{p,q\},q,\ r,\max\{r,s\}),\ (q<r) \vee (q=r \wedge p \le s), \\ \overleftrightarrow{Tr}(\max\{p,q\},q,\ r,\min\{r,s\}),\ (\ q<r) \vee (q=r \wedge p > s). \end{cases} \end{aligned} \tag{10}$$

$$\beta \boxdot \overleftrightarrow{Tr}(a,b,c,d) = \overleftrightarrow{Tr}(\beta{\cdot}a, \beta{\cdot}b, \beta{\cdot}c, \beta{\cdot}d). \tag{11}$$

It is very easy to check that for any pair $\left(\overleftrightarrow{Tr}(a,b,c,d), \overleftrightarrow{Tr}(e,f,g,h)\right) \in \mathbb{K}^+_{Tr} \times \mathbb{K}^+_{Tr} \cup \mathbb{K}^-_{Tr} \times \mathbb{K}^-_{Tr}$, we have:

$$\overleftrightarrow{Tr}(a,b,\ c,d) \boxplus \overleftrightarrow{Tr}(e,f,\ g,h) = \overleftrightarrow{Tr}(a+e,b+f,\ c+g,d+h). \tag{12}$$

Any unary operator $G : \mathbb{R} \supset \mathbb{A} \longrightarrow \mathbb{R}$ may be extended in the case of TrOFNs. Using Kosiński's approach, we define an extended unary operator $\overleftrightarrow{G} : \mathbb{K}_{Tr} \supset \mathbb{H} \longrightarrow \mathbb{K}$ as follows:

$$\overleftrightarrow{\mathscr{L}}(G(a),G(b),G(c),G(d),\ S_L,E_L) = \overleftrightarrow{G}\left(\overleftrightarrow{Tr}(a,b,c,d)\right) \tag{13}$$

where the starting function and the ending function are given by the formulas:

$$\forall\, y \in [G(a),G(b)]\ S_L(y) = \frac{G^{-1}(y)-a}{b-a}, \tag{14}$$

$$\forall\, y \in [G(c),G(d)]\ E_L(y) = \frac{G^{-1}(y)-d}{c-d}. \tag{15}$$

OFN (12) is TrOFN if and only if its starting function and ending function are linear.

3. Evaluation of Imprecision for Oriented Fuzzy Numbers

Information imprecision we understand as a superposition of ambiguity and indistinctness of information [10]. Ambiguity is comprehended in such a way that neither option is unequivocally recommended. Indistinctness is comprehended in such a way that recommended and unrecommended alternatives are not clearly distinguishable.

The increase in OFN ambiguity implies a higher number of recommended alternatives. This increases the risk of choosing the wrong alternative from among the recommended ones. This can eventually lead to one making a decision that will result in the loss of ex-post chances. The possibility of this happening is called the ambiguity risk. Hence, an increase in the ambiguity of OFN causes an increase in the ambiguity risk. A convenient tool for measuring the oriented fuzzy number ambiguity is an extension of energy measure, defined by the following formula:

$$d\left(\overleftrightarrow{\mathscr{L}}(a,b,c,d,L_L,R_L)\right) = \left|\int_a^d \mu_L(x|a,b,c,d,L_L,R_L)dx\right|. \tag{16}$$

This energy measure is an extension of the energy measure $d \in \left[\mathbb{R}^+_0\right]^{\mathbb{F}}$, defined for fuzzy numbers by de Luca and Termini [59]. In decision analysis, to measure the ambiguity risk is to exploit the energy measure.

A comfortable instrument for measuring the indistinctness of an oriented fuzzy number is the entropy measure $e \in \left[\mathbb{R}_0^+\right]^{\mathbb{F}}$. This measure was also proposed by de Luca and Termini [60] and was then modified by Piasecki [61].

In [62], Kosko described the most widely known kind of entropy measure. Unfortunately, in [44,45] it is shown that Kosko's entropy measure is not convenient for portfolio analysis. For this reason, in [23,34] the tool for measuring the OFN indistinctness is proposed as the entropy measure $e \in \left[\mathbb{R}_0^+\right]^{\mathbb{K}}$, defined by the identity:

$$e\left(\overset{\leftrightarrow}{\mathscr{L}}(a,b,c,d,L_L,R_L)\right) = \left|\int_a^d \min\{\mu_L(x|a,b,c,d,L_L,R_L),\ 1-\mu_L(x|a,b,c,d,L_L,R_L)\}dx\right|. \tag{17}$$

This entropy measure is an extension of the entropy measure $e \in \left[\mathbb{R}_0^+\right]^{\mathbb{F}}$, introduced for FNs by Czogała, Gottwald, and Pedrycz in [63]. In decision analysis, we use the entropy measure as a measure of the indistinctness risk. Imprecision risk consists of both ambiguity and indistinctness risk combined.

For any TrOFN $\overset{\leftrightarrow}{Tr}(a,b,c,d)$, its energy and entropy are determined based on the following equations [34]:

$$d\left(\overset{\leftrightarrow}{Tr}(a,b,c,d)\right) = \frac{1}{2}\cdot|d+c-b-a|, \tag{18}$$

$$e\left(\overset{\leftrightarrow}{Tr}(a,b,c,d)\right) = \frac{1}{4}\cdot|d-c+b-a|. \tag{19}$$

The results can be described in a form convenient for the analysis of an investment portfolio. An asset benefit index is the value of the relative profit of owned capital. This relative profit of the owned asset (for example return rate, discount factor) is a function that is dependent on the profit value and the asset value [34].

The benefit index for a two-assets portfolio is equal to the average of the portfolio component benefit indexes. This model is often used to analyze the effects of portfolio diversification. In our research, the portfolio component benefit indexes are imprecisely estimated. If portfolio component benefit indexes are described by trapezoidal oriented fuzzy numbers $\overset{\leftrightarrow}{\mathscr{K}}, \overset{\leftrightarrow}{\mathscr{L}} \in \mathbb{K}_{Tr}$, then the portfolio benefit index is determined by the function $\varpi : (\mathbb{K}_{Tr})^2 \times [0,1] \to \mathbb{K}_{Tr}$, given by this formula [34]:

$$\varpi\left(\overset{\leftrightarrow}{\mathscr{K}}, \overset{\leftrightarrow}{\mathscr{L}}, \lambda\right) = \left(\lambda \boxdot \overset{\leftrightarrow}{\mathscr{K}}\right) \boxplus \left((1-\lambda) \boxdot \overset{\leftrightarrow}{\mathscr{L}}\right). \tag{20}$$

The method of determining the parameter λ depends on the kind of considered relative benefits [45].

Theorem 2 [34].*For any real number* $\lambda \in [0,1]$, we have:

- *for any pair* $\left(\overset{\leftrightarrow}{\mathscr{K}}, \overset{\leftrightarrow}{\mathscr{L}}\right) \in \left(\mathbb{K}_{Tr}^- \times \mathbb{K}_{Tr}^-\right) \cup \left(\left(\mathbb{K}_{Tr}^+ \cup \mathbb{R}\right) \times \left(\mathbb{K}_{Tr}^+ \cup \mathbb{R}\right)\right)$

$$d\left(\varpi\left(\overset{\leftrightarrow}{\mathscr{K}}, \overset{\leftrightarrow}{\mathscr{L}}, \lambda\right)\right) = \lambda\cdot d\left(\overset{\leftrightarrow}{\mathscr{K}}\right) + (1-\lambda)\cdot d\left(\overset{\leftrightarrow}{\mathscr{L}}\right), \tag{21}$$

$$e\left(\varpi\left(\overset{\leftrightarrow}{\mathscr{K}}, \overset{\leftrightarrow}{\mathscr{L}}, \lambda\right)\right) = \lambda\cdot e\left(\overset{\leftrightarrow}{\mathscr{K}}\right) + (1-\lambda)\cdot e\left(\overset{\leftrightarrow}{\mathscr{L}}\right), \tag{22}$$

- *for any pair* $\left(\overset{\leftrightarrow}{\mathscr{K}}, \overset{\leftrightarrow}{\mathscr{L}}\right) \in \left(\left(\mathbb{K}_{Tr}^+ \cup \mathbb{R}\right) \times \mathbb{K}_{Tr}^-\right)$

$$d\left(\varpi(\vec{\mathcal{K}},\vec{\mathcal{L}},\lambda)\right) \leq \begin{cases} \lambda \cdot d(\vec{\mathcal{K}}) - (1-\lambda) \cdot d\left(\mathcal{C}ore(\vec{\mathcal{L}})\right), \varpi(\vec{\mathcal{K}},\vec{\mathcal{L}},\lambda) \in \mathbb{K}_{Tr}^{+} \cup \mathbb{R}, \\ (1-\lambda) \cdot d(\vec{\mathcal{L}}) - \lambda \cdot d\left(\mathcal{C}ore(\vec{\mathcal{K}})\right), \varpi(\vec{\mathcal{K}},\vec{\mathcal{L}},\lambda) \in \mathbb{K}_{Tr}^{-} \cup \mathbb{R}, \end{cases} \tag{23}$$

- *for any pair* $\left(\overset{\leftrightarrow}{\mathscr{K}},\overset{\leftrightarrow}{\mathscr{L}}\right) \in \left((\mathbb{K}_{Tr}^{+} \cup \mathbb{R}) \times \mathbb{K}_{Tr}^{-}\right) \cup \left(\mathbb{K}_{Tr}^{-} \times (\mathbb{K}_{Tr}^{+} \cup \mathbb{R})\right)$

$$e\left(\varpi\left(\overset{\leftrightarrow}{\mathscr{K}},\overset{\leftrightarrow}{\mathscr{L}},\lambda\right)\right) \leq \min\left\{\lambda e\left(\overset{\leftrightarrow}{\mathscr{K}}\right),(1-\lambda)e\left(\overset{\leftrightarrow}{\mathscr{L}}\right)\right\}. \tag{24}$$

4. Oriented Fuzzy Present Value

According to the uncertainty theory, any unknown future state of affairs is uncertain [1,2]. The uncertainty results from our lack of knowledge about the future state of affairs. We can model uncertainty with some probability. It is possible when we are able to indicate a specific time in the future, when the effect of a considered situation will already be known to the observer [3–9]. This was initially formulated by Kolmogorov [5,6] and is called "Kolmogorov's postulate".

We understand a security as an authorization to receive a future financial income payable to a certain maturity. The value of this income is understood as an anticipated future value of an asset. In our research, we can point to this particular time in the future at which the considered revenue will already be known to the observer. For this reason, the future value is not burdened by Knight's uncertainty [64]. Together with Kolmogorov's postulate, this allows us to conclude that the future value is a random variable.

In [12], a present value (PV) was redefined as a present equivalent of a payment available at a given time in the future. In our research, the estimation of the fuzzy present value is supplemented by a forecast of the closest changes in price. It was shown in [65] that the closest price changes can be forecasted using a prediction table. The use of oriented fuzzy numbers for a portfolio analysis is more useful than the use of fuzzy numbers, as discussed in detail in [34]. For this reason, an imprecise present value may be estimated using oriented fuzzy numbers [21,32]. PV, defined in this way, is called an oriented present value (O-PV). Any O-PV is described by a monotonic sequence $\left(V_s,\ V_f,\ \check{P},\ V_l,\ V_e\right)$ and then it is estimated by a trapezoidal oriented fuzzy number:

$$\overset{\leftrightarrow}{PV} = \overset{\leftrightarrow}{Tr}\left(V_s,\ V_f,\ V_l,\ V_e\right) \tag{25}$$

where the monotonic sequence $\left(V_s,\ V_f,\ \check{P},\ V_l,\ V_e\right)$ is determined in the following way:

- $\check{P}$ is a quoted price.
- $[V_s, V_e] \subset \mathbb{R}^+$ is an interval of all possible values of PV.
- $\left[V_f, V_l\right] \subset [V_s, V_e]$ is the interval of all prices that do not noticeably differ from the quoted price $\check{P}$.

If we anticipate a price increase, then PV is described by a positively oriented trapezoidal fuzzy number. If we anticipate a price decrease, then PV is described by a negatively oriented trapezoidal fuzzy number.

In our research, oriented present value is used for modelling imprecise evaluations of PV. The imprecision consists of ambiguity and indistinctness. We comprehend the ambiguity in such a way that neither value is unequivocally indicated. An indistinctness is comprehended in such a way that values equal to the present value and values different from the present value are not clearly distinguishable. We use the energy and entropy to measure the ambiguity and indistinctness of the O-PV given by $\overset{\leftrightarrow}{PV} = \overset{\leftrightarrow}{Tr}\left(V_s,\ V_f,\ V_l,\ V_e\right)$.

Let the fixed security Ŝ be given. Using Equations (18), (19) and (25) for fixed security Ŝ, we get:

$$d\left(\overleftrightarrow{PV}(\hat{S})\right) = \frac{1}{2}\cdot\left|V_e + V_l - V_f - V_s\right|, \tag{26}$$

$$e\left(\overleftrightarrow{PV}(\hat{S})\right) = \frac{1}{4}\cdot\left|V_e - V_l + V_f - V_s\right|. \tag{27}$$

Examples of trapezoidal O-PVs are the Japanese candle model [21] and the Heikin-Ashi candle model [24].

Example 1 [32]. *We consider a portfolio of company stocks included in WIG20 quoted on the Warsaw Stock Exchange (WSE). Based on a session closing on the WSE on 28 January 2020, for each observed stock we evaluated its oriented present value assessed by trapezoidal oriented fuzzy number, describing its Japanese candle [21]. Table 1 contains these stocks' O-PVs. For each considered stock Ŝ, we notice its quoted price $\check{P}_s$is the initial price on 29 January 2020.*

Table 1. Recorded values of the portfolio π stocks.

Company's Stock	Present Value $\overleftrightarrow{PV}_s$	Quoted Price $\check{P}_s$	Energy Measure	Entropy Measure
ALR	$\overleftrightarrow{Tr}(27.42;\ 27.30;\ 27.00;\ 26.84)$	27.00	0.44	0.07
CCC	$\overleftrightarrow{Tr}(83.35;\ 88.00;\ 88.00;\ 89.65)$	88.00	3.15	1.575
CDR	$\overleftrightarrow{Tr}(271.50;\ 271.50;\ 276.30;276.30)$	277.00	4.8	0
CPS	$\overleftrightarrow{Tr}(26.42;\ 26.60;\ 27.04;\ 27.34)$	27.20	0.68	0.12
DNP	$\overleftrightarrow{Tr}(155.00;\ 155.00;155.10;\ 157.30)$	155.30	1.2	0.55
JSW	$\overleftrightarrow{Tr}(18.60;\ 19.36;\ 20.14;\ 20.14)$	20.32	1.16	0.19
KGH	$\overleftrightarrow{Tr}(91.78;\ 93.60;\ 93.70;\ 94.90)$	94.24	1.61	0.755
LTS	$\overleftrightarrow{Tr}(83.88;\ 83.40;\ 81.16;\ 80.26)$	81.44	2.93	0.345
LPP	$\overleftrightarrow{Tr}(8205.00;\ 8380.00;\ 8395.00;\ 8460.00)$	8385.00	135	60
MBK	$\overleftrightarrow{Tr}(367.00;\ 366.00;359.80;\ 357.00)$	359.00	8.1	0.95
OPL	$\overleftrightarrow{Tr}(7.01;\ 7.05;\ 7.20;\ 7.35)$	7.17	0.245	0.0475
PEO	$\overleftrightarrow{Tr}(97.22;\ 97.70;\ 98.20;\ 98.66)$	98.20	0.97	0.235
PGE	$\overleftrightarrow{Tr}(7.08;\ 7.15;\ 7.30;\ 7.40)$	7.30	0.235	0.0425
PGN	$\overleftrightarrow{Tr}(3.91;\ 3.88;\ 3.86;\ 3.82)$	3.87	0.055	0.0175
PKN	$\overleftrightarrow{Tr}(83.22;\ 83.00;\ 81.62;\ 81.18)$	81.90	1.71	0.165
PKO	$\overleftrightarrow{Tr}(34.59;\ 34.68;\ 34.90;\ 35.26)$	34.93	0.445	0.1125
PLY	$\overleftrightarrow{Tr}(35.82;\ 35.94;\ 36.76;\ 37.20)$	36.70	1.1	0.14
PZU	$\overleftrightarrow{Tr}(40.72;\ 40.73;\ 40.89;\ 41.11)$	40.88	0.275	0.0575
SPL	$\overleftrightarrow{Tr}(276.20;\ 278.00;\ 281.80;283.80)$	287.00	5.7	0.95
TPE	$\overleftrightarrow{Tr}(1.51;\ 1.53;\ 1.56;\ 1.56)$	1.56	0.04	0.005

We calculate the energy and entropy measure of the oriented present value using Equations (26) and (27), respectively.

It is noteworthy that CCC, CDR, CPS, DNP, JWS, KGH, LPP, OPL, PEO, PGE, PKO, PLY, PZU, SPL, and TPE are estimated by positively oriented trapezoidal fuzzy numbers. It means that they are evaluated by positively oriented present values. Hence, we anticipate that their quoted prices will increase. Conversely, ALR, LTS, MBK, PGN, and PKN are estimated by negatively oriented trapezoidal fuzzy numbers, which means that they are evaluated by negatively oriented present values; thus, we anticipate that their quoted prices will decrease.

5. Expected Discount Factor

Let us assume that the time horizon $t > 0$ of an investment is fixed. In this case, the considered asset is determined by two values:

- the anticipated future value (FV), V_t.
- the assessed present value (PV), V_0.

The basic property of benefits from owning this asset is a simple return rate r_t, given by the formula:

$$r_t = \frac{V_t - V_0}{V_0} = \frac{V_t}{V_0} - 1. \tag{28}$$

According to Kolmogorov's postulate [5,6], FV is a random variable $\widetilde{V}_t : \Omega \longrightarrow \mathbb{R}^+$, where Ω is a set of elementary states ω of the financial market. In a classical approach to a return rate estimation, present value is identified with the observed quoted price $\check{P}$. Thus, the return rate is a random variable determined by identity:

$$r_t(\omega) = \frac{\widetilde{V}_t(\omega) - \check{P}}{\check{P}}. \tag{29}$$

Uncertainty risk is a result of a lack of knowledge about the future state of affairs. In the practice of financial market analysis, the uncertainty risk is usually described by the probability distribution of return rate (29). This function may be given by a cumulative distribution function $F_r(\cdot|\bar{r}) : \mathbb{R} \longrightarrow [0,1]$. We assume that the expected return rate $\bar{r}$ of this distribution exists. The expected discount factor (EDF) $\bar{v}$ must therefore also exist and is determined by the dependency:

$$\bar{v} = (1 + \bar{r})^{-1}. \tag{30}$$

To simplify the reading of Equation (30) and further where Equation (30) is used, ω has been omitted.

Taking (28) and (29), we obtain the following formula describing the return rate:

$$r_t = r_t(V_0, \omega) = \frac{\check{P}\cdot(1 + r_t(\omega))}{V_0} - 1. \tag{31}$$

It implies that the expected return rate may be expressed in the following way:

$$\mathscr{R}(V_0) = \int_{-\infty}^{+\infty} \frac{\check{P}\cdot(1+y)}{V_0} - 1dF_r(y|\bar{r}) = \frac{\check{P}\cdot(1+\bar{r})}{V_0} - 1. \tag{32}$$

In this way, we express the return rate as a decreasing unary operator $\mathscr{R} : \mathbb{R}^+ \longrightarrow \mathbb{R}$ transforming PV. If PV is imprecisely estimated by TrOFN (25), then in line with (13), (14) and (15), the imprecise return rate is given as OFN:

$$\begin{aligned} \mathscr{R}(\hat{S}) = \mathscr{R}\left(\overleftrightarrow{PV}(\hat{S})\right) &= \mathscr{R}\left(\overleftrightarrow{Tr}\left(V_s, V_f, V_l, V_e\right)\right) \\ &= \overleftrightarrow{\mathscr{L}}\left(\tfrac{\check{P}\cdot(1+\bar{r})}{V_e} - 1, \tfrac{\check{P}\cdot(1+\bar{r})}{V_l} - 1, \tfrac{\check{P}\cdot(1+\bar{r})}{V_f} - 1, \tfrac{\check{P}\cdot(1+\bar{r})}{V_s} - 1, S_L, E_L\right), \end{aligned} \tag{33}$$

where the reference functions are given by the formulas:

$$\forall\, y \in \left[\frac{\check{P}\cdot(1+\bar{r})}{V_s} - 1, \frac{\check{P}\cdot(1+\bar{r})}{V_f} - 1\right] S_L(r) = \frac{\frac{\check{P}\cdot(1+\bar{r})}{1+r} - V_s}{V_f - V_s}, \tag{34}$$

$$\forall\, y \in \left[\frac{\check{P}\cdot(1+\bar{r})}{V_l} - 1, \frac{\check{P}\cdot(1+\bar{r})}{V_e} - 1\right] E_L(r) = \frac{\frac{\check{P}\cdot(1+\bar{r})}{1+r} - V_e}{V_l - V_e}. \tag{35}$$

It is noteworthy that the above reference functions are not linear. It implies that OFN (33) is not TrOFN. The use of the return rate does not allow us to bypass the complexity of the arithmetic operations of OFN.

In the next step, we consider the discount factor v_t, which is determined by the return rate r_t. From (28), we get:

$$v_t = (1 + r_t)^{-1}. \tag{36}$$

Then, taking (36) and (29) together, we obtain the following random variable describing a discount factor:

$$v_t = r_t(V_0, \omega) = \frac{V_0}{\check{P}\cdot(1 + r_t(\omega))}. \tag{37}$$

This implies that the expected discount factor EDF may be expressed in the following way:

$$\mathscr{V}(V_0) = \int_{-\infty}^{+\infty} \frac{V_0}{\check{P}\cdot(1+y)} dF_r(y|\bar{r}) = \frac{V_0}{\check{P}\cdot(1+\bar{r})} = \left(\frac{\check{P}\cdot(1+\bar{r})}{V_0}\right)^{-1} = \frac{\bar{v}}{\check{P}}\cdot V_0. \tag{38}$$

Using this equation, we determine the imprecise EDF $\mathscr{V} : \mathbb{R}^+ \to \mathbb{R}^+$ as an increasing operator transforming PV. If PV is imprecisely estimated by TrOFN (25), then in line with (13)–(15), the imprecise EDF is given as TrOFN:

$$\overset{\leftrightarrow}{\mathscr{V}}(\hat{S}) = \overset{\leftrightarrow}{\mathscr{V}}\left(\overset{\leftrightarrow}{PV}(\hat{S})\right) = \overset{\leftrightarrow}{\mathscr{L}}\left(\frac{V_s\cdot\bar{v}}{\check{P}}, \frac{V_f\cdot\bar{v}}{\check{P}}, \frac{V_l\cdot\bar{v}}{\check{P}}, \frac{V_e\cdot\bar{v}}{\check{P}}, S_L, E_L\right) \tag{39}$$

where the reference functions are given by the formulas:

$$\forall\, y \in \left[\frac{V_s\cdot\bar{v}}{\check{P}}, \frac{V_f\cdot\bar{v}}{\check{P}}\right] \; S_L(v) = \frac{\frac{v\cdot\check{P}}{\bar{v}} - V_s}{V_f - V_s} = \frac{\check{P}\cdot v - V_s\cdot\bar{v}}{\bar{v}\cdot\left(V_f - V_s\right)}, \tag{40}$$

$$\forall\, y \in \left[\frac{V_l\cdot\bar{v}}{\check{P}}, \frac{e\cdot\bar{v}}{\check{P}}\right] \; E_L(v) = \frac{\frac{v\cdot\check{P}}{\bar{v}} - V_e}{V_l - V_e} = \frac{\check{P}\cdot v - V_e\cdot\bar{v}}{\bar{v}\cdot(V_l - V_e)}. \tag{41}$$

Note that both the above reference functions are linear. This implies that if an imprecise PV is given by TrOFN (25), then the imprecise EDF $\mathscr{V}$ is TrOFN, which is given by the formula:

$$\overset{\leftrightarrow}{\mathscr{V}}(\hat{S}) = \overset{\leftrightarrow}{\mathscr{V}}\left(\overset{\leftrightarrow}{PV}(\hat{S})\right) = \frac{\bar{v}}{\check{P}} \boxdot \overset{\leftrightarrow}{PV}(\hat{S}). \tag{42}$$

In this case, it is clear that in financial analysis the imprecise EDF is more useful than the imprecise expected return rate. If we apply EDF, then the criterion of maximization of the expected return rate is replaced by the criterion of EDF minimalization.

Using (35), (18) and (19), the energy measure and the entropy measure of the expected discount factor EDF $\overset{\leftrightarrow}{\mathscr{V}}(\hat{S})$ are determined by the formulas:

$$d\left(\overset{\leftrightarrow}{\mathscr{V}}(\hat{S})\right) = \left|\frac{\left(V_e + V_l - V_f - V_s\right)\cdot\bar{v}}{2\check{P}}\right|, \tag{43}$$

$$e\left(\overset{\leftrightarrow}{\mathscr{V}}(\hat{S})\right) = \left|\frac{\left(V_e - V_l + V_f - V_s\right)\cdot\bar{v}}{4\check{P}}\right|. \tag{44}$$

Example 2. *All considerations in this paper are run for the quarterly period of the investment time* $t = 1$ *quarter. We researched the stocks of the portfolio* π *presented in Table 1. For an average expected return rate* $\bar{r} = 0.10144$ *from (30), we get an expected discount factor* $\bar{v} = 0.9079$*. Using* *the one-year time series of quotations, for each considered stock $\hat{S}$using (42), we calculate a quarterly expected discount factor O-EDF.*

For example, for the O-EDF of ALR company stock, we get:

$$\overleftrightarrow{\mathscr{V}}(ALR) = \overleftrightarrow{\mathscr{V}}\left(\overleftrightarrow{PV}(ALR)\right) = \frac{\overline{v}}{\check{P}} \boxdot \overleftrightarrow{PV}(ALR) = \frac{0.9079}{27} \boxdot \overleftrightarrow{Tr}(27.42;\ 27.30;\ 27.00;\ 26.84)$$
$$= 0.0336 \boxdot \overleftrightarrow{Tr}(27.42;\ 27.30;\ 27.00;\ 26.84) = \overleftrightarrow{Tr}(0.9220;\ 0.9180;\ 0.9079;\ 0.9025)$$

We calculate the O-EDF in a similar way for other company stocks.

Additionally, using Equations (43) and (44), we determine, respectively, the energy and entropy measure of the O-EDF. All evaluations obtained in this way are presented in Table 2.

Table 2. O-EDF of portfolio π components and their energy and entropy measures.

Company's Stock	OEDF $\overleftrightarrow{V}_s$	Energy Measure	Entropy Measure
ALR	$\overleftrightarrow{Tr}(0.9220;\ 0.9180;\ 0.9079;\ 0.9025)$	0.0148	0.0024
CCC	$\overleftrightarrow{Tr}(0.8599;\ 0.9079;\ 0.9079;\ 0.9249)$	0.0325	0.0163
CDR	$\overleftrightarrow{Tr}(0.8899;\ 0.8899;\ 0.9056;\ 0.9056)$	0.0157	0.0000
CPS	$\overleftrightarrow{Tr}(0.8819;\ 0.8879;\ 0.9026;\ 0.9126)$	0.0227	0.0040
DNP	$\overleftrightarrow{Tr}(0.9062;\ 0.9062;\ 0.9067;\ 0.9196)$	0.0070	0.0032
JSW	$\overleftrightarrow{Tr}(0.8311;\ 0.8650;\ 0.8999;\ 0.8999)$	0.0518	0.0085
KGH	$\overleftrightarrow{Tr}(0.8842;\ 0.9017;\ 0.9027;\ 0.9143)$	0.0155	0.0073
LTS	$\overleftrightarrow{Tr}(0.9351;\ 0.9298;\ 0.9048;\ 0.8948)$	0.0327	0.0039
LPP	$\overleftrightarrow{Tr}(0.8884;\ 0.9074;\ 0.9090;\ 0.9160)$	0.0146	0.0065
MBK	$\overleftrightarrow{Tr}(0.9281;\ 0.9256;\ 0.9099;\ 0.9028)$	0.0205	0.0024
OPL	$\overleftrightarrow{Tr}(0.8876;\ 0.8927;\ 0.9117;\ 0.9307)$	0.0310	0.0060
PEO	$\overleftrightarrow{Tr}(0.8988;\ 0.9033;\ 0.9079;\ 0.9122)$	0.0090	0.0022
PGE	$\overleftrightarrow{Tr}(0.8805;\ 0.8892;\ 0.9079;\ 0.9203)$	0.0292	0.0053
PGN	$\overleftrightarrow{Tr}(0.9173;\ 0.9103;\ 0.9056;\ 0.8962)$	0.0129	0.0041
PKN	$\overleftrightarrow{Tr}(0.9225;\ 0.9201;\ 0.9048;\ 0.8999)$	0.0190	0.0018
PKO	$\overleftrightarrow{Tr}(0.8991;\ 0.9014;\ 0.9071;\ 0.9165)$	0.0116	0.0029
PLY	$\overleftrightarrow{Tr}(0.8861;\ 0.8891;\ 0.9094;\ 0.9203)$	0.0272	0.0035
PZU	$\overleftrightarrow{Tr}(0.9044;\ 09046;\ 0.9081;\ 0.9130)$	0.0061	0.0013
SPL	$\overleftrightarrow{Tr}(0.8737;\ 0.8794;\ 0.8915;\ 0.8978)$	0.0180	0.0030
TPE	$\overleftrightarrow{Tr}(0.8788;\ 0.8904;\ 0.9079;\ 0.9079)$	0.0233	0.0029

Note that the O-EDF of a security described in this way is a TrOFN with an identical orientation to the O-PV used for its estimation.

6. Expected Discount Factors for Portfolio

By a financial portfolio, we mean an arbitrary, finite set of assets. Any asset is treated as a fixed security in a long position. On the other hand, any portfolio is also a security. Let us consider the case of a multi-asset portfolio π^*, built of assets Y_i . We describe this portfolio as the set $\pi^* = \{Y_i\ :\ i = 1, 2, \ldots, n\}$. Any asset Y_i is determined by a block of n_i of stocks $\hat{S}_i$ quoted at the price $\check{P}_i \in \mathbb{R}^+$. Any security is characterized by its imprecise PV evaluated by TrOFN:

$$\overleftrightarrow{Tr}\left(V_s^{(i)}, V_f^{(i)}, V_l^{(i)}, V_e^{(i)}\right). \tag{45}$$

This means that:

$$\overleftrightarrow{PV}(Y_i\ ) = n_i \boxdot \overleftrightarrow{PV}(\hat{S}_i) \tag{46}$$

where the quoted value is equal to:

$$M_i = n_i \cdot \check{P}_i \tag{47}$$

and by its EDF $\overline{v}_i$ determined by (30). Taking into account all of the above, we evaluate any asset Y_i by its imprecise EDF

$$\overleftrightarrow{\mathscr{V}}(Y_i) = \overleftrightarrow{\mathscr{V}}(\hat{S}_i) = \overleftrightarrow{Tr}\left(D_s^{(i)}, D_f^{(i)}, D_l^{(i)}, D_e^{(i)}\right). \tag{48}$$

Example 3. *We construct a portfolio π consisting of the stocks presented in Table 1. Table 3 contains information on the number of stocks of a given company in the portfolio, the value of the current block of these stocks, and their price. For example, for ALR we get, from (46):*

$$\overleftrightarrow{PV}(Y_1) = 170 \boxdot \overleftrightarrow{PV}(ALR) = 170 \boxdot \overleftrightarrow{Tr}(27.42;\ 27.30;\ 27.00;\ 26.84)$$
$$= \overleftrightarrow{Tr}(4661.40;\ 4641.00;\ 4590.00;\ 4562.80)$$

and from (47), we get: $M_1 = 170 \cdot 27.00 = 4590.00$. We do the same with the other assets.

Table 3. Recorded values of the portfolio π stocks.

Assets	Company's Stock	Number of Stocks in Block	Present Value $\overleftrightarrow{PV}_s$	Price M_i
Y_1	ALR	170	$\overleftrightarrow{Tr}(4661.40;\ 4641.00;\ 4590.00;\ 4562.80)$	4590.00
Y_2	CCC	10	$\overleftrightarrow{Tr}(833.50;\ 880.00;\ 880.00;\ 896.50)$	880.00
Y_3	CDR	17	$\overleftrightarrow{Tr}(4615.50;\ 4615.50;\ 4697.10;4697.10)$	4709.00
Y_4	CPS	50	$\overleftrightarrow{Tr}(1321.00;\ 1330.00;\ 1352.00;\ 1367.00)$	1360.00
Y_5	DNP	5	$\overleftrightarrow{Tr}(775.00;\ 775.00;775.50;\ 786.50)$	776.50
Y_6	JSW	200	$\overleftrightarrow{Tr}(3720.00;\ 3872.00;\ 4028.00;\ 4028.00)$	4064.00
Y_7	KGH	8	$\overleftrightarrow{Tr}(734.24;\ 748.80;\ 749.60;\ 759.20)$	769.92
Y_8	LTS	50	$\overleftrightarrow{Tr}(4194.00;\ 4170.00;\ 4058.00;\ 4013.00)$	4072.00
Y_9	LPP	1	$\overleftrightarrow{Tr}(8205.00;\ 8380.00;\ 8395.00;\ 8460.00)$	8385.00
Y_{10}	MBK	25	$\overleftrightarrow{Tr}(9175.00;\ 9150.00;8995.00;\ 8925.00)$	8975.00
Y_{11}	OPL	100	$\overleftrightarrow{Tr}(701.00;\ 705.00;\ 720.00;\ 735.00)$	717.00
Y_{12}	PEO	10	$\overleftrightarrow{Tr}(972.20;\ 977.00;\ 982.00;\ 986.60)$	982.00
Y_{13}	PGE	100	$\overleftrightarrow{Tr}(708.00;\ 715.00;\ 730.00;\ 740.00)$	730.00
Y_{14}	PGN	1200	$\overleftrightarrow{Tr}(4692.00;\ 4656.00;\ 4632.00;\ 4584.00)$	4644.00
Y_{15}	PKN	50	$\overleftrightarrow{Tr}(4161.00;\ 4150.00;\ 4081.00;\ 4059.00)$	4095.00
Y_{16}	PKO	30	$\overleftrightarrow{Tr}(1037.70;\ 1040.40;\ 1047.00;\ 1057.80)$	1047.90
Y_{17}	PLY	60	$\overleftrightarrow{Tr}(2149.20;\ 2156.40;\ 2205.60;\ 2232.00)$	2202.00
Y_{18}	PZU	25	$\overleftrightarrow{Tr}(1018.00;\ 1018.25;\ 1022.25;\ 1027.75)$	1022.00
Y_{19}	SPL	10	$\overleftrightarrow{Tr}(2762.00;\ 2780.00;\ 2818.00;2838.00)$	2870.00
Y_{20}	TPE	200	$\overleftrightarrow{Tr}(302.00;\ 306.00;\ 312.00;\ 312.00)$	312.00

The present value of the portfolio is the sum of the present values of its components. The components' PVs are estimated by trapezoidal oriented fuzzy numbers. The addition of TrOFN is not associative. Additionally, multiple addition depends on the order of the components. For this reason, a portfolio's present value, given as the sum of its components' PVs, is not clearly defined. Therefore, calculating the portfolio PV, we order the portfolio components. We use a method of ordering the assets proposed and justified in [22]. At the outset, we distinguish the portfolio of rising assets $\pi^+ = \left\{Y_i \in \pi^* :\ \overleftrightarrow{PV}_i \in \mathbb{K}_{Tr}^+\right\}$ and the portfolio of falling assets $\pi^- = \pi^* \backslash \pi^+$. Then, using (29), we calculate the PV of portfolio

π^+, denoted by the symbol $\overset{\leftrightarrow}{PV}^{+}$, and the PV of portfolio π^-, denoted by the symbol $\overset{\leftrightarrow}{PV}^{+}$. Thus, we get:

$$\overset{\leftrightarrow}{PV}^{+} = \overset{\leftrightarrow}{Tr}\left(V_s^{(+)}, V_f^{(+)}, V_l^{(+)}, V_e^{(+)}\right) = \overset{\leftrightarrow}{Tr}\left(\sum_{Y_i \in \pi^+} V_s^{(i)}, \sum_{Y_i \in \pi^+} V_f^{(i)}, \sum_{Y_i \in \pi^+} V_l^{(i)}, \sum_{Y_i \in \pi^+} V_e^{(i)}\right), \tag{49}$$

$$\overset{\leftrightarrow}{PV}^{-} = \overset{\leftrightarrow}{Tr}\left(V_s^{(-)}, V_f^{(-)}, V_l^{(-)}, V_e^{(-)}\right) = \overset{\leftrightarrow}{Tr}\left(\sum_{Y_i \in \pi^-} V_s^{(i)}, \sum_{Y_i \in \pi^-} V_f^{(i)}, \sum_{Y_i \in \pi^-} V_l^{(i)}, \sum_{Y_i \in \pi^-} V_e^{(i)}\right). \tag{50}$$

Finally, we calculate the PV of portfolio π^*, denoted by the symbol $\overset{\leftrightarrow}{PV}^{*}$. We then get:

$$\overset{\leftrightarrow}{PV}^{*} = \overset{\leftrightarrow}{PV}^{+} \boxplus \overset{\leftrightarrow}{PV}^{-} = \overset{\leftrightarrow}{Tr}\left(V_s^{(+)}, V_f^{(+)}, V_l^{(+)}, V_e^{(+)}\right) \boxplus \overset{\leftrightarrow}{Tr}\left(V_s^{(-)}, V_f^{(-)}, V_l^{(-)}, V_e^{(-)}\right) = \overset{\leftrightarrow}{Tr}\left(V_s^{(*)}, V_f^{(*)}, V_l^{(*)}, V_e^{(*)}\right). \tag{51}$$

Example 4. *For the portfolio determined in Example 3, we have the portfolio of rising assets:*

$$\pi^+ = \{Y_2, Y_3, Y_4, Y_5, Y_6, Y_7, Y_9, Y_{11}, Y_{12}, Y_{13}, Y_{16}, Y_{17}, Y_{18}, Y_{19}, Y_{20}\}$$

and the portfolio of falling assets:

$$\pi^- = \{Y_1, Y_8, Y_{10}, Y_{14}, Y_{15}\}.$$

Using (49), (50) and (51), we get the O-PV of portfolios π^+, π^-, and π^*, respectively:

$$\overset{\leftrightarrow}{PV}^{+} = \overset{\leftrightarrow}{Tr}(29854.34,\ 30299.35,\ 30714.05,\ 30923.45),$$

$$\overset{\leftrightarrow}{PV}^{-} = \overset{\leftrightarrow}{Tr}(26883.40,\ 26767.00,\ 26356.00,\ 26143.80),$$

$$\overset{\leftrightarrow}{PV}^{*} = \overset{\leftrightarrow}{PV}^{+} \boxplus \overset{\leftrightarrow}{PV}^{-} = \overset{\leftrightarrow}{Tr}(56737.74,\ 57066.35,\ 57070.05,\ 57070.05).$$

Now we can start calculating the EDFs of the considered portfolios. The values of portfolios π^+, π^-, and π^* are calculated in the following way, respectively:

$$M^+ = \sum_{Y_i \in \pi^+} M_i,\ M^- = \sum_{Y_i \in \pi^-} M_i,\ M^* = M^+ + M^-. \tag{52}$$

The share q_i^+ of the asset $Y_i \in \pi^+$ in the portfolio π^+ and the share q_i^- of the asset $Y_i \in \pi^-$ in the portfolio π^- are given by the formulas:

$$q_i^+ = \frac{M_i}{M^+},\ q_i^- = \frac{M_i}{M^-}. \tag{53}$$

The share q^+ of portfolio π^+ in the portfolio π^* and the share q^- of portfolio π^- in the portfolio π^* are given by the formulas:

$$q^+ = \frac{M^+}{M^*},\ q^- = \frac{M^-}{M^*}. \tag{54}$$

The EDF $\overline{v}^+$ of portfolio π^+, the EDF $\overline{v}^-$ of portfolio π^-, and the EDF $\overline{v}^*$ of portfolio π^* are calculated as follows:

$$\overline{v}^+ = \left(\sum_{Y_i \in \pi^+} \frac{q_i^+}{\overline{v}_i}\right)^{-1},\ \overline{v}^- = \left(\sum_{Y_i \in \pi^-} \frac{q_i^-}{\overline{v}_i}\right)^{-1},\ \overline{v}^* = \left(\frac{q^+}{\overline{v}^+} + \frac{q^-}{\overline{v}^-}\right)^{-1}. \tag{55}$$

Due to the results obtained in [21,34,44] and (12), we find that:

- the imprecise EDF $\overset{\leftrightarrow}{\mathscr{V}}^{+}$ of portfolio π^{+} is given by the formula:

$$\overset{\leftrightarrow}{\mathscr{V}}^{+}=\overset{\leftrightarrow}{Tr}\left(D_s^{(+)},D_f^{(+)},D_l^{(+)},D_e^{(+)}\right)=\overline{v}^{+}\boxdot\left(\boxplus_{Y_i\in\pi^{+}}\left(\frac{q_i^{+}}{\overline{v}_i}\boxdot\overset{\leftrightarrow}{\mathscr{V}}(Y_i)\right)\right)=\overline{Tr}\left(\sum_{Y_i\in\pi^{+}}\frac{\overline{v}^{+}\cdot q_i^{(+)}}{\overline{v}_i}\cdot D_s^{(i)},\sum_{Y_i\in\pi^{+}}\frac{\overline{v}^{+}\cdot q_i^{(+)}}{\overline{v}_i}\cdot D_f^{(i)},\sum_{Y_i\in\pi^{+}}\frac{\overline{v}^{+}\cdot q_i^{(+)}}{\overline{v}_i}\cdot D_l^{(i)},\sum_{Y_i\in\pi^{+}}\frac{\overline{v}^{+}\cdot q_i^{(+)}}{\overline{v}_i}\cdot D_e^{(i)}\right). \quad (56)$$

- the imprecise EDF $\overset{\leftrightarrow}{\mathscr{V}}^{-}$ of portfolio π^{-} is given by the formula:

$$\overset{\leftrightarrow}{\mathscr{V}}^{-}=\overset{\leftrightarrow}{Tr}\left(D_s^{(-)},D_f^{(-)},D_l^{(-)},D_e^{(-)}\right)=\overline{v}^{-}\boxdot\left(-\left(\frac{q_i^{-}}{\overline{v}_i}\boxdot\overset{\leftrightarrow}{\mathscr{V}}(Y_i)\right)\right)=\overline{Tr}\left(\sum_{Y_i\in\pi^{-}}\frac{\overline{v}^{-}\cdot q_i^{(-)}}{\overline{v}_i}\cdot D_s^{(i)},\sum_{Y_i\in\pi^{-}}\frac{\overline{v}^{-}\cdot q_i^{(-)}}{\overline{v}_i}\cdot D_f^{(i)},\sum_{Y_i\in\pi^{-}}\frac{\overline{v}^{-}\cdot q_i^{(-)}}{\overline{v}_i}\cdot D_l^{(i)},\sum_{Y_i\in\pi^{-}}\frac{\overline{v}^{-}\cdot q_i^{(-)}}{\overline{v}_i}\cdot D_e^{(i)}\right), \quad (57)$$

- the imprecise EDF $\overset{\leftrightarrow}{\mathscr{V}}^{*}$ of portfolio π^{*} is given by the formula:

$$\overset{\leftrightarrow}{\mathscr{V}}^{*}=\overset{\leftrightarrow}{Tr}\left(D_s^{(*)},D_f^{(*)},D_l^{(*)},D_e^{(*)}\right)=\left(\frac{\overline{v}^{*}\cdot q^{+}}{\overline{v}^{+}}\boxdot\overset{\leftrightarrow}{\mathscr{V}}^{+}\right)\boxplus\left(\frac{\overline{v}^{*}\cdot q^{-}}{\overline{v}^{-}}\boxdot\overset{\leftrightarrow}{\mathscr{V}}^{-}\right). \quad (58)$$

Then, using (56), (21) and (22), and mathematical induction, we find that the energy measure and the entropy measure of EDF $\overset{\leftrightarrow}{\mathscr{V}}^{+}$ are determined by the formulas:

$$d\left(\overset{\leftrightarrow}{\mathscr{V}}^{+}\right)=\sum_{Y_i\in\pi^{+}}\frac{\overline{v}^{+}\cdot q_i^{(+)}}{\overline{v}_i}\cdot d\left(\overset{\leftrightarrow}{\mathscr{V}}(Y_i)\right), \quad (59)$$

$$e\left(\overset{\leftrightarrow}{\mathscr{V}}^{+}\right)=\sum_{Y_i\in\pi^{+}}\frac{\overline{v}^{+}\cdot q_i^{(+)}}{\overline{v}_i}\cdot e\left(\overset{\leftrightarrow}{\mathscr{V}}(Y_i)\right). \quad (60)$$

Similarly, using (57), (21) and (22), and mathematical induction, we find that the energy measure and the entropy measure of EDF $\overset{\leftrightarrow}{\mathscr{V}}^{-}$ are determined by the formulas:

$$d\left(\overset{\leftrightarrow}{\mathscr{V}}^{-}\right)=\sum_{Y_i\in\pi^{-}}\frac{\overline{v}^{-}\cdot q_i^{(-)}}{\overline{v}_i}\cdot d\left(\overset{\leftrightarrow}{\mathscr{V}}(Y_i)\right), \quad (61)$$

$$e\left(\overset{\leftrightarrow}{\mathscr{V}}^{-}\right)=\sum_{Y_i\in\pi^{-}}\frac{\overline{v}^{-}\cdot q_i^{(-)}}{\overline{v}_i}\cdot e\left(\overset{\leftrightarrow}{\mathscr{V}}(Y_i)\right). \quad (62)$$

From (58), (23) and (24), the energy measure and entropy measure of EDF $\overset{\leftrightarrow}{\mathscr{V}}^{*}$ of portfolio π^{*} meet the following conditions:

$$d(\overleftrightarrow{V}^{*})\leq\begin{cases}\dfrac{\overline{v}^{*}\cdot q^{+}}{\overline{v}^{+}}\cdot d(\overleftrightarrow{V}^{+})-\dfrac{\overline{v}^{*}\cdot q^{-}}{\overline{v}^{-}}\cdot d\left(\mathcal{C}ore(\overleftrightarrow{V}^{-})\right),\overleftrightarrow{V}^{*}\in\mathbb{K}_{Tr}^{+}\cup\mathbb{R},\\ \dfrac{\overline{v}^{*}\cdot q^{-}}{\overline{v}^{-}}\cdot d(\overleftrightarrow{V}^{-})-\dfrac{\overline{v}^{*}\cdot q^{+}}{\overline{v}^{+}}\cdot d\left(\mathcal{C}ore(\overleftrightarrow{V}^{+})\right),\overleftrightarrow{V}^{*}\in\mathbb{K}_{Tr}^{-}\cup\mathbb{R},\end{cases} \quad (63)$$

$$e\left(\overset{\leftrightarrow}{\mathscr{V}}^{*}\right)\leq\min\left\{\frac{\overline{v}^{*}\cdot q^{+}}{\overline{v}^{+}}\cdot e\left(\overset{\leftrightarrow}{\mathscr{V}}^{+}\right),\frac{\overline{v}^{*}\cdot q^{-}}{\overline{v}^{-}}\cdot e\left(\overset{\leftrightarrow}{\mathscr{V}}^{-}\right)\right\}. \quad (64)$$

Example 5. *For the portfolio used in Example 3, we use (52) to calculate the quoted values of portfolios π^{+}, π^{-}, and π^{*} and we get $M^{+}=30{,}827.32$, $M^{-}=26{,}376.00$, and $M^{*}=57{,}203.32$. Table 4 contains information on the share q_i^{+} of the asset $Y_i\in\pi^{+}$ in the portfolio π^{+} and the share q_i^{-} of the asset $Y_i\in\pi^{-}$ in the portfolio π^{-}. We obtained these values using (53).*

Table 4. The share q_i^+ of the asset $Y_i \in \pi^+$ in the portfolio π^+ and the share q_i^- of the asset $Y_i \in \pi^-$ in the portfolio π^-.

Company	Share of the Asset in the Portfolio
ALR	0.1740
CCC	0.0286
CDR	0.1528
CPS	0.0441
DNP	0.0252
JSW	0.1318
KGH	0.0250
LTS	0.1544
LPP	0.2720
MBK	0.3403
OPL	0.0233
PEO	0.0319
PGE	0.0237
PGN	0.1761
PKN	0.1553
PKO	0.0340
PLY	0.0714
PZU	0.0332
SPL	0.0931
TPE	0.0101

From (54), we calculate the share $q^+ = 0.5389$ of portfolio π^+ in the portfolio π^* and the share $q^- = 0.4611$ of portfolio π^- in the portfolio π^*. Using (55), we obtain the EDF $\overline{v}^+ = 0.9077$ of portfolio π^+, the EDF $\overline{v}^- = 0.9078$ of portfolio π^-, and the EDF $\overline{v}^* = 0.90775$ of portfolio π^*. Utilising (56), (57) and (58), we calculate the imprecise expected discount factors of portfolios π^+, π^- and π^*, respectively:

$$\overset{\leftrightarrow}{\mathcal{V}}^{+} = \overset{\leftrightarrow}{Tr}(0.8797, 0.8927, 0.9050, 0.9112),$$

$$\overset{\leftrightarrow}{\mathcal{V}}^{-} = \overset{\leftrightarrow}{Tr}(0.9253, 0.9214, 0.9072, 0.8999),$$

$$\overset{\leftrightarrow}{\mathcal{V}}^{*} = \overset{\leftrightarrow}{Tr}(0.9008, 0.9060, 0.9060, 0.9060),$$

The energy measure of EDF $\overset{\leftrightarrow}{\mathcal{V}}^{*}$ of portfolio π^* equals $d\left(\overset{\leftrightarrow}{\mathcal{V}}^{*}\right) = 0.0027$. On the other hand, using (59) and (61) we get $d\left(\overset{\leftrightarrow}{\mathcal{V}}^{+}\right) = 0.0218$ and $d\left(\overset{\leftrightarrow}{\mathcal{V}}^{-}\right) = 0.0198$ as well as $d\left(\mathcal{C}ore(\overset{\leftrightarrow}{\mathcal{V}}^{-})\right) = 0.0142$, which is $d\left(\overset{\leftrightarrow}{\mathcal{V}}^{*}\right) \leq 0.0053$, which means that the condition (63) is satisfied. Additionally, we get $d\left(\overset{\leftrightarrow}{\mathcal{V}}^{*}\right) = 0.0027 \leq 0.0053 \leq min\left\{d\left(\overset{\leftrightarrow}{\mathcal{V}}^{+}\right), d\left(\overset{\leftrightarrow}{\mathcal{V}}^{-}\right)\right\}$. Similarly, the entropy measure of EDF $\overset{\leftrightarrow}{\mathcal{V}}^{*}$ of portfolio π^* is $e\left(\overset{\leftrightarrow}{\mathcal{V}}^{*}\right) = 0.0013$. On the other hand, using (60) and (62) we get $e\left(\overset{\leftrightarrow}{\mathcal{V}}^{+}\right) = 0.0048$ and $e\left(\overset{\leftrightarrow}{\mathcal{V}}^{*}\right) = 0.0028$, meaning that $e\left(\overset{\leftrightarrow}{\mathcal{V}}^{*}\right) = 0.0013 \leq min\left\{e\left(\overset{\leftrightarrow}{\mathcal{V}}^{+}\right), e\left(\overset{\leftrightarrow}{\mathcal{V}}^{-}\right)\right\}$, which means that the condition (64) is also satisfied. It is worth stressing that in [25] it was indicated that linear portfolio analysis is not possible for the considered portfolio π^*.

The above results allow us to conclude that portfolio diversification reduces uncertainty risk and imprecision risk. Additionally, it is worth noting that different orientations of index profits significantly affect ambiguity and indistinctness of portfolio profit index.

Note that if index profits are described by TrOFNs with different orientations then portfolio diversification significantly reduces the ambiguity of the portfolio profit index and if the index profits are described by TrOFNs with the same orientations, then portfolio diversification only averages the ambiguity of the portfolio profit index. Similarly, if index profits are described by TrOFNs with different orientations, then portfolio diversification significantly reduces the indistinctness of the portfolio profit index and if index profits are described by TrOFNs with the same orientations, then portfolio diversification only averages the indistinctness of the portfolio profit index [23].

In order to make it easier for the reader to use the presented method, we present it in a shortened form using a simple algorithm:

We consider a portfolio of company stocks.

STEP 1. Based on a session closing on, for example, the Warsaw Stock Exchange, on a fixed day, for each observed stock we assess its O-PV equal to TrOFN, describing its Japanese candle and we calculate the energy and entropy measure of the oriented present value.

STEP 2. For each considered stock $\hat{S}$, we note its quoted price $\check{P}_s$ as the initial price on the next day.

STEP 3. For each considered stocks $\hat{S}$, we calculate a quarterly expected discount factor O-EDF.

STEP 4. We construct a portfolio π consisting of the stocks from STEP 1. We determine the number of stocks of a given company in the portfolio and calculate the value of the current block of these stocks and their price M_i.

STEP 5. For the portfolio from STEP 4, we describe the portfolio of the rising assets π^+ and the portfolio of the falling assets π^-. We calculate the O-PV of portfolio π^+, π^-, and π^*, i.e., $\overleftrightarrow{PV}^{+}$, $\overleftrightarrow{PV}^{-}$, and $\overleftrightarrow{PV}^{*} = \overleftrightarrow{PV}^{+} \boxplus \overleftrightarrow{PV}^{-}$.

STEP 6. We calculate the values of the portfolios π^+, π^-, and π^*, i.e., $M^+ = \sum_{Y_i \in \pi^+} M_i$, $M^- = \sum_{Y_i \in \pi^-} M_i$ and $M^* = M^+ + M^-$, the share $q_i^+ = \frac{M_i}{M^+}$ of portfolio π^+ in portfolio π^* and the share $q_i^- = \frac{M_i}{M^-}$ of portfolio π^- in portfolio π^*, the share $q^+ = \frac{M^+}{M^*}$ of portfolio π^+ in portfolio π^* and the share $q^- = \frac{M^-}{M^*}$ of portfolio π^- in portfolio π^*.

STEP 7. We calculate the EDF $\overline{v}^+ = \left(\sum_{Y_i \in \pi^+} \frac{q_i^+}{\overline{v}_i}\right)^{-1}$ of portfolio π^+, the EDF $\overline{v}^- = \left(\sum_{Y_i \in \pi^-} \frac{q_i^-}{\overline{v}_i}\right)^{-1}$ of portfolio π^-, and the EDF $\overline{v}^* =$ of portfolio π^*.

STEP 8. We calculate the imprecise EDF $\overleftrightarrow{\mathscr{V}}^{+} = \overline{v}^+ \boxdot \left(\boxed{+}_{Y_i \in \pi^+} \left(\frac{q_i^+}{\overline{v}_i} \boxdot \overleftrightarrow{\mathscr{V}}(Y_i)\right)\right)$ of portfolio π^+, the imprecise EDF $\overleftrightarrow{\mathscr{V}}^{-} = \overline{v}^- \boxdot \left(-\left(\frac{q_i^-}{\overline{v}_i} \boxdot \overleftrightarrow{\mathscr{V}}(Y_i)\right)\right)$ of portfolio π^-, and the imprecise EDF $\overleftrightarrow{\mathscr{V}}^{*} = \left(\frac{\overline{v}^* \cdot q^+}{\overline{v}^+} \boxdot \overleftrightarrow{\mathscr{V}}^{+}\right) \boxplus \left(\frac{\overline{v}^* \cdot q^-}{\overline{v}^-} \boxdot \overleftrightarrow{\mathscr{V}}^{*}\right)$ of portfolio π^* and we calculate the energy and entropy measures of these imprecise EDFs.

7. Conclusions

Statements describing particular definitions and facts influence the state of knowledge. Linguistic variables (variables with values that are actually words or sentences of natural language) are used to formulate statements. After Knight, imprecise information is identified with imperfect information. In the imprecision of information, we often distinguish ambiguity and indistinctness [10,66]. The ambiguity of information can be defined as a lack of unequivocal distinction of recommended options among many considered alternatives. Indistinctness of information is defined as a lack of unequivocal distinction between an information and its contradiction. Such a formal model of imprecision is a fuzzy set membership function. In our paper, imprecise information about a given financial instrument is presented with the use of O-EDF $\overset{\leftrightarrow}{\mathscr{V}} \in \mathbb{K}$. In this situation, the membership function of O-EDF $\overset{\leftrightarrow}{\mathscr{V}} \in \mathbb{K}$ models the assessment imprecision of a financial instrument.

A rate of return is a function of a future value, which is uncertain. This is due to an investor's lack of certain knowledge about the future. It means that no investor is sure of their future profits or losses. An increase of uncertainty may increase the risk of making an incorrect financial decision. In this paper, we evaluate the uncertainty risk using the oriented expected discount factor. An increase in ambiguity means an increase in the amount of recommended alternative information about the state of affairs. The increase in the ambiguity of O-EDF $\overset{\leftrightarrow}{\mathscr{V}} \in \mathbb{K}$ suggests a higher number of alternative recommendations to choose from. This increases the risk of making an incorrect choice from among the recommended alternatives.

This may imply making a decision that will result in less than the maximum profit, which can be understood as a loss of certainty. Ambiguity and uncertainty have the same depiction of the risks they cause. We will assess both risks using different methods because the effects of a decision made both under the risk of ambiguity and under the risk of uncertainty are the same, but their reasons are different. The ambiguity risk loading the oriented fuzzy expected factor $\overset{\leftrightarrow}{\mathscr{V}}$ is appraised by the energy measure $d\left(\overset{\leftrightarrow}{\mathscr{V}}\right)$. An increase in the indistinctness of $\overset{\leftrightarrow}{\mathscr{V}}$ means that it is more difficult to distinguish between recommended and unrecommended decisions. It implies an increase in the indistinctness risk, that is, the possibility of making an unrecommended decision. The indistinctness risk of O-EDF $\overset{\leftrightarrow}{\mathscr{V}}$ is appraised by the entropy measure $e\left(\overset{\leftrightarrow}{\mathscr{V}}\right)$. Imprecision risk includes ambiguity and indistinctness risk. Hence, an increase in uncertainty risk or in imprecision risk worsens the conditions for making a decision. Risk assessment related to uncertainty as to the state of affairs plays a very important role in the decision-making process. Piasecki [67] shown that there exists a possibility of limiting the uncertainty risk by increasing the imprecision risk. Taking indistinctness into account allows us to reject those decision options that have a low cognitive value of the collected information, even if they are attractive from the point of view of the relationship between the expected profits and the assessment of the uncertainty risk.

In our paper, we examined the imprecise security evaluation in the context of uncertainty. The main objective of this article was to analyze the possibility of managing the risk of a portfolio of multiple assets. The portfolio was built using imprecise information describing the present value of the component assets. These imprecise present values were represented by TrOFNs. Relationships between the imprecision risk burdening portfolio components and the same risk burdening multi–assets portfolios have been described. We showed that the proposed portfolio analysis can be fully used for portfolios π^{+} of rising securities and π^{-} of falling ones. This allows the investor to manage the portfolio risk since only rising securities can receive a BUY or ACCUMULATE recommendation, while only falling securities can receive a SELL or REDUCE recommendation. We showed that the portfolio diversification can lower uncertainty risk and imprecision risk. All conclusions are compatible with financial practice and theory. Results obtained with the use of the imprecise expected discount factor were applied as input data for the robo-advice systems described in [32]. In further studies, we will compare the methods using trapezoidal fuzzy numbers and trapezoidal oriented fuzzy numbers in portfolio analysis, and we will show that the use of trapezoidal oriented fuzzy numbers is more useful. The aim of further research will also be to investigate the relationship between oriented fuzzy numbers, trapezoidal fuzzy numbers, and intuitionistic fuzzy numbers.

Funding: This research received no external funding.

Institutional Review Board Statement: Not applicable.

Informed Consent Statement: Not applicable.

Conflicts of Interest: The author declares no conflict of interest.

References

1. Von Mises, L. *The Ultimate Foundation of Economic Science an Essay on Method*; D. Van Nostrand Company, Inc.: Princeton, NJ, USA, 1962.
2. Kaplan, S.; Barish, N.N. Decision-Making Allowing Uncertainty of Future Investment Opportunities. *Manag. Sci.* **1967**, *13*, B569–B577. [CrossRef]
3. Caplan, B. Probability, common sense, and realism: A reply to Hulsmann and Block. *Q. J. Austrian Econ.* **2001**, *4*, 69–86.
4. Czerwiński, Z. Enumerative induction and the theory of games. *Studia Log.* **1960**, *10*, 24–36. [CrossRef]
5. Kolmogorov, A.N. *Grundbegriffe der Wahrscheinlichkeitsrechnung*; Julius Springer: Berlin/Heidelberg, Germany, 1933. [CrossRef]
6. Kolmogorov, A.N. *Foundations of the Theory of Probability*; Chelsea Publishing Company: New York, NY, USA, 1956.
7. Von Lambalgen, M. Randomness and foundations of probability: Von Mises' axiomatization of random sequences. *Inst. Math. Stat. Lect. Notes-Monogr. Ser.* **1996**, *30*, 347–367.
8. Von Mises, R. *Probability, Statistics and Truth*; The Macmillan Company: New York, NY, USA, 1957.
9. Sadowski, W. Forecasting and decision making. In *Quantitative Wirtschafts-und Unternehmensforschung*; Henn, R., Schips, B., Stähly, P., Eds.; Springer: Berlin/Heidelberg, Germany, 1980. [CrossRef]
10. Klir, G.J. Developments in uncertainty-based information. *Adv. Comput.* **1993**, *36*, 255–332. [CrossRef]
11. Zadeh, L. Fuzzy sets. *Inf. Control.* **1965**, *8*, 338–353. [CrossRef]
12. Piasecki, K. Basis of financial arithmetic from the viewpoint of the utility theory. *Oper. Res. Decis.* **2012**, *22*, 37–53. [CrossRef]
13. Piasecki, K. Revision of the Kosiński's Theory of Ordered Fuzzy Numbers. *Axioms* **2018**, *7*, 16. [CrossRef]
14. Kacprzak, D.; Kosiński, W. Optimizing Firm Inventory Costs as a Fuzzy Problem. *Stud. Logic Gramm. Rhetor* **2014**, *37*, 89–105. [CrossRef]
15. Prokopowicz, P.; Czerniak, J.; Mikołajewski, D.; Apiecionek, Ł.; Slezak, D. *Theory and Applications of Ordered Fuzzy Number. Tribute to Professor Witold Kosiński*; Studies in Fuzziness and Soft Computing, 356; Springer: Berlin/Heidelberg, Germany, 2017. [CrossRef]
16. Kosiński, W.K.; Kosiński, W.; Kościeński, K. Ordered fuzzy numbers approach to an investment project evaluation. *Manag. Prod. Eng. Rev.* **2013**, *4*, 50–62. [CrossRef]
17. Piasecki, K.; Łyczkowska-Hanćkowiak, A. On Present Value Evaluation under the Impact of Behavioural Factors Using Oriented Fuzzy Numbers. *Symmetry* **2021**, *13*, 468. [CrossRef]
18. Piasecki, K. Expected return rate determined as oriented fuzzy number. In Proceedings of the 35th International Conference Mathematical Methods in Economics Conference, Hradec Králové, Czech Republic, 13–15 September 2017; pp. 561–565.
19. Łyczkowska-Hanćkowiak, A.; Piasecki, K. Two-assets portfolio with trapezoidal oriented fuzzy present values. In Proceedings of the 36th International Conference Mathematical Methods in Economics Conference, Jindřichův Hradec, Czech Republic, 12–14 September 2018; pp. 306–311.
20. Łyczkowska-Hanćkowiak, A. Sharpe's Ratio for Oriented Fuzzy Discount Factor. *Mathematics* **2019**, *7*, 272. [CrossRef]
21. Piasecki, K.; Łyczkowska-Hanćkowiak, A. Representation of Japanese Candlesticks by Oriented Fuzzy Numbers. *Econometrics* **2020**, *8*, 1. [CrossRef]
22. Łyczkowska-Hanćkowiak, A.; Piasecki, K. The Present Value of a Portfolio of Assets with Present Values Determined by Trapezoidal Ordered Fuzzy Number. *Oper. Res. Decis.* **2018**, *28*, 41–56. [CrossRef]
23. Piasecki, K.; Łyczkowska-Hanćkowiak, A. Imprecision Indexes of Oriented Fuzzy Numbers. In Proceedings of the International Workshop on Intuitionistic Fuzzy Sets and General Nets—IWIFSGN 2019, Warsaw, Poland, 24–25 October 2019; Atanassov, K.T., Atanassova, V., Kacprzyk, J., Kałuszko, A., Krawczak, M., Owsiński, J.W., Sotirov, S.S., Sotirova, E., Szmidt, E., Zadrożny, S., Eds.; Artificial Intelligence and Soft Computing 1308; Springer: Cham, Switzerland, 2021; pp. 112–124. [CrossRef]
24. Piasecki, K.; Łyczkowska-Hanćkowiak, A. Heikin-Ashi technique with use of oriented fuzzy numbers. In Proceedings of the International Workshop on Intuitionistic Fuzzy Sets and General Nets IWIFSGN—IWIFSGN 2020, Warsaw, Poland, 10–11 December 2020; Atanassov, K.T., Atanassova, V., Kacprzyk, J., Kałuszko, A., Krawczak, M., Owsiński, J.W., Sotirov, S.S., Sotirova, E., Szmidt, E., Zadrożny, S., Eds.; Artificial Intelligence and Soft Computing. (accepted for publication).
25. Łyczkowska-Hanćkowiak, A.; Piasecki, K. Portfolio discount factor evaluated by oriented fuzzy numbers. In Proceedings of the 39th International Conference Mathematical Methods in Economics MME 2021, Prague, Czech Republic, 8–10 September 2021; Fejfar, J., Hruška, M., Eds.; Czech University of Life Sciences Prague: Praha-Suchdol, Czech Republic, 2021; pp. 299–304.
26. Roszkowska, E.; Kacprzak, D. The fuzzy SAW and fuzzy TOPSIS procedures based on ordered fuzzy numbers. *Inf. Sci.* **2016**, *369*, 564–584. [CrossRef]
27. Rudnik, K.; Kacprzak, D. Fuzzy TOPSIS method with ordered fuzzy numbers for flow control in a manufacturing system. *Appl. Soft Comput.* **2017**, *52*, 1020–1041. [CrossRef]
28. Piasecki, K.; Roszkowska, E. On application of ordered fuzzy numbers in ranking linguistically evaluated negotiation offers. *Adv. Fuzzy Syst.* **2018**, *2018*. [CrossRef]
29. Kacprzak, D. A doubly extended TOPSIS method for group decision making based on ordered fuzzy numbers. *Expert Syst. Appl.* **2018**, *116*, 243–254. [CrossRef]

30. Piasecki, K.; Roszkowska, E.; Łyczkowska-Hanćkowiak, A. Simple Additive Weighting Method Equipped with Fuzzy Ranking of Evaluated Alternatives. *Symmetry* **2019**, *11*, 482. [CrossRef]
31. Piasecki, K.; Roszkowska, E.; Łyczkowska-Hanćkowiak, A. Impact of the Orientation of the Ordered Fuzzy Assessment on the Simple Additive Weighted Method. *Symmetry* **2019**, *11*, 1104. [CrossRef]
32. Łyczkowska-Hanćkowiak, A. On Application Oriented Fuzzy Numbers for Imprecise Investment Recommendations. *Symmetry* **2020**, *12*, 1672. [CrossRef]
33. Piasecki, K.; Roszkowska, E.; Wachowicz, T.; Filipowicz-Chomko, M.; Łyczkowska-Hanćkowiak, A. Fuzzy Representation of Principal's Preferences in Inspire Negotiation Support System. *Entropy* **2021**, *23*, 981. [CrossRef]
34. Piasecki, K.; Łyczkowska-Hanćkowiak, A. Oriented Fuzzy Numbers vs. Fuzzy Numbers. *Mathematics* **2021**, *9*, 523. [CrossRef]
35. Markowitz, H.S.M. Portfolio Selection. *J. Financ.* **1952**, *7*, 77–91. [CrossRef]
36. Li, X.; Quin, Z.; Kar, S. Mean-Variance-Skewness model for portfolio selections with fuzzy returns. *Eur. J. Oper. Res.* **2010**, *202*, 239–247. [CrossRef]
37. Quin, Z.; Wen, M.; Gu, C. Mean-absolute deviation portfolio selection model with fuzzy returns. *Iran. J. Fuzzy Syst.* **2011**, *8*, 61–75. [CrossRef]
38. Tsaur, R. Fuzzy portfolio model with different investor risk attitudes. *Eur. J. Oper. Res.* **2013**, *227*, 385–390. [CrossRef]
39. Tanaka, H.; Guo, P.; Turksen, B. Portfolio selection based on fuzzy probabilities and possibility distributions. *Fuzzy Sets Syst.* **2000**, *111*, 387–397. [CrossRef]
40. Duan, L.; Stahlecker, P. A portfolio selection model using fuzzy returns. *Fuzzy Optim. Decis. Mak.* **2011**, *10*, 167–191. [CrossRef]
41. Guo, H.; Sun, B.; Karimi, H.R.; Ge, Y.; Jin, W. Fuzzy investment portfolio selection models based on interval analysis approach. *Math. Probl. Eng.* **2012**, *2012*. [CrossRef]
42. Gupta, P.; Mittal, G.; Saxena, A. Asset portfolio optimization using fuzzy mathematical programming. *Inf. Sci.* **2008**, *178*, 1734–1755. [CrossRef]
43. Gupta, P.; Mittal, G.; Mehlawat, M.K. Multiobjective expected value model for portfolio selection in fuzzy environment. *Optim. Lett.* **2013**, *7*, 1765–1791. [CrossRef]
44. Piasecki, K.; Siwek, J. Two-Asset Portfolio with Triangular Fuzzy Present Values—An Alternative Approach. In *Contemporary Trends in Accounting, Finance and Financial Institutions. Springer Proceedings in Business and Economics*; Choudhry, T., Mizerka, J., Eds.; Springer: Cham, Switzerland, 2018. [CrossRef]
45. Piasecki, K.; Siwek, J. Multi-asset portfolio with trapezoidal fuzzy present values. *Stat. Rev.* **2018**, *LXV*, 183–199. [CrossRef]
46. Buckley, I.J. The fuzzy mathematics of finance. *Fuzzy Sets Syst.* **1987**, *21*, 257–273. [CrossRef]
47. Ward, T.L. Discounted fuzzy cash flow analysis. In Proceedings of the 1985 Fall Industrial Engineering Conference Proceedings; 1985; pp. 476–481.
48. Greenhut, J.G.; Norman, G.; Temponi, C.T. Towards a fuzzy theory of oligopolistic competition. In Proceedings of the IEEE Proceedings of ISUMA-NAFIPS, College Park, MD, USA, 17–20 September 1995; pp. 286–291. [CrossRef]
49. Sheen, J.N. Fuzzy economical analysis of cogeneration project. In Proceedings of the 2004 IEEE Region 10 Conference TENCON 2004, Chiang Mai, Thailand, 21–24 November 2004; Volume 3, pp. 311–314. [CrossRef]
50. Huang, X. Risk curve and fuzzy portfolio selection. *Int. J. Prod. Econ.* **2007**, *106*, 1102–1112. [CrossRef]
51. Tsao, C.T. Assessing the probabilistic fuzzy Net Present Value for a capital, Investment choice using fuzzy arithmetic. *J. Chin. Inst. Ind. Eng.* **2005**, *22*, 106–118. [CrossRef]
52. Calzi, M.L. Towards a general setting for the fuzzy mathematics of finance. *Fuzzy Sets Syst.* **1990**, *35*, 265–280. [CrossRef]
53. Piasecki, K. Behavioural present value. *SSRN Electron. J.* **2011**, *1*. [CrossRef]
54. Piasecki, K.; Łyczkowska-Hanćkowiak, A. On Approximation of Any Ordered Fuzzy Number by A Trapezoidal Ordered Fuzzy Number. *Symmetry* **2018**, *10*, 526. [CrossRef]
55. Dubois, D.; Prade, H. Operations on fuzzy numbers. *Int. J. Syst. Sci.* **1978**, *9*, 613–629. [CrossRef]
56. Delgado, M.; Vila, M.A.; Voxman, W. On a canonical representation of fuzzy numbers. *Fuzzy Sets Syst.* **1998**, *93*, 125–135. [CrossRef]
57. Goetschel, R.; Voxman, W. Elementary fuzzy calculus. *Fuzzy Sets Syst.* **1986**, *18*, 31–43. [CrossRef]
58. Kosiński, W.; Prokopowicz, P.; Ślęzak, D. Fuzzy numbers with algebraic operations: Algorithmic approach. In *Proc. IIS'2002*; Klopotek, M., Wierzchoń, S.T., Michalewicz, M., Eds.; Physica: Heidelberg, Germany; Sopot, Poland, 2002; pp. 311–320.
59. De Luca, A.; Termini, S. Entropy and energy measures of fuzzy sets. In *Advances in Fuzzy Set Theory and Applications*; Gupta, M.M., Ragade, R.K., Yager, R.R., Eds.; North-Holland: Amsterdam, The Netherlands, 1979; pp. 321–338.
60. De Luca, A.; Termini, S. A definition of a nonprobabilistic entropy in the settings of fuzzy set theory. *Inform. Control* **1972**, *20*, 301–312. [CrossRef]
61. Piasecki, K. Some remarks on axiomatic definition of entropy measure. *J. Intell. Fuzzy Syst.* **2017**, *33*, 1945–1952. [CrossRef]
62. Kosko, B. Fuzzy entropy and conditioning. *Inf. Sci.* **1986**, *40*, 165–174. [CrossRef]
63. Pedrycz, W.; Gottwald, S.; Czogała, E. Measures of fuzziness and operations with fuzzy sets. *Stochastica* **1982**, *6*, 187–205.
64. Knight, F.H. *Risk, Uncertainty, and Profit, Hart, Schaffner & Marx*; Houghton Mifflin Company: Boston, MA, USA, 1921.

65. Piasecki, K.; Stasiak, M.D. The Forex Trading System for Speculation with Constant Magnitude of Unit Return. *Mathematics* **2019**, *7*, 623. [CrossRef]
66. Stirling, W.C. *Satisficing Games and Decision Making*; Cambridge University Press: Cambridge, UK, 2003. [CrossRef]
67. Piasecki, K. Effectiveness of securities with fuzzy probabilistic return. *Oper. Res. Decis.* **2011**, *21*, 65–78.

Article

The Role of Metaheuristics as Solutions Generators

Hanane El Raoui [1,2,3], Marcelino Cabrera-Cuevas [4] and David A. Pelta [2,*]

1 TICLab, ESIN—Ecole Supérieure d'Informatique et du Numérique, International University of Rabat, Rabat 11100, Morocco; hanane.elraoui@uir.ac.ma
2 Department of Computer Science and Artificial Intelligence, Universidad de Granada, 18014 Granada, Spain
3 Modeling and Mathematical Structures Laboratory, Sidi Mohamed Ben Abdellah University, Fez 30000, Morocco
4 Department of Languages and Information Systems, Universidad de Granada, 18014 Granada, Spain; mcabrera@ugr.es
* Correspondence: dpelta@ugr.es

Abstract: Optimization problems are ubiquitous nowadays. Many times, their corresponding computational models necessarily leave out of consideration several characteristics and features of the real world, so trying to obtain the optimum solution can not be enough for a problem solving point of view. The aim of this paper is to illustrate the role of metaheuristics as solutions' generators in a basic problem solving framework. Metaheuristics become relevant in two modes: firstly because every run (in the case of population based techniques) allows to obtain a set of potentially good solutions, and secondly, if a reference solution is available, one can set up a new optimization problem that allows to obtain solutions with similar quality in the objectives space but maximally different structure in the design space. Once a set of solutions is obtained, an example of an a posteriori analysis to rank them according with decision maker's preferences is shown. All the problem solving framework steps, emphasizing the role of metaheuristics are illustrated with a dynamic version of the tourist trip design problem (for the first mode), and with a perishable food distribution problem (for the second one). These examples clearly show the benefits of the problem solving framework proposed. The potential role of the symmetry concept is also explored.

Keywords: metaheuristics; optimization; tourist trip design; vehicle routing

Citation: Raoui, H.E.; Cabrera-Cuevas, M.; Pelta, D.A. The Role of Metaheuristics as Solutions Generators. *Symmetry* **2021**, *13*, 2034. https://doi.org/10.3390/sym13112034

Academic Editors: Kóczy T. László, István A. Harmati and José Carlos R. Alcantud

Received: 19 August 2021
Accepted: 21 October 2021
Published: 28 October 2021

Publisher's Note: MDPI stays neutral with regard to jurisdictional claims in published maps and institutional affiliations.

1. Introduction

In most 'real world' decision-making and/or optimization problems, countless decision criteria and objectives can not be evidently included in their computational model formulation; or, if included, they may increase the complexity of such models up to a point where they can not be exactly solved. Final decisions (i.e., selecting a solution to deploy) are often taken based not only on modeled objectives, but also on potentially subjective decision makers goals, biases, and preferences [1].

Many mathematical optimization algorithms are devoted to finding single optimal solutions to single-objective problems or, in the best case, to determine sets of non-inferior solutions to multi-objective formulations [2–4]. In any case, it should never be forgotten that we are, hopefully, obtaining the optimum solution for the model.

Assuming the existence of unmodeled (or hard to model) goals and parameters implies that non-conventional solving approaches are needed, not only to search the decision space for optimal solutions, but also to explore the decision region for alternative solutions with good quality. Such solutions can be sub-optimal for the model point of view, but valuable from additional perspectives.

It is here where the role of metaheuristics as solutions' generators becomes relevant in two senses. Firstly, because several runs (a single one, in the case of population based techniques) allow to obtain a set of potentially good solutions. Secondly, let us suppose a reference solution x^* for a problem P is available. Then, it is possible to define a new optimization problem P' where the objective is to find a solution x that maximize the difference with x^* subject to $f(x^*) - f(x) \leq \gamma$, being f the objective function from problem P and γ the maximum variation in quality allowed. In other words, the aim in P' is to obtain solutions with similar quality but maximally different structure. This last approach is the so-called 'Modeling to Generate Alternatives' (MGA) approach proposed by [4–6].

Setting apart this MGA approach, the consideration of metaheuristics in the sense posed here is not being explored (to the best of our knowledge). Most of the metaheuristics publications emphasize their role as "solvers" trying to find the best possible solution for the problem at hand usually without making further analysis.

In this context, the aim of this contribution, is to illustrate the role of metaheuristics as solutions' generators in a basic problem solving framework that considers the problem formulation, its resolution, and the analysis of solutions beyond their objective function value.

Two examples are considered. The first problem is a dynamic version of the tourist trip design problem. In the static version, we have a set of points of interest (POI), each one having a visit time and a level of interest. The aim is to select a subset of points that maximizes the level of interest subject to certain available time. The problem is usually modeled as a team orienteering problem [7].

In the dynamic version, the level of interest of a POI depends on the time that it is being visited, thus adding an extra level of complexity. We will describe the many characteristics associated with a solution and how difficult it is to include them *a priori* in a model, thus requiring a different solving strategy.

The second one, is a perishable food distribution problem (basically, a vehicle routing problem) [8]. Originally modeled as a many objectives' problem, we describe a solving strategy that first solves a sub-problem and then, using the obtained solution, sets a new optimization problem with a fuzzy constraint that allows to obtain new solutions with similar quality but maximally different structure (as in the MGA approach).

The paper is structured as follows. In Section 2 the proposed basic problem solving framework is presented, together with an existing approach to rank solutions according to the user's preference. Next, Sections 3 and 4 present the two examples commented before. For each one, every step of the solving framework is illustrated with a specific dataset, and the corresponding analyses of results are presented.

In each example, a metaheuristic is used to generate a set of solutions. We should highlight here that the specific details of the implemented methods are omitted, as they are not relevant in this context: we do not want to make comparisons against other methods over a set of well-defined objectives.

Finally, Section 5 is devoted to conclusions and further work.

2. Basic Problem Solving Framework

The role of metaheuristics as solutions' generators is best seen when integrated in the following problem solving framework. Although every step is clear enough, we describe the mains aspects considered here.

Let us depart from an optimization problem P. The following steps are required to solve it.

1. **Problem description**: determine a set of solutions' features for problem P, $\mathcal{F} = \{f_1, f_2, \ldots, f_n\}$. Such features are used to assess the quality of a solution;
2. **Model formulation**: Define the subset of features to be used in P optimization model $\mathcal{F}' = \{f_i, f_j, \ldots, f_t\}$. For every feature, either a maximization or minimization goal should be established;
3. **Problem Solving**: "Solve" the problem P. Depending on the solver algorithm:
    (a) An optimum solution s^* (or at least a reference one) is obtained. From such solution, generate new ones with similar quality but as different as possible;
    (b) A set of solutions is obtained;

4. **Analysis of Solutions**: For every solution, calculate the corresponding values for the features in $\mathcal{F} - \mathcal{F}'$;
5. **Ranking**: Rank the solutions according to the user's preferences.

Let us illustrate these steps with the well known travelling salesmen problem (TSP). Given a solution (a route), features like the distance travelled, time of the route, fuel consumption, average length of the inter-city paths, the length of the longest path, and so on can be measured.

In a basic TSP formulation, a single goal is defined: distance minimization.

It is in Step 3 where metaheuristics comes into play. Suppose we solve the problem using a genetic algorithm, so it is easy to obtain a set of routes (Step 3.b). For example, the best m solutions from the final population are kept.

Then, in Step 4, those m solutions can be organized in a matrix M as:

	$\mathcal{F} - \mathcal{F}'$		$\mathcal{F}'$	
	f_1	f_2	$\dots$	f_n
s_1	M_{11}	M_{12}	$\dots$	M_{1n}
s_2	M_{21}	M_{22}	$\dots$	M_{2n}
s_m	M_{m1}	M_{m2}	$\dots$	M_{mn}

where M_{ij} is the value achieved by solution s_i on feature f_j. Those values M_{ij} with $j \in \mathcal{F}'$ are given as the output of the optimization process, while those for $j \in \mathcal{F} - \mathcal{F}'$ are calculated after the optimization.

Finally, in Step 5 the values in every M row can be combined (for example using some aggregation function [9]) to obtain a score q_i for every solution s_i. The relevance of the objectives for the user can be expressed through a set of weights. Sorting according to q_i allows to rank the solutions.

Many options are available for this step. Here, we will use the approach in [10] that allows to select a solution of interest from a set of solutions. The user states the preferences, just providing a linear ordering of the objectives. A brief summary of the approach follows.

2.1. Ranking Solutions According to the User's Preferences

Let us depart from the solutions' matrix M. Additionally, suppose the user provides a set of weights $W = \{w_1, w_2, \dots, w_n\}$ where if the feature f_i is more relevant than f_j for the decision maker, then $w_i \geq w_j$. It should hold that $\sum_{j=1}^{m} w_j = 1$. Under these assumptions, a score q_i for the solution s_i can be calculated as $\sum_{j=1}^{m} w_j \times x_{ij}$.

However, this approach has a problem. Let us suppose we have just three criteria and the given preference order is f_2, f_1, f_3. Then, we need the weights in such a way that $w_2 \geq w_1 \geq w_3$ with $w_1 + w_2 + w_3 = 1$. As the reader may notice, there are infinite values for w_i that verifies both conditions, and every possible set of values will give a different score for the alternative.

Instead of a single score value, the authors in [10] proposed to calculate an interval of the potential scores that a solution can attain and then sort the solutions in terms of their intervals.

The main steps are detailed below.

1. Solutions' matrix normalization: normalization is needed because different measurements should be combined;
2. Intervals calculation: the user's preferences are given as an ordinal relation among features denoted as $f_1 \succeq_p f_2 \succeq_p \dots \succeq_p f_n$. The symbol $\succeq_p$ is to be read as "at least as preferred to". This implies that the weights are ordered as $w_1 \geq w_2 \geq \dots \geq w_m$.
All the potential scores that a solution s_i can attain are included in the interval denoted as $I_i = [L_i, U_i]$ with L_i, U_i are, respectively, the minimum, and the maximum scores (obtained through solving two simple linear programming problems. See [10] fo details);

3. Select a reference solution s^*: such solution s^* and its corresponding interval $I^* = [L^*, U^*]$ is the one with the greatest lower bound $\forall\, i, L^* \geq L_i$, thus there is no solution that always scores better than s^*;
4. Compare solutions against s^*: every solution s_i is compared against s^* using a possibility function that measures the possibility degree of a solution being better than another using their corresponding intervals.
   Let us have two solutions s_X, s_Y with their corresponding non-negative intervals $X = [x_l, x_r], Y = [y_l, y_r]$ and $x_l, x_r, y_l, y_r \in R_0^+$. The possibility degree of s_X being greater than s_Y is stated in terms of $P(X \geq Y)$ as in [10] and considering a user with a neutral attitude is defined as follows [11]:
   (a) if $X \cap Y = \emptyset$ (intervals do not overlap)
   $$P(X \geq Y) = \begin{cases} 0 & x_r \leq y_l \\ 1 & x_l \geq y_r \end{cases}$$
   (b) if $X \cap Y \neq \emptyset$ (intervals overlap)
   $$P(X \geq Y) = \frac{x_r - y_l}{x_r - x_l + y_r - y_l}$$
5. Ranking of Solutions: for every solution, s_i the value $P(s_i \geq s^*)$ is calculated. Then, they are are sorted based on such possibility degree values.

3. The Time-Dependant Tourist Trip Design Problem

The tourist trip design problem (TTDP) involves two essential aspects: the selection of points of interest suitable to the tourist's preferences, and the generation of itineraries to visit these points (all or a subset of them). Although the first aspect is usually related to recommendation systems, which have to identify user preferences, the second is related to decision and optimization issues.

In general, TTDP models are based on the well-known *orienteering problem* (OP), *team orienteering problem* (TOP) and their variants for route design ([12–16].

3.1. Problem Description

The tourist trip design problem with time-dependent scores (TTDP-TDS) starts from a set N of nodes or points of interest (POIs). Each node $i \in N$ has an associated interest or score S_i and several recommendation factors f_{it} to weight the interest according to the time at which the node is visited. The travel time from node i to node j is given by t_{ij}.

On the other hand, not all nodes in the set can be visited due to there is a maximum available time (T_{max}). There is a node n_{ini} defined as the starting and ending point of the route (for example, the location of a hotel).

A solution to the problem will be a permutation of a subset of nodes (POIs), i.e., the ordered list of POIs to visit. Over such a list, it is easy to imagine the many features that could be calculated.

3.1.1. Features

- I, Overall interest;
- $\#POIs$, Number of POIs in the route;
- T_{visit}, Time spent visiting POIs;
- T_{travel}, Total walking time between POIs on the route;
- T_{route}, Overall trip duration: $T_{visit} + T_{travel}$;
- E, Route "efficiency": $100 \times (1 - T_{visit}/t_{route})$, the percentage of time in the POIs with respect to the trip time;
- Average, maximum. and minimum time spent in POIs;
- Average, maximum and minimum travel time between POIs;
- Average number of POIs visited in every time period;

- Percentage of POIs visited at the best time of the day.

Considering all of these features in a single model is, if not impossible, very difficult. Moreover, different optimization problems can be defined: Max I, Max $\#POIs$, (Max I, Min $\#POIs$), (Max I, Max E), etc.

In this contribution, we depart from the following model.

3.1.2. Parameters

- N—Node set (the POIs);
- n_{ini}—Index of the start/end node of the route;
- T—Number of time periods;
- S_i—Interest of node i;
- t_{ij}—Travel time between node i and node j;
- f_{it}—Recommendation factor to visit node i in period t;
- v_i—Time required to visit node i (service time);
- T_{max}—Maximum time available for the route.

3.1.3. Decision Variables

The following variables are considered:

- y_{it}, a binary variable. If the node i is visited in time period t, then $y_{it} = 1$, and 0 otherwise;
- π, a permutation of k elements (the length of route) where $\pi(i)$ represents the node that is visited at position i of the route.

3.1.4. Objective Function

$$Max\ I = \sum_{i \in N \setminus \{n_{ini}\}} \sum_{t \in T} S_i \cdot f_{it} \cdot y_{it} \tag{1}$$

subject to:

$$T_{visit} + T_{travel} \leq T_{max} \tag{2}$$

where

$$T_{visit} = \sum_{i \in N \setminus \{n_{ini}\}} \sum_{t \in T} v_i \cdot y_{it} \tag{3}$$

$$T_{travel} = t_{ini,\pi(1)} + \left(\sum_{i=1}^{k-1} t_{\pi(i),\pi(i+1)} \right) + t_{\pi(k),ini} \tag{4}$$

Although a visit to a point can span two different time periods, here the contribution of such point to the overall interest is determined by the period at which its visit starts.

3.2. Problem Solving and Generation of Solutions

We describe next the basic details of the metaheuristic used, the dataset and the way the set of solutions is generated.

3.2.1. Metaheuristic

A crossover-less evolutionary algorithm (EA) is used to solve this problem.

Each individual of the initial population is generated using a GRASP-like procedure. Initially, an initial POI is randomly selected. Subsequently, a list of candidate POIs is constructed, an element from the list is randomly chosen and then added to the solution. The process is repeated as long as the maximum time available is not exceeded.

In each generation, from the current population of *popSize* individuals, K parents are chosen by roulette wheel selection and incorporated into an intermediate population. Using one of the several mutation operators available, a child solution is generated from each parent. If the new solution is feasible and better than the parent, it is incorporated into the intermediate population.

Finally, the intermediate population is sorted, and the best *popSize* elements are taken to replace the original population.

The proposed algorithm always operates with feasible solutions. Therefore, non-feasible solutions obtained by mutation are discarded.

3.2.2. Dataset

A dataset with the 50 most visited POIs of Granada city, Spain is used. The distribution of these POIs is shown in Figure 1 and the dataset is available at http://18.156.111.23/TTDP-TDS/web/ (accessed on 1 August 2021).

Figure 1. Map of Granada and set of points of interest.

Four time slots ($T = 4$) and a value $T_{max} = 360$ are considered. Each time slot has a duration of 90 min.

3.2.3. Generation of Solutions

The generation of solutions is done running the EA several times. More specifically, we made 40 independent runs of the EA with different parameters. Every run deals with $popSize = 50$ solutions and 100 generations. After every run, we kept the best solution, thus obtaining a set of 40 solutions.

3.3. *Analysis of Solutions' Diversity*

In order to explore the relation between the interest and the diversity of the solutions, we will use the Jaccard's similarity coefficient to assess how similar two solutions A, B (considered as two sets of POIs) are. The coefficient is defined as:

$$J(A,B) = \frac{|A \cap B|}{|A \cup B|} \tag{5}$$

It is easy to see that if routes A and B contain the same POIs, $A = B = A \cap B = A \cup B$, then $J(A,B) = 1$. On the other hand, if routes A and B do not share any element at all, $|A \cap B| = 0$, so $J(A,B) = 0$.

From the set of 40 solutions $S = \{s_1, s_2, \ldots, s_{40}\}$, we select the best one in terms of interest (s^*) and compute $J(s^*, s_i)\ \forall\, i$.

The scatter plot in Figure 2 displays the relation between similarity and interest difference for every $s_i \in S$ and s^*.

Solutions 5, 12, 14, 5 have the same interest that s^* but a different structure, with a Jaccard coefficient ≤ 0.8. There are other three solutions (3,4, 10) with a minor difference in cost (less than 2 units) and rather different from s^*, ($J < 0.7$). If the user may allow routes with less interest (up to 6%) then all the solutions in the range of interest difference between 2 and 4 are available.

Figure 2. Similarity vs. Interest difference for every solution against s^*. The lower the values in both axis, the better the solution is.

3.4. *Users Profiles and Ranking*

Now, having those 40 solutions, we calculate the values of several additional features (beyond the route interest I) which are:

- I, Route Interest;
- $\#POIs$, Number of visited POIs;
- E, Route efficiency;
- T_{travel}, total walking time.

The corresponding values for every solution are shown in Table A1 (Appendix B), while Table 1 summarizes the values of the features over the 40 solutions. We can observe that a wide variety of values are available. There are solutions ranging from 11 to 15 POIs, with an interest which is (in the worst case) 18% lower than the best solution. Regarding the travel time, values from 25 min to almost an hour of walking between POIs can be considered and looking at the efficiency, it is clear that the solutions are very effective at reducing the travel time while maximizing the visit time.

Table 1. Summary of values for every feature considered, over the generated 40 solutions.

	Min	Max	Mean	Std. Dev.
Number of POIs	11	15	13.33	1.57
Interest	74.00	90.00	83.01	4.57
Travel Time	24.78	54.78	37.19	6.36
Efficiency	84.77	92.60	89.18	1.60

Now, let us define the following three profiles.

1. **Guided tour**: A tourist wanting to visit as many of the city's most interesting POIs as possible while maximizing the time available. To do so, we rank the solutions taking the interest of the solution as the main criterion, ordering the rest of features as $I \succeq_p \#POIs \succeq_p E \succeq_p T_{travel}$;
2. **Casual visit**: A visitor, who is in the city for another reason, wants to visit the most interesting places in the city but wants to have as much free time as possible during

this visit. In this case, we maximize the Effectiveness of the visit and order the rest of the criteria by $E \succeq_p I \succeq_p \#POIs \succeq_p T_{travel}$;

3. **Large groups or people with reduced mobility**: Due to the size of the groups or the mobility problems of the visitors, it is necessary to keep the journeys to be made as short as possible. In this last profile we minimize travel time between POIs T_{travel}, and we set the order of the rest of the criteria by $T_{travel} \succeq_p I \succeq_p \#POIs \succeq_p E$

For every profile, the procedure described in Section 2.1 is applied. Figure 3 shows the best 10 solutions of the corresponding rankings. The values for the reference solution (s^*) are indicated in the top row, and the columns are ordered in terms of the criteria importance. Values in red cells are worse than the reference value, while the green ones are better.

Regarding Profile 1, the reference solution $s^* = s_9$ is the top ranked alternative. Solution 4 has a slightly lower value of interest while attaining the same values in the rest of features. Nevertheless, this solution is quite different from s^*, having $J(s^*, s_4) = 0.58$.

Solutions 5, 12, and 14 have the same value of interest, but they are slightly worse in the rest of the factors.

	Interest	POIs	Efficiency	Travel time
Ref. Solution	90,00	15	90,57%	33,83
9	0,00	0	0,00%	0,00
4	-1,50	0	0,00%	0,00
14	0,00	-1	-0,41%	1,09
12	0,00	-1	-1,24%	4,39
1	-3,50	0	-0,69%	1,64
5	0,00	0	-2,96%	10,70
16	-4,00	0	-0,82%	2,69
6	-3,50	0	-1,46%	5,25
17	-4,00	0	-1,34%	4,77
8	-3,50	0	-1,77%	5,27

(**a**) Profile 1

	Efficiency	Interest	POIs	Travel time
Ref. Solution	90,57%	90,00	15	33,83
9	0,00%	0,00	0	0,00
4	0,00%	-1,50	0	0,00
37	2,03%	-10,00	-3	-9,05
1	-0,69%	-3,50	0	1,64
39	0,90%	-6,00	-3	-5,41
14	-0,41%	0,00	-1	1,09
11	-0,30%	-7,50	0	0,64
16	-0,82%	-4,00	0	2,69
6	-1,46%	-3,50	0	5,25
17	-1,34%	-4,00	0	4,77

(**b**) Profile 2

	Travel time	Interest	POIs	Efficiency
Ref. Solution	33,83	90,00	15	90,57%
37	-9,05	-10,00	-3	2,03%
39	-5,41	-6,00	-3	0,90%
27	-4,36	-8,00	-3	0,35%
9	0,00	0,00	0	0,00%
4	0,00	-1,50	0	0,00%
26	-5,33	-12,00	-4	0,62%
11	0,64	-7,50	0	-0,30%
14	1,09	0,00	-1	-0,41%
1	1,64	-3,50	0	-0,69%
30	-3,13	-8,00	-4	0,00%

(**c**) Profile 3

Figure 3. Best 15 solutions for the defined user profiles.

Regarding Profile 2, again the reference solution $s^* = s_9$ is the top ranked alternative followed by s_4. Then, solution s_{37} appears (which was not among the top 15 solutions for Profile 1). This solution has better efficiency and travel time than the reference solution, but is ranked in third place due to the interest (which is 10 units less than the reference). Solution s_{39} is also interesting because it is better in efficiency and travel time, while just having 6 units less of interest. In turn, the route uses 12 POIs instead of 15 which is a remarkable point.

Finally, for Profile 3, we observe that the reference solution is ranked in the fourth place. Solutions 37, 39, and 27 are top ranked, providing improvements in travel time (shorter displacements between POIs) and greater efficiency. All of them had 12 POIs which nevertheless allowed to reach up to 88% of the reference interest (90 units).

So, Profiles 1 and 2 share the top two solutions. The third solution from Profile 2 does not even appear for Profile 1. None of the top 3 solutions from Profile 3 are available in Profile 1 and the third solution from Profile 3 does not appear in Profile 2.

These results clearly highlight the need of having a set of solutions to choose from instead of concentrating the efforts in obtaining one single best solution.

4. The Perishable Food Distribution Problem

Perishable food distribution has become increasingly tricky due to the features of these goods: perishables have a short shelf life, and their quality deteriorates over time during the distribution process until it is consumed. Therefore, ensuring food quality has become a major priority that must be met if customers are to be satisfied.

In a real-world setting, the distribution of perishable food entails a set of complex specifications that might be difficult to incorporate into mathematical models. Such parameters are frequently conflicting and, in some cases, unquantifiable [2,17,18].

A well-designed distribution plan that prioritizes short transit times and distances can help to preserve products quality while reducing waste. However, improving the sustainability of perishables distribution network necessitates balancing several competing goals, such as minimizing the travel costs (e.g., fuel consumption costs, refrigeration costs, the damage cost), maximize the average freshness, meeting consumers requirements to ensure their satisfaction (e.g., on-time delivery, good service level), and reducing environmental effect.

4.1. Problem Description

The problem is represented as a direct graph denoted by $G = (V, A)$, being $V = \{0, 1, \ldots, n+1\}$ the set of vertices, nodes 0 and $n+1$ represent the depot, and the remaining nodes $C = \{1, \ldots, n\}$ are the customers to serve. Each customer i should be served within a specific time slot defined by $[e_i, l_i]$.

The problem is modeled under the assumption of fleet homogeneity: all the refrigerated trucks have the same refrigeration characteristics, the same load capacity, and the same constant velocity.

When considering solutions for the perishable food distribution problem, several features can be defined.

4.1.1. Features

- Fixed, transportation and refrigeration costs;
- Damage cost;
- Average freshness;
- Quality of Service (or Service level);
- Total Tardiness.

From this set of available objectives, we deal with the following model formulation.

4.1.2. Parameters

- $C_{(i,j)}$: the transportation cost from node i to j;
- F: fixed cost associated to a vehicle;
- $T_{(i,j)}$: the travel time from i to j;
- $d_{(i,j)}$: the distance between node i and j;
- C_e: the cost per unit time for the refrigeration during the transportation process;
- U_i: the unloading time at customer i;
- C'_e: the unit refrigeration cost during the unloading;
- S_i: the necessary time to serve customer i;
- q_i: the demand of customer i;
- Q: the loading capacity of trucks;

- K: the fleet of trucks;
- $[e_i, l_i]$: the time window of customer i.

4.1.3. Decision Variables

- $x^k_{(i,j)}$: a 0–1 decision variable, equal to 1 in case the truck k travels on arc (i, j), 0 otherwise;
- y^k_i: a 0–1 decision variable, equal to 1 in case the customer i is served by the truck k, 0 otherwise;
- t^k_i: a continuous decision variable representing the starting service time at customer i using the truck k.

4.1.4. Objective Function

The objective is to minimize the travel cost, calculated from: C_1, the fixed costs for using a vehicle denoted; C_2, the transportation costs and C_3 the refrigeration costs.

- **C_1, The fixed costs:** represent the vehicle's maintenance and depreciation costs.

$$C_1 = F \sum_{k \in K} y^k_0 \tag{6}$$

where k is the number of refrigerated trucks available and F represents the fixed cost related to a vehicle;

- **C_2, The transportation costs**: are proportional to the vehicle mileage. We consider only the fuel consumption cost and express it as:

$$C_2 = \sum_{k \in K} \sum_{i \in V} \sum_{j \in V} C_{(i,j)} x^k_{(i,j)} d_{(i,j)} \tag{7}$$

- **C_3, The refrigeration costs**: are calculated as

$$C_3 = C_{31} + C_{32} \tag{8}$$

where C_{31} are the refrigeration costs of the transportation process:

$$C_{31} = C_e \sum_{k \in K} \sum_{i \in V} \sum_{j \in V} x^k_{(i,j)} T_{(i,j)} \tag{9}$$

and C_{32} is the cost of energy supplied during the unloading:

$$C_{32} = C'_e \sum_{k \in K} \sum_{i \in V} y^k_i U_i \tag{10}$$

Considering the objectives described above, the problem can be formulated as a mixed integer program (MIP) giving by the following:

$$\text{Min } C_1 + C_2 + C_3 \tag{11}$$

subject to:

$$x^k_{(i,i)} = 0, \quad \forall i \in V, \quad \forall k \in K \tag{12}$$

$$x^k_{(0,n+1)} = 1, \quad \forall k \in K \tag{13}$$

$$\sum_{k \in K} y^k_i = 1, \quad \forall i \in C \tag{14}$$

$$\sum_{i \in V} x^k_{(i,j)} = y^k_j, \quad \forall j \in C, \quad \forall k \in K \tag{15}$$

$$\sum_{j \in V} x^k_{(i,j)} = y^k_i, \quad \forall i \in C, \quad \forall k \in K \tag{16}$$

$$\sum_{j \in v} x^k_{(0,j)} \leq 1, \quad \forall k \in K \tag{17}$$

$$\sum_{i \in v} x^k_{(i,n+1)} \leq 1, \quad \forall k \in K \tag{18}$$

$$\sum_{i \in C} D_i y^k_i \leq Q, \quad \forall k \in K \tag{19}$$

$$t_i + S_i + T_{(i,j)} x^k_{(i,j)} - M(1 - x^k_{(i,j)}) \leq t_j, \quad \forall i,j \in V, \quad \forall k \in K \tag{20}$$

$$e_i \leq t^k_i \leq l_i, \quad \forall i \in C, \quad \forall k \in K \tag{21}$$

$$\sum_{k \in K} y^k_0 \leq K \tag{22}$$

$$\sum_{j \in V} x^k_{(i,j)} \leq y^k_i, \quad \forall i \in V, \quad \forall k \in K \tag{23}$$

The objective (11) corresponds to the total cost minimization. Constraint (12) avoids going from a point to itself. Constraint (13) avoids going from the depot to node $n+1$ which represent the depot. Constraint (14) states that only one vehicle visits each customer exactly once. Constraint (15) and (16) ensure that each vehicle that arrives at a customer must depart for another destination. The vehicle departs from the depot 0, and returns to the depot node $n+1$ according to the Constraint (17) and (18). Constraint (19) guarantees the respect of vehicle loading capacity. The connection between the start service at a specific customer and the next one is established by Constraint (20). Constraint (21) is the time window constraint. Constraint (22) guarantees that the number of trucks departing from the depot do not exceed the available fleet of trucks K. Constraint (23) states that the vehicle must traverse the arc (i, j) in order to serve the customer j.

4.2. Problem Solving and Generation of Solutions

We describe next the basic details of the metaheuristic used, the dataset and the way the set of solutions is generated.

4.2.1. Metaheuristic

To solve the proposed problem, we use a basic implementation of a general variable neighborhood search GVNS metaheuristic. In GVNS, variable neighborhood descent (VND) is used for local search. When designing a GVNS, one should take the following decisions: the number of neighborhood structures, the order in which they will be explored, how the initial feasible solution is generated, the acceptance criterion and the stopping conditions. These elements define the configuration of the GVNS.

GVNS starts by an initialization phase in which a feasible solution is generated. The best solution is set as the first feasible solution. After selecting the neighborhood structures for the shaking stage N_s, and for local search N_k, the stopping condition is then chosen. The stopping condition corresponds to a number of iterations M that will be set in the computational experience. The shaking process, the local search, and the move decision are repeated until completing all neighborhood structures ($S = S_{max}$). In the shaking phase a solution X' is generated randomly at the Sth neighborhood of$X^*(X' \in N_s(X^*))$. Then, the local search is performed to find a better solution X'' from X' using the N_K neighborhood structures.

4.2.2. Dataset

The proposed approach is evaluated through a dataset of 50 customers in Granada, Spain. The customers' time window takes a random value in the interval $[0, 240]$ with an interval width ≥ 45. For customer's target time, we assume that it is the midpoint of the corresponding time window. The dataset is available at http://18.156.111.23/PER-FOOD/DataPerishableFood.csv (accessed on 1 August 2021).

The refrigerated vehicles used to deliver products have a fixed cost of 25 €, and the fuel consumption is estimated to 3 €/km. The other parameters are: $P = 20$ €/unit, $\partial_1 = 0.002$, $\partial_2 = 0.003$, $C_e = 0.03$ €/unit, $C_e' = 0.04$ €/unit and $\delta = 0.8$.

4.2.3. Generation of Solutions

Although multiple runs of GVNS may allow to obtain a set of solutions, we illustrate here a different approach inspired in the so called 'Modeling to Generate Alternatives' (MGA) strategy [5,6]. MGA tries to generate a set of solutions that are quantifiable good across all the modeled goals, while remaining as different as possible from each other (in the decision space).

Let us assume that the best solution obtained for the problem is X^*, with $Z^* = F(X^*)$ as the objective value. To generate alternative solutions X that are maximally different from X^*, the following problem is addressed:

$$Min\ Jaccard(X, X^*) \tag{24}$$

Subject to:

$$X \in D \tag{25}$$

$$|F(X) - Z^*| \leq_f T \tag{26}$$

where $Jaccard(x_1, x_2)$ is the similarity coefficient used in the previous example. In the original MGA approach, the problem is stated as maximizing a dissimilarity function. The parameter T is the tolerance threshold related to the optimal objective value Z^* and should be defined by the decision maker.

We define the Jaccard's coefficient between two routing solutions as the ratio of the number of shared arcs to the number of total arcs used in both solutions. Let $y_{ijk} = 1$ if arc (i, j) from vertex i to vertex j is used by any vehicle in solution r_k, and $y_{ijk} = 0$ otherwise.

$$Jaccard(p, q) = \frac{\sum_{i=0}^{n}\sum_{j=0}^{n} y_{ijp} \cdot y_{ijq}}{\sum_{i=0}^{n}\sum_{j=0}^{n} sign(y_{ijp} + y_{ijq})} \tag{27}$$

where $y_{ijp} \cdot y_{ijq} = 1$ iff arc (i, j) is used by both solutions, and $sign(y_{ijp} + y_{ijq}) = 1$ if any of the solutions use it. If solutions p and q are the same, the sum in the numerator will equal the sum in the denominator, and therefore $Jaccard(p, q) = 1$. On the other hand, if they are two completely different solutions with no arc in common, the numerator will equal 0, and then $Jaccard(p, q) = 0$

In second place, we need to manage the fuzzy constraint (26) (which is crisp in the original MGA approach). The membership function that represents the satisfaction degree of the fuzzy constraint is is defined as:

$$\mu(\tau) = \begin{cases} 1 & ;\tau < T \\ 1 - \frac{\tau - T}{\Delta^t} & ;T \leq \tau < T' \\ 0 & ;\tau \geq T' \end{cases} \tag{28}$$

where $T' = T + \Delta^t$ is the endurable distance threshold, $\tau = |F(X) - Z^*|$ is the cost difference between the reference solution and the obtained alternative.

Using the concept of $\alpha - cuts$ and the parametric approach [19], the fuzzy constraint is transformed into:

$$|F(X) - Z^*| \leq T + \Delta^t(1 - \alpha) \tag{29}$$

Now, the GVNS parameters, we use the following values: $\delta_1 = 0, \delta_2 = 1, \Lambda = 2, \mu = 1$. The stopping criteria M which correspond to the number of iterations, is fixed to 10. The other parameters are: $P = 20$ €/Unit, $\partial_1 = 0.002$, $\partial_2 = 0.003$, $C_e = 0.03$ €/unit, $C_e' = 0.04$ €/unit and $\delta = 0.8$.

To generate a set of alternative solutions, we start by finding a reference solution X^* by running 10 times the GVNS metaheuristic. The best solution obtained (according to Equation (12)) is considered as the reference solution X^*.

Then, using X^*, the new problem considering as objective the Equation (24) subject to the constraints (25) and (29) is solved for $\alpha \in [0, 0.2, 0.4, 0.6, 0.8, 1]$. For each value of α, a tailored version of GVNS is run 10 times.

At the end, 62 different solutions are obtained and the corresponding values for the following features are calculated: the damage cost, the average freshness, the service level, and the tardiness. The mathematical definition of these criteria can be found in Appendix A.

Table A2 (in the Appendix B) shows the results obtained for every value of α. It should be noted that several solutions per GVNS were obtained and just the different ones are reported.

4.3. Analysis of Solutions' Diversity

Table 2 summarizes the values of the features over the 62 solutions. It is clear that a wide variety of alternatives are available to choose from.

Table 2. Summary of values for every feature considered, over the generated 62 solutions.

	Min	**Max**	**Mean**	**Std.Dev**
Travel Cost	1167.85	1252.36	1215.43	24.97
Total damage	7530.24	8024.43	7709.90	109.80
Average freshness	20.12	22.39	21.11	0.51
Tardiness	69,137.18	216,830.61	134,649.45	36,362.44
Service level	15.96	20.31	18.10	1.07

As we explicitly generate the set of solutions minimizing the similarity with X^*, it is interesting to observe the relation between the Jaccard coefficient, and the differences in the calculated values for every solution. This information is shown in Figure 4 (for visualization purposes, just a subset of solutions are shown in every case). The Y axis reflects the difference as percentage.

When considering the distance (which is correlated with the travel cost), Figure 4a shows that all the solutions are worst in this feature. This is not surprising, as X^* has the lowest distance. There are several interesting solutions, like 10, 5, 6, and 9, that had a quite similar distance but a Jaccard similarity below 0.9. Solution 28 is also interesting because it has a low level of similarity (less than 0.45), while attaining a quite similar distance value (less than 2% of difference).

With an increment in the distance between 2% and 5%, up to eight solutions can be found with a Jaccard value between $[0.4, 0.62]$.

For the average freshness and service level, the situation is more interesting, because there are better and worse solutions than the reference one. Focusing in the average freshness (Figure 4b), those solutions with a positive value in Y are better than the reference solution. Solution 15 allows to improve up to 4% in the criteria while having a similarity value of 0.7. Other improvements between 1% and 2% can be obtained with quite different routes (solutions 47, 25, 52).

The results in terms of the service level are shown in Figure 4c. Here, solution 59 provides more than a 10% of improvement with a moderate value of similarity (less than 0.65). Several solutions providing improvements higher than 6% and similarity below 0.5 are available. In this case, even solution 37 may be interesting, as it has a minor decrement in service level but very low level of similarity.

Finally, Figure 4d shows the solution in terms of the Tardiness criteria. Here, all the solutions are better than the reference one. It is clear here that tardiness more similar than the reference solution can not be attained. Great improvements can be obtained, but with quite different solutions (look at the low values for similarity).

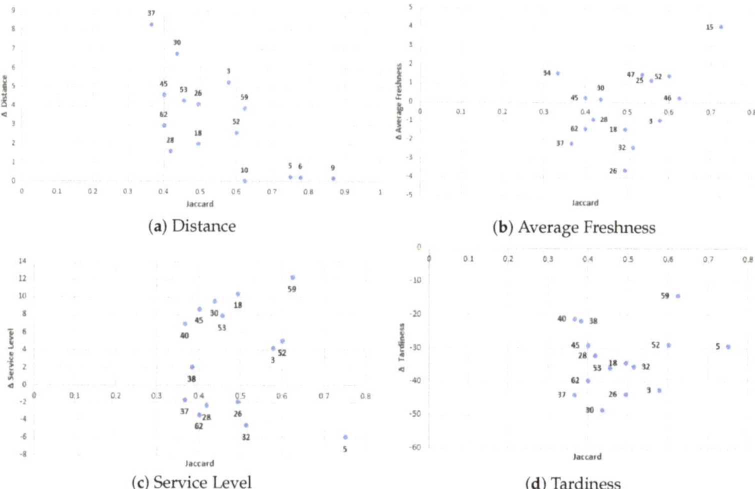

(**a**) Distance (**b**) Average Freshness

(c) Service Level (**d**) Tardiness

Figure 4. Variation of individual criterion and Jaccard's coefficient with respect to the reference solution. Just a subset of solutions is shown.

4.4. User Profiles and Ranking

The previous analysis is useful for detecting good solutions when analysed from just two points of view: similarity vs. a single criterion.

Here, we define three user's profiles, representing the preferences of a decision maker. As stated before, the preferences are indicated through a linear ordering of the criteria that is described below.

- Economic-centric (E-c): Travel Cost = Total Damage $\succeq_p$ Average freshness $\succeq_p$ Tardiness $\succeq_p$Service level;
- Product-centric (P-c): Average freshness $\succeq_p$ Travel Cost $\succeq_p$ Total Damage$\succeq_p$ Tardiness $\succeq_p$ Service level;
- Customer Satisfaction-centric (C-c): Tardiness = Service level $\succeq_p$ Travel Cost $\succeq_p$ Total Damage $\succeq_p$ Average freshness.

where the symbol $\succeq_p$ should be read as "at least as preferred to".

For every profile, the procedure described in Section 2.1 to rank the solutions is applied. The best 15 solutions under each user profile are shown in Table 3–5.

Every solution is compared against the reference solution in terms of the following criteria: travel cost, total damage, average freshness, service level, and tardiness. Values in red cells are worse than the reference value, while the green ones are better.

Table 3. Ranking of solutions under the Economic−centric profile.

	Travel Cost	Total Damage	Average Freshness	Tardiness	Service Level
Reference Solution	**1168.4**	**8024.42**	**21.52**	**134,158.6**	**17.45**
18	1.72	−3.64	−1.46	−34.27	10.44
30	5.78	−3.29	0.17	−48.47	9.57
26	3.46	−4.90	−3.64	−43.70	−1.87
53	3.54	−4.78	−6.35	−35.80	7.88
3	4.50	−3.31	−0.99	−42.35	4.25
62	2.60	−4.22	−1.41	−39.59	−3.35
5	0.19	−5.60	−4.46	−29.38	−5.88
28	1.46	−3.61	−0.95	−32.17	−2.33
45	3.94	−3.00	0.24	−29.08	8.69
52	2.21	−2.07	1.39	−28.91	5.09
37	7.06	−4.11	−2.21	−44.00	−1.67
10	−0.04	−5.13	−6.21	−14.61	2.23
9	0.17	−3.18	−1.48	−17.28	2.45
6	0.14	−4.61	−4.48	−15.06	2.22
59	3.24	−4.77	−3.61	−14.21	12.28

Table 4. Ranking of solutions under the product-centric profile.

	Average Freshness	Travel Cost	Total Damage	Tardiness	Service Level
Reference Solution	**21.52**	**1168.4**	**8024.42**	**134,158.60**	**17.45**
30	0.17	5.78	−3.29	−48.47	9.57
52	1.39	2.21	−2.07	−28.91	5.09
3	−0.99	4.50	−3.31	−42.35	4.25
62	−1.41	2.60	−4.22	−39.59	−3.35
45	0.24	3.94	−3.00	−29.08	8.69
37	−2.21	7.06	−4.11	−44.00	−1.67
28	−0.95	1.46	−3.61	−32.17	−2.33
18	−1.46	1.72	−3.64	−34.27	10.44
26	−3.64	3.46	−4.90	−43.70	−1.87
47	1.47	6.07	−3.06	−18.42	4.74
32	−2.41	2.93	−2.79	−35.33	−4.58
15	4.01	2.54	−0.45	−1.24	9.68
34	1.53	6.61	−3.40	−12.16	4.29
25	1.16	5.57	−2.27	−14.17	7.58
46	0.23	3.58	−3.47	−15.53	3.11

Table 5. Ranking of solutions under the customer-centric profile.

	Tardiness	Service Level	Travel Cost	Total Damage	Average Freshness
Reference Solution	**134,158.60**	**17.45**	**1168.4**	**8024.42**	**21.52**
30	−48.47	9.57	5.78	−3.29	0.17
3	−42.35	4.25	4.50	−3.31	−0.99
26	−43.70	−1.87	3.46	−4.90	−3.64
18	−34.27	10.44	1.72	−3.64	−1.46
37	−44.00	−1.67	7.06	−4.11	−2.21
53	−35.80	7.88	3.54	−4.78	−6.35
62	−39.59	−3.35	2.60	−4.22	−1.41
45	−29.08	8.69	3.94	−3.00	0.24
52	−28.91	5.09	2.21	−2.07	1.39
32	−35.33	−4.58	2.93	−2.79	−2.41
28	−32.17	−2.33	1.46	−3.61	−0.95
5	−29.38	−5.88	0.19	−5.60	−4.46
40	−21.31	6.99	4.11	−4.01	−2.64
38	−21.90	2.07	3.57	−4.42	−3.11
59	−14.21	12.28	3.24	−4.77	−3.61

We can observe that the proposed approach allowed to generate a variety of alternative solutions, each of which shows a different behavior for each criterion. This diversity allows the decision maker (DM) to select the most appropriate solution, considering different perspectives and depending on his preference. Regarding the E-c profile, solution 18 is the top-ranked alternative. This solution is quite different from the reference solution X^* having a Jaccard coefficient $J(X^*, s_18) = 0.49$. s_{18} performs better than X^* in terms of service level, tardiness, and damage. On the other hand, has approximately the same travel cost as the reference solution. Solution 30 is slightly worse in terms of travel cost. However, it allows improvement in all the remaining criteria. Regarding the P-c profile, solution 30 is now the best one. This solution has a worse travel cost than the reference solution, but it is better in the remaining criteria. Solution 52 appears in the second place. Note that it was ranked in the 10th position for the E-c profile. Regarding the C-c profile, solution 30 retain the first place by improving significantly all the criteria, with a slight increase in travel cost compared to X^*.

It should be noted that solutions 26 and 30 are among the TOP-3 best alternatives for both the E-c and C-C profile. Another interesting alternative is solution 3 which maintains its place in the TOP-3 for the P-C and C-c profile. Solution 18 is the best alternative for a DM that have economic focus, and retain the third position in terms of service level improvement. However, this solution does not have interest if the focus is on the product freshness or customer satisfaction.

It is highly remarkable that the solution X^* does not appear among the top 15 solutions for all the profiles, which illustrate the ability of our proposed approach in generating performant and quite dissimilar solutions that will serve the decision maker to select the appropriate solution from his perspective.

5. Conclusions

The computational models of relevant optimization problems necessarily leave out of consideration several characteristics and features of the real world. So trying to obtain the optimum solution can not be enough for a problem solving point of view. Moreover, it is doubtful that a single solution would be able to meet all the real specifications. Therefore, it is desirable for the decision maker to have a *set of solutions* that are good enough for the modeled objective, but can perform also well when it comes to other unmodeled criteria.

In the context of a basic problem solving framework, we illustrated the role of metaheuristics as solutions' generators. In the tourist trip design problem, independent runs of an evolutionary algorithm allowed to obtain the set. In the case of perishable food distribution, a more complex strategy was used, inspired from the MGA approach.

Considering a set of solutions and different user profiles, allows to properly rank the available solutions. Using a similarity measure, such as the Jaccard's coefficient, also allowed to better understand the relation between the reference solution and the other ones, not only in the objectives space, but also in the decision space.

This opportunity to consider different perspectives will grant a decision better suited to the real context.

As future work, we will further explore MGA-like approaches, where some measurement of similarity between solutions is needed. Here, the concept of symmetry appears to be crucial as potentially symmetric solutions in the design space can be equal in the objectives space. In other words, different ways to attain the same results can be explored with a proper usage of the symmetry/anti-symmetry ideas.

Finally, connecting the solutions' analysis with multicriteria decision-making approaches will be also explored.

Author Contributions: Investigation, H.E.R. and D.A.P.; methodology, H.E.R. and M.C.-C.; software, H.E.R. and M.C.-C.; writing—original draft, H.E.R., M.C.-C. and D.A.P.; funding acquisition, D.A.P. All authors have read and agreed to the published version of the manuscript.

Funding: This research is supported by project PID2020-112754GB-I00 from MCINAEI/10.13039/501100011033.

Institutional Review Board Statement: Not applicable.

Acknowledgments: Authors acknowledge the support from J.L. Verdegay during the preparation and writing of the manuscript.

Conflicts of Interest: The authors declare no conflict of interest.

Appendix A. Additional Quality Criteria for the Perishable Food Distribution Problem

The alternative solutions generated will be evaluated for a set of criteria relevant to perishable food distribution. It is expected that they can help the decision maker in selecting the most appropriate solution according to his or her preferences. In this work, we consider the four criteria described below:

Appendix A.1. Damage Cost

The quality of perishable foods degrades with the progression of time and temperature fluctuations during transportation and handling. Damage costs are incurred when product quality falls below a certain threshold. The following feature can be used to express the quality of refrigerated products: $D_t = D_0 e^{-\partial t}$, where D_t, and D_0 are, respectively, the quality of product at time t and from the depot 0. The parameter ∂ is the product's spoilage rate, which is assumed to increase as the temperature rises (Ref. [20]). As a result, we distinguish between damage cost during distribution C_{dmg}, and the damage cost during the unloading C'_{dmg} due to the temperature changes (∂ varies also). The damage cost C_{dmg} is expressed as follows:

$$C_{dmg} = \sum_{k \in K} \sum_{i \in C} y_i^k P q_i \left(1 - e^{-\partial_1 \left(t_i^k - t_0^k\right)}\right) \tag{A1}$$

where the coefficient ∂_1 reflects the spoilage rate of product when the vehicle is locked, t_i^k the arrival time of vehicle k at customer i, t_0^k the departure time of vehicle k from the depot,and y_i^k is 0–1 variable taking the value 1 in case the vehicle k is servicing the customer i, and 0 otherwise.

The damage cost during the unloading C_{42} is defined as:

$$C'_{dmg} = \sum_{k \in K} \sum_{i \in C} y_i^k P q_{in} \left(1 - e^{-\partial_2 S_i}\right) \tag{A2}$$

with q_{in} the remaining quantity of product after servicing the customer i, the necessary time to serve is customer i is S_i, and ∂_2 is the spoilage rate when the vehicle is opened.

Appendix A.2. Average Freshness

The average freshness, denoted by AVGF, is defined as [21]:

$$AVGF = \frac{\sum_{k \in K} \sum_{i \in C} y_i^k q_i e^{-\partial_1 \left(t_i^k - t_0^k\right)}}{\sum_{i \in C} q_i} \tag{A3}$$

Appendix A.3. Service Level

This criterion reflects the fact that customers may want to obtain their products at a certain time so that they can begin planning meals, for example. We use the target time as an indicator to determine the service level using the following function:

$$SL_i(t_i) = \begin{cases} 0 & t_i < e_i \\ f(t_i) & e_i \leq t_i < T_g^i \\ 1 & t_i = T_g^i \\ g(t_i) & T_g^i < t_i \leq l_i \\ 0 & t_i > l_i \end{cases} \tag{A4}$$

$SL_i(t_i)$ represents the service level ensured for customer i if we deliver his demand at time t_i. T_g^i is the target time, e_i and l_i are, respectively, the lower, and upper bounds of the time window. The function f is a non-decreasing function, while g is a decreasing function defined as follows:

$$f(t_i) = \frac{t_i - e_i}{T_g^i - e_i} \tag{A5}$$

$$g(t_i) = \frac{l_i - t_i}{l_i - T_g^i} \tag{A6}$$

A good alternative solution is the one that maximizes the following function:

$$SL = \sum_{i \in C} SL_i(t_i) \tag{A7}$$

Appendix A.4. Total Tardiness

We assume that we have a pre-defined set of customers' priorities and they must be served according to their priority level. To consider priority indexes, we define a precedence matrix P, where $P_{ij} = 1$ indicates that the customer i should be supplied before the customer j, and $P_{ij} = 0$ if customer i might be supplied after customer j.

A good solution is the one that minimize the tardiness as much as possible. Tardiness is increased when a lower-priority customer is served before a higher-priority customer, and can be expressed using the following formula:

$$\sum_{i \in C} \sum_{j \in C} t_j - t_i \tag{A8}$$

With t_j, t_i are the arrival times at customer j and i, respectively.

Appendix B. Sets of Solutions

The set of solutions for both problems, together with the calculated values for all the features, are provided here. Table A1 shows the 40 solutions for the time-dependant tourist trip design problem, and Table A2 shows the 62 solutions for the perishable food distribution problem grouped in terms of every α value

Table A1. Set of solutions for the dynamic tourist trip design problem. X^* is the reference solution.

Solution	Interest	# POIs	Efficiency	Travel Time
X^*	90.00	15	90.57%	33.83
1	86.50	15	89.88%	35.47
2	86.00	15	88.33%	41.60
3	88.00	15	88.40%	41.33
4	88.50	15	90.57%	33.83
5	90.00	15	87.61%	44.53
6	86.50	15	89.12%	39.08
7	81.00	15	87.03%	46.18
8	86.50	15	88.80%	39.10
9	90.00	15	90.57%	33.83
10	88.00	14	86.57%	48.08
11	82.50	15	90.28%	34.47
12	90.00	14	89.33%	38.22
13	74.50	14	87.54%	44.12
14	90.00	14	90.16%	34.92
15	84.50	15	84.77%	54.78
16	86.00	15	89.76%	36.52
17	86.00	15	89.24%	38.60
18	86.50	15	88.50%	40.30
19	86.00	15	86.42%	47.93
20	80.00	14	86.46%	46.98
21	82.00	13	88.43%	39.92
22	82.00	12	90.05%	32.03
23	84.00	12	89.83%	33.95
24	82.00	12	90.10%	33.50
25	84.00	13	89.58%	33.72
26	78.00	11	91.19%	28.50
27	82.00	12	90.92%	29.47
28	78.00	12	90.17%	31.08
29	86.00	14	88.32%	39.03
30	82.00	11	90.57%	30.70
31	74.00	12	86.71%	44.43
32	79.50	12	90.13%	31.75
33	80.00	11	90.53%	30.85
34	80.00	11	90.51%	30.92
35	78.00	11	88.28%	39.83
36	76.00	12	88.70%	35.68
37	80.00	12	92.60%	24.78
38	74.00	11	89.84%	33.93
39	84.00	12	91.48%	28.42
40	78.00	12	90.10%	35.15

Table A2. Set of solutions for the perishable food distribution problem in terms of every α value. X^* is the reference solution.

		Jaccard Coeff	Travel Cost	Total Damage	Average Freshness	Tardiness	Service Level
	X^*		**1168.40**	**8024.42**	**21.52**	**134,158.60**	**17.45**
	1	0.47	1168.35	8024.43	21.53	134158.60	17.45
	2	0.45	1218.24	7871.47	21.53	160,106.88	18.81
	3	0.58	1220.88	7759.02	21.31	77,340.90	18.19
	4	0.44	1211.40	7642.49	20.80	150,318.80	16.34
	5	0.75	1170.63	7574.83	20.57	94,747.56	16.42
	6	0.78	1170.01	7654.86	20.56	113,959.93	17.84
	7	0.45	1183.76	7593.93	20.96	172,670.60	19.66
	8	0.58	1184.17	7647.87	20.92	134,514.11	20.10
	9	0.87	1170.28	7769.21	21.21	110,974.37	17.88
	10	0.62	1167.85	7613.17	20.19	114,563.32	17.84
	11	0.70	1172.65	7665.73	20.62	129,750.23	17.06
$\alpha = 1$	**12**	0.62	1196.22	7989.30	21.64	129,427.45	18.24
	13	0.47	1242.50	7719.73	21.16	145,819.23	17.91
	14	0.44	1224.65	7530.24	20.12	179,372.30	17.95
	15	0.72	1198.04	7988.70	22.39	132,493.70	19.14
	16	0.51	1205.16	7774.37	21.70	163,477.79	20.08
	17	0.62	1236.82	7581.91	20.45	145,892.33	18.78
	18	0.49	1188.44	7732.66	21.21	88,179.91	19.27
	19	0.42	1235.29	7753.68	21.27	121,543.30	15.96
	20	0.56	1244.36	7745.72	21.81	189,232.09	17.86
	21	0.33	1239.06	7594.26	20.71	188,649.18	18.59
	22	0.26	1243.11	7713.25	21.52	149,117.63	18.95
	23	0.45	1228.98	7703.10	21.26	124,985.42	17.27
	24	0.62	1236.82	7581.91	20.45	161,864.14	18.78
$\alpha = 0.8$	**25**	0.56	1233.45	7842.50	21.78	115,142.75	18.77
	26	0.49	1208.76	7631.62	20.74	75,526.48	17.13
	27	0.56	1244.36	7745.72	21.81	216,830.61	15.80
	28	0.42	1185.46	7734.76	21.32	90,998.06	17.04
	29	0.53	1187.79	7670.50	20.70	115,110.98	18.45
	30	0.44	1235.87	7760.68	21.56	69,137.18	19.12
	31	0.42	1214.54	7647.11	20.82	155,266.90	16.94
$\alpha = 0.6$	**32**	0.51	1202.58	7800.86	21.01	86,761.52	16.65
	33	0.62	1236.82	7581.91	20.45	203,346.27	18.78
	34	0.33	1245.61	7751.62	21.86	117,838.20	18.20
	35	0.47	1200.82	7735.57	21.40	130,465.05	16.65
	36	0.38	1244.21	7543.49	20.62	113774.55	17.98
	37	0.37	1250.88	7694.87	21.05	75,133.26	17.16
	38	0.38	1210.05	7669.48	20.86	104,778.07	17.81
$\alpha = 0.4$	**39**	0.58	1240.43	7729.76	21.06	179,606.29	16.77
	40	0.37	1216.35	7702.60	20.96	105,564.80	18.67
	41	0.62	1235.22	7830.84	21.72	119,825.69	18.02
	42	0.47	1243.45	7605.22	21.20	175925.44	19.00
	43	0.44	1211.40	7642.49	20.80	141,176.06	16.34
	44	0.44	1224.65	7530.24	20.12	163,449.29	17.95
$\alpha = 0.2$	**45**	0.40	1214.36	7784.08	21.58	95,139.87	18.97
	46	0.62	1210.16	7746.02	21.58	113,320.22	17.99
	47	0.53	1239.28	7778.74	21.84	109,442.38	18.28
	48	0.58	1240.43	7729.76	21.06	179,606.29	16.77
	49	0.51	1205.16	7774.37	21.70	163,477.79	20.08
	50	0.49	1230.69	7570.80	20.39	176,408.72	18.37
	51	0.75	1170.63	7574.83	20.57	139,394.36	16.42
	52	0.60	1194.15	7858.23	21.83	95,379.98	18.34
	53	0.45	1209.71	7640.73	20.16	86,129.95	18.83
	54	0.53	1247.40	7721.07	21.20	187,736.49	17.34
$\alpha = 0$	**55**	0.56	1247.56	7696.99	21.08	182,320.04	18.63
	56	0.49	1252.36	7717.77	21.18	195,880.30	17.59
	57	0.47	1216.26	7955.55	21.68	126,305.39	17.12
	58	0.47	1198.73	7679.40	20.83	158,445.86	18.63
	59	0.62	1206.17	7641.84	20.75	115,100.21	19.60
	60	0.40	1208.93	7632.71	20.95	148,361.60	20.31
	61	0.40	1235.61	7747.33	21.33	125,947.14	19.89
	62	0.40	1198.73	7686.03	21.22	81,051.91	16.87

References

1. Gunalay, Y.; Yeomans, J.S. Simulation-optimization techniques for modelling to generate alternatives in waste management planning. *J. Appl. Oper. Res.* **2011**, *3*, 23–35.
2. Janssen, J.; Krol, M.; Schielen, R.; Hoekstra, A. The effect of modelling quantified expert knowledge and uncertainty information on model based decision making. *Environ. Sci. Policy* **2010**, *13*, 229–238. [CrossRef]
3. Walker, W.E.; Harremoës, P.; Rotmans, J.; Van Der Sluijs, J.P.; Van Asselt, M.B.; Janssen, P.; Krayer von Krauss, M.P. Defining uncertainty: A conceptual basis for uncertainty management in model-based decision support. *Integr. Assess.* **2003**, *4*, 5–17. [CrossRef]
4. Brill, E.D., Jr.; Chang, S.Y.; Hopkins, L.D. Modeling to generate alternatives: The HSJ approach and an illustration using a problem in land use planning. *Manag. Sci.* **1982**, *28*, 221–235. [CrossRef]
5. Loughlin, D.H.; Ranjithan, S.R.; Brill, E.D., Jr; Baugh, J.W., Jr. Genetic algorithm approaches for addressing unmodeled objectives in optimization problems. *Eng. Optim.* **2001**, *33*, 549–569. [CrossRef]
6. Zechman, E.M.; Ranjithan, S.R. An evolutionary algorithm to generate alternatives (EAGA) for engineering optimization problems. *Eng. Optim.* **2004**, *36*, 539–553. [CrossRef]
7. Gavalas, D.; Konstantopoulos, C.; Mastakas, K.; Pantziou, G. A survey on algorithmic approaches for solving tourist trip design problems. *J. Heuristics* **2014**, *20*, 291–328. [CrossRef]
8. Utama, D.M.; Dewi, S.K.; Wahid, A.; Santoso, I. The vehicle routing problem for perishable goods: A systematic review. *Cogent Eng.* **2020**, *7*, 1816148. [CrossRef]
9. Beliakov, G.; Pradera, A.; Calvo, T. *Aggregation Functions: A Guide for Practitioners;* Studies in Fuzziness and Soft Computing; Springer: Berlin/Heidelberg, Germany, 2007. [CrossRef]
10. Torres, M.; Pelta, D.A.; Lamata, M.T.; Yager, R.R. An approach to identify solutions of interest from multi and many-objective optimization problems. *Neural Comput. Appl.* **2021**, *33*, 2471–2481. [CrossRef]
11. Liu, F.; Pan, L.H.; Liu, Z.L.; Peng, Y.N. On possibility-degree formulae for ranking interval numbers. *Soft Comput.* **2018**, *22*, 2557–2565. [CrossRef]
12. Schilde, M.; Doerner, K.F.; Hartl, R.F.; Kiechle, G. Metaheuristics for the bi-objective orienteering problem. *Swarm Intell.* **2009**, *3*, 179–201. [CrossRef]
13. Souffriau, W.; Vansteenwegen, P.; Vertommen, J.; Berghe, G.V.; Oudheusden, D.V. A personalized tourist trip design algorithm for mobile tourist guides. *Appl. Artif. Intell.* **2008**, *22*, 964–985. [CrossRef]
14. Deitch, R.; Ladany, S.P. The one-period bus touring problem: Solved by an effective heuristic for the orienteering tour problem and improvement algorithm. *Eur. J. Oper. Res.* **2000**, *127*, 69–77. [CrossRef]
15. Garcia, A.; Vansteenwegen, P.; Arbelaitz, O.; Souffriau, W.; Linaza, M.T. Integrating public transportation in personalised electronic tourist guides. *Comput. Oper. Res.* **2013**, *40*, 758–774. [CrossRef]
16. Vansteenwegen, P.; Souffriau, W.; Berghe, G.V.; Oudheusden, D.V. The City Trip Planner: An expert system for tourists. *Expert Syst. Appl.* **2011**, *38*, 6540–6546. [CrossRef]
17. Brugnach, M.; Tagg, A.; Keil, F.; de Lange, W.J. Uncertainty matters: computer models at the science–policy interface. *Water Resour. Manag.* **2007**, *21*, 1075–1090. [CrossRef]
18. Matthies, M.; Giupponi, C.; Ostendorf, B. Environmental Decision Support Systems: Current Issues, Methods and Tools. *Environ. Model. Softw.* **2007**, *22*, 123–127.
19. Verdegay, J. Fuzzy Mathematical Programming. In *Fuzzy Information and Decision Processes;* Gupta, M.M., Sanchez, E., Eds.; North-Holland: New York, NY, USA, 1982.
20. Wang, S.; Tao, F.; Shi, Y.; Wen, H. Optimization of vehicle routing problem with time windows for cold chain logistics based on carbon tax. *Sustainability* **2017**, *9*, 694. [CrossRef]
21. Wang, X.; Wang, M.; Ruan, J.; Li, Y. Multi-objective optimization for delivering perishable products with mixed time windows. *Adv. Prod. Eng. Manag.* **2018**, *13*, 321–332. [CrossRef]

Article

A Novel Hybrid Method for KPI Anomaly Detection Based on VAE and SVDD

Yun Zhao, Xiuguo Zhang *, Zijing Shang and Zhiying Cao *

School of Information Science and Technology, Dalian Maritime University, Dalian 116026, China; zhao_yun@dlmu.edu.cn (Y.Z.); shangzj@dlmu.edu.cn (Z.S.)
* Correspondence: zhangxg@dlmu.edu.cn (X.Z.); czysophy@dlmu.edu.cn (Z.C.)

Abstract: Key performance indicator (KPI) anomaly detection is the underlying core technology in Artificial Intelligence for IT operations (AIOps). It has an important impact on subsequent anomaly location and root cause analysis. Variational auto-encoder (VAE) is a symmetry network structure composed of encoder and decoder, which has attracted extensive attention because of its ability to capture complex KPI data features and better detection results. However, VAE is not well applied to the modeling of KPI time series data and it is often necessary to set the threshold to obtain more accurate results. In response to these problems, this paper proposes a novel hybrid method for KPI anomaly detection based on VAE and support vector data description (SVDD). This method consists of two modules: a VAE reconstructor and SVDD anomaly detector. In the VAE reconstruction module, firstly, bi-directional long short-term memory (BiLSTM) is used to replace the traditional feedforward neural network in VAE to capture the time correlation of sequences; then, batch normalization is used at the output of the encoder to prevent the disappearance of *KL* (Kullback–Leibler) divergence, which prevents ignoring latent variables to reconstruct data directly. Finally, exponentially weighted moving average (EWMA) is used to smooth the reconstruction error, which reduces false positives and false negatives during the detection process. In the SVDD anomaly detection module, smoothed reconstruction errors are introduced into the SVDD for training to determine the threshold of adaptively anomaly detection. Experimental results on the public dataset show that this method has a better detection effect than baseline methods.

Keywords: key performance indicator (KPI); anomaly detection; variational auto-encoder (VAE); support vector data description (SVDD)

Citation: Zhao, Y.; Zhang, X.; Shang, Z.; Cao, Z. A Novel Hybrid Method for KPI Anomaly Detection Based on VAE and SVDD. *Symmetry* **2021**, *13*, 2104. https://doi.org/10.3390/sym13112104

Academic Editors: Kóczy T. László and István A. Harmati

Received: 11 September 2021
Accepted: 26 October 2021
Published: 5 November 2021

1. Introduction

In recent years, with the development of technologies, such as machine learning and deep learning, the concept of Artificial Intelligence for IT operations (AIOps) has been proposed. AIOps combines Artificial Intelligence (AI) with operation and maintenance (O and M) to automatically monitor and manage IT services, and improve O and M efficiency. KPI (key performance indicator) anomaly detection is an underlying core technology of intelligent operation and maintenance. Most of the key technologies of intelligent operation and maintenance depend on the results of KPI anomaly detection [1]. In order to provide an efficient and reliable service, KPIs must be monitored in real time to detect anomalies on time. It is necessary for those KPI fluctuations with relatively short durations that must also be accurately monitored to avoid future economic losses. KPI data is a time series data with specific meaning, obtained through periodic sampling in the format of (timestamp, value). KPIs can be roughly divided into two types: service KPIs and machine KPIs. Service KPIs can reflect the scale and quality of web services, such as web page response time, web page visits, number of connection errors, etc. Machine KPIs can reflect the health status of machines (servers, routers, and switches), such as CPU utilization, memory utilization, disk IO, network card throughput, etc. In addition, KPIs also show diversity in shape characteristics, which can be roughly divided into periodic KPIs, stable

KPIs, and continuously fluctuating KPIs. In the actual scenes, the occurrence frequency of anomalies is very low, which leads to extremely unbalanced data samples. Due to the complexity of the business system, it will be constantly updated and upgraded, resulting in the diversity of anomaly types. Because of these characteristics, the precision and recall of existing anomaly detection algorithms are not high, and there are a lot of false positives and false negatives. This not only increases the workload of operation and maintenance personnel, but also makes them unable to find abnormal KPIs timely and accurately.

At present, a series of KPI anomaly detection methods have been proposed by the academia and industry, and these methods are gradually changing from statistical methods to machine learning methods. Deep learning is a subset of machine learning that can automatically learn features from data to achieve good performance and flexibility. As a powerful symmetrical neural network, deep generative models have been widely used in the field of anomaly detection. The learning goal of deep generation models is to narrow the gap between the restored data and the original data as much as possible. Based on the idea that normal data patterns occur frequently and anomalies rarely occur, the "compression restore" process will find the main data patterns instead of restoring the abnormal patterns. Anomaly detection needs to learn the normal pattern of data, so generation models are very suitable. Among them, the representative algorithms are variational auto-encoder (VAE) [2] and generative adversarial network (GAN) [3].

VAE is an unsupervised generative network model, which is composed of encoder and decoder. The encoder maps the input data X to latent variable Z, and the decoder maps the latent variable Z back to X. Generally, the encoder and decoder are the same and share network parameters, so this architecture is called symmetrical [4]. VAE and GAN learn the distribution of normal data, while abnormal data cannot fit this distribution. Anomaly detection is based on the asymmetry of normal data and abnormal data distribution. VAE obtains the distribution of data by variational inference. GAN directly uses the generator to simulate the distribution of data, and the discriminator determines whether the distribution simulated by the generator is good or bad. VAE is less difficult to train and more robust to noise than GAN, so it is more suitable for KPI anomaly detection. However, KPI anomaly detection methods based on VAE still have the following problems:

(1) VAE is not well suited for time series modeling. Previous VAE-based KPI anomaly detection methods [5,6] regard time series as sliding windows, ignoring the time relationship between sliding windows in the encoding process. In order to solve this problem, researchers combine LSTM [7] and VAE. Specifically, LSTM is used to replace the feedforward neural network in VAE, which can extract the characteristics, such as time dependence and correlation between data [8,9]. However, when VAE combines with the strong autoregressive decoder (LSTM), *KL* (Kullback–Leibler) divergence will disappear [10]. Because of the autoregressive of decoder, latent variables in VAE are often ignored and data is reconstructed directly. At this time, the approximate posterior is close to the prior, which causes the *KL* divergence term in the loss function to be reduced to 0. Some studies [10–13] have tried to solve this problem before, but additional parameters or training processes need to be added.

(2) VAE needs to set the threshold for anomaly detection. VAE detects anomalies by comparing the reconstruction results with the original inputs, that is, reconstruction errors. To some extent, the reconstruction error represents an instantaneous measure of anomaly degree. If a threshold is set directly on the reconstruction error, it will lead to a large number of false positives and false negatives. Moreover, for a large number of different types of KPIs, it is difficult to set a unified threshold for reconstruction errors. Early VAE-based anomaly detection studies [5,14] often ignored the importance of threshold selection. Some studies [14,15] adjusted the threshold through cross-validation. However, anomalous samples are rare, and establishing a sufficiently large validation set is a luxury. Other attempts [5,16] only evaluate the best performance of models in the test set, which makes it difficult to reproduce

the results in practical application. Therefore, anomaly detection models need to determine the threshold automatically.

In response to the above-mentioned problems, this paper applies VAE and SVDD to KPI anomaly detection. We use the public data set collected from the real operation and maintenance environment to prove the effectiveness of this method. The main contributions of this paper are summarized as follows:

(1) In order to better capture the time correlation of the KPI time series, the encoder and decoder of the VAE are designed as BiLSTM [17]. Compared with LSTM, BiLSTM processes sequences in both positive and negative directions. Its advantage lies in considering not only past KPI data, but also future KPI data.
(2) It focuses on the problem of the disappearance of *KL* divergence in the loss function during model training, avoiding the strong autoregressive decoder to ignore latent variables and directly reconstruct the data. In this paper, batch normalization [18] is used at the output of the encoder to make the *KL* divergence have a lower bound greater than zero. This method can effectively prevent the disappearance of *KL* divergence without introducing any new model components or modifying targets.
(3) Due to the unpredictability of system behavior, normal behavior can also lead to sharp error peaks. In this paper, EWMA [19] is used to smooth the reconstruction error to suppress frequent error peaks. Simultaneously, the effect of eliminating short-term trends and retaining long-term trends can be achieved, which will minimize false positives and false negatives in the detection process.
(4) In order to solve the threshold adaptation problem of KPI anomaly detection, smoothed reconstruction errors are put into the SVDD [20] for training. The threshold determined by the SVDD has good adaptability and improves the performance of anomaly detection.

2. Related Work

At present, there are few anomaly detection methods for KPI, but, as a kind of time series data, many time series anomaly detection methods are worthy of reference. The existing studies in this section are divided into three categories: traditional statistical methods, supervised machine learning methods, and unsupervised machine learning methods.

The method based on traditional statistics is the earliest method to study time series. The general idea of this method is to make some assumptions about the distribution of data, and then use the statistical inference method to find the anomalies under this assumption. For example, the well-known $3-\sigma$ [21] criterion assumes that data follow a normal distribution, and, if some values exceed 3 standard deviations, they can be considered outliers. With the development of technology, the ARIMA [22,23] and Holt–Winters [24] methods are proposed. Both of these algorithms use a predictive idea to fit the law of time series. Then, prediction results are compared with actual time series, and anomalies are determined by setting a threshold for prediction errors. However, anomaly detection methods based on traditional statistics usually have simple assumptions about time series. Moreover, experts are required to select detectors for given time series and fine-tune the parameters of detectors based on the training data. Therefore, it does not apply to complex monitoring indicator data in actual O and M scenes.

The method based on supervised machine learning can avoid parameters adjustment in traditional statistical algorithms. Among them, the EGADS [25] framework developed by Yahoo and the Opprentice [26] framework developed by then Tsinghua Netman Laboratory are very representative. EGADS and Opprentice are supervised ensemble learning methods. These two methods use anomaly scores output by various traditional anomaly detection algorithms as features, and use user feedback as labels to train anomaly classifiers, which have achieved good results in KPI anomaly detection. However, supervised methods rely heavily on good manual labeling, which is usually not feasible in practical applications. In addition, the ensemble learning classifier based on multi anomaly detectors also faces some problems, such as a large amount of calculation, imbalance of positive and negative samples, among others. Therefore, unsupervised learning methods have become the main research direction of KPI anomaly detection.

The method based on unsupervised machine learning determines the "normal area" by using a single class label (normal KPI samples). Then, by comparing the difference between the KPI observation value and "normal area", we can infer the abnormal degree of data. Since normal samples are far more numerous than abnormal samples in anomaly detection, the model can still be trained even without labels. However, traditional unsupervised machine learning methods need to spend a lot of time to extract the features of data for anomaly detection, such as OCSVM [27], K-means [28], GMM [29], etc. Since deep generative models can automatically capture complex features from data and have higher accuracy, they have received extensive attention. Donut [5] was the first unsupervised model that applied a deep generative model to KPI anomaly detection. Donut puts forward innovations, such as M-ELBO, MCMC iteration, and missing value zero fillings on the basis of VAE, which has excellent performance on periodic KPIs. Subsequently, Buzz [6] solved the problem that was difficult for donut as it handles more complex data distribution. It measures the distance of data distributions and generates distributions through the Wasserstein distance. In fact, Buzz optimizes the likelihood evidence lower bound of a variant VAE by adversarial training. Buzz has a better detection effect on aperiodic KPIs. However, the KPI anomaly detection method based on VAE does not consider the time dependence of data, which limits its applicability to time series. Although LSTM-VAE [8] solves this problem, it will encounter the problem of *KL* divergence disappearing during training. In addition, the VAE judges anomalies by means of reconstruction, with manual determination of thresholds that have poor adaptability. Some recent studies [30,31] have achieved good results by using models to automatically determine thresholds.

This paper uses the public benchmark data set, with related research works of anomaly detection under the same data set that are introduced as follows. KPI-TSAD [32] is a time series deep learning model based on convolution and long short-term memory (LSTM) neural network, and uses a variational auto-encoder (VAE) oversampling model to solve the imbalanced classification problem. Although the method based on supervised learning has achieved good performance in anomaly detection, it needs a lot of labeled data for training. LSTM-based VAE-GAN [9] regards the long short-term memory (LSTM) network as the encoder, generator, and discriminator of VAE-GAN, and jointly trains the encoder, generator, and discriminator. In the anomaly detection stage, anomalies are detected based on reconstruction errors and discrimination results. However, it needs to accumulate certain data to adjust the threshold of the abnormal score. PAD [33] is a method for robust prediction and unsupervised anomaly detection. The prediction block (LSTM) obtains a clean input from the time series reconstructed by the VAE, making it robust to anomalies and noise. At the same time, because LSTM helps to maintain a long-term sequence pattern, VAE performs better in anomaly detection. ALSR [34] is a machine learning scheme for continuous interval KPI anomaly detection. The anomaly detection scheme is optimized by using the different characteristics of abnormal points in the continuous anomaly interval, so that it has better detection accuracy. FluxEV [35] mainly improves SPOT [36], which is only sensitive to extreme values and therefore cannot detect local fluctuations. The method of moment estimation is used to optimize maximum likelihood estimation in SPOT to improve computational efficiency. This paper mainly aims at solving the problems of KPI anomaly detection based on VAE. However, the works discussed above are different from the problems addressed in this paper. For example, KPI-TSAD [32] solves the problem of data imbalance through VAE. LSTM-based VAE-GAN [9] aims to resolve the problem of errors in the mapping of GAN from real-time space to potential space. PAD [33] considers two aspects: state prediction and anomaly detection. ALSR [34] mainly focuses on anomaly detection in continuous intervals. FluxEV [35] focuses on improving computational efficiency.

3. Anomaly Detection Method

3.1. Method Flow

The problem to be solved in this paper is how to detect anomalies in KPI time series data. In order to solve this problem, a novel hybrid anomaly detection method is proposed. The method flow is shown in Figure 1, which mainly includes data preprocessing, a VAE-based reconstruction module, and SVDD-based anomaly detection module. In the training stage, data preprocessing was firstly carried out, that is, missing value filling and data normalization were performed on the original KPI time series. Then, the VAE reconstruction was carried out, that is, the BiLSTM-VAE model was trained and batch normalization was used to prevent the disappearance of the KL divergence. Finally, SVDD anomaly detection was carried out, and reconstruction errors smoothed by EWMA were put into the SVDD for training. The center a and radius R of the SVDD hypersphere were calculated, and the radius R is the threshold of anomaly detection. In the test stage, the test data were preprocessed and input into the trained BiLSTM-VAE model to obtain the reconstructed test data. If the smoothed reconstruction error was less than or equal to the threshold R, it would be judged as normal. If it was greater than the threshold R, it would be judged as abnormal.

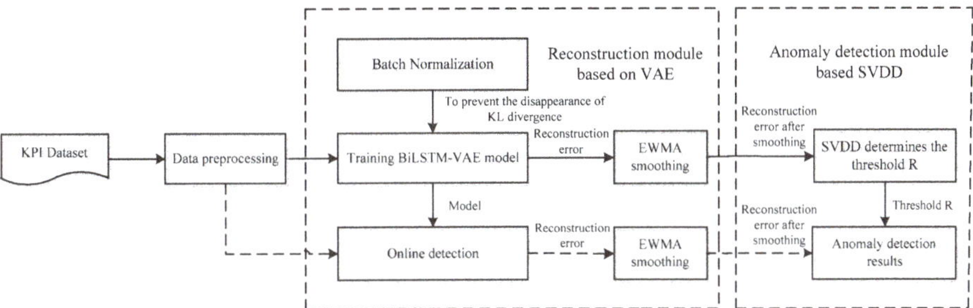

Figure 1. Method flow chart.

3.2. Data Preprocessing

3.2.1. Missing Value Processing

In the real scene, there may be a small number of missing reports or noise data may be deleted in the original KPI data, resulting in the loss of values in the data. Supplementing appropriate data is helpful for subsequent model training. When the number of missing values is small, the effects of nearest neighbor interpolation, linear interpolation, and cubic polynomial interpolation are similar [37]. There are fewer missing values in the KPI dataset used in this paper. After measuring speed and simplicity, the linear interpolation method was selected to deal with missing values. First, the slope was calculated according to the data before and after the missing value, and then the missing KPI data was supplemented according to the slope. Figure 2 shows the interpolation method. If the data x_k was lost, the slope would be calculated as follows:

$$b = \frac{x_{k+1} - x_{k-1}}{t_{k+1} - t_{k-1}}, \tag{1}$$

Next, the missing KPI data x_k was calculated based on the slope b:

$$x_k = x_{k-1} + b \times (t_k - t_{k-1}), \tag{2}$$

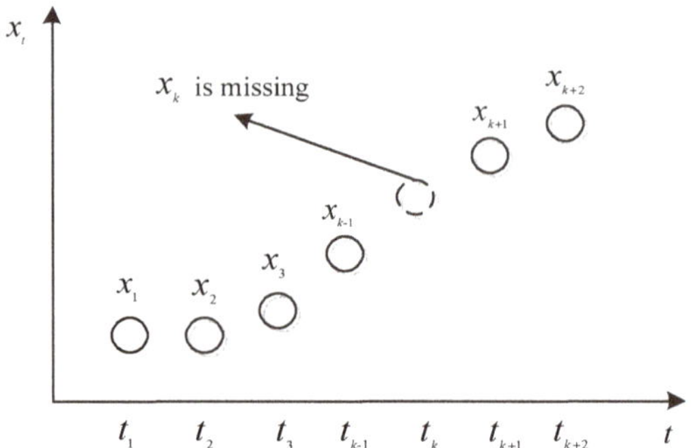

Figure 2. Schematic diagram of the interpolation method.

3.2.2. Data Standardization

In order to eliminate the dimensional influence between indicators, data standardization is needed. After data standardization, all indicators are in the same order of magnitude, which is suitable for comprehensive comparative evaluation. In addition, it can reduce the training time of the model and make the training process converge as soon as possible. This paper normalizes the KPI data, and the data was mapped to the range of 0–1. The normalization formula is as follows:

$$x^* = \frac{x - \min}{\max - \min}, \tag{3}$$

3.3. Reconstruction Module Based on VAE

3.3.1. BiLSTM-VAE Model

In physics, symmetry has a more profound meaning, which refers to invariance under certain transformations. In the VAE, the data is invariant in time and space after encoding and decoding operations, so it just conforms to the concept of symmetry.

The encoder of the VAE is used to learn the distribution of training data and generate the compressed value of training data, and the decoder reconstructs the compressed data. The basic idea is to use a deep neural network to model two complex probability density functions: posteriori probability distribution and conditional probability distribution. The neural network fitting $x \rightarrow z$ is called the inference network $q_\varphi(z \mid x)$, as shown in Formula (4). The neural network fitting $z \rightarrow x$ is called the generative network $p_\theta(x \mid z)$, as shown in Formula (5).

$$z \sim Enc(x) = q_\varphi(z \mid x), \tag{4}$$

$$x \sim Dec(z) = p_\theta(x \mid z), \tag{5}$$

KPI data belongs to time series. Using the memory function of LSTM [7], LSTM network units are introduced into the VAE network to replace traditional neural units in the inference network and generation network. The time dependence and correlation of input data can be learned, which is helpful to extract appropriate features in the hidden layer and reconstruct the input sequence. In this paper, BiLSTM [17] is used as the encoder and decoder of VAE. Compared with LSTM, the advantage of BiLSTM is not only to consider the past KPI data, but also to consider the future KPI data. Figure 3 shows the network structure of BiLSTM-VAE.

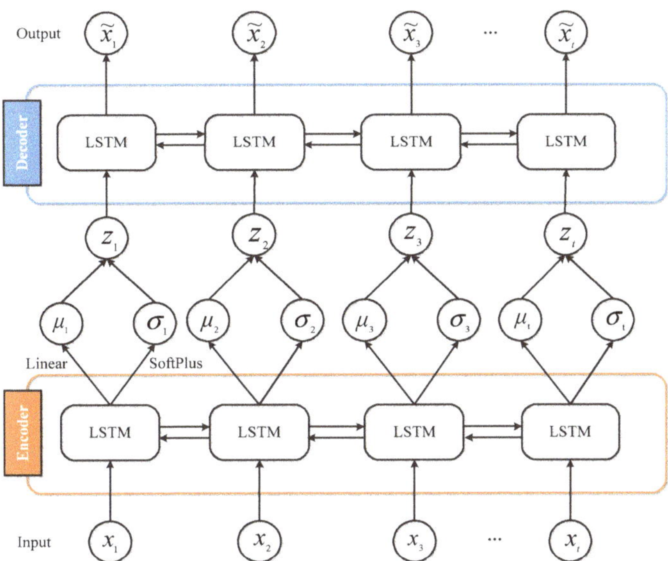

Figure 3. BiLSTM-VAE network structure.

Firstly, time series were divided into sub-sequences corresponding to input variables by the sliding window with a certain step size. Each input sample of the encoder was a vector of a specific size, expressed as $x = \{x_1, x_2, x_3, \cdots, x_t\}$. Then, the encoder encoded input variables into latent variables through the inference network. It was assumed that the true posterior of latent variables z obeys the standard Gaussian distribution (standard normal distribution), i.e., $p_\theta(z) = N(0, I)$. According to the description of reference [2], the standard normal distribution can simulate any distribution through a sufficiently complex function. It can be proved by the inverse transformation sampling theorem. $F(x)$ is a cumulative distribution function; U is a standard normal distribution variable between 0 and 1; and $F^{-1}(U)$ is a sample of the target distribution. Then, the following formula can be obtained:

$$\begin{aligned} &P(F^{-1}(U) \le x) \\ &= P(U \le F(x)) \ , \\ &= F(x) \end{aligned} \tag{6}$$

In short, we can use a normal distribution to obtain a complex distribution through the $D(z)$ function of the decoder to output $\tilde{x}$. In this way, $\tilde{x}$ and x have the same probability distribution and content. Therefore, this assumption was reasonable. Specifically, given a real sample x_t, we assumed that there is a distribution $p_\theta(z \mid x_t)$ exclusively belonging to x_t, and further assumed that this distribution is (independent and multivariate) normal distribution. If $p_\theta(z \mid x_t)$ belongs exclusively to x_t, it is reasonable to say that z sampled from this distribution should be restored to x_t. Since the distributions assumed above are normal distributions, it was necessary to obtain the corresponding variance and mean. Then, a z_t was sampled from this exclusive distribution and $\tilde{x}$ was obtained through a decoder $\tilde{x}_t = D(z_t)$.

The approximate posterior distribution $q_\varphi(z \mid x)$ also obeys the Gaussian distribution $N(\mu, \sigma I)$, where μ and σ are the mean and variance of the Gaussian distribution. It is not difficult to see that the function of the encoder is to generate the mean μ and variance σ through two networks. The encoder was parameterized through BiLSTM with an activation function to generate hidden state sequences in both directions, forward $\rightarrow$ and backward $\leftarrow$. The final encoder hidden states of both passes were concatenated with each other to produce the vector $h_t = [\overrightarrow{h}_t, \overleftarrow{h}_t]$. μ and σ were derived from the final encoder hidden state h_t using two fully connected layers with linear and Softplus activations, respectively.

Latent variables were obtained by reparameterization, that is, $z = \mu + \sigma \odot \varepsilon$. Among them, $\varepsilon \sim N(0, I)$ is an auxiliary noise variable, and $\odot$ represents the product at the element level. Evidently, noise will increase the difficulty of reconstruction. Resampling essentially adds "Gaussian noise" to the encoder result (the mean value), so that the decoder result can be robust to noise. Another encoder result (the variance) was used to dynamically adjust the intensity of noise. Intuitively, when the decoder is not well trained, it will appropriately reduce the noise to make the fitting easier (the reconstruction error becomes smaller). On the contrary, if the decoder is well trained, the noise will increase, making the fitting more difficult (the reconstruction error becomes larger).

Finally, the decoder decoded latent variables back to the original data space through the generation network $p_\theta(x \mid z)$, so as to obtain reconstructed data samples.

In VAE, the parameters of the network were optimized by maximizing the lower bound of evidence $ELBO_{vae}$, as shown in Formula (7):

$$ELBO_{vae} = E_{q_\varphi(z|x)}[\log\, p_\theta(x \mid z)] - D_{KL}(q_\varphi(z \mid x) \mid p_\theta(z)), \tag{7}$$

where the first term represents the reconstruction term, and $E_{q_\varphi(Z|X)}$ is the logarithmic likelihood estimate of the posterior probability of x. The second term represents the regularization term, which measures the gap between approximate posterior $q_\varphi(z \mid x)$ and true posterior $p_\theta(z)$ by *KL* divergence. The goal of optimization is to maximize the likelihood function of generated data and minimize the *KL* divergence between the approximate posterior distribution and the true posterior distribution. In short, on the one hand, the output was fitted to the input as much as possible. On the other hand, the noise was appropriately increased through the *KL* divergence to prevent over-fitting.

3.3.2. Batch Normalization Prevents the Disappearance of *KL* Divergence

When VAE is used with a powerful autoregressive decoder (LSTM), *KL* divergence often disappears. This is generally believed due to the strong autoregression of the decoder, that is, the generated network $p_\theta(x \mid z)$ is too strong. This will cause the model to abandon the use of the approximate posterior of encoder and directly use the latent variables of the model. At the same time, the *KL* divergence term will quickly decrease to 0, that is, prior and approximate posterior are equal. In addition, the reparameter operation will introduce noise during training. When it has high noise, latent variables are difficult to be used, so the VAE ignores latent variables and carries out the reconstruction independently. When the *KL* divergence is 0, the encoder outputs a constant vector. The use of VAE usually focuses on its ability to construct coding vectors unsupervised. Therefore, the problem of the *KL* divergence disappearance must be solved when applying VAE.

Based on the batch normalized-VAE (BN-VAE) method proposed in reference [18], this paper solves the problem of *KL* disappearance. This method has good results in language modeling, text classification, and dialog generation. This paper applies it to the field of anomaly detection for the first time. The core idea of BN-VAE is to apply BN to the mean vector output by the encoder, so as to ensure that the expected lower limit of *KL* divergence distribution is positive, as shown in Figure 4.

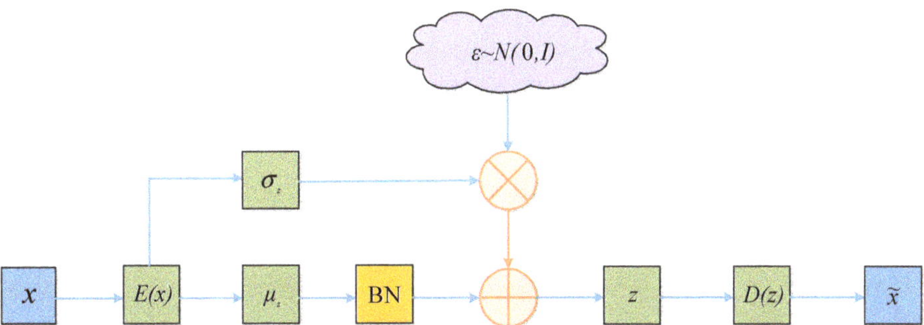

Figure 4. Adding BN to VAE prevents the disappearance of the *KL* divergence.

In order to explain how BN was associated with *KL* divergence, the *KL* divergence term formula is given first:

$$KL = \frac{1}{2}\sum_{i=1}^{d} \mu_i^2 + \sigma_i^2 - \log\sigma_i^2 - 1, \tag{8}$$

In the above formula, d is the dimension of latent variables. μ_i and σ_i are the mean and standard deviation of posterior distribution of the i-th dimension of latent variables, respectively. In the actual calculation, we often used batch training, so the above formula was further calculated during the training process:

$$\begin{aligned} KL &= \frac{1}{2b}\sum_{j=1}^{b}\sum_{i=1}^{d} (\mu_{i,j}^2 + \sigma_{i,j}^2 - \log\sigma_{i,j}^2 - 1) \\ &= \frac{1}{2}\sum_{i=1}^{d} (\frac{\sum_{j=1}^{b}\mu_{i,j}^2}{b} + \frac{\sum_{j=1}^{b}\sigma_{i,j}^2}{b} - \frac{\sum_{j=1}^{b}\log\sigma_{i,j}^2}{b} - 1) \end{aligned}, \tag{9}$$

In the above formula, b is the size of batch. When b is large enough, the *KL* term will approximate the average value of *KL* of the whole data set. Thus, we limited the distribution of *KL* in the data set by limiting the distribution of mean and variance. In this way, *KL* was equivalent to the distribution of posterior distribution parameters of latent variables. Therefore, the above formula can be expressed as follows:

$$\begin{aligned} E[KL] &= \frac{1}{2}\sum_{i=1}^{d} (Var[\mu_i] + E^2[\mu_i] + E^2[\sigma_i^2] - E[\log\sigma_i^2] - 1) \\ &\geq \frac{1}{2}\sum_{i=1}^{d} (Var[\mu_i] + E^2[\mu_i]) \end{aligned}, \tag{10}$$

In the above formula, $E^2[\sigma_i^2] - E[\log\sigma_i^2] - 1 \geq 0$ can be derived from $e^x \geq x+1$, so the inequality holds. Through this transformation, it is not difficult to realize that batch normalization can be used to constrain the distribution of mean. The mean value in the posterior distribution was performed as follows:

$$\hat{\mu}_i = \gamma\frac{\mu_i - \mu_{Bi}}{\sigma_{Bi}} + \beta, \tag{11}$$

In the above formula, $\hat{\mu}_i$ is the mean value of μ_i transformed by the BN layer. μ_{Bi} and σ_{Bi} represent the mean and standard deviation of μ_i. β and γ are parameters in batch normalization, which can control the variance and mean value of μ_i distribution, respectively. Finally, Formula (12) was obtained by replacing μ_i in KL formula:

$$\begin{aligned} E[KL] &\geq \frac{1}{2}\sum_{i=1}^{d}\left(Var[\mu_i] + E^2[\mu_i]\right) \\ &= \frac{d}{2}\left(\gamma^2 + \beta^2\right) \end{aligned}, \tag{12}$$

Therefore, as long as β and γ are well controlled (mainly γ is fixed to a certain constant), the KL divergence term can have a positive lower bound. In this way, KL divergence and BN were cleverly linked to avoid the disappearance of KL divergence.

3.3.3. EWMA Smoothing Reconstruction Errors

The difference sequence $d = | \tilde{x} - x |$ can be obtained by comparing reconstructed KPI sequence with the original sequence. However, the original difference sequence represents an instantaneous measure of the predictability of the current input. Nevertheless, in many practical applications, the underlying system is inherently unpredictable. In this case, predictable change usually means meaningless behavior. This is seen, for example, in the latency of HTTP requests for websites. Although the latency is usually low, it is not uncommon for random jumps to reach the peak corresponding to anomaly scores. In fact, abnormal observations usually occur continuously, and it is acceptable to trigger an alarm in a short time. Setting thresholds directly on original difference sequences will lead to many false positives. Therefore, this paper uses EWMA [19] to smooth the difference sequence to suppress the frequently occurring error peaks. System behavior is usually not perfectly predictable, and normal behavior can also cause sharp peaks in error values [38]. At time k, the smoothed sequence e_k was obtained according to the original difference sequence d_k. The calculation process is shown in Formula (13):

$$e_k = \alpha d_k + (1 - \alpha)e_{k-1}, \tag{13}$$

In the above formula, $\alpha(0 < \alpha < 1)$ is the weight coefficient of EWMA for the historical measurement value. The closer its value is to 1, the lower weight for the past measurement value. α determines the ability of EWMA to track sudden changes in actual data, namely timeliness. With the increase in α, the timeliness of EWMA is stronger; otherwise, it is weaker. EWMA also shows a certain ability to absorb instantaneous bursts. By controlling α, short-term fluctuations are eliminated and long-term development trends are retained, providing a smooth form of sequences.

3.4. *Anomaly Detection Module Based on SVDD*

SVDD [20] is an algorithm that can describe the target data in a hypersphere, which can contain as many data points as possible. It can be described as: if only one class can be judged, then the smallest hypersphere needs to be found through SVDD to include all the data of this class. When the hypersphere is used to identify new data, if the data fall within the hypersphere, the data are considered to belong to this class. Otherwise, the data do not belong to this class.

When training the SVDD classifier, this paper inputs the reconstruction error of normal data into the SVDD for training to determine the threshold. In the test phase, the reconstruction error of abnormal data was greater than that of normal data, so it exceeded the threshold to realize anomaly detection. However, different KPI curves correspond to different reconstruction error curves. If a fixed threshold is set based on human experience, a large number of false positives and false negatives will be caused. Therefore, this paper inputs the reconstruction error into SVDD for training to determine the threshold, which can adaptively set different thresholds for different KPIs.

The goal of the SVDD is to find support vectors and use them to construct a minimal closed hypersphere that contains all or most of the target training samples. In this paper, target training samples are the smoothed reconstruction error of normal KPI, expressed as $e = \{e_1, e_2, e_t, \cdots, e_n\}$, where n is the number of samples. The sample was distributed in a ball with center a and radius R, i.e., $\|e_t - a\|^2 \leq R^2$. By introducing the slack variable ξ_t, it was allowed that some samples were no longer in the ball, that is, $\|e_t - a\|^2 \leq R^2 + \xi_t$. The training objective was to minimize the value of radius R and slack variable ξ_t, so the objective function was expressed as Formula (14):

$$\min F(R, a, \xi_t) = R^2 + C \sum_{t=1}^{n} \xi_t$$
$$s.t. \begin{cases} \|e_t - a\|^2 \leq R^2 + \xi_t, (t = 1, 2, \cdots, n) \\ \xi_t \geq 0 \end{cases}, \tag{14}$$

Among them, ξ_t is the slack variable, which is used to measure a small amount of abnormal data outside the hypersphere. C is the penalty coefficient used to control the volume of the hypersphere, and its value ranges from 0 to 1. The slack variable ξ_t prevented the model from being "destroyed" by individual extreme data points. In short, if most data points are in a small area and only a few abnormal data are far away from them, the model prefers to regard those few data points as anomalies. To avoid the model making excessive sacrifices to cater to few data points, the model tolerated some data points that did not meet the rigid constraints and gave them some elasticity. C adjusted the influence of the slack variable. Generally speaking, the slack space is given to those data points that need slack. If C is large, the loss caused by the slack variable in the loss function is large. Then, the slack variable will be reduced during training. In this way, the model does not tolerate those outliers and just wants to include them. On the contrary, if C is small, the model will give outliers greater elasticity, so that they can not be included.

In order to make the training process easier to understand, the hypersphere was visualized in two-dimensional and three-dimensional space respectively, as shown in Figures 5 and 6. The hypersphere corresponds to a curve in two-dimensional space and a sphere in three-dimensional space. Under normal circumstances, the data will not show spherical distribution, so the Gaussian kernel function method was used to improve the expression ability of model. Figures 5a and 6a show the contour distance visualization of hypersphere in two-dimensional and three-dimensional space, respectively. Figures 5b and 6b show the decision boundary visualization of hypersphere in two-dimensional and three-dimensional space, respectively.

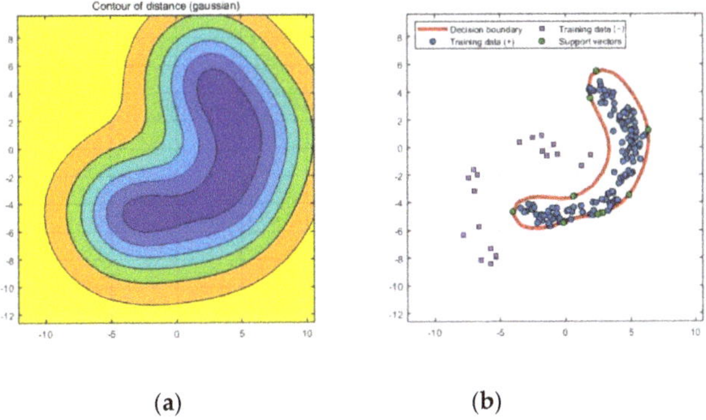

(a) (b)

Figure 5. Visualization of the hypersphere in two-dimensional space: (**a**) Contour of distance; and (**b**) Decision boundary.

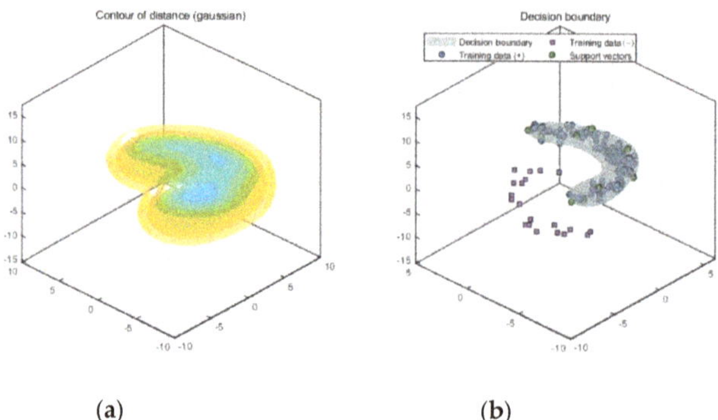

(a) (b)

Figure 6. Visualization of the hypersphere in three-dimensional space: (**a**) Contour of distance; and (**b**) Decision boundary.

The optimization problem was solved by the Lagrange multiplier method, and the following Lagrange function was obtained:

$$L(R, a, \xi_t) = R^2 + C\sum_{t=1}^{n} \xi_t - \sum_{t=1}^{n} \lambda_t \left[R^2 + \xi_t - \|e_t - a\|^2\right] - \sum_{t=1}^{n} \beta_t \xi_t, \tag{15}$$

where λ_t and β_t are Lagrange multipliers. The distance from e_t to a is recorded as $g(e_t)$, and the calculation formula is as follows:

$$g(e_t) = \|e_t - a\| = \sqrt{(e_t, e_t) - 2\sum_{i=1}^{n} \lambda_i (e_i, e_t) + \sum_{i=1}^{n}\sum_{j=1}^{n} \lambda_i \lambda_j (e_i, e_j)}, \tag{16}$$

The calculation formula of radius R is as follows:

$$R = \sqrt{(e_s, e_s) - 2\sum_{i=1}^{n} \lambda_i (e_i, e_s) + \sum_{i=1}^{n}\sum_{j=1}^{n} \lambda_i \lambda_j (e_i, e_j)}, \tag{17}$$

Among them, e_s is a support vector on the sphere of the hypersphere. e_i and e_j are any two samples input to the SVDD. In addition, in order to make the samples linearly separable in the feature space, it was necessary to map samples from the original space to the high-dimensional feature space by using a kernel function. In this paper, a Gaussian kernel function is used to map samples from original space to appropriate feature space. The expression of Gaussian kernel function is:

$$K_{Gauss}(e_i, e_j) = \exp(-\|e_i - e_j\|^2 / s^2), \tag{18}$$

where s is the Gaussian kernel parameter.

In the anomaly detection stage, $g(e_{test}) > R$ indicates that the distance from e_{test} to a is greater than R, then x_{test} is the abnormal KPI data. $g(e_{test}) \leq R$ indicates that the distance from e_{test} to a is less than or equal to R, then x_{test} is the normal KPI data. The process of anomaly detection is shown in Figure 7.

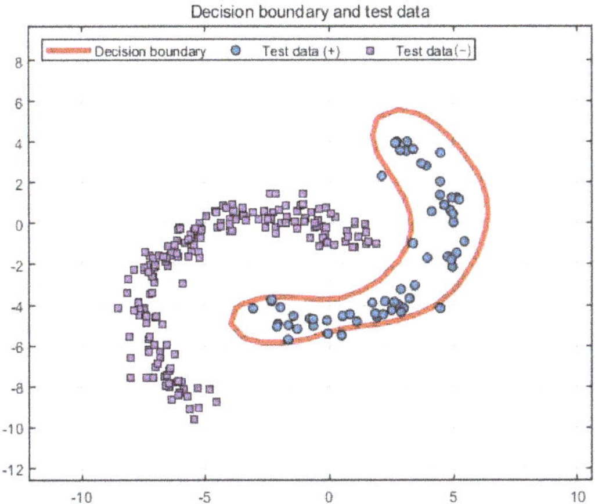

Figure 7. Anomaly detection based on SVDD.

4. Experimental Procedure

The experimental environment of this paper was Windows 10 (64-bit) operating system. The hardware configuration was Intel (R) Core (TM) i7-8700CPU@3.20 GHz 16 G RBM and 237 G solid state drive. The development language was python3.6, the development framework was Keras, and the back-end engine was TensorFlow.

4.1. Dataset

The KPI dataset used in this paper was published by the AIOps challenge competition (http://iops.ai/competition_detail/?competition_id=5&flag=1 (accessed on 10 September 2021)), which provides the KPI desensitization time series with anomaly labels. The data were collected from the real operation and maintenance environment of top Internet companies, such as Sogou, Tencent, eBay, Baidu, and Alibaba, and the sampling interval was 1 min. We randomly selected two KPIs to verify the proposed method in this paper. As shown in Table 1, the ratio of normal samples to abnormal samples in the data set was obviously very uneven, and abnormal samples account for less than 10% of the total number of samples. Figure 8 shows the visualization effect of two KPIs. It can be observed that KPIs showed a certain degree of periodicity and trend.

Table 1. Dataset details.

Dataset	KPI 1	KPI 2
Total points	128,562	129,035
Anomaly points	10,550/8.21%	7666/5.94%
Missing points	3233/0.02%	2755/0.02%
Duration	91 days	91 days
Sample Frequency	1412.77	1417.97

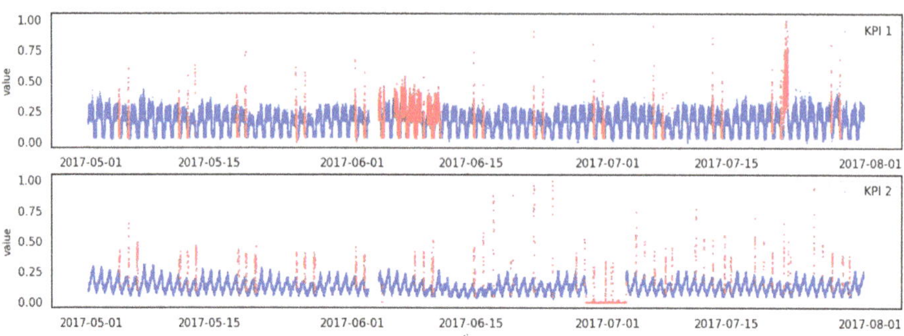

Figure 8. Two KPIs from real production environments.

4.2. Evaluation Metrics

In the anomaly detection module, this paper determines the threshold through SVDD. When the reconstruction error was greater than the threshold, the point was judged as an anomaly. Operation and maintenance personnel usually only care about whether the anomaly detection algorithm can detect a continuous anomaly interval, rather than detecting each anomaly point in the anomaly interval. Therefore, the evaluation of this paper adopts the strategy described in the literature [5]. If the anomaly detection algorithm made a judgment fast enough (before the maximum allowable delay) after the beginning of anomalies, it was considered to have successfully detected the whole anomaly segment. The alarm delay was the time difference between first anomaly point and first detection point in the anomaly segment. If the anomaly detection algorithm did not issue any alarm before the maximum allowable delay, even if the anomaly detection algorithm detected the anomaly, we considered that the algorithm failed to successfully detect the anomaly segment. Figure 9 shows the anomaly detection results with an alarm delay of 1 min (1 grid). The first line represents the real labeled data, including 10 consecutive time points and 2 anomaly intervals. The second line represents the output results of the anomaly detection method. The third line represents the anomaly detection results corrected according to the alarm delay. For the first anomaly interval, if the anomaly detection method found an anomaly within the longest delay alarm, it was considered that the whole anomaly interval was successfully detected. For the second anomaly interval, it was considered that the anomaly interval was not successfully detected, because the detection result exceeded the alarm delay.

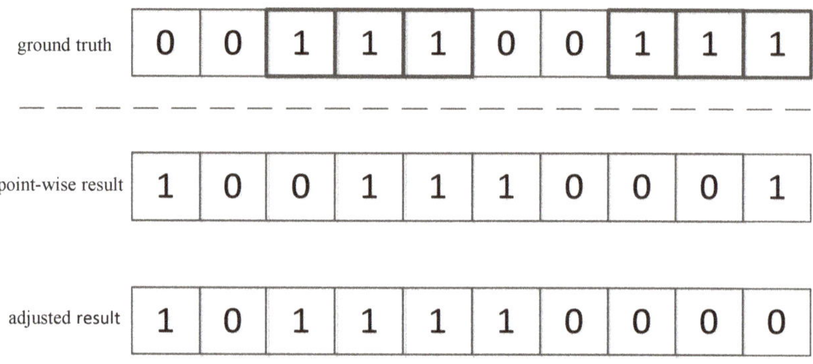

Figure 9. Description of anomaly detection policies.

Therefore, the anomaly detection of time series can be regarded as a classification problem. In this paper, *Precision*, *Recall*, and *F1-score* are used to evaluate the performance of detection. *Precision* represents the proportion of correct prediction being positive in relation to the total prediction being positive. *Recall* represents the proportion of correct prediction being positive in relation to the total actual being positive. *F1-score* is the weighted harmonic average of *Precision* and *Recall*. The calculation formula is as follows:

$$Precision = \frac{TP}{TP + FP}, \tag{19}$$

$$Recall = \frac{TP}{TP + FN}, \tag{20}$$

$$F1\text{-}score = \frac{2 \times Precision \times Recall}{Precision + Recall}, \tag{21}$$

where TP is the number of anomaly points correctly detected; FP is the number of normal points incorrectly identified as anomaly points; and FN is the number of anomaly points incorrectly identified as normal points.

4.3. Experimental Parameter Setting

In the experiment, each KPI time series was divided into training set and test set by 8:2. After the many repeated experiments that were conducted in the context of this paper, the final hyperparameters are shown in Table 2 on the premise of balancing the time-consuming and detection effect.

Table 2. Main hyperparameter settings.

Hyperparameter Name	Hyperparameter Value
Batch size	256
Number of iterations	100
Optimizer	Adam
Learning rate	0.0005
LSTM unit size	128
Latent variable dimension	10
Sliding window length	12
Alarm delay	7
Penalty coefficient of SVDD	0.25
Gaussian kernel parameter of SVDD	9

In the process of adjusting the parameters, we found that the sliding window length W and latent variable dimension K had a great influence on the results of the anomaly detection. Too short sliding windows could not obtain the relationship between adjacent points. Too long sliding windows relied too much on historical information and lacked sensitivity to current values. Latent variables represent all the important information needed to contain the original data point. The representation ability of potential space varies with the dimension of latent variables. Therefore, this paper tests the best *F1-score* of algorithm under different sliding window lengths and latent variable dimensions, as shown in Figure 10. It can be seen that, when the sliding window length W = 12 and latent variable dimension K = 10, the *F1-score* reaches the optimal value. In addition, when SVDD determined the threshold, the selection of the penalty coefficient and kernel parameters also had an important impact on the effect of anomaly detection. In this paper, the accuracy of anomaly detection is used as the fitness function, and the penalty coefficient C = 0.25 and the Gaussian kernel parameter s = 9 are obtained by the particle swarm optimization algorithm [39].

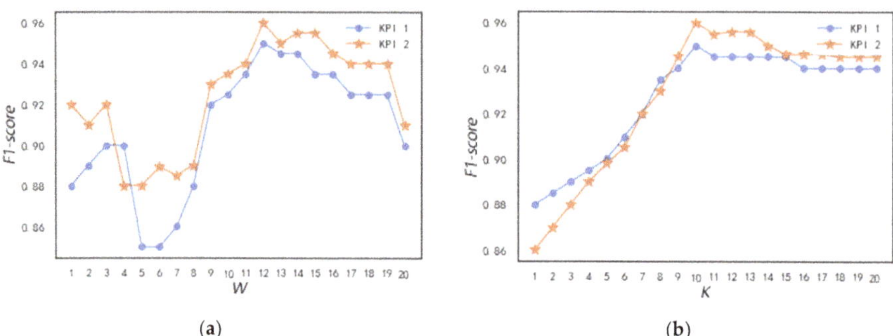

(a) (b)

Figure 10. The best *F1-score* under different sliding window lengths and latent variable dimensions: (**a**) Different sliding window lengths; and (**b**) Different latent variable dimensions.

4.4. Experimental Results of Anomaly Detection

In this paper, normal data are used to train BiLSTM-VAE model and the distribution of normal data is learned. During the test phase, the model did not reconstruct the anomalous data well, because of the different distributions of the anomalous data from the normal data. To visually observe this, we plotted the reconstruction effect of two KPIs on a partial test set. As shown in Figure 11, normal samples of two KPIs can be reconstructed well, while abnormal samples cannot be reconstructed well, resulting in higher reconstruction errors. Figure 12 shows original the reconstruction errors of two KPIs. It can be seen that reconstruction errors of normal points are closer to 0, while abnormal points will lead to the error peak. However, setting a fixed threshold directly on the original reconstruction error threshold will not only lead to a large number of false positives and false negatives, but also to the need to adjust the threshold manually. In addition, it is unrealistic to set a unified threshold for different KPIs, which may lead to poor adaptability. Therefore, this paper uses EWMA to smooth the reconstruction error and SVDD to adaptively determine the threshold. Figure 13 shows smoothed reconstruction errors of two KPIs, and the red dotted line is the threshold determined by SVDD. It can be seen from the figure that the threshold of KPI 1 is 0.04, and the threshold of KPI 2 is 0.018. The errors of normal points are lower than the threshold, and the errors of abnormal points are higher than the threshold. In summary, the method in this paper can accurately detect anomalies and adaptively determine the optimal threshold for each KPI.

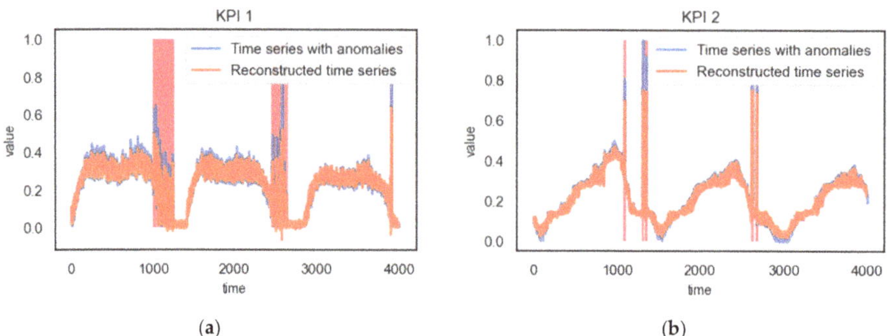

(a) (b)

Figure 11. The reconstruction effect of two KPIs: (**a**) KPI 1; and (**b**) KPI 2.

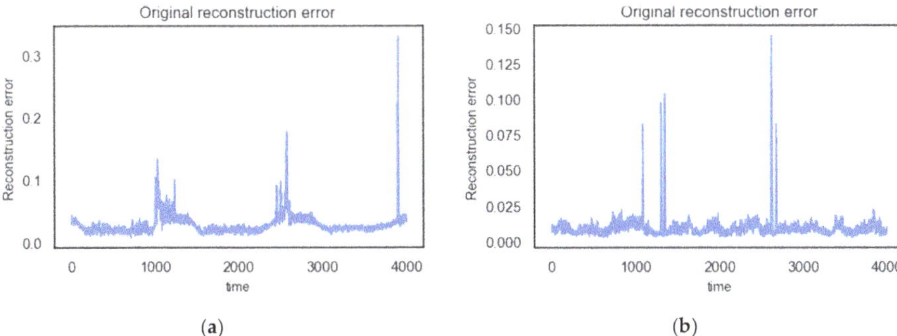

Figure 12. Original reconstruction errors of two KPIs: (**a**) KPI 1; and (**b**) KPI 2.

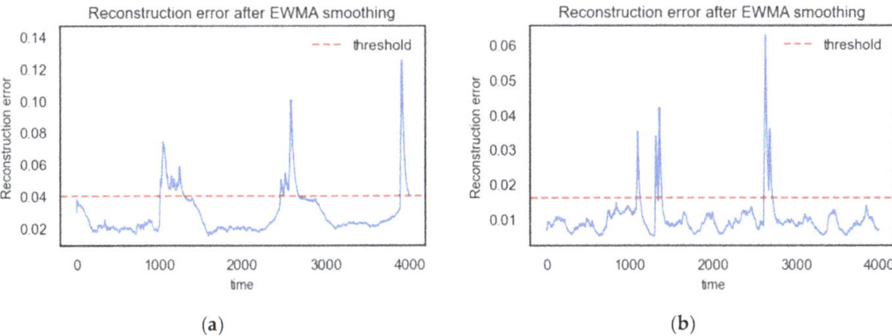

Figure 13. Smoothed reconstruction errors of two KPIs: (**a**) KPI 1; and (**b**) KPI 2.

4.5. Comparative Experiment and Analysis

In order to verify the effectiveness of this method, we selected three methods as baseline to test the detection effect of each method on the KPI data set. These methods included the VAE-based anomaly detection method Donut [5] and LSTM-VAE [8]. In fact, VAE was part of our approach. In order to make a fair comparison, the hyperparameters of Donut and LSTM-VAE were the same as the method in this paper. In addition, we also compared the supervised learning method Opprentice [26] as the most competitive method of non-deep learning:

- Opprentice [26] is an ensemble supervised algorithm that uses random forest classifiers. Its principal concept is to use more than ten different types of detectors to extract hundreds of abnormal features. Then, using the manually labeled data and anomaly features, the anomaly detection problem can be transformed into a supervised classification problem in machine learning. The extracted features are used as the input of machine learning algorithm. The points on the KPI curve are divided into normal points and abnormal points through a classification algorithm, so as to realize anomaly detection.
- Donut [5] is an unsupervised anomaly detection algorithm based on VAE. Through the improved variational lower bound and Markov chain Monte Carlo interpolation technology, the algorithm can be used without labels. Donut applies a sliding window on the KPI to obtain the sub-sequence, and tries to identify the normal pattern. Then, anomalies are determined by reconstruction probability. In fact, it selects a threshold for each KPI.
- LSTM-VAE [8] combines LSTM and VAE to make it more suitable for time series modeling. Specifically, it replaces the feedforward neural network in VAE with LSTM. LSTM-VAE fuses sequences and reconstructs their expected distribution by

introducing a schedule based variational a priori. In the anomaly detection phase, it uses an anomaly score based on reconstruction probability and a state-based threshold.

Table 3 shows the best *Precision*, *Recall*, and *F1-score* of various anomaly detection methods on two KPIs. Opprentice based on machine learning performed worse than VAE based on deep learning, and it needed labels for training, which was difficult to achieve in the actual scene. Donut made a series of improvements based on VAE so that it can train without labels. Since Donut treats time series as sliding windows and does not process time information, Donut's performance was poor when anomaly detection relies on time information. LSTM-VAE can extract the time correlation of sequence better than Donut, so a better detection effect was obtained. The method in this paper is VAE-SVDD—VAE uses BiLSTM as encoder and decoder; batch normalization avoids the disappearance of the *KL* divergence; EWMA smooths the original reconstruction error; and SVDD adaptively determines the threshold. Through the above improvements, VAE-SVDD had higher *Precision*, *Recall*, and *F1-score* compared to other baseline methods.

Table 3. Experimental results of various anomaly detection methods.

Method	KPI 1			KPI 2		
	Precision	*Recall*	*F1-Score*	*Precision*	*Recall*	*F1-Score*
Opprentice	0.72	0.66	0.69	0.78	0.70	0.74
Donut	0.83	0.76	0.79	0.86	0.83	0.84
LSTM-VAE	0.91	0.84	0.87	0.90	0.85	0.87
VAE-SVDD	0.95	0.96	0.95	0.97	0.96	0.96

VAE-SVDD uses VAE to reconstruct KPI data and uses SVDD to train the reconstruction error again. It is necessary to compare the complexity of VAE-SVDD with other detection methods. This paper records the anomaly detection time of various anomaly detection methods. Table 4 shows the average duration of 5 anomaly detections performed by each method on the test set (containing 26,358 data). The Opprentice method is based on a random forest, and it can improve efficiency through parallelization, so the detection time is short. Donut based on VAE is more complex than the Opprentice method, resulting in slightly longer detection time. LSTM-VAE integrates LSTM and VAE, which evidently leads to a longer detection time. In VAE-SVDD, the encoder and decoder of VAE were designed as BiLSTM, and the threshold was determined by SVDD. However, these optimization mechanisms also increased the detection time. Although the detection time of VAE-SVDD was slightly longer than other methods, the detection was more accurate. Therefore, it is still acceptable in the actual situation.

Table 4. Average detection time.

Evaluation Index	Opprentice	Donut	LSTM-VAE	VAE-SVDD
Detection time (s)	34.5	46.3	53.8	65.2

4.6. Effects of Different Components

4.6.1. Time Correlation

In this paper, BiLSTM network is used as the encoder and decoder of VAE, which can better capture the time correlation of sequence data. We compared the distribution of latent variables between VAE and BiLSTM-VAE to prove that time correlation has a positive effect. In order to verify this time correlation, part of the test set containing anomalies was selected. At the same time, the hours of timestamp were extracted as labels, that is, there were 24 types of labels.

In order to facilitate visualization, we used a principal component analysis (PCA) [40] and t-distributed Stochastic Neighbor Embedding (t-SNE) [41] to reduce the dimension of latent variables to 2. Latent variables can be regarded as the characteristic representation of data. PCA replaces original 10 features with a smaller number of 2 features. New features are the linear combination of old features. These linear combinations maximize the sample variance and try to make new features irrelevant to each other. The mapping from old features to new features captures the inherent variability in the data. In addition, it has the advantage of being quick and easy to implement. However, the nonlinear correlation between samples may be lost after linear dimension reduction using PCA. In contrast, t-SNE is a nonlinear dimension reduction method. It converts the similarity of data points into joint probability and optimizes the *KL* error between low-dimensional data and high-dimensional data. t-SNE dimension reduction can not only maintain the difference of data, but also maintain the local structure of data. However, the results of t-SNE have a certain degree of randomness, rather than the consistency of PCA results. Therefore, we combined the two methods to visualize latent variables after dimension reduction. It can be more reasonable to prove that our method better captures the potential pattern of data, that is, time correlation.

Figure 14 shows the two-dimensional visualization effect of VAE latent variables on KPI 1. Figure 15 shows the two-dimensional visualization effect of BilSTM-VAE latent variables on KPI 1. Figure 16 shows the two-dimensional visualization effect of VAE latent variables on KPI 2. Figure 17 shows the two-dimensional visualization effect of BilSTM-VAE latent variables on KPI 2. All subgraphs (a) are the visualization effect of the PCA on latent variables after dimension reduction. All subgraphs (b) are the visualization effect of t-SNE on latent variables after dimension reduction. Evidently, the latent variables distribution of BiLSTM-VAE is more regular than VAE. It was proved that BiLSTM-VAE captures the time correlation of sequences better than VAE. It can be observed from the figure that the latent variables of the time-aligned sequence are roughly in the same area, and the anomaly moment will show a large deviation (such as 8 o'clock). The literature [5] explains this effect for the first time, which is called time gradient.

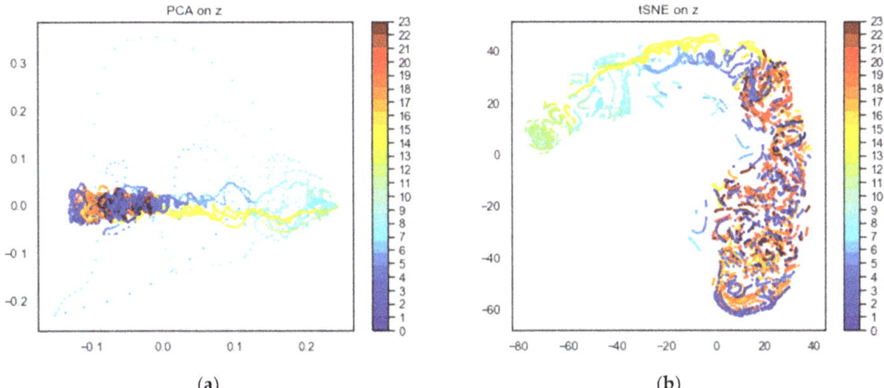

Figure 14. Two-dimensional visualization of the VAE latent variables for KPI 1: (**a**) PCA on z; and (**b**) t-SNE on z.

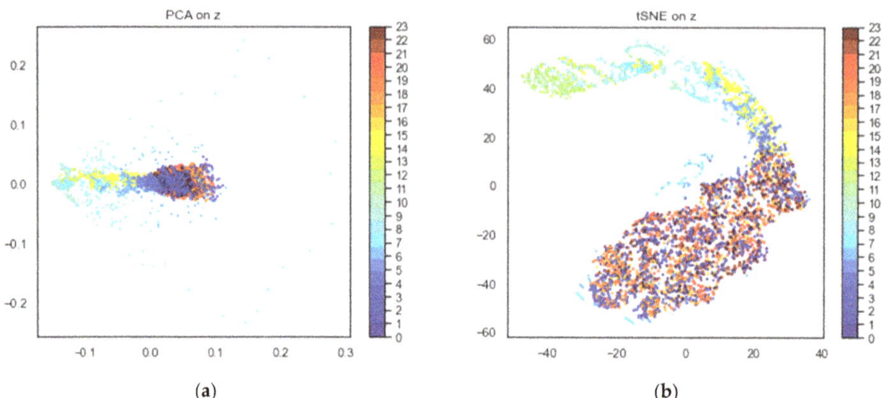

(**a**) (**b**)

Figure 15. Two-dimensional visualization of the BiLSTM-VAE latent variables for KPI 1: (**a**) PCA on z; and (**b**) t-SNE on z.

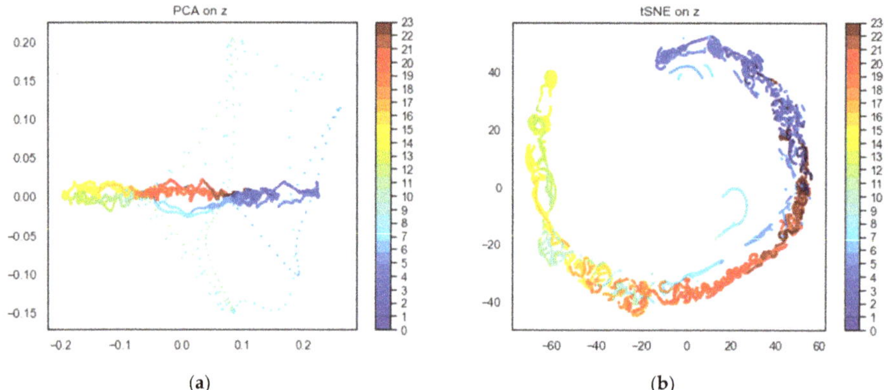

(**a**) (**b**)

Figure 16. Two-dimensional visualization of the VAE latent variables for KPI 2: (**a**) PCA on z; and (**b**) t-SNE on z.

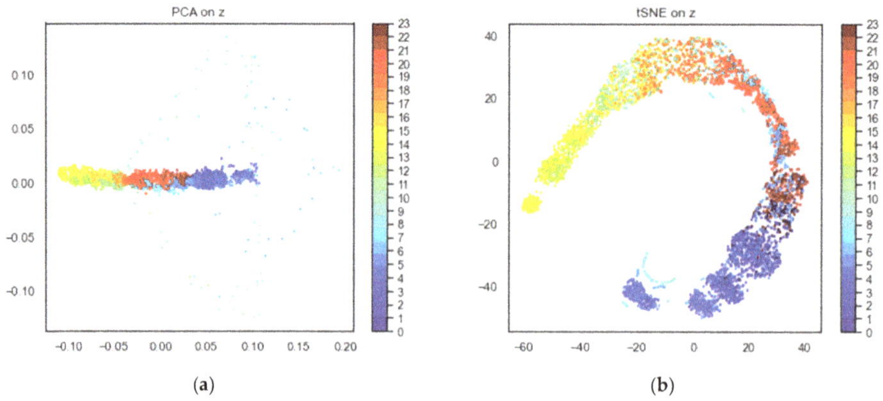

(**a**) (**b**)

Figure 17. Two-dimensional visualization of the BiLSTM-VAE latent variables for KPI 2: (**a**) PCA on z; and (**b**) t-SNE on z.

4.6.2. Batch Normalization

We used batch normalization to prevent the disappearance of the *KL* divergence during model training. In order to highlight the effect of batch normalization, we visualized the loss and accuracy of the model during training. As shown in Figure 18a, the accuracy of the model with BN (97%) is significantly higher than that of the model without BN (87%). From the perspective of training speed, the model with BN is already very close to the final convergence at the 22nd iteration. The model without BN is close to the final convergence at the 42nd time, indicating that the model with BN is faster. As shown in Figure 18b, the model with BN is lower than the model without BN, regardless of the loss of training set or test set. In conclusion, BN can accelerate the training speed of the model and even play a positive role in improving accuracy and reducing loss.

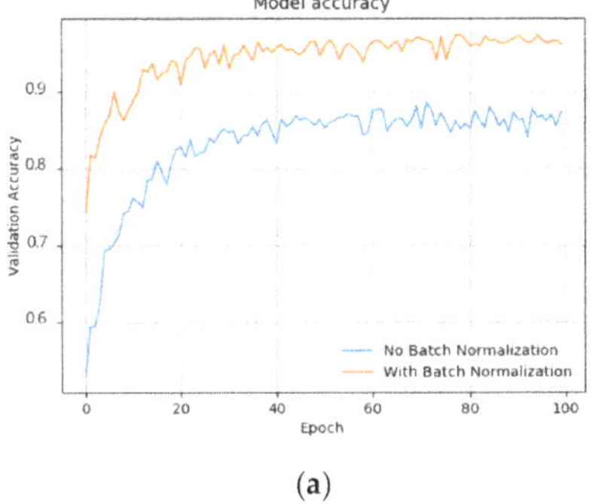

(a)

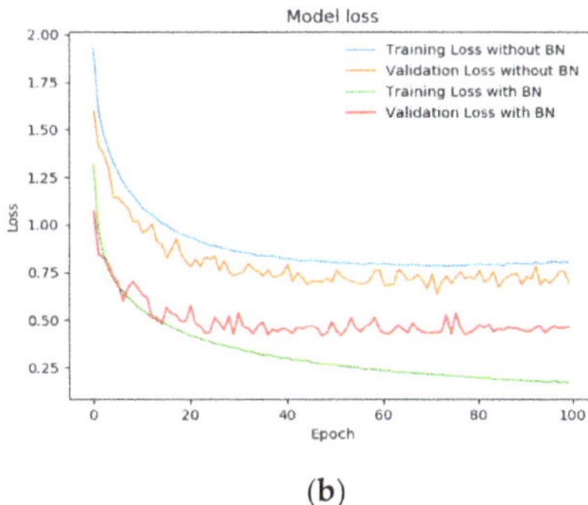

(b)

Figure 18. Comparative effects of batch normalization are introduced during training: (**a**) Model accuracy; and (**b**) Model loss.

4.6.3. EWMA Smoothing

We used EWMA to smooth the error sequence, which reduces false positives and false negatives during the detection process. To prove that the smoothing operation can improve the effectiveness of anomaly detection, we compared the detection effects of no smoothing and EWMA smoothing on two KPIs, as shown in Table 5. As can be seen from the table, *Precision*, *Recall*, and *F1-score* all improved greatly after EWMA smoothing.

Table 5. EWMA smoothing effect.

Dataset	Method	*Precision*	*Recall*	*F1-Score*
KPI 1	No smoothing	0.88	0.85	0.86
	EWMA smoothing	0.95	0.96	0.95
KPI 2	No smoothing	0.91	0.88	0.89
	EWMA smoothing	0.97	0.96	0.96

4.6.4. Adaptive Threshold

We used SVDD to determine the threshold of anomaly detection, so that different thresholds can be set adaptively for different KPIs. Figure 19 shows the PRC curve and ROC curve of the two threshold methods, proving that the adaptive threshold has a better effect than the fixed threshold. The larger the area under the curve is, the better the model effect is. Compared with the traditional fixed threshold, the adaptive threshold not only has a higher precision rate and recall rate, but also produces a higher true positive rate (TPR) under the same false positive rate (FPR).

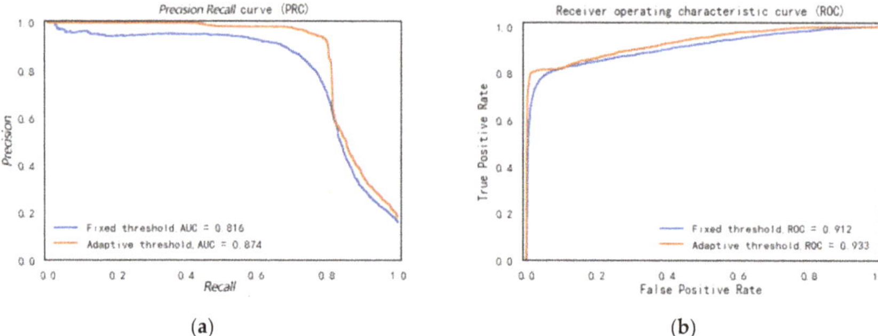

(**a**) (**b**)

Figure 19. Comparison of the adaptive threshold and fixed threshold: (**a**) PRC; and (**b**) ROC.

5. Conclusions

In this paper, a novel KPI anomaly detection method is proposed by combining VAE and SVDD. In this method, firstly, the encoder and decoder of VAE were designed as BiLSTM to capture the time dependence of data. Then, batch normalization was used on the output of the encoder to prevent the *KL* divergence from disappearing. In addition, EWMA was used to smooth reconstruction errors to eliminate accidental error peaks. Finally, smoothed reconstruction error sequences were put into the SVDD for training to determine the threshold of anomaly detection adaptively. In the experiment, the appropriate sliding window length and latent variable dimension were selected. The visualization effect of latent variables showed that time-aligned sequences are in the same region of latent variables space, and the model can better capture the time correlation of sequences. Batch normalization can speed up training and reduce loss. The reconstruction error after smoothing can reduce false positives and false negatives in the detection to some extent. Compared with the fixed threshold, the adaptive threshold has more flexibility and a better effect. The comparison result with current advanced baseline methods shows that the method in this paper has a better detection effect. Moreover, although the method in this paper is applied to KPI's univariate time series, it is also applicable to multivariate time series. The adaptive threshold can be applied not only to reconstruction errors, but also effectively to prediction errors.

In the future, we will continue our work focusing on the following two aspects:

(1) The linear interpolation method is too simple. When there are many missing values, some errors may be caused. Next, we will explore interpolation methods that can handle both linear and nonlinear data, such as modeling interpolation.

(2) The duration of anomaly detection is important. Next, we can improve the VAE-SVDD model structure and adjust parameters to obtain better performance.

Author Contributions: Y.Z., conceptualization, methodology, investigation, validation, formal analysis, writing—original draft, and writing—review and editing; X.Z., supervision, validation, formal analysis, and writing—review and editing; Z.C., supervision, validation, formal analysis, and writing—review and editing; Z.S., supervision, validation, formal analysis, and writing—review and editing. All authors have read and agreed to the published version of the manuscript.

Funding: This work was partly supported by the National Key R & D Program of China (Grant No. 2018YFB1601502) and the LiaoNing Revitalization Talents Program (Grant No. XLYC1902071).

Data Availability Statement: The data used to support the findings of this study are available from the corresponding author upon request.

Acknowledgments: The authors would like to thank all anonymous reviewers and editors for their helpful suggestions for the improvement of this paper.

Conflicts of Interest: The authors declare that they have no known competing financial interest or personal relationship that could appear to influence the work reported in this paper.

References

1. Pei, D.; Zhang, S.; Pei, C. Intelligent operation and maintenance based on machine learning. *Commun. CCF* **2017**, *13*, 68–72.
2. Kingma, D.P.; Welling, M. Auto-encoding variational bayes. In Proceedings of the 2nd International Conference on Learning Representations (ICLR), Banff, AB, Canada, 14–16 April 2014.
3. Goodfellow, I.J.; Pouget-Abadie, J.; Mirza, M.; Xu, B.; Warde-Farley, D.; Ozair, S.; Courville, A.; Bengio, Y. Generative Adversarial Nets. In Proceedings of the 28th Conference on Neural Information Processing Systems (NIPS), Montreal, QC, Canada, 8–13 December 2014; pp. 2672–2680.
4. Huang, Y.; Li, Y.; Liu, Y.; Jing, R.; Li, M. A Multiple Comprehensive Analysis of scATAC-seq Based on Auto-Encoder and Matrix Decomposition. *Symmetry* **2021**, *13*, 1467. [CrossRef]
5. Xu, H.; Feng, Y.; Chen, J.; Wang, Z.; Qiao, H.; Chen, W. Unsupervised Anomaly Detection via Variational Auto-Encoder for Seasonal KPIs in Web Applications. In Proceedings of the 27th World Wide Web (WWW) Conference, Lyon, France, 23–27 April 2018; pp. 187–196.
6. Chen, W.; Xu, H.; Li, Z.; Pei, D.; Chen, J.; Qiao, H. Unsupervised Anomaly Detection for Intricate KPIs via Adversarial Training of VAE. In Proceedings of the IEEE Conference on Computer Communications (IEEE INFOCOM), Paris, France, 29 April–2 May 2019; pp. 1891–1899.
7. Hochreiter, S.; Jürgen, S. Long Short-Term Memory. *Neural Comput.* **1997**, *9*, 1735–1780. [CrossRef] [PubMed]
8. Daehyung, P.; Hoshi, Y.; Kemp, C.C. A Multimodal Anomaly Detector for Robot-Assisted Feeding Using an LSTM-based Variational Autoencoder. *IEEE Robot. Autom. Lett.* **2017**, *3*, 1544–1551.
9. Niu, Z.; Yu, K.; Wu, X. LSTM-Based VAE-GAN for Time-Series Anomaly Detection. *Sensors* **2020**, *20*, 3738. [CrossRef]
10. Bowman, S.R.; Vilnis, L.; Vinyals, O.; Dai, A.M.; Bengio, S. Generating Sentences from a Continuous Space. In Proceedings of the 20th Conference on Computational Natural Language Learning (CoNLL), Berlin, Germany, 11–12 August 2016; pp. 10–21.
11. Kingma, D.P.; Salimans, T.; Jozefowicz, R.; Chen, X.; Sutskever, I.; Welling, M. Improving Variational Inference with Inverse Autoregressive Flow. In Proceedings of the 30th Conference on Neural Information Processing Systems (NIPS), Barcelona, Spain, 5–10 December 2016.
12. Xu, J.; Durrett, G. Spherical Latent Spaces for Stable Variational Autoencoders. In Proceedings of the Conference on Empirical Methods in Natural Language Processing, Brussels, Belgium, 31 October–4 November 2018; pp. 4503–4513.
13. Yang, Z.; Hu, Z.; Salakhutdinov, R.; Berg-Kirkpatrick, T. Improved Variational Autoencoders for Text Modeling using Dilated Convolutions. In Proceedings of the 34th International Conference on Machine Learning (ICML), Sydney, Australia, 6–11 August 2017; pp. 3881–3890.
14. An, J.; Cho, S. Variational autoencoder based anomaly detection using reconstruction probability. *Spec. Lect. IE* **2015**, *2*, 1–18.
15. Suh, S.; Chae, D.H.; Kang, H.G.; Choi, S. Echo-state conditional variational autoencoder for anomaly detection. In Proceedings of the International Joint Conference on Neural Networks (IJCNN), Vancouver, BC, Canada, 24–29 July 2016; pp. 1015–1022.
16. Zong, B.; Song, Q.; Min, M.R.; Cheng, W.; Lumezanu, C.; Cho, D.; Chen, H. Deep autoencoding Gaussian mixture model for unsupervised anomaly detection. In Proceedings of the 6th International Conference on Learning Representations (ICLR), Vancouver, BC, Canada, 30 April–3 May 2018.
17. Graves, A.; Schmidhuber, J. Framewise phoneme classification with bidirectional LSTM and other neural network architectures. *Neural Netw.* **2005**, *18*, 602–610. [CrossRef]
18. Zhu, Q.L.; Bi, W.; Liu, X.J.; Ma, X.Y.; Li, X.L.; Wu, D.P. A Batch Normalized Inference Network Keeps the KL Vanishing Away. In Proceedings of the 58th Annual Meeting of the Association-for-Computational-Linguistics (ACL), Seattle, WA, USA, 5–10 July 2020; pp. 2636–2649.
19. Hunter, J. The Exponentially Weighted Moving Average. *J. Qual. Technol.* **1986**, *18*, 19–25. [CrossRef]
20. Tax, D.M.; Duin, R.P. Support Vector Data Description. *Mach. Learn.* **2004**, *54*, 45–66. [CrossRef]
21. Pukelsheim, F. The three sigma rule. *Am. Stat.* **1994**, *48*, 88–91.
22. Yaacob, A.H.; Tan, I.K.; Chien, S.F.; Tan, H.K. ARIMA Based Network Anomaly Detection. In Proceedings of the Second International Conference on Communication Software and Networks (ICCSN), Singapore, 26–28 February 2010; pp. 205–209.
23. Yu, Q.; Jibin, L.; Jiang, L.R. An improved ARIMA-based traffic anomaly detection algorithm for wireless sensor networks. *Int. J. Distrib. Sens. Netw.* **2016**, *12*, 9653230. [CrossRef]
24. Kalekar, P. Time series forecasting using Holt-Winters exponential smoothing. *Kanwal Rekhi Sch. Inf. Technol.* **2004**, *4329008*, 1–13.
25. Laptev, N.; Amizadeh, S.; Flint, I. Generic and Scalable Framework for Automated Time-series Anomaly Detection. In Proceedings of the 21st ACM SIGKDD Conference on Knowledge Discovery and Data Mining (KDD), Sydney, Australia, 10–13 August 2015; pp. 1939–1947.
26. Liu, D.P.; Zhao, Y.J.; Xu, H.W.; Sun, Y.Q.; Pei, D.; Luo, J.; Jing, X.W.; Feng, M. Opprentice: Towards Practical and Automatic Anomaly Detection Through Machine Learning. In Proceedings of the ACM Internet Measurement Conference(IMC), Tokyo, Japan, 28–30 October 2015; pp. 211–224.
27. Erfani, S.M.; Rajasegarar, S.; Karunasekera, S.; Leckie, C. High-dimensional and large-scale anomaly detection using a linear one-class SVM with deep learning. *Pattern Recognit.* **2016**, *58*, 121–134. [CrossRef]
28. Wazid, M.; Das, A.K. An Efficient Hybrid Anomaly Detection Scheme Using K-Means Clustering for Wireless Sensor Networks. *Wirel. Pers. Commun.* **2016**, *90*, 1971–2000. [CrossRef]

29. Laxhammar, R.; Falkman, G.; Sviestins, E. Anomaly detection in sea traffic—A comparison of the Gaussian Mixture Model and the Kernel Density Estimator. In Proceedings of the 12th International Conference on Information Fusion(FUSION), Seattle, WA, USA, 6–9 July 2009; pp. 756–763.
30. Wang, X.H.; Du, Y.; Lin, S.J.; Cui, P.; Shen, Y.T.; Yang, Y.P. adVAE: A self-adversarial variational autoencoder with Gaussian anomaly prior knowledge for anomaly detection. *Knowl.-Based Syst.* **2020**, *190*, 105187. [CrossRef]
31. Luo, P.; Wang, B.H.; Li, T.Y.; Tian, J.W. ADS-B anomaly data detection model based on VAE-SVDD. *Comput. Secur.* **2021**, *104*, 102213. [CrossRef]
32. Qiu, J.; Du, Q.F.; Qian, C.S. KPI-TSAD: A Time-Series Anomaly Detector for KPI Monitoring in Cloud Applications. *Symmetry* **2019**, *11*, 1350. [CrossRef]
33. Chen, R.Q.; Shi, G.H.; Zhao, W.L.; Liang, C.H. A Joint Model for IT Operation Series Prediction and Anomaly Detection. *Neurocomputing* **2021**, *448*, 130–139. [CrossRef]
34. Wang, J.Y.; Jing, Y.H.; Qi, Q.; Feng, T.T.; Liao, J.X. ALSR: An adaptive label screening and relearning approach for interval-oriented anomaly detection. *Expert Syst. Appl.* **2019**, *136*, 94–104. [CrossRef]
35. Li, J.; Di, S.; Shen, Y.; Chen, L. FluxEV: A Fast and Effective Unsupervised Framework for Time-Series Anomaly Detection. In Proceedings of the Fourteenth International Conference on Web Search and Data Mining (WSDM), Virtual Event, Israel, 8–12 March 2021; pp. 824–832.
36. Siffer, A.; Fouque, P.A.; Termier, A.; Largouet, C. Anomaly Detection in Streams with Extreme Value Theory. In Proceedings of the 23rd International Conference on Knowledge Discovery and Data Mining (KDD), Halifax, NS, Canada, 13–17 August 2017; pp. 1067–1075.
37. Yang, D. Influences of different interpolation methods on GPS time series. *Gnss World China* **2019**, *44*, 66–69.
38. Hundman, K.; Constantinou, V.; Laporte, C.; Colwell, I.; Soderstrom, T. Detecting Spacecraft Anomalies Using LSTMs and Nonparametric Dynamic Thresholding. In Proceedings of the 24th ACM SIGKDD Conference on Knowledge Discovery and Data Mining (KDD), London, UK, 19–23 August 2018; pp. 387–395.
39. Robinson, J.; Rahmat-Samii, Y. Particle swarm optimization in electromagnetics. *IEEE Trans. Antennas Propag.* **2004**, *52*, 397–407. [CrossRef]
40. Bro, R.; Smilde, A.K. Principal Component Analysis. *J. Mark. Res.* **2014**, *6*, 2812–2831. [CrossRef]
41. Laurens, V.D.M.; Hinton, G. Visualizing Data using t-SNE. *J. Mach. Learn. Res.* **2008**, *9*, 2579–2605.

MDPI

Article

Functional Evaluation Using Fuzzy FMEA for a Non-Invasive Measurer for Methane and Carbone Dioxide

Lidilia Cruz-Rivero [1,*], María Leonor Méndez-Hernández [1], Carlos Eusebio Mar-Orozco [2], Alberto A. Aguilar-Lasserre [3], Alfonso Barbosa-Moreno [2] and Josué Sánchez-Escobar [1]

1 Instituto Tecnológico Superior de Tantoyuca, National Technological Institute of Mexico, Veracruz 92100, Mexico; draleonormendez@gmail.com (M.L.M.-H.); josue_se@outlook.es (J.S.-E.)
2 Instituto Tecnológico de Ciudad Madero, National Technological Institute of Mexico, Veracruz 92100, Mexico; carlos.mar.orozco@gmail.com (C.E.M.-O.); a.barbosa.moreno@gmail.com (A.B.-M.)
3 Instituto Tecnológico de Orizaba, National Technological Institute of Mexico, Veracruz 92100, Mexico; albertoaal@hotmail.com
* Correspondence: lilirivero@gmail.com; Tel.: +52-(789)-100-5673

Abstract: This paper combines the use of two tools: Failure Mode and Effect Analysis (FMEA) and Fuzzy Logic (FL), to evaluate the functionality of a quantifier prototype of Methane gas (CH_4) and Carbon Dioxide (CO_2), developed specifically to measure the emissions generated by cattle. Unlike previously reported models for the same purpose, this device reduces damage to the integrity of the animal and does not interfere with the activities of livestock in their development medium. FMEA and FL are used to validate the device's functionality, which involves identifying possible failure modes that represent a more significant impact on the operation and prevent the prototype from fulfilling the function for which it was created. As a result, this document presents the development of an intelligent fuzzy system type Mamdani, supported in the Fuzzy Inference System Toolbox of MatLabR2018b®, for generating a risk priority index. A Fuzzy FMEA model was obtained to validate the prototype for measuring Methane and Carbon Dioxide emissions, which allows considering this prototype as a reliable alternative for the reliable measurement of these gases. This study was necessary as a complementary part in the validation of the design of the prototype quantifier of CH_4 and CO_2 emissions. The methods used (classic FMEA and Fuzzy FMEA) to evaluate the RPN show asymmetric graphs due to data disparity. Values in the classical method are mostly lower than the Mamdani model results due to the description of the criteria with which it is evaluated.

Keywords: FMEA; Fuzzy Logic; greenhouse gases

Citation: Cruz-Rivero, L.; Méndez-Hernández, M.L.; Mar-Orozco, C.E.; Aguilar-Lasserre, A.A.; Barbosa-Moreno, A.; Sánchez-Escobar, J. Functional Evaluation Using Fuzzy FMEA for a Non-Invasive Measurer for Methane and Carbone Dioxide. *Symmetry* **2022**, *14*, 421. https://doi.org/10.3390/sym14020421

Academic Editor: Basil Papadopoulos

Received: 25 December 2021
Accepted: 16 February 2022
Published: 20 February 2022

1. Introduction

There are methodologies applied to security and reliability engineering, which include fuzzy FMEA, fuzzy Bayesian networks, fuzzy Markov chains, to name a few. When developing prototypes, it is not always possible to obtain fully reliable data due to the unavailability of primary observations and the consequent scarcity of data on the failure of their components [1]. To handle such situations, fuzzy set theory has been used successfully in approaches to security and reliability evaluation under conditions of uncertainty.

The use of Fuzzy Logic to help decision-making was applied in health in conjunction with the Failure Mode and Effects Analysis (FMEA) to improve decision-making. This was achieved by improving the way patients were processed before the study, the assessment, and treatment assignment; the use of Fuzzy Logic in the application of the FMEA was considered better than its conventional use [2]. The use of software to manage the risks of failure in medical equipment was developed around the use of the FMEA; it is described that the classic use of the FMEA is not enough to achieve an accurate analysis, so, in this case, additionally the use of the Fuzzy Logic was made to replace how the three main parameters of the FMEA are evaluated [3]. FMEA has been applied extensively in the reliability engineering domain. Risk Priority Number (RPN), which is the product of Occurrence (O), Severity (S), and Detection (D) of a failure, is the most important measure used in FMEA for prioritizing risk, and they integrate fuzzy belief structure and grey relational projection method (GRPM) to avoid the use of traditional RPN [4].

According to work "An extension to fuzzy developed failure mode and effect analysis FDFMEA application for aircraft landing system," the use of FMEA methodology and Fuzzy Logic was implemented for the evaluation of commercial aircraft in Iran, as described in this work; the main focus of the study was the analysis of the components of the landing systems since a large number of failures were recorded annually, in addition to the results of the Fuzzy Logic model also evaluated by the conventional method to offer a comparison of the results, the defuzzification method applied to the Fuzzy Logic model was Center of Gravity and for the construction of the model five participating experts to describe according to their knowledge the elements of evaluation of Severity, Occurrence, and Detection, and with the results, it was possible to establish measures and prioritize failures [5].

In the analysis of a lathe-type polishing machine, the fuzzy approach with the FMEA was also used, in this case, to improve reliability, because one of the most important parts of the entire polishing system presented a series of failures, and it was considered to individually evaluate the components of this part of the machine. As in other works, a model of Fuzzy Logic was used, and some of the important data that are mentioned in work around the use of the FMEA are that conventional use considers the evaluation of RPN almost equally in the evaluation criteria of Severity, Occurrence, and Detection; since these are pre-established, the assessment of the RPN is inadequate, instead they consider that each criterion should be based around the context of this problem with different weights and that these should also be established at the discretion of the experts in the field, so the use of Fuzzy Logic was chosen to improve this part of the methodology [6].

Greenhouse Gas (GHG) emissions are an important factor contributing to current climate change; it is said that of the sources that contribute to the increase in emissions due to livestock activities [7], the breeding of cattle is pointed out as one of the activities of this sector that contributes the most in emissions, although in the absence of reliable monitoring methods there is imprecision in terms of the data reported worldwide, and due to this different works have been done to quantify the emissions of livestock, through the creation of measurement devices to determine more precisely to what extent livestock influences emissions; however, according to this work reporting on this type of device, these have not been suitable for: the comfort and integrity of the animal and neither to measure in the natural conditions where cattle usually develop.

A device capable of belching CO_2 and CH_4 gas emissions from cattle was developed since, at the time in Mexico, there were no reports about the development of this type of device, and due to the cattle ranching activity present in several of the states, one of them was the state of Veracruz. The only countries that documented the development of this type of device were the United States, Colombia, and Argentina [8].

The prototype was statistically analyzed to verify that the proposed design is functional [9]. This device presents an alternative to measure GHG without harming the animal; however, it is necessary to carry out functionality studies, so an FMEA is used to evaluate the possibilities of failures, to explain the possible causes and the effects they can represent.

The application of the FMEA in conjunction with the Artificial Intelligence technique called Fuzzy Logic (FL), is proposed to ensure the correct functioning of the prototype through a functional evaluation, which implies identifying possible failure modes that represent a more significant impact on the operation and prevent the prototype from fulfilling the function for which it was created. The FL, in this case, is used to address the imprecision of the evaluation criteria of the classic FMEA and adapt them to the specific problem and facilitate the evaluation process commonly used in this tool. In this case, the prototype quantifier of gas emissions was put into operation in a group of cows previously selected to carry out a series of measurement tests; during the realization of these tests, different failures were observed in the equipment, and possible failure modes were considered, creating uncertainty about the correct functioning of the other elements that make up the device and the possibility that these may fail at a certain time and affect the measurements.

Due to the behaviors that cattle perform, especially in natural development conditions, several parts of the device are exposed to situations that represent a certain degree of risk for the proper functioning of the device, and the possibility of risk to failure in some elements was unknown before the performance of the tests of the quantifier. During the trials carried out, they resulted in damage: sensors, connectors, and the support structure that holds the device to the cow; these failures caused mostly inaccuracy in the measurements as verified after repairing some of these failures.

This work aims to know and assess the risk of possible failure modes, causes, and effects of a device that quantifies carbon dioxide (CO_2) and methane (CH_4) emissions from cattle; through the use of the Failure Mode and Effect Analysis (FMEA) and Fuzzy Logic (LD) methodology.

Currently, the application of the FMEA methodology has been extended in different areas and adapted to specific needs. However, it is generally recognized that there are four types of FMEA: system, design, process, and service [10]. Failure mode and effects analysis (FMEA) is a typical prevention reliability analysis method that has an inherent advantage in improving systems and processes. However, traditional FMEA also contains some deficiencies in rating risks, weighing risk factors, and ranking failure modes [11].

To overcome the inherent drawbacks of the conventional FMEA method and the uncertainty regarding the experts' evaluation, fuzzy methods, for example, triangular/trapezoidal and fuzzy sets, have been developed [12].

As mentioned by [2,13], combining the FMEA method with fuzzy theory provides a more efficient tool than the classic FMEA method in the presence of imprecise information and uncertainty. Fuzzy Logic could be used to reduce the drawback in assessing and prioritizing failures of traditional FMEA with more certainty.

Fuzzy Logic is a technique that allows us to represent and manipulate some variables' imprecise and uncertain nature. However, its use is not limited only to these variables but also to those that are known and precise to a certain degree; the technique allows us to make these types of variables coexist mainly in cases of decision making and the development of intelligent systems, cases where imprecision and uncertainty modeling is necessary and also the inclusion of known variables for a joint analysis.

In a work related to fuzzy FMEA assessment of hydroelectric earth dam failure modes, the rule bases for the two stages and the membership functions were obtained through in-depth interviews with a focus group composed of experts. The fuzzification process assured more consistency to the RPN calculation, treated the imprecision, and provided fair management value to prioritize actions and improve monitoring processes [14], which is the goal of this proposed work.

The use of analysis methods such as classic FMEA to assess the safety and reliability of a system relies heavily on knowledge of component failure. Any uncertainty that arises in the probability of component failure will impact the results [1]. On the other hand, the lack of availability of failure data would introduce uncertainty in the analysis results in this sense. To reduce this uncertainty, it is proposed to use fuzzy FMEA.

2. Materials and Methods

2.1. Prototype Description

The prototype to be evaluated fulfills the primary function of estimating the emissions of gases that cattle generate and exhale, specifically those of CO_2 and CH_4 gas, the main components in this device are: an emitter and a receiver; in turn, the subcomponents are: a battery, a support structure that adjusts to the characteristics of the bovine tube, two sensors that quantify gas emissions, straps that attach the support structure and other components to the head of the animal and the receiver which is connected to the computer to record the information of the measurements. The following image shows the device as well as the installation in cattle (Figure 1).

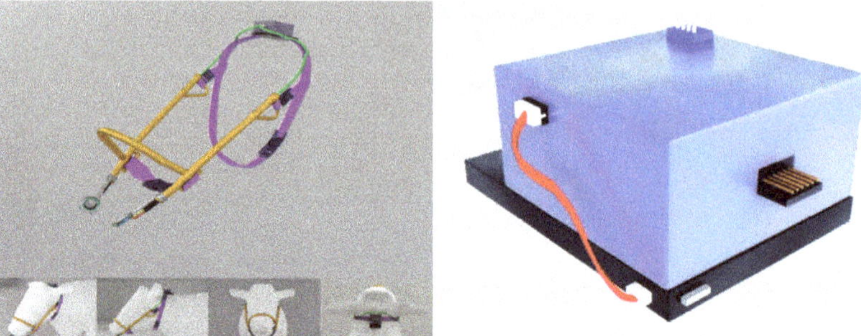

Figure 1. Gas emission quantifier for cattle. Source: Retrieved from Autodesk 3ds Max, self-made content.

Components associated with major failures and representing a negative impact on the operation of the quantifier were strategically selected (Table 1).

Table 1. Elements selected for the development of the fault study Receiver Sender.

Receiver	Sender
Emitter-battery connection	Receiver-computer connection
Emitter-sensor connection	Receiver DIP
Issuer's DIP	On-screen light trimmer potentiometer
Battery	XBee Coordinator Module
Straps	
Support structure	
Sensors	
Sensor holder	
XBee Router Module	

The proposed framework for dealing with fuzzy FMEA is illustrated in Figure 2. The framework comprises three main phases, including (1) prototype components under study; (2) classic FMEA development; (3) determining the NPR through a Mamdani-type Fuzzy Logic model; (4) fuzzification process; and (5) defuzzification.

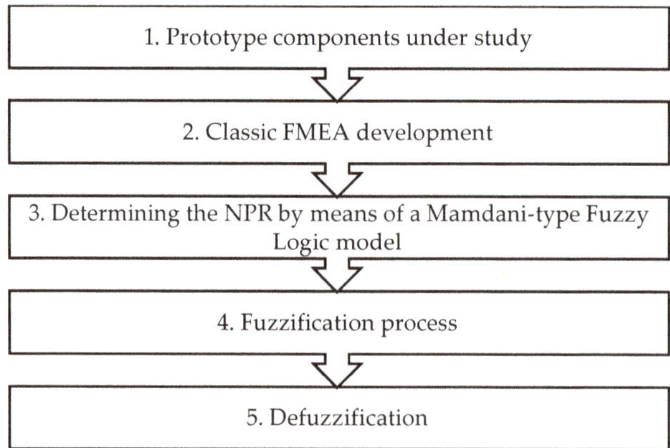

Figure 2. The methodology flowchart.

2.2. Classic FMEA Development

For the development of the FMEA, those components associated with important faults and that represent a great negative impact on the operation of the quantifier was selected. For the selection of the necessary elements to carry out the prototype failure risk study, first, a group of specialists participating in the design of the device was the ones who, through brainstorming, managed to generate a list of both potential components and subcomponents for the study, after an analysis based on the relationship of components and their importance in the operation, it was possible to prioritize the said list, the resulting elements that will be used for the development of the study are shown below (Table 2).

Table 2. Classic FMEA.

Main Component	Subcomponent	No	Failure Mode	Causes of Failure	Failure Effects
Emitter	Battery-emitter line	1	Unplugging connectors	Line exposed to the outside of the emitter stuck with objects in the environment.	Emitter and sensors out of operation due to lack of electrical power supply, interruption of measurements.
		2	False in connectors	Non-fixed plug type connectors, contact of the line with the animal's body or objects in the environment.	Electric current and intermittent emitter operation, uncalibrated sensor, discontinuous and incorrect data sending.
		3	Short circuit	Connectors discovered outdoors, the humidity of the environment, or animal fluids.	Partial or total damage to internal components of the emitter and/or battery.
Emitter	Emitter-sensor line	4	Unplugging connectors	Line exposed to the outside of the emitter, inadequate line length, line stuck with objects in the environment.	Sensors out of operation due to lack of electrical power supply.
		5	False in connectors	Non-fixed plug-type connectors, contact of the line with the animal's body or objects in the environment.	Electric current and intermittent sensor operation, sensor uncalibrated, incorrect and discontinuous measurements.
		6	Short circuit	Connectors discovered outdoors, ambient humidity, or animal fluids.	Partial or total damage to the sensors and/or internal controllers of the emitter.
Emitter	Emitter DIP	7	Inactive electric current function	Switch exposed to the elements, deactivated by contact with the animal's body or objects in the environment.	The passage of electric current from the emitter to each of the sensors is not allowed; the sensors will not make measurements.
		8	Inactive ground current function	Switch exposed to the elements, deactivated by contact with the animal's body or objects in the environment.	Emitter components and sensors are vulnerable to damage from electrical surges.
		9	Inactive TX radio frequency function	Switch exposed to the elements, deactivated by contact with the animal's body or objects in the environment.	The sensors perform measurements, but the XBee Coordinator module does not transmit the data to the receiver for data recording.
Emitter		10	Inactive TX radio frequency function	Switch exposed to the elements, deactivated by contact with the animal's body or objects in the environment.	The signal of the measurements is transmitted from the XBee Coordinator module, but communication with the XBee Router module of the receiver is not achieved.
		11	Fractured switches	Exposed switch location, sudden movements of the animal, or improper handling of the operator.	Difficulty manipulating the functions of the switches in the emitter.
		12	Short circuit	Connectors discovered outdoors, ambient humidity, or animal fluids.	Instability of functions, damage of emitter components.

Table 2. *Cont.*

Main Component	Subcomponent	No	Failure Mode	Causes of Failure	Failure Effects
Emitter	Battery	13	Thermal leakage	The inadequate protective case against high or low temperatures.	Poor battery performance, interruption of measurements, incomplete measurement periods.
		14	Short circuit	Lack of tightness, broken protective sheath, moisture filtration from the environment, or animal fluids.	Damage to emitter box components and/or sensors.
Emitter	Straps	15	Loose safety clasps	Animal struggle, poor operator fit, or poorly resistant plastic material.	Instability of the structure at the animal head or loose structure of the animal head.
		16	Breaking of security bands	Friction wear, moisture deterioration, sudden movement of the animal, or low resistance of the material.	Instability of the structure at the head of the animal or loose structure of the head of the animal.
Emitter	Support structure	17	Folded metal arms	Flexible material, animal struggle, or operator mismanagement.	Improper position of the sensor to the animal's tube, low sensor measurement range, unreliable gas estimates with a high level of variation.
		18	Desoldered arm joints	Poor welding work, poorly resistant welding, or overexertion of the structure.	Instability of the structure at the head of the animal or loose structure of the head of the animal.
Emitter	Sensors	19	Low measurement sensitivity	Obstruction of the sensor by food debris or mucus from the animal, poor posture, and strong wind currents.	Unreliable gas estimates with a high level of variation.
		20	Short circuit	The printed circuit of the discovered sensor, ambient humidity, animal mucus, or food debris.	Instability of electric current in sensors, discontinuous measurement lapses, the partial or total damage to the sensor.
		21	Led measurement indicator without operating.	Led desoldering of the printed circuit, damaged by moisture or melted by shocks.	Difficulty identifying sensor malfunction in real-time.
Emitter	Sensor holder	22	Bent structure	Slightly rigid structural material, sudden movements of the animal's tube, and/or the wrong fit in the animal's tube.	Improper position of the sensor to the animal's tube, low sensor measurement range, unreliable gas estimates with a high level of variation.
		23	Loose Assembly Snaps	Sudden movements of the animal's trunk and obstruction with objects in the environment.	Instability of sensors to the animal's trunk.
Receiver	XBee Modules (Router and Coordinator)	24	Communication signal loss	Unfavorable topographical conditions, interference from other signals, or low range capability of the modules.	Gas measurement data loss during signal interruption between XBee modules.
Receiver	Receiver DIP	25	Inactive electric current function	Operator mishandling or fractured switch.	The passage of electric current from the computer is not allowed, difficulty manipulating the functions of the switches.
		26	Inactive grounding function	Operator mishandling or fractured switch.	In case of short circuit damage of the receiver components, difficulty manipulating the functions of the switches.
		27	Inactive TX radio frequency function	Operator mishandling or fractured switch.	The receiver does not send back the communication signal to the transmitter, having difficulty manipulating the switches' functions.
		28	Inactive TX radio frequency function	Operator mishandling or fractured switch.	The receiver does not receive the signal from the transmitter, difficulty manipulating the functions of the switches on the receiver.

Table 2. *Cont.*

Main Component	Subcomponent	No	Failure Mode	Causes of Failure	Failure Effects
Receiver	Trimmer LCD Screen Potentiometer	29	Incorrect light regulation	The sweeping regulator, presence of moisture or internal dirt.	Difficulty manipulating the lighting intensity on the receiver screen.
Receiver	Receiver-computer connection	30	Incorrect light regulation	Error in software drivers, operator mishandling.	Interruption in the transfer of information, loss of data in the logger software.

2.3. *NPR through a Mamdani-Type Fuzzy Logic Model3*

For the elaboration of the fuzzy FMEA model, the MatLabR2018b® LD toolbox was used. This toolbox made it possible to define a fuzzy system through dialogs and windows that facilitated data entry. A set of functions was available to analyze the behavior of these systems.

In the adaptation of the criteria and the RPN, it is also necessary to provide the linguistic or categorical values that serve as a reference to use as values that allow the Mamdani type model to assess rules; it is based on these values that the ranges of values will be considered representative by the specialists, which in fuzzification is where the categorical criteria will be modeled with these scales, geometric figures, and the criteria proposed for evaluation by specialists are those shown in the following table (Table 3).

Table 3. Adequacy of parameters for the assessment of the Risk Priority Number.

Linguistic Value	Severity	Linguistic Value	Occurrence	Linguistic Value	Detection
Very Low	It does not represent significant affectation; the operation of the prototype will be almost normal.	Remote	Just once for each measurement period.	Very High	In most tests are identified, detailed inspections are not necessary.
Low	It represents minimal affectation; the prototype will work with slight affectations.	Low	Approximately up to three times for each measurement period.	High	They are almost always detected in tests; no detailed inspections are necessary.
Medium	The affectations are significant; the device will work, the information generated will be unreliable.	Regular	Approximately up to five times per measurement period.	Medium	Sometimes they are detected during testing; basic inspection is necessary.
High	The affectations are severe; sometimes, it will not work, the information generated will not be reliable.	High	Approximately up to seven times for each measurement period.	Low	Hardly detected during testing, a detailed inspection is necessary.
Very High	The affectations will not allow the operation of the prototype.	Very High	More than seven for each measurement period.	Very Low	They are seldom detected during testing; inspection with special methods is necessary.

The Fuzzy Logic model is proposed to evaluate the risk of possible failures in the prototype; the construction is done around the variables used by the FMEA. The process for obtaining the risk or RPN is carried out through inference rules based on human knowledge.

Model Architecture

For this case study, the general architecture of the Pure Type Fuzzy System or Mamdani was used [15]. Mamdani fuzzy models do not require mathematical models of the system to control and are obtained from fuzzy rules or fuzzy conditional statements [16], like the one presented in this work. The structure of the fuzzy model poses input and output variables, with the inputs for this model being the evaluation criteria Severity, Occurrence, and Detection and output being the RPN indicator.

The development of the Mamdani model consists of three important processes [15]: fuzzification, the introduction of inference rules, and the defuzzification of the outputs. The following diagram represents the structure of this type of model applied to the case (Figure 3).

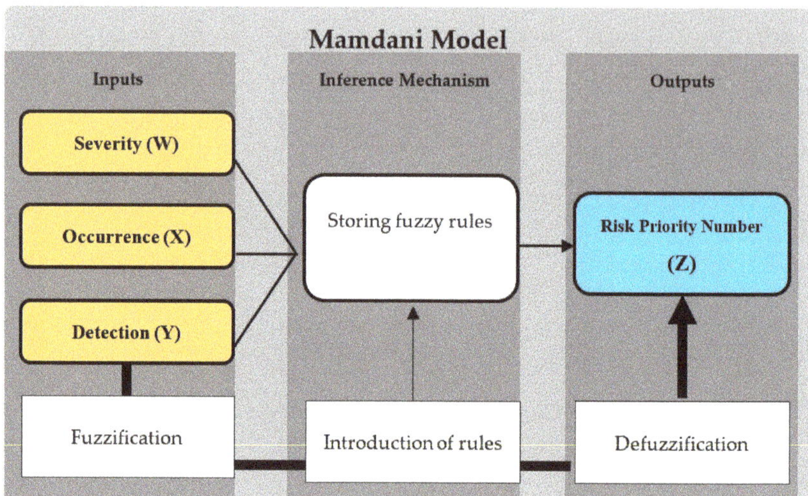

Figure 3. General diagram of the Mamdani Type Fuzzy Logic model for evaluating the quantifying prototype Source: Own elaboration.

2.4. Fuzzification Process

The fuzzification process is carried out for the input and output variables of the model, these being those of the FMEA: Severity, Occurrence, Detection and the RPN, for each variable, the geometric figures that best fit the membership function to the type of variable that is modeled are assigned, the approximate ranges of the variables are also established and are associated with linguistic values, being those that will represent the fuzzy sets of the Mamdani model.

The input variables for this model are Severity (W) (Table 4, Figure 4), Occurrence (X), and Detection (Y). Based on these and from the resulting fuzzy sets, the inference rules that determine the corresponding output scenarios will be established; these variables in the model represent the evaluation parameters with which it is sought to analyze the risk or RPN (Z) of each of the rulings raised in the FMEA.

Table 4. Characteristics of the fuzzy sets of the input variable Severity (W).

Fuzzy Sets	Fuzzy Range	Geometric Figures Values
Very Low	0–2	[0, 0, 1, 2]
Low	1–4	[1, 2, 3, 4]
Medium	3–6	[3, 4, 5, 6]
High	5–8	[5, 6, 7, 8]
Very High	7–10	[7, 8, 10, 10]

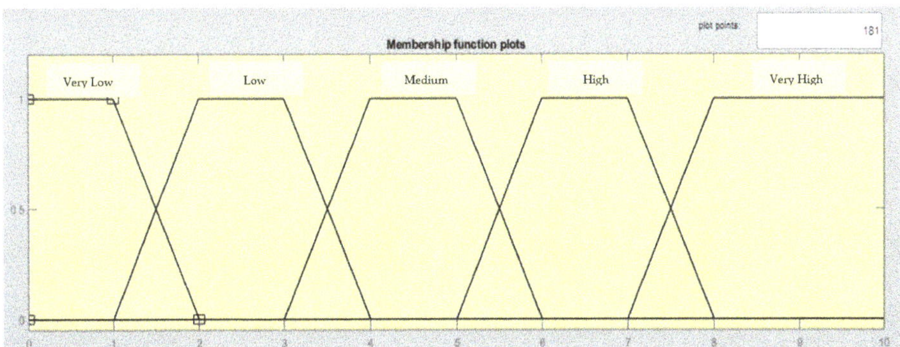

Figure 4. Fuzzy Sets Severity Variable (Retrieved from MATLAB R2018b, Screenshot of Membership Function Editor window, self-made content).

Equation (1). Severity Fuzzy Set "Very Low" (W).

$$\mu_{Very\ Low}(Severity) = \begin{cases} 1; & W \le 1 \\ 1 - \frac{W-1}{2-1}; & 1 < W < 2 \\ 0; & 2 \le W \end{cases} \quad (1)$$

Equation (2). Severity Fuzzy Set "Low" (W).

$$\mu_{Low}(Severity) = \begin{cases} 0; & W \le 1 \\ 1 - \frac{2-W}{2-1}; & 1 < W < 2 \\ 1; & 2 \le W \le 3 \\ 1 - \frac{W-3}{4-3}; & 3 < W < 4 \\ 0; & 4 \le W \end{cases} \quad (2)$$

Equation (3). Severity Fuzzy Set "Medium" (W).

$$\mu_{Medium}(Severity) = \begin{cases} 0; & W \le 3 \\ 1 - \frac{4-W}{4-3}; & 3 < W < 4 \\ 1; & 4 \le W \le 5 \\ 1 - \frac{W-5}{6-5}; & 5 < W < 6 \\ 0; & 6 \le W \end{cases} \quad (3)$$

Equation (4). Severity Fuzzy Set "High" (W).

$$\mu_{High}(Severity) = \begin{cases} 0; & W \le 5 \\ 1 - \frac{6-W}{6-5}; & 5 < W < 6 \\ 1; & 6 \le W \le 7 \\ 1 - \frac{W-7}{8-7}; & 7 < W < 8 \\ 0; & 8 \le W \end{cases} \quad (4)$$

Equation (5). Severity Fuzzy Set "Very High" (W).

$$\mu_{Very\ High}(Severidad) = \begin{cases} 0; & W \le 7 \\ 1 - \frac{8-W}{8-7}; & 7 < W < 8 \\ 1; & 8 \le W \end{cases} \quad (5)$$

In this model, the Occurrence was raised in considering the measurement periods, that is, to establish a reference metric on how often failures may occur in the prototype. The time in which it was used for the gas measurement studies in the cattle was taken into account, which in this case was per week: 5 days of measurement and two days of rest, and for each day of measurement, two tests were taken so that in a measurement period the prototype was used on ten occasions, in this context it is that the scale was established regarding the failures that occurred and from this, it was possible to relate the frequency to categorical values that qualify the occurrence. As a result, the following fuzzy sets are proposed with the representative ranges of each geometric figure (Table 5).

Table 5. Characteristics of the fuzzy sets of the input variable Occurrence (X).

Fuzzy Sets	Fuzzy Range	Geometric Figures Values
Remote	0–2	[0, 0, 1, 2]
Low	1–4	[1, 2, 3, 4]
Regular	3–6	[3, 4, 5, 6]
High	5–8	[5, 6, 7, 8]
Very High	7–10	[7, 8, 10, 10]

The detection variable represents the ease with which the prototype operator can identify the failure modes during the tests and/or after the tests, the scale established as in the conventional way reaches up to 10, for this case the sets associated with such a scale are the following (Table 6).

Table 6. Characteristics of the fuzzy sets of the input variable Detection (Y).

Fuzzy Sets	Fuzzy Range	Geometric Figures Values
Very High	0–2	[0, 0, 1, 2]
High	1–4	[1, 2, 3, 4]
Medium	3–6	[3, 4, 5, 6]
Low	5–8	[5, 6, 7, 8]
Remote	7–10	[7, 8, 10, 10]

3. Results

3.1. Risk Priority Number

For the RPN output variable, it was considered appropriate to use seven fuzzy sets associated with linguistic terms and denoting a category that allows evaluating the intensity of the variable (Table 7), the scale that is used for the sets, as well as that used in the classic FMEA, is up to 1000, to make a comparison or make symmetry analysis between the results of the Mamdani model and those of the conventional method (Figure 5).

Table 7. Characteristics of the fuzzy sets of the output variable Risk Priority Number (Z).

Fuzzy Sets	Fuzzy Range	Geometric Figures Values
Very Low	0–250	[0, 125, 250]
Low	125–375	[125, 250, 375]
Significant	250–500	[250, 375, 500]
Middle	375–625	[375, 500, 625]
Predominant	500–750	[500, 625, 750]
High	625–875	[625, 750, 875]
Very high	750–1000	[750, 875, 1000]

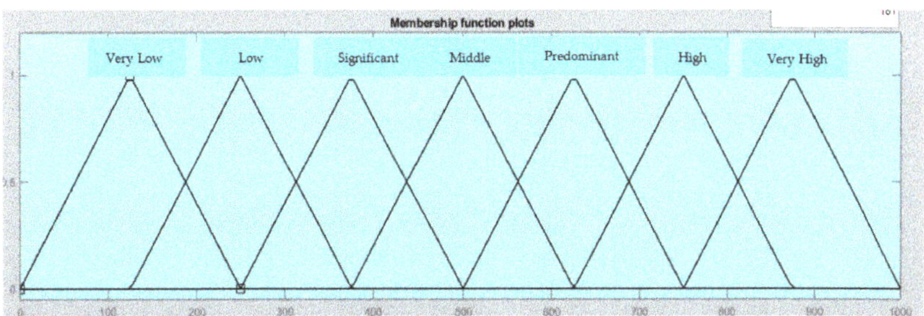

Figure 5. Fuzzy sets of the Risk Priority Number variable; Source: Retrieved from MATLAB R2018b, Screenshot of Membership Function Editor window, self-made content.

Equation (6). RPN Fuzzy Set "Very Low" (Z).

$$\mu_{Very\ Low}(RPN) = \begin{cases} 0; & Z \leq 0 \\ 1 - \frac{125-Z}{125-0}; & 0 < Z \leq 125 \\ 1 - \frac{Z-125}{250-125}; & 125 < Z < 250 \\ 0; & 250 \leq Z \end{cases} \quad (6)$$

Equation (7). RPN Fuzzy Set "Low" (Z).

$$\mu_{Low}(RPN) = \begin{cases} 0; & Z \leq 125 \\ 1 - \frac{250-Z}{250-125}; & 125 < Z \leq 250 \\ 1 - \frac{Z-250}{375-250}; & 250 < Z < 375 \\ 0; & 375 \leq Z \end{cases} \quad (7)$$

Equation (8). RPN Fuzzy Set "Significant" (Z).

$$\mu_{Significant}(RPN) = \begin{cases} 0; & Z \leq 250 \\ 1 - \frac{375-Z}{375-250}; & 250 < Z \leq 375 \\ 1 - \frac{Z-375}{250-375}; & 375 < Z < 500 \\ 0; & 500 \leq Z \end{cases} \quad (8)$$

Equation (9). RPN Fuzzy Set "Middle" (Z).

$$\mu_{Middle}(NPR) = \begin{cases} 0; & Z \leq 375 \\ 1 - \frac{375-Z}{375-250}; & 375 < Z \leq 500 \\ 1 - \frac{Z-375}{250-375}; & 500 < Z < 625 \\ 0; & 625 \leq Z \end{cases} \quad (9)$$

Equation (10). RPN Fuzzy Set "Predominant" (Z).

$$\mu_{Predominant}(NPR) = \begin{cases} 0; & Z \leq 500 \\ 1 - \frac{625-Z}{625-500}; & 500 < Z \leq 625 \\ 1 - \frac{Z-625}{750-625}; & 625 < Z < 750 \\ 0; & 750 \leq Z \end{cases} \quad (10)$$

Equation (11). RPN Fuzzy Set "High" (Z).

$$\mu_{Alto}(RPN) = \begin{cases} 0; & Z \leq 625 \\ 1 - \frac{750-Z}{750-625}; & 625 < Z \leq 750 \\ 1 - \frac{Z-750}{875-750}; & 750 < Z < 875 \\ 0; & 875 \leq Z \end{cases} \quad (11)$$

Equation (12). RPN Fuzzy Set "Very High" (Z).

$$\mu_{Very\ High}(NPR) = \begin{cases} 0; & Z \leq 750 \\ 1 - \frac{875-Z}{875-750}; & 750 < Z \leq 875 \\ 1 - \frac{Z-875}{1000-875}; & 875 < Z < 1000 \\ 0; & 1000 \leq Z \end{cases} \tag{12}$$

The inference rules used for this Mamdani type model were considered according to the number of input variables and their sets, all possible combinations that represent the scenarios that can be presented given the conditions that are established were considered; as a result, 125 inference rules were formulated, the outputs were defined by specialists, to assign in each case the corresponding categorical rating of the RPN. These rules were introduced into the MATLAB model, as shown in the following figure (Figure 6).

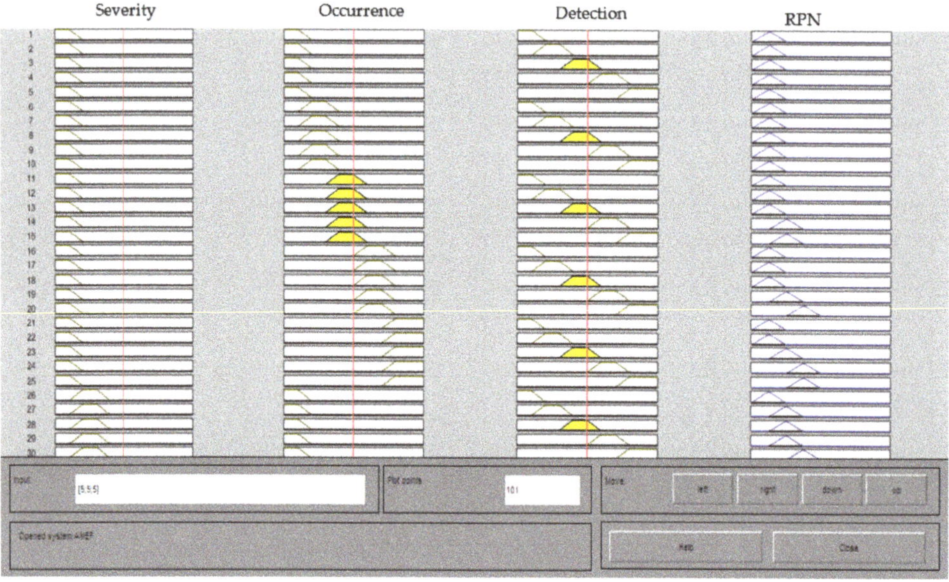

Figure 6. Mamdani Type Fuzzy Logic Model Inference Rules for Evaluating the Quantifying Prototype Source: Retrieved from MATLAB R2018b, Rule Viewer window screenshot, self-made content.

3.2. Defuzzification Process

The values that correspond to the sets of the RPN output variable are obtained numerically, according to the case of the rule activated in the Mamdani model when evaluating the different failure modes of the FMEA. The potential failure modes that were identified in the prototype were 30. Each failure mode was assessed using the defuzzification method above, and the values obtained from RPN are as shown below (Table 8).

Table 8. Defuzzification of the RPN variable from the Mamdani model was applied to the quantifying prototype.

Failure Mode	RPN Fuzzy Model
1	625
2	250
3	625
4	625
5	375
6	500
7	500
8	375
9	375
10	250
11	125
12	375
13	125
14	625
15	125
16	375
17	375
18	125
19	750
20	500
21	125
22	500
23	125
24	125
25	500
26	375
27	375
28	250
29	125
30	250

As shown in the table above, the RPN values obtained from the Mamdani model are repeated between some failure modes. This is because the values provided by the "Center of Gravity" defuzzification method have not been manipulated, but left by default in the Mamdani model, and in this way only one rule is activated per failure mode as the case may be, and although the sets that make up the rules are several of these coincide since in the output variable you only have seven sets, but if instead of leaving the default values that allow only one rule to be activated when the input sets are provided and the input values of each variable will be manipulated, more rules could be activated at the same time so that the value in the output may be different from the seven that are repeated. This would occur when the provided values enter the regions of intersection between sets, and what this causes is that other rules are activated and the result is different depending on these rules and the method of defuzzification used.

For this case, the default values have been left since the evaluation was intended to be carried out around using the linguistic terms of input so that only the necessary rule will be activated.

3.3. Response Surfaces

In this type of application, the use of the response surfaces that result from the Mamdani model of MATLAB are not a direct resource to obtain the type of result that is sought, which is to obtain the level of risk in a particular way in the sets of the output variable. However, these surfaces allow us to visualize and understand the logic that exists in the graphical representation of the RPN. That means the risk assessment is possible once the fuzzy rules have been established as a precedent in the inference base of the system.

Therefore, a detailed analysis of the surfaces is not necessary in this model as in other application cases.

4. Discussion

With the defuzzification method "Center of Gravity", it was possible to obtain the RPN results for each mode of failure of the FMEA, as shown in the following figure (Figure 7). The numbering of the "failure modes" corresponds to that presented in the format of Table 7.

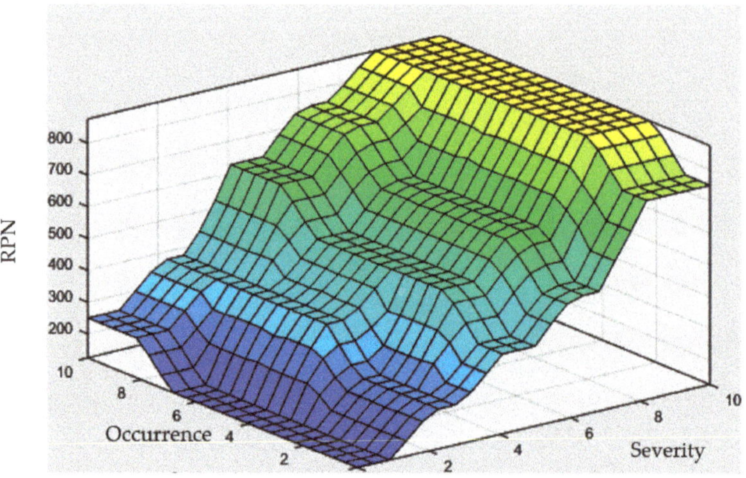

Figure 7. Response Surface Occurrence vs. Severity of the Risk Assessment Model to Failure Source: MATLAB R2018b software capture, self-developed content.

The response surfaces generated from the model are shown in Figures 8 and 9. In them, you can see how the level of risk increases or decreases depending on the input criteria Severity, Occurrence, and Detection, when these tend to take a certain value on the lower axes of the three-dimensional plane.

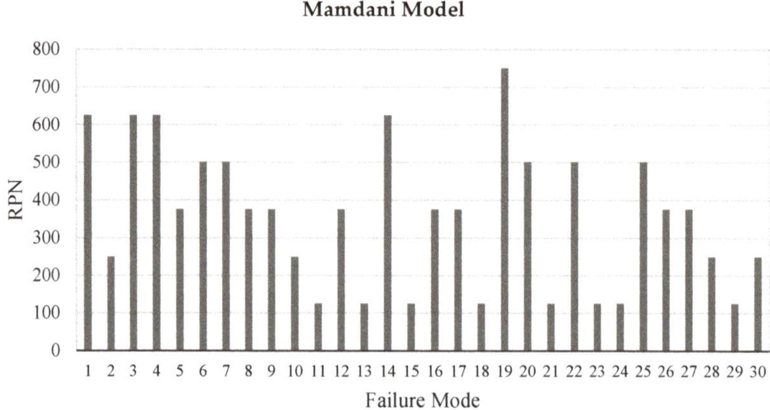

Figure 8. Representation of the Risk Priority Number by the Fuzzy Logic model.

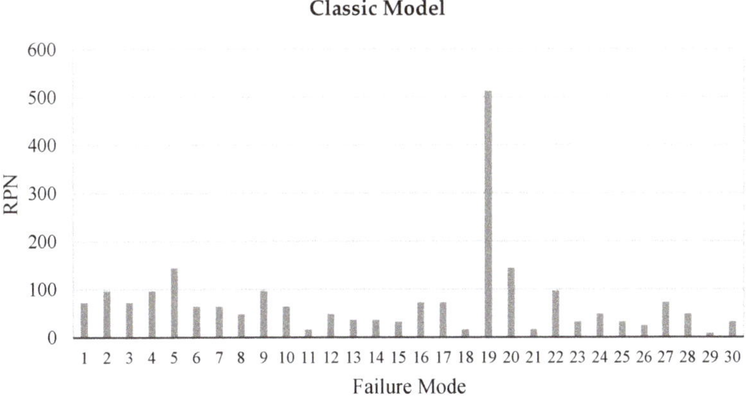

Figure 9. Representation of the Risk Priority Number by the classic model.

The results obtained from the methods used (classic FMEA and Fuzzy FMEA) to evaluate the RPN show asymmetric graphs. In the classical method, the values are mostly lower compared to the results of the Mamdani model; this is due to the description of the criteria with which it is evaluated.

There are two main reasons why the results vary significantly. The evaluation criteria in the Fuzzy Logic model for the FMEA have been appropriate to the context of the problem so that both linguistic and numerical values have a very different meaning in the evaluation.

In the classical method, the criteria used do not fit or comprehensively represent the context of what is being evaluated; obtaining the RPN is given through the multiplication of the evaluation criteria (Severity x Occurrence x Detection) in this method, not always the failures with greater severity are the priority.

It sometimes happens that failures with less severity but with higher value of occurrence or detection when multiplied, give a higher RPN value and take higher priority; that is, the priority depends in the first instance on the highest RPN value based on the evaluation parameters, without considering their relevance and only when the values are repeated, then the individual criteria assessment is deemed to be determined according to their significance which should go first.

Severity is addressed first because it relates to the effects of failures; Detection is used over the Occurrence because it depends on the client, which is more important than just failure frequencies (Stamatis, 2003); however, due to this classic method of obtaining, priority is given to failures with a higher number of RPN than to those with greater Severity.

The method used in the Mamdani model to obtain the RPN is inference through rules, and these are established according to the knowledge and reasoning of specialists about the level or intensity of the evaluation criteria; in this method, the highest RPN is assigned based on the relevance suggested in the literature of the evaluation criteria, in the order: Severity, Detection, and Occurrence, therefore the values obtained from the Mamdani model are considered with greater validity, and because the description of each criterion has been appropriate to the specific situation of the problem, therefore the results are more representative than in a conventional way.

The relevant fundamentals of the classic FMEA are described and later the use of fuzzy set theory is proposed as an alternative to reduce the uncertainty that can be generated by the NPR calculated with the traditional method, hence the disparity of data mentioned is made between Figures 8 and 9. The review shows the context in which the technique may be more appropriate and highlights the potential usefulness of fuzzy set theory to address uncertainty specifically in this case study, in software engineering, security and reliability.

There are methodologies applied to security and reliability engineering, which include fuzzy FMEA [1]. When prototypes are developed, it is not always possible to obtain reliable data; this is because they are recently created and, in most cases, due to the lack of availability of primary observations and the scarcity of data on the failure of their components. It is also difficult to establish risk parameters. To handle such situations, fuzzy set theory is an alternative for the evaluation of security and reliability under conditions of uncertainty, such as what is presented in this work.

5. Conclusions

The description of recommended actions for this case was established for failure modes whose RPN values were equivalent to those of the fuzzy "Significant" set, whose value was equal to or greater than 250. The FMEA methodology aims at continuous improvement, so it is necessary to carry it out periodically to identify failures and establish actions, in this case being a prototype, and it is essential to develop improvement actions for most failures to ensure operation, so it was determined to establish as a reference a low RPN for the context of the problem, to address most of the prototype's shortcomings and then re-evaluate the results.

It can be seen that in the classical method, the values are mostly lower compared to the results of the Mamdani model; this is due to the description of the criteria with which it is evaluated, and it can be said that there are two main reasons why the results vary significantly. One is that the evaluation criteria in the Fuzzy Logic model have been appropriate to the context of the problem so that both linguistic and numerical values have a very different meaning in the evaluation, while in the classical method the criteria used do not fit or represent in their entirety the context of what is evaluated.

The implementation of the Fuzzy FMEA allowed us to collect, order, and evaluate the information of the prototype concerning the operational failures of the different components; it can be said that the design of the studied quantifier is better in some important characteristics compared to other prototypes designed for the same purpose, however; the operating uncertainty conditions to which the quantifier is subjected affect the precision. Specifically, they are the characteristics related to the configuration or arrangement of the sensors to the tube of the animal that needs to be modified so that the sensors are no longer affected since, by their design and location, they suffer rubbing with solid surfaces that damage them. In general terms, these are the aspects that need to be improved to achieve greater accuracy.

Author Contributions: L.C.-R.: conceptualization, writing—original draft preparation, methodology and research; A.A.A.-L.: software and research; C.E.M.-O.: formal analysis and investigation; J.S.-E.: writing—original draft preparation; A.B.-M.: review and editing; M.L.M.-H.: supervision and project administration. All authors have read and agreed to the published version of the manuscript.

Funding: This research received no external funding

Institutional Review Board Statement: Not applicable

Informed Consent Statement: Not applicable

Acknowledgments: To TECNM/ Instituto Tecnológico Superior de Tantoyuca, Francisco Osorio, and Michelle Castillo for the support.

Conflicts of Interest: The authors declare no conflict of interest.

References

1. Kabir, S.; Papadopoulos, Y. A Review of Applications of Fuzzy Sets to Safety and Reliability Engineering. *Int. J. Approx. Reason.* **2018**, *100*, 29–55. [CrossRef]
2. Chanamool, N.; Naenna, T. Fuzzy FMEA Application to Improve Decision-Making Process in an Emergency Department. *Appl. Soft Comput.* **2016**, *43*, 441–453. [CrossRef]
3. Batbayar, K.; Takacs, M.; Kozlovszky, M. *Medical Device Software Risk Assessment Using FMEA and Fuzzy Linguistic Approach: Case Study*; Institute of Electrical and Electronics Engineers (IEEE): Piscataway, NJ, USA, 2016; pp. 197–202. [CrossRef]
4. Li, Z.; Chen, L. A Novel Evidential FMEA Method by Integrating Fuzzy Belief Structure and Grey Relational Projection Method. *Eng. Appl. Artif. Intell.* **2019**, *77*, 136–147. [CrossRef]

5. Yazdi, M.; Daneshvar, S.; Setareh, H. An Extension to Fuzzy Developed Failure Mode and Effects Analysis (FDFMEA) Application for Aircraft Landing System. *Saf. Sci.* **2017**, *98*, 113–123. [CrossRef]
6. Zhou, Y.; Xia, J.; Zhong, Y.; Pang, J. An Improved FMEA Method Based on the Linguistic Weighted Geometric Operator and Fuzzy Priority. *Qual. Eng.* **2016**, *28*, 491–498. [CrossRef]
7. Grossi, G.; Goglio, P.; Vitali, A.; Williams, A. Livestock and Climate Change: Impact of Livestock on Climate and Mitigation Strategies. *Anim. Front.* **2019**, *9*, 69–76. [CrossRef] [PubMed]
8. Silva-Martínez, K.L.; Cruz-Rivero, L.; Arrieta-González, A.; Purroy-Vasquez, R. Non-Invasive Measurer for Methane and Carbone Dioxide Emissions in Bovine Cattle through TRIZ. *Agron. Res.* **2020**, *18*, 1018–1026. [CrossRef]
9. Cruz-Rivero, L.; Mateo-Diaz, N.F.; Purroy-Vasquez, R.; Angeles-Herrera, D.; Osorio-Cruz, F. Statistical Analysis for a Non-Invasive Methane Gas and Carbon Dioxide Measurer for Ruminants. In Proceedings of the 2020 IEEE International Conference on Engineering Veracruz (ICEV), Boca del Rio, Mexico, 26–29 October 2020; Institute of Electrical and Electronics Engineers (IEEE): Piscataway, NJ, USA, 2020; pp. 1–7. [CrossRef]
10. Stamatis, D. *Failure Mode and Effect Analysis: FMEA from Theory to Execution*; American Society for Quality: Milwaukee, WI, USA, 2003.
11. Jin, C.; Ran, Y.; Zhang, G. Interval-Valued Q-Rung Orthopair Fuzzy FMEA Application to Improve Risk Evaluation Process of Tool Changing Manipulator. *Appl. Soft Comput.* **2021**, *104*, 107192. [CrossRef]
12. Boral, S.; Howard, I.; Chaturvedi, S.K.; McKee, K.; Naikan, V. An Integrated Approach for Fuzzy Failure Modes and Effects Analysis Using Fuzzy AHP and Fuzzy MAIRCA. *Eng. Fail. Anal.* **2020**, *108*, 104195. [CrossRef]
13. Yeganeh, A.; Heravi, M.Y.; Razavian, S.B.; Behzadian, K.; Shariatmadar, H. Applying a New Systematic Fuzzy FMEA Technique for Risk Management in Light Steel Frame Systems. *J. Asian Arch. Build. Eng.* **2021**, *20*, 1–22. [CrossRef]
14. Ribas, J.R.; Severo, J.C.R.; Guimarães, L.F.; Perpetuo, K.P.C. A Fuzzy FMEA Assessment of Hydroelectric Earth Dam Failure Modes: A Case Study in Central Brazil. *Energy Rep.* **2021**, *7*, 4412–4424. [CrossRef]
15. Mamdani, E.H. Application of Fuzzy Logic to Approximate Reasoning Using Linguistic Synthesis. *IEEE Trans. Comput.* **1977**, *C-26*, 1182–1191. [CrossRef]
16. Castillo, O.; Aguilar, L.T. Fuzzy Control Synthesis for Systems with Discontinuous Friction. In *Type-2 Fuzzy Logic in Control of Nonsmooth Systems: Studies in Fuzziness and Soft Computing*; Springer Science and Business Media: Cham, Switzerland, 2018; Volume 373, pp. 73–83. [CrossRef]

 MDPI

Article

Genetic-Algorithm-Inspired Difficulty Adjustment for Proof-of-Work Blockchains

Zi Hau Chin [1], Timothy Tzen Vun Yap [2,*] and Ian Kim Teck Tan [3]

1 School of Information Technology, Monash University Malaysia, Jalan Lagoon Selatan, Bandar Sunway, Subang Jaya 47500, Malaysia; zi.chin@monash.edu
2 Faculty of Computing and Informatics, Multimedia University, Persiaran Multimedia, Cyberjaya 63100, Malaysia
3 School of Mathematical and Computer Sciences, Heriot-Watt University Malaysia, 1, Jalan Venna P5/2, Precinct 5, Putrajaya 62200, Malaysia; i.tan@hw.ac.uk
* Correspondence: timothy@mmu.edu.my

Abstract: In blockchains, the principle of proof-of-work (PoW) is used to compute a complex mathematical problem. The computation complexity is governed by the difficulty, adjusted periodically to control the rate at which new blocks are created. The network hash rate determines this, a phenomenon of symmetry, as the difficulty also increases when the hash rate increases. If the hash rate grows or declines exponentially, the block creation interval cannot be maintained. A genetic algorithm (GA) is proposed as an additional mechanism to the existing difficulty adjustment algorithm for optimizing the blockchain parameters. The study was conducted with four scenarios in mind, including a default scenario that simulates a regular blockchain. All the scenarios with the GA were able to achieve a lower standard deviation of the average block time and difficulty compared to the default blockchain network without GA. The scenario of a fixed difficulty adjustment interval with GA was able to reduce the standard deviation of the average block time by 80.1%, from 497.1 to 98.9, and achieved a moderate median block propagation time of 6.81 s and a stale block rate of 6.67%.

Keywords: blockchain; difficulty adjustment; genetic algorithm; proof-of-work

Citation: Chin, Z.H.; Yap, T.T.V.; Tan, I.K.T. Genetic-Algorithm-Inspired Difficulty Adjustment for Proof-of-Work Blockchains. *Symmetry* **2022**, *14*, 609. https://doi.org/10.3390/sym14030609

Academic Editor: Kóczy T. László

Received: 16 February 2022
Accepted: 16 March 2022
Published: 18 March 2022

1. Introduction

The first prominent application in this field was conceptualized in a paper published by a pseudonymous author, or authors, named Satoshi Nakamoto. The paper is titled "A peer-to-peer electronic cash system" [1] and it proposed a framework for a decentralized electronic currency called Bitcoin. The framework actualized the research that was previously proposed by Haber and Stornetta [2], and was then coupled with a patent on the Merkle tree [3]. It was then implemented and proven to be a workable decentralized electronic currency that is immutable with limited circulation, similar to a sovereign-backed currency but without a centralized authority.

The core of the decentralization is the method for validating each transaction, where anyone can contribute processing resources to assist in validating the transactions in the electronic currency network. As it is decentralized, the process requires consensus between the nodes (miners) that are validating the transactions. The overall process is called proof-of-work (PoW) [4]. To obtain the consensus, the Bitcoin network conducts voting among the miners approximately every 10 min on the state of the Bitcoin network, to consolidate the validated transactions into a block (block time). In order to maintain the period between voting, that is the 10 min interval or the block time, the miners agree among themselves to adjust a variable called the difficulty, which is directly related to the overall computational resources that are available in the Bitcoin network. With more computational resources, the difficulty is set higher, and with less, it is set lower.

Bitcoin has grown in worldwide acceptance, and since its introduction, other electronic currencies (cryptocurrencies) have emerged. As the miners are rewarded for their effort in solving a mathematical puzzle (mining), the miners will tend to provide their computation resources to the most profitable cryptocurrency. This causes "coin-hopping", where miners switch to the more profitable cryptocurrency [5]. As the miners switch their resources to mine a different cryptocurrency, the cryptocurrency they departed from will still have the same difficulty settings, until the next consensus is conducted. This can cause an issue with attempting to maintain the block time, due to a lack of computation resources for mining with high difficulty settings. Bitcoin's difficulty adjustment algorithm is therefore susceptible to this drawback [6].

In PoW blockchains, the phenomenon of symmetry is observed, as difficulty also increases when the hash rate increases [7]. The block time can only be maintained by the difficulty adjustment algorithm if the overall computation resources (known as the hash rate) are constant. In the event of an increase in hash rate, the block time decreases and is adjusted through a retargeting mechanism regarding the difficulty, to maintain this equilibrium. However, as the retargeting takes place only after 2016 blocks in Bitcoin, the equilibrium is not maintained for a considerable length of time. Further, the network is not able to retain the block time if the hash rate rises or falls exponentially. In this paper, a genetic algorithm (GA) is introduced into the difficulty adjustment protocol to reduce the effect of the symmetry of the hash rate on the block time due to fluctuating hash rates in the Bitcoin network.

Proof-of-Work (PoW)

Dywork and Noar introduced PoW as a concept to address the issue of junk mail and administering access to shared resources [8], but the term "proof-of-work" was conceived by Jakobsson and Juels [9]. In order to access a shared resource, a feasible function that is reasonably hard must be computed as a requirement, and this serves to fend off malicious utilization of the shared resource.

Bitcoin implements PoW to provide resilience and security. A PoW process known as "mining" creates new Bitcoins, where the user or "miner" attempts to find a solution for a mathematical problem (PoW algorithm) through a particular piece of software. The target (T) is the threshold set for the block hash computed (to be less than) by the miner, for the candidate block to be valid. The difficulty (D) is a metric that indicates how hard it is to find a hash that is smaller than a specific target. As it is difficult to discover a block hash that is smaller than an already tiny value, a lower T will result in a larger D.

The new target, T_{i+1}, is calculated by multiplying T by the actual time it took to mine 2016 blocks and dividing it by the expected time, which is 20,160 min, as shown by the following equation [10]:

$$T_{i+1} = T * \frac{\sum_{i=1}^{2016} X_i}{20160 \text{ min}} \tag{1}$$

where D is calculated by multiplying the target of the genesis block (g_T) with the current target (c_T), given by

$$D = \frac{g_T}{c_T} \tag{2}$$

The block time (B), which is the expected time taken to mine a block in Bitcoin, is approximately 10 min. A retargeting mechanism will automatically and independently ensure that the block time B is as close as possible to the expected 10 min [4]. In this case, T is periodically and dynamically adjusted to meet the expected B of 10 min. Whenever the block time falls below 10 min due to an increasing hash rate, T is lowered (increasing the difficulty) during the adjustment, and vice versa. In addition, a limit is also imposed on the adjustment to prevent drastic changes to D, as shown in Algorithm 1.

However, PoW does not respond or react well when the hash rate experiences sudden changes. This was evident in some blockchain networks where a rapid shift in hash rate was experienced when capable powerful mining hardware for other networks was repurposed specifically for these networks. Since Bitcoin only retargets once every 2016 blocks (approximately 2 weeks), mining is performed at an extremely slow pace until the next retargeting event occurs, when enough blocks are created. The difficulty adjustment algorithm acts as a feedback controller, where the difficulty is the input, and it is manipulated towards the desired block time based on the actual time taken to mine a block (measured output). This reactive approach has a few vital shortcomings [11]:

1. The difficulty adjustment may overshoot or undershoot, thus causing the block time to oscillate significantly.
2. Cryptocurrencies are susceptible to "coin-hopping" or "pool-hopping" attacks, where miners choose to only mine a specific cryptocurrency when it is profitable, and switch to another when it is not.

As a means of mitigating these issues, a GA is introduced into the difficulty adjustment protocol to regulate the variation of the parameters (i.e., block time, retargeting interval, etc.). We show that it is possible to achieve a more dynamic retargeting mechanism that is able to meet the network objectives.

Algorithm 1 Target adjustment limit

```
Set targetTimeSpan = expected time taken to mine a block (s) × difficulty readjustment interval
Set totalInterval = actual time taken to mine N blocks
if totalInterval < targetTimeSpan then
   totalInterval = targetTimeSpan / 4
end if
if totalInterval > targetTimeSpan then
   totalInterval = targetTimeSpan × 4
end if
```

2. Literature Review

Several approaches have been introduced to improve the difficulty adjustment protocol, and in the process reduce the block time fluctuation. Bissias, Thibodeau and Levine proposed a proactive difficulty adjustment algorithm known as Bonded Mining (BM) in response to the relatively reactive nature of typical difficulty adjustment algorithms in PoW [11]. In BM, miners are required to commit to a hash rate which is financially bound to "bonds", where the committed hash rate in turn adjusts the difficulty of the network. As the miners are bound to their commitments, they are required to follow through with them even when it is no longer profitable to do so. However, the commitments only last until the next block is created, and if the miners choose to "deviate from their commitment", they suffer a fine that is equal to their divergence. In evaluating BM, block time stability simulations were carried out. The simulations compared the BM difficulty adjustment algorithm to the one used in Bitcoin Cash (BCH). The results showed that for BCH, the resulting block times diverged significantly from the intended time whenever the hash rate fluctuated, with the lowest block time reaching approximately 250 s and the highest reaching around 1500 s, with a recovery period of at least a day for the correction. Oscillation of the resulting block times was also observed, although the hash rate was maintained. With BM, a relatively lower amplitude in the oscillation and divergence of the block time was maintained, and no oscillation was observed when the hash rate was maintained, resulting in a block time that was closer to the intended block time.

Noda, Okumura and Hashimoto examined the behavior of the winning rate in place of the difficulty, and they found that the winning rate was "mathematically more traceable" [12]. Let W represent the winning rate, which is the probability that a block hash found by a miner is smaller than the target. H represents the hash rate, the total number of hash attempts per time unit. Based on observations, the average block time (B^*) can be calculated as $1/(W \times H)$. The winning rate can be adjusted to achieve a B^* of 10 min. Noda et al. concluded that Bitcoin's difficulty adjustment mechanism made it difficult to maintain steady block creation compared with Bitcoin Cash (BCH). The Bitcoin difficulty adjustment algorithm performed poorly on average, as the winning rate differed from the predicted outcome, and only 63.7% of the required 12,096 blocks were produced. BCH's difficulty adjustment algorithm, on the other hand, was able to create new blocks on a regular basis since the winning rate was modified once every block, based on the simple moving average block time of the preceding 144 blocks since August 2017 [13]. A total of 12,049 blocks were constructed, which was 99.6% of the expected 12,096 blocks, despite the winning rate fluctuating significantly. Bitcoin's difficulty adjustment algorithm produced a greater mean block time and mean standard deviation when comparing block times. When the hash rate varied, Bitcoin's difficulty adjustment algorithm was unable to modify the winning rate to the correct value.

In the case of soft computing approaches for PoW-based blockchains, Zhang and Ma suggested a difficulty adjustment algorithm with a two-layer neural network [14]. The difficulty in Ethereum was adjusted according to Algorithm 2. In order to forecast the state of the blockchain, different variations of previous actual times taken to mine a block (T_k) served as the input features. A two-layer neural network was utilized to perceive distinct patterns based on the obtained variance of T_k. For simplicity and easy computation, a simple neural network with a single hidden layer was chosen. Based on the actual data collected from Ethereum for comparison between the proposed algorithm and its initial complexity modification algorithm, improvements in the nominal hash rate were simulated. During the training phase, a Monte Carlo simulation was performed. Each sample started with a hash rate of 1.46×10^{14} hash/s. For typical and atypical changes, the scale of hash rate variations was from -60% to $+60\%$ of the starting hash rate. The factor affecting the accuracy was the number of blocks mined after the hash rate change, since the accuracy of the neural network steadily improved as time elapsed after the abrupt hash rate change. A sudden hash rate shift was manufactured by injecting an extra 20% hash rate into the mining pool at block height 50,000, which was then removed at block height 100,000. An additional 40% of the hash rate was also inserted into the mining pool at block heights 150,000 and 200,000, then removed at block heights 155,000 and 250,000, respectively. The proposed neural-network-based approach maintained the quick updating and low volatility of the block difficulty in simulations based on real data. The suggested method was better at detecting irregularities and dealing with irregular occurrences such as hash rate surges or drops that only last a short time. However, when the hash rate suddenly increased or decreased, the approach tended to delay changing the difficulty by gradually increasing or reducing the difficulty to the predicted value over a longer period, rather than instantaneously, to guarantee that it was not a malicious assault. This slowed down the difficulty adjustment reaction time in exchange for a smoother and more stable adjustment, resulting in a longer time to achieve the intended difficulty, and stabilized the time needed to mine a block.

Algorithm 2 Ethereum's difficulty adjustment algorithm

New difficulty = *parent block's difficulty* + floor(*parent block's difficulty*/1024)
if *current block's timestamp - parent block's timestamp* < 9 **then**
  New difficulty = *new difficulty* $\times$ 1
else
  New difficulty = *new difficulty* $\times -1$
end if

Zheng et al. proposed a linear-prediction-based difficulty adjustment method for Ethereum to address the present difficulty adjustment algorithm's drawbacks, such as its lack of versatility and sensitivity to hash rate fluctuations [15]. They defined a new term, $PT_n = \frac{d_n}{r_n}$, where d_n is the difficulty and r_n is the hash rate at the n_th block. Despite having a one-block delay compared to the real PT_n, the linear predictor was accurate and obtained a low mean squared error (MSE). The fundamental reason for this was that the PT_n fluctuated with the hash rate, which was precipitated by miners and constantly changing. As a result, the linear prediction could only take the previous PT_n as the primary input and alter it according to the expected trend. The concept of linear prediction was to anticipate present and future values using previously observed values, but only the actual time taken to mine a block could be observed in an actual blockchain. Thus an additional computation step was required to calculate PT_n. The authors proposed two methods to obtain the PT_n value: (i) using the smoothed actual time taken to mine a block or (ii) using the integrated actual time taken to mine a block. The results showed that the linear-prediction-based difficulty adjustment algorithm with integrated actual time taken to mine a block was able to obtain a lower deviation. Nonetheless, while the prediction was capable of generating a value that was close to the actual value, in some cases, such as sudden change in hash rate, the predicted value was considerably higher or lower than the actual value.

In Zhang and Ma's proposal, the difficulty adjustment algorithm determined whether the actual time taken to mine a block was experiencing no change, normal change or abnormal change. When the neural network detected an abnormal change, the algorithm was implemented to adjust the difficulty [14]. In contrast, Zheng et al. suggested using a linear predictor to adjust the difficulty accordingly [15].

In this investigation, we implemented GA to suppress the fluctuations in the average block time and difficulty, and to enable faster adjustment of the difficulty. This was achieved by adjusting the expected time taken to mine a block and the difficulty adjustment interval. The implementation could react faster as it was no longer needed to wait for the next difficulty adjustment to set the accurate difficulty value. A difficulty adjustment could be scheduled immediately after detecting significant deviations from the default time to mine a block.

3. Methods

3.1. Observation of Blockchain Behaviors with Reparameterization

For the initial investigation, a total of 24 virtual machines (VMs) with equivalent specifications were used to simulate and observe the Bitcoin blockchain network for different parameter settings. The VMs were divided into 3 separate blockchain network groups, consisting of 8 VMs per group. Each VM acted as a full node, running the Bitcoin Core software and a miner at the same time. The parameters for the deployed blockchain networks are shown in Table 1. The 3 different blockchain network groups, referred to as Coin A, Coin B and Coin C, were as follows:

- Coin A, the first blockchain network group, was representative of an actual Bitcoin network.
- For Coin B, the default values of 10 min and 1 MB were used for the block time and block size, respectively. The difficulty adjustment interval was set to 60 blocks, and thus the difficulty would be readjusted once for every 60 blocks mined. The value of 60 for the difficulty adjustment interval was selected based on the data obtained by Friedenbach [16].
- For Coin C, the block time and block size were set at 1 min and 1 MB, respectively, while the difficulty adjustment interval was set to within a range of 1 to 20,000.

Data for the observation were collected for 150 days for both Coin A and Coin B. For Coin C, due to the nature of the setup, data were collected for 200 days. The three blockchain networks were allowed to run for a considerable lead time, so that the block time and difficulty reached a steady state.

Table 1. Differences between 3 groups of VMs.

	Coin A	Coin B	Coin C
Block interval	10 min (default)	10 min	1 min
Block size	1 MB (default)	1 MB	1 MB
Difficulty adjustment interval (block)	2016 (default)	60	Variable within constraints
Number of history blocks	2016 (default)	60	Same as difficulty adjustment

3.1.1. Coin B

As shown in Table 2, we were able to obtain a lower standard deviation, decreased by 63.18%, with Coin B compared to Coin A. Additionally, the maximum value of the daily average block time for Coin B was 14.08, which was relatively close to the expected value of 10, whereas it was 27.91 for Coin A.

Table 2. Statistical data for daily average actual time taken to mine a block.

	Coin A	Coin B	Coin C
Min	2.00	3.16	0.63
Max	27.91	14.08	3.43
Mean	10.60	10.48	1.11
Median	10.15	10.33	0.99
Standard Deviation	3.54	1.25	0.44

On commencement of the experiment, it was observed that the obtained block time deviated from the expected time and it took 17 days or 2 sequences of difficulty adjustment before it reached the expected block time of 10 min, as shown in Figure 1.During the experiment, a sudden decrease in hash rate was simulated by lowering the hash rate of each miner. We decreased the overall hash rate by 50% on Day 71, which was 200 blocks away from the next difficulty adjustment at that time. As observed in Figure 2, there was a steep increase in the block time after the decrease in hash rate up to Day 73, lasting for 3 days for Coin A. This was due to the fact that the difficulty adjustment algorithm was unable to react rapidly by adjusting the difficulty in response to the sudden drop in hash rate. The block time recovered back to around the expected time of 10 min on Day 74 because the difficulty adjustment occurred immediately before the end of Day 73. Following this, we increased the hash rate back to its value before the drop on day 75. From that time, we observed a decrease in block time due to the increase in hash rate. Since the hash rate was increased at the beginning of the 2016-block cycle for difficulty adjustment, it took about 7 days of mining before the difficulty adjustment took place. On Day 82, the block time was back to within the expected range. The Coin A blockchain network showed relatively smooth changes in difficulty due to the fact that the difficulty was only readjusted once every 2016 blocks. In total, the difficulty was readjusted only nine times over the course of the investigation.

On the other hand, Coin B was able to reach the expected block time of 10 min right away, after the two sequences of difficulty adjustments at the beginning. Since the difficulty readjusted once every 60 blocks instead of every 2016 blocks, only about 24 h was needed, as shown in Figure 3. Moreover, although the obtained block time increased to 13.77 min with the same occurrence of a decreased hash rate on Day 71 as in the previous experiment, we observed that Coin B was able to stabilize and obtain a block time of 10.20 min on the next day, as shown in Figure 4. In addition, the increase in hash rate on Day 75 only decreased the obtained block time to 7.59 min for one day, and it returned to around the expected block time on the following day. Figure 5 shows the recorded difficulty of the Coin B network. Throughout the experiment, the difficulty for Coin B readjusted a total of 200 times within 148 days. Furthermore, as shown in Table 3, Coin B was able to obtain a lower standard deviation compared to Coin A, even though it fluctuated a great deal in comparison, as shown in Figure 5. The network was able to react to the changes in hash rate by readjusting the difficulty, due to the shorter difficulty interval.

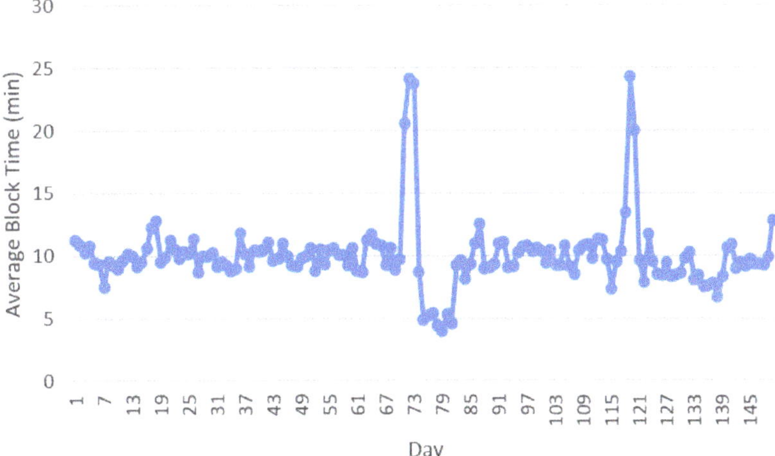

Figure 1. Average block time for Coin A.

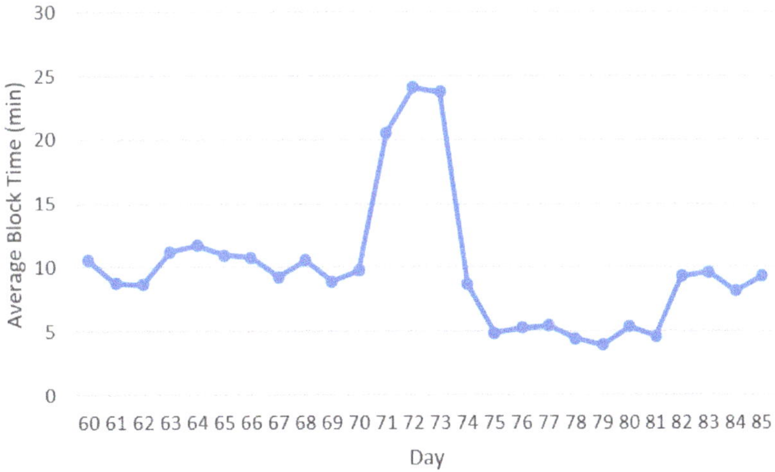

Figure 2. Average block time for Coin A from Day 60 to Day 85.

Table 3. Statistical data for difficulty.

	Coin A	Coin B	Coin C
Min	1.00	1.00	0.00
Max	6.62	4.66	17.23
Mean	2.78	2.86	0.24
Median	2.60	2.92	0.23
Standard Deviation	1.07	0.60	0.14

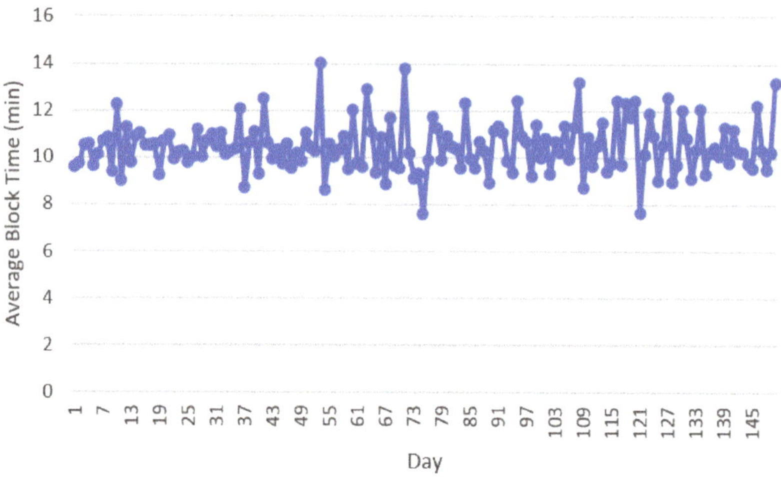

Figure 3. Average block time for Coin B.

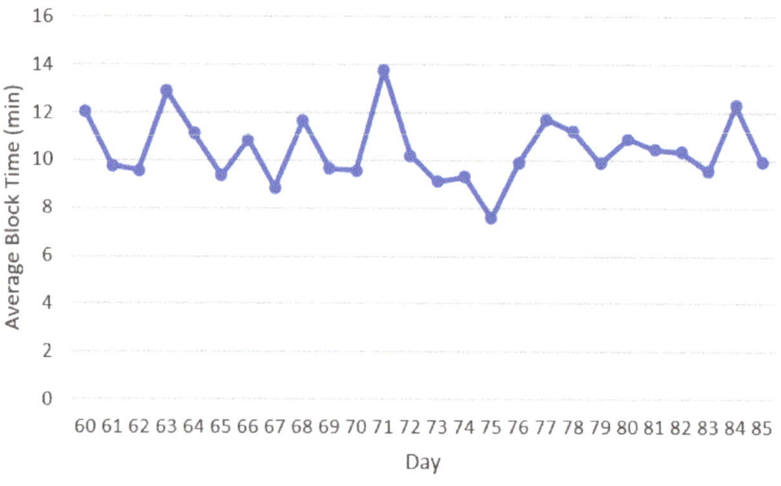

Figure 4. Average block time for Coin B from Day 60 to Day 85.

3.1.2. Coin C

Reparameterization of the blockchain PoW for Coin C was achieved manually by changing the parameters once when at least 10,000 blocks had been mined (varying depending on the difficulty adjustment interval). Unlike the other experimental setups, this blockchain network was set to a block interval of 1 min, with varying difficulty adjustment intervals ranging from 1 block to 20,000 blocks. A block interval of 1 min was deployed instead of the default 10 min to reduce the length of the experiment. Moreover, the initial and minimum difficulty was set to 0.00024414, unlike in previous experiments where it was set to 1 to ensure that the miners were able to mine a block every minute. This was because the miners were unable to mine a block every minute with a minimum difficulty of 1, even when the maximum hash rate was available. The value of 0.00024414 was taken from Dogecoin, as Dogecoin has the same block interval of 1 min. The available hash rate was set to the maximum value throughout the experiment.

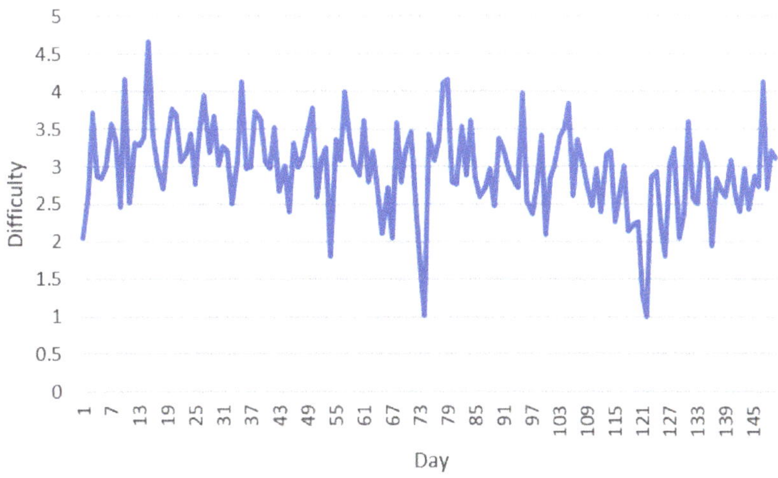

Figure 5. Difficulty graph for Coin B (sampled daily).

Tables 4 and 5 record the statistical data for the obtained block time and difficulty for Coin C, with a difficulty adjustment interval of 1 to 20,000. We modified the difficulty adjustment interval and then waited for 10,000 blocks to be mined before setting a new difficulty adjustment interval (Tables 4 and 5). A difficulty interval of 1 indicates that the difficulty will be adjusted once a block is mined, whereas a difficulty interval of 20,000 indicates that the difficulty will be adjusted after 20,000 blocks are mined. From Figure 6, when the difficulty was readjusted once for every block mined, we observed that the difficulty fluctuated widely, with a standard deviation of 2.40 and a standard deviation of 0.47 for the obtained block time. Moreover, the standard deviation of the obtained block time of 0.47 was the second-highest value. Considering that only the actual time taken to mine a block for the previous block was referred to when adjusting the difficulty, it tended to overshoot or undershoot. Thus, the actual block time deviated from the expected value and was unable to maintain the expected time, as observed in the left-most part of Figure 7.

Table 4. Statistical data for block time for Coin C.

Difficulty Adjustment Interval	Min	Max	Mean	Median	Standard Deviation
1	1.03	2.72	2.34	2.50	0.47
100	0.98	1.04	1.01	1.01	0.01
200	0.89	1.09	1.00	0.99	0.06
400	0.95	1.09	1.02	1.02	0.04
800	0.88	1.06	0.98	1.00	0.06
1000	0.82	1.13	1.01	1.04	0.10
1500	0.81	1.36	1.03	0.98	0.16
2000	0.91	1.47	1.03	0.96	0.18
4000	0.85	1.52	1.02	1.00	0.15
8000	0.78	1.88	1.08	0.98	0.29
10,000	0.74	2.24	1.02	0.91	0.32
20,000	0.63	3.43	1.12	0.96	0.58

Table 5. Statistical data for difficulty for Coin C.

Difficulty Adjustment Interval	Min	Max	Mean	Median	Standard Deviation
1	0.00	17.24	0.73	0.43	2.40
100	0.15	1.17	0.26	0.26	0.08
200	0.13	0.32	0.26	0.27	0.04
400	0.15	0.27	0.23	0.24	0.03
800	0.11	0.27	0.22	0.25	0.05
1000	0.16	0.27	0.21	0.20	0.03
1500	0.19	0.30	0.25	0.28	0.04
2000	0.24	0.30	0.28	0.29	0.02
4000	0.25	0.30	0.28	0.29	0.02
8000	0.23	0.30	0.25	0.24	0.02
10,000	0.21	0.26	0.22	0.23	0.02
20,000	0.17	0.24	0.20	0.20	0.01

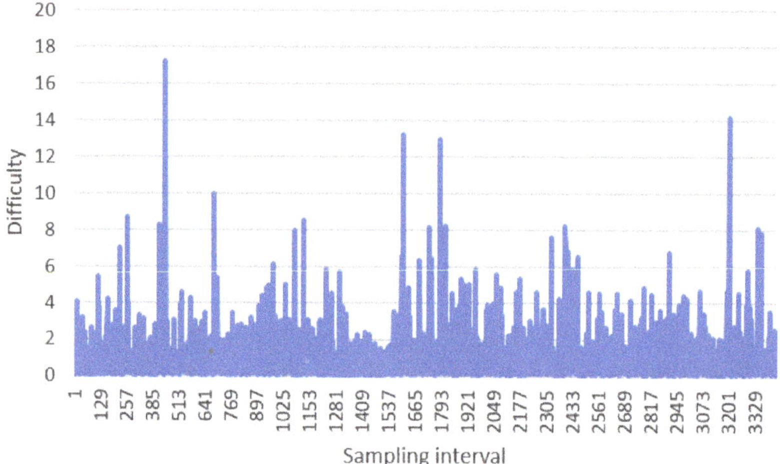

Figure 6. Difficulty graph for Coin C with difficulty adjustment interval of 1 block (Day 1 to Day 10). Data were recorded for every block; the graph only shows the difficulty when there was an adjustment.

In our next experiment, the difficulty adjustment interval was set to 100 blocks and the block time and difficulty history were recorded as shown in Figures 8 and 9, respectively. Compared to a difficulty adjustment interval of 1, a decrease in the deviation of the difficulty and the block time was observed. Moreover, the obtained block time deviated minimally from the expected block time, thus maintaining the obtained block time as close to the expected value as possible with the lowest standard deviation. Nonetheless, an even lower standard deviation of 0.04 for the difficulty was obtained with a difficulty adjustment interval of 200 blocks, as recorded in Table 5, despite an increase in the standard deviation of the block time. As observed from Figures 10 and 11, the obtained block time deviated slightly from the expected value, but the deviation of the obtained difficulty decreased as the difficulty only readjusted once every 200 blocks.

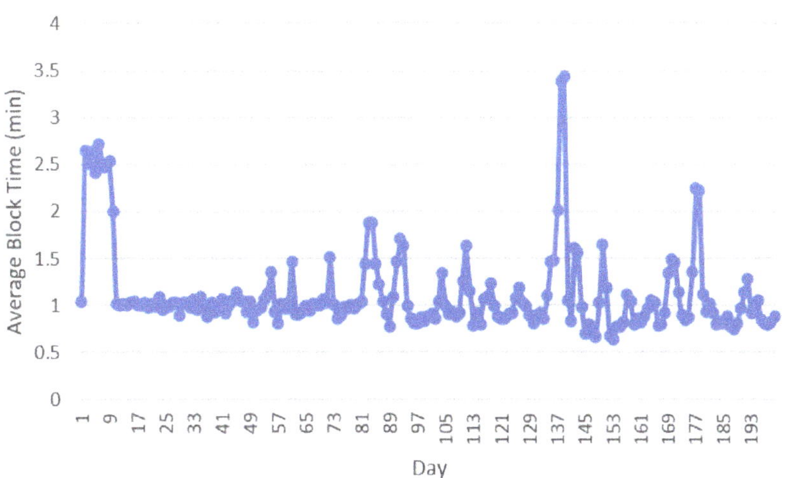

Figure 7. Average block time for Coin C.

For Coin C, we were able to observe that as the difficulty adjustment interval increased, the standard deviation of the block time increased, while the standard deviation of the difficulty decreased. Although the standard deviation of the difficulty was lowest when the difficulty adjustment interval was equal to 20,000, this value gave the highest standard deviation for the block time. This was a consequence of the difficulty not adjusting frequently. If there was a sudden increase or decrease in the hash rate with a difficulty adjustment interval of 20,000, the difficulty was unable to readjust in response. For example, when the total hash rate decreased by 50% while the difficulty remained the same immediately after the difficulty adjustment, the block time increased to 20 min instead of the expected 10 min. With a block time of 20 min, in the worst-case scenario, it would be approximately 278 days before the next difficulty adjustment. Therefore, there must be a balance between the block time and the difficulty.

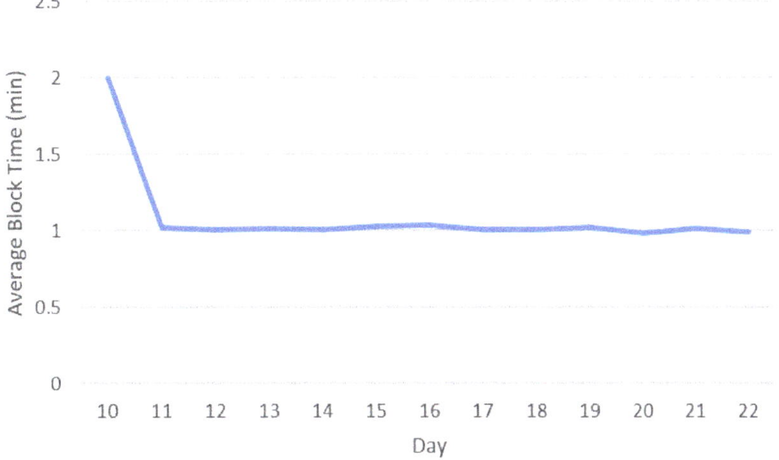

Figure 8. Average block time for Coin C with difficulty adjustment interval of 100 blocks.

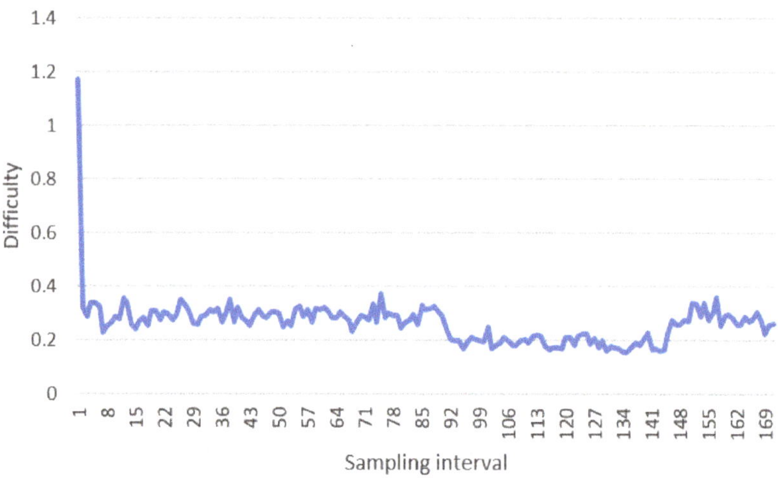

Figure 9. Difficulty graph for Coin C with difficulty adjustment interval of 100 blocks (Day 10 to Day 22). Data were recorded for every block; the graph only shows the difficulty when there was an adjustment.

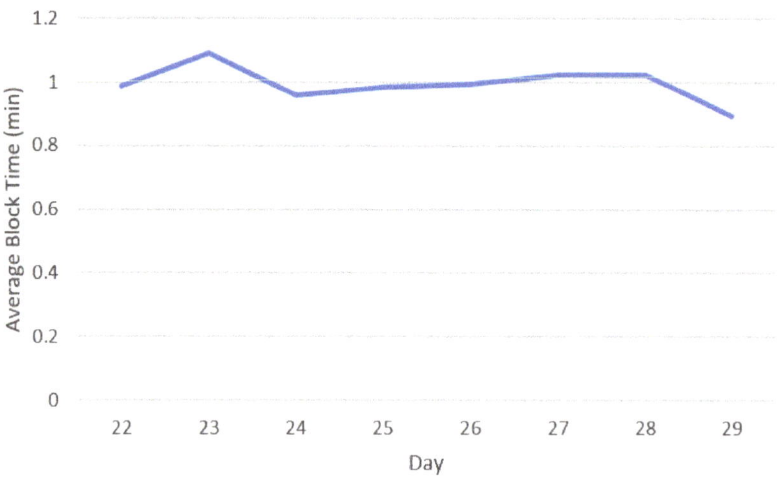

Figure 10. Average block time for Coin C with difficulty adjustment interval of 200 blocks.

3.2. Optimization of Parameters with Genetic Algorithm

As the main contribution of this research, GA is proposed as a reparameterization mechanism for the PoW protocol, to remedy the inherent trade-offs in the difficulty adjustment in PoW. Through simulations, the key factors that dramatically influence the performance of the PoW protocol are determined, together with their suitability regarding the GA optimization. The proposed method is not intended to be a replacement for the Bitcoin difficulty adjustment mechanism. In fact, it serves the same purpose, i.e., to govern the pace at which Bitcoins are issued. The GA helps by determining the best settings to use in order to control the length of time it takes to mine a block. The total actual time needed to mine the previous N blocks is used as the deterministic seed in the GA, to maintain consistency across all nodes. Otherwise, nodes may arrive at different parameters and the state consensus would be lost. In this study, the variables considered for optimization are the block interval (s) and the difficulty adjustment interval (number of blocks).

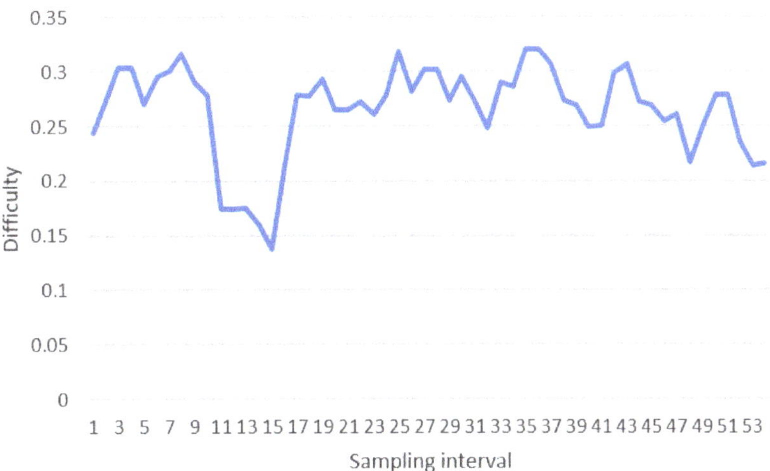

Figure 11. Difficulty graph for Coin C with difficulty adjustment interval of 200 blocks (Day 22 to Day 29). Data were recorded for every block; the graph only shows the difficulty when there was an adjustment.

The block interval requires a minimum value of 1 s and a maximum value of 600 s. The difficulty adjustment interval dictates the number of blocks mined before the difficulty readjustment. In Bitcoin, the difficulty readjusts once every 2016 blocks (approximately 14 days). In this study, the difficulty adjustment interval could be set as low as retargeting once every block up to retargeting once every 4032 blocks. Because there are at least two optimization variables to consider, we utilized the non-dominated sorting genetic algorithm II (NSGA-II) [17], which is a multi-objective GA.

A Java-based architecture for multi-objective optimization with metaheuristics, jMetal 6.0, was used as a basis and modified to suit our purpose [18]. A version of NSGA-II known as "parallel NSGA-II" was chosen for our experiment as it benefits from a multi-core processor to perform the feature evaluations of in parallel. At the time, there was only one option for assessing a solution list in parallel, called "synchronous parallelism". This suggests that this parallel algorithm's action was similar to that of a sequential algorithm. The computation of the output, however, was impaired because the algorithm alternated between parallel and sequential, and hence it did not scale well. However, it was considered more than adequate for our use case.

The simulations were performed using an adaptation of SimBlock [19], an event-driven blockchain network simulator. During the simulation, the default starting values for the block interval and the difficulty adjustment interval for the experiments were 600 s and 2016 blocks, respectively. The GA was set to run throughout the whole simulation, as it detected whether the latest N blocks were mined too fast or too slow compared to the expected time for mining N blocks. It approximated the state of mining for each set of optimization variables, where each set of optimization variables was applied to mining 10,000 blocks. The set of optimization variables were evaluated based on two objectives:

1. Standard deviation of average block time.
2. Standard deviation of difficulty.

The goal was to control the variations in the difficulty and average block time, in addition to achieving faster adjustment with the utilization of the GA, and the objective functions were chosen based on these criteria. Studies were performed to investigate the effectiveness of the objective functions. In our simulations with SimBlock, every simulation was furnished with the same set of hash rates, ending when a total of 10,000 blocks were generated. The overall flow of the simulation is described in Algorithm 3. Firstly, the network was constructed based on the number of nodes, the degree distribution and the region. The latency, download and upload bandwidths of the node were determined by the region in which it was located. A check was subsequently conducted for each block mined to decide whether the average actual time taken to mine N blocks was too fast or too slow. The GA was activated if the average actual time taken to mine the N blocks was 25% slower or faster than the planned time. For the GA, the fitness score was generated for selection after simulated mining of 10,000 blocks. Next, crossover and mutation were continued until convergence was achieved and the best solution was obtained. The best solution was then applied to the network, and new blocks were mined with the new parameters thereafter. The GA had to wait until the difficulty was adjusted before it was able to optimize again. This was to ensure that the previously found best solution was given ample time to affect the network.

Algorithm 3 Overall flow of simulation

```
start
Construct network based on the number of node, distribution of degree and region
repeat
  Generate block
  if actual time taken to mine the latest N blocks are too fast or too slow then
    GA optimization
  end if
  if current block height == difficulty adjustment interval then
    adjust difficulty
  end if
until current block height == 10,000
end
```

Table 6 shows the hyperparameters for the GA applied in the simulations, while Table 7 shows the parameters used for our blockchain simulation. For the hyperparameters of the GA, simulations were performed to obtain the optimal hyperparameters for our studies. On observation, the results from these hyperparameters were comparable to results obtained with higher population sizes and max generation at shorter execution times. Due to the nature of both optimization variables, the differences in some values, such as 3103 blocks and 3105 blocks, were negligible. Although a higher population size and max generation can cover more possible solutions, it significantly increased the time needed to complete the optimization, and the improvement in the obtained objectives was minimal. With these hyperparameters, the average run time of the GA for one iteration in our simulation was 165.74 s or 2.7 min, and the GA was run several times within an iteration of our simulations. Since mining of blocks was ongoing while the GA was optimizing, the faster the GA finished, the better the optimization.

Table 6. Hyperparameters of the GA.

Parameter	Value
Population size	200
Max generation	50
Crossover probability	90%
Crossover distribution index	20
Mutation probability	50%
Mutation distribution index	20

Table 7. Blockchain parameters for simulation.

Parameter	Value
Number of nodes	9000
Average of hash rate	80,093,787 TH/s
Average block size	1 MB
End block height	10,000
Geographical distribution	Based on the Bitcoin network
Bandwidth and latency	Based on data collected from 6 regions

4. Results and Discussion

The parameters shown in Table 7 were collected based on the data from the Bitcoin network from June 2019 to December 2019. However, in this case, the number of nodes refers to the reachable public listening nodes at that time and not the total number of full nodes in the Bitcoin network.

The simulations were performed for four scenarios with a runtime of 20 iterations for each scenario:

1. Default;
2. Fixed block interval;
3. Fixed difficulty adjustment interval;
4. Variable block and difficulty adjustment intervals.

Each iteration simulated the mining of 10,000 blocks for all the scenarios, and at intervals of 1500 blocks, the available hash rate was increased. For Scenario 1, the simulation used fixed default values for both optimization variables and ran without GA, i.e., there were no modifications to the optimization variables to represent a Bitcoin blockchain network. Since there were no other experiments using similar methods to the best of our knowledge, comparisons were made with the Bitcoin network. We set one of the optimization variables in Scenarios 2 and 3 to the default values, while allowing the GA to adjust the other optimization variable within the range defined. In addition, in Scenario 4, the GA was able to modify all the optimization variables. Table 8 records the outcomes of the simulations.

As shown in Table 9, in Scenario 2 the GA was activated 3.4 times on average, with a minimum of 3 times and a maximum of 7 times. For Scenario 3, the GA was triggered 5.3 times on average, with a minimum of 5 times and a maximum of 7 times, which was slightly higher in terms of the minimum and average but with the same maximum as in Scenario 2. In contrast, the GA was triggered 7.9 times on average in one iteration for Scenario 4. The minimum number was 6 times and the maximum was 12 times.

Table 8. Average of Objective 1, Objective 2, median block propagation time (t_{MBP}) and stale block rate (r_s) for 20 iterations. Objective 1: standard deviation of average block time. Objective 2: standard deviation of difficulty.

Scenarios	Objective 1	Objective 2	t_{MBP} (s)	r_s (%)
Without GA (default)	497.10	2.83×10^{15}	6.40	1.12
GA (fixed block interval)	376.80	1.92×10^{15}	6.44	1.80
GA (fixed difficulty adjustment interval)	98.90	2.18×10^{14}	6.81	6.67
GA (variable block and difficulty adjustment intervals)	102.75	2.86×10^{14}	7.02	32.04

Table 9. The number of times GA ran within an iteration.

Scenarios	Min	Average	Max
GA (fixed block interval)	3	3.4	7
GA (fixed difficulty adjustment interval)	5	5.3	7
GA (variable block and difficulty adjustment intervals)	6	7.9	12

4.1. Fixed Block Interval

For this scenario, the block interval was fixed to 10 min, only allowing the GA to optimize the difficulty adjustment interval within a range of 1 to 4032 blocks. The value obtained for Objective 1 in Scenario 2 was 376.80, a 24.2% decrease compared to the value of 497.10 obtained if GA was not applied (Table 8). Additionally, the value obtained for Objective 2 showed a decrease of 32.15%, from 2.83×10^{15} to 1.92×10^{15}. During each simulation, a sudden rise in hash rate was implemented by increasing the hash rate immediately after the first six blocks were mined. For all iterations, the hash rate increased by 655.43%. For all 20 iterations, it was noted that after the GA had run for the first time, the difficulty adjustment interval was still optimized to 4006. The subsequent GA run then refined the difficulty adjustment interval to at least 3300 blocks. Values ranged from 566 to 4030 for the remaining tailored difficulty adjustment intervals, but the difficulty adjustment period was more than 3000 blocks for 90 percent of the time.

On the other hand, t_{MBP} and r_s showed small increases of 0.63% and 60.71%, respectively. When the difficulty adjustment interval was low, an improvement in the stale block rate was seen. However, as the difficulty adjustment interval increased from 800 blocks to 20,000 blocks, the stale block rate only increased by 1.44%. Experiments were conducted to study the effects of different difficulty adjustment intervals on t_{MBP} and r_s. Additional simulations were conducted with SimBlock but without GA, and the results are reported in Table 10. The findings show that the difficulty adjustment interval influenced r_s, in one direction or the other. When the difficulty adjustment interval was 1 block, where the difficulty changed after a block was mined, r_s was largest. By increasing the difficulty adjustment interval to just 10 blocks, the obtained value of r_s decreased to 0.41%, an approximately 90.16% decrease. Nevertheless, we noted an increase in r_s as the difficulty adjustment interval increased, although the value of r_s obtained with a difficulty adjustment interval of 4000 blocks was still lower than for the difficulty adjustment interval of 1 block. This was caused by the low difficulty adjustment interval (1 block), as a low difficulty adjustment interval has the tendency to overshoot or undershoot. On the other hand, t_{MBP} was highest with a difficulty adjustment interval of 1 block, whereas with difficulty adjustment intervals from 10 blocks to 4000 blocks, a slight increase in t_{MBP} was observed as the difficulty adjustment interval increased. However, the obtained t_{MBP} value was lowest when the difficulty adjustment interval was 100 blocks.

Table 10. Effect of difficulty adjustment interval on t_{MBP} and r_s.

Difficulty Adjustment Interval (Blocks)	t_{MBP} (s)	r_s (%)
1	6.62	4.17
10	6.34	0.41
100	5.84	0.47
1000	6.39	0.89
4000	6.44	1.82

4.2. Fixed Difficulty Adjustment Interval

The difficulty adjustment interval was fixed at 2016 blocks, and the block interval started at 600 s but was optimized by the GA within a range from 1 to 600 s. As shown in Table 8, even lower values for Objective 1 and Objective 2 were obtained than in the two previous simulations. The decreases were 80.10% and 92.29%, respectively, compared to when GA was not applied. An identical occurrence was observed in this scenario where, for all the 20 iterations, the block interval was optimized to 78 s after the first GA run. Throughout the simulations, the minimum block interval was 4 s while the maximum was 152 s. We observed that the GA seemed to favor a lower block interval, mainly due to the objectives. A lower block interval reduced the mean of the actual time it takes to mine a block, while lowering the standard deviation. In addition, since a lower difficulty was required so that the desired actual time taken to mine a block could be reached, a lower block interval caused the mean and standard deviation of the difficulty to decrease. A block interval of less than 10 s in the actual Bitcoin environment is implausible, since the median block propagation time of Bitcoin was measured at 8.7 s [20].

It takes time for information from a freshly mined block to propagate from the miner to the remaining nodes. A higher block interval can ensure that the newly mined block is able to propagate to a majority of the nodes. Stale blocks are blocks that have been mined but are no longer part of the longest chain. They occur when more than one miner concurrently manages to mine a valid block. There is a temporary fork, where each node in the network sees a separate block tip every time this event occurs. The stale block rate increases when the block interval decreases [21], even more so if the block interval is lower than the median block propagation time. The probability of the nodes generating a stale block rises proportionally with the block interval and the time passed until a node in the network learns of the new block. Based on a new analysis by Neudecker [22], the median block propagation time in the real Bitcoin network has been reduced to less than 1 s for most of the time, with enhancements to the block propagation time after the implementation of a Bitcoin enhancement protocol (BIP) such as BIP 0152 in 2016. This does not mean, however, that the block interval should be set to a very low value such as 1 s. For a block interval of 1 s, the difficulty is very low, and thus it is too easy for a miner to mine a block. Therefore, this increases the likelihood of miners successfully mining a block concurrently, increasing the numbers of stale blocks.

In this scenario, t_{MBP} and r_s were increased by approximately 6.4% and 495%, respectively, compared to when the GA was not used. Nevertheless, we performed some simulations with low block intervals and a fixed difficulty adjustment interval of 2016. It was observed that the lower the block interval, the higher the value of r_s, as shown in Table 11. Moreover, with a block interval of 1 s, the value of r_s obtained was a huge 1144.65%. This translated to approximately 11.4 stale blocks produced per mined block. Interestingly, the t_{MBP} was also affected by the block interval, increasing with an increasing block interval. The obtained value of t_{MBP} was highest when the block interval was 100 s.

Table 11. Effect of block interval on t_{MBP} and r_s.

Block Interval (s)	t_{MBP} (s)	r_s (%)
1	6.59×10^{-3}	1144.65
10	4.05	119.00
100	7.15	11.28
300	6.59	3.88

4.3. Variable Block and Difficulty Adjustment Intervals

In Scenario 4, Objectives 1 and 2 achieved a decrease of 79.33% (102.75 vs. 497.1) and 89.89% (2.86×10^{14} vs. 2.83×10^{15}), respectively. For the block interval and difficulty adjustment interval, the range of applied values was 1 s to 190 s and 5 blocks to 4027 blocks, respectively. In contrast, the t_{MBP} value increased slightly by 9.6%, from 6.40 s to 7.02 s. However, r_s increased significantly from 1.12% to 32.04%, which was an increase of 2760.71%.

The huge increase in r_s for Scenario 4 was due to the fact that the variable block interval could be as low as 1 s. The increase in t_{MBP} was believed to be due to the fact that some available bandwidths were utilized to propagate stale blocks, thus causing a slight increase in the block propagation time. However, Objectives 1 and 2 for Scenario 4 increased by 3.89% and 31.19%, respectively, compared to Scenario 3. This was mainly due to the recorded minimum value of 1 s for the block interval being even lower than the block interval of 4 s for Scenario 3, where the GA was only able to optimize in terms of the block interval, thus contributing to the higher value of r_s.

Figures 12 and 13 show the recorded difficulty history for Default and Scenario 4. From Figure 12, without GA, the difficulty was unable to reach the expected value before the mining ended, even after five difficulty adjustments. With GA (Figure 13), the blockchain was quicker to reach the intended difficulty, reaching it just after the third difficulty adjustment, and it was comparatively stable.

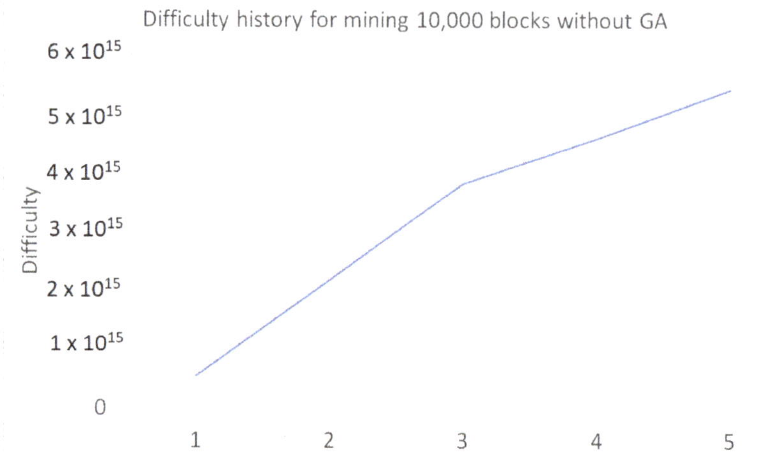

Figure 12. Difficulty history (Default).

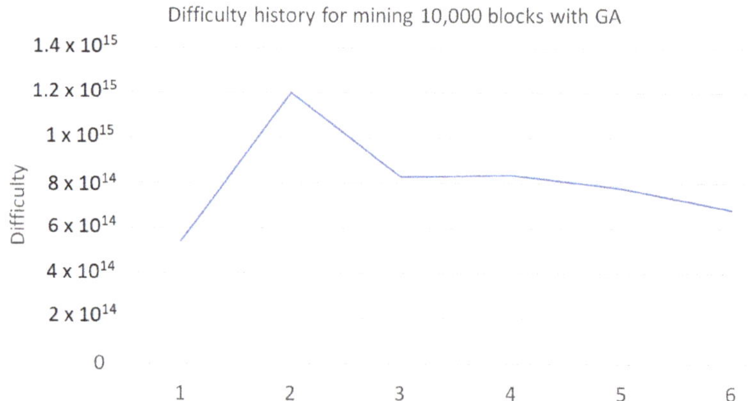

Figure 13. Difficulty history (Scenario 4).

4.4. Application Considerations

The average elapsed GA time (population size = 200 and maximum generation = 50) was 165.74 s or 2.7 min during our simulations, with parallel processing enabled, using 10 threads out of a total of 16 threads. A rise in central processing unit (CPU) usage was observed during this period, with a maximum of around 90% with an Intel Xeon E5-2650 v2 processor (2.6 GHz, 8 cores and 16 threads). If the GA was applied on an actual Bitcoin blockchain network, the machines running the nodes may suffer some loss of processing power, and this varies between different hardware specifications. One response is to detect an anomaly or abnormality in the block time. When the average actual block time is greater or smaller than a predetermined threshold, the GA runs in place of the default parameters. Furthermore, to ensure consistency across all the nodes, the total actual time taken to mine the previous N blocks is used as the determinant seed in the GA, otherwise, the nodes may arrive at different parameters and state consensus would be lost.

Although lower standard deviations for average block time and difficulty were obtained, increases in t_{MBP} and r_s were found. Although a stale block is not directly harmful and does not cause major problems on the network, there are a few ways in which it can impact the network slightly, such as causing poor propagation of the network [23]. In addition, double spending is one of the potential problems caused by stale blocks. On 27 January 2020, a USD 3 double-spend from a stale block occurred, which was the first stale block found since 16 October 2019. Because of the low value involved, it was very unlikely to be a targeted attack. In addition, the distributed aspect of a blockchain ensures that the success of an attack depends on managing 51% of the network's mining hash rate, making these types of attacks nearly impossible. The growing r_s also impacts t_{MBP}, as a certain bandwidth is lost when propagating stale blocks, thereby increasing the t_{MBP} whenever r_s is high.

Moreover, the GA was observed to be tending towards a block interval that was as low as possible. This was caused by a low block interval resulting in a low standard deviation of the average block time, which was one of the objective functions. A low block interval such as 1 s gives rise to a high r_s, as time is needed to propagate the block. As seen in one study, the longer the network propagation time, the more frequently miners were mining on top of old blocks, hence increasing the stale block rate [24]. It is worth looking at defining a new range for the optimization variables. Block intervals of 1 s or 2 s are not suggested, and therefore these numbers might be removed from the range for better results. In order to increase the GA's performance, new optimization variables and goal functions may be considered. For example, additional objective functions such as the block propagation time and stale block rate allow the GA to produce better optimization by not focusing solely on low block intervals and difficulty adjustment intervals (to obtain low average block times and difficulties), as they should also decrease the median block propagation time and stale block rate at the same time. However, this could also have an adverse effect, as these objective functions may interfere with the original intention of optimizing the block and difficulty adjustment intervals. On the other hand, decoupling the sliding window from the difficulty adjustment interval for the optimization variables should assist in improving the performance for low difficulty adjustment intervals. Alternatively, the timing of when to activate the GA for optimization and alternative strategies to prevent the GA from continuously optimizing could be examined. This will be the subject of future investigation.

5. Conclusions

A GA was proposed as an optimization approach for the difficulty adjustment intervals of a PoW blockchain. The aim of integrating the GA was to ensure that, by tuning the block and difficulty intervals, the blockchain could respond quickly to any sudden occurrence such as a large decrease or increase in hash rate. Using an evolutionary approach, the GA was expected to evolve to identify suitable intervals for changing the difficulty rates, in order to minimize the standard deviation of the average block time, defined as the time to generate one block. The GA optimized two variables (block interval and difficulty adjustment interval) based on the two objective functions (the standard deviations of average block time and difficulty). The optimal combination of variables was chosen and the new block mined was based on the new parameters.

The suggested difficulty adjustment technique aimed to be reliable enough to reduce the standard deviation of difficulty variations, resulting in minimal volatility. The purpose was to produce equal and consistent difficulty outputs from each chain in the network, while keeping the computation simple. However, issues such as when to activate the GA for optimization and how to prevent the GA from continuously optimizing could also be investigated.

Author Contributions: Conceptualization, T.T.V.Y. and I.K.T.T.; methodology, Z.H.C.; software, Z.H.C.; validation, Z.H.C., T.T.V.Y. and I.K.T.T.; formal analysis, Z.H.C.; investigation, Z.H.C.; resources, T.T.V.Y. and I.K.T.T.; data curation, Z.H.C.; writing—original draft preparation, Z.H.C.; writing—review and editing, Z.H.C., T.T.V.Y. and I.K.T.T.; visualization, Z.H.C.; supervision, T.T.V.Y.

and I.K.T.T.; project administration, T.T.V.Y. and I.K.T.T.; funding acquisition, T.T.V.Y. and I.K.T.T. All authors have read and agreed to the published version of the manuscript.

Funding: This research was funded by the Ministry of Higher Education, Malaysia, under the Fundamental Research Grant Scheme, with grant number FRGS/1/2018/ICT02/MMU/03/6.

Conflicts of Interest: The authors declare no conflict of interest. The funders had no role in the design of the study; in the collection, analyses, or interpretation of data; in the writing of the manuscript, or in the decision to publish the results.

Abbreviations

The following abbreviations are used in this manuscript:

PoW	Proof-of-work
GA	Genetic algorithm
BM	Bonded Mining
BCH	Bitcoin Cash
MSE	Mean squared error
VM	Virtual machine
t_{MBP}	Median block propagation time
r_s	Stale block rate

References

1. Nakamoto, S. Bitcoin: A Peer-to-Peer Electronic Cash System. 2008. Available online: https://bitcoin.org/bitcoin.pdf (accessed on 9 November 2021).
2. Haber, S.; Stornetta, W.S. How to time-stamp a digital document. *J. Cryptol.* **1991**, *3*, 99–111. [CrossRef]
3. Bayer, D.; Haber, S.; Stornetta, W.S. Improving the Efficiency and Reliability of Digital Time-Stamping. In *Sequences II*; Springer: New York, NY, USA, 1993; pp. 329–334.
4. Chin, Z.H.; Yap, T.T.V.; Tan, I.K.T. On the trade-offs of Proof-of-Work algorithms in blockchains. In *Computational Science and Technology*; Alfred, R., Lim, Y., Haviluddin, H., On, C.K., Eds.; Springer: Singapore, 2020; pp. 575–584.
5. Meshkov, D.; Chepurnoy, A.; Jansen, M. Short Paper: Revisiting Difficulty Control for Blockchain Systems. In *Data Privacy Management, Cryptocurrencies and Blockchain Technology*; Garcia-Alfaro, J., Navarro-Arribas, G., Hartenstein, H., Herrera-Joancomartí, J., Eds.; Springer International Publishing: Cham, Switzerland, 2017; pp. 429–436.
6. Voshmgir, S. *Token Economy: How Blockchains and Smart Contracts Revolutionize the Economy*; Shermin Voshmgir, BlockchainHub: Berlin, Germany, 2019.
7. Jung, H.; Lee, H. ECCPoW: Error-correction code based proof-of-work for ASIC resistance. *Symmetry* **2020**, *12*, 988. [CrossRef]
8. Dwork, C.; Naor, M. Pricing via Processing or Combatting Junk Mail. In Proceedings of the 12th Annual International Cryptology Conference on Advances in Cryptology—CRYPTO '92, Santa Barbara, CA, USA, 16–20 August 1992; Springer: London, UK; pp. 139–147.
9. Jakobsson, M.; Juels, A. Proofs of Work and Bread Pudding Protocols. In Proceedings of the IFIP TC6/TC11 Joint Working Conference on Secure Information Networks: Communications and Multimedia Security—CMS '99, Leuven, Belgium, 20–21 September 1999; pp. 258–272.
10. Antonopoulos, A.M. *Mastering Bitcoin: Programming the Open Blockchain*, 2nd ed.; O'Reilly Media, Inc.: Sebastopol, CA, USA, 2017.
11. Bissias, G.; Thibodeau, D.; Levine, B.N. Bonded Mining: Difficulty Adjustment by Miner Commitment. In *Data Privacy Management, Cryptocurrencies and Blockchain Technology*; Pérez-Solà, C., Navarro-Arribas, G., Biryukov, A., Garcia-Alfaro, J., Eds.; Springer International Publishing: Cham, Switzerland, 2019; pp. 372–390.
12. Noda, S.; Okumura, K.; Hashimoto, Y. An Economic Analysis of Difficulty Adjustment Algorithms in Proof-of-Work Blockchain Systems. 2019. Available online: https://papers.ssrn.com/sol3/papers.cfm?abstract_id=3410460 (accessed on 10 November 2021).
13. Aggarwal, V.; Tan, Y. A Structural Analysis of Bitcoin Cash's Emergency Difficulty Adjustment Algorithm. 2019. Available online: https://papers.ssrn.com/sol3/papers.cfm?abstract_id=3383739 (accessed on 10 November 2021)
14. Zhang, S.; Ma, X. A General Difficulty Control Algorithm for Proof-of-Work Based Blockchains. In Proceedings of the ICASSP 2020—2020 IEEE International Conference on Acoustics, Speech and Signal Processing (ICASSP), Barcelona, Spain, 4–8 May 2020; pp. 3077–3081. [CrossRef]
15. Zheng, K.; Zhang, S.; Ma, X. Difficulty Prediction for Proof-of-Work Based Blockchains. In Proceedings of the 2020 IEEE 21st International Workshop on Signal Processing Advances in Wireless Communications (SPAWC), Atlanta, GA, USA, 26–29 May 2020; pp. 1–5. [CrossRef]
16. Friedenbach, M. Fast(er) Difficulty Adjustment for Secure Sidechains. Available online: https://scalingbitcoin.org/transcript/milan2016/fast-difficulty-adjustment (accessed on 12 November 2021).

17. Deb, K.; Pratap, A.; Agarwal, S.; Meyarivan, T. A fast and elitist multiobjective genetic algorithm: NSGA-II. *IEEE Trans. Evol. Comput.* **2002**, *6*, 182–197. [CrossRef]
18. Nebro, A.J.; Durillo, J.J.; Vergne, M. Redesigning the JMetal Multi-Objective Optimization Framework. In Proceedings of the Companion Publication of the 2015 Annual Conference on Genetic and Evolutionary Computation—GECCO Companion '15, Madrid, Spain, 11–15 July 2015; Association for Computing Machinery: New York, NY, USA; pp. 1093–1100. [CrossRef]
19. Chin, Z.H.; Yap, T.T.V.; Tan, I.K.T. Simulating Difficulty Adjustment in Blockchain with SimBlock. In Proceedings of the 2nd ACM International Symposium on Blockchain and Secure Critical Infrastructure—BSCI '20, , Taipei, Taiwan, 6 October 2020; Association for Computing Machinery: New York, NY, USA; pp. 192–197. [CrossRef]
20. Croman, K.; Decker, C.; Eyal, I.; Gencer, A.E.; Juels, A.; Kosba, A.; Miller, A.; Saxena, P.; Shi, E.; Gün Sirer, E.; et al. On Scaling Decentralized Blockchains. In *Financial Cryptography and Data Security*; Clark, J., Meiklejohn, S., Ryan, P.Y.A., Wallach, D., Brenner, M., Rohloff, K., Eds.; Springer: Berlin/Heidelberg, Germany, 2016; pp. 106–125.
21. Gervais, A.; Ritzdorf, H.; Karame, G.O.; Capkun, S. Tampering with the Delivery of Blocks and Transactions in Bitcoin. In Proceedings of the 22nd ACM SIGSAC Conference on Computer and Communications Security—CCS '15, Denver, CO, USA, 12–16 October 2015; ACM: New York, NY, USA; pp. 692–705. [CrossRef]
22. Neudecker, T. *Characterization of the Bitcoin Peer-to-Peer Network (2015–2018)*; Technical Report 1; Karlsruher Institut für Technologie (KIT): Karlsruhe, Germany, 2019. [CrossRef]
23. Chipolina, S. Bitcoin Blockchain Sees Two Stale Blocks in One Day. Available online: https://decrypt.co/43380/bitcoin-blockchain-sees-two-stale-blocks-in-one-day (accessed on 17 November 2021).
24. Rahmadika, S.; Siwan, N.; Lee, K.; Kweka, B.; Rhee, K.H. The Dilemma of Parameterizing Propagation Time in Blockchain P2P Network. *J. Inf. Process. Syst.* **2020**, *16*, 699–717. [CrossRef]

MDPI

Article

A New Approach to COVID-19 Detection: An ANN Proposal Optimized through Tree-Seed Algorithm

Muhammet Fatih Aslan [1,*], Kadir Sabanci [1] and Ewa Ropelewska [2]

[1] Department of Electrical and Electronics Engineering, Karamanoglu Mehmetbey University, Karaman 70100, Turkey; kadirsabanci@kmu.edu.tr

[2] Fruit and Vegetable Storage and Processing Department, The National Institute of Horticultural Research, Konstytucji 3 Maja 1/3, 96-100 Skierniewice, Poland; ewa.ropelewska@inhort.pl

* Correspondence: mfatihaslan@kmu.edu.tr; Tel.: +90-541-438-5285

Abstract: Coronavirus disease (COVID-19), which affects the whole world, continues to spread. This disease has infected and killed millions of people worldwide. To limit the rate of spread of the disease, early detection should be provided and then the infected person should be quarantined. This paper proposes a Deep Learning-based application for early and accurate diagnosis of COVID-19. Compared to other studies, this application's biggest difference and contribution are that it uses Tree Seed Algorithm (TSA)-optimized Artificial Neural Networks (ANN) to classify deep architectural features. Previous studies generally use fully connected layers for end-to-end learning classification. However, this study proves that even relatively simple AlexNet features can be classified more accurately with the TSA-ANN structure. The proposed hybrid model provides diagnosis with 98.54% accuracy for COVID-19 disease, which shows asymmetric distribution on Computed Tomography (CT) images. As a result, it is shown that using the proposed classification strategy, the features of end-to-end architectures can be classified more accurately.

Keywords: ANN; AlexNet; COVID-19; transfer learning; TSA

Citation: Aslan, M.F.; Sabanci, K.; Ropelewska, E. A New Approach to COVID-19 Detection: An ANN Proposal Optimized through Tree-Seed Algorithm. *Symmetry* **2022**, *14*, 1310. https://doi.org/10.3390/sym14071310

Academic Editors: Kóczy T. László and István A. Harmati

Received: 28 May 2022
Accepted: 22 June 2022
Published: 24 June 2022

1. Introduction

The SARS-CoV-2 virus, which emerged in 2019, affected the whole world and is still an ongoing problem all over the world [1,2]. Although it is proclaimed by the World Health Organization (WHO) that the mortality rate of the virus is lower than other coronaviruses, the high spread rate of the virus obligated the WHO to declare the virus, also called COVID-19, as a pandemic on 11 March 2020 [3–5]. Due to the contagious effect of COVID-19, curfews were imposed in many countries and social life was restricted. Because of this restriction, people spent more time at home and used social media platforms more intensively [6]. Worldwide, from 8 December 2020 to the present (10 June 2022), more than 530 million people have been infected and 6.3 million deaths have occurred due to COVID-19 [7]. Fever, dry cough, loss of appetite, and fatigue are the most common symptoms of COVID-19. In some cases it is possible to encounter liver injury, septic shock, and pneumonia [8].

The main diagnostic approaches in COVID-19 today are generally real-time Reverse Transcription Polymerase Chain Reaction (rRT-PCR), chest Computed Tomography (CT) imaging, X-ray imaging, etc. [9]. The rRT-PCR is a method that reveals the presence of a specific genetic sequence in a pathogen which can be a virus. The most important advantage of this method is the ability to create an almost instant result. Therefore, the rRT-PCR test is the most-used diagnostic method to detect COVID-19 [10]. However, the percentage of rRT-PCR positives in throat swab samples is reported as roughly 30–60% because of limitations in sample collection, kit performance and transportation [11,12].

The most common alternative to rRT-PCR is chest Computed Tomography (CT) imaging. CT images of different angles of the chest area are used for COVID-19 diagnosis. This method has better sensitivity in COVID-19 diagnosis than rRT-PCR tests [9,12]. Unlike the rRT-PCR test, CT imaging does not need any extra equipment over hospital equipment. Although the CT imaging method has many advantages as mentioned above, the main disadvantage of the CT imaging method is that professional personnel are required for the interpretation of CT images [13]. Although this need can be easily met under normal conditions, the heavy workload of specialist doctors can cause problems in pandemic conditions. All healthcare professionals are also under a heavy workload, as the COVID-19 pandemic is causing significant constraints on healthcare systems around the world [14].

Recently, Artificial Intelligence (AI), which represents a symmetrical imitation of the human brain, is becoming more capable in many areas including medical imaging tasks such as CT imaging, MR imaging, and X-ray imaging [13]. In literature, AI has been employed for various tasks such as bone age determination, COVID-19 diagnosis, abnormal problems in the chest or tuberculosis detection etc. [15–17]. Deep Learning (DL), which is the trend application of artificial intelligence today, is now successfully used in many medical diagnostic applications due to the huge amount of data available. One of the most important features of DL is that it processes big data efficiently. It also eliminates the need to manually extract image features. In this respect, it provides superiority to Machine Learning (ML) methods such as Support Vector Machines (SVM) and Artificial Neural Networks (ANN). In DL, hierarchical feature extraction is performed by deriving high-level features from lower-level features.

Researchers have conducted numerous studies in the past to diagnose COVID-19 based on Convolutional Neural Networks (CNN) to take advantage of DL. Most of these works use popular CNN architectures such as AlexNet, ResNet, Xception, etc. Wu et al. [18] presented a ResNet50 architecture-based algorithm to identify COVID-19 patients, and their accuracy rate was 76%. Ardakani et al. [19] presented a DL-based application using CT scan images to diagnose COVID-19. They benchmarked different CNN models trained on CT scan images with each other, and finally calculated that ResNet-101 and Xception models had 99.51% and 99.02% accuracy rates, respectively. These models had better accuracy than the other CNN models. Jaiswal et al. [20] developed a transfer learning-based application using the pre-trained Densenet 201 architecture. This CNN module classified COVID-19 and non-COVID-19 data with 96.21% accuracy. Wang et al. [21] stated that a clinical diagnosis can be achieved before pathogenic testing with AI techniques, and in this context, they conducted a study that analyzed changes in CT images of infected patients. For this, they modified the pre-trained Inception model. The overall accuracy was 89.66%. Sethy and Behera [22] extracted features using a large number of different pre-trained CNN models such as ResNet50, ResNet101, InceptionV3, GoogleNet, and VGG16. The authors then provided the classification with SVM. They stated that the ResNet-SVM structure provides high accuracy of 95.33%. Deng et al. [23] used five CNN models, including Xception, ResNet50, etc., to diagnose COVID-19 from chest X-ray and CT images. They achieved accuracies of 84% and 75% for Chest X-ray and CT scan images, respectively. Narin et al. [24] used CNN models, ResNet50, Inception-ResNetV2, InceptionV3, ResNet152, and ResNet101 for diagnosis of COVID-19-infected patients using chest X-ray images. The ResNet50 model gave the most successful classification accuracy according to the performance values obtained using five-fold cross-validation. Aslan et al. [13] performed a novel hybrid model to classify chest X-ray images as COVID-19, Viral Pneumonia, or Normal. For hybrid architecture, they combined modified AlexNet and Bidirectional Long Short-Term Memory (BiLSTM). At the end of the study, the authors stated that the hybrid architecture achieved 98.702% classification success. Mukherjee et al. [25] designed a CNN—tailored Deep Neural Network (DNN) that can collectively train and test both CT scans and chest X-ray images. In practice, the authors provided an overall accuracy of 96.28%. Aslan et al. [26] first performed lung segmentation with ANN for the detection of COVID-19 with CT images. They classified features extracted from segmented lung images with deep CNN models using different machine learning methods. They determined the parameters of each machine learning model by hyperparameter optimization. At the end of the study, the authors stated that the highest accuracy was obtained with the DenseNet201-SVM structure, with an accuracy of 96.29%. Serte et al. [27] used a generative adversarial network (GAN) in addition to data augmentation to increase the number of CT samples in a small dataset. They then fed the images generated with data augmentation and GAN into different CNN models and performed COVID-19 detection. At the end of the study, 89% diagnostic accuracy was achieved.

It is possible to observe that previously made DL-based methods often have high classification success in detecting COVID-19. These studies are promising in terms of early COVID-19 detection, but many hyperparameters which belong to the DL network may increase also the success rate of the detection of infection. In literature, Artificial Bee Colony (ABC) [28], Bayesian Optimization [29], Tree Seed Algorithm (TSA) [30], and other optimization methods are used in various fields. These optimization techniques can tune hyperparameters in the deep network for a stronger prediction. However, for the detection of COVID-19, few studies have used optimization methods to improve current success. Ucar and Korkmaz [31] used Bayesian optimization to fine-tune the hyperparameters of the SqueezeNet CNN architecture, and as a result, they achieved a classification success of 98.3%. Nour et al. [32] designed a CNN model with five convolution layers for COVID-19 diagnosis. The deep-network features obtained with this CNN model were used to feed the ML algorithms, the k-nearest neighbor, SVM, and decision tree. The hyperparameters of the ML methods have been optimized using the Bayes optimization algorithm. As a result, the highest classification accuracy of 98.97% was obtained with SVM. Elaziz et al. [33] applied a modified Bayesian optimization algorithm together with an ML for the selection of useful features from CT images. Toğaçar et al. [34] used MobileNetV2 and SqueezeNet architectures to detect COVID-19. The most effective features were determined with the Social-Mimic Optimization method. Then, the features were fused and classified with the SVM algorithm, and their accuracy rate was 99.27%.

In this study, a DL-based COVID-19 detection system using CT-scans is proposed. A publicly available SARS-CoV-2 CT scan dataset [35,36] is used in this study. The dataset includes 1230 and 1252 CT scans of uninfected and infected patients, respectively. The application of COVID-19 detection can be addressed in two steps. The first step performs feature extraction and classification by modifying the pre-designed AlexNet architecture, as in many previous studies. The second step gives the extracted features to the optimized ANN classifier to improve the results in the first step. The ANN is trained by Tree Seed Algorithm (TSA) optimization method. In order to determine the best structure, various transfer functions and hidden layer neuron numbers are examined. The results prove that optimized ANN increases classification success.

The contributions of this study are listed as follows:

- A high-accuracy diagnosis of COVID-19 has been performed automatically.
- To improve the classification performance of end-to-end architectures, ANN is applied instead of fully connected layers.
- For a high classification performance, ANN is optimized by the TSA method.
- The proposed method can increase the diagnostic accuracy of previous studies using the CNN model.
- The applied experimental work outperforms many previous studies.

The remainder of the paper is organized as follows. In Section 2, the dataset and suggested methods are explained in detail. Section 3 contains the results of the proposed method. Section 4 discusses the proposed method and makes comparisons with previous studies. Finally, Section 5 concludes the work overall and provides information about future work.

2. Materials and Methods

This section provides detailed information about the dataset, TSA optimization, mAlexNet architecture, and the proposed hybrid model.

2.1. SARS-CoV-2 Ct-Scan Dataset

In this study, a publicly available dataset called SARS-CoV-2 Ct-Scan Dataset [35,36] is used due to having large numbers of data and the ability to compare previous studies done with this dataset. Five researchers from Lancaster University United Kingdom and Public Hospital of the Government Employees of Sao Paulo created this dataset with 1252 infected patients' CT scans and 1230 uninfected patients' CT scans in May 2020. There are a totally of 2482 CT scans in the dataset. The data in this dataset were generated by collecting asymmetrical CT scans of real patients in hospitals in Sao Paulo, Brazil. CT scans of two patients with positive and negative diagnosis of COVID-19 are shown in Figure 1.

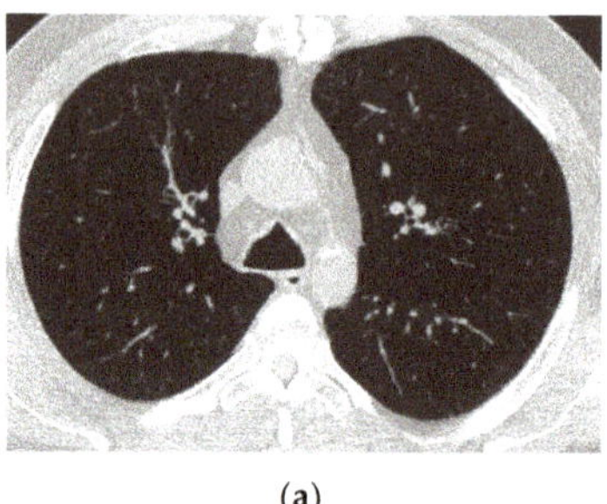
(a)

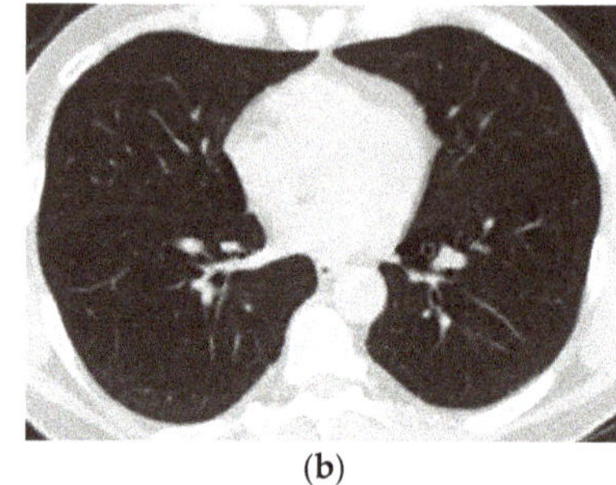
(b)

Figure 1. Positive and negative images in the SARS-CoV-2 Ct-Scan dataset. (**a**) CT scan of a patient infected with COVID-19; (**b**) CT scan of a patient not infected with COVID-19.

2.2. The Tree Seed Algorithm (TSA)

The TSA algorithm is based on natural phenomena between trees and their spread seeds. The land where trees are grown is assumed as search space. Each tree is a solution candidate for the problem to be optimized. The seeds are produced by the trees in order to grow a new tree, which is going to be a candidate for the solution of the problem. The purpose of the optimization process is to generate new coordinates of the seeds using the available information. Two equations used for this purpose are presented in Equations (1) and (2) [30].

$$S_{ij} = T_{ij} + \propto_{ij} \times (B_j - T_{rj}) \tag{1}$$

$$S_{ij} = T_{ij} + \propto_{ij} \times (T_{ij} - T_{rj}) \tag{2}$$

In Equations (1) and (2), S_{ij} represents the jth dimension of ith seed and T_{ij} is the jth dimension of the ith tree. B_j is the jth dimension of the best tree ever found. T_{rj} is the jth dimension of the rth tree, which is randomly selected from the population. $\propto$ is the scaling factor which is randomly produced between 1 and -1. In seed production, two equations are presented, thus there has to be a criterion that determines which equation will be used. This is controlled by the search tendency (ST) parameters in the range of 0 to 1. While the higher value of ST makes the algorithm condense on local solutions, the lower ST value forces it to make a global search. Although the number of seeds produced by a single tree was completely random, TSA's performance analysis determined that TSA performed best when the number of seeds produced by each tree was between 10% and 25% of the population size [30].

In the initial of the algorithm, the optimization parameters are created randomly in a range of specific upper and lower bounds for each parameter. Then, for each dimension of each seed of each tree in the population, Equation (1) or Equation (2) is randomly applied and new seeds are created. The termination criteria are checked each iteration. These criteria may be the number of iteration limits, error threshold or no change in error, etc. When termination criteria are met, the best result in the population is reported. The flowchart of TSA is presented in Figure 2.

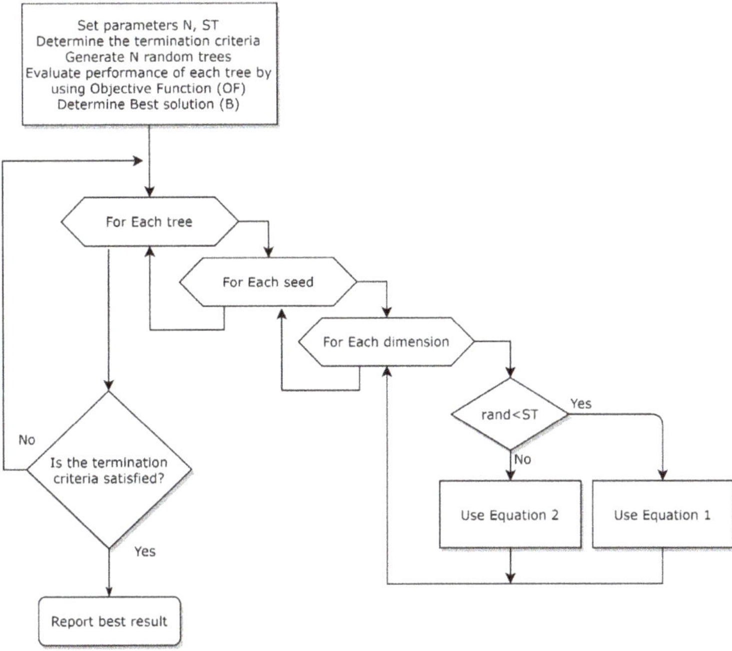

Figure 2. Flow diagram of Tree Seed Algorithm.

2.3. Proposed mAlexNet Architecture

Due to the high performance of CNNs in various image recognition applications, they are highly preferred [37,38]. In this study, to classify the CT-scan images, a modified AlexNet (mAlexNet) created by using transfer learning is used. AlexNet is made up of 25 layers together with a convolution layer, Rectified Linear Unit (ReLU), fully connected (fc) layer, normalization layer, pooling layer, etc. The three layers at the end of the AlexNet model have been slightly modified (fine-tuning) to be compatible with existing study inputs. In order to distinguish whether a CT image shows COVID-19 or not, these final three layers are removed. Other parameters of the pre-trained AlexNet architecture are retained. Instead of the removed layers, new layers suitable for this study are added, as in Figure 3. This new architecture is called the modified AlexNet (mAlexNet). In the fc8 layer of pre-trained AlexNet, the number of the neurons is 1000. The number of features used for classification in our application is 25. Therefore, the number of neurons in the fc8 layer is changed to 25 in the mAlexNet architecture. In Figure 3, the mAlexNet structure is presented. Table 1 shows the layer parameters and training options of mAlexNet. When the training options are examined, it can be realized that the Mini Batch parameter, which provides the training data to be divided into smaller parts, is 40. The optimization algorithm Stochastic Gradient Descent with Momentum (SGDM) is applied to reduce the training error. Parameters of the SGDM algorithm are also shown in Table 1.

$$\theta_{l+1} = \theta_l - \alpha \nabla E(\theta_l) + \gamma(\theta_l - \theta_{l-1}) \tag{3}$$

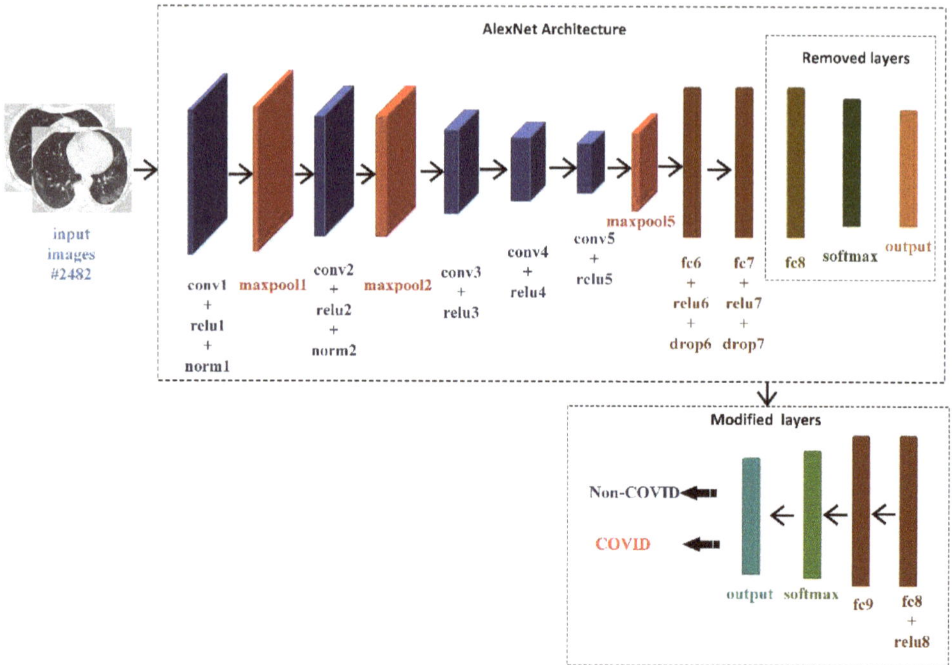

Figure 3. mAlexnet model.

Table 1. mAlexnet layer parameters and training options.

Layer Name	Size	Filter Size	Stride	Padding	Output Channel	Activation Function
conv1	55 × 55	11 × 11	4	0	96	relu
maxpool1	27 × 27	3 × 3	2	0	96	-
conv2	27 × 27	5 × 5	1	2	256	relu
maxpool2	13 × 13	3 × 3	2	0	256	-
conv3	13 × 13	3 × 3	1	1	384	relu
conv4	13 × 13	3 × 3	1	1	384	relu
conv5	13 × 13	3 × 3	1	1	256	relu
maxpool5	6 × 6	3 × 3	2	0	256	-
fc6	-	-	-	-	4096	relu
fc7	-	-	-	-	4096	relu
fc8	-	-	-	-	25	relu
fc9	-	-	-	-	2	softmax

Training Options

Optimization Alg.	Maximum Epoch	Mini Batch Size	Initial Learning Rate (α)	Momentum (γ)
SGDM	25	40	0.001	0.95

With SGDM, the weights of the network are updated according to the estimation error. Equation (3) is used to update the weights. With this equation, the weights are updated according to the loss function ($E(\theta_l)$). In order for the error value to decrease after each update, the loss function is moved in the direction of the negative gradient. The speed of this movement depends on the learning rate (α). The contribution of the current weight value to the weight value in the previous iteration is determined by the Momentum (γ) coefficient. The values of α and γ parameters in Equation (3) are shown in Table 1.

2.4. Proposed TSA-ANN Model

This section discusses the classification of mAlexNet features using TSA-optimized ANN (TSA-ANN). The mAlexNet architecture described in the previous section is combined with the TSA-ANN structure as in Figure 4. 25; features extracted via mAlexNet are given to the TSA-ANN structure for classification.

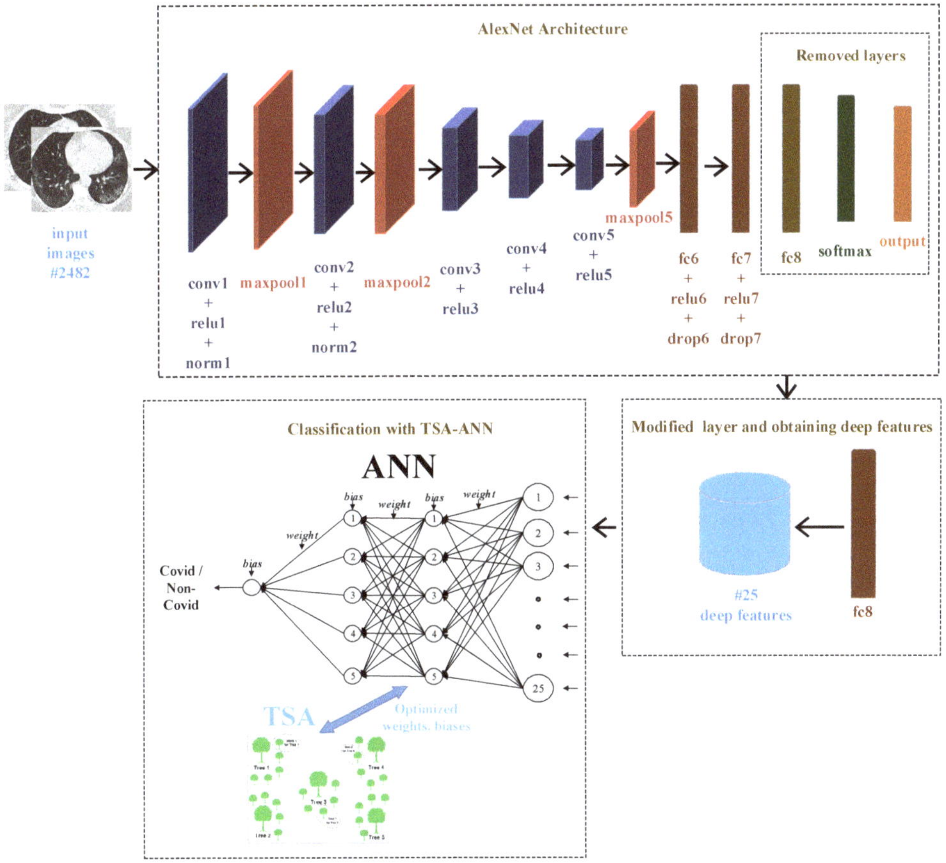

Figure 4. mAlexNet—TSA-ANN model.

The proposed ANN modsel consists of an input layer, two hidden layers, and an output layer. In the input layer, there are 25 neurons. It is also the same as the size of the input features. The number of neurons in both hidden layers is determined to be five by the trial and error method. The output layer has a single neuron.

The activation functions used in the neural network are Hyperbolic Tangent Sigmoid and Logarithmic Sigmoid functions. Since the range of the output layer is [0 1], the activation function of the output layer is determined to be a Logarithmic Sigmoid function. The activation functions of hidden layers are determined to be Hyperbolic Tangent Sigmoid functions by the trial and error method. The Logarithmic Sigmoid and Hyperbolic Tangent Sigmoid functions are presented in Equations (4) and (5), respectively.

$$AF(x) = \frac{1}{1+e^{-x}} = logsig(x) \tag{4}$$

$$AF(x) = \frac{e^{x}-e^{-x}}{e^{x}+e^{-x}} = tansig(x) \tag{5}$$

The applied input data are transferred to each neuron in the first hidden layer by multiplying the corresponded coefficient $W_{(k,i,j)}$. The (k,i,j) indexes indicate the connection coefficient in the kth layer, between ith data and jth neuron. For each neuron, there is also a bias value represented by $B_{(k,j)}$ symbol. Then, the sum all of the inputs to a neuron is applied to an Activation Function (AF). The result of this function is the value of the neuron. This value of the jth neuron in the kth layer $V_{(k,j)}$ can be calculated by the following formula:

$$V_{(k,j)} = AF\left(\sum_{i=1}^{N_{k-1}} V_{(k-1,i)} W_{(k,i,j)} + B_{(k,j)}\right) \tag{6}$$

where the N_{k-1} is the number of neurons in the $(k-1)$th layer. As seen from the formula, in each layer, there are $N_{k-1} * N_k$ W parameters and N_k B parameters. Thus, the total number of parameters that need to be optimized is $(N_{k-1}+1) * N_k$. In Table 2, the number of neurons, activation functions, and parameter count of each layer are presented.

Table 2. Neural network properties.

Layers	Number of Neurons	Activation Function	Parameter Count
Layer 1 (Input Layer)	25	-	-
Layer 2 (1st Hidden Layer)	5	Hyperbolic Tangent Sigmoid	130
Layer 3 (2st Hidden Layer)	5	Hyperbolic Tangent Sigmoid	30
Layer 4 (Output Layer)	1	Logarithmic Sigmoid	6

As presented in Table 2, the weight and bias values of each layer after the input layer of the ANN are optimized with TSA. This optimization process can be thought of as the training phase of the network. For example, 25 × 5 + 5 = 130 parameters for Layer 2, 5 × 5 + 5 = 30 parameters for Layer 3 and 5 × 1 + 1 = 6 parameters for the output layer should be optimized. This optimization process aims to minimize the error value. Therefore, the objective function is related to the error between the target value and the predicted value. A total of 166 network parameters need to be optimized for the training of the network. During this period, the training dataset is used. The Mean Absolute Error (MAE) (see Equation (7)) for the training dataset generated by the ANN with network parameters is used as the objective function of TSA optimization. The n, y, and $\hat{y}$ in Equation (7) represent the number of data, the actual output value, and the estimated output value, respectively. During the optimization, the error value is reduced in each iteration. After the optimization is completed, the ANN is updated with the determined parameters. This ANN is called a trained ANN from now on. The performance of the trained ANN is determined by using a test dataset. These results are presented in detail in Section 3.

$$MAE = \frac{1}{\mathrm{n}} \sum |y - \hat{y}| \tag{7}$$

3. Results

The performance of both architectures expressed above is calculated by using a test dataset. Both architectures are created with the architectural parameters specified in Table 1. Training and testing of the methods developed within the scope of this application are carried out on a laptop computer with Intel Core i7-7700HG CPU, NVIDIA GeForce GTX 1050 4 GB, 16 GB RAM. Firstly, 80% of the COVID-19 dataset is used in the training of the mAlexNet structure in Figure 3. The training graph is successfully obtained as seen in Figure 5. Then, the performance of mAlexNet is tested with the remaining 20% of the dataset, called the test dataset.

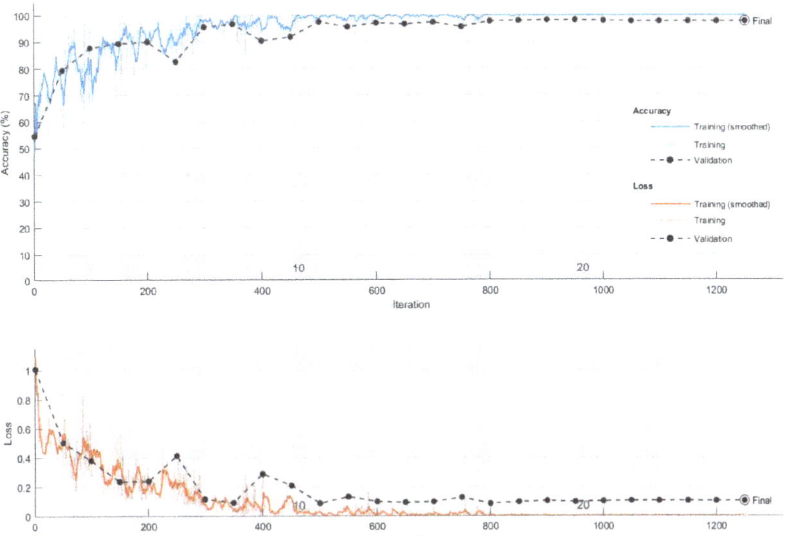

Figure 5. Training graphics of mAlexNet model.

Figure 6 shows the confusion matrix of mAlexNet and TSA-ANN implementations. Figure 6a is the confusion matrix obtained by classifying test data with mAlexNet. Figure 6b is the confusion matrix obtained as a result of the application that enables classification of mAlexNet features with the optimized ANN (TSA-ANN) structure. The parameters used for the optimization algorithm in the TSA-ANN application are as follows: population size: 50, number of iterations: 1000, and search tendency: 0.1. When both confusion matrices are examined, it is seen that COVID-19 infected scans with asymmetrical patterns are more successfully diagnosed with the TSA-ANN structure. Additionally, various performance metrics such as accuracy, specificity, MCC, F1-score, recall, and precision are calculated and presented. For each metric, formulations are given between Equations (8)–(13) [13,39,40]. These metrics prove the robustness and unbiasedness of the classification performance obtained as a result of the proposed method. Table 3 shows the results for these performance metrics. According to Table 3, both applications are successful in detecting COVID-19 infection using CT scans. But the performance of the hybrid TSA-ANN added mAlexNet model is better than the single mAlexNet structure in terms of accuracy. The accuracy rates of models are 97.92% and 98.54% for mAlexNet and mAlexNet + TSA-ANN, respectively. Additionally, Precision, Sensitivity, F1-Score, MCC, and Specificity indicators are also better in hybrid architecture, as seen in Table 3. It is possible to conclude that deep architecture using the TSA-ANN classification has better performance and unbiased classification.

$$\text{Accuracy} = \frac{tp + tn}{tp + fp + tn + fn} \times 100 \tag{8}$$

$$\text{Sensitivity} = \frac{tp}{tp + fn} \tag{9}$$

$$\text{Specificity} = \frac{tn}{tn + fp} \tag{10}$$

$$\text{Precision} = \frac{tp}{tp + fp} \tag{11}$$

$$\text{F1-score} = \frac{2tp}{2tp + fp + fn} \tag{12}$$

$$\text{MCC} = \frac{(tp * tn) - (fn * fp)}{\sqrt{(tp + fn) * (tn + fp) * (tp + fp) * (tn + fn)}} \quad (13)$$

tp: True Positive
tn: True Negative
fp: False Positive
fn: False Negative

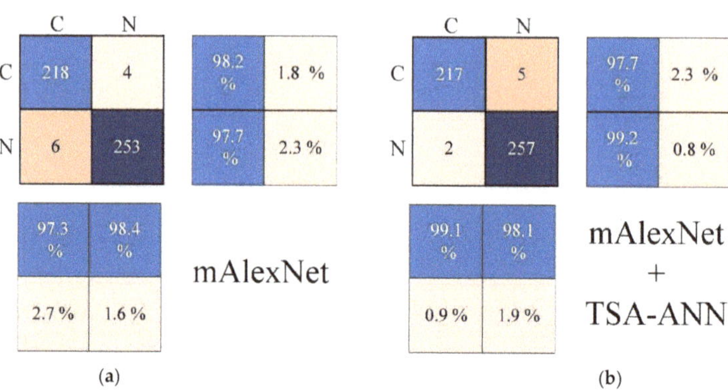

Figure 6. Confusion matrices of proposed models. (**a**) mAlexNet; (**b**) mAlexNet + TSA-ANN.

Table 3. Performance metrics of the proposed models.

Model	Accuracy	Sensitivity	Specificity	Precision	F1-Score	MCC
mAlexNet	97.92	0.9820	0.9768	0.9732	0.9776	0.9582
mAlexNet + TSA-ANN	98.54	0.9775	0.9923	0.9909	0.9841	0.9708

To prove the robustness of the proposed method, the same application is also performed on a different dataset. For this, the COVID-19 Radiology database [17] is preferred. Only the COVID-19 and Normal classes in this dataset are used. However, the data numbers in this dataset are unbalanced. The number of COVID-19 classes is 219, while the Normal (non-COVID) class number is 1341. Therefore, first, data augmentation is applied for the class with few data, resulting in a total number of images of 2436. 80% of all data is allocated as training and 20% as testing. Figure 7 shows the confusion matrices obtained after applying the proposed method on the COVID-19 Radiology database. Figure 7a is provided by mAlexNet and average accuracy is 99.38%. Figure 7b is provided by mAlexNet-TSA-ANN with an average accuracy of 99.59%. The results show that the proposed method is also effective in different datasets.

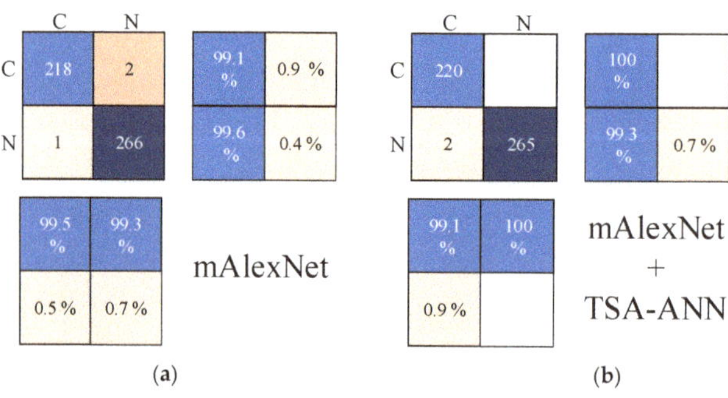

Figure 7. Confusion matrices of proposed models on COVID-19 Radiology database. (**a**) mAlexNet; (**b**) mAlexNet + TSA-ANN.

4. Discussion

Many DL-based diagnostic methods developed so far have directly used CNN models. However, different techniques can be used to improve the performance obtained with a CNN model. The aim of this study is to show that the mAlexNet–TSA-ANN hybrid structure is highly effective in detecting COVID-19. Moreover, this experiment was implemented on AlexNet, which is simpler than other CNN models. The results showed that even a fine-tuned AlexNet (mAlexNet) combined with the TSA-ANN method outperforms many studies. Some of the previous studies suggesting different DL-based methods are shown in Table 4. According to Table 4, it is seen that the proposed mAlexNet–TSA-ANN structure is superior to previous studies using the same dataset.

Table 4. Benchmarking of the proposed mAlexNet-TSA-ANN with previous studies.

Study	Method	Accuracy (%)
Soares et al. [41]	xDNN	97.38%
Özkaya et al. [42]	CNN + SVM	94.03%
Tetila et al. [43]	Inception-Resnet-v2	98.4%
Panwar et al. [44]	Color Visualization (Grad-CAM)	95%
Wang et al. [45]	Contrastive Learning	90.83 ± 0.93
Jaiswal et al. [20]	DenseNet201	96.25%
Öztürk et al. [46]	WOA-MLP	88.06%
Silva et al. [47]	EfficientNet	98.50%
Yazdani et al. [48]	Attentional Convolutional Network	92%
Proposed approach	**mAlexNet—TSA-ANN**	**98.54%**

Although the proposed method is superior to previous studies in terms of accuracy, it has some limitations. The most important limitation is that the developed optimization approach slows down the training process of the network. Only the mAlexNet structure realizes faster training. Another limitation is that the TSA optimization method used also needs parameter optimization. In addition, the results obtained are valid only for the datasets used. Its success on other different datasets is unpredictable. For a general success, a study should be done that includes all different datasets.

5. Conclusions

Early detection of COVID-19 disease is crucial due to the high rate of spread among humans. This study uses computed tomography (CT) images to quickly and accurately diagnose COVID-19. Both proposed models include the AlexNet architecture. In the first model, mAlexNet architecture is created and classification is performed. In the second model, 25 features extracted from each image with mAlexNet are given to the TSA-optimized ANN for classification. The different aspect of this study compared to other studies is the use of the TSA-ANN-based hybrid model. The high accuracy achieved with the hybrid architecture shows that the model is a powerful classifier. By feeding the trained hybrid architecture with CT images, COVID-19 can be detected quickly. In future studies, to increase the success of the system, the lungs will be determined from the CT images by applying semantic segmentation. Then, the number of images will be increased with the obtained lung images through data enhancement methods. It is planned to achieve higher accuracy by testing different architectures with these images.

Author Contributions: M.F.A.: Investigation and writing—reviewing and editing. K.S.: Investigation, methodology, supervision. E.R.: Investigation and supervision, editing. All authors have read and agreed to the published version of the manuscript.

Funding: This research received no external funding.

Institutional Review Board Statement: Not applicable.

Informed Consent Statement: Not applicable.

Data Availability Statement: The dataset is cited in the manuscript. The data access link is https://www.kaggle.com/datasets/plameneduardo/sarscov2-ctscan-dataset (accessed on 10 May 2022).

Conflicts of Interest: The authors declare no conflict of interest.

References

1. Green, A. Li wenliang. *Lancet* **2020**, *395*, 682. [CrossRef]
2. Khan, A.I.; Shah, J.L.; Bhat, M.M. Coronet: A deep neural network for detection and diagnosis of COVID-19 from chest X-ray images. *Comput. Methods Programs Biomed.* **2020**, *196*, 105581. [CrossRef]
3. Mahase, E. Coronavirus: COVID-19 has killed more people than SARS and MERS combined, despite lower case fatality rate. *BMJ* **2020**, *368*, m641. [CrossRef]
4. Spinelli, A.; Pellino, G. COVID-19 pandemic: Perspectives on an unfolding crisis. *Br. J. Surg.* **2020**, *107*, 785–787. [CrossRef]
5. WHO. *Coronavirus Disease 2019 (COVID-19) Situation Report–51, 11 March 2020*; WHO: Geneva, Switzerland, 2020.
6. Thakur, N.; Han, C.Y. An Exploratory Study of Tweets about the SARS-CoV-2 Omicron Variant: Insights from Sentiment Analysis, Language Interpretation, Source Tracking, Type Classification, and Embedded URL Detection. *Preprints* **2022**, 2022050238. [CrossRef]
7. JH. COVID-19 Map—Johns Hopkins Coronavirus Resource Center. Available online: https://coronavirus.jhu.edu/map.html (accessed on 10 June 2022).
8. Guan, W.-J.; Ni, Z.-Y.; Hu, Y.; Liang, W.-H.; Ou, C.-Q.; He, J.-X.; Liu, L.; Shan, H.; Lei, C.-L.; Hui, D.S. Clinical characteristics of coronavirus disease 2019 in China. *N. Engl. J. Med.* **2020**, *382*, 1708–1720. [CrossRef]
9. Jiang, Y.; Chen, H.; Loew, M.; Ko, H. COVID-19 CT Image Synthesis with a Conditional Generative Adversarial Network. *arXiv* **2020**, arXiv:2007.14638. [CrossRef]
10. Jawerth, N. How is the COVID-19 virus detected using real time RT-PCR. *IAEA Bull.* **2020**, *61*, 8–11.
11. Ai, T.; Yang, Z.; Hou, H.; Zhan, C.; Chen, C.; Lv, W.; Tao, Q.; Sun, Z.; Xia, L. Correlation of chest CT and RT-PCR testing in coronavirus disease 2019 (COVID-19) in China: A report of 1014 cases. *Radiology* **2020**, *296*, 200642. [CrossRef] [PubMed]
12. Fang, Y.; Zhang, H.; Xie, J.; Lin, M.; Ying, L.; Pang, P.; Ji, W. Sensitivity of chest CT for COVID-19: Comparison to RT-PCR. *Radiology* **2020**, *296*, 200432. [CrossRef]
13. Aslan, M.F.; Unlersen, M.F.; Sabanci, K.; Durdu, A. CNN-based transfer learning-BiLSTM network: A novel approach for COVID-19 infection detection. *Appl. Soft Comput.* **2020**, *98*, 106912. [CrossRef] [PubMed]
14. Barello, S.; Palamenghi, L.; Graffigna, G. Stressors and Resources for Healthcare Professionals during the COVID-19 Pandemic: Lesson Learned From Italy. *Front. Psychol.* **2020**, *11*, 2179. [CrossRef] [PubMed]
15. Chen, X.; Li, J.; Zhang, Y.; Lu, Y.; Liu, S. Automatic feature extraction in X-ray image based on deep learning approach for determination of bone age. *Future Gener. Comput. Syst.* **2020**, *110*, 795–801. [CrossRef]
16. Kieu, P.N.; Tran, H.S.; Le, T.H.; Le, T.; Nguyen, T.T. Applying multi-CNNs model for detecting abnormal problem on chest X-ray images. In Proceedings of the 2018 10th International Conference on Knowledge and Systems Engineering (KSE), Ho Chi Minh City, Vietnam, 1–3 November 2018; pp. 300–305.
17. Chowdhury, M.E.; Rahman, T.; Khandakar, A.; Mazhar, R.; Kadir, M.A.; Mahbub, Z.B.; Islam, K.R.; Khan, M.S.; Iqbal, A.; Al Emadi, N. Can AI help in screening viral and COVID-19 pneumonia? *IEEE Access* **2020**, *8*, 132665–132676. [CrossRef]
18. Wu, X.; Hui, H.; Niu, M.; Li, L.; Wang, L.; He, B.; Yang, X.; Li, L.; Li, H.; Tian, J. Deep learning-based multi-view fusion model for screening 2019 novel coronavirus pneumonia: A multicentre study. *Eur. J. Radiol.* **2020**, *128*, 109041. [CrossRef]
19. Ardakani, A.A.; Kanafi, A.R.; Acharya, U.R.; Khadem, N.; Mohammadi, A. Application of deep learning technique to manage COVID-19 in routine clinical practice using CT images: Results of 10 convolutional neural networks. *Comput. Biol. Med.* **2020**, *121*, 103795. [CrossRef]
20. Jaiswal, A.; Gianchandani, N.; Singh, D.; Kumar, V.; Kaur, M. Classification of the COVID-19 infected patients using DenseNet201 based deep transfer learning. *J. Biomol. Struct. Dyn.* **2021**, *39*, 5682–5689. [CrossRef]
21. Wang, S.; Kang, B.; Ma, J.; Zeng, X.; Xiao, M.; Guo, J.; Cai, M.; Yang, J.; Li, Y.; Meng, X.; et al. A deep learning algorithm using CT images to screen for Corona virus disease (COVID-19). *Eur. Radiol.* **2021**, *31*, 6096–6104. [CrossRef]
22. Sethy, P.K.; Behera, S.K. Detection of Coronavirus Disease (COVID-19) Based on Deep Features. *Preprints* **2020**, 2020030300. [CrossRef]
23. Deng, X.; Shao, H.; Shi, L.; Wang, X.; Xie, T. A classification–detection approach of COVID-19 based on chest X-ray and CT by using keras pre-trained deep learning models. *Comput. Model. Eng. Sci.* **2020**, *125*, 579–596. [CrossRef]
24. Narin, A.; Kaya, C.; Pamuk, Z. Automatic detection of coronavirus disease (COVID-19) using X-ray images and deep convolutional neural networks. *arXiv* **2020**, arXiv:2003.10849. [CrossRef] [PubMed]
25. Mukherjee, H.; Ghosh, S.; Dhar, A.; Obaidullah, S.M.; Santosh, K.; Roy, K. Deep neural network to detect COVID-19: One architecture for both CT Scans and Chest X-rays. *Appl. Intell.* **2021**, *51*, 2777–2789. [CrossRef]
26. Aslan, M.F.; Sabanci, K.; Durdu, A.; Unlersen, M.F. COVID-19 diagnosis using state-of-the-art CNN architecture features and Bayesian Optimization. *Comput. Biol. Med.* **2022**, *142*, 105244. [CrossRef] [PubMed]
27. Serte, S.; Dirik, M.A.; Al-Turjman, F. Deep Learning Models for COVID-19 Detection. *Sustainability* **2022**, *14*, 5820. [CrossRef]

28. Karaboga, D.; Basturk, B. A powerful and efficient algorithm for numerical function optimization: Artificial bee colony (ABC) algorithm. *J. Glob. Optim.* **2007**, *39*, 459–471. [CrossRef]
29. Frazier, P.I. A tutorial on bayesian optimization. *arXiv* **2018**, arXiv:1807.02811.
30. Kiran, M.S. TSA: Tree-seed algorithm for continuous optimization. *Expert Syst. Appl.* **2015**, *42*, 6686–6698. [CrossRef]
31. Ucar, F.; Korkmaz, D. COVIDiagnosis-Net: Deep Bayes-SqueezeNet based Diagnostic of the Coronavirus Disease 2019 (COVID-19) from X-ray Images. *Med. Hypotheses* **2020**, *140*, 109761. [CrossRef]
32. Nour, M.; Cömert, Z.; Polat, K. A novel medical diagnosis model for COVID-19 infection detection based on deep features and Bayesian optimization. *Appl. Soft Comput.* **2020**, *97*, 106580. [CrossRef]
33. Elaziz, M.A.; Hosny, K.M.; Salah, A.; Darwish, M.M.; Lu, S.; Sahlol, A.T. New machine learning method for image-based diagnosis of COVID-19. *PLoS ONE* **2020**, *15*, e0235187. [CrossRef]
34. Toğaçar, M.; Ergen, B.; Cömert, Z. COVID-19 detection using deep learning models to exploit Social Mimic Optimization and structured chest X-ray images using fuzzy color and stacking approaches. *Comput. Biol. Med.* **2020**, *121*, 103805. [CrossRef] [PubMed]
35. Angelov, P.; Almeida Soares, E. SARS-CoV-2 CT-Scan Dataset. Available online: https://www.kaggle.com/plameneduardo/sarscov2-ctscan-dataset (accessed on 10 June 2022).
36. Angelov, P.; Almeida Soares, E. Explainable-by-Design Approach for COVID-19 Classification via CT-Scan. *medRxiv* **2020**. Available online: https://www.medrxiv.org/content/10.1101/2020.04.24.20078584v1 (accessed on 10 June 2022).
37. Krizhevsky, A.; Sutskever, I.; Hinton, G.E. Imagenet classification with deep convolutional neural networks. *Commun. ACM* **2017**, *60*, 84–90. [CrossRef]
38. Simonyan, K.; Zisserman, A. Very deep convolutional networks for large-scale image recognition. *arXiv* **2014**, arXiv:1409.1556.
39. Aslan, M.F.; Sabanci, K.; Durdu, A. A CNN-based novel solution for determining the survival status of heart failure patients with clinical record data: Numeric to image. *Biomed. Signal Process. Control* **2021**, *68*, 102716. [CrossRef]
40. Sabanci, K.; Aslan, M.F.; Ropelewska, E.; Unlersen, M.F. A convolutional neural network-based comparative study for pepper seed classification: Analysis of selected deep features with support vector machine. *J. Food Process Eng.* **2021**, *45*, e13955. [CrossRef]
41. Soares, E.; Angelov, P.; Biaso, S.; Froes, M.H.; Abe, D.K. SARS-CoV-2 CT-Scan Dataset: A Large Dataset of Real Patients CT Scans for SARS-CoV-2 Identification. *medRxiv* **2020**. Available online: https://www.medrxiv.org/content/10.1101/2020.04.24.20078584v3 (accessed on 10 June 2022).
42. Özkaya, U.; Öztürk, Ş.; Budak, S.; Melgani, F.; Polat, K. Classification of COVID-19 in Chest CT Images using Convolutional Support Vector Machines. *arXiv* **2020**, arXiv:2011.05746.
43. Tetila, E.; Bressem, K.; Astolfi, G.; Sant'Ana, D.A.; Pache, M.C.; Pistori, H. System for quantitative diagnosis of COVID-19-associated pneumonia based on superpixels with deep learning and chest CT. *Res. Sq.* **2020**. [CrossRef]
44. Panwar, H.; Gupta, P.K.; Siddiqui, M.K.; Morales-Menendez, R.; Bhardwaj, P.; Singh, V. A deep learning and grad-CAM based color visualization approach for fast detection of COVID-19 cases using chest X-ray and CT-Scan images. *Chaos Solitons Fractals* **2020**, *140*, 110190. [CrossRef]
45. Wang, Z.; Liu, Q.; Dou, Q. Contrastive cross-site learning with redesigned net for COVID-19 CT classification. *IEEE J. Biomed. Health Inform.* **2020**, *24*, 2806–2813. [CrossRef]
46. Öztürk, Ş.; Yigit, E.; Özkaya, U. Fused Deep Features Based Classification Framework for COVID-19 Classification with Optimized MLP. *Selcuk Univ. J. Eng. Sci. Technol.* **2020**, *8*, 15–27.
47. Silva, P.; Luz, E.; Silva, G.; Moreira, G.; Silva, R.; Lucio, D.; Menotti, D. COVID-19 detection in CT images with deep learning: A voting-based scheme and cross-datasets analysis. *Inform. Med. Unlocked* **2020**, *20*, 100427. [CrossRef] [PubMed]
48. Yazdani, S.; Minaee, S.; Kafieh, R.; Saeedizadeh, N.; Sonka, M. COVID CT-net: Predicting COVID-19 from chest CT images using attentional convolutional network. *arXiv* **2020**, arXiv:2009.05096.

Article

Utilizing Language Models to Expand Vision-Based Commonsense Knowledge Graphs

Navid Rezaei [1] and Marek Z. Reformat [1,2,*]

[1] Department of Electrical and Computer Engineering, University of Alberta, Edmonton, AB T6G 1H9, Canada
[2] Information Technology Institute, University of Social Sciences, 90-113 Łódź, Poland
* Correspondence: reformat@ualberta.ca

Abstract: The introduction and ever-growing size of the transformer deep-learning architecture have had a tremendous impact not only in the field of natural language processing but also in other fields. The transformer-based language models have contributed to a renewed interest in commonsense knowledge due to the abilities of deep learning models. Recent literature has focused on analyzing commonsense embedded within the pre-trained parameters of these models and embedding missing commonsense using knowledge graphs and fine-tuning. We base our current work on the empirically proven language understanding of very large transformer-based language models to expand a limited commonsense knowledge graph, initially generated only on visual data. The few-shot-prompted pre-trained language models can learn the context of an initial knowledge graph with less bias than language models fine-tuned on a large initial corpus. It is also shown that these models can offer new concepts that are added to the vision-based knowledge graph. This two-step approach of vision mining and language model prompts results in the auto-generation of a commonsense knowledge graph well equipped with physical commonsense, which is human commonsense gained by interacting with the physical world. To prompt the language models, we adapted the chain-of-thought method of prompting. To the best of our knowledge, it is a novel contribution to the domain of the generation of commonsense knowledge, which can result in a five-fold cost reduction compared to the state-of-the-art. Another contribution is assigning fuzzy linguistic terms to the generated triples. The process is end to end in the context of knowledge graphs. It means the triples are verbalized to natural language, and after being processed, the results are converted back to triples and added to the commonsense knowledge graph.

Keywords: commonsense; knowledge graph; linguistic terms; language models; deep learning

Citation: Rezaei, N.; Reformat, M.Z. Utilizing Language Models to Expand Vision-Based Commonsense Knowledge Graphs. *Symmetry* **2022**, *14*, 1715. https://doi.org/10.3390/sym14081715

Academic Editors: Kóczy T. László and István A. Harmati

Received: 22 June 2022
Accepted: 11 August 2022
Published: 17 August 2022

1. Introduction

There has been a renewed interest in commonsense as a stepping stone toward achieving human-level intelligence. Some of the new research has shown how important commonsense knowledge graphs can be in training artificial intelligence (AI) models, which exhibit commonsense [1,2].

Commonsensical concepts should be symmetric to any changes in their representation. In the case of an ideal commonsense knowledge graph and an ideal language model, transforming concepts between the two representations of knowledge should not change their meaning. By an ideal language model, we mean a language model that is sufficiently large and capable that can understand language and all the concepts within. At the same time, an ideal commonsense knowledge graph is a knowledge graph that contains all correct commonsensical concepts.

The knowledge-symmetric transformation depends on the architecture of the language model and the knowledge graph, both of which are not ideal. These issues make deriving a transformation process that symmetrically maps knowledge from one to the other challenging and impractical. To compensate for this, we introduce a prompting methodology based on questions and answers to extract from the language model the knowledge missing in the knowledge graph. In that way, the symmetry of concepts is preserved by mapping them between two knowledge storage paradigms.

The main building blocks of knowledge graphs used to represent commonsense knowledge are subjects, predicates, and objects. Subjects and objects are other words for the nodes of the graph. The tail of a relationship is called an object, and the head is called a subject. The directed edge connecting the two is called a predicate. Knowledge graphs are directed heterogeneous graphs in some sense.

Artificial intelligence (AI) models are reported to have limited commonsense abilities [3,4]. Acquiring commonsense by AI systems can make the sample efficient in adapting to new environments, as proposed in [5]. Commonsense knowledge graphs can help AI systems both explicitly and implicitly: explicitly by querying the commonsense knowledge graph itself, or implicitly by knowledge transfer methods, such as the fine-tuning of language models as reported in [1]. This is similar to how a BERT model is fine-tuned on a SQuAD dataset for reading comprehension [6]. In addition, expressing commonsense knowledge in the symbolic format can help with commonsense knowledge explainability and the vetting process.

Using only vision to generate commonsense knowledge as proposed in [7,8] has its advantages and disadvantages. By processing images and videos, we can perceive visual cues that are not usually written or spoken about, but they make our common understanding of how physical entities exist and interact. On the other hand, fine-tuned vision-based deep learning models are limited to the concept and relation vocabulary that they are trained on, which is usually limited, and are not usually capable of understanding the intricacies of natural language. An ideal self-supervised vision model, which can absorb and learn all the visual interactions, could theoretically suffice. However, the current vision models have shortcomings that we believe can be addressed via the utilization of language models. For example, scene graph generation models are limited regarding the number of detected relationship types. Increasing the number of relationship types does not provide a satisfying solution, as there is still a bias toward the frequently seen relationships in the supervised training [9].

In this paper, we explore and use the extra knowledge that language models offer to expand on the limited auto-generated vision-based commonsense knowledge graphs. We chose to use few-shot learning in larger transformer-architecture-based language models, as larger models have shown to perform well on language benchmarks without requiring further fine-tuning on a specific task. We experiment with not only adding new concepts to the vision-based commonsense knowledge graph but also new types of relationships with fuzzy-style linguistic weights.

1.1. Commonsense Definition

Having a good definition of commonsense is imperative to better understand and discuss the work and results. Commonsense is simple, as almost everyone knows it, and is challenging, as no one often talks or writes about it.

Yann LeCun, an inventor of convolutional neural networks, believes that a collection of models of the world that represents what is likely, plausible, or impossible makes our commonsense [3]. John McCarthy classifies human commonsense into two categories of knowledge and ability. The commonsense ability is the action based on the gained commonsense knowledge [10].

Commonsense knowledge is inherently uncertain and context-dependent. The degree of correctness of commonsense knowledge depends on the common group of observers. For instance, the people who live in the northern hemisphere know July to be a hot summer month, while the people in the southern hemisphere observe it as a colder winter month.

Commonsense can also be classified into different topics, such as physical interactions, order of events, and social dynamics. In this paper, we mainly focus on physical commonsense, such as the usage of an object and its relative location, compared to other objects.

In a nutshell, commonsense knowledge graphs are graphs that represent facts and relations between them that are characteristic of real-world scenarios and situations. Such graphs focus on elements and aspects related to everyday activities, arrangements of things, and normal/natural circumstances, such as *flower in vase*, *tree has trunk*, *food on plate*, *shoe is less likely made of metal*, or *arm is most likely to be able to move, bend and be strong*.

Such facts seem very obvious and normal/natural for a human being, but this knowledge is not easy to be acquired by a machine. The gap in the processes of learning that type of information is filled out by techniques and methods linked to collecting and representing commonsense knowledge.

1.2. Contributions

The goal is to utilize transformer-based language models to expand vision-based commonsense knowledge graphs. In this paper, we propose an extension of the methodology for constructing a commonsense knowledge graph proposed in [7,8] with a technique based on questions and answers prompting very large language models. The new technique addresses generating prompts that are used as inputs to the language models. First, a prompt is entered into the model. Then, the obtained response that contains facts/information is added to expand a knowledge graph. The method is illustrated in Figure 1.

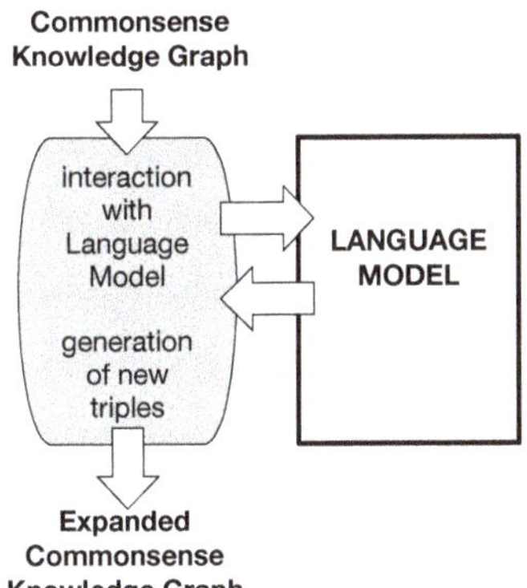

Figure 1. Expansion of a vision-based commonsense knowledge graph with relevant but new information.

In particular, the contributions of the paper are as follows:

- A multi-modal methodology for constructing commonsense knowledge graphs;
- A process of generating question/answer-based prompts for language models based on triples extracted from an existing commonsense knowledge graph, or based on the input from users;
- An expansion of the standard structure of knowledge graphs by introducing an approach to add degrees of likeliness as indicators of the 'strength' of triples that are added to commonsense knowledge graphs; the degrees are expressed with linguistic terms , such as **more likely**, **less likely**;
- An evaluation process based on Amazon Mechanical Turk.

2. Related Work

The work presented in this paper falls into the category of tasks that focus on completion and expansion of a commonsense knowledge base. There is related literature that addresses the methods and tools to achieve these goals.

2.1. Expansion of Knowledge Bases

The work presented in [11,12] focuses on link prediction between known entities within a graph. These methods are not able to expand beyond the current conceptual knowledge in the graph. They are more suited toward finding possible relations between currently known concepts.

Recent works have tried to use language models, especially transformers [13], to achieve better results in the tasks of the completion and expansion of the knowledge base. The authors of [14] use language models to construct knowledge graphs: they assume to have a subject and object and then use the language model to predict an appropriate relationship between them.

Ref. [15] indicates that using the next token prediction capability of pre-trained language models, one can use them as a factual knowledge base, e.g., to find the birthplace of a specific person. Among the language models analyzed, the largest transformer-based language model, BERT-Large [16], performed better than others. This paper confirms the overall consensus in the research community that the larger the language models become, the more capable they become.

There has been recent works to train very large generic transformer-based language models, such as GPT-3 by OpenAI [17], Meta's OPT-175B [18], and Google's PaLM [19]. There is a common consensus among all recent findings that larger language models can potentially be more capable of performing diverse tasks. Additionally, they do not need costly fine-tuning and data collection. Yet, providing appropriate prompts to language models can be challenging.

Processes of generating prompts are a subject of recent research publications. Prompts serve as input to large language models [20] and are used to reduce the amount of data required for fine-tuning [21]. By prompt, we mean a set of tokens and a short text that constitute the input to the model. Prompts could have different purposes, such as providing context, tone, or a sample of expected responses. They are part of the few-shot-learning process and are usually used instead of fine-tuning a language model. While prompts benefit the overall performance, their design does not follow a specific rule. Some even call the process 'prompt engineering.' Question and answering tasks are improved by few-example prompts when using large language models [17]. Extra chain-of-thought language prompts that contain reasoning steps are shown to improve more complex tasks related to arithmetics and commonsense [20]. The chain-of-thought process helps find missing parts of knowledge [22]. The work is similar to unsupervised data creation [23]. However, questions and answers used in this paper serve as prompts to foundation models. They are not used directly on text for reading comprehension.

2.2. *Construction and Expansion of Commonsense Knowledge*

A body of literature [1,2] focuses on annotating commonsense knowledge graphs to train language models for predicting commonsense information based on the given subject and predicate. The human-annotated knowledge graphs are typically in the size of millions and cover social interactions, events, and entity commonsense. The ATOMIC-COMET work [1] is based on manually creating a commonsense knowledge graph. This graph is used to train a small language model, such as GPT-2, on the human-annotated data. Our approach is different. We focus on generating a commonsense knowledge graph automatically rather than manually. The method comprises two phases, the first based on vision and the second enriching the results using language. The manual generation of commonsense knowledge graphs can become costly, as shown in [21]. Our approach seems to be more similar to [2], where GPT-3 is utilized to generate commonsense knowledge graphs. The proposed method is different in multiple ways. One is that we use a two-step method, where the feed for GPT-3 is provided by visual data, while [2] uses human-annotated data. Moreover, [2] only generates the most probable results, while we generate both highly probable and less probable results. Our approach has a cost of roughly one-fifth of the method described in [2] when considering the linguistic generation part and using the same GPT-3 model size. The reduced cost is because our prompt method accommodates the generation of $N = 5$ triples in one pass. Another difference is that [2] is proposed to only find an object given the subject and predicate, while our approach works in both ways and can suggest an appropriate subject given the predicate and object.

Several papers have experimented with the new very large transformers, such as GPT-3 [2]. This work focuses on prompting GPT-3 with some annotated commonsense triples, then extracting GPT-3's commonsense and adding it to a graph. It assumes pre-defined predicates and does not explicitly discuss the weight of the triples. The process takes subjects and predicates and uses causal language models to predict the most suitable objects. The method we introduce in this paper can also predict the triples' subjects.

It was discussed in [24,25] that training a model on commonsense knowledge base completion (CKBC) task suffers from low-coverage training data. Therefore, training on specific data results in the model's over-fitting and reduces its performance on novel data. Based on these observations, we focused our efforts on generically trained language models, which are large enough to accommodate few-shot learning.

A few recent works report on generating knowledge graphs from visual data. As an example, the NEIL method [26] extracts object relationships in images and results in 10,000 triples using 10 types of predicates.

One of the main physical commonsense knowledge graphs is ConceptNet [27]. As much as it can be helpful and treated as a reference, it has some drawbacks that our work can potentially resolve in the future. First, ConceptNet is mainly human annotated and cannot be continuously and cost-effectively updated. Our work suggests a methodology to continuously and automatically update the missing commonsense knowledge. Second, ConceptNet has a limited predicate related to location–the vague *AtLocation* predicate. Our method is able to enrich the commonsense knowledge with more fine-grained relations, such as *Above*, *Below*, and others. Third, ConceptNet is limited in terms of its predicate types, too. Our approach can enrich ConceptNet with new types of predicates, such as *NotIsA* or *CanEat*. An essential weakness of ConceptNet is its lack of context. For example, finding a desk in a classroom is more probable than in a bar. Our approach can potentially expand and enrich ConceptNet with weighted contextual relations. Moreover, it can be done automatically if part of ConceptNet is used as a seed commonsense knowledge graph.

3. Image-Based Construction of Commonsense Knowledge Graph

In our previous works, we introduced methodologies to generate a commonsense knowledge graph, called *world-perceiving knowledge graph* (*WpKG*), by only using visual data [7,8]. Like human infants who gain commonsense details about their physical world before they learn to express them in language, the introduced process focuses on deducing commonsense knowledge by observing many images.

The *WpKG* paper [7] introduces a methodology to auto-generate commonsense using deep learning models to perform object detection and relation prediction. The final *WpKG* graph has 7000 triples using 50 predicate and 150 entity types. [8] expands on the previous work to generate contextual and weighted commonsense knowledge graph, *C-WpKG*, in 93 contexts using state-of-the-art object and relation detection models. In the following sub-sections, we describe the process of reaching these results.

3.1. Extraction of Scene Graphs

The first step in the process is to analyze each image individually by detecting the existing objects and extracting possible relations between the objects in the image. The resulting graph representing objects in images as nodes and their relationships as edges is called a scene graph.

A convolutional neural network (CNN) model, such as Faster-RCNN [28], is used to detect the objects. To produce image features, ResNeXt-101-FPN CNN model [29] is utilized, which is needed for the region proposal network (RPN) of the Faster-RCNN model. The output of the pre-trained object detection model includes objects in the image, together with their bounding boxes and class scores.

To predict relations between the objects and generate a scene graph for each image, the MOTIFS model [30] unbiased by the Causal-TDE method [9] is used. Then, the scene graph for each image is generated based on the object features and relations between them.

3.2. Fusion of Scene Graphs

Regularly observed phenomena make up collective commonsense knowledge. Similarly, we aggregate the scene graphs extracted from the images into a single knowledge graph that comprises possible commonsense relations. To differentiate between relationships to know if a phenomenon is a one-time event or a typical one, we assign weights to the links representing the relations. Different methods of assigning weights to the observations are investigated. Among them, a probability-based approach is selected. It correlates the most with human commonsense during human evaluations. This weight assignment method follows Equation (1).

$$w_{t_i} = \sum_{j=1}^{|D_T|} \delta(t_i, t_j) \cdot P(t_j) \tag{1}$$

where w_{t_i} is a weight of the t_i triple, $\delta(\cdot)$ is Kronecker delta function, $P(t_j)$ represents the probability of detecting each instance of triple t_j, which is made of a subject (s), predicate (p), and object (o). The weights are also normalized by $\max\{w_{t_i} : t_i \in D_T\}$. The list of all detected triples is represented by D_T.

Variations of the same method have been shown to work in context-free and contextual scenarios. In this paper, we only focus on context-free visual commonsense knowledge.

4. Expanding Knowledge Graph Using Language Model

The automatic construction of commonsense knowledge graphs requires retrieving commonsense knowledge. It seems natural—also for a human being—to start that process by analyzing images and pictures representing real-world situations. Yet, to further increase commonsense knowledge and expand knowledge graphs, other sources of information are required and beneficial. One of them is verbal, textual information.

Therefore, to diversify information embedded in vision-based commonsense knowledge graphs and further expand them, we propose a human-like method of assimilating commonsense knowledge using linguistic-based data sources.

4.1. Methodology

The proposed method is intuitive and straightforward. It starts with interaction with a language model using short texts created based on the commonsense knowledge graph to be expanded. Then, the obtained results, i.e., the retrieved pieces of information and facts, are added to the graph as triples. The overview of the process is illustrated in Figure 2. It shows the *WpKG* as a graph from which some triples are extracted. The information from these triples is used to instantiate prompt templates (Section 4.3) that represent training data for a language model. The instantiated prompts are entered into the model. As a result, the obtained pieces of information are converted into new triples. These new triples are added to the *WpKG*, leading to its expansion.

4.2. Language Models

Larger language models, such as GPT-3, have shown promising results on diverse benchmarks with only a few examples of each task. The results are sometimes even comparable with smaller language models, which are fine-tuned on a large corpus of data. Recent research has shown the usefulness and effectiveness of large language models in automatic commonsense knowledge generation [2]. In this paper, we utilize different versions of a large language model, called GPT-3 [17], to expand vision-based commonsense knowledge graphs.

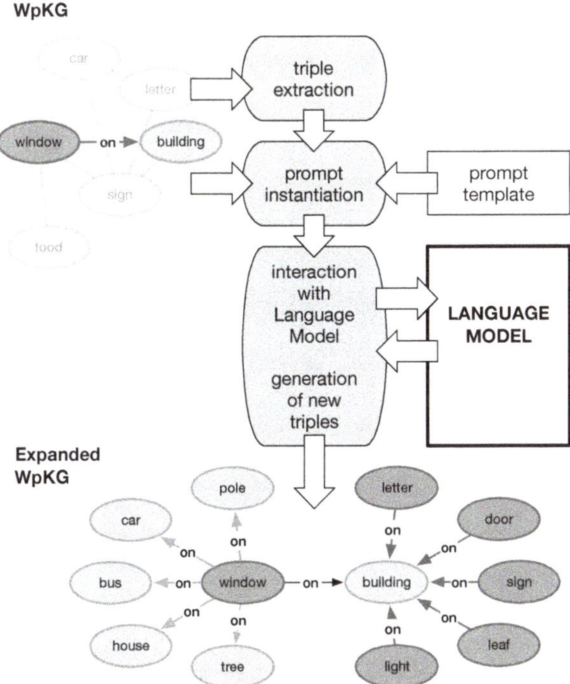

Figure 2. Process of expanding a graph using language model.

GPT-3 is a causal language model with almost the same yet larger architecture as previous iterations of the same model (GPT and GPT-2). The goal of a causal model is to predict the next token given the previous tokens. The language model assigns a probability to all the tokens to decide which one could happen next.

Choosing the highest probability next token may not be the best option, given the task. In this paper, we use nucleus (top-p) sampling to generate the text responses [31] and also adjust the temperature of the sampling to reach better results.

By reducing the temperature, we basically increase the likelihood of high-probability next tokens and reduce the likelihood of low-probability next tokens. This setting results in more deterministic next tokens to be chosen when selecting the next token randomly. The temperature is implemented as a coefficient inside the softmax function. Empirically, we observed that lower temperature works better for simpler cases, while the higher temperature can work for more complex cases that need diverse results, e.g., finding objects that are less likely to exist given a subject and a predicate.

In nucleus (top-p) sampling, instead of sampling from all the tokens, the algorithm chooses from the set of tokens that their cumulative probability of occurrence next is smaller than a given probability p. In our experiments, we keep the p value equal to one to choose from the most diverse vocabulary possible.

4.3. Language Model Prompts

Retrieving information from GPT-3 involves prompting the model with a few examples that serve as a few-shot learning training data. The content and the structure of the responses depend on these prompts. Therefore, experimentation with different prompts to achieve the desired structure is necessary. Formally, the examples that define the structure and content of an interaction with a language model are called *prompts*.

The purpose of a prompt is to 'show' the model how to interpret and respond to an input text appended to the prompt, which in our case is a question. For example, one wants to retrieve a piece of information about the most common items found on a table in a conference room. In such a case, the following prompt is constructed and used:

prompt:	*Q: What can be found on table in bar? Name five.*
	A: bottle, class, cup, napkin, fork.
	Q: What can be found on table in conference room? Name five.
GPT-3 response:	**paper, glass, laptop, phone, box.**

This example is a simple explanation of the role of the prompt. As it can be seen, the first part of the prompt—*Q* and *A*—is one-shot training data and 'teaches' the model that for a type of question like Q, a proper response looks like *A*. After that, the 'real' question *Q: What can be found on table in conference room?* is asked. Then, finally, the model responds with five items it 'thinks' represents the most suitable response.

Sometimes, one example is not enough, and multiple examples need to be provided to serve as few-shot learning training data. Empirically, we find that explaining the task and a well-defined question format help the model respond better.

To achieve more accurate results, we also utilize the chain-of-thought prompting method introduced in [20] for the fuzzy and the predicate expansion cases, described in Sections 5.2 and 5.3, respectively. In each example answer in the prompt, we hand-craft a reasoning that can help narrow down to the correct response. The model learns to generate a similar pattern and, as a result, generates a reasoning before answering the asked question.

5. Expansion of Commonsense Graph

To illustrate the benefits of using a language model for expanding the *WpKG*, we extract information from GPT-3 to construct different triples. It shows how versatile the interaction with the model can be and how different results are obtained. The presented utilization of GPT-3 involves the following scenarios:

- Asking for subjects and objects for given relations using a basic prompt template;
- Asking for the most and least likely subjects and objects for given relations to construct fuzzy triples;
- Asking for the most and least likely objects with novel relations given by a user.

Expanding the existing graph means 'asking' the language model to provide answers that contain the most suitable pieces of information that are directly added to the graph as nodes—subjects and objects—and relations that link the existing nodes to the newly added ones.

The questions are prepared based on templates that are initialized with facts/information obtained from the *WpKG* or from a user. Three sets of templates are constructed, one for each type of defined-above scenarios.

5.1. Simple Triples

In the beginning, a straightforward scenario that involves adding simple triples, i.e., triples that are not associated with degrees of strength of relations between subjects and objects, is presented. In such a case, GPT-3 is asked questions that result in retrieving from the model facts that are interpreted as subjects or as objects. It means that the questions are of the format $\langle ?s, relation_X, object_X \rangle$ when subjects are asked for, or $\langle subject_X, relation_X, ?o \rangle$ when objects are asked for. The retrieved subjects and objects are added as triples with the $relation_X$ to the *WpKG*.

In a nutshell, the process—for a single $relation_X$—is as follows:

- Extract five triples with the $relation_X$ from the *WpKG*.
- Select randomly one triple from the set of five, say, triple *k*; it is used in the process of customization of a prompt template for the $relation_X$.

   - Extract a set of five most popular objects Obj_k fitting ⟨$subject_k$, $relation_X$, -⟩ from the *WpKG*.
   - Extract a set of five most popular subjects Sub_k fitting ⟨ -, $relation_X$, $object_k$⟩ from the *WpKG*.
   - Audit the instantiated prompt and make changes if necessary.
- For each extracted triple ⟨$subject_i$, $relation_X$, $object_i$⟩:
   - Put $subject_i$ and $relation_X$ into the question template and append to the prompt.
   - Put the prompt to the language model to initiate the text generation.
   - Extract the five new objects Obj_{LM} from the generated text.
   - Add five new triples ⟨$subject_i$, $relation_X$, -⟩ with objects from Obj_{LM} to *WpKG*.
   - Put $relation_X$ and $object_i$ into the question template and append to the prompt.
   - Put the prompt to the language model to initiate the text generation.
   - Extract the five new subjects Sub_{LM} from the generated text.
   - Add five new triples ⟨-, $relation_i$, $object_X$⟩ with subjects from Sub_{LM} to *WpKG*.

As it is described above, the process of asking GPT-3 involves the instantiation of prompt templates. For the simple triples case, the prompt templates for asking for both *objects* and *subjects* are shown in Table 1. Following the aforementioned process, it can be seen that the prompts are filled out with facts/information obtained originally from *WpKG*, and the same initialization is used for prompting GPT-3 for all other *objects* or *subjects* obtained from the randomly selected $relation_X$s. Depending on the predicate $relation_X$, different variations of the prompt templates are created to result in meaningful questions and answers.

Table 1. Sample template for simple triple.

	SIMPLE_TEMPLATE_A for ⟨subject⟩
prompt:	*Answer with five items separated with comma.* *Q: What is* ⟨$relation_X$⟩ ⟨$object_k$⟩*? Name five.* *A: elements of* Sub_k *Q: What is* ⟨$relation_X$⟩ ⟨$object_i$⟩*? Name five.*
	SIMPLE_TEMPLATE_B for ⟨object⟩
prompt:	*Answer with five items separated with comma.* *Q: What* ⟨$subject_k$⟩ *can be* ⟨$relation_X$⟩*? Name five.* *A: elements of* Obj_k *Q: What* ⟨$subject_i$⟩ *can be* ⟨$relation_X$⟩*? Name five.*

The templates from Table 1 are used with five different relations: *behind, in, has, on,* and *watching*. The instantiated prompt templates, together with the results of querying GPT-3 for the relation *on*, are shown in Table 2 for extracting *subjects*, and in Table 3 for extracting *objects*.

It can be seen that, for example, selecting $object_A$ = *plate*, we obtain the following triples: ⟨*food, on, plate*⟩, ⟨*drink, on, plate*⟩, ⟨*utensils, on, plate*⟩, ⟨*napkin, on, plate*⟩, and ⟨*tablecloth, on, plate*⟩, Table 2. Similarly selecting $subject_A$ = *hair*, we obtain the triples such as ⟨*hair, on, head*⟩, ⟨*hair, on, beard*⟩, ⟨*hair, on, eyebrows*⟩, ⟨*hair, on, eyelashes*⟩, and ⟨*hair, on, pubic*⟩ (Table 3). Another example, this time in a graphical form, that shows an expansion of the triple ⟨*window, on, building*⟩ is illustrated in Figure 3. Besides the original triple, the figure includes its extension on both *subject* and *object* sides.

Table 2. Query and results for ⟨-, *on*, -⟩ for *subject*.

user:		*Answer with five items separated with comma.* *Q: What is on building? Name five.* *A: letter, door, sign, leaf, light.* *Q: What is* $\langle relation_A \rangle$ $\langle object_A \rangle$*? Name five.*
where:		$relation_A = on$ $object_A$ = *{ building, sign, man, plate, head }*
GPT-3 responses for		
	building:	$subject_A \in$ *{letter, door, sign, leaf, light}*
	sign:	$subject_A \in$ *{words, letters, numbers, shapes, colors}*
	man:	$subject_A \in$ *{shirt, pants, belt, shoes, socks}*
	plate:	$subject_A \in$ *{food, drink, utensils, napkin, tablecloth}*
	head:	$subject_A \in$ *{hair, hat, ear, eyebrow, eyelash }*

Table 3. Query and results for ⟨-, *on*, -⟩ for *object*.

user:		*Answer with five items separated with comma.* *Q: What window can be on? Name five.* *A: pole, car, bus, house, tree.* *Q: What* $\langle subject_B \rangle$ *can be* $\langle relation_B \rangle$*? Name five.*
where:		$subject_B$ = *{ window, letter, hat, food, hair }* $relation_B = on$
GPT-3 responses for		
	window:	$object_B \in$ *{pole, car, bus, house, tree}*
	letter:	$object_B \in$ *{A, B, C, D, E }*
	hat:	$object_B \in$ *{baseball, cowboy, graduation, party, winter}*
	food:	$object_B \in$ *{apple, banana, orange, pear, grape}*
	hair:	$object_B \in$ *{head, beard, eyebrows, eyelashes, pubic}*

Of course, not all obtained *subjects* and *objects* are correct, especially in the case of asking for *objects*. For example, triples generated for the subject *letter*, Table 3, are quite inferior. A human-wise evaluation was performed; see Section 6.3 for details.

In the prompts, we chose the *What* question word, as it is generic enough to result in diverse types of results. However, a more fine-tuned selection of the question word may result in more relevant results, as suggested in [23].

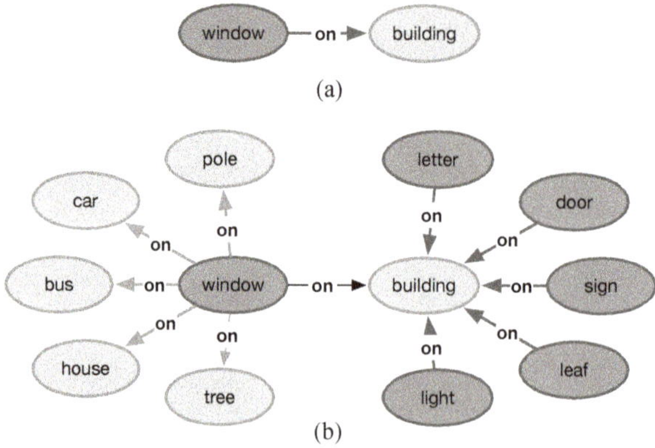

Figure 3. Expanded *WpKG*—simple triples: original triple (**a**); and after its extension (**b**).

We utilized the largest GPT-3 model, with 175 billion parameters, for the experiments. We started with a softmax temperature of 0.0 to obtain more deterministic results. However, we observed that the model sometimes shies away from generating text with this temperature setting and immediately generates an end token. To fix the problem, we increased the temperature to 0.7 and then to 1.0 to increase the chances for a good response.

5.2. Fuzzy Triples with Linguistic Terms

The remarkable abilities of GPT-3 can be utilized to extract *subjects* and *objects* when the triples need to be labeled with the degrees of the plausibility of their occurrence. Triples with such information can be added to the *WpKG* when the prompt, and its question-and-answer parts, used to query GPT-3 are constructed/designed in a specific way. The prompt templates presented in the previous section have to be modified.

To invoke responses from GPT-3 that give a quantifiable assessment of relation strength, the prompts should be more verbal to contextualize interaction with the model. The experiments with multiple approaches have led to the prompts that are the same, even if GPT-3 is asked to provide facts related to a variety of topics.

Due to the fact that two degrees of relation strength are considered, two prompts are designed and used: one for generating triples that represent high likeliness and one for building triples that are of low likeliness. Both of them are shown in Table 4. A quick look at them indicates that the prompts refer to quite different domains/topics—the questions are related to window and number. Yet, they work very well with the relations we use as examples—the same as for the simple triples in Section 5.1.

Another interesting 'feature' of these prompts is the very little need for instantiation. Only the last questions, Q_S for *subjects* and Q_O for *objects*, Table 4, are initialized to reflect the relations of interest.

Table 4. Template for fuzzy triple with linguistic terms.

	FUZZY_TEMPLATE_X for the linguistic term most likely
prompt:	*Answer with five items separated with comma.* Q: *What* **most likely** *has window? Name five.* A: *Window is usually used to see through. Therefore, train, building, house, car, bus.* Q: *What number can* **most likely** *be on? Name five.* A: *Number is made of digits and can be written on different things for information. Therefore, train, sidewalk, track, street, building.* Q_S: *What is* **most likely**$\langle relation_X\rangle$ $\langle object_X\rangle$*? Name five.* Q_O: *What does/is* $\langle subject_X\rangle$ **most likely** *be/-* $\langle relation_X\rangle$*? Name five.*
	FUZZY_TEMPLATE_Y for the linguistic term less likely
prompt:	*Answer with five items separated with comma.* Q: *What* **less likely** *has window? Name five.* A: *Window is usually used to see through. Therefore, hat, drawer, vase, basket, box.* Q: *What number can* **less likely** *be on? Name five.* A: *Number is made of digits and can be written on different things for information. Therefore, window, people, rock, tree, jacket.* Q_S: *What is* **less likely**$\langle relation_Y\rangle$ $\langle object_Y\rangle$*? Name five.* Q_O: *What does/is* $\langle subject_Y\rangle$ **less likely** *be/-* $\langle relation_Y\rangle$*? Name five.*

As an example of using the prompt templates, the results for a $relation_X = relation_Y = on$ are included. Please note that different question templates are developed to fit various types of relations. The obtained *subjects* and *objects* are in Tables 5 and 6 for the linguistic terms **most likely** and **less likely**, respectively.

Again, not all obtained *subjects* and *objects* are correct. For example, triples ⟨*hat*, (**most likely**) *on*, -⟩, Table 5, or ⟨*hat*, (**less likely**) *on*, *person*⟩, ⟨*food*, (**less likely**) *on*, *stove*⟩, Table 6, are quite inferior. As before, there is also a graphical representation in Figure 4 of the addition of new triples with the relation *on* that have *building* as their *object*. It can be seen that the most likely *subjects* are quite reasonable, while the less likely *subjects* are a bit odd. A human-wise evaluation is performed; see Section 6.3 for details.

For the **most likely** case, the softmax temperature starts at 0.0 and increases to 0.7 and 1.0, in the case that no text is generated. For the **less likely** case, we observe better results if the initial temperature is set to 0.7 and increases to 1.0 if needed.

Table 5. Query and results for ⟨-, **most likely** *on*, -⟩ for *object*.

user:	*Answer with five items separated with comma.* *Q: What* **most likely** *has window? Name five.* *A: Window is usually used to see through.* *Therefore, train, building, house, car, bus.* *Q: What number can* **most likely** *be on? Name five.* *A: Number is made of digits and can be written on different things for information.* *Therefore, train, sidewalk, track, street, building.* Q_O: *What* ⟨$subject_X$⟩ *can* **most likely** *be* ⟨$relation_X$⟩? *Name five.*
where:	$subject_X$ = { *window, letter, hat, food, hair* } $relation_X$ = *on*
GPT-3 responses for	
window:	$object_X \in$ {*train, building, house, car, bus*}
letter:	$object_X \in$ {*train, sidewalk, track, street, building*}
hat:	$object_X \in$ {*baseball cap, fedora, beanie, cowboy hat, sun hat*}
food:	$object_X \in$ {*apple, banana, orange, grape, strawberry*}
hair:	$object_X \in$ {*person, animal, doll, toy, statue*}

Table 6. Query and results for ⟨-, **less likely** *on*, -⟩ for *object*.

user:	*Answer with five items separated with comma.* *Q: What* **less likely** *has window? Name five.* *A: Window is usually used to see through.* *Therefore, train, building, house, car, bus.* *Q: What number can* **less likely** *be on? Name five.* *A: Number is made of digits and can be written on different things for information.* *Therefore, train, sidewalk, track, street, building.* Q_O: *What* ⟨$subject_Y$⟩ *can* **less likely** *be* ⟨$relation_Y$⟩? *Name five.*
where:	$subject_Y$ = { *window, letter, hat, food, hair* } $relation_Y$ = *on*
GPT-3 responses	
window:	$object_Y \in$ {*number, people, rock, tree, jacket*}
letter:	$object_Y \in$ {*number, people, rock, tree, jacket*}
hat:	$object_Y \in$ {*window, book, cat, person, wall*}
food:	$object_Y \in$ {*sink, counter, stove, refrigerator, table*}
hair :	$object_Y \in$ {*shoulder, leg, foot, arm, hand*}

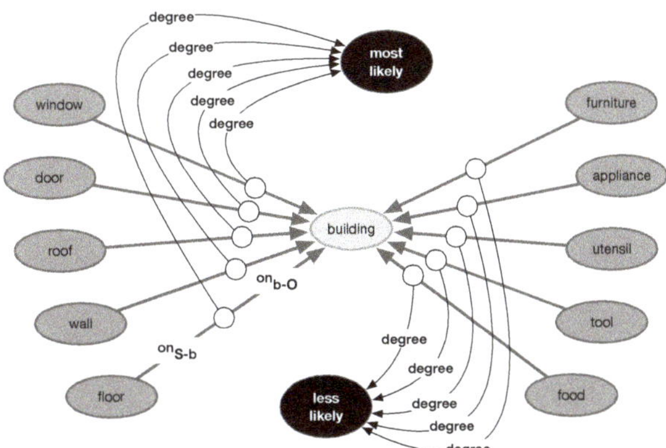

Figure 4. Expanded WpKG—triples with linguistic terms.

5.3. Fuzzy Triples with Novel User-Provided Relations

The last scenario focuses on the generation of new triples that contain novel relations provided by a user. It means the user gives relations that do not exist in the initial vision-based knowledge graph. We selected three novel relations: *used for*, *made of*, and *has property*. We opted for *triples with linguistic terms* and their respective prompts instead of the *simple triples* scenario, as more information about triples is obtained. The prompt templates used here are included in Table 4.

The results obtained for a $subject_X$ = *arm* and the user provided $relation_X \in$ *{used for, made of, has property}* are included in Table 7 for the fuzzy term **most likely**, and in Table 8 for the fuzzy term **less likely**. Graphically, the generated triples for $subject_X$ = *shoe* are in Figure 5. As in the previous cases, not all triples—constructed based on the obtained sets of objects—are satisfactory. The human evaluation results are presented in Section 6.3.

Table 7. Query and results for ⟨-, (**most likely**) *used for/made of/has property*, -⟩ for *object*.

user:	*Answer with five items separated with comma.* *Q: What* **most likely** *has window? Name five.* *A: Window is usually used to see through.* *Therefore, train, building, house, car, bus.* *Q: What number can* **most likely** *be on? Name five.* *A: Number is made of digits and can be written on different things for information.* *Therefore, train, sidewalk, track, street, building.* Q_O*: What is* ⟨$subject_X$⟩ **most likely** ⟨$relation_X$⟩*? Name five.*
where:	$subject_X$ = *arm* $relation_X \in$ **{used for, made of, has property}**
GPT-3 responses for	
used for:	$object_X \in$ *{lifting, carrying, pushing, pulling, holding}*
made of:	$object_X \in$ *{human, animal, plastic, metal, wood}*
has property:	$object_X \in$ *{to move, to bend, to be strong, to be flexible, to grip}*

Table 8. Query and results for ⟨-, *(***less likely***) used for*, -⟩ for *object*.

user:		*Answer with five items separated with comma.* *Q: What* **less likely** *has window? Name five.* *A: Window is usually used to see through.* *Therefore, train, building, house, car, bus.* *Q: What number can* **less likely** *be on? Name five.* *A: Number is made of digits and can be written on different things for information.* *Therefore, train, sidewalk, track, street, building.* Q_O*: What is* ⟨*subject*$_Y$⟩ **less likely** ⟨*relation*$_Y$⟩*? Name five.*
where:		*subject*$_X$ = *arm* *relation*$_X$ ∈ **{used for, made of, has property}**
GPT-3 responses for		
	used for:	*object*$_X$ ∈ *{hat, drawer, vase, basket, box}*
	made of:	*object*$_X$ ∈ *{metal, plastic, glass, wood, fabric}*
	has property:	*object*$_X$ ∈ *{number, window, glass, bottle, box}*

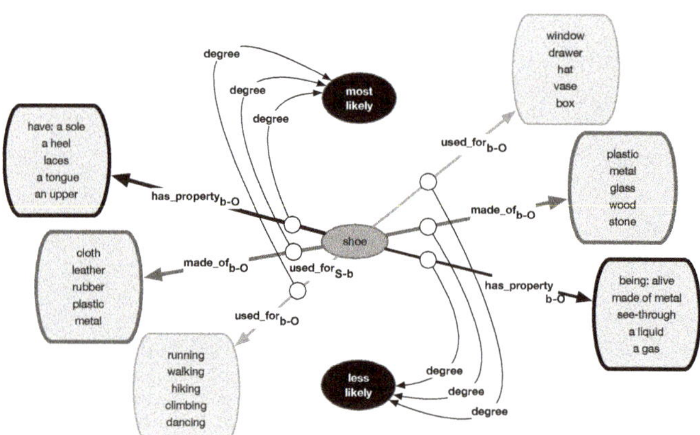

Figure 5. Expanded *WpKG*–fuzzy triples with *shoe* as their *subject* and user-provided relations *has_property, made_of, used_for*.

6. Discussion

The presented method for expanding existing commonsense knowledge graphs represents an example of a new approach to constructing knowledge graphs in a specific domain using very large language models and prompts. It can be said that these techniques are in their infancy; therefore, there are a number of aspects that need to be investigated regarding the approach itself as well as evaluation of the obtained results.

6.1. Vision-Based Commonsense Graph

Similar to how toddlers learn about their environment, our approach is based on two steps. First, we generate commonsense knowledge using vision models and then expand it using language models.

The evaluation of the weighted commonsense knowledge graph generated using only visual data is presented in Table 9 from our previous work [7,8]. Three different approaches for determining the weights (strengths) of relations are proposed and evaluated. Depending on the weighting mechanism, the accuracy of the generated commonsense triples ranges from 87.6% to 93%. Among these, the DPbM (detection probability-based method) correlates highly with human commonsense, while other methods still show good results.

Table 9. Human evaluation of the three weighting mechanisms defined in [8]. Three reviewers were given top 100 triples from each restaurant and classroom contextual commonsense knowledge graphs (total of 600 evaluations per method). Alpha is Krippendorff's Alpha [32] measuring consensus among evaluators.

Weighing Schema	Accept	Reject	N/A	Accuracy (%)	Alpha
DPbM	560	22	18	**93.0**	**0.78**
ROM	526	60	14	87.6	0.63
WOM	538	51	11	89.7	0.72

6.2. Preliminary Experiments with Language Models

The high accuracy obtained using automatic vision-based weighted commonsense knowledge generation does come with some specific challenges of its own. For example, the concept and relation vocabulary is limited only to the dictionary provided to the underlying models during the supervised training of the vision models. Adding a new vocabulary requires several time-consuming and costly tasks. They include human annotation on images to label objects and relations between them and then the fine-tuning of models for object detection and scene graph generation. Even if we accept the time and cost of adding a new vocabulary, it is shown in [9] that there is a bias toward the most common relationship type. It prevents the process from effectively going beyond specific vocabulary.

To address the issue of limited vocabulary, we have investigated using language models to extend the initial vision-based commonsense knowledge graph. We opted to use very large language models, such as GPT-3, for two main reasons. One is their capability to offer new concepts beyond the known ones with acceptable precision. The other reason is the flexibility and time/cost saving of using prompts instead of fine-tuning, which usually requires large amounts of costly human-annotated data.

Our experimental results support the overfitting statement explained in [24,25] stating that training on specific data reduces performance on novel data. We initially experimented with comparing one-shot-prompted 175-billion-parameter unsupervised-trained GPT-3 versus variations of smaller language models fine-tuned on an initial 5000-triple vision-based commonsense knowledge graph. Although the GPT-3 result accuracy was lower than a fine-tuned language model, the novelty of the vocabulary offered was much better. GPT-3 with 175 billion parameters predicted 15 times more vocabulary than the RoBERTa-large model with 355 million parameters.

6.3. Evaluation of Commonsense Knowledge Graph

To the best of our knowledge, there is limited benchmark data or a well-established method suitable for evaluating constructed commonsense knowledge graphs, especially when there are mostly novel generated concepts. There are benchmarks introduced in works such as [33], but are more related to knowledge base completion rather than expansion to new concepts. For mostly novel concepts, human evaluation of the results seems to be the preferred method, mainly in generative model scenarios, as performed in [2].

In this work, the process applied to assess the quality of the constructed commonsense knowledge graph is fully based on human evaluation using Amazon MTurk annotators. Amazon Mechanical Turk https://www.mturk.com (accessed on 12 August 2022) (MTurk) is a crowd-sourcing marketplace that provides, among multiple services, assistance in data annotation tasks. Three sets of validation tasks are performed for simple triples (Section 5.1), fuzzy triples (Section 5.2), and fuzzy triples with user-provided relations (Section 5.3).

The evaluation results are shown in Table 10 for only the new triples that did not exist in the original commonsense knowledge graph. As it can be seen, the results are encouraging. To gain some insight into the evaluation process and to better understand the evaluation results, it should be stressed that MTurk controls who is involved in the evaluation task. To increase the confidence in results, each triple is evaluated by three independent annotators.

To make the evaluation task easier and more intuitive for the annotators, we generated sentences from triples. Based on each predicate, a manual pattern is introduced. Once a sentence is generated using a fixed pattern, it is passed through an off-the-shelf grammar correction module to fix obvious errors. The sentences are then manually vetted to make sure they are grammatically correct and are based on the original triples.

In the description given, the annotators were asked to assume visual commonsense when encountering any of these statements. For example, in the case of *It is likely to see cloud behind cow.*, we asked them to imagine that they are in a field and they see cows. Then it makes sense to see clouds behind the cows.

Some examples of the triples and their evaluation scores are presented:

- *Shoe is used for running.* –> Correct with 0.95 confidence.
- *Shoe is not likely to be alive.* –> Incorrect with 0.95 confidence.
- *Shoe is not usually made of stone.* –> Correct with 0.65 confidence.

As we can see in the examples, finding a well-understood and easy-to-annotate verbalization of triples can affect the result. For example, in the case of *Shoe is not likely to be alive.*, the statement makes sense based on our understanding; however, it was not the case with the three annotators.

Table 10. Results of human evaluation of generated triples. Overall, *Likely* and *Unlikely* columns show the accuracies regarding total triples, most-likely triples, and less-likely triples, respectively. *N* represents the number of triples evaluated in each case.

Triple Type	*N*	Overall Accuracy	*Likely* Accuracy	*Unlikely* Accuracy
Simple	122	72.95%	N/A	N/A
with Linguistic Terms	287	67.94%	68.09%	67.81%
with New Relations	148	72.97%	66.22%	79.73%

A few examples are analyzed under Table 11 to understand the obtained results better. Triples without linguistic terms are called *Simple*. Triples *with Linguistic Terms* contain two terms, **most likely** and **less likely**. Triples *with New Relations* refer to triples with linguistic terms generated with predicates that do not exist in the initial commonsense knowledge graph. For brevity, the initial parts of the prompts are removed. Only the last part of the prompt (question) is kept. The process of generating triples *with Linguistic Terms* and *with New Relation* uses the chain-of-thought prompting methods, shown in Sections 5.2 and 5.3, while *Simple* triples are generated using a simple question and answering prompting method, shown in Section 5.1.

The obtained results are compared with the results found in similar works. TransOMCS paper [34] reports an overall accuracy of 56% while focusing on the automatic mining of commonsense knowledge from linguistic graphs. The results in TransOMCS are based on 100 randomly selected tuples from the overall results set, which five Amazon mTurk workers evaluated. Another comparable work focuses on symbolic knowledge distillation from large language models, mostly about commonsense social relations, without relationship weights [2]. This work reports a human-evaluated correctness percentage of 73.3% when GPT-3 is used with prompts to complete a knowledge graph. The reported value is close to the comparable case of *Simple* triples as shown in Table 10. The approach used in [2] requires text completion for every subject and predicate to generate each triple. On the other hand, our approach uses prompts that generate $N = 5$ new concepts during a single run. It results in roughly one-fifth of the cost when both methods use the same model.

Table 11. Examples: two correct and one incorrect for each type of generated triple. Correct parts of the response are in teal color, while the incorrect parts are in red color.

Simple	with Linguistic Terms	with New Relations
Q: What is in room? Name five. A: bed, table, chair, lamp, computer.	Q: What is most likely to be in bowl? Name five. A: Bowl is a container and can hold different things. Therefore, rice, cereal, milk, soup, fruit.	Q: What is shoe most likely used for? Name five. A: Shoe is used to protect feet. Therefore, running, walking, hiking, climbing, dancing.
Q: What does face have? Name five. A: eyes, nose, mouth, ears, chin.	Q: Where is flower less likely in? Name five. A: Flower is a natural thing and needs soil, water and sunlight to grow. Therefore, sink, oven, book, pencil, phone.	Q: What property does sidewalk most likely have? Name five. A: Sidewalk is a path for pedestrians. Therefore, it is usually level, made of concrete, has a curb, is smooth, and is wide.
Q: What food can be on? Name five. A: apple, banana, orange, pear, grape.	Q: What hat can most likely be on? Name five. A: Hat is worn on the head and can be made of different materials. Therefore, baseball cap, fedora, beanie, cowboy hat, sun hat.	Q: What property does arm less likely have? Name five. A: Arm is a body part and usually has skin, muscle, bone, and hair. Therefore, number, window, glass, bottle, box.

To further demonstrate the scalability of the proposed method, we generated 1905 triples with linguistic terms. Triples with 13 different predicate types from our vision-based commonsense knowledge graph were used for the generation purpose. There are 1075 triples with the linguistic term **less likely** and 830 with the term **more likely**.

All the triples were evaluated using three Amazon mTurk annotators on the Amazon SageMaker platform. The human evaluations of **more likely** triples resulted in higher accuracy of 72.15%, while the **less likely** triples resulted in an accuracy of 62.1%. We only considered triples with at least 95% evaluation confidence among the three annotators (662 triples). The evaluation of triples with different predicate types and linguistic terms resulted in different accuracies, as shown in Figure 6. This scaling experiment shows that the generated dataset size can expand from the initial hundreds of triples to thousands and beyond.

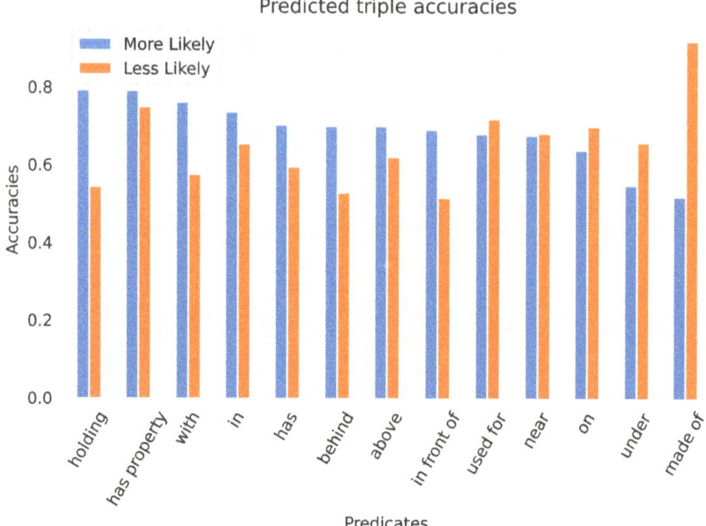

Figure 6. Human (mTurk) annotation accuracy of different predicate types and linguistic terms.

7. Conclusions

There is a growing interest and a need for collecting and storing knowledge that represents information about real-world scenarios and things and activities of everyday life. That type of information—named commonsense—becomes essential when one wants to build autonomous systems that exist around us and assist us in daily duties.

The commonsense knowledge is present in different visual and verbal forms and is learned via observations, experiences, and interaction with others.

A simple attempt to address extracting commonsense knowledge and representing it as a graph is presented here. The previous work [7] showed a method of analyzing images and constructing a commonsense knowledge graph via the fusion of multiple scene graphs extracted from images.

This paper, perceived as a continuation of the work on images, presents a methodology for expanding existing commonsense graphs with facts retrieved from language models. The development of very large language models opens an opportunity to use them for multiple tasks involving retrieving pieces of information and facts in various domains. This capability of the models was utilized here to pull out commonsense information that is easily added to the existing knowledge graphs. Specific prompts and their templates were constructed to retrieve related information. This information was transformed into triples and added to the commonsense graph. Three different types of new triples were considered: simple ones, fuzzy ones with linguistic terms describing degrees of their likeliness, and ones with specific relations provided by the user.

A validation process of new triples was designed and executed—the Amazon service called Mechanical Turk was utilized. The obtained evaluations confirmed the usefulness of the proposed methodology for expanding commonsense graphs.

At the same time, more work is needed to construct prompts that improve the correctness of retrieved information and create triples with more subtle degrees of likeliness. Additionally, more investigation regarding the suitability of different language models is mandated. In this paper, we used the chain-of-thought prompting method [20]. While this prompting method leads to good results, it seems interesting and important to investigate other prompt methods, such as [35], to see if better and more accurate results are achievable.

Author Contributions: Conceptualization, N.R. and M.Z.R.; methodology, N.R. and M.Z.R.; software, N.R. and M.Z.R.; validation, N.R. and M.Z.R.; formal analysis, N.R. and M.Z.R.; investigation, N.R. and M.Z.R.; resources, N.R. and M.Z.R.; data curation, N.R. and M.Z.R.; writing—original draft preparation, N.R. and M.Z.R.; writing—review and editing, N.R. and M.Z.R.; visualization, N.R. and M.Z.R.; supervision, M.Z.R.; project administration, N.R. and M.Z.R.; funding acquisition, M.Z.R. All authors have read and agreed to the published version of the manuscript.

Funding: This research was funded by the Autonomous Systems Initiative (ASI), Alberta, Canada.

Data Availability Statement: The code, data, and experiments are going to be available under the following repository: https://github.com/navidre/lm_vision_kg_expansion (accessed on 10 August 2022).

Conflicts of Interest: The authors declare no conflict of interest.

Abbreviations

The following abbreviations are used in this paper:

WpKG	World-Perceiving Knowledge Graph
C-WpKG	Contextual World-Perceiving Knowledge Graph
GPT	Generative Pre-Trained Transformer
BERT	Bidirectional Encoder Representations from Transformers
AI	Artificial Intelligence
DPbM	Detection Probability-Based Method
ROM	Relative Occurrence Method
WOM	Weighted Occurrence Method
VG	Visual Genome

References

1. Hwang, J.D.; Bhagavatula, C.; Bras, R.L.; Da, J.; Sakaguchi, K.; Bosselut, A.; Choi, Y. COMET-ATOMIC 2020: On Symbolic and Neural Commonsense Knowledge Graphs. In Proceedings of the AAAI, Virtual Conference, 2–9 February 2021.
2. West, P.; Bhagavatula, C.; Hessel, J.; Hwang, J.D.; Jiang, L.; Bras, R.L.; Lu, X.; Welleck, S.; Choi, Y. Symbolic knowledge distillation: From general language models to commonsense models. *arXiv* **2021**, arXiv:2110.07178.
3. LeCun, Y. A Path Towards Autonomous Machine Intelligence Version 0.9.2. Available online: https://openreview.net/pdf?id=BZ5a1r-kVsf (accessed on 27 June 2022).
4. Choi, Y. The Curious Case of Commonsense Intelligence. *J. Am. Acad. Arts Sci.* **2022**, *151*, 139–155. [CrossRef]
5. Bommasani, R.; Hudson, D.A.; Adeli, E.; Altman, R.; Arora, S.; von Arx, S.; Bernstein, M.S.; Bohg, J.; Bosselut, A.; Brunskill, E.; et al. On the opportunities and risks of foundation models. *arXiv* **2021**, arXiv:2108.07258.
6. Rajpurkar, P.; Zhang, J.; Lopyrev, K.; Liang, P. SQuAD: 100,000+ Questions for Machine Comprehension of Text. In Proceedings of the EMNLP, Austin, TX, USA, 1–5 November 2016.
7. Rezaei, N.; Reformat, M.Z.; Yager, R.R. Image-Based World-perceiving Knowledge Graph (WpKG) with Imprecision. *Inf. Process. Manag. Uncertain Knowl. Based Syst.* **2020**, *1237*, 415–428.
8. Rezaei, N.; Reformat, M.Z.; Yager, R.R. Generating Contextual Weighted Commonsense Knowledge Graphs. In Proceedings of the International Conference on Information Processing and Management of Uncertainty in Knowledge-Based Systems, Milan, Italy, 11–15 July 2022; Springer: Berlin/Heidelberg, Germany, 2022; pp. 593–606.
9. Tang, K.; Niu, Y.; Huang, J.; Shi, J.; Zhang, H. Unbiased Scene Graph Generation From Biased Training. In Proceedings of the 2020 IEEE/CVF Conference on Computer Vision and Pattern Recognition (CVPR), Seattle, WA, USA, 13–19 June 2020; pp. 3713–3722.
10. McCarthy, J. *Formalizing Common Sense*; Intellect Books: Bristol, UK, 1990; Volume 5.
11. Bordes, A.; Usunier, N.; Garcia-Duran, A.; Weston, J.; Yakhnenko, O. Translating Embeddings for Modeling Multi-relational Data. In *Proceedings of the Advances in Neural Information Processing Systems*; Burges, C.J., Bottou, L., Welling, M., Ghahramani, Z., Weinberger, K.Q., Eds.; Curran Associates, Inc.: Red Hook, NY, USA, 2013; Volume 26.
12. Yang, B.; Yih, W.; He, X.; Gao, J.; Deng, L. Embedding Entities and Relations for Learning and Inference in Knowledge Bases. In Proceedings of the 3rd International Conference on Learning Representations, ICLR 2015, San Diego, CA, USA, 7–9 May 2015; Conference Track Proceedings; Bengio, Y., LeCun, Y., Eds., 2015.
13. Vaswani, A.; Shazeer, N.; Parmar, N.; Uszkoreit, J.; Jones, L.; Gomez, A.N.; Kaiser, L.u.; Polosukhin, I. Attention is All you Need. In Proceedings of the Advances in Neural Information Processing Systems, Long Beach, CA, USA, 4–9 December 2017; Guyon, I., Luxburg, U.V., Bengio, S., Wallach, H., Fergus, R., Vishwanathan, S., Garnett, R., Eds.; Curran Associates, Inc.: Red Hook, NY, USA, 2017; Volume 30.
14. Wang, C.; Liu, X.; Song, D.X. Language Models are Open Knowledge Graphs. *arXiv* **2020**, arXiv:2010.11967.

15. Petroni, F.; Rocktäschel, T.; Riedel, S.; Lewis, P.; Bakhtin, A.; Wu, Y.; Miller, A. Language Models as Knowledge Bases? In Proceedings of the 2019 Conference on Empirical Methods in Natural Language Processing and the 9th International Joint Conference on Natural Language Processing (EMNLP-IJCNLP), Hong Kong, China, 3–7 November 2019; pp. 2463–2473.
16. Devlin, J.; Chang, M.W.; Lee, K.; Toutanova, K. BERT: Pre-training of Deep Bidirectional Transformers for Language Understanding. In Proceedings of the 2019 Conference of the North American Chapter of the Association for Computational Linguistics: Human Language Technologies, Minneapolis, MN, USA, 2–7 June 2019; Volume 1 (Long and Short Papers); Association for Computational Linguistics: Minneapolis, SP, USA, 2019; pp. 4171–4186. [CrossRef]
17. Brown, T.; Mann, B.; Ryder, N.; Subbiah, M.; Kaplan, J.D.; Dhariwal, P.; Neelakantan, A.; Shyam, P.; Sastry, G.; Askell, A.; et al. Language Models are Few-Shot Learners. In Proceedings of the Advances in Neural Information Processing Systems, Online, 6–12 December 2020; Larochelle, H., Ranzato, M., Hadsell, R., Balcan, M., Lin, H., Eds.; Curran Associates, Inc.: Red Hook, NY, USA, 2020; Volume 33, pp. 1877–1901.
18. Zhang, S.; Roller, S.; Goyal, N.; Artetxe, M.; Chen, M.; Chen, S.; Dewan, C.; Diab, M.; Li, X.; Lin, X.V.; et al. Opt: Open pre-trained transformer language models. *arXiv* **2022**, arXiv:2205.01068.
19. Chowdhery, A.; Narang, S.; Devlin, J.; Bosma, M.; Mishra, G.; Roberts, A.; Barham, P.; Chung, H.W.; Sutton, C.; Gehrmann, S.; et al. Palm: Scaling language modeling with pathways. *arXiv* **2022**, arXiv:2204.02311.
20. Wei, J.; Wang, X.; Schuurmans, D.; Bosma, M.; Chi, E.; Le, Q.; Zhou, D. Chain of thought prompting elicits reasoning in large language models. *arXiv* **2022**, arXiv:2201.11903.
21. Rezaei, N.; Reformat, M.Z. Super-Prompting: Utilizing Model-Independent Contextual Data to Reduce Data Annotation Required in Visual Commonsense Tasks. *arXiv* **2022**, arXiv:2204.11922.
22. Khot, T.; Sabharwal, A.; Clark, P. What's Missing: A Knowledge Gap Guided Approach for Multi-hop Question Answering. In Proceedings of the 2019 Conference on Empirical Methods in Natural Language Processing and the 9th International Joint Conference on Natural Language Processing (EMNLP-IJCNLP), Hong Kong, China, 3–7 November 2019; pp. 2814–2828.
23. Fabbri, A.R.; Ng, P.; Wang, Z.; Nallapati, R.; Xiang, B. Template-Based Question Generation from Retrieved Sentences for Improved Unsupervised Question Answering. In Proceedings of the 58th Annual Meeting of the Association for Computational Linguistics, Online, 5–10 July 2020; pp. 4508–4513.
24. Jastrzębski, S.; Bahdanau, D.; Hosseini, S.; Noukhovitch, M.; Bengio, Y.; Cheung, J. Commonsense mining as knowledge base completion? A study on the impact of novelty. In Proceedings of the Workshop on Generalization in the Age of Deep Learning, Munich, Germany, 8–14 September 2018; Association for Computational Linguistics: New Orleans, LO, USA, 2018; pp. 8–16. [CrossRef]
25. Davison, J.; Feldman, J.; Rush, A.M. Commonsense knowledge mining from pretrained models. In Proceedings of the 2019 Conference on Empirical Methods in Natural Language Processing and the 9th International Joint Conference on Natural Language Processing (EMNLP-IJCNLP), Hong Kong, China, 3–7 November 2019; pp. 1173–1178.
26. Chen, X.; Shrivastava, A.; Gupta, A. NEIL: Extracting Visual Knowledge from Web Data. In Proceedings of the IEEE International Conference on Computer Vision 2013, Sydney, Australia, 1–8 December 2013; pp. 1409–1416.
27. Speer, R.; Chin, J.; Havasi, C. Conceptnet 5.5: An open multilingual graph of general knowledge. In Proceedings of the Thirty-first AAAI Conference on Artificial Intelligence, San Francisco, CA, USA, 4–9 February 2017; pp. 4444–4451.
28. Ren, S.; He, K.; Girshick, R.B.; Sun, J. Faster R-CNN: Towards Real-Time Object Detection with Region Proposal Networks. *IEEE Trans. Pattern Anal. Mach. Intell.* **2015**, *39*, 1137–1149. [CrossRef] [PubMed]
29. Xie, S.; Girshick, R.; Dollár, P.; Tu, Z.; He, K. Aggregated residual transformations for deep neural networks. In Proceedings of the IEEE Conference on Computer Vision and Pattern Recognition, Honolulu, HI, USA, 21–26 July 2017; pp. 1492–1500.
30. Zellers, R.; Yatskar, M.; Thomson, S.; Choi, Y. Neural Motifs: Scene Graph Parsing with Global Context. In Proceedings of the 2018 IEEE/CVF Conference on Computer Vision and Pattern Recognition, Salt Lake City, UT, USA, 18–23 June 2018; pp. 5831–5840.
31. Holtzman, A.; Buys, J.; Du, L.; Forbes, M.; Choi, Y. The Curious Case of Neural Text Degeneration. In Proceedings of the International Conference on Learning Representations, New Orleans, LA, USA, 6–9 May 2019.
32. Hayes, A.F.; Krippendorff, K. Answering the Call for a Standard Reliability Measure for Coding Data. *Commun. Methods Meas.* **2007**, *1*, 77–89. [CrossRef]
33. Li, X.; Taheri, A.; Tu, L.; Gimpel, K. Commonsense Knowledge Base Completion. In Proceedings of the 54th Annual Meeting of the Association for Computational Linguistics, Berlin, Germany, 7–12 August 2016; (Volume 1: Long Papers); Association for Computational Linguistics: Berlin, Germany, 2016; pp. 1445–1455. [CrossRef]
34. Zhang, H.; Khashabi, D.; Song, Y.; Roth, D. TransOMCS: From linguistic graphs to commonsense knowledge. In Proceedings of the Twenty-Ninth International Conference on International Joint Conferences on Artificial Intelligence, Yokohama, Japan, 7–15 January 2021; pp. 4004–4010.
35. Wang, X.; Wei, J.; Schuurmans, D.; Le, Q.; Chi, E.; Zhou, D. Self-consistency improves chain of thought reasoning in language models. *arXiv* **2022**, arXiv:2203.11171.

MDPI
St. Alban-Anlage 66
4052 Basel
Switzerland
Tel. +41 61 683 77 34
Fax +41 61 302 89 18
www.mdpi.com

Symmetry Editorial Office
E-mail: symmetry@mdpi.com
www.mdpi.com/journal/symmetry

www.ingramcontent.com/pod-product-compliance
Lightning Source LLC
LaVergne TN
LVHW071622170726
843515LV00009B/2453

9783036561554